T0139095

HANDBOOK OF ROBUST LOW-RANK AND SPARSE MATRIX DECOMPOSITION

Applications in Image and Video Processing

HANDBOOK OF ROBUST LOW-RANK AND SPARSE MATRIX DECOMPOSITION

Applications in Image and Video Processing

edited by

Thierry Bouwmans
Université de La Rochelle, France

Necdet Serhat Aybat
Université de La Rochelle, France

El-hadi Zahzah
Pennsylvania State University, University Park, USA

CRC Press
Taylor & Francis Group
Boca Raton London New York

CRC Press is an imprint of the
Taylor & Francis Group, an **informa** business

CRC Press
Taylor & Francis Group
6000 Broken Sound Parkway NW, Suite 300
Boca Raton, FL 33487-2742

Printed on acid-free paper
Version Date: 20160411

International Standard Book Number-13: 978-1-4987-2462-3 (Hardback)

Visit the Taylor & Francis Web site at
http://www.taylorandfrancis.com

and the CRC Press Web site at
http://www.crcpress.com

Contents

Part IV: Applications in Image and Video Processing

Part V: Applications in Background/Foreground Separation for Video Surveillance

Preface

Robust subspace learning and tracking by decomposition into low-rank and sparse matrices provide a suitable framework for computer vision applications. Thus, decomposition into low-rank and sparse matrices has been developed in different formulation problems such as robust principal component analysis, robust non-negative matrix factorization, robust matrix completion, subspace tracking, and low-rank minimization. These different approaches differ from the decomposition, the corresponding optimization problem, and the solvers. The optimization problem can be NP-hard in its original formulation, and it can be convex or not follow the constraints and the loss functions used. Thus, the key challenges concern the design of efficient relaxed models and solvers, which have to have as few iterations as possible, and be as efficient as possible.

As the advances in these different problem formulations are fundamental for computer vision applications, this field has witnessed a number of significant publications since the papers of Cands et al., and Chandrasekharan et al. in 2009. A representative example concerns the background/foreground separation in video surveillance. Up to now, many efforts have been made to develop methods that perform well visually with reduced computational cost. However, no algorithm has emerged that is able to simultaneously address all of the key challenges that accompany real-world videos. Thus, effective decompositions for robustness to deal with both real-life scenes with fixed cameras and mobile devices have recently been developed. Another feature of computer vision applications is that the decomposition has to be computed in real-time and low memory requirements. Algorithms have to be designed to meet these requirements.

In this context, this handbook solicited contributions to address this wide range of robust low-rank and sparse matrix decompositions for applications in image and video processing. Thus, it groups the works of the leading teams in this field over recent years. By incorporating both existing and new ideas, this handbook gives a complete overview of the concepts, theories, algorithms, and applications related to robust low-rank and sparse matrix decompositions. First, an introduction to robust principal component analysis via low-rank and sparse matrices decomposition for beginners is provided by surveying the different decompositions, loss functions, optimization problems, and solvers. Furthermore, leading methods and algorithms for robust low-rank and sparse matrix decompositions are presented. Moreover, an accompanying website[1] is provided. This website contains the list of chapters, their abstracts, and links to some software demonstrations. It allows the reader to have quick access to the main resources and codes in the field. Finally, with this handbook, we aim to bring a one-stop solution, i.e., access to a number of different decompositions, algorithms, implementations, and benchmarking techniques in a single volume.

The handbook consists of five parts. **Part I** presents an overall introduction to robust PCA via decomposition into low-rank and sparse matrices. Chapter 1 provides a first complete survey of the decomposition into low-rank and sparse matrices. Furthermore, the

[1] `https://sites.google.com/site/lowranksparsedecomposition/`

authors provide an accompanying website: the DLAM Website.[2] This website contains a full list of the references in the field, links to available datasets and codes. In each case, the list is regularly updated and classified according to the sections of this chapter. Chapter 2 gives a detailed review of algorithms for stable PCA. Chapter 3 investigates dual smoothing and value function techniques for variational matrix decomposition. Thus, the authors review some recent techniques in convex optimization and contribute several novel results. A distinguishing feature of Chapter 3 is the emphasis on a range of optimization formulations of the RPCA problem. When a few columns of the data matrix are generated by mechanisms different from the rest of the columns, the existence of these outlying columns tends to destroy the low-rank structure of the data matrix. Thus, Chapter 4 presents a low-rank and block-sparse matrix decomposition insensitive to column/row outliers. Chapter 5 focuses on the control of the sparsity in robust PCA.

Part II concerns robust matrix factorization/completion problems. Chapter 6 unifies nuclear norm and bilinear factorization for low-rank matrix decomposition. The authors present very convincing results in the several applications, such as background/foreground separation, structure from motion, face reconstruction, and motion estimation between photometric stereo sequences. Chapter 7 describes a robust non-negative matrix factorization under separability assumption. The algorithm called RobustXray is used for background/-foreground separation when illumination changes occur. Chapter 8 provides nonconvex approaches and efficient algorithms for robust matrix completion. The authors provide several results for image/video recovery and removing shadows from faces. Chapter 9 develops a factorized robust matrix completion. Results on video background subtraction show that this approach is robust against several challenges such as illumination changes and dynamic backgrounds.

Part III focuses on robust online subspace estimation, learning, and tracking. Chapter 10 develops online robust algorithms for robust PCA. Thus, the authors study the problem of sequentially recovering a sparse vector and a vector from a low-dimensional subspace from knowledge of their sum. Two main approaches are then presented: Recursive Projected Compressed Sensing (ReProCS) and Modified-PCP. A full evaluation is provided for background/foreground separation against state-of-the-art RPCA algorithms. Chapter 11 provides incremental methods for robust local subspace estimation. Furthermore, the authors generalize their model from a single low-rank subspace with a sparse set of possibly-large deviations, to a low-dimensional manifold with the same type of deviations. Thus, local subspace models and endogenous sparse representations are proposed to obtain a robust approximation of the backgrounds component of a video sequence captured by a non-stationary camera. Finally, a transform invariant incremental RPCA algorithm is described. Chapter 12 presents a Robust Orthonormal Subspace Learning called ROSL for efficient low-rank recovery. Different from convex methods using the nuclear norm, ROSL utilizes a novel rank measure on the low-rank matrix that imposes the group sparsity of its coefficients under orthonormal subspace. The authors present several applications such as in background/foreground separation and removing shadows from faces. Chapter 13 presents a unified view of nonconvex heuristic approaches. Then, the authors propose two non-convex models, i.e., l_p-norm heuristic recovery (pHR) and log-sum heuristic recovery (LHR) for corrupted matrix learning. Experimental results on noisy depth maps fusion for muti-view stereo show the robustness of these two non-convex models.

[2] `https://sites.google.com/site/robustdlam/home`

Part IV concerns applications in image and video processing. Chapter 14 developed a variational approach. The authors evaluated their method on foreground detection in blurred and noisy video, and detection of network anomalies. Chapter 15 recovered low-rank and sparse matrices in the presence of missing and grossly corrupted observations. The authors present results on text removal, background/foreground separation, and face reconstruction. Collaborative filtering and subspace clustering are also investigated. Chapter 16 briefly presents the application of low-rank and sparse matrix decompositions in hyperspectral video processing. Chapter 17 investigates an accelerated dynamic MRI using low-rank plus sparse reconstruction with separation of background and dynamic components.

Part V presents resources and applications in background/foreground separation for video surveillance. Chapter 18 describes the LRSLibrary, which provides a collection of low-rank and sparse decomposition algorithms in MATLAB®. The library was designed for background/foreground separation in videos, but it can also be used or adapted for other computer vision. Currently the LRSLibrary contains a total of 72 matrix-based and tensor-based algorithms. The LRSLibrary was tested successfully in MATLAB R2013b in both the x86 and x64 versions. Chapter 19 develops a Dynamic Mode Decomposition (DMD) for Robust PCA. The DMD decomposition yields oscillatory time components of the video frames that have contextual implications. Furthermore, the authors present a multi-resolution DMD (MRDMD) that allows them to separate components that are happening on different time scales. Chapter 20 provides three algorithms for stochastic RPCA applied to background/-foreground separation. First, Markov Random Fields (MRF) are used to take into account the spatial constraints of the foreground objects. Then, multiple features and dynamic feature selection are added to improve the detection in the case of highly dynamic backgrounds. Finally, the authors present a depth-extended version which is robust in the presence of camouflage in color. Chapter 21 presents a Bayesian sparse estimation applied to background/foreground separation.

The handbook is intended to be a reference for researchers and developers in industries, as well as graduate students, who are interested in low-rank and sparse matrix decomposition applied to computer vision. Particularly, the application in image and video processing are presented, such as in image analysis, image denoising, motion saliency detection, video coding, key frame extraction, hyperspectral video processing and background/foreground separation. Thus, it can be suggested as a reading text for teaching graduate courses in subjects such as computer vision, image and video processing, real-time architecture, machine learning, and data mining. The editors of this handbook would like to acknowledge, with their sincere gratitude, the contributors, for their valuable chapters, and the reviewers, for the helpful comments concerning the chapters in this handbook. Particularly, we acknowledge Dr. Yuanqiang (Evan) Dong [3] from UtopiaCompression Corporation [4] for his review of the handbook. We also acknowledge the reviewers of the original handbook proposal for their hepful suggestions. Furthermore, we are very grateful for the help that we have received from Randi Cohen, Hayley Ruggieri, and others at CRC Press during the preparation of this handbook. Finally, we would like to acknowledge Shashi Kumar from Cenveo for his valuable support about the LaTeX issues.

[3] `http://vigir.missouri.edu/~evan/index.htm`

[4] `http://www.utopiacompression.com/`

About the Editors

Thierry Bouwmans (http://sites.google.com/site/thierrybouwmans/) is an Associate Professor at the University of La Rochelle, France. His research interests consist mainly of the detection of moving objects in challenging environments. He has recently authored more than 30 papers in the field of background modeling and foreground detection. These papers investigated, in particular the use of fuzzy concepts, discriminative subspace learning models, and robust PCA. They also developed surveys on mathematical tools used in the field and particularly on decomposition in low-rank plus additive matrices. He has supervised five Ph.D. students in background/foreground separation. He is the creator and administrator of the Background Subtraction Website, and has served as a reviewer for numerous international conferences and journals.

Necdet Serhat Aybat (http://www.ie.psu.edu/AboutUs/FacultyStaff/Faculty/Profile/aybat.html) received his Ph.D. degree in Operations Research from Columbia University, Industrial Engineering and Operations Research Department. Currently, he is an assistant professor in the Industrial and Manufacturing Engineering Department at Pennsylvania State University, USA. His research, supported by the National Science Foundation (NSF), focuses on developing fast first-order algorithms for large-scale convex optimization problems coming from diverse application areas, such as compressed sensing, matrix completion, convex regression, and distributed optimization. In particular, he has devised algorithms, with provable computational complexity, for robust and stable principal component pursuit problems. He supervises Ph.D. students in this field, and actively serves as a reviewer for numerous academic journals and a session organizer for international conferences.

El Hadi Zahzah (http://sites.google.com/site/ezahzah/) is an Associate Professor at the University of La Rochelle, France. He obtained his Ph.D. at Toulouse Research Institute in Information Technology (IRIT) Lab. Since 1993, his research interests have consisted mainly in the spatio-temporal relations and detection of moving objects in challenging environments. He has authored more than 60 papers in the field of fuzzy logic, expert systems, image analysis, spatio-temporal modelization, and background modeling and foreground detection. His recent papers investigated the use of fuzzy concepts, discriminative subspace learning models, and robust Principal Component Analysis (PCA). He also develops surveys on mathematical tools and has supervised seven Ph.D. students.

List of Contributors

Narendra Ahuja, University of Illinois at Urbana-Champaign, USA

Aleksandr Aravkin, IBM T.J. Watson Research Center, USA

Necdet Serhat Aybat, Pennsylvania State University, USA

S. Derin Babacan, Google Inc., USA

Feng Bao, Tsinghua University, China

Stephen Becker, University of Colorado Boulder, USA

Alexandre Bernardino, Instituto Superior Tecnico, Portugal

Thierry Bouwmans, Laboratoire Master of Interior Architecture (MIA), University de La Rochelle, La Rochelle, France

Steven Brunton, University of Washington, USA

Ricardo Cabral, Carnegie Mellon University, USA

Emmanuel Candes, Stanford University, USA

Jen-Mei Chang, California State University, Long Beach, USA

Zhaofu Chen, Northwestern University, Illinois, USA

Hong Cheng, Department of Systems Engineering and Engineering Management, The Chinese University of Hong Kong, China

James Cheng, Department of Computer Science and Engineering, The Chinese University of Hong Kong, China

Joao Paulo Costeira, Instituto Superior Tecnico, Portugal

Qionghai Dai, Department of Automation, Tsinghua University, China

Fernando De La Torre, Component Analyis Laboratory, Carnegie Mellon University, USA

Yue Deng, University of California, San Francisco, USA

Yunlong Feng, Kulak Leuven, ESAT-STADIUS, Belgium

Xing Fu, University of Washington, USA

Torin Gerhart, Western Digital Corporation, USA

Georges B. Giannakis, University of Minnesota, USA

Jacob Grosek, Air Force Research Laboratories, USA

Han Guo, Iowa State University, USA

Sajid Javed, Kyungpook National University, Republic of Korea

Soon Ki Jung, Kyungpook National University, Republic of Korea

Aggelos K. Katsaggelos, Northwestern University, USA

Abhishek Kumar, IBM T.J. Watson Research Center, USA

Jake Nathan Kutz, Department of Applied Mathematics, University of Washington, USA

Qiuwei Li, Colorado School of Mines, USA

Yuanyuan Liu, Department of Computer Science and Engineering, The Chinese University of Hong Kong, China

Hassan Mansour, Mitsubishi Electric Research Laboratories (MERL), USA

Gonzalo Mateos, University of Rochester, USA

Rafael Molina, Universidad de Granada, Spain

Shinichi Nakajima, Technische Universität Berlin, Germany

Arye Nehorai, Department of Electrical and Systems Engineering, Washington University, St. Louis, USA

Ricardo Otazo, New York University School of Medicine, USA

Fatih Porikli, Australian National University/NICTA, Australia

Paul Rodriguez, Pontificia Universidad Catolica del Peru, PUCP, Peru

Fanhua Shang, Department of Computer Science and Engineering, The Chinese University of Hong Kong, China

Xianbiao Shu, University of Illinois at Urbana-Champaign, USA

Vikas Sindhwani, Google Research, New York, USA

Andrews Sobral, Laboratoire L3i, University La Rochelle, France

Daniel K. Sodickson, New York University School of Medicine, USA

Masashi Sugiyama, University of Tokyo, Japan

Johan Suykens, KU Leuven, ESAT-STADIUS, Belgium

Gongguo Tang, Colorado School of Mines, USA

Dong Tian, Mitsubishi Electric Research Laboratories (MERL), USA

Namarata Vaswani, Iowa State University, USA

Anthony Vetro, Mitsubishi Electric Research Laboratories (MERL), USA

Brendt Wohlberg, Los Alamos National Laboratory, USA

Yuning Yang, KU Leuven, ESAT-STADIUS, Belgium

El-Hadi Zahzah, Lab. L3i, University La Rochelle, France

Jichun Zhan, Iowa State University, USA

I

Robust Principal Component Analysis

1

Robust Principal Component Analysis via Decomposition into Low-Rank and Sparse Matrices: An Overview

Thierry Bouwmans
Lab. MIA, Univ. La Rochelle, France

El-Hadi Zahzah
Lab. L3I, Univ. La Rochelle, France

1.1 Introduction

RPCA via decomposition in low-rank and sparse matrices proposed by Candes et al. [20] in 2009 is currently the most investigated RPCA method. In this chapter, we review this method and all these modifications in terms of decomposition, solvers, incremental algorithms, and real-time implementations. These different RPCA methods via decomposition in low-rank and sparse matrices are fundamental in several applications [21]. Indeed, as this decomposition is nonparametric and does not make many assumptions, it is widely applicable to a large scale of problems that include the following:

- **Latent variable model selection:** Chandrasekaran et al. [24] proposed to discover the number of latent components, and to learn a statistical model over the entire collection of variables by only observing samples of a subset of a collection of random variables. The geometric properties of the decompostion of low-rank plus sparse matrices play an important role in this approach [24] [108].

- **Image processing:** Sometimes, it is needed to separate information from noise or outliers in image processing. RPCA framework was applied with success in image analysis [220] such as image denoising [49], image composition [16], image colorization [201], image alignment and rectification [126], multi-focus image [172], and face recognition [185].

- **Video processing:** Numerous authors used the RPCA framework in applications such as action recognition [71], motion estimation [149], motion saliency detection [191], video coding [214] [56] [27] [28], key frame extraction [32], hyperspectral video processing [44], video restoration [78], and in background and foreground separation [3] [124] [128].

- **3D Computer Vision:** Structure from Motion (SfM) refers to the process of automatically generating a 3D structure of an object by its tracked 2D image frames. Practically, the goal is to recover both 3D structure, namely 3D coordinates of scene points, and motion parameters, namely attitude (rotation) and position of the cameras, starting from image point correspondences. Then, finding the full 3D reconstruction of this object can be posed as a low-rank matrix recovery problem [95] [7] [186].

Here, we choose to focus on the application of background/foreground separation, which is a representative application of RPCA, and which has witnessed very numerous papers (more than 200) since 2009 [18]. Applying RPCA via decomposition in low-rank and sparse matrices in video surveillance, the background sequence is modeled by the low-rank subspace that can gradually change over time, while the moving foreground objects constitute the correlated sparse outliers. For example, Fig. 1.1 shows original frames of sequences from the BMC dataset[5] [171] and their decomposition into the low-rank matrix L and sparse matrix S. We can see that L corresponds to the background whereas S corresponds to the foreground. The fourth image shows the foreground mask obtained by thresholding the matrix

[5] http://bmc.iut-auvergne.com/

FIGURE 1.1 RPCA via decomposition into low-rank and sparse matrices in foreground/background separation: Original image (309), low-rank matrix L (background), sparse matrix S (foreground), foreground mask (Sequences from BMC 2012 dataset [171]).

S. So, the different advances in the different problem formulations of the decomposition into low-rank and sparse matrices are fundamental and can be applied to background modeling and foreground detection in video surveillance [18] [17].

Considering all of this, this chapter develops a comprehensive review of the different RPCA methods based on decomposition into low-rank and sparse matrices. The rest of this chapter is organized as follows. Firstly, we provide a preliminary overview and a unified view of RPCA via decomposition into low-rank and sparse matrices in Section 1.2. Then, we review each original method in its section (Section 1.3 to Section 1.17). For each method, we investigate how it is solved, and if incremental and real-time versions are available for real-time applications such as background/foreground separation. Finally, we conclude with promising research directions in Section 1.18.

1.2 Decomposition into Low-Rank and Sparse Matrices

1.2.1 A Preliminary Overview

The aim of this section is to allow the readers a quick preliminary overview of the different RPCA approaches that are reviewed in detail in the different sections of this chapter. These different approaches differ from the decomposition, the corresponding optimization problem, and the solvers. These different approaches can be classified as follows:

1. **RPCA via Principal Component Pursuit (RPCA-PCP):** The first work on RPCA-PCP developed by Candes et al. [20] proposed a convex optimization to address the robust PCA problem. Under minimal assumptions, this approach, called Principal Component Pursuit (PCP), perfectly recovers the low-rank and the sparse matrices. The background sequence is then modeled by a low-rank subspace that can gradually change over time, while the moving foreground objects constitute the correlated sparse outliers. So, Candes et al. [20] showed visual results on foreground detection that demonstrated encouraging performance, but PCP presents several limitations for foreground detection. The first limitation is that the algorithms required to be solved are computationally expensive.

The second limitation is that PCP is a batch method that stacked a number of training frames in the input observation matrix. In real-time applications such as foreground detection, it would be more useful to estimate the low-rank matrix and the sparse matrix in an incremental way quickly when a new frame comes rather than in a batch way. The third limitation is that the spatial and temporal features are lost as each frame is considered as a column vector. The fourth limitation is that PCP imposed the low-rank component being exactly low-rank and the sparse component being exactly sparse but the observations such as in video surveillance are often corrupted by noise affecting every entry of the data matrix. The fifth limitation is that PCP assumed that all entries of the matrix to be recovered are exactly known via observation and that the distribution of corruption should be sparse and random enough without noise. These assumptions are rarely verified in the case of real applications because **(1)** only a fraction of entries of the matrix can be observed in some environments, **(2)** the observation can be corrupted by both impulsive and Gaussian noise, and **(3)** the outliers, i.e., moving objects are spatially localized.

Many efforts have recently been concentrated to develop low-computational algorithms for solving PCP [91] [92] [19] [204] [194] [155] [184] [116] [97] [93] [107] [46]. Other authors investigated incremental algorithms of PCP to update the low-rank and sparse matrix when new data arrives [129] [131] [130] [132]. Real-time implementations [4] [5] [167] have been developed, too.

Moreover, other efforts have addressed problems that appear specifically in real applications, such as:

(a) **Presence of noise:** Noise in the image is due to a poor-quality image source such as images acquired by a web cam or images after compression.

(b) **Quantization of the pixels:** The quantization can induce at most an error of 0.5 in the pixel value.

(c) **Spatial and temporal constraints of the foreground pixels:** Moving objects are localized in a connexed area. Furthermore, moving objects present a continuous motion through the sequence. These two points induce spatial and temporal constraints on the detection.

(d) **Local variations in the background:** Variations in the background may be due to a camera jitter or dynamic backgrounds.

To address (a), Zhou et al. [221] proposed a stable PCP (SPCP) that guarantees stable and accurate recovery in the presence of entry-wise noise. Becker et al. [15] proposed an inequality constrained version of PCP to take into account the quantization error of the pixel's value (b). To address (c), Tang and Nehorai [167] proposed a PCP method via a decomposition that enforces the low-rankedness of one part and the block sparsity of the other part. Wohlberg et al. [183] used a decomposition corresponding to a more general underlying model consisting of a union of low-dimensional subspaces for local variation in the background (d).

Practically, RPCA is generally applied in the pixel domain by using intensity or color features, but other features can be used such as depth [73] and motion (optical flow [150]) features. Furthermore, RPCA can be extended to the measurement domain, rather than the pixel domain, for use in conjunction with compressive sensing [179] [180] [84] [79] [80] [192] [223] [139]. Although experiments show that moving objects can be reliably extracted by using a small amount of measurements, we have limited the investigation and the comparative evaluation in this paper to the pixel domain to compare it with the classical background subtraction methods.

2. **RPCA via Outlier Pursuit (RPCA-OP):** Xu et al. [188] proposed a robust PCA via Outlier Pursuit to obtain a robust decomposition when the outliers corrupted entire columns; that is, every entry is corrupted in some columns. Moreover, Xu et al. [188] proposed a stable OP (SOP) that guarantees stable and accurate recovery in the presence of entry-wise noise.

3. **RPCA via Sparsity Control (RPCA-SpaCtrl):** Mateos and Giannakis [112] [113] proposed a robust PCA where a tunable parameter controls the sparsity of the estimated matrix, and the number of outliers as a by-product.

4. **RPCA via Sparse Corruptions (RPCA-SpaCorr):** Even if the matrix A is exactly the sum of a sparse matrix S and a low-rank matrix L, it may be impossible to identify these components from the sum. For example, the sparse matrix S may be low-rank, or the low-rank matrix L may be sparse. To address this issue, Hsu et al. [70] imposed conditions on the sparse and low-rank components in order to guarantee their identifiability from A.

5. **RPCA via Log-sum heuristic Recovery (RPCA-LHR):** When the matrix has high intrinsic rank structure or the corrupted errors become dense, the convex approaches may not achieve good performances. Then, Deng et al. [35] used the log-sum heuristic recovery to learn the low-rank structure.

6. **RPCA via Iteratively Reweighted Least Squares (IRLS):** Guyon et al. [63] proposed to solve the RPCA problem by using an Iteratively Reweighted Least Squares (IRLS) alternating scheme for matrix low-rank decomposition. Furthermore, spatial constraint can be added in the minimization process to take into account the spatial connexity of pixels [61]. The advantage of IRLS over the classical solvers is its fast convergence and its low computational cost. Furthermore, Guyon et al. [60] improved this scheme by addressing in the minimization the spatial constraints and the temporal sparseness of moving objects.

7. **RPCA via Stochastic Optimization (RPCA-SO):** Goes et al. [45] proposed a robust PCA via a stochastic optimization. Feng et al. [39] developed an online Robust PCA (OR-PCA) that processes one sample per time instance and hence its memory cost is independent of the number of samples, significantly enhancing the computation and storage efficiency. The algorithm is equivalent to a reformulation of the batch RPCA [45]. Therefore, Javed et. al [75] modified OR-PCA via the stochastic optimization method to perform it on background subtraction. An initialization scheme is adopted, which converges the algorithm very fast as compared to original OR-PCA. Therefore, OR-PCA further improved the foreground segmentation using the continuous constraints such as Markov Random Filed (MRF) [77] and using dynamic feature selection [76].

8. **Bayesian RPCA (BRPCA):** Ding et al. [36] proposed a Bayesian framework that infers an approximate representation for the noise statistics while simul-

taneously inferring the low-rank and sparse components. Furthermore, Markov dependency is introduced spatially and temporally between consecutive rows or columns corresponding to image frames. This method has been improved in a variational Bayesian framework [12] and a factorized variational Bayesian framework [2]. In a similar manner, Zhao et al. [216] developed a generative RPCA model under the Bayesian framework by modeling data noise as a mixture of Gaussians (MoG).

9. **Approximated RPCA:** Zhou and Tao [218] proposed an approximated low-rank and sparse matrix decomposition. This method, called Go Decomposition, (GoDec) produces an approximated decomposition of the data matrix whose RPCA exact decomposition does not exist due to the additive noise, the predefined rank on the low-rank matrix, and the predefined cardinality of the sparse matrix. GoDec is significantly accelerated by using bilateral random projection. Furthermore, Zhou and Tao [218] proposed a Semi-Soft GoDec that adopts soft thresholding to the entries of S, instead of GoDec, which imposes hard thresholding to both the singular values of the low-rank part L and the entries of the sparse part S.

10. **Sparse Additive Matrix Factorization:** Nakajima et al. [117] [118] developed a framework called Sparse Additive Matrix Factorization (SAMF). The aim of SAMF is to handle various types of sparse noise such as row-wise and column-wise sparsity, in addition to element-wise sparsity (spiky noise) and low-rank sparsity (low-dimensional). Furthermore, their arbitrary additive combination is also allowed. In the original robust PCA [20], row-wise and column-wise sparsity can capture noise observed only in the case when some sensors are broken or their outputs are unreliable. SAMF, due to its flexibility in sparsity design, incorporates side information more efficiently. In background/foreground separation, Nakajima et al. [117] [118] induced the sparsity in SAMF using image segmentation.

11. **Variational Bayesian Sparse Estimator:** Chen et al. [30] proposed a variational Bayesian Sparse Estimator (VBSE) based algorithm for the estimation of the sparse component of an outlier-corrupted low-rank matrix, when linearly transformed composite data are observed. It is a generalization of the original robust PCA [20]. VBSE can achieve background/foreground separation in blurred and noisy video sequences.

Table 1.1 shows an overview of the different RPCA methods via Principal Component Pursuit. The first column indicates the different problem formulations and the second column shows the different categories of each problem formulation. The third column indicates the different methods of each category and the corresponding acronym is indicated in the first parentheses. The fourth column gives the name of the authors and the date of the related publication. The previous surveys in the field are indicated in bold and the reader can refer to them for more references on the corresponding category or sub-category.

TABLE 1.1 Robust Principal Component Analysis: A complete overview.

Categories	Sub-categories	Authors - Dates
Principal Component Pursuit (**Survey Bouwmans and Zahzah [18]**)	PCP	Candes et al. (2009) [20]
	Stable PCP	Zhou et al. (2010) [221]
	Quantized PCP	Becker et al. (2011) [15]
	Block-based PCP	Tang and Nehorai (2011) [167]
	Local PCP	Wohlberg et al. (2012) [183]
Outlier Pursuit	OP	Xu et al. (2010) [188]
	SOP	Xu et al. (2010) [188]
Sparsity Control	SpaCtrl	Mateos et Giannakis (2010) [112]
	SpaCorr	Hsu et al. (2011) [70]
Non-Convex Heuristic Recovery	l_p-HR (pHR)	Deng et al. (2012) [33]
	Log-sum HR (LHR)	Deng et al. (2012) [35]
Iteratively Reweighted Least Square	IRLS	Guyon et al. (2012) [63]
	Spatial IRLS	Guyon et al. (2012) [61]
	Spatio-temporal IRLS	Guyon et al. (2012) [60]
Stochastic Optimization	RPCA-SO	Goes et al. (2014) [45]
	OR-RPCA	Feng et al. (2013) [39]
	OR-RPCA with MRF	Javed et al. (2014) [77]
	OR-RPCA with dynamic feature selection	Javed et al. (2014) [76]
Bayesian RPCA	Bayesian RPCA (BRPCA)	Ding et al. (2011) [36]
	Variational Bayesian RPCA (VBRPCA)	Babacan et al. (2012) [12]
	Factorized Variational Bayesian RPCA (FVBRPCA)	Aicher (2013) [2]
	Bayesian RPCA with MoG noise(MoG-BRPCA)	Zhao et al. (2014) [216]
Approximated RPCA	GoDec	Zhou and Tao (2011) [218]
	Semi-Soft GoDec	Zhou and Tao (2011) [218]
Sparse Additive Matrix Factorization	SAMF	Nakajima et al. (2012) [117]
Variational Bayesian Sparse Estimator	VBSE	Chen et al. (2014) [30]

1.2.2 A Unified View of RPCA via Decomposition in Low-Rank and Sparse Matrices

We present in different tables quick comparisons that concern the different key characteristics of the different RPCAs via decomposition into low-rank and sparse matrices. In this idea, Table 1.2, Table 1.3 and Table 1.4 show an overview of the different RPCA via decomposition in low-rank and sparse matrices in terms of minimization, constraints, and convexity to allow us to define this unified view that we call Decomposition into Low-rank and Sparse Matrices (DLSM).

Notations

To allow the readers an easy comparison, we homogenized all the different notations found throughout the papers as follows:

1. **Matrices:** For the common matrices, A stands for the observation matrix, L is the low-rank matrix, S is the unconstrained (residual) matrix or sparse matrix, and E is the noise matrix. I is the identity matrix. For the specific matrices, the notations are given in the section of the corresponding method.

2. **Indices:** m and n are the number of columns and rows of the observed data matrix A. In the case of background/foreground separation, m corresponds to the number of pixels in a frame, and n corresponds to the number of frames in the sequence. n is usually taken to 200 due to computational and memory limitations. i and j stand for the current indices of the matrix. r is the estimated or fixed rank of the matrix L. p stands for the pth largest value in a truncated matrix.

3. **Norms:** The different norms used in the paper for vectors and matrices can be classified as follows:

   - **Vector l^α-norm with $0 \leq \alpha \leq 2$:** $||V||_{l^0}$ is the l^0-norm of the vector V, and it corresponds to the number of non-zero entries. $||V||_{l^1} = \sum_i v_i$ is the l^1-norm of the vector V, and it corresponds to the sum of the entries. $||V||_{l^2} = \sqrt{\sum_i (v_i)^2}$ is the l^2-norm of the vector V, and it corresponds to the Euclidean distance [210].

   - **Matrix l_α-norm with $0 \leq \alpha \leq 2$:** $||M||_{l_0}$ is the l_0-norm of the matrix M, and it corresponds to the number of non-zero entries [210]. $||M||_{l_1} = \sum_{i,j} |M_{ij}|$ is the l_1-norm of the matrix M [210], and it corresponds to the Manhattan distance [50]. $||M||_{l_2} = \sqrt{\sum_{i,j} M_{i,j}^2}$ is the l_2-norm of the matrix M, also known as the Frobenius norm.

   - **Matrix l_∞-norm:** $||M||_{l_\infty} = \max_{ij} |M_{ij}|$ [209] is the l_∞-norm of the matrix M, which allows us to capture the quantization error of the observed value of the pixel [15].

   - **Matrix $l_{\alpha,\beta}$-norm with $0 \leq \alpha, \beta \leq 2$:** $||M||_{l_{\alpha,\beta}}$ is the $l_{\alpha,\beta}$-mixed norm of the matrix M, and it corresponds to the l^β-norm of the vector formed by taking the l^α-norms of the columns of the underlying matrix. α and β are in

TABLE 1.2 RPCA via decomposition into low-rank and sparse matrices: An homogeneous overview (Part I).

Methods	Decomposition	Minimization	Constraints	Convexity
PCP **Candes et al. [20]**	$A = L + S$	$\min_{L,S} \|\|L\|\|_* + \lambda \|\|S\|\|_{l_1}$	$A - L - S = 0$	Yes
Modified PCP (Fixed Rank) Leow et al. [86]	$A = L + S$	$\min_{L,S} \|\|L\|\|_* + \lambda \|\|S\|\|_{l_1}$	$rank(L) = known \ r$	Yes
Modified PCP (Nuclear Norm Free) Yuan et al. [203]	$A = u1^T + S$	$\min_u \|\|A - u1^T\|\|_{l_1}$	$rank(u1^T) = 1$	Yes
Modified PCP (Capped Norm) Sun et al. [161]	$A = L + S$	$\min_{L,S} rank(L) + \lambda \|\|S\|\|_{l_0}$	$\|\|A - L - S\|\|_F^2 \leq \sigma^2$	No
Modified PCP (Inductive) Bao et al. [13]	$A = PA + S$	$\min_{P,S} \|\|P\|\|_* + \lambda \|\|S\|\|_{l_1}$	$A - PA - S = 0$	Yes
Modified PCP (Partial Subspace Knowledge) Zhan and Vaswani [205]	$A = L + S$	$\min_{L,S} \|\|L\|\|_* + \lambda \|\|S\|\|_{l_1}$	$L + P_{\Gamma\perp} S = P_{\Gamma\perp} A$	Yes
p,q-PCP (Schattern-p norm, l_q norm) Wang et al. [175]	$A = L + S$	$\min_{L,S} \|\|L\|\|_{S_p}^p + \lambda \|\|S\|\|_{l_q}$	$A = L + S$	No
Modified p,q-PCP (Schatten-p norm, l_q Seminorm) Shao et al. [154]	$A = L + S$	$\min_{L,S} \|\|L\|\|_{S_p}^p + \lambda \|\|S\|\|_{L_q}^q$	$A = L + S$	No
Modified PCP (2D-PCA) Sun et al. [162]	$A = L + S$	$\min_{U,V} \frac{1}{T} \sum_{i=1}^{T} \|\|A_i - U \Sigma_i V^T\|\|_F^2$	$U^T U = I_{r \times r}, V^T V = I_{c \times c}$	No
Modified PCP (Rank-N Soft Constraint) Oh [120]	$A = L + S$	$\min_{L,S} \sum_{i=N+1}^{\min(m,n)} \|\sigma_{L,i}\| + \lambda \|\|S\|\|_{l_1}$	$A = L + S$	Yes
Modified PCP (JVFSD-RPCA) Wen et al. [182]	$A = L + S$	$\min_{L,S} \|\|L\|\|_* + \lambda \|\|S\|\|_{l_1}$	$A = L + S$	Yes
Modified PCP (NSMP) Wang and Feng [176]	$A = L + S$	$\min_{L,S} \lambda \|\|L\|\|_* + \mu \|\|S\|\|_2$	$A = L + S$	Yes
Modified PCP (WNSMP) Wang and Feng [176]	$A = L + S$	$\min_{L,S} \lambda \|\|\omega(L)\|\|_* + \mu \|\|\omega^{-1}(S)\|\|_2$	$A = L + S$	Yes
Modified PCP (Implicit Regularizers) He et al. [68]	$A = L + S$	$\min_{L,S} \lambda \|\|L\|\|_* + \varphi(S)$	$A = L + S$	Yes

TABLE 1.3 RPCA via decomposition into low-rank and sparse matrices: An homogeneous overview (Part II).

Methods	Decomposition	Minimization	Constraints	Convexity
SPCP **Zhou et al. [221]**	$A = L + S + E$	$\min_{L,S} \|\|L\|\|_* + \lambda \|\|S\|\|_{l_1}$	$\|\|A - L - S\|\|_F < \delta$	Yes
Modified SPCP (Bilateral Projection) Zhou and Tao [219]	$A = UV + S + E$	$\min_{U,V,S} \lambda \|\|S\|\|_{l_1} + \|\|A - UV - S\|\|_F^2$	$rank(U) = rank(S) \leq r$	Yes
Modified SPCP (Nuclear Norm Free) Yuan et al. [203]	$A = U1^T + S + E$	$\min_{S \in \mathbf{R}^{m \times n}, u \in \mathbf{R}^m} \|\|S\|\|_{l_1} + \frac{\mu}{2} \|\|A - u1^T - S\|\|_F^2$	$rank(U1^T) = 1$	Yes
Modified SPCP (Nuclear Norm Free for blur in video) Yuan et al. [203]	$A = U1^T + S + E$	$\min_{S \in \mathbf{R}^{m \times n}, u \in \mathbf{R}^m} \|\|S\|\|_{l_1} + \frac{\mu}{2} \|\|A - H(u1^T + S)\|\|_F^2$	$rank(U1^T) = 1$	Yes
Modified SPCP (Undercomplete Dictionary) Sprechman et al. [160]	$A = UT + S + E$	$\min_{U,T,S} \lambda \|\|S\|\|_1 + \frac{\lambda_1}{2} (\|\|U\|\|_F^2 + \|\|T\|\|_F^2)$ $+ \frac{1}{2} \|\|A - UT - S)\|\|_F^2$	$rank(UT) \leq r$	Yes
Variational SPCP (Huber penalty) Aravkin et al. [6]	$A = L + S + E$	$\min \Phi(L, S)$	$\rho(L + S - Y) < \epsilon$	Yes
Modified SPCP (Three Term Low-rank Optimization) Oreifej et al. [124]	$A = L + S + E$	$\min_{A,L,S} \lambda \|\|L\|\|_* + \lambda_1 \|\|f_\pi(S)\|\| + \lambda_2 \|\|E\|\|_F^2$	$A = L + S + E$	Yes
QPCP **Becker et al. [15]**	$A = L + S$	$\min_{L,S} \|\|L\|\|_* + \lambda \|\|S\|\|_{l_1}$	$\|\|A - L - S\|\|_\infty < 0.5$	Yes
BPCP **Tang and Nehorai [167]**	$A = L + S$	$\min_{L,S} \|\|L\|\|_* + \kappa(1 - \lambda) \|\|L\|\|_{l_{2,1}} + \kappa \lambda \|\|S\|\|_{l_{2,1}}$	$A - L - S = 0$	Yes
LPCP **Wohlberg et al. [183]**	$A = AU + S$	$\min_{U,S} \alpha \|\|U\|\|_{l_1} + \beta \|\|U\|\|_{l_{2,1}} + \beta \|\|S\|\|_{l_1}$	$A - AU + S = 0$	Yes

TABLE 1.4 RPCA via decomposition into low-rank and sparse matrices: An homogeneous overview (Part III).

Methods	Decomposition	Minimization	Constraints	Convexity
OP Xu et al. [188]	$A = L + S$	$\min_{L,S} \Vert L\Vert_* + \lambda \Vert S\Vert_{l_{1,2}}$	$A - L - S = 0$	Yes
SOP Xu et al. [188]	$A = L + S + E$	$\min_{L,S} \Vert L\Vert_* + \lambda \Vert S\Vert_{l_{1,2}}$	$\Vert A - L - S\Vert_F < \delta$	Yes
Sparsity Control Mateos et Giannakis [112] [113]	$A = M + U^T P + S + E$	$\min_{U,S} \Vert X + 1_N M^T - PU^T - S\Vert_F^2 + \lambda \Vert S\Vert_{l_{2(r)}}$	$UU^T = I_q$	Yes
Sparse Corruptions (case 1) Hsu et al. [70]	$A = L + S$	$\min_{L,S} \Vert L\Vert_* + \lambda \Vert S\Vert_{l_1}$	$\Vert A - L - S\Vert_1 \leq \epsilon_1$ $\Vert A - L - S\Vert_* \leq \epsilon_*$ $\Vert L\Vert_\infty \leq b$	Yes
Sparse Corruptions (case 2) Hsu et al. [70]	$A = L + S$	$\min_{L,S} \Vert L\Vert_* + \lambda \Vert S\Vert_{l_1} + \frac{1}{2\mu} \Vert A - L - S\Vert_{l_2}^2$	$\Vert A - L - S\Vert_1 \leq \epsilon_1$ $\Vert A - L - S\Vert_* \leq \epsilon_*$ $\Vert A - S\Vert_\infty \leq b$	Yes
pHR Deng [33]	$A = L + S$	$\min_{\hat{X} \in \hat{D}} \frac{1}{2}(\Vert Diag(Y)\Vert_{f_{l_p}} + \Vert Diag(Z)\Vert_{f_{l_p}}) + \lambda \Vert S\Vert_{f_{l_p}}$	$\hat{X} = \{Y, Z, L, S\}$ $\hat{D} = \left\{(Y, Z, L, S) : \begin{pmatrix} Y & L \\ L^T & Z \end{pmatrix} \geq 0, (L, S) \in C\right\}$	No
LHR Deng et al. [35]	$A = L + S$	$\min_{\hat{X} \in \hat{D}} \frac{1}{2}(\Vert Diag(Y)\Vert_L + \Vert Diag(Z)\Vert_L) + \lambda \Vert S\Vert_L$	$\hat{X} = \{Y, Z, L, S\}$ $\hat{D} = \left\{(Y, Z, L, S) : \begin{pmatrix} Y & L \\ L^T & Z \end{pmatrix} \geq 0, (L, S) \in C\right\}$	No
IRLS Guyon et al. [63]	$A = UV + S$	$\min_{U \in \mathbb{R}^{n \times p}, V \in \mathbb{R}^{p \times m}} \mu \Vert UV\Vert_* + \Vert (A - UV) \circ W_1\Vert_{l_{2,1}}$	$A - UV - S = 0$	Yes
SO Goes et al. [45]	$A = LR + S + E$	$\min_{L \in \mathbb{R}^{n \times p}, R \in \mathbb{R}^{n \times r}} \frac{1}{2} \Vert A - LR^T - S\Vert_F^2 + \frac{\lambda_1}{2}(\Vert L\Vert_F^2 + \Vert R\Vert_F^2 + \lambda_2 \Vert S\Vert_1$	$A - LR - S = 0$	Yes
BRPCA Ding et al. [36]	$A = D(ZG)W_2 + BX + E$	$-log\ (p(\Theta \mid A, H))$	Distribution constraints	-
VBRPCA Babacan et al. [12]	$A = DB^T + S + E$	$p(A, D, B, S)$	Distribution constraints	-
MoG-RPCA Zhao et al. [216]	$A = L + S$	min KL divergence	MOG distribution constraints for S	-
GoDec Zhou and Tao [218]	$A = L + S + E$	$\min_{L,S} \Vert A - L - S\Vert_F^2$	$rank(L) \leq e, card(S) \leq k$	No
Semi-Soft GoDec Zhou and Tao [218]	$A = L + S + E$	$\min_{L,S} \Vert A - L - S\Vert_F^2$	$rank(L) \leq e, card(S) \leq \tau$ τ is a soft threshold	No
SAMF Nakajima et al. [117]	$A = \sum_{k=0}^{K} S + E$			
VBSE Chen et al. [118]	$A = UV^T + RS + E$	$\rho(A, U, V, S, \gamma, \alpha, \beta)$	Distribution constraints	-

the interval $[0, 2]$. For example, $||M||_{l_{2,0}}$ corresponds to the number of non-zero columns of the matrix M [210]. $||M||_{l_{2,1}}$ forces spatial homogeneous fitting in the matrix M [60], and it is suitable in the presence of column outliers or noise [167] [60] [67]. $||M||_{l_{2,1}}$ is equal to $\sum_j ||M_{:j}||_{l^2}$ [210]. The influence of α and β on the matrices L and S was studied in [60].

- **Matrix L_α-seminorm with $0 < \alpha \leq 2$:** $||M||_{L_\alpha} = (\sum_{i,j} |M_{ij}|^\alpha)^{1/\alpha}$ is the L_α-seminorm of the matrix M [154]. The L_1-seminorm is equivalent to the l_1-norm.

- **Matrix L_α-quasi-norm with $0 < \alpha < 1$:** L_α-quasi-norm is defined by $L_\alpha(M) = \sum_{i=1}^{m}(M_i^2 + \mu)^{\frac{1}{\alpha}}$ [65] [151].

- **Matrix Frobenius norm:** $||M||_F = \sqrt{\sum_{i,j} M_{i,j}^2}$ is the Frobenius norm [210]. The Frobenius norm is sometimes also called the Euclidean norm, which may cause confusion with the vector l^2-norm, which is also sometimes known as the Euclidean norm.

- **Matrix nuclear norm:** $||M||_*$ is the nuclear norm of the matrix M, and it corresponds to the sum of its singular values [210]. The nuclear norm is the l^1-norm applied on the vector composed with the singular values of the matrix [35].

- **Matrix dual norm:** $||.||_d$ is the d dual norm of any norm $||.||_{norm}$ previously defined; that is, $norm \in \{l_\alpha, l_\infty, l_{\alpha,\beta}, L_\alpha, F, *\}$. For example, the dual norm of the nuclear norm is $||.||_2$, called the spectral norm, which corresponds to the largest singular value of the matrix [176].

- **Matrix Schatten-α norm with $0 < \alpha \leq 2$:** The Schatten-α norm $||M||_{S_\alpha} = (\sum_{k=1}^{min(m,n)}(\sigma_k(M))^\alpha)^{1/\alpha}$, where $\sigma_k(M)$, denotes the kth singular values of M, can also be used as a surrogate for the nuclear norm as in [175] [154]. The Schatten-1-norm is equivalent to the nuclear norm.

- **Matrix Log-sum norm:** The Log-sum norm $||M||_L$ is defined as $\sum_{ij} log\ (|M_ij| + \delta)$, with $\delta > 0$ as a small regularization constant [35].

- **Matrix l_α-concave-norm with $0 < \alpha < 1$:** The l_α-concave-norm of a matrix M is defined by $f_{l_\alpha}(M) = \sum_{ij} |M_{ij}|^\alpha$ [33].

4. **Loss functions:** Several loss functions can be used to enforce the low-rank or sparse constraints (on L and S, respectively), and minimize the error E. Some loss functions are defined on the previous norms, such as: l_0-loss function ($||.||_{l_0}$), l_1-loss function ($||.||_{l_1}$), l_2-loss function ($||.||_{l_2}$), nuclear norm function, Frobenius loss function, and Log-sum heuristic function [35].

   Other loss functions can be used, such as Least Squares (LS) loss function ($||.||_F^2$), Huber loss function [6], M-estimator-based loss functions [68], and the generalized fused Lasso loss function [187]. Most of the time, proxy loss functions are

TABLE 1.5 Loss functions used for the low rank and sparse constraints, and to minimize the error in the RPCA framework.

Constraints	Original loss function	Surrogate loss functions
Low-rank	$rank(.)$ [20]	**PCP/SPCP/QPCP/BPCP/LPCP/OP**
L		nuclear norm [20]
		Modified PCP
		capped nuclear norm [161]
		Schatten-α norm [175] [154], Rank-N [120]
		LHR
		Log-sum heuristic [34]
Sparsity	l_0-norm [20]	**PCP/Modified PCP**
S		l_1-norm [20], capped l_1-norm [161]
		l_α-norm [175], L_α-seminorm [154]
		dual norm [176], M-estimator [68]
		SPCP
		Generalized fused Lasso [187]
		BPCP/LPCP
		$l_{2,1}$-norm [167] [183]
		OP
		$l_{1,2}$-norm [107]
		LHR
		Log-sum heuristic [34]
Error	**PCP**	**Modified PCP**
E	Equality ($\|\|A-L-S\|\|_F = 0$)	Inequality ($\|\|A-L-S\|\|_F^2 \leq \sigma^2$)
	Frobenius norm [20]	Frobenius norm [161]
		Modified SPCP
		Inequality ($\rho(A-L-S) \leq \epsilon$):
		Huber penalty [6]
		QPCP
		Inequality ($\|\|\|A-L-S\|\|_{l_\infty} \leq 0.5$)
		l_∞-norm [15]
	SPCP	**SPCP**
	Inequality ($\|\|A-L-S\|\|_F < \delta$)	Equality ($\|\|A-L-S-E\|\|_F = 0$)
	Frobenius norm [221]	+ Frobenius norm [124] on E

used as surrogates for the original loss function ($rank(.)$ loss function for the low-rank constraint and l_0-loss function for the sparsity constraint) to obtain a solvable problem. Table 1.5 shows an overview of the different loss functions used in the RPCA framework. The Frobenius norm loss funntion is a valid proxy for nuclear norm loss function, but it fails to recover the low-rank matrix without rank estimate [157]. The l_1-loss function may be suboptimal, since the l_1-norm is a loose approximation of the l_0-norm and often leads to an over-penalized problem. The l_2-loss function is sensitive to outliers and missing data. The Least Squares (LS) loss function is known to be very sensitive to outliers, too [113].

Decomposition into Low-rank and Sparse Matrices (DLSM): From this homogenized overview, we can see that all the decompositions in the different methods can be considered in a unified view that we call Decomposition into Low-rank and Sparse Matrices (DLSM). Thus, all the decompositions can be written in a general formulation as follows:

$$A = \sum_{k=1}^{K} M_k \tag{1.1}$$

with $K \in \{2, 3\}$. The matrix A can be corrupted by element-wise outliers/noise, column or row wise outliers/noise or missing data [110] as can be seen in Figure 1.2. The characteristics of the matrices M_k are as follows:

- The first matrix $M_1 = L$ is a low-rank matrix. In some decompositions, L is decomposed as a product of two matrices UV obtained by bilateral projection [219].

- The second matrix M_2 is a sparse matrix S. This second matrix can be decomposed as follows: **(1)** a sum of two matrices, S_1 and S_2, which correspond to the

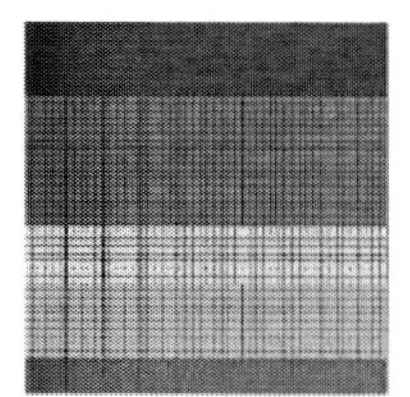

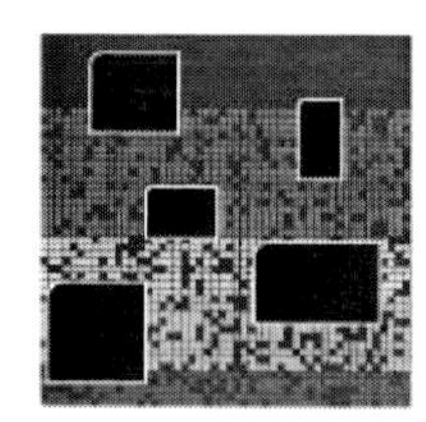

FIGURE 1.2 Illustration of different types of corruption for the matrix A: a) A without noise, b) A with element-wise outliers/noise, and c) A with both element-wise outliers/noise and missing data. Images from the slides of Ma et al. [110]

foreground and the dynamic backgrounds [22], and **(2)** a product of two matrices RT where R is an unconstrained matrix and T is a sparse matrix [30] [31] [29].

- The third matrix M_3 is generally the noise matrix E. The noise can be modeled by a Gaussian, a mixture of Gaussians (MoG), or a Laplacian distribution. M_3 can capture turbulences in the case of background/foreground separation [124].

Practically, the decomposition is $A = L + S$ when $K = 2$. In the case of $K = 3$, the decomposition is $A = L + S + E$. This decomposition is called "stable" decomposition as it separates the outliers in S and the noise in E.

The corresponding minimization problem of Equation 1.1 with $K = 2$ is formulated as follows:

$$\min_{L,S} \ \lambda_1 f_{low}(L) + \lambda_2 f_{sparse}(S) \quad \text{subj} \quad C \tag{1.2}$$

where λ_1 and λ_2 are regularization parameters. C is a constraint on A, L, and S. $f_{low}(L)$ is a loss function that constrains the matrix L to be low rank like the following ones: $rank(.)$ and $||.||_*$. $f_{sparse}(S)$ is a loss function which constrains the matrix S to be sparse like the following ones: $||.||_{l_0}$ and $||.||_{l_1}$. This minimization problem could be NP-hard, and convex or not, following the constraints and the loss functions used. Practically, when the problem is NP hard or not convex, the constraints are relaxed and the loss functions are changed to obtain a tractable and convex problem. For example, the original formulation in RPCA [20] used the $rank(.)$ and the l_0-norm as loss functions for L and S, respectively as shown in Equation 1.12. As this problem is NP-hard, this formulation is relaxed with the nuclear norm and the l_1-norm as shown in Equation 1.13. To minimize confusion, the models that minimize rank functions and nuclear norms are named as the original model and the relaxed model, respectively.

Thus, the corresponding minimization problem of Equation 1.2 can be formulated with norms to be convex and solvable as follows:

$$\min_{L,S} \ \lambda_1 ||L||^{p_1}_{norm_1} + \lambda_2 ||S||^{p_2}_{norm_2} \quad \text{subj} \quad C_1 \tag{1.3}$$

where λ_1 and λ_2 are regularization parameters. p_1 and p_2 are taken in the set $\{1, 2\}$. $||.||_{norm_1}$ and $||.||_{norm_2}$ could be any norm of the following set of norms: l_α-norm, l_∞, $l_{\alpha,\beta}$ mixed norm, L_α-seminorm, Frobenius norm, nuclear norm, dual norm, and Schatten norm. $||.||_{norm_1}$ and $||.||_{norm_2}$ are taken to enforce the low-rank and sparsity constraints of L and S, respectively. The constraint C_1 is generally based on **(1)** an equality such as $||A - L - S||^{p_0}_{norm_0} = 0$ or $rank(L) = r$, or **(2)** an inequality such as $||A - L - S||^{p_0}_{norm_0} \leq q$ or $rank(L) \leq r$. $||.||_{norm_0}$ is a norm taken in the set of norms previously defined. Moreover, the minimization problem formulated in Equation 1.3 can be written in its Lagrangian form as follows:

$$\min_{L,S} \frac{\lambda_0}{2}||A - L - S||^{p_0}_{norm_0} + \lambda_1||L||^{p_1}_{norm_1} + \lambda_2||S||^{p_2}_{norm_2} \quad \text{subj} \quad C_2 \tag{1.4}$$

where λ_0 is a regularization parameter. C_2 is the constraint similar to the constraint C_1.

Similarly, for the stable version, the minimization problem of Equation 1.1 with $K = 3$ is formulated as follows:

$$\min_{L,S} \lambda_1 f_{low}(L) + \lambda_2 f_{sparse}(S) + \lambda_3 f_{noise}(E) \quad \text{subj} \quad C \tag{1.5}$$

where λ_3 is a regularization parameter. C is a constraint on A, L, S, and E. $f_{noise}(E)$ is a function that constrains E.Thus, the corresponding minimization problem of Equation 1.5 can be formulated with norms to be convex and solvable as follows:

$$\min_{L,S} \lambda_1||L||^{p_1}_{norm_1} + \lambda_2||S||^{p_2}_{norm_2} + \lambda_3||E||^{p_3}_{norm_3} \quad \text{subj} \quad C_1 \tag{1.6}$$

where d p_3 is taken in the set $\{1, 2\}$. $||.||_{norm_3}$ could be any previous norms. Then, the minimization problem formulated in Equation 1.6 can be written in its Lagrangian form as follows:

$$\min_{L,S} \frac{\lambda_0}{2}||A - L - S - E||^{p_0}_{norm_0} + \lambda_1||L||^{p_1}_{norm_1} + \lambda_2||S||^{p_2}_{norm_2}\lambda_3||E||^{p_3}_{norm_3} \quad \text{subj} \quad C_2 \tag{1.7}$$

where λ_0 is the regularization parameter. C_2 is the constraint similar to the constraint C_1.

Algorithms for solving the decompositions: Algorithms called solvers are then used to solve the minimization problem in its original form or in its Lagrangian form. Furthermore, instead of directly solving the original convex optimizations, some authors use their strongly convex approximations in order to design efficient algorithms. Zhang et al. [208] proved that these strongly convex programmings guarantee the exact low-rank matrix recovery as well. Moreover, solvers have different characteristics in terms of complexities: complexity per iteration, complexity to reach an accuracy of ϵ precision (ϵ-optimal solution), and convergence rate complexity following the number of iterations. The key challenges related to the solvers are the following [109]: **1)** to make the iterations as few as possible (choice of the solvers), and **2)** make the iterations as efficient as possible (choice of SVD algorithm). Practically, the solvers can be broadly classified into two categories [29] as follows:

- **Regularization-based approaches:** The decomposition is formulated as regularized fitting, where the regularizers are convex surrogates for rank and sparsity. Examples of algorithms in this category include the following solvers: Singular Value Thresholding (SVT) [19], the Accelerated Proximal Gradient (APG) [92], and the Augmented Lagrange Multipliers (ALM) [91]. All the solvers for the different problem formulations are grouped in Table 1.6, Table 1.7, Table 1.8, and Table 1.9.

- **Statistical inference-based approaches:** Hierarchical statistical models are used to model the data-generation process. Prior distributions are selected to capture the low-rank and sparse properties of the respective terms. The joint distribution can be determined from the priors and conditional distributions. Posterior distributions of the unknowns are approximated using Bayesian inference approaches. Representative algorithms in this category can be found in [36] [12] [2] [216].

TABLE 1.6 Solvers for RPCA-PCP: An overview of their complexity per iteration at running time O_{iter}, their complexity O_{pre} to reach an accuracy of ϵ precision, and their convergence rate O_{conv} for T iterations. "Unknown" stands for not indicated by the authors.

Solvers for PCP	Complexity
Basic solvers	
Singular Value Threshold (SVT[1]3) Cai et al. [19]	$O_{iter}(mnmin(mn))$, O_{pre}=unknown, O_{conv}=unknown
Iterative Thresholding (IT) Wright et al. [184]	$O_{iter}(mnmin(mn))$, $O_{pre}(\sqrt{L/\epsilon})$, $O_{conv} = 1/T^2$
Accelerated Proximal Gradient (APG)[13]) Lin et al. [92]	$O_{iter}(mnmin(mn))$, $O_{pre}(\sqrt{1/\epsilon})$, $O_{conv}(1/T^2)$ Full SVD
Dual Method (DM)[13]) Lin et al. [92]	$O_{iter}(rmn)$, $O_{pre}(\sqrt{1/\epsilon})$, $O_{conv}(1/T^2)$ Partial SVD
Exacted Augmented Lagrangian Method (EALM) (EALM[13]) Lin et al. [91]	$O_{iter}(mnmin(mn))$, O_{pre}=unknown, $O_{conv}(1/\mu_T)$ Full SVD
Inexact Augmented Lagrangian Method (IALM) (IALM[13]) Lin et al. [91]	$O_{iter}(rmn)$, O_{pre}=unknown, $O_{conv}(1/\mu_T)$ Partial SVD, Linear Time SVD [198] Limited Memory SVD (LMSVD[14]) [98] Symmetric Low-Rank Product-Gauss-Newton [99]
Alternating Direction Method (ADM) (LRSD[15]) Yuan and Yang [204]	$O_{iter}(mnmin(mn))$, O_{pre}=unknown, O_{conv}=unknown
Symmetric Alternating Direction Method (SADM[16]) (SADAL) Ma [107], Goldfarb et al. [46]	O_{iter}=Unknown, $O_{pre}(1/\epsilon)$, O_{conv}=Unknown
Non-Convex Splitting ADM (NCSADM) Chartrand [26]	O_{iter}=Unknown, O_{pre}=Unknown, O_{conv}=Unknown
Variant of Douglas-Rachford Splitting Method (VDRSM) Zhang and Liu [211]	O_{iter}=Unknown, O_{pre}=Unknown, O_{conv}=Unknown
Proximity Point Algorithm (PPA) Zhu et al. [222]	O_{iter}=Unknown, O_{pre}=Unknown, O_{conv}=Unknown
Proximal Iteratively Reweighted Algorithm (PIRA) Wang et al. [175] (5)	O_{iter}=Unknown, O_{pre}=Unknown, O_{conv}=Unknown
Alternating Rectified Gradient Method (ARGM) (l_1-ARG) Kim et al. [82]	O_{iter}=Unknown, O_{pre}=Unknown, O_{conv}=Unknown
Parallel Direction Method of Multipliers (PDMM) Wang et al. [174]	O_{iter}=Unknown, O_{pre}=Unknown, $O_{conv}(1/T)$
Generalized Accelerated Proximal Gradient (GAPG) He et al. [68]	O_{iter}=Unknown, O_{pre}=Unknown, O_{conv}=Unknown
Improved alternating direction method (IADM) Chai et al. [23]	O_{iter}=Unknown, O_{pre}=Unknown, O_{conv}=Unknown
Linearized solvers	
Linearized Augmented Lagrangian Method (LALM) Yang and Yuan [194]	$O_{iter}(mnmin(mn))$, O_{pre}=Unknown, O_{conv}=Unknown
Linearized Alternating Direction Method (LADM) Yang and Yuan [194]	$O_{iter}(mnmin(mn))$, O_{pre}=Unknown, O_{conv}=Unknown
LADM with Adaptive Penalty (LADMAP[16]) Lin et al. 2011 [93]	$O_{iter}(rmn)$, $O_{pre}(1/\epsilon)$, O_{conv}=Unknown Accelerated version
Linearized Symmetric ADM (LSADM[16]) (ALM) Ma [107], Goldfarb et al. [46]	O_{iter}=Unknown, $O_{pre}(1/\epsilon)$, O_{conv}=Unknown
Fast Linearized Symmetric ADM (Fast-LSADM[16]) (FALM) Ma [107], Goldfarb et al. [46]	O_{iter}=Unknown, $O_{pre}(\sqrt{1/\epsilon})$, O_{conv}=Unknown
Linearized Alternating Direction Method (LADM) (LMaFit[17]) Shen et al. [155]	$O_{iter}(rmn)$, $O_{pre}(1/\epsilon)$, O_{conv}=Unknown

[13]http://perception.csl.uiuc.edu/matrix-rank/samplecode.html

[14]http://www.caam.rice.edu/ yzhang/LMSVD/lmsvd.html

[15]http://math.nju.edu.cn/ jfyang/LRSD/index.html

[16]Available on request by email to the corresponding author

[17]http://lmafit.blogs.rice.edu/

TABLE 1.7 Solvers for RPCA-PCP: An overview of their complexity per iteration at running time O_{iter}, their complexity O_{pre} to reach an accuracy of ϵ precision, and their convergence rate O_{conv} for T iterations. "Unknown" stands for not indicated by the authors.

Solvers for PCP	Complexity
Fast solvers	
Randomized Projection for ALM (RPALM) Mu et al. [116]	$O_{iter}(pmn)$, O_{pre}=Unknown, O_{conv}=Unknown
l_1 filtering (LF[18]) Liu et al. [97] [96]	$O_{iter}(r^2(m+n))$, O_{pre}=Unknown, O_{conv}=Unknown
Block Lanczos with Warm Start Lin and Wei [94]	less than $O_{iter}(pmn)$, O_{pre}=Unknown, O_{conv}=Unknown Partial SVD
Exact Fast RCPA (EFRPCA) Abdel-Hakim and El-Saban [1]	$O_{iter}(mk^2)$ with $k \ll n$, O_{pre}=Unknown, O_{conv}=Unknown Full SVD
Inexact Fast RPCA (IFRPCA) Abdel-Hakim and El-Saban [1]	$O_{iter}(mk^2)$ with $k \ll n$, O_{pre}=Unknown, O_{conv}=Unknown Partial SVD
Matrix Tri-Factorization (MTF) Liu et al. [102]	$O_{iter}(n^3+(r^3+r^2n+mn^2+rn^2))$ O_{pre}=Unknown, O_{conv}=Unknown
Fast Tri-Factorization(FTF) Liu et al. [101]	$O_{iter}(r^3+r^2(m+n)+r^2m+rmn)$ O_{pre}=Unknown, O_{conv}=Unknown
PRoximal Iterative SMoothing Algorithm (PRISMA) Orabona et al. [123]	$O_{iter}(nm)$, $O_{pre}(\log(\epsilon)/\epsilon)$, O_{conv}=Unknown
Fast Alterning Minimization (FAM[19]) Rodriguez and Wohlberg [144]	O_{iter}=Unknown, O_{pre}=Unknown, O_{conv}=Unknown
Fast ADM of Multipliers (FADMM) Yang and Wang [199]	O_{iter}=Unknown, O_{pre}=Unknown, O_{conv}=Unknown
Fast ADM with Smoothing Technique (FADM-ST) Yang [197]	O_{iter}=Unknown, O_{pre}=Unknown, O_{conv}=Unknown
Online solvers	
Online Alternating Direction Method (OADM) Wang and Banerjee [173]	O_{iter}=Unknown, O_{pre}=Unknown, $O_{conv}(1/T)$
Non-convex solvers	
Difference of Convex (DC) Sun et al. [161]	O_{iter}=Unknown, O_{pre}=Unknown, O_{conv}=Unknown
Fast Alternating Difference of Convex (FADC) Sun et al. [161]	O_{iter}=Unknown, O_{pre}=Unknown, O_{conv}=Unknown
Non-convex Alternating Projections(AltProj) Netrapalli et al. [119]	$O_{iter}(r^2mn)$, $O_{pre}(\log(1/\epsilon))$, O_{conv}=Unknown
2D solvers	
Iterative method for Bi-directional Decomposition (IMBD) Sun et al. [162]	O_{iter}=Unknown, O_{pre}=Unknown, O_{conv}=Unknown

[18] Available on request by email to the corresponding author

[19] https://sites.google.com/a/istec.net/prodrig/Home/en/pubs

TABLE 1.8 Solvers for RPCA-SPCP: An overview of their complexity per iteration at running time O_{iter}, their complexity O_{pre} to reach an accuracy of ϵ precision, and their convergence rate O_{conv} for T iterations. "Unknown" stands for not indicated by the authors.

Methods	Solvers	Complexity
RPCA via SPCP Zhou et al. [221]	Alternating Splitting Augmented Lagrangian Method (ASALM[20] Tao and Yuan [168]	O_{iter}=Unknown, O_{pre}=Unknown, O_{conv}=Unknown
	Variational ASALM (VASALM[20]) Tao and Yuan [168]	O_{iter}=Unknown, O_{pre}=Unknown, O_{conv}=Unknown
	Parallel ASALM (PSALM[20]) Tao and Yuan [168]	O_{iter}=Unknown, O_{pre}=Unknown, O_{conv}=Unknown
	Non Smooth Augmented Lagrangian Algorithm (NSA[21]) Aybat et al. [8]	O_{iter}=Unknown, O_{pre}=Unknown, O_{conv}=Unknown
	First-order Augmented Lagrangian for Composite norms (FALC[21]) Aybat et al. [9]	O_{iter}=Unknown, O_{conv}=Unknown, $O_{pre}(1/\epsilon)$
	Augmented Lagragian method for Conic Convex (ALCC[21]) Aybat et al. [10]	O_{iter}=Unknown, O_{conv}=Unknown, $O_{pre}(\log(1/\epsilon))$
	Partially Smooth Proximal Gradient (PSPG[21]) Aybat et al. [11]	O_{iter}=Unknown, O_{conv}=Unknown, $O_{pre}(1/\epsilon)$
	Alternating Direction Method - Increasing Penalty (ADMIP[21]) Aybat et al. [115]	O_{iter}=Unknown, O_{conv}=Unknown, O_{pre}=Unknown
	Inexact Alternating Minimization - Matrix Manifolds (IAM-MM[22]) Hintermüller and Wu [69]	$O_{iter}(rmn)$,O_{conv}=Unknown, O_{pre}=Unknown
RPCA via Modified SPCP Zhou and Tao [219]	Greedy Bilateral Smoothing (GreBsmo[23]) Zhou and Tao [219]	$O_{iter}(max(\|\Omega\| r^2, mnr^3)$, O_{conv}=Unknown, O_{pre}=Unknown
	Bilinear Generalized Approximate Message Passing (BiG-AMP[24]) Parker and Schniter [125]	$O_{iter}(mn + nl + ml)$, O_{conv}=Unknown, O_{pre}=Unknown

[20] Available on request by email to the corresponding author

[21] http://www2.ie.psu.edu/aybat/codes.html

[22] http://www.uni-graz.at/imawww/ifb/r2pcp/index.html

[23] https://sites.google.com/site/godecomposition/GreBsmo.zip

[24] http://www2.ece.ohio-state.edu/ schniter/BiGAMP/BiGAMP.html

TABLE 1.9 Solvers for RPCA (except for PCP and SPCP): An overview of their complexity per iteration at running time O_{iter}, their complexity O_{pre} to reach an accuracy of ϵ precision, and their convergence rate O_{conv} for T iterations. "Unknown" stands for not indicated by the authors.

Methods	Solvers	Complexity
RPCA via Quantized PCP (RPCA-QPCP) Becker et al. [15]	Templates for First-Order Conic Solvers (TFOCS[25]) Becker et al. [15]	$O_{iter}(mlogn)$, O_{conv}=Unknown, $O_{pre}(1/\epsilon)$
RPCA via Block-based PCP (RPCA-BPCP) (RPCA-LBD) Tang and Nehorai [167]	Augmented Lagrangian Method (ALM) Tang and Nehorai [167]	$O_{iter}(mnmin(mn))$, O_{conv}=Unknown, O_{pre}=Unknown
RPCA via Local PCP (RPCA-LPCP) Wohlberg et al. [183]	Split Bregman Algorithm (SBA) Goldstein and Osher [47]	O_{iter}=Unknown, O_{conv}=Unknown, O_{pre}=Unknown
RPCA via Outlier Pursuit (RPCA-OP[26]) Xu et al. [188]	Singular Value Threshold (SVT) Cai et al. [19]	$O_{iter}(mnmin(mn))$, O_{conv}=Unknown, O_{pre}=Unknown
RPCA via Stable Outlier Pursuit(RPCA-SOP[26]) Xu et al. [188]	Singular Value Threshold (SVT) Cai et al. [19]	$O_{iter}(mnmin(mn))$, O_{conv}=Unknown, O_{pre}=Unknown
RPCA with Sparsity Control (RPCA-SpaCtrl) Mateos and Giannakis [112] [113]	Alternating Minimization (AM) Zhou et al. [224]	O_{iter}=Unknown, O_{conv}=Unknown, O_{pre}=Unknown
RPCA via Sparse Corruptions (RPCA-SpaCorr) Hsu et al. [70]	- -	O_{iter}=Unknown, O_{conv}=Unknown, O_{pre}=Unknown
RPCA via Log-sum Heuristic Recovery (RPCA-LHR) Deng et al. [35]	Majorization-Minimization (MM) Fazel [38], Lange et al. [85]	O_{iter}=Unknown, O_{conv}=Unknown, O_{pre}=Unknown
Bayesian RPCA (B-RPCA[27]) Ding et al. [36]	Markov chain Monte Carlo (MCMC) Robert and Cassela [142]	$O_{iter}(r(m+n)+mn)$ O_{conv}=Unknown, O_{pre}=Unknown
	Variational Bayesian Inference (VB) Beal [14]	O_{iter}=Unknown, O_{conv}=Unknow, O_{pre}=Unknow
Variational Bayesian RPCA (VB-RPCA[28]) Babacan et al. [12]	Approximate Bayesian Inference (AB) Beal [14]	$O_{iter}(min(n^3, r^3) + min(m^3, r^3))$ O_{conv}=Unknown, O_{pre}=Unknown
Approximated RPCA (A-RPCA) (GoDec[29]) Zhou and Tao [218]	Naive GoDec Zhou and Tao [218]	Linear convergence
	Fast Godec via Bilateral Random Projection Zhou and Tao [218]	Linear convergence

[25] http://cvxr.com/tfocs/

[26] http://guppy.mpe.nus.edu.sg/ mpexuh/publication.html

[27] http://people.ee.duke.edu/ lcarin/BCS.html

[28] http://www.dbabacan.info/software.html

[29] http://sites.google.com/site/godecomposition/code

Adequation for the background/foreground separation: For each RPCA method, we investigated its adequation with the application of background/foreground separation in their corresponding section in terms of following criteria: **(1)** its robustness to noise, **(2)** its spatial and temporal constraints, **(3)** the existence of an incremental version, and **(4)** the existence of a real-time implementation. Table 1.10 shows an overview of these issues and the following observations can be made:

1. **Robustness to noise:** Noise is due to a poor quality image source such as images acquired by a web cam or images after compression. It affects the entries of the matrix A. In each problem formulation, assumptions are made to assure the exact recovery of the decomposition. PCP assuming that all entries of the matrix to be recovered are exactly known via the observation and that the distribution of corruption should be sparse and random enough without noise. These assumptions are rarely verified in the case of real applications because only a fraction of entries of the matrix can be observed and the observation can be corrupted by both impulsive and Gaussian noise. The robustness of PCP can be improved by taking into account entry-wise noise in SPCP, quantization error in QPCP, and the presence of outliers in entire columns in BPCP. The other methods address sparsity control, recovery guarantees, or the entry-wise noise, too.

2. **Spatial and temporal constraints:** *Spatial constraints* of the foreground object are addressed by BPCP [167] [58], LBPCP [183], BRPCA [36], and IRLS [63] [61] [60]. *Temporal constraints* are addressed by RPCA with dense optical flow [43], RPCA with consistent optical flow [72], RPCA with smoothness and arbitrariness constraints [57] and BRPCA [36]. Only RPCA with smoothness and arbitrariness constraints [57], BRPCA [36], and spatio-temporal IRLS [60] address both the spatial and temporal constraints. The different strategies used to take into account the spatial and/or temporal coherence can be classified as follows:

   - For the ***regularization-based approaches***, the main strategies to take into account the *spatial coherence* consist of using a mixed norm ($||.||_{2,1}$ [167] [58] [63] [61] [60]) on the matrices L and/or S, using a structured sparsity norm [100] on the matrix S, and adding a term in the minimization problem such as a Total Variation penalty [61] [60] [57] [22] or a gradient [63] [61] [60] [183] on the matrix S. For the *temporal coherence*, optical flow is used as in [43] [72]. Thus, the motion information can be used in an adaptive λ [43] [100], which is a function of the motion consistency to ensure that all the changes caused by the foreground motion will be entirely transfered to the matrix S. Thus, the minimization problem in Equation 1.2 can be extended to:

$$\min_{L,S} \ \lambda_1 f_{low}(L) + \lambda_2 f_{sparse}(\Pi(S)) + \lambda_3 f_{back}(L) + \lambda_4 f_{fore}(S) \quad \text{subj} \quad C \tag{1.8}$$

     where λ_3 and λ_4 are regularization parameters. $f_{back}(L)$ and $f_{fore}(S)$ are loss functions that allow the minimization to take into account the characteristics of the background and the foreground, respectively. $f_{back}(L)$ can be a mixed norm. $f_{fore}(S)$ can be the gradient or the Total Variation on S. The function $\Pi()$ allows us to add a confidence map [124] [159] on S. C_1 contains the constraints that are both on the recovery and the spatial/temporal aspects, such as $W \circ A = W \circ (L + S + E)$ [193] [202] [159]. Note that the first part of Equation 1.8 with $f_{low}(L)$ and $f_{sparse}(S)$ concerns the decomposition into low-rank and sparse matrices, and the second part with $f_{back}(L)$ and $f_{fore}(S)$ concerns the application to background/foreground separation. Thus, the minimization problem can be formulated as follows:

$$\begin{aligned} &\min_{L,S} \ \lambda_1 ||L||_{norm_1}^{p_1} + \lambda_2 ||S||_{norm_2}^{p_2} \\ &+\lambda_3 ||L||_{l_{2,1}} + \delta_1 ||grad(S)||_{l_1} + \delta_2 TV(S) + \delta_3 \Omega(S) \quad \text{subj} \quad C_2 \end{aligned} \tag{1.9}$$

where δ_1, δ_2, and δ_3 are regularization parameters. $norm_2$ is usually taken to force spatial homogeneous fitting in the matrix S, that is, for example, the norm $l_{2,1}$ with $p_2 = 1$ [167] [58] [63] [61] [60]. $||grad(S)||_1$, $TV(S)$, and $\Omega(S)$ are a gradient [63] [61] [60] [183], a total variation [61] [60] [57] [22], and a structured norm [100] applied on the matrix S, respectively. Furthermore, S can be processed with a linear operator that weights its entries according to their confidence of corresponding to a moving object such that the most probable elements are unchanged and the least are set to zero [124]. Note that the term $\lambda_3 ||L||_{l_{2,1}}$ ensures the recovered L has exactly zero columns corresponding to the outliers. C_2 is the constraint similar to the constraint C_1.

- For ***the statistical inference-based approaches***, Markov Random Fields (MRF) are used to extract temporally and spatially localized moving objects as in BRPCA [36]. Statistical total variation can also be used as in the approach based on smoothness and arbitrariness constraints [57].

3. **Incremental algorithms:** Incremental algorithms are needed to update the low-rank and additive matrices when new data arrives. Several incremental algorithms can be found in the literature as follows: PCP [129] [131] [130] [132] [51] [181] [147] [190], SPCP [160], RPCA-SpaCtrl [112] [113], Approximated RPCA [28]). Thus, the decomposition can be written as follows:

$$A_t = L_t + S_t + E_t \tag{1.10}$$

where t is the indice for the time. L_t, S_t, E_t are determined from L_{t-1}, S_{t-1}, E_{t-1} and the current observation.

4. **Real-time implementations:** As background/foreground separation needs to be achieved in real-time, several strategies have been developed and are generally based on submatrices computation [127] or GPU implementations [4] [5]. Real-time implementations can be found for PCP [4] [5] [127] [55] and for SPCP [111].

All these key challenges need to be addressed in the different RPCA methods via decompositions in low-rank and sparse matrices so that they are applied adequately to real-time applications such as background modeling and foreground detection in video taken by a static camera.In the next sections, we reviewed this method and all these modifications in terms of decomposition, solvers, incremental algorithms, and real time implementations.

TABLE 1.10 Decomposition into Low-rank and Sparse Matrices: Adequation for background/foreground separation

Methods	Robustness	Spatial and temporal constraints/Incremental/Real time
PCP Candes et al. [20]	Sparse noise	Temporal [43]/ RR-PCP[6] [129]/CAQR[10] [4] [5] Temporal [72] / ReProCS[6] [130] [133]/ Real-Time PCP[11] [127] Spatial and temporal [57]/Support-Predicted Modified-CS[6] [130]/LRSRR [55] -/Support-Predicted Modified-CS RR-PCP[6] [131]/- -/Automated ReProCS[7] [132]/- -/Practical ReProCS (Prac-ReProCS[8]) [51]/- -/Incremental Low-Rank (iLR) Algorithm [181]/- -/Incremental PCP (incPCP[9]) [147]/-
SPCP Zhou et al. [221]	Entry-wise noise	-/Fast Trainable Encoders [160]/DFC-PROJ[12] [111] -/-/DFC-PROJ-ENS[12] [111] -/-/DFC-NYS[12] [111] -/-/DFC-NYS-ENS[12] [111]
QPCP Becker et al. [15]	Quantization error	-/-/-
BPCP Tang and Nehorai [167]	Outliers in entire columns	Spatial [167] [58]/-/-
LPCP Wohlberg et al. [183]	Multimodal backgrounds	Spatial [183]/-/-
OP Xu et al. [188]	Outliers in entire columns	-/-/-
SOP Xu et al. [188]	Outliers in entire columns Entry-wise noise	-/-/-
Sparsity control Mateos et Giannakis [112] [113]	Sparsity control	-/Yes/-
Sparse Corruptions (case 1) Hsu et al. [70]	Recovery guarantees	-/-/-
Sparse Corruptions (case 2) Hsu et al. [70]	Recovery guarantees	-/-/-
LHR Deng et al. [34]	High intrinsic rank structure Dense outliers	-/-/-
ILRS Guyon et al. [63]	Entry-wise noise	Spatial [61], Spatial and temporal [60]/-/-
BRPCA Ding et al. [36]	Entry-wise noise	Spatial and temporal [36]/-/-
VBRPCA Babacan et al. [12]	Entry-wise noise	-/-/-
GoDec Zhou and Tao [218]	Entry-wise noise	-/-/-
Semi-Soft GoDec Zhou and Tao [218]	Entry-wise noise	-/ILRSD [28]/-
SAMF Nakajima et al. [117]	Entry-wise, row-wise, column-wise noise	Spatial [117]/-/-
VBSE Chen et al. [118]	Gaussian noise	-/-/-

[6] http://home.engineering.iastate.edu/ chenlu/ReProCS/ReProCS.htm

[7] http://home.engineering.iastate.edu/ chenlu/ReProCS/ReProCs-auto/results.htm

[8] http://www.ece.iastate.edu/ hanguo/PracReProCS.html

[9] https://sites.google.com/a/istec.net/prodrig/Home/en/pubs/incpcp

[10] Available on request by email to the corresponding author

[11] http://www.nari.ee.ethz.ch/commth/research/downloads/

[12] http://www.cs.berkeley.edu/ ameet/dfc/

1.3 Principal Component Pursuit

RPCA via PCP proposed by Candes et al. [20] in 2009 is currently the most investigated method. In the following sub-sections, we reviewed this method and all these modifications in terms of decomposition, solvers, incremental algorithms, and real-time implementations. Table 1.11 and Table 1.12 show an overview of the Principal Component Pursuit methods and their key characteristics.

1.3.1 Principle

Candes et al. [20] [184] have proposed a convex optimization to address the robust PCA problem. The observation matrix A is assumed to be represented as:

$$A = L + S \tag{1.11}$$

where L is a low-rank matrix and S must be a sparse matrix with a small fraction of nonzero entries. The straightforward formulation is to use l_0-norm to minimize the energy function:

$$\min_{L,S} \ rank(L) + \lambda ||S||_{l_0} \quad \text{subj} \quad A - L - S = 0 \tag{1.12}$$

where $\lambda > 0$ is an arbitrary balanced parameter. But this problem is NP-hard, and a typical solution might involve a search with combinatorial complexity. This research seeks to solve for L with the following optimization problem:

$$\min_{L,S} \ ||L||_* + \lambda ||S||_{l_1} \quad \text{subj} \quad A - L - S = 0 \tag{1.13}$$

where $||.||_*$ and $||.||_1$ are the nuclear norm (which is the l_1-norm of singular value) and l_1-norm, respectively, and $\lambda > 0$ is an arbitrary balanced parameter. Usually, $\lambda = \frac{1}{\sqrt{max(m,n)}}$. Under these minimal assumptions, this approach, called a Principal Component Pursuit (PCP) solution, perfectly recovers the low-rank and the sparse matrices.

Candes et al. [221] showed results on face images and background modeling that demonstrated encouraging performance. The low-rank minimization concerning L offers a suitable framework for background modeling due to the correlation between frames. So, minimizing L and S implies that the background is approximated by a low-rank subspace that can gradually change over time, while the moving foreground objects constitute the correlated sparse outliers that are contained in S. To obtain the foreground mask, S needs to be thresholded. The threshold is determined experimentally. $rank(L)$ influences the number of modes of the background that can be represented by L: If $rank(L)$ is too high, the model will incorporate the moving objects in its representation; if the $rank(L)$ is to low, the model tends to be uni-modal and then the multi-modality that appears in dynamic backgrounds will be not captured. The quality of the background/foreground separation is directly related to the assumption of the low rank and sparsity of the background and foreground, respectively. In this case, the best separation is then obtained only when the optimization algorithm has converged.

Essentially, the nuclear-norm term corresponds to low-frequency components along the temporal while the l_1 norm describes the high-frequency components. However, the low-frequency components can leak into extracted background images for areas that are dominated by moving objects. The leakage, as ghost artifacts that appear in extracted backgrounds, cannot be well-handled by adjusting the weights between the two regularization parameters. Practically, RPCA-PCP present several other limitations developed in Section 1.2.1 and an overview of their solutions is given in the following sections.

TABLE 1.11 Principal Component Pursuit: A complete overview (Part 1).

Categories	Authors - Dates
Decomposition	
1) original PCP	Candes et al. (2009) [20]
2) modified-PCP (Fixed Rank)	Leow et al. (2013) [86]
3) modified-PCP (Nuclear Norm Free)	Yuan et al. (2013) [203]
4) modified-PCP (Capped Norms)	Sun et al. (2013) [161]
5) modified-PCP (Inductive)	Bao et al. (2012) [13]
6) modified-PCP (Partial Subspace Knowledge)	Zhan and Vaswani (2014) [205]
7) p,q-PCP (Schattern-p Norm, l_q Norm Regularization)	Wang et al. (2014) [175]
8) modified p,q-PCP (Schattern-p Norm, l_q Seminorm)	Shao et al. (2014) [154]
9) modified PCP (2D-PCA)	Sun et al. (2013) [162]
10) modified PCP (Rank-N Soft Constraint)	Oh (2012) [120]
11) Joint Video Frame Set Division RPCA (JVFSD-RPCA)	Wen (2014) [182]
12) Nuclear norm and Spectral norm Minimization Problem (NSMP)	Wang and Feng [176]
13) Weighted function NSMP (WNSMP)	Wang and Feng [176]
14) Implicit Regularizers (IR)	He et al. (2013) [68]
15) Random Learning (RL)	Rahmani and Atia (2015) [138]
16) Shape Constraint (SC)	Yang et al. (2015) [200]
17) Generalized Fused Lasso regularization (GFL)	Xin et al. (2015) [187]
18) Double Nuclear Norm-Based Matrix Decomposition (DNMD)	Zhang et al. (2015) [207]
19) Self-paced Matrix Factorization (SPMF)	Zhao et al. (2015) [215]
20) K-Sparsity Prior (K-SP)	Karl and Osendorfer (2015) [81]
Solvers	
1) Basic solvers	
Singular Values Decomposition (SVT)	Cai et al. (2008) [19]
Iterative Thresholding (IT)	Wright et al. (2009) [184]
Accelerated Proximal Gradient (APG)	Lin et al.(2009) [92]
Dual Method (DM)	Lin et al.(2009) [92]
Exacted Augmented Lagrangian Method (EALM)	Lin et al. (2009) [91]
Inexact Augmented Lagrangian Method (IALM)	Lin et al. (2009) [91]
Alternating Direction Method (ADM)	Yuan and Yang (2009) [204]
Symmetric Alternating Direction Method (SADM)	Goldfarb et al. (2010) [46]
Non-Convex Splitting ADM (NCSADM)	Chartrand (2012) [26]
Douglas-Rachford Splitting Method (DRSM)	Gandy and Yamada (2010] [42]
Variant of Douglas-Rachford Splitting Method (VDRSM)	Zhang and Liu (2013) [211]
Proximity Point Algorithm (PPA)	Zhu et al. (2014) [222]
Proximal Iteratively Reweighted Algorithm (PIRA)	Wang et al. (2014) [175]
Alternating Rectified Gradient Method (ARGM)	Kim et al. (2014) [82]
Parallel Direction Method of Multipliers (PDMM)	Wang et al. (2014) [174]
Generalized Singular Value Thresholding (GSVT)	Lu et al. (2014) [106]
Generalized Accelerated Proximal Gradient (GAPG)	He et al. (2013) [68]
Improved Alternating Direction Method (IADM)	Chai et al. (2013) [23]
Optimal Singular Values Shrinkage (OptShrink)	Moore et al. (2014) [114]
Iterative Thresholding with Primal-Dual Method (IT-PDM)	Fan et al. (2014) [37]
2) Linearized solvers	
Linearized ADM (LADM)	Yang and Yuan (2011) [194]
Linearized ADM with Adaptive Penalty (LADMAP)	Lin et al. (2011) [93]
Linearized Symmetric ADM (LSADM)	Goldfarb et al. (2010) [46]
Fast Linearized Symmetric ADM (Fast-LSADM)	Goldfarb et al. (2010) [46]
Linearized IAD Contraction Methods (LIADCM)	Gu et al. (2013) [48]
3) Fast solvers	
Randomized Projection for ALM (RPALM)	Mu et al. (2011) [116]
l_1 filtering (LF)	Liu et al. (2011) [97]
Block Lanczos with Warm Start	Lin and Wei (2010) [94]
Exact Fast Robust Principal Component Analysis (EFRPCA)	Abdel-Hakim and El-Saban (2012) [1]
Inexact Fast Robust Principal Component Analysis (IFRPCA)	Abdel-Hakim and El-Saban (2012) [1]
Matrix Tri-Factorization (MTF)	Liu et al. (2013) [102]
Fast Tri-Factorization (FTF)	Liu et al. (2013) [101]
PRoximal Iterative SMoothing Algorithm (PRISMA)	Orabona et al. (2012) [123]
Fast Alterning Minimization (FAM)	Rodriguez and Wohlberg (2013) [144]
Fast Alternating Direction Method of Multipliers (FADMM)	Yang and Wang (2014) [199]
Fast Alternating Direction Method with Smoothing Technique (FADM-ST)	Yang (2014) [197]
Fast Randomized Singular Value Thresholding (FRSVT)	Oh et al. (2015) [121]
4) Online solvers	
Online Alternating Direction Method (OADM)	Wang and Banerjee (2013) [173]
5) Non-convex solvers	
Difference of Convex (DC)	Sun et al. (2013) [161]
Fast Alternating Difference of Convex (FADC)	Sun et al. (2013) [161]
Non-convex Alternating Projections(AltProj)	Netrapalli et al. (2014) [119]
Iterative Shrinkage-Thresholding/Reweighted Algorithm (ISTRA)	Zhong et al. (2015) [217]
6) 2D solvers	
Iterative Method for Bi-directional Decomposition (IMBD)	Sun et al. (2013) [162]

TABLE 1.12 Principal Component Pursuit: A complete overview (Part 2).

Categories	Authors - Dates
Incremental Algorithms	
Recursive Robust PCP (RR-PCP)	Qiu and N. Vaswani (2010) [129]
Recursive Project Compressive Sensing (ReProCS)	Qiu and Vaswani (2011 [130]
Support-Predicted Modified-CS RR-PCP)	Qiu and Vaswani (2011) [131]
Support-Predicted Modified-CS)	Qiu and Vaswani (2011) [130]
Automated ReProCS	Qiu and Vaswani (2012) [132]
Practical ReProCS (Prac-ReProCS)	Guo et al. (2013) [51]
Incremental Low-Rank (iLR) Algorithm	Wei et al. (2014) [181]
Incremental PCP (incPCP)	Rodriguez and Wohlberg (2014) [147]
Online RPCA (ORPCA)	Xu (2014) [190]
Real-Time Implementations	
CAQR	Anderson et al. (2010) [4]
Real-Time PCP	Pope et al. (2011) [127]
LR Submatrix Recovery/Reconstruction (LRSRR)	Guo et al. (2014) [55]
Multi-Features Algorithms	
Multi-Features Algorithm (MFA)	Gan et al. (2013) [41]
Multi-Task RPCA (MTRPCA)	Wang and Wan (2014) [177]
Spatial-Temporal Algorithms	
Dense Optical Flow	Gao et al. (2012) [43]
Consistent Optical Flow	Huang et al. (2013) [72]
Total Variation (TV) Regularizer	Guo et al. (2014) [57]
Compressive Sensing Algorithms	
Sparsity Reconstruction for Compressive Sensing (SpaRCS)	Waters et al. (2011) [179]
SpaRCS with Partial Support Knowledge (SpaRCS-PSK)	Zonoobi and Kassim (2013) [223]
Adaptive Reconstruction Compressive Sensing (ARCS)	Yang et al. (2013) [192]
LRSD for Compressive Sensing (LRSDCS)	Jiang et al. (2014) [84]
Recursive Low-rank and Sparse Decomposition (rLSDR)	Li and Qi (2014) [89]
Optimal PCP Solutions	
Minimum Description Length (MDL)	Ramirez and Shapiro (2012) [141]
Saliency Measure	Gao et al. (2012) [43]
SVD Algorithms	
Full SVD	-
Partial SVD	-
Linear Time SVD	Yang and An (2013) [198]
smaller-scale SVD	Zhang and Tian (2013) [212]
block-SVD	Chai et al. (2013) [23]
Limited Memory SVD (LMSVD)	Liu et al. (2013) [98]
Symmetric Low-Rank Product-Gauss-Newton (SLRPGN)	Liu et al. (2014) [99]

1.3.2 Algorithms for Solving PCP

Several algorithms called solvers have been proposed for solving PCP. An overview of these solvers as well as their complexity (when they are available) are grouped in Table 1.6. For an $m \times n$ input matrix A with estimated rank r, the complexity per iteration at running time is formulated as $O_{iter}(f_{iter}(m, n, r, ..))$ where $f_{iter}()$ is a function. The complexity to reach an accuracy of ϵ precision (ϵ-optimal solution) is formulated as $O_{pre}(f_{pre}(\epsilon))$ where $f_{pre}()$ is a function. The convergence rate is formulated as $O_{conv}(f_{conv}(T))$, where $f_{conv}()$ is a function of T, which is the number of iterations. All these algorithms require solving the following type of subproblem in each iteration:

$$\min_{L,S} \quad \eta ||L||^{p_1}_{norm_1} + \lambda ||S||^{p_2}_{norm_2} \tag{1.14}$$

The above problem can have a closed form solution or not, following the application. So, several solvers can be found in the the litterature:

- **Basic solvers:** When the problem is supposed to have a closed-form solution, PCP can be reformulated as a semidefinite program and then be solved by standard interior point methods [25]. However, interior point methods have difficulty in handling large matrices because the complexity of computing the step direction is $O((mnmin(m, n))^2)$, where $m \times n$ is the size of the data matrix. If $m = n$, then the complexity is $O(n^6)$. So the generic interior point solvers are too limited for many real applications where the number of data are very large. To overcome the scalability issue, only the first-order information can be used. Cai et al. [19] showed that this technique, called Singular Value Thresholding (SVT), can be used to minimize the nuclear norm for matrix completion. As the matrix recovery problem in Equation (1.13) needs to minimize a combination of both the l_1-norm and the nuclear norm, Wright et al. [184] adopted an iterative thresholding technique (IT) to solve it and obtained similar convergence and scalability properties to interior point methods. However, the iterative thresholding scheme converges extremely slowly with $O_{pre} = \sqrt{L/\epsilon}$ where L is the Lipschitz constant of the gradient of the objective function. To alleviate this slow convergence, Lin et al. [92] proposed two algorithms: the accelerated proximal gradient (APG) algorithm and the gradient-ascent algorithm applied to the dual of the problem in Equation (1.13). However, these algorithms are all the same too slow for real application. More recently, Lin et al. [91] proposed two algorithms based on augmented Lagrange multipliers (ALM). The first algorithm is called the Exact ALM (EALM) method, and has a Q-linear convergence speed, while the APG is in theory only sub-linear. The second algorithm is an improvement of the EALM that is called the Inexact ALM (IALM) method, which converges practically as fast as the EALM, but the required number of partial SVDs is significantly less. The IALM is at least five times faster than APG, and its precision is also higher [91]. IALM was evaluated in the application of background/foreground separation by Yang [196]. However, the direct application of ALM treats Equation (1.13) as a generic minimization problem and ignores its separable structure emerging in both the objective function and the constraint [91]. Hence, the variables S and L are minimized simultaneously. Yuan and Yang [204] proposed to alleviate this by the Alternating Direction Method (ADM), which minimizes the variables L and S serially. ADM achieves it with less computation cost than ALM. Recently, Chartrand [26] proposed a non-convex splitting version of the ADM [91], called NCSADM. This non-convex generalization of [91] produces a sparser model that is better able to separate moving objects and stationary objects. Furthermore, this splitting algorithm maintains the background model while removing substantial noise, more so than the convex regularization does. The ALM neglects the separable structure in both the objective function and the constraint. Thus, Zhang and Liu [211] proposed a variant of the Douglas-Rachford splitting method (VDRSM) for accomplishing recovery in the case of illumination changes and dynamic backgrounds. In a similar way, Zhu et al. [222] proposed a Proximity Point Algorithm (PPA) based on the Douglas-Rachford splitting method. The convex optimization problem is solved by canceling the constraint of the variables, and the proximity operators of the objective function are computed alternately. The new algorithm can exactly recover the low-rank and sparse simultaneously, and it is proved to be convergent. Another approach developed by Chai et al. [23] is an improved alternating direction method (IADM) algorithm with a block-based SVD approach. Experimental results [23] on the I2R dataset [87] show that IADM outperforms SVT [19], APG [92], IALM [91], and ADM [204] with less computation time.

- **Linearized solvers:** When the resulting subproblems do not have closed-form solutions, Yang and Yuan [194] proposed to linearize these subproblems such that closed-form solutions of these linearized subproblems can be easily derived. Global convergence of these Linearized ALM (LALM) and ADM (LADM) algorithms are established under standard assumptions. Recently, Lin et al. [93] improved the convergence for the Linearized Alternating Direction Method with an Adaptive Penalty (LADMAP). They proved the global convergence of LADM and applied it to solve Low-Rank Representation (LRR). Furthermore, the fast version of LADMAP reduces the complexity $O(mnmin(m,n))$ of the original LADM-based method to $O(rmn)$, where r is the rank of the matrix to recover, which is supposed to be smaller than m and n. Guyon et al. [59] evaluated LADMAP in the application of background/foreground separation. In a similar way, Ma [107] and Goldfarb et al. [46] proposed a Linearized Symmetric Alternating Direction Method (LSADM) for minimizing the sum of two convex functions. This method requires at most $O(1/\epsilon)$ iterations to obtain an ϵ-optimal solution, and its fast version, called Fast-LSADM, requires at most $O(1/\sqrt{\epsilon})$ with little change in the computational effort required at each iteration. Experimental results on background/foreground separation are provided by Guyon et al. [62].

- **Fast solvers:** All the previous solvers require computing SVDs for some matrices, resulting in $O(mnmin(m,n))$ complexity. Although partial SVDs are used to reduce the complexity to $O(rmn)$, such a complexity is still high for large data sets. Therefore, recent research focuses on the reduction of the complexity by avoiding computation of SVD. Shen et al. [155] presented a method where the low-rank matrix is decomposed in a product of two-low rank matrices and then minimized over the two matrices alternatively. Although they do not require nuclear norm minimization and thus the computation of SVD, the convergence of the algorithm is not guaranteed as the problem is non-convex. Furthermore, both the matrix-matrix multiplication and QR decomposition-based rank estimation technique require $O(rmn)$ complexity. So, this method does not essentially reduce the complexity. In another way, Mu et al. [116] reduced the problem scale by random projections (linear or bilinear projection) but different random projections may lead to radically different results. Furthermore, additional constraints to the problem slow down the convergence. The complexity of this method is $O(pmn)$ where $p \times m$ is the size of the random projection matrix with $p \ll m$, $p \ll n$ and $p > r$. So, this method is still not linear complex with respect to the matrix size. Its convergence needs more iterations than IALM but it requires less computation time. In another way, Liu et al. [97] [96] proposed an algorithm, called l_1-filtering, for exactly solving PCP with an $O(r^2(m+n))$ complexity. This method is a truly linear cost method to solve PCP problems when the date size is very large while the target rank is small. Moreover, l_1-filtering is highly parallelizable. It is the first algorithm that can exactly solve a nuclear norm minimization problem in linear time. Numerical experiments [97] [96] show the great performance of l_1-filtering in speed compared to the previous algorithms for solving PCP. In another way, Orabona et al. [123] proposed an optimization algorithm called the PRoximal Iterative SMoothing Algorithm (PRISMA) which is decomposed into three parts: a smooth part, a simple non-smooth Lipschitz part, and a simple non-smooth non-Lipschitz part. Furthermore, a time variant smoothing strategy is used to obtain a guarantee that does not depend on knowing in advance the total number of iterations nor a bound on the domain. Numerical experiments [123] show that PRISMA required less iterations than Fast-LSADM [46]. Another approach developed by Rodriguez and Wohlberg [144] is able to compute a sparse approximation even after the first outer loop, (taking approximately 12 seconds for a $640 \times 480 \times 400$ color test video), which is approximately an order of magnitude faster than IALM [91] with the same accuracy. Yang and Wang [199] proposed a Fast Alternating Direction Method of Multipliers(FADMM) algorithms, which slightly outperforms IALM [91] and ADM [204] in terms of computation times. Yang [197] improved FADMM by using a smoothing technique, which is used to smooth the non-smooth terms in the objective function.

- **Online solvers:** The previous solvers are mainly batch ones, but online algorithms are better adapted for real-time application. So, Wang and Banerjee [173] proposed an efficient online learning algorithm named online ADM (OADM), which can solve online convex optimization under linear constraints where the objective could be nonsmooth.

- **Non-convex solvers:** Sun et al. [161] developed, for the non-convex RPCA formulation of RPCA with capped norms, two algorithms: Difference of Convex (DC) and Fast Alternating Difference of Convex (FADC). DC programming treats a non-convex function as the difference of two convex functions, and then iteratively solves it on the basis of the combination of the first convex part and the linear approximation of the second convex part. Numerical measurements [161] demonstrate that the DC approach performs better than both IALM [91] and NSA [8] in terms of the low-rank and sparsity. In another way, Netrapalli et al. [119] proposed a Non-convex Alternating Projections algorithm (AltProj) to solve a non-convex formulation of RPCA. The overall complexity of AltProj is $O(r^2 mn\log(1/\epsilon))$. This is drastically lower than the best-known bound of $O(m^2n/\epsilon)$ on the number of iterations required by convex methods, and just a factor r away from the complexity of naive PCA. AltProj is around 19 times faster than IALM [91]. Moreover, visually, the background extraction seems to be of better accuracy.

- **2D solvers:** Sun et al. [162] [164] developed an iterative algorithm for robust 2D-PCA via alternating optimization, which learns the projection matrices by bidirectional decomposition. To further speed up the iteration, Sun et al. [162] [164] proposed an alternating greedy approach or l_0-norm regularization, minimizing over the low-dimensional feature matrix and the sparse error matrix.

1.3.3 Algorithms for Incremental PCP

PCP is an offline method that treats each image frame as a column vector of the matrix A. In real-time applications such as foreground detection, it would be more useful to estimate the sparse matrix in an incremental way quickly as a new frame comes, rather than in a batch way. Furthermore, the sparsity structure may change slowly or in a correlated way, which may result in a low-rank sparse matrix. In this case, PCP assumption will not be satisfied and S can't be separated from L. Moreover, the principal directions can change over time. So, the rank of the matrix L will keep increasing over time, thus making PCP infeasible after a time. This last issue can be solved by not using all frames of the sequences. So, several algorithms for incremental algorithms are available in literature.

1. **Recursive Projected Compressive Sensing (ReProCS):** To address the first two issues, Qiu and Vaswani [129] proposed an online approach called Recursive Robust PCP (RR-PCP) in [129], and Recursive Projected Compressive Sensing (ReProCS) in [130] [133] [136] [135]. The aim of ReProCS is to causally keep updating the sparse matrix S_t at each time, and keep updating the principal directions. The t-th column of A, A_t, is the data acquired at time t and can be decomposed as follows:

$$A_t = L_t + S_t = [UI]\,[x_t S_t]^t \qquad (1.15)$$

where $x_t = U^T L_t$ and the matrix U is an unknown $m \times m$ orthonormal matrix. The support of S_t changes slowly over time. Let N_t denote the support of x_t, which is assumed piecewise constant over time, and given an initial estimate of $P_t = (U)_{N_t} = \hat{P}_t$, Qiu and Vaswani [129] solved for the sparse component S_t by finding the orthogonal complement matrix $\hat{P}_{t,\perp}$, and then using the projection M_t onto $\hat{P}_{t,\perp}$, denoted by y_t:

$$y_t = \hat{P}_{t,\perp}^T M_t = \hat{P}_{t,\perp}^T L_t + \hat{P}_{t,\perp}^T S_t \tag{1.16}$$

to solve S_t. The low-rank component is closed to zero if $\hat{P}_t \approx P_t$, otherwise new directions are added. Furthermore, recent estimates of $L_t = A_t - S_t$ are stored and used to update P_t. Confirming the first results obtained in [206] [104], a correctness result for ReProCS is given by Lois and Vawani [103]. However, ReProCS requires the support x_t to be fixed and quite small for a given support size S_t, but this does often not hold. So, ReProCS could not handle large outliers' support sizes.

2. **Support-Predicted Modified-CS:** Qiu and Vaswani [131] address the large outliers' support sizes by using time-correlation of the outliers. This method, called Support-Predicted Modified-CS RR-PCP [131] and Support-Predicted Modified-CS [130], is also an incremental algorithm and outperforms the ReProCS. However, this algorithm is only adapted for specific situations where there are only one or two moving objects that remain in scene. But this is not applicable to real videos where multiple and time-varying numbers of objects can enter or leave the scene. Moreover, it requires knowledge of foreground motion.

3. **Automated Recursive Projected CS (A-ReProCS):** To address the limitation of the Support-Predicted Modified-CS, Qiu and Vaswani [132] proposed a method called automated Recursive Projected CS (A-ReProCS) that ensures robustness when there are many nonzero foreground pixels; that is, there are many moving objects or large moving objects. Furthermore, A-ReProCS outperforms the previous incremental algorithms when foreground pixels are correlated spatially or temporally and when the foreground intensity is quite similar to the background one.

4. **ReProCS with cluster-PCA (ReProCS-cPCA):** To address the fact that the structure that we require is that L_t is in a low-dimensional subspace and the eigenvalues of its covariance matrix are "clustered," Qiu and Vaswani [134] [136] introduced a Recursive Projected Compressive Sensing with cluster-PCA (ReProCS-cPCA). Under mild assumptions, ReProCS-cPCA with high probability can exactly recover the support set of S_t at all times. Furthermore, the reconstruction errors of both S_t and L_t are upper bounded by a time-invariant and small value.

5. **Practical ReProCS (Prac-ReProCS):** Guo et al. [51] [53] [54] [52] designed a practically usable modification of the theoretical ReProCS algorithm. This practical ReProCS (Prac-ReProCS) requires much fewer parameters that can be set without any model knowledge, and it exploits practically valid assumptions such as denseness for L_t, slow subspace change for L_t, and correlated support change of S_t.

6. **Incremental Low-Rank (iLR) Algorithm:** Wei et al. [181] proposed an incremental low-rank matrix decomposition algorithm that maintains a clean background matrix adaptive to dynamic changes with both effectiveness and efficiency guarantees. Instead of a batch RPCA, which requires a large number of video frames (usually 200 frames) for each time period, 15 frames only are required with iLR. The iLR algorithm is about 9 times faster than a batch RPCA.

7. **Incremental PCP (incPCP):** Rodriguez and Wohlberg [147] [145] proposed an incremental PCP, which processes one frame at a time. Obtaining similar results to batch PCP algorithms, it has an extremely low memory footprint and a computational complexity that allows real-time processing. Furthermore, incPCP is also able to quickly adapt to changes in the background. A MATLAB-only implementation of this algorithm [146] running on a standard laptop (Intel i7- 2670QM quad-core, 6GB RAM, 2.2 GHz) can process color videos of size 640 and 1920 at a rate of 8 and 1.5 frames per second, respectively. On the same hardware, an ANSI-C implementation [146] can deliver a rate of 49.6 and 7.2 frames per second for grayscale videos of size 640 and 1920, respectively. Furthermore, Rodriguez and Wohlberg [148] developed a translational and rotational jitter invariant, incPCP.

1.3.4 Methods for Real-Time Implementation of PCP

Despite the efforts to reduce the time complexity, the corresponding algorithms have prohibitive computational time for real applications such as foreground detection. The main computation in PCP is the singular value decomposition of the large matrix A. Instead of computing a large SVD on the CPU, Anderson et al. [4] [5] proposed an implementation of the communication-avoiding QR (CAQR) factorization that runs entirely on the GPU. This implementation achieved $30\times$ speedup over the implementation on CPU using Intel Math Kernel Library (Intel MKL).

In another way, Pope et al. [127] proposed a variety of methods that significantly reduce the computational time of the ALM algorithm. These methods can be classified as follows:

- **Reducing the computation time of SVD:** The computation of the SVD is reduced using the Power method [127] that enables users to compute the singular values in a sequential manner, and to stop the procedure when a singular value is smaller than a threshold. The use of the Power method by itself results in $4.32\times$ lower runtime. Furthermore, the gain is improved by a factor of $2.02\times$ speedup if the procedure is stopped when the singular value is smaller than the threshold. If the rank of L is fixed and the Power SVD is stopped when the number of singular value is equal to $rank(L)$, the additional speedup is 17.35.
- **Seeding the PCP algorithm:** PCP operates on matrices consisting of blocks of contiguous frames acquired with a fixed camera. So, the low-rank matrix does not change significantly from one block to the next. Thus, Pope et al. [127] use the low-rank component obtained by the ALM algorithm from the previous block as a starting point for the next block. This method allows an additional speedup of 7.73.
- **Partioning into subproblems:** Pope et al. [127] proposed to partition the matrix A into P smaller submatrices. The idea is to combine the solutions of the P corresponding PCP subproblems to recover the solution of the full matrix A

at lower computational complexity.

In this way, Pope et al. [127] demonstrated that the PCP algorithm can be in fact suitable for real-time foreground/background separation for video-surveillance applications using off-the-shelf hardware.

In a similar manner, Guo et al. [55] proposed a low rank matrix recovery scheme, which splits the original RPCA into two small ones: a low-rank submatrix recovery and a low-rank submatrix reconstruction problem. This method showed a speedup of the ALM algorithm by more than 365 times compared to a C implementation with less requirements of both time and space. In addition, this method significantly cuts the computational load for decomposing the remaining frames.

1.3.5 Methods for Finding the Optimal PCP Solution

PCP recovers the true underlying low-rank matrix when a large portion of the measured matrix is either missing or arbitrarily corrupted. However, in the absence of a true underlying signal L and the deviation S, it is not clear how to choose a value of λ that produces a good approximation of the given data A for a given application. A typical approach would involve some cross-validation step to select λ to maximize the final results of the application. The issue with cross-validation in this situation is that the best model is selected indirectly in terms of the final results, which can depend in unexpected ways on later stages in the data processing chain of the application. Instead, Ramirez and Sapiro [140] [141] addressed this issue via the Minimum Description Length (MDL) principle [66] and so proposed an MDL-based low-rank model selection. The principle is to select the best low-rank approximation by means of a direct measure of the intrinsic ability of the resulting model to capture the desired regularity from the data. To obtain the family of models $\mathbf{M}$ corresponding to all possible low-rank approximation of A, Ramirez and Sapiro [140] [141] applied the RPCA decomposition for a decreasing sequence of values of λ, $\{\lambda_t : t = 1, 2, 3, ...\}$, obtaining a corresponding sequence of decomposition $\{(L_t, S_t), t = 1, 2, 3, ...\}$. This sequence is obtained via a simple modification of the ALM algorithm [91] to allow warm restarts; that is, where the initial ALM iterate for computing (L_t, S_t) is (L_{t-1}, S_{t-1}). Finally, Ramirez and Sapiro [140] [141] select the pair $(L_{\hat{t}}, S_{\hat{t}})$,$\hat{t} = \arg\min_t \{MDL(L_t) + MDL(S_t)\}$ where $MDL(L_t)+MDL(S_t) = MDL(A|M)$ denoted the description length in bits of A under the description provided by a given model $M \in \mathbf{M}$. Experimental results show that the best λ is not the one determined by the theory in Candes et al. [20].

Another approach was developed by Gao et al. [43] and consists of a two-pass RPCA process. The first-pass RPCA done on block resolution detects region with salient motion. Then, a saliency measure in each area is computed and permits users to adapt the value of λ following the motion in the second-pass RPCA. Experimental results show that this block-sparse RPCA outperforms the original PCP [20] and the ReProCS [130].

1.3.6 Modified PCP

In the literature, there are several modifications that concern the improvements of the original PCP, and they can be classified as follows:

1. **Fixed rank:** Leow et al. [86] proposed a fixed-rank algorithm for solving background recovering problems because low-rank matrices have known ranks. The decomposition involves the same model as PCP in Equation 1.11 but the minimization problem differs by the constraint as follows:

$$\min_{L,S} ||L||_* + \lambda||S||_{l_1} \quad \text{subj} \quad rank(L) = r \tag{1.17}$$

where r is the rank of the matrix L and r is known.

2. **Nuclear norm free:** Another variant of PCP was formulated by Yuan et al. [203] who proposed a nuclear-norm-free algorithm to avoid the SVD computation. The low-rank matrix is thus represented as $u1^T$ where $u \in \mathbf{R}^m$ and 1 denotes the vector $\mathbf{R}^n$. Accordingly, a noiseless decomposition is formulated as follows:

$$A = u1^T + S \tag{1.18}$$

Then, the corresponding minimization problem is the following one:

$$\min_{u} ||A - u1^T||_{l_1} \tag{1.19}$$

Note that the closed-form solution of Equation 1.19 is given by the median of the entries of the matrix A. In other words, the background is extracted as the median at each pixel location of all frames of a surveillance video. As no iteration is required at all to obtain the solution of 1.19, its computation for solving should be significantly cheaper than any iterative schemes for solving Equation 1.13 numerically. Furthermore, this model extracts the background more accurately than the original PCP. Moreover, Yuan et al. [203] developed a noise and a blur and noise nuclear-norm-free models of SPCP as well. In a similar way, Yang et al. [195] proposed a nonconvex model for background/foreground separation that can incorporate both the nuclear-norm-free model and the use of nonconvex regularizers.

3. **Capped norms:** In another way, Sun et al. [161] presented a nonconvex formulation using the capped norms for matrices and vectors, which are the surrogates of the rank function and the l_0-norm, and are called capped nuclear norm and the capped l_1-norm, respectively. The minimization problem is formulated as follows:

$$\min_{L,S} rank(L) + \lambda||S||_{l_0} \quad \text{subj} \quad ||A - L - S||_F^2 \leq \sigma^2 \tag{1.20}$$

where σ^2 is the level of Gaussian noise. The capped nuclear norm is then:

$$\frac{1}{\theta_1}\left[||L||_* + \sum_{i=1}^{p} max(\sigma_i(L) - \theta_1), 0)\right] \tag{1.21}$$

and the capped l_1-norm is formulated as follows:

$$\frac{1}{\theta_2}\left[||S||_{l_1} + \sum_{i=1}^{p} max(S_{ij}) - \theta_2), 0)\right] \tag{1.22}$$

for some small parameters θ_1 and θ_2. If all the singular values of L are greater than θ_1 and all the absolute values of elements in S are greater than θ_2, then the approximation will become equality. The smaller θ_1 and θ_2 are, the more accurate the capped norm approximation will be. The recovery precision is controled via

θ_1 and θ_2. By carefully choosing θ_1 and θ_2, L and S are more accurately determined than with the nuclear norm and the l_1-norm approximation. This capped formulation can be solved via two algorithms. One is based on the Difference of Convex functions (DC) framework and the other tries to solve the sub-problems via a greedy approach. Experimental results [161] show better performance for the capped formulation of PCP than the original PCP [20] and SPCP [221] on the I2R dataset [87].

4. **Inductive approach:** Bao et al. [13] proposed the following decomposition:

$$A = PA + S \tag{1.23}$$

where $P \in \mathbf{R}^{n \times n}$ is the low-rank projection matrix. The related optimization problem is formulated as follows:

$$\min_{P,S} ||P||_* + \lambda ||S||_{l_1} \quad \text{subj} \quad A - PA - S = 0 \tag{1.24}$$

This is solved by IALM [91]. Furthermore, Bao et al. [13] developed an inductive version that requires less computational cost in processing new samples.

5. **Partial Subspace Knowledge:** Zhan and Vaswani [205] proposed a modified-PCP with partial subspace knowledge. They supposed that a partial estimate of the column subspace of the low-rank matrix L is available. This information is used to improve the PCP solution, i.e., allow recovery under weaker assumptions. So, the modified PCP requires significantly weaker incoherence assumptions than PCP, when the available subspace knowledge is accurate. The corresponding optimization problem is written as follows:

$$\min_{L,S} ||L||_* + \lambda ||S||_{l_1} \quad \text{subj} \quad L + P_{\Gamma^\perp} S = P_{\Gamma^\perp} A \tag{1.25}$$

where $P_{\Gamma^\perp}$ is a projection matrix, Γ is a linear space of matrices with column span equal to that of the columns of S, and $\Gamma^\perp$ is the orthogonal complement. Zhan and Vaswani [205] applied with success their modified-PCP to the background-foreground separation problem, in which the subspace spanned by the background images is not fixed but changes over time and the changes are gradual.

6. **Schatten-p,l_q-PCP (p,q-PCP):** The norms introduced by Candes et al. [20] are not tight approximations, which may deviate the solution from the authentic one. Thus, Wang et al. [175] considered a non-convex relaxation, which consists of a Schatten-p norm and an l_q-norm with $0 < p, q \leq 1$ that strengthen the low rank and sparsity, respectively. The Schatten-p norm ($||.||_{S_p}$) is a popular non-convex surrogate of the rank function. Thus, the miminization problem is the following one:

$$\min_{L,S} ||L||_{S_p}^p + \lambda ||S||_{l_q} \quad \text{subj} \quad A - L - S = 0 \tag{1.26}$$

By replacing the Schatten-p norm and an l_q-norm by their expression, the minimization problem can be written as follows:

$$\min_{L,S} \lambda_1 \sum_{i=1}^{min(m,n)} (\sigma_i(L))^p + \lambda_2 \sum_{i=1}^{m} \sum_{j=1}^{n} |S_{ij}|^q \quad (1.27)$$

where $\sigma_i(L)$ denotes the ith singular values of L. When $p = q = 1$, p,q-PCP degenerates into the original convex PCP. Smaller values of p and q help p,q-PCP to well approximate the original formulation of RPCA. The solver used is a Proximal Iteratively Reweighted Algorithm (PIRA) based on the alternating direction method of multipliers, where in each iteration the underlying objective function is linearized to have a closed-form solution. Experimental results [175] on the I2R dataset [87] show better performance for p,q-PCP (in its stable formulation) in comparison to the original SPCP [221] and SPCP solved via NSA [8].

7. **Modified Schatten-p,l_q-PCP:** Shao et al. [154] proposed a similar approach to p,q-PCP [175] but they used the L_q-seminorm as a surrogate to the l_1-norm instead of the l_q-norm. Thus, the miminization problem is the following one:

$$\min_{L,S} ||L||_{S_p}^p + \lambda ||S||_{L_q}^q \quad \text{subj} \quad A - L - S = 0 \quad (1.28)$$

Furthermore, Shao et al. [154] used two different solvers based on the ALM and the APG methods as well as efficient root-finder strategies.

8. **2D-PCA:** To take into account the two-dimensional spatial information, Sun et al. [162] extracted a distinguished feature matrix for image representation, instead of matrix-to-vector conversion. Thus, the miminization problem is the following one:

$$\min_{U,V,S} \lambda ||S||_{l_0} + \frac{1}{2} ||A - U\Sigma V^T - S||_F^2 \quad \text{subj} \quad UU^T = I, VV^T = I \quad (1.29)$$

where $U\Sigma V^T = L$. As distinct from l_1-norm relaxation, Sun et al. [162] developed an iterative method to solve Equation (1.28) efficiently via alternating optimization, by a specific greedy algorithm for the l_0-norm regularization. So, a robust 2D-PCA model by sparse regularization is then solved via an alternating optimization algorithm. Results on dynamic backgrounds from the I2R dataset [87] show the effectiveness of the Robust 2D-PCA (R2DPCA), compared with the conventional 2D-PCA [83] and PCP solved via IALM [91].

9. **Rank-N Soft Constraint:** Oh [120] proposed an RPCA with Rank-N Soft Constraint (RNSC) based on the observation that the matrix A should be rank N without corruption and noise. So, the decomposition is formulated as estimating sparse error matrix and minimizing rank of low-rank matrix consisting of N principal components associated to the N largest singular largest values. Thus, the miminization problem with rank-N soft constraint is the following one:

$$\min_{L,S} \sum_{i=N+1}^{min(m,n)} ||\sigma_i(L)|| + \lambda ||S||_{l_1} \quad \text{subj} \quad A - L - S = 0 \quad (1.30)$$

where $\sigma_i(L)$ represents the ith singular value of the low-rank matrix L, and N is a contraint parameter for rank-N. Minimizing a partial sum of singular values can minimize the rank of the matrix L and satisfy rank-N constraint. Subsequently, Oh [120] applied the RPCA with Rank-1 Soft Constraint on the edge images for moving objects detection under global illumination changes.

10. **JVFSD-RPCA:** Wen et al. [182] reconstructed the input video data and aimed to make the foreground pixels not only sparse in space but also sparse in "time" by using a Joint Video Frame Set Division and RPCA-based (JVFSD-RPCA) method. In addition, they used the motion as a priori knowledge. The proposed method consists of two phases. In the first phase, a Lower Boundbased Within-Class Maximum Division (LBWCMD) method divided the video frame set into several subsets. In this way, the successive frames are assigned to different subsets in which the foregrounds are located at the scene randomly. In the second phase, each subset with the frames is increased with a small quantity of motion. This method shows robustness in the case of dynamic backgrounds.

11. **NSMP/WNSMP:** Wang and Feng [176] improved the RPCA method to find a new model to separate the background and foreground and reflect the correlation between them as well. For this, they proposed a "low-rank + dual" model and they used the reweighted dual function norm instead of the normal norms so as to get a better and faster model. So, the original minimization problem is improved by a nuclear norm and spectral norm minimization problem (NSMP). Thus, the minimization problem with the dual norm is the following one:

$$\min_{L,S} \ \lambda||L||_* + \mu||S||_2 \quad \text{subj} \quad A - L - S = 0 \tag{1.31}$$

where the spectral norm $||.||_2$ is the dual norm of the nuclear norm, and it corresponds to the largest singular value of the matrix [176]. As the nuclear norm regularized is not a perfect approximation of the rank function, Wang and Feng [176] proposed a weighted function nuclear norm and spectral norm minimization problem (WNSMP) with the corresponding minimization problem:

$$\min_{L,S} \ \lambda||\omega(L)||_* + \mu||\omega^{-1}(S)||_2 \quad \text{subj} \quad A - L - S = 0 \tag{1.32}$$

where $\omega()$ denotes the weighted function that directly adds the weights onto the singular values of the matrix, and, for any matrix X, weighted function norm is defined as follows: $||\omega(X)||_* = \sum_{i=1}^{min(m,n)} \omega_i \sigma_i(X)$ and $||\omega^{-1}(X)||_2 = \max_i \frac{1}{\omega_i}\sigma_i(X)$w where $\sigma_i(X)$ represents the ith singular value of the matrix X. Although this minimization problem with the weighted function nuclear norm is nonconvex, fortunately it has a closed-form solution due to the special choice of the value of weights, and it is also a better approximation of the rank function. NSMP and WNSMP show more robustness on the I2R dataset [87] than RPCA solved IALM [20] and GoDec [12].

12. **Implicit Regularizers:** He et al. [68] proposed a robust framework for low-rank matrix recovery via implicit regularizers of robust M-estimators (Huber, Welsch, l_1-l_2) and their minimizer functions. Based on the additive form of halfquadratic optimization, proximity operators of implicit regularizers are developed such that both low-rank structure and corrupted errors can be alternately recovered. Thus,

the minimization problem with implicit regularizers is formulated as follows:

$$\min_{L,S} \lambda||(L)||_* + \varphi(S) \quad \text{subj} \quad A - L - S = 0 \tag{1.33}$$

where the implicit regularizer $\varphi(y)$ is defined as the dual potential function of a robust loss function $\phi(x)$ where $\phi(x) = \min_y \frac{1}{2}||x - y||_2^2 + \varphi(x)$. If $\phi(x)$ is the Huber M-estimator, the implicit regularizer $\varphi(y)$ becomes $\mu\lambda||(L)||_{l_1}$. When the M-estimator $\phi(x)$ is the Welsch M-estimator, the minimization problem becomes the sample-based maximum correntropy problem. Compared with the mean square error, the model in Equation (1.31) is more robust to outliers due to M-estimation. Experimental results [68] on the I2R dataset [87] show that the Welsch M-estimator outperforms the Huber-estimator and the l_1-l_2-estimator.

1.4 Stable Principal Component Pursuit

PCP is limited to the low-rank component being exactly low-rank and the sparse component being exactly sparse, but the observations in real applications are often corrupted by noise affecting every entry of the data matrix. Therefore, Zhou et al. [221] proposed a stable PCP (SPCP) that guarantees stable and accurate recovery in the presence of entry-wise noise. In Chapter 2, Aybat reviewed this method and all these modifications in terms of decomposition, solvers, incremental algorithms, and real, time implementations.

1.5 Quantization-Based Principal Component Pursuit

Becker et al. [15] proposed an inequality-constrained version of RPCA proposed by Candes et al. [20] to take into account the quantization error of the pixel's value. Indeed, each pixel has a value between $0, 1, 2, \ldots, 255$. This value is the quantized version of the real value, which is between $[0, 255]$. So, the idea is to apply RPCA to the real observations instead of applying it to the quantized ones. Indeed, it is unlikely that the quantized observation can be split nicely into a low-rank and sparse component. So, Becker et al. [15] supposed that $L + S$ is not exactly equal to A, but rather that $L + S$ agrees with A up to the precision of the quantization. The quantization can induce at most an error of 0.5 in the pixel value. This measurement model assumes that the observation matrix A is represented as follows:

$$A = L + S + Q \tag{1.34}$$

where Q is the error of the quantization. Then, the objective function is the same as the equality version in Equation (1.13), but instead of the constraints $L+S = A$, the constraints are $||A - L - S||_{l_\infty} \leq 0.5$. So, the quantization-based PCP (QPCP) is formulated as follows:

$$\min_{L,S} ||L||_* + \lambda||S||_{l_1} \quad \text{subj} \quad ||A - L - S||_{l_\infty} \leq 0.5 \tag{1.35}$$

The l_∞-norm allows us to capture the quantization error of the observed value of the pixel.

Algorithms for solving QPCP: Becker et al. [15] used a general framework for solving this convex cone problem, called Templates for First-Order Conic Solvers (TFOCS). First, this approach determines a conic formulation of the problem and then its dual. Then, Becker et al. [15] applied smoothing and solved the problem using an optimal first-order method. This approach allows us to solve the problem in compressed sensing.

1.6 Block-Based Principal Component Pursuit

Tang and Nehorai [167] proposed a block-based PCP (BPCP) that enforces the low-rankedness of one part and the block sparsity of the other part. This decomposition involves the same model as PCP in Equation (1.11); that is, $A = L+S$, where L is the low-rank component but S is a block-sparse component. The low-rank matrix L and the block-sparsity matrix S can be recovered by the following optimization problem [218]:

$$\min_{L,S} \quad ||L||_* + \kappa(1-\lambda)||L||_{l_{2,1}} + \kappa\lambda||S||_{l_{2,1}}$$

$$\text{subj} \quad A - L - S = 0 \tag{1.36}$$

where $||.||_*$ and $||.||_{l_{2,1}}$ are the nuclear norm and the $l_{2,1}$-norm, respectively. The $l_{2,1}$-norm corresponds to the l^1-norm of the vector formed by taking the l^2-norms of the columns of the underlying matrix. The term $\kappa(1-\lambda)||L||_{l_{2,1}}$ ensures that the recovered matrix L has exact zero columns corresponding to the outliers. In order to eliminate ambiguity, the columns of the low-rank matrix L corresponding to the outlier columns are assumed to be zeros. BPCP was evaluated by Guyon et al. [64] [58] in the application of background/foreground separation.

Algorithm for solving BPCP: Tang and Nehorai [167] designed an efficient algorithm to solve the convex problem in Equation (1.36) based on the ALM method. This algorithm allows us to decompose the matrix A in low-rank and block-sparse matrices with respect to the $||.||_{l_{2,1}}$ and the extra term $\kappa(1-\lambda)||L||_{l_{2,1}}$.

1.7 Local Principal Component Pursuit

PCP is highly effective, but the underlying model is not appropriate when the data are not modeled well by a single low-dimensional subspace. Wohlberg et al. [183] proposed a decomposition corresponding to a more general underlying model consisting of a union of low-dimensional subspaces.

$$A = AU + S \tag{1.37}$$

This idea can be implemented as the following problem:

$$\min_{U,S} \ \alpha||U||_{l_1} + \beta||U||_{l_{2,1}} + \beta||S||_{l_1} \quad \text{subj} \quad A - AU - S = 0 \tag{1.38}$$

The explicit notion of low rank, and its nuclear norm proxy is replaced by representability of a matrix as a sparse representation on itself. The $l_{2,1}$-norm encourages rows of U to be zero, but does not discourage nonzero values among the entries of a nonzero row. The l_1-norm encourages zero values within each nonzero row of S.

To better handle noisy data, Wohlberg et al. [183] modified Equation (1.38) with a penalized form and added a Total Variation penalty on the sparse deviations for contiguous regions as follows:

$$\min_{U,S} \ \frac{1}{2}||A - DU - S||_{l_2}^2 + \alpha||U||_{l_1}$$

$$+\beta||U||_{l_{2,1}} + \beta||S||_{l_1} + \delta||grad(S)||_{l^1} \quad \text{subj} \quad A - DU - S = 0 \tag{1.39}$$

where the dictionary D is derived from the data A by mean subtraction and scaling, and $grad(S)$ is a vector valued discretization of the 3D gradient of S. An appropriate sparse U can be viewed as generating a locally low-dimensional approximation DU of $A - S$. When the dictionary is simply the data (i.e., $D = A$), the sparse deviations (or outliers) S are also the deviations of the dictionary D, so constructing the locally low-dimensional approximation as $(D-S)U$, implying an adaptive dictionary $D-S$, should allow U to be even sparser.

Algorithm for solving LPCP: Wohlberg et al. [183] proposed to solve Equation (1.38) using the Split Bregman Algorithm (SBA) [47]. Adding terms relaxing the equality constraints of each quantity and its auxiliary variable, Wohlberg et al. [183] introduced Bregman variables in Equation (1.38). So, the problem is split into an alternating minimization of five subproblems. Two subproblems are l_2 problems that are solved by techniques for solving linear systems such as conjugate gradient. The other three subproblems are solved very cheaply using shrinkage, i.e., generalized shrinkage and soft shrinkage.

1.8 Outlier Pursuit

Xu et al. [188] proposed a robust PCA via Outlier Pursuit (OP) to obtain a robust decomposition when the outliers corrupted entire columns; that is, every entry is corrupted in some columns. This method involves the nuclear norm minimization and recovers the correct column space of the uncorrupted matrix, rather than the exact matrix itself. The decomposition involves the same model as PCP in Equation (1.11), that is, $A = L + S$. A straightforward formulation is to use l_0-norm to minimize the energy function:

$$\min_{L,S} \ rank(L) + \lambda ||S||_{0,c} \quad \text{subj} \quad A - L - S = 0 \tag{1.40}$$

where $||S||_{0,c}$ stands for the number of nonzero columns of a matrix, and it is equivalent to $||S||_{l_{2,0}}$ which corresponds to the number of non-zero columns as well [210]. $\lambda > 0$ is an arbitrary balanced parameter. But because this problem is NP-hard, typical solution: might involve a search with combinatorial complexity. This research seeks to solve for L with the following optimization problem:

$$\min_{L,S} \ ||L||_* + \lambda ||S||_{l_{1,2}} \quad \text{subj} \quad A - L - S = 0 \tag{1.41}$$

where $||.||_*$ and $||.||_{1,2}$ are the nuclear norm and the $l_{1,2}$-norm, respectively. The $l_{1,2}$-norm corresponds to the l_2-norm of the vector formed by taking the l_1-norms of the columns of the underlying matrix. $\lambda > 0$ is an arbitrary balanced parameter. Adapting the OP algorithm to the noisy case, that is, $A = L + S + E$, Xu et al. [188] proposed a robust PCA via Stable Outlier Pursuit (SOP):

$$\min_{L,S} \ ||L||_* + \lambda ||S||_{l_{1,2}} \quad \text{subj} \quad ||A - L - S)||_F < \delta \tag{1.42}$$

where S is supported on at most γn columns and $\lambda = \frac{3}{7\sqrt{\gamma n}}$.

Algorithm for solving OP and SOP: Xu et al. [188] used the Singular Value Threshold (SVT) algorithm to solve these two minimization problems.

1.9 Sparsity Control

Mateos and Giannakis [112] [113] proposed a robust PCA using a bilinear decomposition with Sparsity Control (RPCA-SpaCtrl). The decomposition involves the following model:

$$A = M + PU^T + S + E \tag{1.43}$$

where M is the mean matrix, the matrix U has orthogonal columns, P are the principal components matrix, S is the outliers' matrix, and E is a zero-mean matrix. The percentage of outliers determines the degree of sparsity in S. The criterion for controlling outlier sparsity is to seek the relaxed estimation:

$$\min_{U,S} ||X + 1_N M^T - PU^T - S||_F^2 + \lambda ||S||_{l_{2(r)}} \quad \text{subj} \quad UU^T = I_q \tag{1.44}$$

where $||S||_{l_{2(r)}} = \sum_{i=1}^{m \times n} ||S_i||_{l^2}$ is the row-wise l^2-norm. The non-differentiable l^2-norm regularization term controls rows-wise sparsity on the estimator of S. The sparsity is then also controlled by the parameter λ. To optimize Equation (1.44), Mateos and Giannakis [112] [113] used an alternating minimization algorithm [224].

Algorithm for incremental SpaCtrl: An incremental version of Equation (1.44) is obtained using the Exponentially Weighted Least Squares (EWLS) estimator as follows:

$$\min_{U,S} \sum_{i=1}^{m \times n} \beta^{(m \times n)-i} \left[||X_n + m - U^T T_i - S_i||_{l_2}^2 + \lambda ||S_i||_{l_2} \right] \tag{1.45}$$

where β is a learning rate between 0 and 1. So, the entire history of data is incorporated in the online estimation process. Whenever $\beta < 1$, past data are exponentially discarded, thus enabling operation in nonstationary backgrounds. Towards deriving a real-time, computationally efficient, and recursive solver of Equation (1.45), an AM scheme is adopted in which iteration k coincides with the time scale $i = 1, 2, ...$ of data acquisition. Experimental results [113] show that RPCA-SpaCtrl with $\lambda = 9.69 \times 10^{-4}$ presents better performance than the naive PCA [122] and RSL [169] with less time computation.

1.10 Sparse Corruption

Even if the matrix A is exactly the sum of a sparse matrix S and a low-rank matrix L, it may be impossible to identify these components from the sum. For example, the sparse matrix S may be low-rank, or the low-rank matrix L may be sparse. So, Hsu et al. [70] imposed conditions on the sparse and low-rank components in order to guarantee their identifiability from A. This method requires that S not be too dense in any single row or column, and that the singular vectors of L not be too sparse. The level of denseness and sparseness are considered jointly in the conditions in order to obtain the weakest possible conditions. This decomposition RPCA with Sparse Corruption (RPCA-SpaCorr) involves the same model as PCP in Equation (1.11), that is, $A = L + S$. Then, Hsu et al. [70] proposed two convex optimizations. The first is the following constraints formulation:

$$\min_{L,S} ||L||_* + \lambda ||S||_{l_1} \quad \text{subj} \quad ||A - L - S||_{l_1} \leq \epsilon_1$$
$$\text{and} \quad ||A - L - S||_* \leq \epsilon_* \tag{1.46}$$

where $\lambda > 0$, $\epsilon_1 \geq 0$ and $\epsilon_* \geq 0$. The second is the regularized formulation:

$$\min_{L,S} ||L||_* + \lambda ||S||_{l_1} + \frac{1}{2\mu} ||A - L - S||_{l_2}^2$$
$$\text{subj} \quad ||A - L - S||_{l_1} \leq \epsilon_1 \quad \text{and} \quad ||A - L - S||_* \leq \epsilon_* \tag{1.47}$$

where $\mu > 0$ is the regularization parameter. Hsu et al. [70] added a constraint to control $||L||_{l_\infty}$. That is, $||L||_\infty \leq b$ is added in Equation (1.46) and $||A - S||_{l_\infty} \leq b$ is added in Equation (1.47). The parameter b is a natural bound for L and is typically 510 for image processing. Hsu et al. [70] determined two identifiability conditions that guarantee the recovery. The first measures the maximum number of non-zero entries in any row or column of S. The second one measures the sparseness of the singular vectors L. Then, a mild strengthening of these measures is achieved for the recovery guarantees.

1.11 Log-Sum Heuristic Recovery

When the matrix has high intrinsic rank structure or the corrupted errors become dense, the convex approaches may not achieve good performances. Thus, Deng et al. [35] used the Log-sum Heuristic Recovery (LHR) to learn the low-rank structure. The decomposition involves the same model as PCP in Equation (1.11), that is, $A = L + S$. Although the objective in Equation (1.13) involves the nuclear norm and the l_1-norm, it is based on the l_1 heuristic since the nuclear norm can be regarded as a specific case of l_1-norm [35]. Replacing the nuclear norm by its l_1-norm formulation, the problem can be solved as follows:

$$\min_{\hat{X} \in \hat{D}} \frac{1}{2}(||diag(Y)||_{l_1} + ||diag(Z)||_{l_1}) + \lambda ||E||_{l_1} \tag{1.48}$$

where $\hat{X} = \{Y, Z, L, S\}$ and

$$\hat{D} = \left\{ (Y, Z, L, S) : \begin{pmatrix} Y & L \\ L^T & Z \end{pmatrix} \geq 0, (L, S) \in C \right\}$$

$(L, S) \in C$ stands for convex constraint. Y and Z are both symmetric and positive definite. $\geq$ represents semi-positive definite. The convex problem with two norms in Equation (1.13) has been converted to an optimization only with l_1-norm and therefore it is called l_1-heuristic. Next, Deng et al. [35] used the logsum term to represent the sparsity of signals and obtained the Log-sum Heuristic Recovery (LHR) model:

$$\min_{\hat{X} \in \hat{D}} \frac{1}{2}(||diag(Y)||_L + ||diag(Z)||_L) + \lambda ||E||_L \tag{1.49}$$

where $||X||_L = \sum_{ij} log\ (|X_ij| + \delta)$ with $\delta > 0$ is a small regularization constant. This model is non-convex but the convex upper bound can be easily defined. LHR can remove much denser errors from the corrupted matrix as compared to PCP.

Algorithm for solving LHR: Deng et al. [35] used the majorization-minimization (MM) [38] [85] algorithm that replaces the hard problem by a sequence of easier ones. It proceeds in an Expectation Maximization (EM)-like fashion by repeating two steps of majorization and minimization in an iterative way. During the majorization step, it constructs the convex upper bound of the non-convex objective. In the minimization step, it minimizes the upper bound.

1.12 Iteratively Reweighted Least Squares Minimization

Guyon et al. [63] proposed the decomposition solved via IRLS with the following model:

$$A = L + S = UV + S \tag{1.50}$$

where U is a low-rank matrix corresponding to the background model plus noise, and V reconstructs L by linear combination. S corresponds to the moving objects. The model involves the error reconstruction determined by the following constraints:

$$\min_{U \in \mathbb{R}^{n \times p}, V \in \mathbb{R}^{p \times m}} \mu ||UV||_* + ||(A - UV) \circ W_1||_{l_{\alpha,\beta}} \quad (1.51)$$

where $||.||_*$ denote the nuclear norm and $||.||_{l_{\alpha,\beta}}$ is a mixed norm. W_1, which is a weighted matrix, is iteratively computed and aims to enforce the fit exclusively on a guessed background region. A function $\Phi(.)$ smoothes the error like spatial median filtering and transforms the error to obtain a suitable weighted mask for regression:

$$W = \Phi(A - UV), \Phi(x) = e^{-\gamma TV(A-UV)} \quad (1.52)$$

By including local penalty as a constraint in RPCA, it explicitly increases local coherence of the sparse component as filled/plain shapes for moving objects. Furthermore, the decomposition is split into two parts. The first part tracks 1-Rank decomposition since the first eigen-vector is strongly dominant in video surveillance. For the mixed norm, Guyon et al. [63] used $||.||_{l_{2,1}}$ instead of the usual $||.||_{l_{1,1}}$ because it forces spatial homogeneous fitting. Thus, the SVD algorithm can be seen as an iterative regression and then the IRLS algorithm is used. So, Guyon et al. [63] increased local coherence of the error for moving objects by including local penalty as a constraint in the decomposition. Using the same approach, Guyon et al. [61] added spatial constraint in the minimization based on the gradient, and Guyon et al. [60] proposed a spatio-temporal version. Another variant of RPCA via IRLS has been developed by Lu et al [105].

1.13 Stochastic Optimization

Feng et al. [39] proposed an Online Robust PCA (OR-PCA) algorithm. The main idea is to develop a stochastic optimization algorithm to minimize the empirical cost function, which processes one sample per time instance in an online manner. The coefficients that correspond to noise and the basis are optimized in an alternative manner. The coefficients that correspond to noise and the basis are optimized in an alternative manner. The low-dimensional subspace, called a low-rank matric basis, is first initialized randomly and then updated after every frame per time instance. Moreover, OR-PCA decomposes the nuclear norm of the objective function of the traditional PCP algorithms into an explicit product of two low-rank matrices, i.e., basis and coefficients. The main function in OR-PCA is formulated as:

$$\min_{L \in \mathbb{R}^{n \times p}, R \in \mathbb{R}^{n \times r}} \frac{1}{2} ||A - LR^T - S||_F^2 + \frac{\lambda_1}{2} (||L||_F^2 + ||R||_F^2) + \lambda_2 ||S||_{l_1} \quad (1.53)$$

where R is a coefficient matrix. λ_1 controls the basis and coefficients for the low-rank matrix, whereas λ_2 controls the sparsity pattern, which can be tuned according to video analysis. In addition, basis and coefficient depend on the value of rank. In the case of video background modeling, no visual results [39] have been found using this technique. Therefore, Javed et. al [75] modified OR-PCA via the stochastic optimization method for background subtraction applications. An initialization scheme is adopted, which converges the algorithm very fast as compared to the original OR-PCA.

In order to perform OR-PCA, a number of video frames are first initialized as a low-dimensional basis, then stochastic optimization is performed on each input frame to separate the low-rank and sparse component. As compared to conventional RPCA via PCP-based schemes, no batch optimizations are needed, therefore OR-PCA is applicable for real-time processing. In addition, global pre-processing steps such as Laplacian and Gaussian images

TABLE 1.13 Bayesian RPCA: A complete overview.

Bayesian RPCA	Categories	Authors - Dates
Decompositions	1) Original BRPCA	Ding et al. (2011) [36]
	2) Variational BRPCA (VBRPCA)	Babacan et al. (2011) [12]
	3) Factorized Variational BRPCA (FVBRPCA)	Aicher (2013) [2]
	4) MOG-RPCA	Zhao et al. (2014) [216]
Solvers	Markov chain Monte Carlo (MCMC)	Robert and Cassela (2004) [142]
	Variational Bayesian Inference (VB)	Beal (2003) [14]
	Approximate Bayesian Inference (AB)	Beal (2003) [14]

are introduced in modified OR-PCA, which increases the detection rate. Using these modifications in the original scheme, both memory cost and computational time are decreased, since the idea is based on to process one single frame per time instance, but the method shows some weak performance when large variations in the background scenes occuring such as waving trees and moving water surface.

Therefore, Javed et al. [77] further improved the foreground segmentation using the continuous constraints, such as Markov Random Field (MRF). OR-PCA via image decomposition using an initialization scheme including continuous MRF with tuned parameters shows a drastic improvement in the experimental results, especially in case the highly dynamic backgrounds. In this work, a good parameter range is provided according to different background scenarios. A huge amount of experimental results [77] are provided, which shows a very nice potential for its real-time applicability. This scheme was improved with dynamic feature selection [76] and a depth-extended version with spatiotemporal constraints [73].

1.14 Bayesian RPCA

Bayesian Robust Principal Component Analysis approaches have also been investigated for RPCA and used a Bayesian framework in the decomposition into low-rank and sparse matrices. Ding et al. [36] modeled the singular values of L and the entries of S with beta-Bernoulli priors, and used a Markov chain Monte Carlo (MCMC) sampling scheme to perform inference. This method, called Bayesian RPCA (BRPCA), needs many sampling iterations, always hampering its practical use. In a similar approach, Babacan et al. [12] adopted the automatic relevance determination (ARD) approach to model both L and S, and utilized the Variational Bayes (VB) method to do inference. This method, called Variational Bayesian RPCA (VBRPCA), is more computationally efficient. However, these three methods assume a certain noise prior (a sparse noise plus a dense noise), which cannot always effectively model the diverse types of noises occurring in practice. To address this problem, Zhao et al. [216] proposed a generative RPCA model under the Bayesian framework by modeling data noise as a mixture of Gaussians (MoG). Table 1.13 shows an overview of the Bayesian Robust Principal Component Analysis methods.

1.14.1 Bayesian RPCA

Ding et al. [36] proposed a Bayesian Robust PCA (BRPCA). Assuming that the observed data matrix A can be decomposed in three matrices as in SPCP [221], the Bayesian model is then as follows:

$$A = D(ZG)W_2 + B \times X + E \tag{1.54}$$

where $D \in \mathbf{R}^{n \times r}$, $W_2 \in \mathbf{R}^{r \times m}$, and $G \in \mathbf{R}^{r \times r}$ are diagonal matrices and $X \in \mathbf{R}^{n \times m}$. The diagonal matrix Z has binary entries along the diagonal, and the binary matrix $B \in \{0, 1^{n \times m}\}$ is sparse. r defines the largest possible rank that may be inferred for L, and r is set to a large value. The low-rank, sparse, and noise component are obtained as follows.

- **Low-rank Component:** The low-rank component is modeled as $L = D(ZG)W_2$. This is similar to SVD except for the extra diagonal matrix Z with diagonal elements $z_{k,k} \in 0, 1$ for $k = 1, ..., r$. The product ZG is a diagonal matrix, too. The use of Z decouples the rank learning and the singular value learning. r is chosen large and then the diagonal entries of ZG are sparse. The binary diagonal matrix Z is modeled as follows:

$$z_{k,k} \sim Bernoulli(p_k) \tag{1.55}$$

$$p_k \sim Beta(\alpha_0, \beta_0), k = 1, ...r \tag{1.56}$$

with $\alpha_0 > 0$ and $\beta_0 > 0$. The parameters α_0 and β_0 imposed the sparseness of the diagonal of Z. α_0 and β_0 are set, respectively, to $1/K$ and $(K-1)/K$. Each diagonal entry in G, denoted as $g_{k,k}$ for $k = 1, ..., r$, is obtained from a normal-gamma distribution:

$$g_{k,k} \sim \mathcal{N}(0, \tau^{-1})k = 1, ...r. \tag{1.57}$$

$$\tau \sim Gamma(a_0, b_0) \tag{1.58}$$

with $a_0 > 0$ and $b_0 > 0$. a_0 and b_0 are set to 10^{-7}. The column of matrices D and W_2 are obtained from normal distribution:

$$d_k \sim \mathcal{N}(0, (1/N)I_N)k = 1, ...K. \tag{1.59}$$

$$w_{2,m} \sim \mathcal{N}(0, (1/K)I_K)m = 1, ...M. \tag{1.60}$$

with I_N as the $N \times N$ identity matrix. The decomposition can be rewritten as follows:

$$l_m = D(ZG)w_m = \sum_{k=1}^{K} z_{k,k} g_{k,k} w_{k,m} d_k, m = 1, ..., M \tag{1.61}$$

So, each column of L is the weighted sum of the dictionary elements in D, and K is the size of the dictionary. The weights $z_{k,k}g_{k,k}{}_{k=1:K}$ determine the dictionary elements that are active to construct L. The weights $w_{k,m}{}_{k=1:K}$ determine the importance of the selected dictionary elements for the representation of the mth column of L.

- **Sparse Component:** The sparse component is modeled as $S = B \times X$, where B is a binary matrix. This decomposition separates the learning sparseness from the learning of values. Each column of B is modeled as follows:

$$b_m \sim \prod_{n=1}^{N} Bernoulli(\pi_n), m = 1, ..., M \tag{1.62}$$

$$\pi_n \sim Beta(\alpha_1, \beta_1), n = 1, ...N. \tag{1.63}$$

The sparseness prior is made with the parameters α_1 and β_1. α_1 and β_1 are set, respectively, to $1/N$ and $(N-1)/N$. The columns of X are obtained from a normal-gamma distribution:

$$x_m \sim \mathcal{N}(0, v^{-1}I_N), m = 1, ..., M \tag{1.64}$$

$$v \sim Gamma(c_0, d_0). \tag{1.65}$$

with $c_0 > 0$ and $d_0 > 0$. c_0 and d_0 are set to 10^{-6}. Ding et al. [36] address the dependency of the sparse component in time and space with a Markov structure.

If the parent node of $I_t(i,j)$, noted $I_{t-1}(i,j)$, is non-zero, its child node is also non-zero with a high probability. To introduce spatial dependence, Ding et al. [36] define the state of $F_t(i,j)$ as follows:

$$S(F_t(i,j)) = active \;\; if \;\; ||N(F_t(i,j)||_0 \geq \rho \tag{1.66}$$

$$S(F_t(i,j)) = inactive \;\; otherwise \tag{1.67}$$

where $\rho = 5$, which imposes that a node is active if the sparse component contains at least 5 non-zero members in its neighborhood defined by $N(F_t(i,j)) = \}F_(k,l) : |k-i| \leq 1, |l-j| \leq 1\}$. Then, a child node depends on its parent node in time and on its neighbors in space. Markov dependency is then imposed by modifying Equations (1.62) and (1.63) as follows.

$$b_t \sim \prod_{n=1}^{N} Bernoulli(\pi_{nt}), t = 1, ..., M \tag{1.68}$$

$$\pi_{nt} \sim Beta(\alpha_H, \beta_H) \;\; if \;\; S(b_{n,t-1}) = active$$
$$\text{with } \; n = 1, ...N, t = 2, ..., M. \tag{1.69}$$

$$\pi_{nt} \sim Beta(\alpha_L, \beta_L) \;\; if \;\; S(b_{n,t-1}) = inactive \tag{1.70}$$
$$\text{with } \; n = 1, ...N, t = 2, ..., M.$$

where H and L indicate the high and low states in the Markov model and α_H,α_H, β_L, and β_H are set to assume that the sparseness will be propagated along time with high probability. For $t = 1$, Equations (1.62) and (1.63) are used, since there are no parent nodes for the first frame.

- **Noise Component:** The noise is modeled by a Gaussian distribution as follows:

$$e_{n,m} \sim \mathcal{N}(0, \gamma_m^{-1}), \;\; \text{with } \; n = 1, ..., N \tag{1.71}$$

$$\gamma_m \sim Gamma(e_0, f_0) \;\; \text{for } \; m = 1, ..., M, \tag{1.72}$$

with $e_{n,m}$ as the entry at row n and column m of E. c_0 and d_0 are set to 10^{-6}.

Then, the posterior density function of the BRPCA is as follows:

$$\begin{aligned} -log\ (p(\Theta|A,H)) &= \frac{\tau}{2}||G||_F^2 - log\ ([f_{BB}(Z;H))] \\ &+\frac{N}{2}\sum_{k=1}^{r}||d_k||_{l^2}^2 + \frac{1}{2}\sum_{m=1}^{M}||w_m||_{l^2}^2 + \frac{v}{2}||X||_F^2 \\ &-log\ [f_{BB}(B;H)] + \frac{1}{2}||Y-L-S||_F^2 \\ &-log\ [Gamma(\tau|H)Gamma(v|H)Gamma(\gamma|H)] \\ &+constant \end{aligned} \tag{1.73}$$

where Θ represents all model parameters, $f_{BB}(.|H)$ represents the beta-Bernoulli prior, and $H = \{\alpha_0, \alpha_1, \beta_0, \beta_1, a_0, b_0, c_0, d_0, e_0, f_0\}$ are model hyper-parameters.

Algorithms for solving BRPCA: Ding et al. [36] proposed approximating the posterior density function in Equation (1.73) with two algorithms:

- **Markov chain Monte Carlo (MCMC) analysis implemented with Gibbs sampler [142]:** The posterior distribution is approximated by a set of samples, collected by iteratively drawing each random variable from its conditional posterior distribution, given the most recent values of all the other parameters.

- **Variational Bayesian inference (VB) [14]:** A set of distributions $q(\Theta)$ allow us to approximate the true posterior distributions $p(\Theta|A)$, and uses a lower bound to approximate the true log-likelihood of the model $log\ (p(A|\Theta)$. The algorithm iteratively updates $q(\Theta)$ so that the lower bound approaches $log\ (p(A|\Theta)$.

The computational complexity of MCMC and VB iteration is approximatively the same. The VB solution may find a local-optimal solution that may be not be the global-optimal best solution. Ding et al. [36] found that MCMC works quite effectively in practice.

Relation to PCP and SPCP: For the low-rank component, Ding et al. [36] employed a Gaussian prior to obtain a constraint on Frobenius norm $||G||_F^2$ with a beta-Bernoulli distribution, to address the sparseness of singular value and to obtain a small number of non-zero singular values, while PCP employs an l_1-norm that is relaxed to an l_1-norm when solving the problem in a convex way. For the sparse component, the constraint on the Frobenius matrix norm $||X||_F^2$ and the beta-Bernoulli distribution are used to impose sparseness instead of the l_1-norm in PCP. The error term $(2\mu)^{-1}||A-L-S||_F^2$ in SPCP [184] corresponds to the Gaussian prior placed on the measurement noise in Equation (1.71). For solving the problem, the main difference is that BRPCA uses numerical methods to estimate a distribution for the unknown parameters, whereas optimization-based methods effectively search a single solution that minimizes a functional analogous to $-log\ (p(\Theta|A,H))$.

1.14.2 Variational Bayesian Robust Principal Component Analysis

Babacan et al. [12] proposed a Variational Bayesian Robust PCA (VBRPCA). Assuming that the observed data matrix A can be decomposed in three matrices as in SPCP [221], the variational Bayesian model is then as follows:

$$A = DB^T + S + E \tag{1.74}$$

where DB^T is the low-rank component with $D \in \mathbf{m} \times \mathbf{r}$ and $B \in \mathbf{r} \times \mathbf{n}$, S is the sparse component with arbitrarily large coefficients, and E is the dense error matrix with relatively smaller coefficients. The low-rank, sparse, and noise component are obtained as follows. The low-rank component L is then given by DB^T. So, L is the sum of outer-products of the columns of D and B, that is,

$$L = \sum_{i=1}^{k} d_{.i} b_{.i}^T \tag{1.75}$$

where $k \geq r$. $d_{.i}$ and $d_{i.}$ denote the ith column and row of D, respectively. To impose column sparsity in D and B, such that most columns in D and B are set equal to zero, the columns are defined with Gaussians priors as follows:

$$p(D|\gamma) = \prod_{i=1}^{k} \mathcal{N}(d_{.i}|0, \sigma_i I) \tag{1.76}$$

$$p(B|\gamma) = \prod_{i=1}^{k} \mathcal{N}(b_{.i}|0, \sigma_i I) \tag{1.77}$$

where σ_i is the variance. Most of the variances are very small values during inference to reduce the rank of the estimate. Then, the following conditional distribution for the observations are obtained:

$$p(A|D, B, S, \beta) = \mathcal{N}(A|DB^T + S, \gamma^{-1}I) \tag{1.78}$$

$$= exp[\frac{\beta}{2}||A - DB^T - S||_F^2 \tag{1.79}$$

where β is a uniform hyperprior. The modeling of the sparse component S is done by using independent Gaussian priors on its coefficients S_{ij} as follows:

$$p(E|\alpha) = \prod_{i=1}^{m}\prod_{j=1}^{n} \mathcal{N}(S_{ij}|0, \alpha_{ij}^{-1}) \tag{1.80}$$

where $\alpha = \{\alpha_{ij}\}$, α_{ij} is the precision of the Gaussian on the $(i, j)^{th}$ coefficient and $p(\alpha_{ij})$=const $\forall i, j$. Finally, the joint distribution is expressed as follows:

$$p(A, D, B, S, \gamma, \alpha, \beta)$$

$$= p(A|D, B, S, \beta)p(D|\gamma)p(B|\gamma)p(S|\alpha)p(\gamma)p(\alpha)p(\beta) \tag{1.81}$$

where $p(\gamma_i) = \frac{1}{\gamma_i}^{a+1} exp(\frac{-b}{\gamma_i})$ and $p(\beta)$ is a constant, assuming that the noise precision has a uniform prior.

Algorithm for solving VBRPCA: The exact full-Bayesian inference using joint distributions in Equation (1.81) is intractable because $p(y)$ can't be computed by marginalizing all variables. Therefore, Babacan et al. [12] used an inference procedure based on mean-field variational Bayes. The aim is to compute posterior distribution approximations by minimizing the Kullback-Leibler divergence in an alternating way for each variable. Letting $z = (D, D, S, \gamma, \alpha, \beta)$, the posterior approximation $q(z_k)$ of each variable $z_k \in z$ is then determined as follows:

$$log\ (q(z_k)) = \langle log\ (p(A, z))\rangle_{\frac{z}{z_k}} + const \tag{1.82}$$

where $\frac{z}{z_k}$ is the set z without z_k. The distribution $p(A, z)$ is the joint probability distribution given in Equation (1.81). The posterior factorization $q(z) = \prod q(z_k)$ is used such that the posterior distribution of each unknown is estimated by holding the others fixed using their most recent distributions. Thus, the expectations of all parameters in the joint distribution are taken with respect to their most recent distributions, and the result is normalized to find the approximate posterior distribution. Since all distributions are in the conjugate exponential family, the form of each posterior approximation is easily determined.

1.14.3 Factorized Variational Bayesian RPCA

Aicher [2] proposed a Factorized Variational Bayesian RPCA. This model is slightly different from BRPCA [36] and VBRPCA [12] in how sparse noise is modeled and incorporated, as well as in the use of variational Bayes instead of MCMC.

$$A = UV^T + Z^* \circ B + E \tag{1.83}$$

where $\circ$ denotes the Hadamard element-wise multiplication. The low-rank matrix is $L = UV^T$ and U is restricted to be an $n \times r$ matrix and V to be an $r \times m$ matrix so that the rank of L less than or equal to r. The sparse matrix is $S = Z^* \circ B$, while B is set to be a sparse binary matrix, and Z^* is without constraint. For numerical reasons, Z^* is treated as a very diffuse Gaussian matrix. To induce sparsity in S, a prior on B is selected such that it is sparse. E is a small Gaussian noise term and the prior on its variance is small compared with the variance of Z^*. Instead of solving Equation 1.83, it is more numerically convenient to solve the following problem:

$$A = UV^T + Z^* \circ B + E \circ (1 - B) \tag{1.84}$$

To infer U,V,B,Z, and E, Aicher [2] approximated the posterior distribution with a factorizable distribution. This variational approach selects the distribution q closest to the posterior in the sense of Kullback-Leibler (KL) divergence. By parameterizing q, Aicher [2] converted the inference scheme back into an objective maximization problem. After selecting a distribution to approximate the posterior, the expectations of U,V,B,Z, and E are taken to estimate them. Experimental results [2] show that FVBRPCA performs slightly better than RPCA solved via IALM [91], VBRPCA [12], and GoDec [218].

1.14.4 Bayesian RPCA with MoG Noise

Zhao et al. [216] developed a generative RPCA model under the Bayesian framework by modeling data noise as a mixture of Gaussians (MoG). The MoG is a universal estimator to continuous distributions and thus MoG-BRPCA is able to fit a wide range of noises such as Laplacian, Gaussian, sparse noises, and any combinations of them.

1.15 Approximated RPCA

1.15.1 GoDec

Zhou and Tao [218] proposed a randomized low-rank and sparse-matrix decomposition called "Go Decomposition" (GoDec). GoDec estimates the low-rank part L and the sparse part S by using the same decomposition as SPCP [221]:

$$A = L + S + E \tag{1.85}$$

To solve the problem in Equation (1.85), GoDec alternatively assigns the low-rank approximation of $A - S$ of L and the sparse approximation to $A - L$ to S. This approximated decomposition problem seeks to solve the minimization of the following decomposition error:

$$\min_{L,S} \; ||A - L - S||_F^2 \quad \text{subj} \quad rank(L) \leq r, card(S) \leq k. \tag{1.86}$$

Algorithm for solving GoDec: The optimization problem in Equation (1.86) is solved by alternatively solving the two following subproblems:

$$L_t = arg \min_{rank(L) \leq e} \; ||A - L - S_{t-1}||_F^2 \tag{1.87}$$

$$S_t = arg \min_{card(S) \leq e} \; ||A - L_t - S||_F^2 \tag{1.88}$$

Although both subproblems have nonconvex constraints, their global solutions L_t and S_t exist. Indeed, these subproblems can be solved by updating L_t via singular-value hard thresholding of $A - S_{t-1}$, and updating S_t via entry-wise hard thresholding of $A - L_t$, respectively, as follows:

$$L_t = \sum_{i=1}^{r} \lambda_i U_i V_i^T \;\; with \;\; SVD(A - S_{t-1}) = UGV^T \tag{1.89}$$

$$S_t = P_\Omega(A - L_t) \text{ with } \Omega : |(A - L_t)_{i,j\in\Omega}| \neq 0$$
$$\text{and } \geq \left|(A - L_t)_{i,j\in\bar{\Omega}}\right|, |\Omega| \geq k \quad (1.90)$$

where $P_\Omega(.)$ is defined as the projection of the matrix on the observed entries following the sampling set Ω. The main computation time is due to the computation of the SVD for $A - S_{t-1}$ in the updating L_t sequence. To significantly reduce the time cost, Zhou and Tao [218] replaced the SVD with a Bilateral Random Projection (BRP)-based low-rank approximation.

1.15.2 Semi-Soft GoDec

Zhou and Tao [218] proposed a Semi-Soft GoDec, which adopts soft thresholding to the entries of S, instead of GoDec, which imposes hard thresholding to both the singular values of the low-rank part L and the entries of the sparse part S. This improvement has two main advantages: 1) the parameter k in constraint $card(S) \leq k$ is automatically determined by a soft-threshold τ, thus avoiding the situation when k is chosen too large and some part of noise E is leaked into S; 2) the time cost is substantially smaller than the ordinary GoDec. For example, the background modeling experiments can be accomplished with a speed 4 times faster than ordinary GoDec, while the error is kept the same or even smaller. The approximated decomposition problem seeks to solve the minimization of the following decomposition error:

$$\min_{L,S} ||A - L - S||_F^2 \quad \text{subj} \quad rank(L) \leq e, card(S) \leq \tau. \quad (1.91)$$

where τ is the soft threshold. Chen et al. [28] proposed to use Semi-Soft GoDec for video coding in the existing standard codecs H.264/AVC and HEVC via background/foreground separation. For this, Chen et al. [28] developed an extension of the Semi-Soft GoDec that is able to perform LRSD on new matrix columns with a given low-rank structure, which is called incremental low-rank and sparse decomposition (ILRSD).

1.16 Sparse Additive Matrix Factorization

Nakajima et al. [117] [118] extented the original robust PCA [20] by proposing a unified framework called Sparse Additive Matrix Factorization (SAMF). Instead of RPCA, which only copes with element-wise sparsity (spiky noise) and low-rank sparsity (low-dimensional matrix), SAMF handles various types of sparse noise, such as row-wise and column-wise sparsity. Thus, the decomposition is written as follows:

$$A = \sum_{k=0}^{K} S + E \quad (1.92)$$

where K is the number of sparse matrices. $K = 2$ in the original RPCA [20] in which the element-wise sparse term is added to the low-rank term. For background/foreground separation, the low-rank term and the element-wise sparse term capture the static background and the moving foreground, respectively. Nakajima et al. [117] [118] relied on the natural assumption that a pixel segment with similar intensity values in an image tends to belong to the same object. Thus, Nakajima et al. [117] [118] adopted a segment-wise sparse term, where the matrix is constructed using a precomputed over-segmented. Experimental results [117] [118] on the CAVIAR dataset [40] show that SAMF based on image segmentation (sSAMF) outperforms PCP via IALM [20], which correponds to 'LE'-SAMF in [117] [118].

Algorithm for solving SAMF: First, Nakajima et al. [117] [118] reduced the partial SAMF problem to the standard MF problem, which can be solved analytically. Then, Nakajima et al. [117] [118] derived an iterative algorithm called the mean update (MU) for the variational Bayesian approximation to SAMF, which gives the global optimal solution for a large subset of parameters in each step.

1.17 Variational Bayesian Sparse Estimator

Chen et al. [30] [31] [29] proposed a generalization of the original RPCA [20], where a linear transformation through the use of a known measurement matrix is applied to the outlier corrupted data. The aim is to estimate the outlier amplitudes given the transformed observation. This approach, called variational Bayesian Sparse Estimator (VBSE), can achieve background/foreground separation in blurred and noisy video sequences. Thus, the decomposition is written as follows:

$$A = L + RS + E \tag{1.93}$$

where R models the linear transformation performed on the data. The aim is to obtain accurate estimates for the sparse term S and the low-rank term L, given the noise-corrupted observation A. Although S is sparse, the multiplication with a wide matrix R has an effect of compression, and hence the product RE is not necessarily sparse. Then, Chen et al. [30] [31] modeled the lowk-rank part as follows:

$$||L||_* = \min_{U,V} \frac{1}{2}||U||_F^2 + ||V||_F^2 \quad \text{subj} \quad L = UV^T \tag{1.94}$$

With this relaxation and parametrization, Chen et al. [30] [31] obtained the following optimization problem:

$$\min_{U,V,S} \frac{1}{2}||A - UV^T - RS||_F^2 + \lambda_*(||U||_F^2 + ||V||_F^2) + \lambda_1||E||_{l_1} \quad \text{subj} \quad L = UV^T \tag{1.95}$$

where λ_* and λ_1 are regularization parameters. To enforce column sparsity in U and V, the columns of U and V are modeled with Gaussian priors of precision. Then, Chen et al. [30] [31] incorporated conjugate Gamma hyperprior on the precisions. The sparse part S is modeled by setting the entries be independent of each other, and their amplitudes are modeled by zero-mean Gaussian distributions with independent precisions. For the noise part E, Gaussian priors with zero mean are used to model the dense observation noise. By combining these different stages in a hierarchical Bayesian model, a joint distribution of the observation and all the unknowns is expressed as follows:

$$\rho(A, U, V, S, \gamma, \alpha, \beta) \tag{1.96}$$

where γ and α are hyperparameters and β is the noise precision. To solve VBSE, Chen et al. [30] [31] used an an approximate Bayesian inference. Experimental results [30] [31] on the CAVIAR dataset [40] show that VBSE outperforms PCP solved via APG [92] and PCA solved via IALM [20].

1.18 Conclusion

In this chapter, we have first presented a full review of recent advances in RPCA via Decomposition in Low-rank and Sparse Matrices, which we called DLSM. We evaluated their adequation to a representive application in computer vision (background/foreground separation) by investigating how these methods are solved and if incremental algorithms and real-time implementations can be achieved. In conclusion, this review highlights the following points:

- Decomposition into low-rank plus sparse matrices offers a suitable framework for background/foreground separation. Furthermore, several DLSM models can outperform state-of-the-art models in background/foreground separation, such as the MOG.

- The main disadvantages of the DLSM models is that their original version used batch algorithms that often need too-expensive computation time to reach real-time requirements. Thus, many efforts have been made to reach real-time performance and to develop incremental algorithms as in the works of Rodriguez et al. [144] [146] [143] [145], and Vaswani et al. [129] [136] [52].

- As images are stored in vectors that are oftenly exploited as is, DLSM models in their original version lose the spatial and temporal constraints. Thus, it is more suitable **1)** to add the use of Markov Random Fields [77], **2)** to use structured norms aiming to preserve the spatial structures of images while being insensitive to outliers and missing data [189] [153] [163], or **3)** to formulate RPCA in two dimensions (2D-RPCA) rather than via image-to-vector conversion, which enables the preservation of the image spatial information with reduced computational time [163].

Future research may concern less computational SVD algorithms such as LMSVD [98] for batch algorithms, and DLSM models that would be both incremental and real-time to reach the performance of the state-of-the-art algorithms [152] [156] in terms of computation time and memory requirements. Finally, DLSM models show a suitable potential for background modeling and foreground detection in video surveillance [79] [80]. Furthermore, DLSM can be extended to the measurement domain, rather than the pixel domain, for use in conjunction with compressive sensing. Moreover, other research may concern the extension of DLSM in a tensor-wise way to fully exploit spatial and temporal constraints [165] [166] [170] [88] [90] [213] with incremental algorithms [158] [137] [74].

References

1. A. Abdel-Hakim and M. El-Saban. FRPCA: fast robust principal component analysis. *International Conference on Pattern Recognition, ICPR 2012*, November 2012.
2. C. Aicher. A variational Bayes approach to robust principal component analysis. *REU 2013*, 2013.
3. Z. An. Video background modeling based on optimization algorithms of robust pca. *Thesis*, February 2014.
4. M. Anderson, G. Ballard, J. Demme, and K. Keutzer. Communication-avoiding QR decomposition for GPUs. *Technical Report, ECCS*, 2010.
5. M. Anderson, G. Ballard, J. Demme, and K. Keutzer. Communication-avoiding QR decomposition for GPUs. *IEEE International Parallel and Distributed Processing Symposium, IPDPS 2011*, 2011.
6. A. Aravkin, S. Becker, V. Cevher, and P. Olsen. A variational approach to stable principal component pursuit. *Conference on Uncertainty in Artificial Intelligence, UAI 2014*, July 2014.

7. F. Arrigoni, B. Rossi, and A. Fusiello. Robust and efficient camera motion synchronization via matrix decomposition. *International Conference on Image Processing, ICIAP 2015*, September 2015.
8. N. Aybat, D. Goldfarb, and G. Iyengar. Fast first-order methods for stable principal component pursuit. *Preprint*, 2011.
9. N. Aybat, D. Goldfarb, and G. Iyengar. Efficient algorithms for robust and stable principal component pursuit. *Preprint*, November 2012.
10. N. Aybat and G. Iyengar. A unified approach for minimizing composite norms. *Preprint*, August 2012.
11. N. Aybat and G. Iyengar. An alternating direction method with increasing penalty for stable principal component pursuit. *Computational Optimization and Applications*, 2014.
12. S. Babacan, M. Luessi, R. Molina, and A. Katsaggelos. Sparse Bayesian methods for low-rank matrix estimation. *IEEE Transactions on Signal Processing*, 60(8):3964–3977, 2012.
13. B. Bao, G. Liu, C. Xu, and S. Yan. Inductive robust principal component analysis. *IEEE Transactions on Image Processing*, pages 3794–3800, August 2012.
14. M. Beal. Variational algorithms for approximate Bayesian inference. *PhD Thesis, University of London*, 2003.
15. S. Becker, E. Candes, and M. Grant. TFOCS: flexible first-order methods for rank minimization. *Low-rank Matrix Optimization Symposium, SIAM Conference on Optimization*, 2011.
16. A. Bhardwaj and S. Raman. Robust PCA-based solution to image composition using augmented lagrange multiplier (alm). *Visual Computer*, March 2015.
17. T. Bouwmans. Traditional and recent approaches in background modeling for foreground detection: An overview. *Computer Science Review*, 11:31–66, May 2014.
18. T. Bouwmans and E. Zahzah. Robust PCA via principal component pursuit: A review for a comparative evaluation in video surveillance. *Special Isssue on Background Models Challenge, Computer Vision and Image Understanding, CVIU 2014*, 2014.
19. J. Cai, E. Candes, and Z. Shen. A singular value thresholding algorithm for matrix completion. *International Journal of ACM*, May 2008.
20. E. Candes, X. Li, Y. Ma, and J. Wright. Robust principal component analysis? *International Journal of ACM*, 58(3), May 2011.
21. E. Candes and M. Soltanolkotabi. Discussion of latent variable graphical model selection via convex optimization. *Annals of Statistics*, 40(4), 2012.
22. X. Cao, L. Yang, and X. Guo. Total variation regularized RPCA for irregularly moving object detection under dynamic background. *IEEE Transactions on Cybernetics*, April 2015.
23. Y. Chai, S. Xu, and H. Yin. An improved ADM algorithm for RPCA optimization problem. *Chinese Control Conference, CCC 2013*, pages 4769–4880, July 2013.
24. V. Chandrasekaran, P. Parillo, and A. Willsky. Latent variable graphical model selection via convex optimization. *Annals of Statistics*, 40(4):1935–1967, 2012.
25. V. Chandrasekharan, S. Sanghavi, P. Parillo, and A. Wilsky. Rank-sparsity incoherence for matrix decomposition. *Preprint*, 2009.
26. R. Chartrand. Non convex splitting for regularized low-rank and sparse decomposition. *IEEE Transactions on Signal Processing*, 2012.
27. C. Chen, J. Cai, W. Lin, and G. Shi. Surveillance video coding via low-rank and sparse decomposition. *ACM international conference on Multimedia*, pages 713–716, 2012.
28. C. Chen, J. Cai, W. Lin, and G. Shi. Incremental low-rank and sparse decomposition for

compressing videos captured by fixed cameras. *Journal of Visual Communication and Image Representation*, December 2014.
29. Z. Chen. Multidimensional signal processing for sparse and low-rank problems. *Thesis, Northwestern University, USA*, June 2014.
30. Z. Chen, S. Babacan, R. Molina, and A. Katsaggelos. Variational Bayesian methods for multimedia problems. *IEEE Transaction on Multimedia*, 2014.
31. Z. Chen, R. Molina, and A. Katsaggelos. A variational approach for sparse component estimation and low-rank matrix recovery. *Journal of Communication*, 8(9), September 2013.
32. C. Dang, A. Moghadam, and H. Radha. RPCA-KFE: key frame extraction for consumer video based robust principal component analysis. *Preprint*, May 2014.
33. Y. Deng. Sparse structure for visual information sensing: Theory and algorithms. *Thesis*, 2014.
34. Y. Deng, Q. Dai, R. Liu, and Z. Zhang. Low-rank structure learning via log-sum heuristic recovery. *Preprint*, 2012.
35. Y. Deng, Q. Dai, R. Liu, Z. Zhang, and S. Hu. Low-rank structure learning via nonconvex heuristic recovery. *IEEE Transactions on Neural Networks And Learning Systems*, 24(3), March 2013.
36. X. Ding, L. He, and L. Carin. Bayesian robust principal component analysis. *IEEE Transaction on Image Processing*, 2011.
37. R. Fan, H. Wang, and H. Zhang. A new analysis of the iterative threshold algorithm for RPCA by primal-dual method. *Advanced Materials Research*, pages 989–994, July 2014.
38. M. Fazel. Matrix rank minimization with applications. *PhD Thesis, Stanford University*, March 2002.
39. J. Feng, H. Xu, and S. Yan. Online robust pca via stochastic optimization. *NIPS 2013*, 2013.
40. R. Fisher. CAVIAR: context aware vision using image-based active recognition. *http://homepages.inf.ed.ac.uk/rbf/CAVIAR/*, 2005.
41. C. Gan, Y. Wang, and X. Wang. Multi-feature robust principal component analysis for video moving object segmentation. *Journal of Image and Graphics*, 18(9), 2013.
42. S. Gandy and I. Yamada. Convex optimization techniques for the efficient recovery of a sparsely corrupted low-rank matrix. *Journal of Math-for-Industry*, 2:147–156, 2010.
43. Z. Gao, L. Cheong, and M. Shan. Block-sparse RPCA for consistent foreground detection. *European Conference on Computer Vision, ECCV 2012*, 2012.
44. T. Gerhart. Convex optimization techniques and their application in hyperspectral video processing. *Thesis*, December 2013.
45. J. Goes, T. Zhang, R. Arora, and G. Lerman. Robust stochastic principal component analysis. *AISTATS 2014*, 2014.
46. D. Goldfarb, S. Ma, and K. Scheinberg. Fast alternating linearization methods for minimizing the sum of two convex functions. *Preprint, Mathematical Programming Series A*, 2010.
47. T. Goldstein and S. Osher. The split Bregman method for l_1-regularized problems. *SIAM Journal of Image Science*, 2(2):323–343, 2009.
48. G. Gu, B. He, and J. Yang. Inexact alternating direction based contraction methods for separable linearly constrained convex programming. *Journal of Optimization Theory and Applications*, December 2013.
49. S. Gu, L. Zhang, W. Zuo, and X. Feng. Weighted nuclear norm minimization with application to image denoising. *Preprint*, March 2014.
50. N. Guan, D. Tao, Z. Luo, and J. Shawe-Taylor. MahNMF: manhattan non-negative

matrix factorization. *Journal of Machine Learning Research*, 2012.
51. H. Guo, C. Qiu, and N. Vaswani. Practical ReProCS for separating sparse and low-dimensional signal sequences from their sum. *Preprint*, October 2013.
52. H. Guo, C. Qiu, and N. Vaswani. An online algorithm for separating sparse and low-dimensional signal sequences from their sum. *IEEE Transactions on Signal Processing*, 2014.
53. H. Guo, C. Qiu, and N. Vaswani. Practical ReProCS for separating sparse and low-dimensional signal sequences from their sum - part 1. *International Conference on Acoustics, Speech, and Signal Processing, ICASSP 2014*, May 2014.
54. H. Guo, C. Qiu, and N. Vaswani. Practical ReProCS for separating sparse and low-dimensional signal sequences from their sum - part 3. *GlobalSIP 2014*, 2014.
55. X. Guo and X. Cao. Speeding up low rank matrix recovery for foreground separation in surveillance videos. *International Conference on Multimedia and Expo, ICME 2014*, 2014.
56. X. Guo, S. Li, and X. Cao. Motion matters: A novel framework for compressing surveillance videos. *ACM International Conference on Multimedia*, October 2013.
57. X. Guo, X. Wang, L. Yang, X. Cao, and Y. Ma. Robust foreground detection using smoothness and arbitrariness constraints. *ECCV 2014*, September 2014.
58. C. Guyon, T. Bouwmans, and E. Zahzah. Foreground detection based on low-rank and block-sparse matrix decomposition. *IEEE International Conference on Image Processing, ICIP 2012*, September 2012.
59. C. Guyon, T. Bouwmans, and E. Zahzah. Foreground detection by robust PCA solved via a linearized alternating direction method. *International Conference on Image Analysis and Recognition, ICIAR 2012*, June 2012.
60. C. Guyon, T. Bouwmans, and E. Zahzah. Foreground detection via robust low rank matrix decomposition including spatio-temporal constraint. *International Workshop on Background Model Challenges, ACCV 2012*, November 2012.
61. C. Guyon, T. Bouwmans, and E. Zahzah. Foreground detection via robust low rank matrix factorization including spatial constraint with iterative reweighted regression. *International Conference on Pattern Recognition, ICPR 2012*, November 2012.
62. C. Guyon, T. Bouwmans, and E. Zahzah. Moving object detection by robust PCA solved via a linearized symmetric alternating direction method. *International Symposium on Visual Computing, ISVC 2012*, July 2012.
63. C. Guyon, T. Bouwmans, and E. Zahzah. Moving object detection via robust low rank matrix decomposition with IRLS scheme. *International Symposium on Visual Computing, ISVC 2012*, pages 665–674, July 2012.
64. C. Guyon, T. Bouwmans, and E. Zahzah. Robust principal component analysis for background subtraction: Systematic evaluation and comparative analysis. *INTECH, Principal Component Analysis, Book 1, Chapter 12*, pages 223–238, March 2012.
65. C. Hage and M. Kleinsteuber. Robust PCA and subspace tracking from incomplete observations using l_0-surrogates. *Optimization and Control*, 2012.
66. T. Hastie, R. Tibshirani, and J. Friedman. *The elements of statistical learning: Data mining, inference and prediction*, 2nd Edition, Springer, February 2009.
67. J. He and Y. Zhang. Adaptive stochastic gradient descent on the Grassmannian for robust low-rank subspace recovery. *Preprint*, December 2014.
68. R. He, T. Tan, and L. Wang. Recovery of corrupted low-rank matrix by implicit regularizers. *IEEE Transaction on Pattern Analysis and Machine Intelligence, PAMI 2013*, September 2013.
69. M. Hintermüller and T. Wu. Robust principal component pursuit via inexact alternating minimization on matrix manifolds. *Journal of Mathematics and Imaging Vision*, 2014.

70. D. Hsu, S. Kakade, and T. Zhang. Robust matrix decomposition with sparse corruptions. *IEEE Transactions on Information Theory*, 57(11):7221–7234, 2011.
71. S. Huang, J. Ye, T. Wang, L. Jiang, X. Wu, and Y. Li. Extracting refined low-rank features of robust PCA for human action recognition. *Arabian Journal for Science and Engineering*, 40(2):1427–1441, March 2015.
72. X. Huang, P. Huang, Y. Cao, and H. Yan. A block-sparse RPCA algorithm for moving object detection based on PCP. *Journal of East China, Jiaotong University*, 5:30–36, October 2013.
73. S. Javed, T. Bouwmans, and S. Jung. Depth extended online RPCA with spatiotemporal constraints for robust background subtraction. *Korea-Japan Workshop on Frontiers of Computer Vision, FCV 2015*, January 2015.
74. S. Javed, T. Bouwmans, and S. Jung. Stochastic decomposition into low rank and sparse tensor for robust background subtraction. *ICDP 2015*, July 2015.
75. S. Javed, S. Oh, J. Heo, and S. Jung. Robust background subtraction via online robust PCA using image decomposition. *International Conference on Research in Adaptive and Convergent System, RACS 2014*, 2014.
76. S. Javed, A. Sobral, T. Bouwmans, and S. Jung. OR-PCA with dynamic feature selection for robust background subtraction. *ACM Symposium On Applied Computing, SAC 2015*, 2015.
77. S. Javed, A. Sobral, S. Oh, T. Bouwmans, and S. Jung. OR-PCA with MRF for robust foreground detection in highly dynamic backgrounds. *Asian conference on computer vision, ACCV 2014*, 2014.
78. H. Ji, S. Huang, Z. Shen, and Y. Xu. Robust video restoration by joint sparse and low rank matrix approximation. *SIAM Journal on Imaging Sciences*, 4(4):1122–1142, January 2011.
79. H. Jiang, W. Deng, and Z. Shen. Surveillance video processing using compressive sensing. *Inverse Problems and Imaging*, 6(2):201–214, 2012.
80. H. Jiang, S. Zhao, Z. Shen, W. Deng, P. Wilford, and R. Cohen. Surveillance video analysis using compressive sensing with low latency. *Preprint*, 2014.
81. M. Karl and C. Osendorfer. Improving approximate RPCA with a K-sparsity prior. *International Conference on Learning Representations, ICLR 2015*, 2015.
82. E. Kim, M. Lee, C. Choi, N. Kwak, and S. Oh. Efficient l_1-norm-based low-rank matrix approximations for large-scale problems using alternating rectified gradient method. *International Conference on Multimedia and Expo, ICME 2014*, 2014.
83. H. Kong, X. Li, L. Wang, E. Teoh, J. Wang, and R. Venkateswarlu. Generalized 2D principal component analysis. *IEEE International Joint Conference on Neural Networks, IJCNN 2005*, 1:108–113, July 2005.
84. A. Kyrillidis and V. Cevher. MATRIX ALPS: accelerated low rank and sparse matrix reconstruction. *IEEE Workshop on Statistical Signal Processing, SSP 2012*, 2012.
85. K. Lange, D. Hunter, and I. Yang. Optimization transfer using surrogate objective functions. *Journal of Computational and Graphical Statistics*, 9:1–59, 2000.
86. W. Leow, Y. Cheng, L. Zhang, T. Sim, and L. Foo. Background recovery by fixed-rank robust principal component analysis. *CAIP 2013*, 2013.
87. L. Li, W. Huang, I. Gu, and Q. Tian. Statistical modeling of complex backgrounds for foreground object detection. *IEEE Transaction on Image Processing*, pages 1459–1472, 2004.
88. L. Li, P. Wang, Q. Hu, and S. Cai. Efficient background modeling based on sparse representation and outlier iterative removal. *IEEE Transactions on Circuits and Systems for Video Technology*, December 2014.
89. S. Li and H. Qi. Recursive low-rank and sparse recovery of surveillance video using

compressed sensing. *International Conference on Distributed Smart Cameras, ICDSC 2014*, 2014.

90. Y. Li, J. Yan, Y. Zhou, and J. Yang. Optimum subspace learning and error correction for tensors. *European Conference on Computer Vision, ECCV 2010*, 2010.
91. Z. Lin, M. Chen, L. Wu, and Y. Ma. The augmented Lagrange multiplier method for exact recovery of corrupted low-rank matrices. *UIUC Technical Report*, November 2009.
92. Z. Lin, A. Ganesh, J. Wright, L. Wu, M. Chen, and Y. Ma. Fast convex optimization algorithms for exact recovery of a corrupted low-rank matrix. *UIUC Technical Report*, August 2009.
93. Z. Lin, R. Liu, and Z. Su. Linearized alternating direction method with adaptive penalty for low-rank representation. *NIPS 2011*, December 2011.
94. Z. Lin and S. Wei. A block Lanczos with warm start technique for accelerating nuclear norm minimization algorithms. *Preprint*, 2010.
95. R. Liu, Z. Lin, and Z. Su. Exactly recovering low-rank matrix in linear time via l_1 filter. *Preprint*, August 2011.
96. R. Liu, Z. Lin, Z. Su, and J. Gao. Linear Yme principal component pursuit and its extensions using l_1 filtering. *Neurocomputing*, 2014.
97. R. Liu, Z. Lin, S. Wei, and Z. Su. Solving principal component pursuit in linear time via l_1 filtering. *International Journal on Computer Vision, IJCV 2011*, 2011.
98. X. Liu, Z. Wen, and Y. Zhang. Limited memory block Krylov subspace optimization for computing dominant singular value decomposition. *Preprint*, 2012.
99. X. Liu, Z. Wen, and Y. Zhang. An efficient Gauss-Newton algorithm for symmetric low-rank product matrix approximations. *Technical Report*, June 2014.
100. X. Liu, G. Zhao, J. Yao, and C. Qi. Background subtraction based on low-rank model and structured sparse decomposition. *IEEE Transactions on Image Processing*, 2015.
101. Y. Liu, L. Jiao, and F. Shang. A fast tri-factorization method for low-rank matrix recovery and completion. *Pattern Recognition, PR 2013*, 46:163–173, January 2012.
102. Y. Liu, L. Jiao, and F. Shang. An efficient matrix factorization based low-rank representation for subspace clustering. *Pattern Recognition, PR 2013*, 46:284–292, January 2013.
103. B. Lois and N. Vaswani. A correctness result for online robust PCA. *Preprint*, 2014.
104. B. Lois, N. Vaswani, and C. Qiu. Performance guarantees for undersampled recursive sparse recovery in large but structured noise. *GlobalSIP 2013*, pages 1061–1064, December 2013.
105. C. Lu, Z. Lin, and S. Yan. Smoothed low rank and sparse matrix recovery by iteratively reweighted least squares minimization. *Preprint*, 2014.
106. C. Lu, C. Zhu, C. Xu, S. Yan, and Z. Lin. Generalized singular value thresholding. *Preprint*, December 2014.
107. S. Ma. Algorithms for sparse and low-rank optimization: Convergence, complexity and applications. *Thesis*, Jun. 2011.
108. S. Ma, L. Xue, and H. Zou. Alternating direction methods for latent variable Gaussian graphical model selection. *Neural Computation*, 25:2172–2198, August 2013.
109. Y. Ma. The pursuit of low-dimensional structures in high-dimensional (visual) data: fast and scalable algorithms. *Workshop on Algorithms for Modern Massive Data Sets, MMDS 2012*, 2012.
110. Y. Ma. Pursuit of low-dimensional structures in high-dimensional visual data. *Plenary talk at the Foundations of Computational Mathematics, FoCM 2014*, December 2014.

111. L. Mackey, A. Talwalkar, and M. Jordan. Divide-and-conquer matrix factorization. *Neural Information Processing Systems, NIPS 2011*, December 2011.
112. G. Mateos and G. Giannakis. Sparsity control for robust principal component analysis. *International Conference on Signals, Systems, and Computers*, November 2010.
113. G. Mateos and G. Giannakis. Robust PCA as bilinear decomposition with outlier-sparsity regularization. *Preprint*, November 2011.
114. B. Moore, R. Nadakuditi, and J. Fessler. Improved robust PCA using low-rank denoising with optimal singular value shrinkage. *IEEE Workshop on Statistical Signal Processing, SSP 2014*, pages 13–16, June 2014.
115. C. Mu, Y. Zhang, J. Wright, and D. Goldfarb. Scalable robust matrix recovery: Frank-wolfe meets proximal methods. *Preprint*, 2014.
116. Y. Mu, J. Dong, X. Yuan, and S. Yan. Accelerated low-rank visual recovery by random projection. *International Conference on Computer Vision, CVPR 2011*, 2011.
117. S. Nakajima, M. Sugiyama, and D. Babacan. Sparse additive matrix factorization for robust PCA and its generalization. *ACML 2012*, November 2012.
118. S. Nakajima, M. Sugiyama, and D. Babacan. Variational Bayesian sparse additive matrix factorization. *Machine Learning*, 92:319–347, 2013.
119. P. Netrapalli, U. Niranjan, S. Sanghavi, A. Anandkumar, and P. Jain. Non-convex robust PCA. *Preprint*, October 2014.
120. T. Oh. A novel low-rank constraint method with the sparsity model for moving object analysis. *Master Thesis, KAIST 2012*, 2012.
121. T. Oh, Y. Matsushita, Y. Tai, and I. Kweon. Fast randomized singular value thresholding for nuclear norm minimization. *IEEE International Conference on Computer Vision and Pattern Recognition, CVPR 2015*, June 2015.
122. N. Oliver, B. Rosario, and A. Pentland. A Bayesian computer vision system for modeling human interactions. *ICVS 1999*, January 1999.
123. F. Orabona, A. Argyriou, and N. Srebro. PRISMA: PRoximal Iterative Smoothing Algorithm. *Optimization and Control*, 2012.
124. O. Oreifej, X. Li, and M. Shah. Simultaneous video stabilization and moving object detection in turbulence. *IEEE Transactions on Pattern Analysis and Machine Intelligence, PAMI 2012*, 2012.
125. T. Parker and P. Schniter. Bilinear generalized approximate message passing (BiG-AMP) for matrix completion. *Asilomar Conference on Signals, Systems, and Computers*, November 2012.
126. Y. Peng, A. Ganesh, J. Wright, W. Xu, and Y. Ma. RASL: Robust Alignment by Sparse and Low-rank decomposition for linearly correlated images. *IEEE Transactions on Pattern Analysis and Machine Intelligence*, 34(11):2233–2246, 2012.
127. G. Pope, M. Baumann, C. Studery, and G. Durisi. Real-time principal component pursuit. *Asilomar Conference on Signals, Systems, Computation*, November 2011.
128. H. Qin, Y. Peng, and X. Li. Foreground extraction of underwater videos via sparse and low-rank matrix decomposition. *Workshop on Computer Vision for Analysis of Underwater Imagery, ICPR 2014*, 2014.
129. C. Qiu and N. Vaswani. Real-time robust principal components pursuit. *International Conference on Communication Control and Computing*, 2010.
130. C. Qiu and N. Vaswani. ReProCS: a missing link between recursive robust PCA and recursive sparse recovery in large but correlated noise. *Preprint*, 2011.
131. C. Qiu and N. Vaswani. Support predicted modified-CS for recursive robust principal components' pursuit. *IEEE International Symposium on Information Theory, ISIT 2011*, 2011.
132. C. Qiu and N. Vaswani. Automated recursive projected CS (ReProCS) for real-time

video layering. *International Conference on Computer Vision and Pattern Recognition, CVPR 2012*, 2012.

133. C. Qiu and N. Vaswani. Recursive sparse recovery in large but structured noise - part 1. *Preprint*, November 2012.
134. C. Qiu and N. Vaswani. Recursive sparse recovery in large but structured noise - part 2. *Preprint*, 2012.
135. C. Qiu, N. Vaswani, B. Lois, and L. Hogben. Recursive robust pca or recursive sparse recovery in large but structured noise. *International Conference on Acoustics, Speech, and Signal Processing, ICASSP 2013*, 2013.
136. C. Qiu, N. Vaswani, B. Lois, and L. Hogben. Recursive robust pca or recursive sparse recovery in large but structured noise. *IEEE Transactions on Information Theory*, 2014.
137. C. Qiu, X. Wu, and H. Xu. Recursive projected sparse matrix recovery (ReProCSMR) with application in real-time video layer separation. *IEEE International Conference on Image Processing, ICIP 2014*, pages 1332–1336, October 2014.
138. M. Rahmani and G. Atia. High dimensional low rank plus sparse matrix decomposition. *Preprint*, February 2015.
139. L. Ramesh and P. Shah. R-SpaRCS : An Algorithm for Foreground-Background Separation of Compressively-Sensed Surveillance Videos. *IEEE International Conference on Advanced Video and Signal based Surveillance, AVSS 2015*, 2015.
140. I. Ramirez and G. Sapiro. Low-rank data modeling via the minimum description length principle. *ICASSP 2012*, 2012.
141. I. Ramirez and G. Sapiro. An MDL framework for sparse coding and dictionary learning. *Preprint*, 2012.
142. C. Robert and G. Casellah. *Monte carlo statistical methods*, 2nd edition, New York: Springer, 2004.
143. P. Rodriguez. Real-time incremental principal component pursuit for video background modeling on the TK1. *GPU Technical Conference, GTC 2015*, March 2015.
144. P. Rodriguez and B. Wohlberg. Fast principal component pursuit via alternating minimization. *IEEE International Conference on Image Processing, ICIP 2013*, September 2013.
145. P. Rodriguez and B. Wohlberg. Incremental principal component pursuit for video background modeling. *IEEE Signal Processing Letters*, 2014.
146. P. Rodriguez and B. Wohlberg. A Matlab implementation of a fast incremental principal component pursuit algorithm for video background modeling. *IEEE International Conference on Image Processing, ICIP 2014*, October 2014.
147. P. Rodriguez and B. Wohlberg. Video background modeling under impulse noise. *IEEE International Conference on Image Processing, ICIP 2014*, October 2014.
148. P. Rodriguez and B. Wohlberg. Translational and rotational jitter invariant incremental principal component pursuit for video background modeling. *IEEE International Conference on Image Processing, ICIP 2015*, 2015.
149. G. Ros, J. Alvarez, and J. Guerrero. Motion estimation via robust decomposition with constrained rank. *Preprint*, October 2014.
150. T. Sakai and H. Kuhara. Separating background and foreground optical flow fields by low-rank and sparse regularization. *IEEE International Conference on Acoustics, Speech and Signal Processing, ICASSP 2015*, April 2015.
151. F. Seidel, C. Hage, and M. Kleinsteuber. pROST - a smoothed Lp-norm robust online subspace tracking method for realtime background subtraction in video. *Special Issue on Background Modeling for Foreground Detection in Real-World Dynamic Scenes, Machine Vision and Applications*, 2013.
152. M. Shah, J. Deng, and B. Woodford. Video background modeling: Recent approaches,

issues and our solutions. *Machine Vision and Applications*, 25(5):1105–1119, July 2014.

153. F. Shang, Y. Liu, H. Tong, J. Cheng, and H. Cheng. Structured low-rank matrix factorization with missing and grossly corrupted observations. *Preprint*, September 2014.
154. W. Shao, Q. Ge, Z. Gan, H. Deng, and H. Li. A generalized robust minimization framework for low-rank matrix recovery. *Mathematical Problems in Engineering*, 2014.
155. Y. Shen, Z. Wen, and Y. Zhang. Augmented Lagrangian alternating direction method for matrix separation based on low-rank factorization. *Preprint*, January 2011.
156. A. Shimada, Y. Nonaka, H. Nagahara, and R. Taniguchi. Video background modeling: Recent approaches, issues and our solutions. *Machine Vision and Applications*, 25(5):1121–1131, July 2014.
157. X. Shu, F. Porikli, and N. Ahuja. Robust orthonormal subspace learning: Efficient recovery of corrupted low-rank matrices. *International Conference on Computer Vision and Pattern Recognition, CVPR 2014*, June 2014.
158. A. Sobral, C. Baker, T. Bouwmans, and E. Zahzah. Incremental and multi-feature tensor subspace learning applied for background modeling and subtraction. *International Conference on Image Analysis and Recognition, ICIAR 2014*, October 2014.
159. A. Sobral, T. Bouwmans, and E. Zahzah. Double-constrained RPCA based on saliency maps for foreground detection in automated maritime surveillance. *ISBC 2015 Workshop conjunction with AVSS 2005*, 2015.
160. P. Sprechmann, A. Bronstein, and G. Sapiro. Learning robust low-rank representations. *Optimization and Control*, 2012.
161. Q. Sun, S. Xiang, and J. Ye. Robust principal component analysis via capped norms. *International Conference on Knowledge Discovery and Data Mining, KDD 2013*, pages 311–319, 2013.
162. Y. Sun, X. Tao, Y. Li, and J. Lu. Robust two-dimensional principal component analysis via alternating optimization. *International Conference on Image Processing, ICIP 2013*, September 2013.
163. Y. Sun, X. Tao, Y. Li, and J. Lu. Robust 2D principal component analysis: A structured sparsity regularized approach. *IEEE Transactions on Image Processing*, pages 2515–2526, August 2015.
164. Y. Sun, X. Tao, Y. Li, and J. Lu. Robust two-dimensional principal component analysis: A structured sparsity regularized approach. *IEEE Transactions on Image Processing*, 2015.
165. H. Tan, B. Cheng, J. Feng, G. Feng, W. Wang, and Y. Zhang. Low-n-rank tensor recovery based on multi-linear augmented Lagrange multiplier method. *Neurocomputing*, January 2013.
166. H. Tan, B. Cheng, J. Feng, G. Feng, and Y. Zhang. Tensor recovery via multi-linear augmented Lagrange multiplier method. *International Conference on Image and Graphics, ICIG 2011*, pages 141–146, August 2011.
167. G. Tang and A. Nehorai. Robust principal component analysis based on low-rank and block-sparse matrix decomposition. *CISS 2011*, 2011.
168. M. Tao and X. Yuan. Recovering low-rank and sparse components of matrices from incomplete and noisy observations. *SIAM Journal on Optimization*, 21(1):57–81, 2011.
169. F. De La Torre and M. Black. A robust principal component analysis for computer vision. *International Conference on Computer Vision*, 2001.
170. L. Tran, C. Navasca, and J. Luo. Video detection anomaly via low-rank and sparse decompositions. *IEEE New York Image Processing Workshop, WNYIPW 2012*,

pages 17–20, November 2012.
171. A. Vacavant, T. Chateau, A. Wilhelm, and L. Lequievre. A benchmark dataset for foreground/background extraction. *International Workshop on Background Models Challenge, ACCV 2012*, November 2012.
172. T. Wan, C. Zhu, and Z. Qin. Multifocus image fusion based on robust principal component analysis. *Pattern Recognition Letters*, 34(9):1001–1008, July 2013.
173. H. Wang and A. Banerjee. Online alternating direction method. *Preprint*, 2013.
174. H. Wang, A. Banerjee, and Z. Luo. Parallel direction method of multipliers. *Preprint*, June 2014.
175. J. Wang, M. Wan, X. Hu, and S. Yan. Image denoising with a unified schattern-p norm and l_q norm regularization. *Journal of Optimization Theory and Applications*, April 2014.
176. S. Wang and X. Feng. Optimization of the regularization in background and foreground modeling. *Journal of Applied Mathematics*, 2014.
177. X. Wang and W. Wan. Motion segmentation via multi-task robust principal component analysis. *Journal of Applied Sciences, Electronics and Information Engineering*, 32(5):473–480, September 2014.
178. Y. Wang, Y. Liu, and L. Wu. Study on background modeling method based on robust principal component analysis. *Annual Conference on Electrical and Control Engineering, ICECE 2011*, pages 6787–6790, September 2011.
179. A. Waters, A. Sankaranarayanan, and R. Baraniuk. SpaRCS: recovering low-rank and sparse matrices from compressive measurements. *Neural Information Processing Systems, NIPS 2011*, December 2011.
180. A. Waters, A. Sankaranarayanan, and R. Baraniuk. SpaRCS: recovering low-rank and sparse matrices from compressive measurements. *Technical Report*, 2011.
181. C. Wei, Y. Huang, Y. Wang, and M. Shih. Background recovery in railroad crossing videos via incremental low-rank matrix decomposition. *Asian Conference on Pattern Recognition, ACPR 2013*, November 2013.
182. J. Wen, Y. Xu, J. Tang, Y. Zhan, Z. Lai, and X. Guo. Joint video frame set division and low-rank decomposition for background subtraction. *IEEE Transactions on Circuits and Systems For Video Technology*, 2014.
183. B. Wohlberg, R. Chartrand, and J. Theiler. Local principal component analysis for nonlinear datasets. *International Conference on Acoustics, Speech, and Signal Processing, ICASSP 2012*, March 2012.
184. J. Wright, Y. Peng, Y. Ma, A. Ganesh, and S. Rao. Robust principal component analysis: Exact recovery of corrupted low-rank matrices by convex optimization. *Neural Information Processing Systems, NIPS 2009*, December 2009.
185. J. Wright, A. Yang, A. Ganesh, S. Sastry, and Y. Ma. Robust face recognition via sparse representation. *IEEE Transactions on Pattern Analysis and Machine Intelligence*, 2009.
186. L. Wu, Y. Wang, Y. Liu, and Y. Wang. Robust structure from motion with affine camera via low-rank matrix recovery. *China Information Sciences*, 56(11):1–10, November 2015.
187. B. Xin, Y. Tian, Y. Wang, and W. Gao. Background subtraction via generalized fused Lasso foreground modeling. *Preprint*, April 2015.
188. H. Xu, C. Caramanis, and S. Sanghavi. Robust PCA via outlier pursuit. *NIPS 2010*, 2010.
189. J. Xu, V. Ithapu, L. Mukherjee, J. Rehg, and V. Singh. GOSUS: Grassmannian Online Subspace Updates with Structured-Sparsity. *International Conference on Computer Vision, ICCV 2013*, September 2013.
190. X. Xu. Online robust principal component analysis for background subtraction: A system

evaluation on Toyota car data. *Master's thesis, University of Illinois, Urbana-Champaign, USA*, 2014.
191. Y. Xue, X. Gu, and X. Cao. Motion saliency detection using low-rank and sparse decomposition. *International Conference on Acoustics, Speech, and Signal Processing, ICASSP 2012*, March 2012.
192. F. Yang, H. Jiang, Z. Shen, W. Deng, and D. Metaxas. Adaptive low rank and sparse decomposition of video using compressive sensing. *ICIP 2013*, 2013.
193. J. Yang, X. Sun, X. Ye, and K. Li. Background extraction from video sequences via motion-assisted matrix completion. *IEEE International Conference on Image Processing, ICIP 2014*, October 2014.
194. J. Yang and X. Yuan. Linearized augmented Lagrangian and alternating direction methods for nuclear norm minimization. *Preprint*, 2011.
195. L. Yang, T. Pong, and X. Chen. Alternating direction method of multipliers for non-convex background/foreground extraction. *Preprint*, June 2015.
196. M. Yang. Background modeling from surveillance video using rank minimization. *Artificial Intelligence and Computational Intelligence, AICI 2012*, pages 769–774, 2012.
197. M. Yang. Smoothing technique and fast alternating direction method for robust PCA. *Chinese Control Conference, CCC 2014*, pages 4782–4785, July 2014.
198. M. Yang and Z. An. Video background modeling using low-rank matrix recovery. *Journal of Nanjing University of Posts and Telecommunications*, April 2013.
199. M. Yang and Y. Wang. Fast alternating direction method of multipliers for robust PCA. *Journal of Nanjing University*, 34(2):83–88, April 2014.
200. X. Yang, X. Gao, D. Tao, X. Li, B. Han, and J. Li. Shape-constrained sparse and low-rank decomposition for auroral substorm detection. *IEEE Transactions on Neural Networks and Learning Systems*, 2015.
201. Q. Yao and J. Kwok. Colorization by patch-based local low-rank matrix completion. *AAAI Conference on Artificial Intelligence*, 2015.
202. X. Ye, J. Yang, X. Sun, K. Li, C. Hou, and Y. Wang. Foreground-background separation from video clips via motion-assisted matrix restoration. *IEEE Transactions on Circuits and Systems for Video Technology*, 2015.
203. X. Yuan. Nuclear-norm-free variational models for background extraction from surveillance video. *Cross-straits Optimization Workshop, COW 2013*, March 2013.
204. X. Yuan and J. Yang. Sparse and low-rank matrix decomposition via alternating direction methods. *Optimization Online*, November 2009.
205. J. Zhan and N. Vaswani. Robust PCA with partial subspace knowledge. *Preprint*, 2014.
206. J. Zhan, N. Vaswani, and C. Qiu. Performance guarantees for ReProCS - correlated low-rank matrix entries case. *Preprint*, 2014.
207. F. Zhang, J. Yang, Y. Tai, and J. Tang. Double nuclear norm-based matrix decomposition for occluded image recovery and background modeling. *IEEE Transactions on Image Processing*, 24(6):1956–1966, June 2015.
208. H. Zhang, J. Cai, L. Cheng, and J. Zhu. Strongly convex programming for exact matrix completion and robust principal component analysis. *Preprint*, January 2012.
209. H. Zhang, Z. Lin, C. Zhang, and E. Chang. Exact recoverability of robust PCA via outlier pursuit with tight recovery bounds. *AAAI Conference on Artificial Intelligence*, 2015.
210. H. Zhang, Z. Lin, C. Zhang, and J. Gao. Relations among some low rank subspace recovery models. *Preprint*, 2014.
211. H. Zhang and L. Liu. Recovering low-rank and sparse components of matrices for object detection. *Electronics Letters*, 49(2), January 2013.
212. S. Zhang and J. Tian. Accelerated algorithms for low-rank matrix recovery. *MIPPR*

2013: Parallel Processing of Images and Optimization and Medical Imaging Processing, October 2013.
213. Z. Zhang, S. Yan, M. Zhao, and F. Li. Bilinear low-rank coding framework and extension for robust image recovery and feature representation. *Knowledge-Based Systems*, 2015.
214. L. Zhao, X. Zhang, Y. Tian, R. Wang, and T. Huang. A background proportion adaptive Lagrange multiplier selection method for surveillance video on high HEVC. *International Conference on Multimedia and Expo, ICME 2013*, July 2013.
215. Q. Zhao, D. Meng, L. Jiang, Q. Xie, Z. Xu, and A. Hauptmann. Self-paced learning for matrix factorization. *AAAI Conference on Artificial Intelligence, AAAI 2015*, January 2015.
216. Q. Zhao, D. Meng, Z. Xu, W. Zuo, and L. Zhang. Robust principal component analysis with complex noise. *ICML 2014*, 2014.
217. X. Zhong, L. Xu, Y. Li, Z. Liu, and E. Chen. A nonconvex relaxation approach for rank minimization problems. *National Conference on Artificial Intelligence, AAAI 2015*, January 2015.
218. T. Zhou and D. Tao. GoDec: randomized low-rank and sparse matrix decomposition in noisy case. *International Conference on Machine Learning, ICML 2011*, 2011.
219. T. Zhou and D. Tao. Greedy bilateral sketch, completion and smoothing for large-scale matrix completion, robust PCA and low-rank approximation. *AISTATS 2013*, 2013.
220. X. Zhou, C. Yang, H. Zhao, and W. Yu. Low-rank modeling and its applications in image analysis. *Preprint*, 2014.
221. Z. Zhou, X. Li, J. Wright, E. Candes, and Y. Ma. Stable principal component pursuit. *IEEE ISIT Proceedings*, pages 1518–1522, June 2010.
222. W. Zhu, S. Shu, and L. Cheng. Proximity point algorithm for low-rank matrix recovery from sparse noise corrupted data. *Applied Mathematics and Mechanics*, 35(2):259–268, February 2014.
223. D. Zonoobi and A. Kassim. Lowrank and sparse matrix reconstruction with partial support knowledge for surveillance video processing. *International Conference on Image Processing, ICIP 2013*, September 2013.
224. H. Zou, T. Hastie, and T. Tibshirani. Sparse principal component analysis. *Journal of Computation and Graphical Statistics*, 15(2):265–286, 2006.

2

Algorithms for Stable PCA

Necdet Serhat Aybat
Penn State University, USA

2.1 Introduction

Principal component analysis (PCA) is an essential tool in applications ranging from image and video processing, to web data analysis and bioinformatics. PCA seeks the best low-dimensional approximation to high-dimensional data. In particular, let the data matrix $A \in \mathbb{R}^{m\times n}$ be of the form $A := L^o + N^o$, where L^o is a *low-rank* matrix, i.e., $\mathbf{rank}(L^o) \ll \min\{m,n\}$, and let N^o be a dense small-magnitude noise matrix. PCA generates a rank-$\bar{r}$ approximation to L^o by solving $\min_L\{\|L - A\| : \mathbf{rank}(L) \leq \bar{r}\}$, which requires computing one singular value decomposition (SVD) of A, where $\|X\|$ denotes the spectral norm of $X \in \mathbb{R}^{m\times n}$, i.e., maximum singular value of X. However, when the observed data is corrupted by *gross errors*, classical PCA becomes impractical because even a single grossly corrupted observation can destroy the underlying low-rank structure in such a way that it cannot be recovered using PCA. To remedy this shortcoming, an extended model called robust PCA (RPCA) was considered by Wright et al. [62], Candès et al. [12], and Chandrasekaran et al. [13] when the gross errors are *sparse*. In this model it is assumed that the data matrix $A \in \mathbb{R}^{m\times n}$ is of the form $A := L^o + S^o$, where L^o is a low-rank matrix as in PCA, and S^o is a sparse gross error matrix, i.e., the number of nonzero elements is small. This model attempts to decompose the observed matrix A as a sum of two matrices: a low-rank approximation to A, and a sparse matrix that captures the gross errors. Under certain incoherence assumptions on L^o, and randomness assumption on the sparsity pattern of S^o, Candès et al. [12] showed that one can recover (L^o, S^o) exactly with very high probability as the unique solution of a convex optimization problem, called *Principal Component Pursuit* (PCP).

On the other hand, the observations in real-life applications are often corrupted by noise, possibly affecting every entry of the data matrix, which can be represented with a dense noise matrix N^o. Hence, suppose that the data matrix A can be decomposed as $A = L^o + S^o + N^o$, where L^o and S^o are low-rank and sparse matrices as in robust PCA. However, introducing a third term in the decomposition renders PCP inapplicable as A is not a sum of low-rank and sparse components anymore. To overcome this issue and achieve stable PCA with noisy observations, Zhou et al. [65] proposed another convex model called *stable* PCP (SPCP), and showed that in the presence of dense noise N^o solving SPCP guarantees stable recovery of L^o with an error bound proportional to the overall noise magnitude $\|N^o\|_F$. In the following sections, after we briefly review the statistical recovery guarantees of SPCP and mention some other alternative models for stable recovery in the presence of entry-wise noise, we mainly focus on reviewing iterative algorithms proposed in the literature to solve the SPCP problem and its variants; and discuss their key characteristics, such as convergence rate, and computational complexity per iteration. First, we go over the notation adopted throughout the chapter.

2.1.1 Notation

Let $\mathbb{R}_+ := \{t \in \mathbb{R} : t \geq 0\}$, $\mathbb{R}_{++} := \mathbb{R}_+ \setminus \{0\}$, and sgn(.) denote the signum function, i.e., given $x \in \mathbb{R}$, $\text{sgn}(x)$ is equal to -1, 0, and 1, if $x < 0$, $x = 0$, or $x > 0$, respectively; moreover, for any $x \in \mathbb{R}^n$ and $X \in \mathbb{R}^{m\times n}$, both $\text{sgn}(x)$ and $\text{sgn}(X)$ operate componentwise. $e_j \in \mathbb{R}^n$ denotes the unit vector with the j-th entry as 1, $\mathbf{0}_n$ and $\mathbf{1}_n$ denote vectors in $\mathbb{R}^n$ with all entries equal to 0 and 1, respectively; similarly, $\mathbf{0}_{m\times n}$ and $\mathbf{1}_{m\times n}$ denote matrices in $\mathbb{R}^{m\times n}$ with all entries equal to 0 and 1, respectively. $\mathbf{I}_n$ represents $n \times n$ identity matrix, and given $\sigma \in \mathbb{R}^n$, $\mathbf{diag}(\sigma)$ represents the $n \times n$ diagonal matrix with its diagonal equal to σ. Moreover, given $\mathcal{X} = \{x^{(i)}\}_{i=1}^n \subset \mathbb{R}^m$, $[x^{(1)}, \ldots, x^{(n)}] \in \mathbb{R}^{m\times n}$ represents the matrix of which columns are the elements of $\mathcal{X}$; and $[x^{(1)}; \ldots; x^{(n)}] \in \mathbb{R}^{mn}$ represents a long vector obtained by vertically stacking the elements of $\mathcal{X}$.

Given $X \in \mathbb{R}^{m\times n}$, the nuclear norm $\|X\|_* := \sum_{i=1}^{\mathbf{rank}(X)} \sigma_i(X)$, and the spectral norm $\|X\| := \max\{\sigma_i(X) : \ i = 1, \ldots, \mathbf{rank}(X)\}$, where $\{\sigma_i(X)\}_{i=1}^{\mathbf{rank}(X)} \subset \mathbb{R}_{++}$ denotes the singular values of X; the ℓ_1-norm $\|X\|_1 := \sum_{i=1}^m \sum_{j=1}^n |X_{ij}|$; the ℓ_∞-norm $\|X\|_\infty := \max\{|X_{ij}| : i = 1, \ldots, m, \ j = 1, \ldots, n\}$; the Frobenius norm $\|X\|_F := \sqrt{\sum_{i=1}^m \sum_{j=1}^n Z_{ij}^2}$; and the ℓ_0-"norm" $\|X\|_0$ denotes the number of *non-zero* components of the matrix X. Let $\Omega \subset \{(i,j) : \ i = 1, \ldots, m, \ j = 1, \ldots, n\}$, and define the projection operator $\pi_\Omega : \mathbb{R}^{m\times n} \to \mathbb{R}^{m\times n}$ as follows:

$$(\pi_\Omega(X))_{ij} = \begin{cases} X_{ij}, (i,j) \in \Omega, \\ 0, \quad \text{otherwise}. \end{cases} \tag{2.1}$$

The operator π_{Ω^c} is defined similarly for the complement set Ω^c. Note that the adjoint operator $\pi_\Omega^* = \pi_\Omega$. Given $X, \bar{X} \in \mathbb{R}^{m\times n}$, $X \odot \bar{X}$ denotes componentwise multiplication. Throughout the chapter, $\mathbb{P}(\mathcal{A})$ denotes the probability of some random event $\mathcal{A}$, and the acronym i.i.d. is used for independent and identically distributed. Given a set $\mathcal{X} \subset \mathbb{R}^{m\times n}$, define its indicator function $\mathcal{I}_\mathcal{X} : \mathbb{R}^{m\times n} \to \mathbb{R} \cup \{+\infty\}$ such that

$$\mathcal{I}_\mathcal{X}(X) := \begin{cases} 0, & \text{if } X \in \mathcal{X}; \\ +\infty, \text{o.w.} \end{cases}$$

Moreover, given a closed convex function $\psi : \mathbb{R}^{m\times n} \to \mathbb{R} \cup \{+\infty\}$, $\text{prox}\,\psi : \mathbb{R}^{m\times n} \to \mathbb{R}^{m\times n}$ denotes the proximal map of ψ, also called prox map of ψ, i.e., $\text{prox}\,\psi(X) := \min_Z\{\psi(Z) + \frac{1}{2}\|Z - X\|_F^2\}$.

2.1.2 Recovery Guarantees

Suppose that the data matrix $A \in \mathbb{R}^{m\times n}$ can be decomposed as

$$A = L^o + S^o + N^o, \tag{2.2}$$

where L^o is a *low-rank* matrix, i.e., $r = \mathbf{rank}(L^o) \ll \min\{m,n\}$, S^o is a *sparse* gross "error" matrix, i.e., $s := \|S^o\|_0 \leq mn$, and N^o is a dense noise matrix such that $\|N^o\|_F \leq \delta$ for some $\delta > 0$. In order to stably recover L^o in the presence of the dense noise matrix N^o, Zhou et al. [65] proposed solving the following convex optimization problem, called *Stable Principal Component Pursuit* (SPCP):

$$(L^*, S^*) \in \operatorname*{argmin}_{L,S\in\mathbb{R}^{m\times n}} \{\|L\|_* + \lambda\|S\|_1 : \ \|L+S-A\|_F \leq \delta\}, \tag{2.3}$$

where $\lambda := \frac{1}{\sqrt{\max\{m,n\}}}$. Note that if there is no noise on observations, i.e., $A = L^o + S^o$ and $N^o = \mathbf{0}_{m\times n}$, then $\delta = 0$ and the SPCP formulation in (2.3) is equivalent to the PCP formulation

$$\min_{L,S\in\mathbb{R}^{m\times n}} \{\|L\|_* + \lambda\|S\|_1 : \ L+S=A\}, \tag{2.4}$$

which is discussed in Chapter 1; hence, PCP is a special case of SPCP.

It is clear that if either L^o is sparse, or S^o is low-rank, then recovering unknown components (L^o, S^o) through solving SPCP is hopeless. On the other hand, let $L^o = U\Sigma V^\top$ denote the SVD of L^o, i.e., $U \in \mathbb{R}^{m\times r}$, $V \in \mathbb{R}^{n\times r}$, and $\Sigma \in \mathbb{R}^{r\times r}$ such that $U^\top U = V^\top V = \mathbf{I}_r$ and $\Sigma = \mathbf{diag}(\sigma)$, where $\sigma \in \mathbb{R}^r_{++}$ denote the singular values of L^o; and suppose that L^o and S^o satisfy the following additional assumptions:

$$\max_{1\leq i\leq r} \|U^T e_i\|_2^2 \leq \frac{\mu r}{m}, \quad \max_{1\leq i\leq r} \|V^T e_i\|_2^2 \leq \frac{\mu r}{n}, \quad \|UV^T\|_\infty \leq \sqrt{\frac{\mu r}{mn}}, \tag{2.5}$$

for some $\mu > 0$, and the set of indices corresponding to non-zero components of S^o is distributed uniformly at random among all the subsets of cardinality s. These additional assumptions make sure that *a)* singular vectors of L^o are not sparse – thus, L^o is not sparse; *b)* S^o is not low-rank with high probability.

Under these assumptions, it is shown in [12] that if $N^o = \mathbf{0}_{m\times n}$, then there exists $c > 0$ such that solving the PCP problem (2.4) *exactly* recovers L^o and S^o, i.e., $(L^*, S^*) = (L^o, S^o)$ when $\delta = 0$ in (2.3), with probability of at least $1 - c\max\{m,n\}^{-10}$, provided

$$\mathbf{rank}(L^o) \leq \rho_r \mu^{-1} \min\{m,n\} \left(\log(\max\{m,n\})\right)^{-2} \quad \text{and} \quad \|S^o\|_0 \leq \rho_s mn \tag{2.6}$$

for some constants ρ_r, $\rho_s > 0$. Comparable to the *exact recovery* property of PCP when $N^o = \mathbf{0}_{m\times n}$, it is shown in [65] that under the assumptions stated above, if $N^o \neq \mathbf{0}_{m\times n}$ such that $\|N^o\|_F \leq \delta$, then the solution (L^*, S^*) to the SPCP problem in (2.3) satisfies $\|L^o - L^*\|_F^2 + \|S^o - S^*\|_F^2 \leq Cmn\delta^2$ for some constant C with *high probability.*

In many applications, some of the entries of A may not be available. Let $\Omega \subset \{(i,j) : i = 1,\ldots,m, \ j = 1,\ldots,n\}$ be the index set of the observable entries of A. Suppose that Ω is chosen uniformly at random among subsets of cardinality pmn for some sufficiently large $p \in (0,1]$, L^o satisfies (2.5), S^o is such that for each $(i,j) \in \Omega$, $\mathbb{P}(S^o_{ij} = 0) = 1 - q$ independently of the others for some $q \in (0,1)$, and $N^o = \mathbf{0}_{m\times n}$. Then it is also shown in [12] that there exits $c > 0$ such that with probability at least $1 - c^{-10}$, solving the following modification of PCP,

$$\min_{L,S\in\mathbb{R}^{m\times n}} \{\|L\|_* + \lambda\|S\|_1 : \ \mathcal{P}L + S - A = \mathbf{0}_{m\times n}\} \tag{2.7}$$

with $\lambda = \frac{1}{\sqrt{p\max\{m,n\}}}$, exactly recovers the true low-rank component L^o provided $\mathbf{rank}(L^o) \leq \rho_r \mu^{-1} \min\{m,n\} \left(\log(\max\{m,n\})\right)^{-2}$ and $q \leq q_s$ for some constants ρ_r, $q_s > 0$. For applications with both missing and noisy observations, i.e., $\Omega^c \neq \emptyset$ and $N^o \neq \mathbf{0}_{m\times n}$, Tao and Yuan [59] proposed recovering the low-rank and sparse components of A by solving

$$\min_{L,S\in\mathbb{R}^{m\times n}} \{\|L\|_* + \lambda\|S\|_1 : \|\mathcal{P}L + S - A\|_F \leq \delta\}. \tag{2.8}$$

2.1.3 Overview of Algorithms

When there is no noise in the observations, i.e., $N^o = \mathbf{0}_{m\times n}$ and $A = L^o + S^o$, there are many efficient algorithms for solving the PCP problems in (2.4) and (2.7); on the other hand, in the presence of dense noise, i.e., $N^o \neq \mathbf{0}_{m\times n}$, surprisingly there are only very few methods that can efficiently deal with (2.8) directly. Table 2.1 shows an overview of the Stable Principal Component Pursuit methods that will be discussed in Section 2.2 and Section 2.3.

TABLE 2.1 Stable Principal Component Pursuit: An Overview of Algorithms.

Algorithms	Problem	Authors - Dates
ASALM & VASALM: multi-block ADMM type alg.	(2.8)	Tao and Yuan (2009) [59]
FALC: augmented Lagrangian alg.	(2.8)	Aybat et al. (2010) [5]
SPG: accelerated Nesterov-type alg. + smoothing	(2.8)	Aybat et al. (2011) [3]
PSPG: accelerated proximal gradient alg. + smoothing	(2.8)	Aybat et al. (2013) [3, 4]
ADMIP: increasing penalty ADMM alg.	(2.8)	Aybat et al. (2013) [3, 6]
Quasi Newton method	(2.55)	Aravkin et al. (2014) [1]
FWP: Frank-Wolfe + projected gradient alg.	(2.57)	Mu et al. (2014) [45]
Two-block ADMM alg. for non-negative SPCP	(2.54)	Huai et al. (2015) [35]
DFC: a parallel randomized matrix approximation alg.		Mackey et al. (2011) [43]
GoDec: alternating minimization alg.	(2.60)	Zhou and Tao (2011) [63]
AMS: inexact alternating minimization alg.	(2.61)	Hintermüller and Wu (2014) [33]
LMaFit: multi-block ADMM alg.	(2.62)	Shen et al. (2011) [56]
GreBsmo: multi-block alternating minimization alg.	(2.63)	Zhou and Tao (2013) [64]
AD-MoM: multi-block ADMM alg.	(2.65)	Mardani et al. (2011) [44]

When Zhou et al. proposed the SPCP problem in [65], instead of directly solving it with a customized algorithm for (2.3), they computed a solution to the problem in Lagrangian form:

$$\min_{L,S\in\mathbb{R}^{m\times n}} \|L\|_* + \lambda\|S\|_1 + \frac{\rho}{2}\|L + S - A\|_F^2. \tag{2.9}$$

for an appropriately chosen dual variable $\rho > 0$ using the accelerated proximal gradient (APG) algorithm in [40]. It follows from convex duality that for any given $\delta > 0$, there exists $\rho(\delta) \geq 0$ such that (2.9) is equivalent to (2.3). On the other hand, it should be emphasized that while $\delta > 0$ is usually readily available, and has some natural physical meaning, corresponding $\rho(\delta)$ is usually hard to guess, and tuning ρ requires cross-validation.

The first tailor-made algorithms for the SPCP problem in (2.8) is proposed by Tao and Yuan [59]: a *multi-block* ADMM algorithm, and its variant. While the former one is significantly faster in the numerical tests conducted in [59], it does not have theoretical convergence guarantees that the latter algorithm has. The first $\mathcal{O}(1/\epsilon)$ sublinear convergence rate result for a first-order algorithm customized to (2.8) is shown in [5]. The algorithm proposed in [5] by Aybat and Iyengar is an inexact augmented Lagrangian algorithm; although it is customized for minimizing constrained composite norm problems, it does not fully exploit the special constraint structure in (2.8). To utilize the structure in constraints, Aybat, Goldfarb, and Ma [4] showed that (2.8) can be equivalently formulated as

$$\min_{L,S\in\mathbb{R}^{m\times n}} \{\|L\|_* + \lambda\|\mathcal{P}S\|_1 : \|L + S - \mathcal{P}A\|_F \leq \delta\}, \tag{2.10}$$

in the sense that if (L^*, S^*) is an optimal solution to (2.10), then $(L^*, \mathcal{P}S^*)$ is an optimal solution to the SPCP problem in (2.8); and exploited this equivalence to develop first-order algorithms for (2.10): an alternating linearization method (ALM) for $\delta = 0$, and an accelerated proximal gradient (APG) method for $\delta > 0$. Later, Aybat and Iyengar developed a variable penalty ADMM algorithm to solve (2.8), which has desirable convergence guarantees unlike the multi-block ADMM in [59]. More recently, Aravkin et al. [1] proposed a "quasi-Newton" method to solve a more general formulation that includes the SPCP problem in (2.8) as a special case; and Mu et al. [45] proposed augmenting the Frank-Wolfe method [21] with additional projected gradient steps to solve a flipped formulation related to (2.8), i.e., $\min\{\|\mathcal{P}L + S - A\|_F : \|L\|_* \leq \tau_L, \|S\|_1 \leq \tau_S\}$ for some $\tau_L, \tau_S > 0$. The theoretical and practical convergence properties of these algorithms will be discussed in more detail in the following section. Moreover, there are also other non-convex variants of the SPCP formulation, and iterative solution methods for them; we will very briefly go over these formulations and algorithms in Section 2.3.

Finally, it should also be emphasized that in order to achieve a fast real-time implementation by exploiting modern multi-core and/or distributed computing capabilities, there are some randomized techniques that divide the stable decomposition problem into smaller decomposition subproblems, which can be solved in parallel by calling one of the decomposition algorithms discussed above as a base solver, and finally the subproblem solutions are combined at the end to generate a decomposition for the whole data. In particular, Mackey et al. [43] proposed a framework of this sort, Divide-Factor-Combine (DFC). DFC *randomly* divides the original matrix factorization task into smaller subproblems, and combines the smaller decompositions using an efficient *randomized matrix approximation* technique. The inherent parallelism of DFC allows for near-linear to superlinear speedups in practice, while the theory provides high-probability recovery guarantees for DFC comparable to those possessed by the base algorithm. Moreover, Mackey et al. [43] were able to reduce recovery error of their decomposition further by using ensemble methods that are known to improve the performance of matrix approximation algorithms [38].

2.1.4 Application: Foreground Extraction from Noisy Videos

Extracting the *almost* still background from a sequence of frames in a noisy video is an important task in video surveillance, and interestingly it can be formulated as an SPCP problem if the moving objects in the foreground only occupy a small fraction of each frame. This problem is difficult due to the presence of slightly changing background in the video, e.g., changing illumination conditions such as flickering lights, and/or slowly moving objects such as escalators or tree leaves. Let $F^{(t)}\mathbb{R}^{n_1 \times n_2}$ denote the t-th video frame, and $f^{(t)} \in \mathbb{R}^R$ is obtained by stacking the columns of $F^{(t)}$, where $R = n1n_2$ is the frame resolution. For now, suppose the background is completely stationary, and there is no measurement noise. Then $f^{(t)} = b + s^{(t)}$, where $b \in \mathbb{R}^R$ denotes the background and $s^{(t)} \in \mathbb{R}^R$ denotes the sparse foreground in the t-th frame. Hence, T-frame video can be encoded as matrix $\mathbb{R}^{R\times T} \ni A = [f^{(t)}, \ldots, f^{(T)}]$ such that $A = b\mathbf{1}^\top + [s^{(1)}, \ldots, s^{(T)}]$, i.e., summation of a rank 1 matrix + sparse matrix.

In real-life surveillance videos, the background is never completely stationary, and there is always measurement noise; therefore, we expect that A can be decomposed into the sum of three matrices $A = L^o + S^o + N^o$, where L^o is low rank and S^o is sparse matrices that represent the background and the foreground, respectively, and N^o is a dense noise matrix. When the background changes periodically, e.g., flickering light, or very slowly, e.g., moving escalators, the matrix L^o, which represents the backgrounds in the frames, should be of low rank due to the high correlation between the video frames. The matrix S^o, which represents the moving foregrounds in the frames, should be sparse since the foreground usually occupies a small portion of each frame. Moreover, suppose that $\Omega \subset \{(i,t) : i = 1, \ldots, R,\ t = 1, \ldots, T\}$ denotes the subset of sensor-frame pairs (i,t) such that the signal recorded by i-th sensor in frame-t is completely corrupted due to malfunction. Hence, for foreground-background separation one can solve (2.8) with an appropriately chosen $\lambda > 0$.

2.2 Algorithms for SPCP

This section focusses on algorithms for solving the SPCP formulations in (2.3), (2.8), and (2.10). In the first subsection, we describe the ADMM method and its variant proposed by Tao and Yuan [59], and discuss their convergence properties. In the second subsection, after we go over basic properties of accelerated proximal gradient (APG) methods, we describe the APG implementations for the SPCP problem proposed in [3, 4]. Next, in the third subsection, we revisit the ADMM algorithms for solving the SPCP problem and describe the increasing penalty ADMM algorithm proposed in [6]. Finally, in the fourth and fifth subsections, we briefly review the convex variational framework proposed by Aravkin et al. [1], and the Frank-Wolfe projected gradient method proposed by Mu et al. [45] for solving a convex problem related to the SPCP formulation.

2.2.1 ADMM for SPCP

In this section, we discuss the first tailor-made algorithms for the SPCP problem: the alternating splitting augmented Lagrangian method (ASALM), and its variant (VASALM) proposed by Tao and Yuan in [59] for solving an equivalent formulation of (2.8):

$$\min_{L,S,N\in\mathbb{R}^{m\times n}} \{\|L\|_* + \lambda\|S\|_1 + \mathcal{I}_{\mathcal{X}}(N) : \ L + S + N = \mathcal{P}A\}, \tag{2.11}$$

where $\mathcal{X} = \{N \in \mathbb{R}^{m\times n} : \ \|\mathcal{P}N\|_F \leq \delta\}$, and $\mathcal{I}_{\mathcal{X}}$ denote the indicator function of the set $\mathcal{X}$. Given a constant penalty parameter $\rho > 0$, and a dual variable $Y \in \mathbb{R}^{m\times n}$, let $\mathcal{L}_\rho$ denote the augmented Lagrangian of (2.11):

$$\begin{aligned} \mathcal{L}_\rho(L, S, N; Y) := &\|L\|_* + \lambda\|S\|_1 + \mathcal{I}_{\mathcal{X}}(N) \\ &- \langle Y, L + S + N - \mathcal{P}A\rangle + \frac{\rho}{2}\|L + S + N - \mathcal{P}A\|_F^2. \end{aligned} \tag{2.12}$$

Convergence to optimality would be an immediate result, if one were to compute the primal-dual iterate sequence using the *method of multipliers* [31, 50]: $(L_k, S_k, N_k) \in \operatorname{argmin} \mathcal{L}_{\rho_k}(L, S, N; Y_k)$ and $Y_{k+1} = Y_k - \rho_k(L_k + S_k + N_k - \mathcal{P}A)$, where $\{\rho_k\}_{k\in\mathbb{Z}_+} \subset \mathbb{R}_{++}$. More precisely, Tao and Yuan [59] gave a convergence proof for *method of multipliers* customized to solve (2.11) when $\{\rho_k\}_{k\in\mathbb{Z}_+} \subset \mathbb{R}_{++}$ is chosen such that $\rho_{k+1} = \kappa\rho_k$ for some $\kappa > 1$, and showed that the primal-dual iterate sequence $\{(L_k, S_k, N_k, Y_k)\}_{k\in\mathbb{Z}_+}$ is bounded, and *any* of its *limit points* is an optimal solution to (2.11). However, minimizing $\mathcal{L}_\rho$ jointly in (L, S, N) is not *practical* at all, and it is almost as hard as solving (2.11). On the other hand, in order to exploit the separability of both the objective and constraint functions, Tao and Yuan considered minimizing the augmented Lagrangian alternatingly in one of the variables: L, S, or N, while fixing the other two variables. Hence, given a constant penalty parameter $\rho > 0$, they proposed constructing the ASALM iterate sequence as follows:

$$N_{k+1} = \operatorname*{argmin}_N \mathcal{L}_\rho(L_k, S_k, N; Y_k), \tag{2.13}$$

$$S_{k+1} = \operatorname*{argmin}_S \mathcal{L}_\rho(L_k, S, N_{k+1}; Y_k), \tag{2.14}$$

$$L_{k+1} = \operatorname*{argmin}_L \mathcal{L}_\rho(L, S_{k+1}, N_{k+1}; Y_k), \tag{2.15}$$

$$Y_{k+1} = Y_k - \rho(L_{k+1} + S_{k+1} + N_{k+1} - \mathcal{P}A). \tag{2.16}$$

Compared to minimizing $\mathcal{L}_\rho$ jointly in (L, S, N), the minimization subproblems in (2.13), (2.14), and (2.15) can be solved very efficiently in practice. Indeed, the N-subproblem is nothing but a Euclidean projection, the S-subproblem is the shrinkage or soft-thresholding [18] operation, which is widely used in sparse signals reconstruction algorithms [15], and the L-subproblem is the matrix-shrinkage operation [32], which is used in algorithms for recovering low-rank matrices from few linear measurements [11, 42]; and the solutions to all these subproblems, stated in the following generic forms for some given $\bar{N}, \bar{S}, \bar{L} \in \mathbb{R}^{m\times n}$, can be computed explicitly in closed forms:

$$\begin{aligned} N_+ &= \operatorname*{argmin}_{N\in\mathcal{X}} \|N - \bar{N}\|_F = \mathrm{prox}\mathcal{I}_{\mathcal{X}}(\bar{N}), \\ &= \min\left\{1, \frac{\delta}{\|\mathcal{P}\bar{N}\|_F}\right\}\mathcal{P}\bar{N} + \pi_{\Omega^c}\left(\bar{N}\right), \end{aligned} \tag{2.17}$$

$$\begin{aligned} S_+ &= \operatorname*{argmin}_{S} \lambda\|S\|_1 + \frac{\rho}{2}\|S - \bar{S}\|_F^2 = \mathrm{prox}_{\frac{\lambda}{\rho}}\|.\|_1(\bar{S}), \\ &= \mathrm{sgn}(\bar{S}) \odot \max\left\{|\bar{S}| - \tfrac{\lambda}{\rho}\mathbf{1}_{m\times n},\ \mathbf{0}_{m\times n}\right\}, \end{aligned} \tag{2.18}$$

$$\begin{aligned} L_+ &= \operatorname*{argmin}_{L} \|L\|_* + \frac{\rho}{2}\|L - \bar{L}\|_F^2 = \mathrm{prox}_{\frac{1}{\rho}}\|.\|_*(\bar{L}), \\ &= U\,\mathbf{diag}\left(\max\left\{\sigma - \tfrac{1}{\rho}\mathbf{1}_r,\ \mathbf{0}_r\right\}\right)V^\top, \end{aligned} \tag{2.19}$$

where $U\,\mathbf{diag}(\sigma)V^\top$ denote the SVD of $\bar{L}$, i.e., $\sigma \in \mathbb{R}^r_{++}$ denote the singular values of $\bar{L}$, $U \in \mathbb{R}^{m\times r}$, and $V \in \mathbb{R}^{n\times r}$ such that $U^\top U = V^\top V = \mathbf{I}_r$ and $r = \mathbf{rank}(\bar{L})$.

Compared to the method of multipliers, per-iteration complexity of ASALM is considerably cheaper; moreover, according to numerical results reported in [59], ASALM iterates converge to an optimal solution on synthetic random test problems. However, currently there is no theory that can support this empirical behavior. Note that ASALM is an ADMM algorithm with *three*-blocks, and in a recent paper Chen et al. [14] have shown that direct extension of ADMM algorithm from a *two*-block to a *multi*-block case is not necessarily convergent – we will discuss this issue and possible remedies in more detail in Section 2.2.3.

Tao and Yuan were able to fix the convergence issue using a slightly modified variant of ASALM (VASALM) without any increase in per-iteration complexity. Given algorithm parameters $\kappa > 2$ and $\rho > 0$, the steps of VASALM are as follows:

$$N_{k+1} = \operatorname*{argmin}_{N} \mathcal{L}_\rho(L_k, S_k, N; Y_k), \tag{2.20}$$

$$S_{k+1} = \operatorname*{argmin}_{S} \mathcal{L}_{\kappa\rho}(L_k, S, N_{k+1}; Y_k) + (\kappa - 1)\langle\tilde{Y}_k - Y_k, S\rangle, \tag{2.21}$$

$$L_{k+1} = \operatorname*{argmin}_{L} \mathcal{L}_{\kappa\rho}(L, S_k, N_{k+1}; Y_k) + (\kappa - 1)\langle\tilde{Y}_k - Y_k, L\rangle, \tag{2.22}$$

where $\tilde{Y}_k = Y_k - \rho(L_k + S_k + N_{k+1} - \mathcal{P}A)$ for all $k \geq 1$, and $\{Y_k\}_{k\in\mathbb{Z}_+}$ is defined as in (2.16). Note that for $\kappa = 1$, VASALM and ASALM iterations are very similar, and the only difference is using S_k in (2.22) instead of S_{k+1}. Let $\mathcal{W}^*$ denote the set of primal-dual optimal solutions to the convex problem in (2.11), i.e., $\mathcal{W}^*$ is the convex set of all saddle points of the Lagrangian $\mathcal{L}_0$; and let $\mathcal{X}^* := \{(L^*, S^*, Y^*) :\ \exists N^* \text{ s.t. } (L^*, S^*, N^*, Y^*) \in \mathcal{W}^*\}$ and $X_k := (L_k, S_k, Y_k)$ for all $k \geq 1$. In [59], it is shown that $\{X_k\}_{k\in\mathbb{Z}_+}$ is Fejèr monotone [9] with respect to the closed convex set $\mathcal{X}^*$; hence, it follows that $\{X_k\}_{k\in\mathbb{Z}_+}$ converges to a point in $\mathcal{X}^*$. Hence, although convergence of ASALM iterate sequence is not theoretically guaranteed, VASALM iterate sequence is shown to be converging with a negligible additional execution. The VASALM steps in (2.21) and (2.22) are only slightly different than those in ASALM; however, it is important to note that while ASALM uses S_{k+1} in (2.15) when computing L_{k+1}, VASALM abandons S_{k+1}, and still uses S_k in (2.22). This modification in VASALM helps prove Fejèr monotonicity of the iterate sequence; but it comes at a cost of significant performance degradation in practice – according to numerical results reported in [59], ASALM performs much better than VASALM: the number of SVD computations for ASALM is almost half of what VASALM computes.

Before concluding this section, we will discuss another ADMM implementation for *stable* PCA. Recall that Zhou et al. [65] computed a solution to the Lagrangian formulation in (2.9) using an APG algorithm proposed in [40]. In [59], Tao and Yuan proposed the extensions of ASALM and VASALM to solve the Lagrangian formulation with missing data:

$$\min_{L,S\in\mathbb{R}^{m\times n}} \left\{\|L\|_* + \lambda\|S\|_1 + \frac{\rho}{2}\|\mathcal{P}N\|_F^2 : \ L + S + N = \mathcal{P}A\right\}. \tag{2.23}$$

Note that if all the data is observable, then (2.23) and (2.9) are equivalent. Moreover, given $\delta > 0$, it follows from convex duality that there exists $\rho(\delta)$ such that (2.23) and (2.11) are equivalent. Extension of ASALM for solving (2.23) is also an ADMM algorithm with *three*-blocks, and like ASALM for (2.11), it also lacks proof of convergence; on the other hand, the same convergence guarantees of VASALM for (2.11) also hold for VASALM for (2.23). Again this convergence assurance comes at a cost of significant practical performance degradation compared to ASALM.

2.2.2 APG for SPCP

In Section 2.2.1, we have seen two ADMM-type methods: ASALM and VASALM. While ASALM works well in practice, it does not have any convergence guarantees; on the other hand, while VASALM iterate sequence is shown to be Fejèr monotone with respect to the primal-dual optimal solution set, hence, converging to an optimal solution, its practical performance is significantly worse than ASALM. Moreover, iteration complexity of VASALM is *not* known. In this section, we discuss how a particular class of first-order methods with *optimal* iteration complexity can be efficiently implemented to compute an ϵ-optimal solution to (2.8) within $\mathcal{O}(1/\epsilon)$ iterations, where each iteration requires computing an SVD as a bottleneck operation, similar to ASALM and VASALM. First, as a preliminary, we briefly go over the basics of the *accelerated proximal gradient* (APG) method, which is the backbone of the algorithms proposed in [3, 4]; and next, we discuss the convergence properties of these algorithms to solve (2.10).

Preliminaries: Convergence of APG and Nesterov Smoothing

Let $\psi : \mathbb{R}^{m\times n} \to \mathbb{R}\cup\{+\infty\}$ and $\phi : \mathbb{R}^{m\times n} \to \mathbb{R}$ be proper, closed, convex functions such that ϕ is differentiable on an open set containing $\mathbf{dom}\,\psi$, and $\operatorname{grad}\phi$ is Lipschitz continuous on $\mathbf{dom}\,\psi$ with constant C_ϕ, i.e., $\|\operatorname{grad}\phi(X) - \operatorname{grad}\phi(\tilde{X})\|_F \leq C_\phi\|X - \tilde{X}\|_F$ for all $X, \tilde{X} \in \mathbf{dom}\,\psi$. Consider the following *composite* convex optimization problem:

$$P^* := \min_{X\in\mathbb{R}^{m\times n}} P(X) := \psi(X) + \phi(X). \tag{2.24}$$

Let $\mathcal{X} \subset \mathbb{R}^{m\times n}$ be a closed convex set; for the case $\psi = \mathcal{I}_{\mathcal{X}}$, i.e., the indicator function of $\mathcal{X}$, Nesterov [47, 48] proposed accelerated proximal gradient algorithms such that the iterate sequence $\{X_k\}_{k\in\mathbb{Z}_+}$ satisfies: $P(X_k) - P^* \leq \mathcal{O}(C_\phi/k^2)$ for all $k \geq 1$; and each iteration requires computing $\operatorname{grad}\phi$ and Euclidean projection(s) onto $\mathcal{X}$. Later, in [49] Nesterov extended the APG algorithm in [48] to solve (2.24) with the same iteration complexity of

$\mathcal{O}(C_\phi/k^2)$ when ψ is a general closed convex function. Each iteration of the method in [49] requires *two* proximal map computations and uses all the iterates from previous iterations, i.e., it is an ∞-memory algorithm; on the other hand, the method in [7] proposed by Beck and Teboulle, and the one in [60] proposed by Tseng extend the algorithm for *constrained smooth minimization* in [46] by Nesterov to solve *composite* convex minimization in (2.24) with $\mathcal{O}(C_\phi/k^2)$ iteration complexity. Moreover, this method requires *one* proximal map computation per iteration and only uses the past iterate from the previous iteration, i.e., 1-memory algorithm. The steps of APG algorithm are displayed in Algorithm 2.1.

Algorithm 2.1 - APG: Accelerated Proximal Gradient algorithm - $\mathbf{APG}(X_0)$

1: **input:** $X_0 \in \mathbb{R}^{m\times n}$
2: $k \leftarrow 0$, $t_0 \leftarrow 1$, $\bar{X}_0 \leftarrow X_0$
3: **while** $k \geq 0$ **do**
4: $\quad Q_k \leftarrow \bar{X}_k - \frac{1}{C_\phi}\operatorname{grad}\phi(\bar{X}_k)$
5: $\quad X_{k+1} \leftarrow \operatorname*{argmin}_X \psi(X) + \frac{C_\phi}{2}\|X - Q_k\|_F^2$
6: $\quad t_{k+1} \leftarrow \left(1 + \sqrt{1+4t_k^2}\right)/2$
7: $\quad \bar{X}_{k+1} \leftarrow X_{k+1} + \left(\frac{t_k - 1}{t_{k+1}}\right)(X_{k+1} - X_k)$
8: $\quad k \leftarrow k+1$
9: **end while**

Let $\{(X_k, \bar{X}_k)\}_{k\in\mathbb{Z}_+}$ denote the APG iterate sequence computed as in Figure 2.1, and $\mathcal{X}^*$ denote the set of optimal solutions to (2.24), i.e., $\mathcal{X}^* = \{X \in \mathbb{R}^{m\times n} : \; P(X) = P^*\}$. In [7, 60], it is shown that for any initial point $X_0 \in \mathbf{dom}\,\psi$, the iterate sequence $\{X_k\}_{k\in\mathbb{Z}_+}$ satisfies

$$0 \leq P(X_k) - P^* \leq \frac{2C_\phi\|X_0 - X^*\|_F^2}{k^2}, \quad \forall\, k \geq 1, \text{ and } \forall X^* \in \mathcal{X}^*. \tag{2.25}$$

Hence, (2.25) implies that for *any* $\epsilon > 0$, APG in Figure 2.1 can compute an ϵ-optimal solution X_ϵ to (2.24), i.e., $P^* \leq P(X_\epsilon) \leq P^* + \epsilon$, within $\mathcal{O}\left(\sqrt{\frac{C_\phi}{\epsilon}}\right)$ APG-iterations.

Consider a special case of a composite convex optimization problem in (2.24):

$$P^* := \min_{L,S\in\mathbb{R}^{m\times n}} P(L,S) := \psi(L,S) + \phi(L), \tag{2.26}$$

where $\psi : \mathbb{R}^{m\times n}\times\mathbb{R}^{m\times n} \to \mathbb{R}\cup\{+\infty\}$, $\phi : \mathbb{R}^{m\times n} \to \mathbb{R}$ are closed convex functions such that $\operatorname{grad}\phi$ is Lipschitz continuous on $\mathbb{R}^{m\times n}$ with constant C_ϕ, i.e., $\|\operatorname{grad}\phi(L) - \operatorname{grad}\phi(\tilde{L})\|_F \leq C_\phi\|L - \tilde{L}\|_F$ for all $L, \tilde{L} \in \mathbb{R}^{m\times n}$. Suppose that Step 5 and Step 7 of the APG algorithm in Figure 2.1 are replaced with (2.27) and (2.28), respectively.

$$(L_{k+1}, S_{k+1}) \leftarrow \operatorname*{argmin}_{(L,S)} \psi(L,S) + \frac{C_\phi}{2}\|L - \bar{L}_k + \frac{1}{C_\phi}\operatorname{grad}\phi(\bar{L}_k)\|_F^2, \tag{2.27}$$

$$\bar{L}_{k+1} \leftarrow L_{k+1} + \left(\frac{t_k - 1}{t_{k+1}}\right)(L_{k+1} - L_k). \tag{2.28}$$

With a slight modification, the proof of Theorem 4.4 in [7] still holds for this case, and implies that the APG algorithm with Steps 5 and 7 replaced with (2.27) and (2.28), respectively, generates an iterate sequence $\{(L_k, S_k)\}_{k\in\mathbb{Z}_+}$ such that

$$P(L_k, S_k) - P^* \leq \frac{2C_\phi\|L_0 - L^*\|_F^2}{k^2}, \quad \forall\, k \geq 1,\ \forall L^* : \exists S^* \text{ s.t. } P(L^*, S^*) = P^*. \tag{2.29}$$

Now, consider the problem in (2.10), which is equivalent to the SPCP problem in (2.8). Since the objective function in (2.10) is the sum of two *non-smooth* functions,

$$f(L) := \|L\|_*, \quad \text{and} \quad g(S) := \lambda \|\mathcal{P}S\|_1,$$

one cannot call the APG algorithm to solve (2.10). On the other hand, using the smoothing technique introduced in [48], one can efficiently implement the APG algorithm on an approximate problem to (2.8). First, the smooth versions of the nuclear norm and ℓ_1-norm are briefly discussed next, and then two different APG implementations using the smoothed versions of these norms are described.

For fixed parameters $\mu > 0$ and $\nu > 0$, the smooth approximations, $f_\mu(.)$ and $g_\nu(.)$, to f and g are defined as follows:

$$f_\mu(L) := \max_{W \in \mathbb{R}^{m\times n}} \left\{ \langle L, W\rangle - \frac{\mu}{2}\|W\|_F^2 : \ \|W\| \le 1 \right\}, \tag{2.30}$$

$$g_\nu(S) := \max_{Z \in \mathbb{R}^{m\times n}} \left\{ \langle \pi_\Omega(S), Z\rangle - \frac{\nu}{2}\|Z\|_F^2 : \ \|Z\|_\infty \le \lambda \right\}. \tag{2.31}$$

Let $W_\mu(L)$ and $Z_\nu(S)$ denote the optimal solutions to (2.30) and (2.31), respectively. For any given $L, S \in \mathbb{R}^{m\times n}$, it is shown in [4] that both solutions can be written in closed form:

$$W_\mu(L) = U \ \mathbf{diag}\left(\min\left\{ \frac{\sigma}{\mu}, \mathbf{1}_r \right\} \right) V^\top, \tag{2.32}$$

$$Z_\nu(S) = \operatorname{sgn}(\mathcal{P}S) \odot \min\left\{ \tfrac{1}{\nu}|\mathcal{P}S|, \ \lambda \mathbf{1}_{m\times n} \right\}, \tag{2.33}$$

where $r = \mathbf{rank}(L)$, and $L = U \ \mathbf{diag}(\sigma) V^\top$ is the singular value decomposition of L with $\sigma \in \mathbb{R}^r_{++}$ denoting the vector of singular values. According to Theorem 1 in [48], both f_μ and g_ν are differentiable, and their gradients can be explicitly written as

$$\operatorname{grad} f_\mu(L) = W_\mu(L), \quad \operatorname{grad} g_\nu(S) = \pi^*_\Omega(Z_\nu(S)) = Z_\nu(S). \tag{2.34}$$

Moreover, both $\operatorname{grad} f_\mu$ and $\operatorname{grad} g_\nu$ are Lipschitz continuous with Lipschitz constants $C_{f_\mu} = 1/\mu$ and $C_{g_\nu} = 1/\nu$, respectively.

Smooth and Partially Smooth Proximal Gradient Algorithms for SPCP

Given an optimization problem $P^* = \min\{P(X) : \ X \in \mathcal{X}\}$ for some function P and set $\mathcal{X}$, in the rest of this section, a point $X_\epsilon \in \mathcal{X}$ is called an ϵ-optimal solution, if $P(X_\epsilon) \le P^* + \epsilon$. Consider the problems approximating (2.10) with a smooth objective function

$$\min_{L,S \in \mathbb{R}^{m\times n}} \{ f_\mu(L) + g_\mu(S) : (L,S) \in \mathcal{X} \}, \tag{2.35}$$

and with a partially smooth objective function

$$\min_{L,S \in \mathbb{R}^{m\times n}} \{ f_\mu(L) + g(S) : (L,S) \in \mathcal{X} \}, \tag{2.36}$$

where

$$\mathcal{X} := \{ (L,S) : \ \|L + S - \mathcal{P}A\|_F \le \delta \}$$

denotes the feasible region of (2.10).

The inexact solutions of (2.35) and (2.36) are closely related to (2.10). Indeed, it is shown in [4] that for $\tau = \frac{1}{2}\min\{m,n\}$, and $\lambda = \frac{1}{\sqrt{\max\{m,n\}}}$, (2.30) and (2.31) imply that

$$f_\mu(L) \le f(L) \le f_\mu(L) + \mu\tau, \quad \forall L \in \mathbb{R}^{m\times n}, \tag{2.37}$$

$$g_\mu(S) \le g(S) \le g_\mu(S) + \mu\tau, \quad \forall S \in \mathbb{R}^{m\times n}. \tag{2.38}$$

Hence, as $\mu \to 0$, f_μ and g_μ uniformly converge to f and g, respectively, i.e., $f_\mu \to f$, and $g_\mu \to g$ uniformly as $\mu \to 0$. According to Theorem 2.1 in [4], for any $\epsilon > 0$, if $(L(\mu), S(\mu))$ is an $\frac{\epsilon}{2}$-optimal solution to the *smooth* problem in (2.35) for $\mu = \frac{\epsilon}{4\tau}$, then $(L(\mu), S(\mu))$ is an ϵ-optimal solution to (2.10). Similarly, with $\mu = \frac{\epsilon}{2\tau}$, an $\frac{\epsilon}{2}$-optimal solution to the *partially smooth* problem in (2.36) is an ϵ-optimal solution to (2.10). In the rest, we state and briefly discuss the Smooth Proximal Gradient (SPG) algorithm (a slightly modified version of Algorithm 1 in [3]), and the Partially Smooth Proximal Gradient (PSPG) algorithm [4] to compute an ϵ-optimal solution to (2.35) or (2.36), respectively, with provable iteration complexity bounds.

Algorithm 2.2 - SPG: Smooth Proximal Gradient algorithm - $\mathbf{SPG}(L_0, S_0, \epsilon)$

1: **input:** $L_0, S_0 \in \mathbb{R}^{m\times n}$, $\epsilon > 0$
2: $\mu \leftarrow \epsilon/(2\min\{m,n\})$
3: $k \leftarrow 0$, $t_0 \leftarrow 1$, $\bar{L}_0 \leftarrow L_0$, $\bar{S}_0 \leftarrow S_0$
4: **while** $k \geq 0$ **do**
5: $[U_k, V_k, \sigma_k] \leftarrow \mathrm{SVD}(\bar{L}_k)$ **Comment** : $\bar{L}_k = U_k\ \mathbf{diag}(\sigma_k)V_k^\top$, and $r_k = \mathbf{rank}(\bar{L}_k)$
6: $Q_k^L \leftarrow U_k\, \mathbf{diag}\left(\max\{\sigma_k - \mu\mathbf{1}_{r_k},\ \mathbf{0}_{r_k}\}\right) V^\top$
7: $Q_k^S \leftarrow \mathrm{sgn}(\mathcal{P}\bar{S}_k) \odot \max\left\{|\mathcal{P}\bar{S}_k| - \mu\lambda\mathbf{1}_{m\times n},\ \mathbf{0}_{m\times n}\right\}$
8: $(L_{k+1}, S_{k+1}) \leftarrow \underset{(L,S)}{\mathrm{argmin}}\left\{\|L - Q_k^L\|_F^2 + \|S - Q_k^L\|_F^2 :\ (L,S) \in \mathcal{X}\right\}$
9: $t_{k+1} \leftarrow \left(1 + \sqrt{1 + 4t_k^2}\right)/2$
10: $\bar{L}_{k+1} \leftarrow L_{k+1} + \left(\frac{t_k - 1}{t_{k+1}}\right)(L_{k+1} - L_k)$
11: $\bar{S}_{k+1} \leftarrow S_{k+1} + \left(\frac{t_k - 1}{t_{k+1}}\right)(S_{k+1} - S_k)$
12: $k \leftarrow k + 1$
13: **end while**

The SPG algorithm is an implementation of the APG algorithm, displayed in Figure 2.1, to solve (2.35). In particular, let ψ and ϕ in (2.24) be set as follows: $\psi(L,S) = \mathcal{I}_{\mathcal{X}}(L,S)$ and $\phi(L,S) = f_\mu(L) + g_\mu(S)$, where $\mu = \frac{\epsilon}{4\tau}$ and $\tau = \frac{1}{2}\min\{m,n\}$. It follows from (2.34) that $\mathrm{grad}\,\phi$ is Lipschitz continuous with constant $1/\mu$; hence, $C_\phi = 1/\mu$. Therefore, Step 4 of Figure 2.1 corresponds to computing $[Q_k^L, Q_k^S] = [\bar{L}_k - \mu\,\mathrm{grad}\,f_\mu(\bar{L}_k),\ \bar{S}_k - \mu\,\mathrm{grad}\,g_\mu(\bar{S}_k)]$. However, first computing $\mathrm{grad}\,f_\mu(\bar{L}_k)$ according to (2.32) and (2.34), and then computing Q_k^L can cause numerical problems. It is easy to check that

$$Q_k^L = U_k\, \mathbf{diag}\left(\max\{\sigma_k - \mu\mathbf{1}_{r_k},\ \mathbf{0}_{r_k}\}\right) V^\top, \tag{2.39}$$

where $\bar{L}_k = U_k\ \mathbf{diag}(\sigma_k)V_k^\top$ denotes the SVD of $\bar{L}_k$ and $r_k = \mathbf{rank}(\bar{L}_k)$. Hence, one can compute Q_k^L directly without computing $\mathrm{grad}\,f_\mu(\bar{L}_k)$ to improve numerical stability. Similarly, (2.33) and (2.34) imply that Q_k^S can also be directly computed without computing $\mathrm{grad}\,g_\mu(\bar{S}_k)$ as follows

$$Q_k^S = \mathrm{sgn}(\mathcal{P}\bar{S}_k) \odot \max\left\{|\mathcal{P}\bar{S}_k| - \mu\lambda\mathbf{1}_{m\times n},\ \mathbf{0}_{m\times n}\right\}.$$

Therefore, Step 4 of Figure 2.1 can be computed as in Step 6 and Step 7 of the SPG algorithm displayed in Algorithm 2.2.

Given $(L_0, S_0) \in \mathcal{X} = \{(L,S) :\ \|L+S-\mathcal{P}A\|_F \leq \delta\}$, e.g., $L_0 = \mathbf{0}$ and $S_0 = \mathcal{P}A$, the SPG algorithm keeps all iterates in $\mathcal{X}$; and according to (2.25), the convergence rate in function values of the SPG iterate sequence $\{(L_k, S_k)\}$ is $\mathcal{O}\left(\frac{\mu^{-1}}{k^2}\right)$. In particular, given $\epsilon > 0$, since $\mu = \frac{\epsilon}{2\min\{m,n\}}$, according to the discussion above, SPG can compute an ϵ-optimal solution to the SPCP problem in (2.10) by computing an $\frac{\epsilon}{2}$-optimal solution to (2.35) within at most $N_{\text{SPG}} = 2\frac{\sqrt{2\min\{m,n\}}}{\epsilon}\left(\|L_0 - L^*(\mu)\|_F^2 + \|S_0 - S^*(\mu)\|_F^2\right)^{\frac{1}{2}}$ iterations, where $(L^*(\mu), S^*(\mu))$ is an arbitrary optimal solution to (2.35). However, the overall computational complexity depends on per-iteration complexity of the SPG algorithm, which is partly determined by the complexity of solving the subproblem in Step 8 in Figure 2.2. Now consider the subproblem in Step 8 of Figure 2.2 in the following generic form for some given $\bar{L}, \bar{S} \in \mathbb{R}^{m\times n}$:

$$(L_+, S_+) = \underset{L,S\in\mathbb{R}^{m\times n}}{\operatorname{argmin}} \left\{\|L - \bar{L}\|_F^2 + \|S - \bar{S}\|_F^2 : \ (L, S) \in \mathcal{X}\right\}. \tag{2.40}$$

Lemma 2.1 in [3] implies that (L_+, S_+) can be computed in closed form as follows:

$$L_+ = \frac{\theta_+}{2(1+\theta_+)}\left(\mathcal{P}A - \bar{S}\right) + \frac{2+\theta_+}{2(1+\theta_+)}\bar{L},$$
$$S_+ = \frac{\theta_+}{2(1+\theta_+)}\left(\mathcal{P}A - \bar{L}\right) + \frac{2+\theta_+}{2(1+\theta_+)}\bar{S},$$
$$\theta_+ = \max\left\{0, \ \frac{\|\bar{L} + \bar{S} - \mathcal{P}A\|_F}{\delta} - 1\right\}.$$

The PSPG algorithm proposed in [4] is also an implementation of the modified APG algorithm, with Steps 5 and 7 in Figure 2.1 replaced by (2.27) and (2.28), to solve (2.36). In particular, let ψ and ϕ in (2.26) be set as follows: $\psi(L,S) = g(S) + \mathcal{I}_{\mathcal{X}}(L,S)$ and $\phi(L,S) = f_\mu(L)$, where $\mu = \frac{\epsilon}{2\tau}$ and $\tau = \frac{1}{2}\min\{m,n\}$. It follows from (2.34) that $\operatorname{grad}\phi$ is Lipschitz continuous with constant $1/\mu$; hence, $C_\phi = 1/\mu$. Each step of the PSPG algorithm, displayed in Algorithm 2.3, can be derived using (2.32) and $C_\phi = 1/\mu$. Hence, using exactly the same argument as when we discussed SPG, one can directly compute Q_k^L according to (2.39) without computing $\operatorname{grad} f_\mu(\bar{L}_k)$ to improve numerical stability. Since PSPG is a variant of APG algorithm, given $(L_0, S_0) \in \mathcal{X}$, e.g., $L_0 = \mathbf{0}$ and $S_0 = \mathcal{P}A$, like the SPG algorithm, the PSPG algorithm also keeps all iterates in $\mathcal{X}$; and according to (2.29), the convergence rate in function values of the PSPG iterate sequence $\{(L_k, S_k)\}$ is $\mathcal{O}\left(\frac{\mu^{-1}}{k^2}\right)$. In particular, given $\epsilon > 0$, since $\mu = \epsilon/\min\{m,n\}$, PSPG can compute an ϵ-optimal solution to the SPCP problem in (2.10) by computing an $\frac{\epsilon}{2}$-optimal solution to (2.36) within at most $N_{\text{PSPG}} = 2\frac{\sqrt{\min\{m,n\}}}{\epsilon}\|L_0 - L^*(\mu)\|_F$ iterations, where $L^*(\mu)$ is the low-rank component of an arbitrary optimal solution $(L^*(\mu), S^*(\mu))$ to (2.36).

The overall computational complexity depends on per-iteration complexity, which is partly determined by the complexity of solving subproblem in Step 7 of the PSPG algorithm in Figure 2.3. Since the form of solution and the complexity of this step were not analyzed before, FISTA [7] has not been previously applied to solve (2.36) and its overall complexity for solving (2.36) was unknown until Aybat, Goldfarb, and Ma [4] showed that the subproblem can be efficiently solved and per-iteration complexity of PSPG is determined by the complexity of an SVD. Now consider the subproblem in Step 7 of Figure 2.3 in the following generic form for some $\rho > 0$ and $\bar{L} \in \mathbb{R}^{m\times n}$:

$$(L_+, S_+) = \underset{L,S\in\mathbb{R}^{m\times n}}{\operatorname{argmin}} \left\{\lambda\|\mathcal{P}S\|_1 + \frac{\rho}{2}\|L - \bar{L}\|_F^2 : \ (L, S) \in \mathcal{X}\right\}. \tag{2.41}$$

According to Lemma 6.1 in [4], (L_+, S_+) can be computed in closed form as follows:

$$S_+ = \operatorname{sgn}(\mathcal{P}A - \bar{L}) \odot \max\left\{|\mathcal{P}A - \bar{L}| - \lambda\left(\frac{1}{\theta_+} + \frac{1}{\rho}\right)\mathbf{1}_{m\times n}, \ \mathbf{0}_{m\times n}\right\} - \pi_{\Omega^c}\left(\bar{L}\right),$$
$$L_+ = \frac{\theta_+}{\rho+\theta_+}\mathcal{P}A - S_+ + \frac{\rho}{\rho+\theta_+}\mathcal{P}\bar{L} + \pi_{\Omega^c}\left(\bar{L}\right),$$

Algorithm 2.3 - PSPG: Partially Smooth Proximal Gradient algorithm - **PSPG(L_0, ϵ)**

1: **input:** $L_0 \in \mathbb{R}^{m\times n}$, $\epsilon > 0$
2: $\mu \leftarrow \epsilon/\min\{m,n\}$
3: $k \leftarrow 0$, $t_0 \leftarrow 1$, $\bar{L}_0 \leftarrow L_0$
4: **while** $k \geq 0$ **do**
5: $[U_k, V_k, \sigma_k] \leftarrow \text{SVD}(\bar{L}_k)$ **Comment :** $\bar{L}_k = U_k\,\mathbf{diag}(\sigma_k)V_k^\top$, and $r_k = \mathbf{rank}(\bar{L}_k)$
6: $Q_k^L \leftarrow U_k\,\mathbf{diag}\left(\max\{\sigma_k - \mu\mathbf{1}_{r_k},\ \mathbf{0}_{r_k}\}\right) V^\top$
7: $(L_{k+1}, S_{k+1}) \leftarrow \underset{(L,S)}{\text{argmin}}\left\{\lambda\|\mathcal{P}S\|_1 + \frac{1}{2\mu}\|L - Q_k^L\|_F^2 :\ (L,S) \in \mathcal{X}\right\}$
8: $t_{k+1} \leftarrow \left(1 + \sqrt{1 + 4t_k^2}\right)/2$
9: $\bar{L}_{k+1} \leftarrow L_{k+1} + \left(\frac{t_k - 1}{t_{k+1}}\right)(L_{k+1} - L_k)$
10: $k \leftarrow k+1$
11: **end while**

---

where $\theta_+ = 0$ if $\|\mathcal{P}A - \bar{L}\|_F \leq \delta$; otherwise, θ_+ is the *unique positive* solution to the nonlinear equation $\Gamma(\theta) = 0$, where Γ is defined as follows

$$\Gamma(\theta) := \|\min\left\{\frac{\lambda}{\theta}, \frac{\rho}{\rho+\theta}|\mathcal{P}A - \bar{L}|\right\}\|_F,$$

i.e., $\{\theta > 0 :\ \Gamma(\theta) = 0\}$ is a singleton, and equal to $\{\theta_+\}$. If $\theta_+ = 0$, then $\frac{1}{\theta_+}$ is interpreted as $+\infty$; hence, $S_+ = -\pi_{\Omega^c}\left(\bar{L}\right)$ and $L_+ = \bar{L}$. Moreover, the proof of Lemma 6.1 in [4] shows that θ_+ can be *efficiently* computed in $\mathcal{O}(|\Omega|\log(|\Omega|))$ time.

In [4], PSPG and ASALM are tested on both randomly generated synthetic problems and surveillance video foreground extraction problems. According to numerical results reported in [4], in the synthetic problems the number of partial SVD computations for PSPG and ASALM are on par; on the other hand, surprisingly the average runtime for ASALM is approximately *twice* the runtime for PSPG. This significant gap in the average solution times is explained by the fact that ASALM required more leading singular value computations than PSPG did per partial SVD.

It should be emphasized that in practice adopting a continuation technique for the smoothing parameter speeds up both SPG and PSPG. In particular, for a given approximation error level $\epsilon > 0$, the smoothing parameter μ is fixed in Figure 2.2 and Figure 2.3 accordingly, i.e., $\mu = \frac{\epsilon}{2\min\{m,n\}}$ for SPG, and $\mu = \frac{\epsilon}{\min\{m,n\}}$ for PSPG, so that APG algorithms can compute an ϵ-optimal solution to (2.8) within $\mathcal{O}(1/\epsilon)$ iterations. However, instead of fixing μ throughout the algorithm, choosing $\{\mu_k\}$ such that $\mu_{k+1} < \mu_k$ and $\mu_k \to \mu$, and setting $\mu = \mu_k$ in the k-th iteration of SPG and PSPG works much better in practice. For instance, given $\kappa \in (0,1)$ and $\mu_1 > \mu$, one continuation strategy, similar to the one adopted in [4], sets $\mu_{k+1} = \kappa\mu_k$ if $k \leq \bar{K}$, and $\mu_k = \mu$ for $k > \bar{K}$, where $\bar{K} = \left\lfloor \log_{\frac{1}{\kappa}}\left(\frac{\mu_1}{\mu}\right)\right\rfloor$.

2.2.3 ADMM Revisited

Recall that ASALM, discussed in Section 2.2.1, is a three-block ADMM algorithm proposed by Tao and Yuan [59] to solve (2.8), and although it does not have any convergence guarantees, it works well in practice; and slightly changing the update rule in ASALM leads to VASALM, of which an iterate sequence converges to an optimal solution; but this comes at the cost of degradation in practical convergence speed when compared to ASALM. Next, in

Section 2.2.2, we have reviewed two APG algorithms, SPG and PSPG, proposed in [3, 4], which have stronger convergence guarantees: an ϵ-optimal solution to the SPCP problem in (2.8) can be computed within $\mathcal{O}(1/\epsilon)$ iterations, and computational complexities per iteration of SPG and PSPG are comparable to the work per iteration required by ASALM and VASALM, which is mainly determined by an SVD computation. On the other hand, it is also important to emphasize that SPG and PSPG iterate sequences do not converge to an optimal solution to the SPCP problem in (2.8). In particular, since within SPG and PSPG the smoothing parameter μ is fixed, depending on the approximation parameter ϵ for solving (2.35) and (2.36), respectively, further iterations after reaching an ϵ-optimal solution in $\mathcal{O}(1/\epsilon)$ iterations does not necessarily improve the solution quality.

In this section, we revisit ADMM algorithms for solving the SPCP problem, and discuss the *variable penalty* ADMM algorithm ADMIP [6] (Alternating Direction Method with Increasing Penalty) proposed in [6] as it possesses all the advantages of the methods discussed in the previous two sections:

a) its iterate sequence converges to an optimal solution of (2.8), like VASALM iterate sequence, while it is much faster than ASALM in practice,

b) when the constant penalty parameter is used as a special case, it can compute an ϵ-optimal solution within $\mathcal{O}(1/\epsilon)$ iterations, of which complexity is determined by an SVD.

Let

$$\mathcal{X} := \{(L, S) \in \mathbb{R}^{m\times n} \times \mathbb{R}^{m\times n} : \ \|\mathcal{P}L + S - A\|_F \leq \delta\} \tag{2.42}$$

denote the feasible set of the SPCP problem in (2.8). Using *partial* variable splitting, i.e., only split the L variables in (2.8), to arrive at the following equivalent problem

$$\min_{X,L,S\in\mathbb{R}^{m\times n}} \{\|X\|_* + \lambda \ \|S\|_1 + \mathbf{1}_{\mathcal{X}}(L, S) : \ X - L = \mathbf{0}_{m\times n}\}. \tag{2.43}$$

Let $\psi_1(L) := \|L\|_*$ and $\psi_2(L, S) := \lambda \ \|S\|_1 + \mathbf{1}_{\mathcal{X}}(L, S)$. In this section, we briefly review ADMIP proposed in [6] to solve the following ADMM formulation of the SPCP problem:

$$\min_{X,L,S\in\mathbb{R}^{m\times n}} \{\psi_1(X) + \psi_2(L, S) : \ X - L = \mathbf{0}_{m\times n}\}. \tag{2.44}$$

Variable Penalty Alternating Direction Method of Multipliers

First as a preliminary to this section, we briefly review variable penalty ADMM and its connections to other convex optimization methods in general. Let $\psi_i : \mathbb{R}^{n_i} \to \mathbb{R} \cup \{+\infty\}$ be a proper, closed, convex function, and $A_i : \mathbb{R}^{m\times n_i}$ for all $i = 1, \ldots, M$ and $b \in \mathbb{R}^m$. Let $n := \sum_{i=1}^M n_i$ and $\mathbb{R}^n \ni x = [x^{(1)}; \ldots; x^{(M)}]$. Consider the following convex optimization problem:

$$P^* = \min_{x\in\mathbb{R}^n} P(x) := \sum_{i=1}^M \psi_i(x^{(i)}) \quad \text{s.t.} \quad \sum_{i=1}^M A_i x^{(i)} = b. \tag{2.45}$$

Given $\rho \geq 0$ and a dual variable $y \in \mathbb{R}^m$, the augmented Lagrangian of (2.45) is given by

$$\mathcal{L}_\rho(x^{(1)}, \ldots, x^{(M)}; y) := \sum_{i=1}^M \psi_i(x^{(i)}) - \langle y, \sum_{i=1}^M A_i x^{(i)} - b\rangle + \frac{\rho}{2}\|\sum_{i=1}^M A_i x^{(i)} - b\|_2^2,$$

[6] In an earlier preprint [3], it was named as `NSA` algorithm.

and $g_\rho(y) := \inf_x \mathcal{L}_\rho(x, y)$ denotes the dual function – note that g_0 is the Lagrangian dual function of (2.45).

The *dual ascent* method [57] sets $\rho = 0$, and starting from arbitrary $y_0 \in \mathbb{R}^m$ computes the iterate $x_k = [x_k^{(1)}; \ldots; x_k^{(M)}]$ as follows: $x_k \in \operatorname{argmin} \mathcal{L}_0(x; y_k)$, $y_{k+1} = y_k - \alpha_k \left(\sum_{i=1}^M A_i x_k^{(i)} - b\right)$, where $\alpha_k > 0$ is an appropriately chosen step size. Since $\rho = 0$, $\{x_k^{(i)}\}_{i=1}^M$ can be computed in parallel. This is simply a sub-gradient method in dual space, and it is guaranteed to converge to an optimal solution under certain assumptions [57]; however, it suffers from a slow convergence rate $\mathcal{O}(1/\sqrt{k})$. Unless $\sum_{i=1}^M A_i x_k^{(i)} - b$ is the same for all $x_k \in \operatorname{argmin} \mathcal{L}_0(x; y_k)$ – this is trivially true if $\min \mathcal{L}_0(x; y_k)$ has a unique solution, and g_0 is not differentiable at y_k; on the other hand, for any $\rho > 0$, it can be shown that g_ρ is the Moreau-Yosida regularization of g_0, i.e., $g_\rho(y) = \max_w g_0(w) - \frac{1}{2\rho}\|w - y\|_2^2$; hence, g_ρ is differentiable with a Lipschitz continuous gradient, of which the Lipschitz constant is $\frac{1}{\rho}$, and $y^* \in \operatorname{argmax}_y g_0(y)$ if and only if $y^* \in \operatorname{argmax}_y g_\rho(y)$ – see [52, 53] for details. Using this relation, Rockafellar [54, 55] established the connection between the *method of multipliers* [31, 50] and the proximal-point algorithm. Therefore, for $\{\rho_k\} \subset \mathbb{R}_{++}$, the *method of multipliers* iterate sequence, generated as $x_k \in \operatorname{argmin}_x \mathcal{L}_{\rho_k}(x; y_k)$ and $y_{k+1} = y_k - \rho_k \left(\sum_{i=1}^M A_i x_k^{(i)} - b\right)$, is shown in [24] to converge with the following rate

$$P^* - g_0(y_k) \leq \frac{\|y_0 - y^*\|_2^2}{2\sum_{i=1}^k \rho_i}, \tag{2.46}$$

where $y^* \in \operatorname{argmax} g_0(y)$. Hence, when $\rho_k = \rho$ for some $\rho > 0$, or $\sup_k \rho_k < \infty$, the method of multipliers iterate sequence converges with $\mathcal{O}(1/k)$ rate, and when $\sup_k \rho_k = +\infty$, it converges with $o(1/k)$ rate. Similarly, it is shown in [54, 55] that under assumptions related to *strong* second-order optimality conditions, the primal and dual iterates converge to an optimal pair *superlinearly* when the penalty parameters $\rho_k \nearrow \infty$, while the rate is only *linear* when $\sup_k \rho_k < \infty$. However, this result has not been extended to *variable penalty* ADMM, which will be described next. It should also be emphasized that since $\rho_k > 0$, the *separability* is lost in the augmented Lagrangian function, and x_k computation in *method of multipliers* cannot be parallelized as in *dual ascent*.

Given $y_1 \in \mathbb{R}^m$ and $\{\rho_k\}_{k \in \mathbb{Z}_+} \subset \mathbb{R}_{++}$, according to the Gauss-Seidel update rule, ADMM iterates $x_k = [x_k^{(1)}; \ldots; x_k^{(M)}]$ are computed as follows:

$$x_{k+1}^{(i)} \in \operatorname*{argmin}_{x^{(i)} \in \mathbb{R}^{n_i}} \mathcal{L}_{\rho_k}(x_{k+1}^{(1)}, \ldots, x_{k+1}^{(i-1)}, x^{(i)}, x_k^{(i+1)}, \ldots, x_k^{(M)}; y_k), \quad i = 1, \ldots, M, \tag{2.47}$$

$$y_{k+1} = y_k - \rho_k \left(\sum_{i=1}^M A_i x_{k+1}^{(i)} - b\right). \tag{2.48}$$

First, we consider the *two-block* case, i.e., $M = 2$. When the penalty parameter sequence is constant, i.e., $\rho_k = \rho$ for some $\rho > 0$, the iterate sequence $\{x_k\}$ converges to an optimal solution to (2.45) – see the recent surveys [10, 20] and [19] for more details; moreover, for this setting $\mathcal{O}(1/k)$ rate of convergence has also been shown in [29]. Although convergence is guaranteed for all $\rho > 0$, the empirical performance is critically dependent on the choice of *constant* penalty parameter ρ – it deteriorates very rapidly if the penalty is set too large or too small [22, 23, 37]. Moreover, it is discussed in [41] that under certain assumptions on the objective function, there exists ρ^*, which optimizes the convergence bounds for the constant penalty ADMM scheme; however, estimating ρ^* is difficult in practice [30]. Therefore, adopting a variable penalty sequence within ADMM can be a fruitful approach to fix the issues caused by picking ρ either too large or too small; however, it is difficult to prove the convergence when penalty parameters change in every iteration, which is also remarked by Boyd et al. [10] (see Section 3.4.1), and the existing results on the convergence properties of variable penalty ADMM is very *limited*. The convergence results of He et al. [27, 28, 30] on variable penalty ADMM algorithms implicitly assume that both ψ_1 and ψ_2 in the objective function are *differentiable*; see Assumption A and the following discussion on page 107 in [27]; therefore, these results do not extend to a *non-smooth* optimization problem in (2.44), i.e., to the *two-block* ADMM formulation of (2.8). On the other hand, the ADMM algorithm in [37] can solve (2.45) with $M = 2$ when both ψ_1 and ψ_2 are *non-smooth* convex functions; however, the convergence proof requires that the penalty sequence $\{\rho_k\}$ increases only *finitely* many times; i.e., $\{\rho_k\}$ is *bounded above* – [27, 30] also assume bounded $\{\rho_k\}$.

Next, consider the *multi-block* case, i.e., $M \geq 3$. Through variable splitting (2.45) can be equivalently written using only *two-blocks* – see [8, 10] for details; however, the number of variables and constraints in the equivalent problem increase significantly. As a remedy, one can consider the multi-block ADMM as described in (2.47) and (2.48); however, it has been recently shown that the direct extension of two-block ADMM to three blocks is not necessarily convergent [14] even for constant penalty, i.e., $\rho_k = \rho$ for some $\rho > 0$. Convergence for multi-block ADMM with Gauss-Seidel updates as in (2.47) and (2.48) with $\rho_k = \rho$ has only been established for very special cases of (2.45), e.g., [34, 61]; in certain cases it works well even though lacking theoretical convergence guarantees, e.g., ASALM [59] for solving (2.11). Note that (2.47) cannot be parallelized; however, replacing (2.47) with

$$x_{k+1}^{(i)} \in \operatorname*{argmin}_{x^{(i)} \in \mathbb{R}^{n_i}} \mathcal{L}_\rho(x_k^{(1)}, \ldots, x_k^{(i-1)}, x^{(i)}, x_k^{(i+1)}, \ldots, x_k^{(M)}; y_k), \quad i = 1, \ldots, M, \tag{2.49}$$

one obtains the multi-block ADMM with Jacobian updates. Although this modification is fully parallelizable, there are examples for which the iterate sequence of ADMM with Jacobian updates diverges even for the *two-block* case, e.g., [26]. Very recently, Deng et al. [17] have shown that appending a proximal term $\tau_i \|x^{(i)} - x_k^{(i)}\|_2^2$ to (2.49) generates a convergent sequence for particular choice of $\{\tau_i\}_{i=1}^M \subset \mathbb{R}_{++}$. Moreover, under certain conditions on $\{A_i\}_{i=1}^M$, $o(1/k)$ convergence rate has been established for the proximal Jacobian ADMM in [17]. On the other hand, for a variable penalty multi-block ADMM algorithm even the convergence of the iterate sequence is not known.

ADMIP: Alternating Direction Method with Increasing Penalty

The ADMIP [7] algorithm [6], displayed in Algorithm 2.4, is an implementation of the ADMM algorithm with Gauss-Seidel updates, i.e., (2.47) and (2.48), to solve (2.43), which is equivalent to (2.8). In particular, for given penalty parameter $\rho > 0$, and dual variable $Y \in \mathbb{R}^{m \times n}$, the augmented Lagrangian function corresponding to (2.44) has the following form:

$$\mathcal{L}_\rho(X, L, S; Y) = \psi_1(X) + \psi_2(L, S) - \langle Y, X - L \rangle + \frac{\rho}{2}\|X - L\|_F^2, \tag{2.50}$$

where $\psi_1(L) := \|L\|_*$, $\psi_2(L, S) := \lambda \, \|S\|_1 + \mathbf{1}_{\mathcal{X}}(L, S)$, and $\mathcal{X}$ is defined in (2.42). Hence, the two subproblems in (2.47) can be written equivalently as:

$$X_{k+1} = \operatorname*{argmin}_{X \in \mathbb{R}^{m \times n}} \|X\|_* + \frac{\rho_k}{2}\|X - (L_k + Y_k/\rho_k)\|_F^2, \tag{2.51}$$

$$(L_{k+1}, S_{k+1}) = \operatorname*{argmin}_{L, S \in \mathbb{R}^{m \times n}} \{\lambda \, \|S\|_1 + \frac{\rho_k}{2}\|L - (X_{k+1} - Y_k/\rho_k)\|_F^2 : \ (L, S) \in \mathcal{X}\}. \tag{2.52}$$

[7] In an earlier preprint [3], it was named as `NSA` algorithm.

Note that the X-subproblem in (2.51) is the matrix-shrinkage operation [32], and can be computed as in (2.19). Now consider (2.52) in the following generic form for some $\rho > 0$ and $\bar{L} \in \mathbb{R}^{m\times n}$:

$$\min_{L,S\in\mathbb{R}^{m\times n}} \{\lambda\|S\|_1 + \frac{\rho}{2}\|L - \bar{L}\|_F^2 : (L, S) \in \mathcal{X}\}. \tag{2.53}$$

It can be easily shown that (2.53) is equivalent to the PSPG subproblem in (2.41); in the sense that if (L_+, S_+) denotes the optimal solution to (2.41), then $(L_+, \mathcal{P}S_+)$ is the optimal solution to (2.53). Hence, Step 8 in Algorithm 2.4 can be computed efficiently in at most $\mathcal{O}(|\Omega|\log(|\Omega|))$ time.

Algorithm 2.4 - ADMIP: Alternating Direction Method with Increasing Penalty - **ADMIP**$(L_0, Y_0, \{\rho_k\})$

1: **input:** $L_0, Y_0 \in \mathbb{R}^{m\times n}$, $\{\rho_k\} \subset \mathbb{R}_{++}$
2: $k \leftarrow 0$
3: **while** $k \geq 0$ **do**
4: $Q_k^X \leftarrow L_k + \frac{1}{\rho_k}Y_k$
5: $[U_k, V_k, \sigma_k] \leftarrow \text{SVD}(Q_k^X)$ **Comment** : $Q_k^X = U_k\ \mathbf{diag}(\sigma_k)V_k^\top$, and $r_k = \mathbf{rank}(Q_k^X)$
6: $X_{k+1} \leftarrow U_k\,\mathbf{diag}\left(\max\{\sigma_k - \frac{1}{\rho_k}\mathbf{1}_{r_k},\ \mathbf{0}_{r_k}\}\right) V^\top$
7: $Q_k^L \leftarrow X_{k+1} - \frac{1}{\rho_k}Y_k$
8: $(L_{k+1}, S_{k+1}) \leftarrow \underset{(L,S)}{\text{argmin}}\{\lambda\|S\|_1 + \frac{\rho_k}{2}\|L - Q_k^L\|_F^2 : (L, S) \in \chi\}$
9: $Y_{k+1} \leftarrow Y_k - \rho_k(X_{k+1} - L_{k+1})$
10: $k \leftarrow k + 1$
11: **end while**

Since ADMIP is a two-block ADMM algorithm, if a *constant penalty* scheme is adopted, i.e., $\rho_k = \rho$ for all k for some $\rho > 0$, then according to [29], (X_k, L_k, S_k, Y_k) converges to an optimal solution with $\mathcal{O}(1/k)$ rate of convergence. Moreover, because the case penalty sequence is not constant, Aybat and Iyengar [6] show that both primal and dual ADMIP iterates converge to an optimal primal-dual solution to the SPCP problem in (2.8) under mild conditions on the penalty parameter sequence. In particular, if $\{\rho_k\}_{k\in\mathbb{Z}_+}$ is a non-decreasing sequence such that $\sum_k \rho_k^{-1} = +\infty$, then $L^* := \lim_k X_k = \lim_k L_k$ and $S^* := \lim_k S_k$ exist; and (L^*, S^*) is an optimal solution to (2.8); and if $\{\rho_k\}_{k\in\mathbb{Z}_+}$ also satisfies $\sum_{k\in\mathbb{Z}_+} \rho_k^{-2} = +\infty$, then $Y^* := \lim_{k\in\mathbb{Z}_+} Y_k$ exists whenever $\|\mathcal{P}A - L^*\|_F \neq \delta$, and (L^*, L^*, S^*, Y^*) is a saddle point of the Lagrangian function $\mathcal{L}_0$ in (2.50). Although the convergence rate is not known for ADMIP, it should be emphasized that even the convergence result in [6] is the first one for a variable penalty ADMM when penalties are *not bounded*, the objective function is *non-smooth*, and its subdifferential is *not* uniformly bounded.

The main advantages of adopting an increasing sequence of penalties are as follows:

1. The algorithm is robust in the sense that there is no need to search for ρ^*, depending on problem parameters, that works well in practice, or that optimizes some convergence bounds as in [41].
2. The algorithm is likely to achieve primal feasibility faster.
3. The complexity of initial (transient) iterations can be controlled through controlling $\{\rho_k\}$. The main computational bottleneck in ADMIP (see Algorithm 2.4) is the SVD computation in Step 5. Since the optimal L^* is of low rank, and $L_k \to L^*$, eventually

the SVD computations are likely to be very efficient. However, since the initial iterates in the transient phase of the algorithm may have large rank, the complexity of the SVD in the initial iterations can be quite large. From Step 7 in Figure 2.4, it follows that one does not need to compute singular values smaller than $1/\rho_k$; hence, starting ADMIP with a *small* $\rho_0 > 0$ will significantly decrease the complexity of initial iterations via adopting *partial* SVD computations.

Furthermore, the numerical experiment results reported in [6] clearly demonstrate that adopting an increasing penalty parameter sequence leads to faster convergence in practice; in fact, the performance of ADMIP dominates the performance of constant penalty ADMM for any fixed penalty term. The numerical experiments also confirm that ADMIP is significantly more robust to changes in problem parameters. In [6], Aybat and Iyengar also compared ADMIP against ASALM on both randomly generated synthetic problems and surveillance video foreground extraction problems. According to numerical results reported in [6], on the synthetic problems ASALM requires about *twice* as many iterations for convergence, while the total runtime for ASALM is considerably larger; on a surveillance video foreground extraction problem, both ADMIP and ASALM were able to recover the foreground and the background fairly accurately with only 60% of the pixels functioning, i.e., $\frac{|\Omega|}{mn} = 0.6$ in (2.8); both iteration count and runtime of ADMIP are smaller than those of ASALM.

Before concluding this section, we would like to give an example of alternative *two-block* ADMM formulation for SPCP. Recently, Huai et al. [35] proposed a *two-block* ADMM algorithm to solve the SPCP problem with non-negativity constraints:

$$\min\{\|L\|_* + \lambda\|S\|_1 : \ \|L + S - A\|_F \leq \delta, \ L \geq \mathbf{0}_{m\times n}\}. \tag{2.54}$$

Let $\chi_1 := \{N \in \mathbb{R}^{m\times n} : \ \|N\|_F \leq \delta\}$, and $\chi_2 := \{X \in \mathbb{R}^{m\times n} : \ X \geq \mathbf{0}_{m\times n}\}$. Then Huai et al. [35] equivalently formulated as follows:

$$\begin{aligned} \min_{L,S,N,X\in\mathbb{R}^{m\times n}} \ & \|L\|_* + \lambda\|S\|_1 + \mathcal{I}_{\chi_1}(N) + \mathcal{I}_{\chi_2}(X) \\ \text{s.t.} \ & \begin{pmatrix} \mathbf{I}_m & \mathbf{I}_m \\ -\mathbf{I}_m & \mathbf{0}_{m\times m} \end{pmatrix} \begin{pmatrix} L \\ S \end{pmatrix} + \begin{pmatrix} \mathbf{I}_m & \mathbf{0}_{m\times m} \\ \mathbf{0}_{m\times m} & \mathbf{I}_m \end{pmatrix} \begin{pmatrix} N \\ X \end{pmatrix} = \begin{pmatrix} A \\ \mathbf{0}_{m\times n} \end{pmatrix}. \end{aligned}$$

Hence, one can solve this formulation using *two-block* ADMM alternatingly minimizing in (L, S) and (N, X) blocks. It is important to note that for large-scale SPCP problems, the memory requirement can be a critical issue for this formulation.

2.2.4 Quasi-Newton Method for SPCP

Letting $A = L^o + S^o + N^o$ such that $\|N^o\|_F \leq \delta$, for stable decomposition of A, Aravkin et al. [1] proposed a convex variational framework, which is accelerated with a "quasi-Newton" method. The proposed model is given in (2.55), and it includes the SPCP problem in (2.8) as a special case:

$$\min \psi(L, S) \quad \text{s.t.} \quad \phi(\mathcal{A}(L, S) - A) \leq \delta, \tag{2.55}$$

where $\mathcal{A} : \mathbb{R}^{m\times n} \times \mathbb{R}^{m\times n} \to \mathbb{R}^{m\times n}$ is a linear operator, $\phi : \mathbb{R}^{m\times n} \to \mathbb{R}$ is a smooth convex function such as $\|.\|_F^2$ or the robust Huber penalty, and ψ can be set to either one of the following functions,

$$\begin{aligned} \psi_{\text{sum}}(L, S) &:= \|L\|_* + \lambda\|S\|_1, \\ \psi_{\max}(L, S) &:= \max\{\|L\|_*, \lambda_{\max}\|S\|_1\}, \end{aligned}$$

and $\lambda, \lambda_{\max} > 0$ are some given function parameters. Note that setting $\psi = \psi_{\text{sum}}$, $\rho(.) = \|.\|_F^2$, and $\mathcal{A}(L, S) = \mathcal{P}L + S$ in (2.55), one obtains (2.8).

This approach offers advantages over the original SPCP formulation in terms of practical parameter selection. Indeed, the authors argue that even though setting $\lambda = \frac{1}{\sqrt{\max\{m,n\}}}$ in (2.3) has theoretical justification, which is discussed in Section 2.1.2, many practical problems may violate the underlying assumptions in (2.5) and (2.6); in those cases one needs to tune λ via cross validation, and selecting $\lambda_{\max}$ in $\psi_{\max}$ might be easier than selecting λ in ψ_{sum}.

Instead of directly solving (2.55), Aravkin et al. [1] used Newton's method to find a root of the value function:

$$v(\tau) := \min_{L,S\in\mathbb{R}^{m\times n}} \phi(\mathcal{A}(L,S)-A)-\delta \quad \text{s.t.} \quad \psi(L,S)\leq \tau, \tag{2.56}$$

i.e., given $\delta > 0$, compute $\tau^* = \tau(\delta)$ such that $v(\tau^*) = 0$. According to results in [2], if the constraint in (2.55) is tight at an optimal solution, then there exists $\tau^* = \tau(\delta)$ such that $v(\tau^*) = 0$ and the corresponding optimal solution to (2.56) is also optimal to (2.55).

For the sake of simplicity, suppose that $\phi(.) = \|.\|_F^2$, which ensures that v is differentiable, and let (L_k, S_k) denote the optimal solution to (2.56) when $\tau = \tau_k$. The next point τ_{k+1} is generated using Newton's method: $\tau_{k+1} = \tau_k - \frac{v(\tau_k)}{v'(\tau_k)}$. Aravkin et al. show that v' can be computed efficiently: $v'(\tau_k) = -\psi^\circ(\mathcal{A}^\top \operatorname{grad}\phi(\mathcal{A}(L_k,S_k)-A))$, and ψ° denotes the polar gauge to ψ. Indeed, according to Corollary 4.2 in [2], ψ°_{sum} and $\psi^\circ_{\max}$ can be written in the following closed forms: $\psi^\circ_{\text{sum}}(L,S) = \max\{\|L\|, \frac{1}{\lambda}\|S\|_\infty\}$, and $\psi^\circ_{\max}(L,S) = \|L\| + \frac{1}{\lambda_{\max}}\|S\|_\infty$. Therefore, given $\tau = \tau_k$ if the optimal solution (L_k, S_k) to (2.56) can be computed efficiently, one can minimize (2.55) efficiently using Newton's method as discussed above. Since Euclidean projections onto $\{(L,S) : \psi(L,S) \leq \tau\}$ can be computed efficiently for both ψ_{sum} and $\psi_{\max}$, in order to solve (2.56) Aravkin et al. adopted a projected "quasi-Newton" method that uses Hessian approximations to compute the next iterate.

According to numerical tests reported in [2], QN-max, the quasi-Newton method running on (2.55) with $\psi = \psi_{\max}$ and $\phi(.) = \|.\|_F^2$, is competitive with the state-of-the-art codes, ASALM [59], PSPG [4], and ADMIP [6]; and performs better on the test setting where an exponential noise matrix E^o is directly added to low rank component L^o to generate the observation matrix, i.e., $A = L^o + E^o$, instead of perturbing L^o with $S^o + N^o$ – note that this experimental setup violates model assumptions discussed in Section 2.1.2. Indeed, since exponential noise has a longer tail than the Gaussian noise, during recovery SPCP algorithms artificially decomposed E^o into sparse component, S^o, and a small magnitude, dense noise component, N^o. The authors also used the synthetic experimental setting of Aybat and Iyengar in [6]; and in those experiments the performance of ADMIP is significantly better than QN-max when high accuracy solutions are targeted.

2.2.5 Frank-Wolfe for SPCP

Letting Ω be the subset of observable indices of the data matrix $A = L^o + S^o + N^o \in \mathbb{R}^{m\times n}$, Mu et al. [45] proposed stably decomposing A through solving

$$\phi^* := \min_{L,S\in\mathbb{R}^{m\times n}} \{\phi(L,S) = \tfrac{1}{2}\|\mathcal{P}L + S - A\|_F^2 : \ \|L\|_* \leq \tau_L, \ \|S\|_1 \leq \tau_S\} \tag{2.57}$$

for some $\tau_L, \tau_S > 0$. Recall that full or partial SVD computation is the main bottleneck for many of the algorithms discussed in Section 2.2; to avoid this expensive computation, Mu et al. [45] adopted the Frank-Wolfe algorithm [21] to solve (2.57), which only requires computing the largest singular value and corresponding left and right singular vectors as opposed to full SVD computation. It is important to note that according to discussion in Section 2.2.4, given $\delta > 0$, there exists $\tau(\delta)$ such that solving (2.57) with $\tau_L = \tau_S = \tau(\delta)$

is equivalent to solving $\min\{\psi_{\max}(L,S): \phi(L,S) \leq \delta\}$ with $\lambda_{\max} = 1$. Before discussing further details of the method proposed in [45], as a preliminary we briefly review the Frank-Wolfe method [21] next.

Frank-Wolfe Method

Let $\phi : \mathbb{R}^n \to \mathbb{R}$ be a convex function such that $\operatorname{grad}\phi$ is Lipschitz continuous with constant L over an open set containing convex, compact set $\mathcal{X}$. Consider the constrained problem

$$\phi^* := \min_{x \in \mathbb{R}^n} \phi(x) \quad \text{s.t.} \quad x \in \mathcal{X}, \tag{2.58}$$

where $D := \max_{x,y\in\mathcal{X}} \|x-y\|_2$. Starting from $x_0 \in \mathcal{X}$, the original Frank-Wolfe (FW) method generates the iterate sequence $\{x_k\}$ as follows: $\bar{x}_k \in \operatorname{argmin}_x \langle \operatorname{grad}\phi(x_k), x\rangle$, and $x_{k+1} = x_k + \alpha_k(\bar{x}_k - x_k)$, where $\alpha_k = \frac{2}{k+2}$. In [45], the authors considered a more general update rule, which includes the original update scheme as its special case. In particular, any $x_k \in \mathcal{X}$ satisfying $\phi(x_k) \leq \phi(x_k + \alpha_k(\bar{x}_k - x_k))$ can be chosen as the next iterate.

Algorithm 2.5 - FW: Frank-Wolfe method - $\mathbf{FW}(x_0)$

1: **input:** $x_0 \in \mathcal{X}$
2: $k \leftarrow 0$
3: **while** $k \geq 0$ **do**
4: $\quad \bar{x}_k \in \operatorname{argmin}\{\langle \operatorname{grad}\phi(x_k), x\rangle : x \in \mathcal{X}\}$
5: $\quad \alpha_k \leftarrow \frac{2}{k+2}$
6: $\quad$ Choose $x_k \in \mathcal{X}$ such that $\phi(x_k) \leq \phi(x_k + \alpha_k(\bar{x}_k - x_k))$
7: $\quad k \leftarrow k+1$
8: **end while**

The $\mathcal{O}(1/k)$ convergence rate in function values for the algorithm FW, displayed in Algorithm 2.5, is first given in [16] – see also [45]. In particular, it can be shown that

$$\phi(x_k) - \min_{x\in\mathcal{X}} \phi(x) \leq \frac{2LD^2}{k+2}. \tag{2.59}$$

However, note that this error bound depends on some problem parameters, L, D, which may be unknown in practice. As a remedy, Jaggi [36] proposed the surrogate measure $d_k := \langle \operatorname{grad}\phi(x_k), x_k - \bar{x}_k\rangle$, which can be computed easily without knowing D and L, and it satisfies $\phi(x_k) - \phi^* \leq d_k = \mathcal{O}(1/k)$ – see Theorem 2 in [36]. Moreover, Mu et al. [45] have refined this result and shown that for any $K \geq 1$, there exists $k' \leq K$ such that $d_{k'} \leq \frac{6LD^2}{K+2}$. Hence, given an $\epsilon > 0$, $d_k \leq \epsilon$ is a practical stopping criterion.

Combining Frank-Wolfe and Projected Gradient for SPCP

Note that when the FW method in Figure 2.5 is implemented on (2.57), the bottleneck steps are to minimize linear functions over norm-balls, which can be written in the following generic form for some $\bar{L}, \bar{S} \in \mathbb{R}^{m\times n}$:

$$\begin{aligned} L_+ &:= -\tau_L \bar{u}\bar{v}^\top \in \operatorname{argmin}\{\langle \bar{L}, L\rangle : \|L\|_* \leq \tau_L\}, \\ S_+ &:= -\tau_S e_i e_j^\top \in \operatorname{argmin}\{\langle \bar{S}, S\rangle : \|S\|_1 \leq \tau_S\}, \end{aligned}$$

where $\bar{u} \in \mathbb{R}^m$ and $\bar{v} \in \mathbb{R}^n$ are the left and right singular vectors corresponding to the largest singular value of $\bar{L}$; $e_i \in \mathbb{R}^m$ and $e_j \in \mathbb{R}^n$ are unit vectors with i-th and j-th components equal to 1, respectively, such that $|\bar{S}_{ij}| = \|\bar{S}\|_\infty$. Hence, computing L_+ is much cheaper than computing the full SVD of $\bar{L}$. On the other hand, as pointed out in [45], one clear disadvantage of the FW method in (2.57) is that at every iteration only one entry of the sparse component is updated. This leads to very slow convergence in practice. Hence, Mu et al. [45] proposed combining FW iterations with an additional projected gradient step in the S-block. In particular, they proposed the FWP method displayed in Figure 2.5, where after FW iterate $(L_{k+1}, \tilde{S}_{k+1})$ is computed, an extra projected gradient step is computed in Step 9 – note that $\mathrm{grad}_S\, \phi(L_{k+1}, \tilde{S}_{k+1}) = \mathcal{P}L_{k+1} + \tilde{S}_{k+1} - A$. According to numerical results in [45], where FWP is tested on video and image processing problems such as foreground extraction from noisy surveillance videos, and shadow and specularity removal from face images, the additional projected gradient step significantly improves the convergence rate observed in practice; moreover, Mu et al. also showed that the FWP iterate sequence, generated as shown in Algorithm 2.6, satisfies $\phi(L_k, S_k) - \phi^* \leq \frac{16(\tau_L^2+\tau_S^2)}{k+2}$.

Algorithm 2.6 - FWP: Frank-Wolfe method with projected gradient step - **FWP**

1: **input:** $L_0 = S_0 = \mathbf{0}_{m\times n}$
2: $k \leftarrow 0$
3: **while** $k \geq 0$ **do**
4: $\quad \bar{L}_k \in \mathrm{argmin}\{\langle \mathcal{P}L_k + S_k - A, L\rangle : \ \|L\|_* \leq \tau_L\}$
5: $\quad \bar{S}_k \in \mathrm{argmin}\{\langle \mathcal{P}L_k + S_k - A, S\rangle : \ \|S\|_1 \leq \tau_S\}$
6: $\quad \alpha_k \leftarrow \frac{2}{k+2}$
7: $\quad L_{k+1} \leftarrow L_k + \alpha_k(\bar{L}_k - L_k)$
8: $\quad \tilde{S}_{k+1} \leftarrow S_k + \alpha_k(\bar{S}_k - S_k)$
9: $\quad S_{k+1} \leftarrow \mathrm{argmin}\{\|S - (\tilde{S}_{k+1} - \mathcal{P}L_{k+1} + \tilde{S}_{k+1} - A)\|_F : \ \|S\|_1 \leq \tau_S\}$
10: $\quad k \leftarrow k+1$
11: **end while**

2.3 Algorithms for Non-Convex Formulations

In Section 2.2, we briefly discussed the algorithms for solving convex optimization problems in (2.8), and related formulations. In the literature, there are several *non-convex* models related to the original SPCP problem. Throughout this section, suppose that the data matrix $A \in \mathbb{R}^{m\times n}$ satisfies (2.2).

In [63] Zhou and Tao proposed enforcing the structure on the desired solution via constraints instead of using $\|.\|_*$ and $\|.\|_1$ as in the SPCP formulation. Indeed, their model ensures low-rank structure on the variable L corresponding to low-rank component, and the sparsity of the variable S corresponding to sparse component by using *rank* and *cardinality constraints*, respectively:

$$\min_{L,S\in\mathbb{R}^{m\times n}} \|L + S - A\|_F^2 \quad \text{s.t.} \quad \mathbf{rank}(L) \leq \bar{r}, \quad \mathbf{card}(S) \leq \bar{s}, \tag{2.60}$$

where $\bar{r}, \bar{s} > 0$ are given model parameters such that $\bar{r} \geq \mathbf{rank}(L^o)$ and $\bar{s} \geq \mathbf{card}(S^o)$. Zhou and Tao [63] proposed the GoDec algorithm to solve this model. GoDec is an *alternating minimization* method that alternatingly minimizes the objective in one variable while fixing the other one. They showed that GoDec iterate sequence converges to a local minima, which may be far away from the global optimum. In the implementation, they speed up the algorithm using bilateral random projections instead of computing an SVD for the L-subproblem – see [63] for details. After generating data matrices according to (2.2), the authors compared GoDec with IALM [40], and they report that GoDec is three times faster than IALM according to numerical results in [63] – IALM [40] is a variable penalty ADMM algorithm for the PCP problem in (2.4).

In [33], Hintermüller and Wu considered a slightly modified version of (2.60):

$$\min_{L,S\in\mathbb{R}^{m\times n}} \|L+S-A\|_F^2 + \frac{\rho}{2}\|L\|_F^2 \quad \text{s.t.} \quad \mathbf{rank}(L) \le \bar{r}, \quad \mathbf{card}(S) \le \bar{s}, \tag{2.61}$$

where $\bar{r}, \bar{s} > 0$ are given model parameters as in (2.60), and $0 \le \rho \ll 1$ is a given regularization parameter. The authors proposed an *inexact alternating minimization method* AMS to solve (2.61), where the iterates L_{k+1} and S_{k+1} are computed as "inexact" solutions to subproblems $\min_L\{\|L+S_k-A\|_F^2 + \rho\|L\|_F^2 : \mathbf{rank}(L) \le \bar{r}\}$ and $\min_S\{\|L_{k+1}+S-A\|_F^2 : \mathbf{card}(S) \le \bar{s}\}$, respectively. In particular, the authors show that when the iterates (L_k, S_k) satisfy certain inexact optimality conditions for subproblems, if the iterate sequence has a limit point such that the constraints are tight at the point, then the limit point satisfies the necessary first-order optimality conditions. In the numerical tests, Hintermüller and Wu compared AMS against the *constant penalty* ADMM algorithm in [12], whose steps are the same as those of IALM [40] with a constant penalty parameter sequence. It is reported that AMS is approximately 3 times faster than the ADMM in [12] for foreground extraction from surveillance videos.

For the case $N^o = \mathbf{0}_{m\times n}$, given $\bar{r}$ such that $\bar{r} \ge \mathbf{rank}(L^o)$ and the set of observable indices Ω, Shen et al. [56] considered using bilateral factorization $L = UV^\top$ to guarantee low-rank structure

$$\min_{U\in\mathbb{R}^{m\times\bar{r}}, V\in\mathbb{R}^{n\times\bar{r}}, L\in\mathbb{R}^{m\times n}} \|\mathcal{P}L - A\|_1 \quad \text{s.t.} \quad UV^\top = L. \tag{2.62}$$

Shen et al. proposed a three-block ADMM algorithm to solve (2.62). Note that (2.62) is a non-convex problem; and the ADMM algorithms in general even with two-blocks is not necessarily convergent for non-convex problems. In [56], it is shown that if the iterate sequence converges, then it converges to a KKT point of (2.62). According to numerical test results reported in [56], the proposed algorithm is around one order of magnitude faster than IALM [40], and the authors argue that this speedup is a result of solving the factorization model free of any SVD computations. However, the effectiveness of the proposed methods is limited to the cases where the sparse matrix does not dominate the low-rank one in magnitude.

Later in [64], Zhou and Tao proposed the following model for stable decomposition when $N^o \ne \mathbf{0}_{m\times n}$:

$$\min_{U,V,S} \lambda\|S\|_1 + \|UV^\top + S - A\|_F^2 \quad \text{s.t.} \quad \mathbf{rank}(U) = \mathbf{rank}(V) \le \bar{r}, \tag{2.63}$$

where $\lambda > 0$ and $\bar{r}$ such that $\bar{r} \ge \mathbf{rank}(L^o)$. Zhou and Tao [64] proposed solving (2.63) using a Greedy Bilateral Smoothing algorithm (GreBsmo), which is based on alternating minimization in three blocks: U, V and S. The iterate sequence generated by alternating minimization methods may converge to a point that is not optimal, even for convex problems when the objective is *non-smooth*; therefore, GreBsmo does not have strong convergence properties like the convex methods discussed in Section 2.2. On the other hand, in [64] the authors report that GreBsmo considerably speeds up the decomposition, performing 30–100 times faster than both Godec [63] and IALM [40] when applied to foreground extraction problems as in Section 2.1.4.

Suppose that $A = R(L^o + S^o) + N^o$, where R is a given matrix, and L^o, S^o, N^o are low-rank, sparse, and dense noise components. Mardani et al. [44] developed a distributed algorithm for stably decomposing A for detecting traffic flow anomalies within Internet protocol networks with a fixed topology, represented as a directed graph. The Lagrangian form of the SPCP problem in (2.9) is a special case of the proposed formulation in [44] when R is the *identity matrix*, and the network consists of a *single* node. Letting $R = \mathbf{I}_m$ and

$A \in \mathbb{R}^{m\times n}$ such that $\bar{r} \geq \mathbf{rank}(L^o)$ for some $\bar{r} > 0$, Mardani et al. exploited the following characterization of $\|.\|_*$ due to Recht et al. [51]:

$$\|L\|_* = \min_{U\in\mathbb{R}^{m\times\bar{r}}, V\in\mathbb{R}^{n\times\bar{r}}} \{\tfrac{1}{2}\|U\|_F^2 + \tfrac{1}{2}\|V\|_F^2 : UV^\top = L\}, \tag{2.64}$$

and proposed the following non-convex problem, which is equivalent to the Lagrangian form of the SPCP problem in (2.9):

$$\min_{U\in\mathbb{R}^{m\times\bar{r}}, V\in\mathbb{R}^{n\times\bar{r}}, S\in\mathbb{R}^{m\times n}} \frac{1}{2}\left(\|U\|_F^2 + \|V\|_F^2\right) + \lambda\|S\|_1 + \frac{\rho}{2}\|UV^\top + S - A\|_F^2. \tag{2.65}$$

To solve (2.65), the authors proposed using a *three-block* ADMM algorithm after variable splitting. In particular, on a single node network, the ADMM formulation considered in [44] is given as

$$\min_{U\in\mathbb{R}^{m\times\bar{r}}, V\in\mathbb{R}^{n\times\bar{r}}, S\in\mathbb{R}^{m\times n}} \left\{\|U\|_F^2 + \|V\|_F^2 + 2\lambda\|S\|_1 + \rho\|UV^\top + \bar{S} - A\|_F^2 : S = \bar{S}\right\}. \tag{2.66}$$

Here, (V, S), U, and $\bar{S}$ denotes the three ADMM blocks in which, alternatingly, the augmented Lagrangian is minimized while the other two blocks are fixed. Unfortunately, there is no convergence guarantees on the ADMM iterate sequence for non-convex problems. Indeed, the iterate sequence may not even have a limit point.

Inspired by the formulation in (2.65), Sprechman et al. [58] proposed combining robust decomposition with dictionary learning. In particular, suppose that $U^o \in \mathbb{R}^{m\times\bar{r}}$ is a known or previously learned *under-complete dictionary* with $\bar{r} \ll m$ *atoms*, and every random observation $a \in \mathbb{R}^m$ has the following representation: $a = U^o v + s + \eta$, where $v \in \mathbb{R}^{\bar{r}}$, $s \in \mathbb{R}^m$ is a sparse vector, and $\eta \in \mathbb{R}^m$ is dense noise vector such that $\|\eta\|_2 \leq \delta$ for some $\delta > 0$. This interpretation brings the SPCP problem close to the context of dictionary learning in the sparse modeling domain. In particular, if U^o was known, then given an observation $a \in \mathbb{R}^m$, Sprechman et al. [58] proposed solving $\min_{v\in\mathbb{R}^{\bar{r}}, s\in\mathbb{R}^m} \|v\|_2^2 + 2\lambda\|s\|_1 + \rho\|U^o v + s - a\|_F^2$. To make this model more realistic, Sprechman et al. [58] also developed an *online* version of their algorithm that learns U^o while approximately decomposing the i.i.d. data samples $\{a_t\}_{t\in\mathbb{Z}_+}$ arriving sequentially.

Recently, Yuan et al. [39] considered a special case where $L^o = u^o\mathbf{1}_n^\top$, i.e., $\mathbb{R}^{m\times n} \ni A = u^o\mathbf{1}^\top + S^o + N^o$. Recall that as discussed in Section 2.1.4, this model has applications in foreground extraction from noisy surveillance videos when the background is constant. Since this model forces the low-rank component to have rank-1, it cannot capture subtle movements in the background such as moving escalators, flickering lights, etc., and uses sparse and noise components to capture these changes in the background. Under constant background assumption, Yuan et al. [39] considered the Lagrangian form of the SPCP problem for an appropriately chosen $\rho > 0$:

$$\min_{S\in\mathbb{R}^{m\times n}, u\in\mathbb{R}^m} \|S\|_1 + \frac{\rho}{2}\|u\mathbf{1}_n^\top + S - A\|_F^2.$$

Even under the simplifying rank-1 assumption, this model has no closed-form solution and needs to be solved iteratively. Moreover, considering that there might be a blur in a noisy video surveillance video, Yuan et al. [39] also proposed the following extension:

$$\min_{S\in\mathbb{R}^{m\times n}, u\in\mathbb{R}^m} \|S\|_1 + \frac{\rho}{2}\|H(u\mathbf{1}_n^\top + S) - A\|_F^2, \tag{2.67}$$

where $H : \mathbb{R}^{m\times n} \to \mathbb{R}^{m\times n}$ is the regular blurring operator – the blur is assumed to be frame-wise [25]. Yuan et al. [39] proposed a three-block ADMM algorithm, which is not necessarily convergent – see Section 2.2.3, to solve (2.67). On the noisy videos tested in [39], the proposed method performed well in terms of runtime compared to methods solving (2.9), which is mainly because the proposed method does not require computing SVD as the $\|.\|_*$ term is dropped from the formulation.

2.4 Conclusion

The data observations in real-life applications are often corrupted by noise, possibly affecting every entry of the data matrix. In this chapter, we reviewed optimization algorithms for decomposing data matrices A that can be decomposed as $A = L^o + S^o + N^o$, where L^o and S^o are low-rank and sparse matrices as in robust PCA, and N^o is a dense noise matrix. When there is no noise in the observations, i.e., $N^o = \mathbf{0}_{m\times n}$ and $A = L^o + S^o$, there are many efficient algorithms for solving the PCP problems in (2.4) and (2.7); on the other hand, in the presence of dense noise, i.e., $N^o \neq \mathbf{0}_{m\times n}$, the PCP model becomes inapplicable as A is not a sum of low-rank and sparse components anymore, and for stable decomposition one should consider the SPCP formulation in (2.8), which has uses in numerous applications arising from background separation from surveillance video, shadow and specularity removal from face images, and video denoising from heavily corrupted data. Surprisingly there is only a few methods that can efficiently deal with (2.8) directly. In the first part of this chapter, we briefly describe ADMM type methods proposed by Tao and Yuan [59], and Aybat and Iyengar [6]; efficient APG implementations proposed by Aybat et al. [3, 4]; and review the convex variational framework proposed by Aravkin et al. [1], and the Frank-Wolfe projected gradient method proposed by Mu et al. [45] for solving a convex problem related to the SPCP formulation. In the second part, we very briefly go over some non-convex variants of the SPCP formulation, and iterative solution methods for them.

References

1. A. Aravkin, S. Becker, V. Cevher, and P. Olsen. A variational approach to stable principal component pursuit. In *Conference on Uncertainty in Artificial Intelligence, UAI2014*, 2014.
2. A. Aravkin, J. Burke, and M. Friedlander. Variational properties of value functions. *SIAM Journal on Optimization*, 23(3):1689–1717, 2013.
3. N. Aybat, D. Goldfarb, and G. Iyengar. Fast first-order methods for stable principal component pursuit. *Published online at arXiv:1105.2126 [math.OC], May 2011*, 2011.
4. N. Aybat, D. Goldfarb, and S. Ma. Efficient algorithms for robust and stable principal component pursuit problems. *Computational Optimization and Applications*, 58(1):1–29, 2014.
5. N. Aybat and G. Iyengar. A unified approach for minimizing composite norms. *Mathematical Programming*, 144(1-2):181–226, 2014.
6. N. Aybat and G. Iyengar. An alternating direction method with increasing penalty for stable principal component pursuit. *Computational Optimization and Applications*, pages 1–34, 2015.
7. A. Beck and M. Teboulle. A fast iterative shrinkage-thresholding algorithm for linear inverse problems. *SIAM Journal on Imaging Sciences*, 2(1):183–202, 2009.
8. D. Bertsekas and J. Tsitsiklis. Partial solutions manual parallel and distributed computation: Numerical methods. *Athena Scientific*, 2003.
9. J. Borwein and Q. Zhu. *Techniques Of Variational Analysis*. Springer, 2005.
10. S. Boyd, N. Parikh, E. Chu, B. Peleato, and J. Eckstein. Distributed optimization and

statistical learning via the alternating direction method of multipliers. *Foundations and Trends® in Machine Learning*, 3(1):1–122, 2011.

11. J. Cai, E. Candès, and Z. Shen. A singular value thresholding algorithm for matrix completion. *SIAM Journal on Optimization*, 20(4):1956–1982, 2010.
12. E. Candès, X. Li, Y. Ma, and J. Wright. Robust principal component analysis? *Journal of the ACM, JACM 2011*, 58(3):11, 2011.
13. V. Chandrasekaran, S. Sanghavi, P. Parrilo, and A. Willsky. Rank-sparsity incoherence for matrix decomposition. *SIAM Journal on Optimization*, 21(2):572–596, 2011.
14. C. Chen, B. He, Y. Ye, and X. Yuan. The direct extension of ADMM for multi-block convex minimization problems is not necessarily convergent. *Mathematical Programming*, pages 1–23, 2014.
15. S. Chen, D. Donoho, and M. Saunders. Atomic decomposition by basis pursuit. *SIAM Journal on Scientific Computing*, 20(1):33–61, 1998.
16. V. Demyanov and A. Rubinov. *Approximate Methods in Optimization Problems.* Modern Analytic and Computational Methods in Science and Mathematics. American Elsevier Publishing Company, Inc., 1970.
17. W. Deng, M. Lai, Z. Peng, and W. Yin. Parallel multi-block ADMM with $o(1/k)$ convergence. *arXiv preprint arXiv:1312.3040*, 2013.
18. D. Donoho and I. Johnstone. Adapting to unknown smoothness via wavelet shrinkage. *Journal of the American Statistical Association*, 90(432):1200–1224, 1995.
19. J. Eckstein and D. Bertsekas. On the Douglas-Rachford splitting method and the proximal point algorithm for maximal monotone operators. *Mathematical Programming*, 55(1-3):293–318, 1992.
20. J. Eckstein and W. Yao. Augmented Lagrangian and alternating direction methods for convex optimization: A tutorial and some illustrative computational results. *RUTCOR Research Reports*, 32, 2012.
21. M. Frank and P. Wolfe. An algorithm for quadratic programming. *Naval Research Logistics Quarterly*, 3(1-2):95–110, 1956.
22. M. Fukushima. Application of the alternating direction method of multipliers to separable convex programming problems. *Computational Optimization and Applications*, 1(1):93–111, 1992.
23. R. Glowinski. *Augmented Lagrangian Methods: Applications to the Numerical Solution of Boundary-Value Problems.* Studies in Mathematics and its Applications. Elsevier Science, 2000.
24. O. Güler. On the convergence of the proximal point algorithm for convex minimization. *SIAM Journal on Control and Optimization*, 29(2):403–419, 1991.
25. P. Hansen, J. Nagy, and D. O'Leary. *Deblurring Images: Matrices, Spectra, and Filtering.* Fundamentals of Algorithms. Society for Industrial and Applied Mathematics, 2006.
26. B. He, L. Hou, and X. Yuan. On full jacobian decomposition of the augmented Lagrangian method for separable convex programming. *SIAM Journal on Optimization*, 2013.
27. B. He, L. Liao, D. Han, and H. Yang. A new inexact alternating directions method for monontone variational inequalities. *Mathematical Programming, Series A*, 92:103–118, 2002.
28. B. He and H. Yang. Some convergence properties of a method of multipliers for linearly constrained monotone variational inequalities. *Operations Research Letters*, 23:151–161, 1998.
29. B. He and H. Yang. On the $o(1/n)$ convergence rate of the *-rachford alternating direction method. *SIAM Journal on Numerical Analysis*, 50(2):700–709, 2012.
30. B. He, H. Yang, and S. Wang. Alternating direction method with self-adaptive penalty parameters for monotone variational inequalities. *Journal of Optimization Theory and Applications*, 106(2):337–356, 2000.

31. M. Hestenes. Multiplier and gradient methods. *Journal of optimization theory and applications*, 4(5):303–320, 1969.
32. N. Higham. Computing a nearest symmetric positive semidefinite matrix. *Linear algebra and its applications*, 103:103–118, 1988.
33. M. Hintermüller and T. Wu. Robust principal component pursuit via inexact alternating minimization on matrix manifolds. *Journal of Mathematical Imaging and Vision*, 51(3):361–377, 2015.
34. M. Hong and Z. Luo. On the linear convergence of the alternating direction method of multipliers. *arXiv preprint arXiv:1208.3922*, 2012.
35. K. Huai, M. Ni, F. Ma, and Z. Yu. A customized proximal point algorithm for stable principal component pursuit with non-negative constraint. *Journal of Inequalities and Applications*, 2015(1), 2015.
36. M. Jaggi. Revisiting Frank-Wolfe: Projection-free sparse convex optimization. In *International Conference on Machine Learning, ICML 2013*, pages 427–435, 2013.
37. S. Kontogiorgis and R. Meyer. A variable-penalty alternating direction method for convex optimization. *Mathematical Programming*, 83:29–53, 1998.
38. S. Kumar, M. Mohri, and A. Talwalkar. Ensemble nystrom method. In *Advances in Neural Information Processing Systems*, pages 1060–1068, 2009.
39. X. Li, M. Ng, and X. Yuan. Median filtering-based methods for static background extraction from surveillance video. *Numerical Linear Algebra with Applications*, 2015.
40. Z. Lin, A. Ganesh, J. Wright, L. Wu, M. Chen, and Y. Ma. Fast convex optimization algorithms for exact recovery of a corrupted low-rank matrix. *Computational Advances in Multi-Sensor Adaptive Processing, CAMSAP 2009*, 61, 2009.
41. P. Lions and B. Mercier. Splitting algorithms for the sum of two nonlinear operators. *SIAM Journal on Numerical Analysis*, 16:964–979, 1979.
42. S. Ma, D. Goldfarb, and L. Chen. Fixed point and bregman iterative methods for matrix rank minimization. *Mathematical Programming*, 128(1-2):321–353, 2011.
43. L. Mackey, M. Jordan, and A. Talwalkar. Divide-and-conquer matrix factorization. In *Advances in Neural Information Processing Systems*, pages 1134–1142, 2011.
44. M. Mardani, G. Mateos, and G. Giannakis. Unveiling anomalies in large-scale networks via sparsity and low rank. In *On Signals, Systems and Computers (ASILOMAR), 2011 Conference Record of the Forty Fifth Asilomar Conference*, pages 403–407. IEEE, 2011.
45. C. Mu, Y. Zhang, J. Wright, and D. Goldfarb. Scalable robust matrix recovery: Frank-Wolfe meets proximal methods. *arXiv preprint*, March 2014.
46. Y. Nesterov. A method for unconstrained convex minimization problem with the rate of convergence $\mathcal{O}(1/k^2)$. *Doklady AN SSSR*, 269(3):543–547, 1983.
47. Y. Nesterov. *Introductory lectures on convex optimization*, volume 87. Springer Science & Business Media, 2004.
48. Y. Nesterov. Smooth minimization of non-smooth functions. *Mathematical programming*, 103(1):127–152, 2005.
49. Y. Nesterov. Gradient methods for minimizing composite functions. *Mathematical Programming*, 140(1):125–161, 2013.
50. M. Powell. A method for nonlinear constraints in minimization problems. In R. Fletcher, editor, *Optimization*, pages 283–298. Academic Press, New York, 1969.
51. B. Recht, M. Fazel, and P. Parrilo. Guaranteed minimum-rank solutions of linear matrix equations via nuclear norm minimization. *SIAM review*, 52(3):471–501, 2010.
52. E. Rockafellar. The multiplier method of Hestenes and Powell applied to convex programming. *Journal of Optimization Theory and applications*, 12(6):555–562, 1973.
53. R. Rockafellar. A dual approach to solving nonlinear programming problems by unconstrained optimization. *Mathematical Programming*, 5(1):354–373, 1973.

54. R. Rockafellar. Augmented Lagrangians and applications of the proximal point algorithm in convex programming. *Mathematics of operations research*, 1(2):97–116, 1976.
55. R. Rockafellar. Monotone operators and the proximal point algorithm. *SIAM Journal on Control and Optimization*, 14(5):877–898, 1976.
56. Y. Shen, Z. Wen, and Y. Zhang. Augmented Lagrangian alternating direction method for matrix separation based on low-rank factorization. *Optimization Methods and Software*, 29(2):239–263, 2014.
57. N. Shor, K. Kiwiel, and A. Ruszcay. *Minimization methods for non-differentiable functions.* Springer-Verlag New York, Inc., 1985.
58. P. Sprechmann, A. Bronstein, and G. Sapiro. Learning robust low-rank representations. *arXiv preprint arXiv:1209.6393*, 2012.
59. M. Tao and X. Yuan. Recovering low-rank and sparse components of matrices from incomplete and noisy observations. *SIAM Journal on Optimization*, 21(1):57–81, 2011.
60. P. Tseng. On accelerated proximal gradient methods for convex-concave optimization. *SIAM Journal on Optimization*, 2008.
61. E. Wei and A. Ozdaglar. On the $o(1/k)$ convergence of asynchronous distributed alternating direction method of multipliers. *arXiv preprint arXiv:1307.8254*, 2013.
62. J. Wright, A. Ganesh, S. Rao, Y. Peng, and Y. Ma. Robust principal component analysis: Exact recovery of corrupted low-rank matrices via convex optimization. In *Advances in neural information processing systems*, pages 2080–2088, 2009.
63. T. Zhou and D. Tao. Godec: Randomized low-rank & sparse matrix decomposition in noisy case. In *International Conference on Machine Learning, Bellevue, WA, USA*, 2011.
64. T. Zhou and D. Tao. Greedy bilateral sketch, completion & smoothing. In *International Conference on Artificial Intelligence and Statistics*, pages 650–658, 2013.
65. Z. Zhou, X. Li, J. Wright, E. Candes, and Y. Ma. Stable principal component pursuit. In *IEEE International Symposium on Information Theory, ISIT 2010*, pages 1518–1522, 2010.

3

Dual Smoothing and Value Function Techniques for Variational Matrix Decomposition

Aleksandr Aravkin
IBM Research, USA

Stephen Becker
University of Colorado Boulder, USA

3.1 Introduction

This chapter reviews some recent techniques in convex optimization and contributes several novel results. We apply these techniques to the robust principal component analysis (RPCA) problem. A distinguishing feature of this chapter, in the context of this handbook, is the emphasis on a range of optimization formulations of the RPCA problem.

Linear superposition is a useful model for many applications, including nonlinear mixing problems. Surprisingly, we can perfectly distinguish multiple elements in a given signal using convex optimization as long as they are concise and look sufficiently different from one another. RPCA is a key example, where we decompose a signal into low rank and sparse components; and *stable principal component pursuit (SPCP)*, where we also seek an explicit noise component within the RPCA decomposition. Applications include alignment of occluded images [70], scene triangulation [97], model selection [32], face recognition, and document indexing [26].

Our model is

$$A = L + S + E \qquad (3.1)$$

where A is the observed matrix, L is a low-rank matrix, S is a sparse matrix, and E is an unstructured nuisance matrix (e.g., a stochastic error term). Sometimes the problem is called RPCA when $E = 0$ and SPCP when $E \neq 0$, but we do not distinguish the terms and in general allow $E \neq 0$.

3.1.1 Variations

The RPCA problem uses regularization on the summands L and S in order to improve the recovery of the solution. In the classic RPCA formulation [27], the observed data are fit exactly, the 1-norm regularizer is applied to S to promote sparsity, while the nuclear norm is applied to L to penalize rank:

$$\min \|\|L\|\|_* + \lambda\|S\|_1 \quad \text{s.t.} \quad A = L + S. \tag{3.2}$$

The 1-norm $\|\cdot\|_1$ and nuclear norm $\|\|\cdot\|\|_*$ are given by $\|S\|_1 = \sum_{i,j} |s_{i,j}|$, $\|\|L\|\|_* = \sum_i \varepsilon_i(L)$, where $\varepsilon(L)$ is the vector of singular values of L. The parameter $\lambda > 0$ controls the relative importance of the low-rank term L vs. the sparse term S. This problem has been analyzed by [26, 33], and it has perfect recovery guarantees under stylized incoherence assumptions. There is even theoretical guidance for selecting a minimax optimal regularization parameter λ [26]. Unfortunately, many practical problems only approximately satisfy the idealized assumptions, and hence, we typically tune RPCA via cross-validation techniques.

The SPCP variant is

$$\begin{aligned} &\underset{L,S}{\text{minimize}} \ \|\|L\|\|_* + \lambda_{\text{sum}}\|S\|_1 \\ &\text{subject to } \|L + S - Y\|_F \leq \varepsilon, \end{aligned} \tag{SPCP$_{\text{sum}}$}$$

where the ε parameter accounts for the unknown perturbations $Y - (L + S)$ in the data not explained by L and S.

To cope with practical tuning issues of SPCP, we propose the following new variant called "max-SPCP":

$$\begin{aligned} &\underset{L,S}{\text{minimize}} \ \max\left(\|\|L\|\|_*, \lambda_{\max}\|S\|_1\right) \\ &\text{subject to } \|L + S - Y\|_F \leq \varepsilon, \end{aligned} \tag{SPCP$_{\max}$}$$

where $\lambda_{\max} > 0$ acts similar to λ_{sum}. Our work shows that this new formulation offers both modeling and computational advantages over (SPCP$_{\text{sum}}$).

Cross-validation with (SPCP$_{\max}$) to estimate $(\lambda_{\max}, \varepsilon)$ is significantly easier than estimating $(\lambda_{\text{sum}}, \varepsilon)$ in (SPCP$_{\text{sum}}$). For example, given an *oracle* that provides an ideal separation $Y \simeq L_{\text{oracle}} + S_{\text{oracle}}$, we can use $\varepsilon = \|L_{\text{oracle}} + S_{\text{oracle}} - Y\|_F$ in both cases. However, while we can estimate $\lambda_{\max} = \|L_{\text{oracle}}\|_* / \|S_{\text{oracle}}\|_1$, it is not clear how to choose λ_{sum} from data. Such cross validation can be performed on a similar dataset, or it could be obtained from a probabilistic model.

It is useful to define $\phi(L, S) = \min \|\|L\|\|_* + \lambda_S\|S\|_1$ as a regularizer on the decision variable (L, S). The formulation (3.2) then tries to find the tuple $(\overline{L}, \overline{S})$ that fits the data perfectly, and is minimal with respect to ϕ.

There are several modeling choices that have been made in order to get to formulation (3.2). First, sparsity-promoting and rank-penalizing penalties were selected, and we will respect these choices in the entire chapter. Second, exactly how the choice of component regularizers come together to make ϕ is important; in (3.2), the component penalties are added with a tradeoff parameter λ, but we shall see that other choices can be made as well. Third, (3.2) assumes that $E = 0$, and this is a very limiting choice. We may want to control data fitting error to a given level ε instead, and this requires choice of penalty ρ to measure the misfit error:

$$\min \phi(L, S) \quad \text{s.t.} \quad \rho(A - L - S) \leq \varepsilon. \tag{3.3}$$

In recent works, ρ is often taken to be the least squares penalty or 2-norm, but in general one can make other choices.

Finally, once ρ and ϕ have been selected, the final choice is the type of regularization formulation one wants to solve. Formulation (3.3) minimizes the regularizer subject to a constraint on the misfit error. Two other common formulations are

$$\min \rho(A - L - S) \quad \text{s.t.} \quad \phi(L, S) \leq \tau, \tag{3.4}$$

which minimizes the error subject to a constraint on the regularizer, and

$$\min \rho(A - L - S) + \lambda\phi(L, S), \tag{3.5}$$

which minimizes the sum of error and regularizer with another tradeoff parameter to balance these goals.

All three formulations can be effectively used, and are equivalent in the sense that solutions match for certain values of parameters λ, τ, and ε. Formulation (3.3) is preferable from a modeling perspective when the misfit level ε is known ahead of time, or can be estimated. However, formulations (3.4) and (3.5) often have fast first-order algorithms available for their solution.

It turns out that we can exploit algorithms for (3.4) to solve (3.3) using the graph of the value function for problem (3.4). This concept was first applied for sparsity optimization by [83], and later extended by [2]. In the next sections, we go over concepts from convex analysis that allow us to explain these ideas.

3.1.2 A Primer on SPCP

The theoretical and algorithmic research on SPCP formulations (and source separation in general) is rapidly evolving. Hence, it is important to set the stage first in terms of the available formulations to highlight our contributions.

To this end, we illustrate (SPCP_{sum}) and ($\text{SPCP}_{\max}$) via different convex formulations. Flipping the objective and the constraints in ($\text{SPCP}_{\max}$) and (SPCP_{sum}), we obtain the following convex programs

$$\begin{aligned} &\underset{L,S}{\text{minimize}} \ \frac{1}{2}\|L + S - A\|_F^2 \\ &\text{s.t.} \quad |||L|||_* + \lambda_{\text{sum}}\|S\|_1 \leq \tau_{\text{sum}} \end{aligned} \tag{flip-SPCP$_{\text{sum}}$}$$

$$\begin{aligned} &\underset{L,S}{\text{minimize}} \ \frac{1}{2}\|L + S - A\|_F^2 \\ &\text{s.t.} \quad \max(|||L|||_*, \lambda_{\max}\|S\|_1) \leq \tau_{\max} \end{aligned} \tag{flip-SPCP$_{\max}$}$$

Note that solutions of (flip-SPCP_{sum}) and (flip-$\text{SPCP}_{\max}$) are implicitly related to the solutions of (SPCP_{sum}) and ($\text{SPCP}_{\max}$) via the Pareto frontier by [2, Theorem 2.1].

For the range of parameters where the constraints are *active*, for any parameter ε there exist corresponding parameters $\tau_{\text{sum}}(\varepsilon)$ and $\tau_{\max}(\varepsilon)$, for which the optimal value of (flip-SPCP_{sum}) and (flip-$\text{SPCP}_{\max}$) is ε, and the corresponding optimal solutions $(\overline{S}_s, \overline{L}_s)$ and $(\overline{S}_m, \overline{L}_m)$ are also optimal for (SPCP_{sum}) and ($\text{SPCP}_{\max}$). The activity requirement means that in the range of parameters of interest, we must have that any solution satisfy $\|\overline{L} + \overline{S} - M\|_F = \varepsilon$, and $|||\overline{L}|||_* + \lambda_{\text{sum}}\|\overline{S}\|_* = \tau_{\text{sum}}$ for (flip-SPCP_{sum}), or $\max\left(|||\overline{L}|||_*, \lambda_{\max}\|\overline{S}\|_1\right) = \tau_{\max}$ for (flip-$\text{SPCP}_{\max}$).

For completeness, we also include the Lagrangian formulation, which is covered by our new algorithm:

$$\underset{L,S}{\text{minimize}} \ \lambda_{\text{L}}|||L|||_* + \lambda_{\text{S}}\|S\|_1 + \frac{1}{2}\|L + S - A\|_F^2 \tag{lag-SPCP}$$

Problems (flip-SPCP_{max}) and (flip-SPCP_{sum}) are easier to solve because they involve a differentiable function, and algorithms based on gradients are easier to accelerate than algorithms based only on non-smooth and constrained terms. The disadvantage of some of these formulations is that it is again not as clear how to tune the parameters. Surprisingly, an algorithm we propose in this paper can solve (SPCP_{max}) and (SPCP_{sum}) using a sequence of flipped problems that specifically exploits the structured relationship. In practice, we will see that better tuning also leads to faster algorithms, e.g., fixing ε ahead of time to an estimated 'noise floor' greatly reduces the amount of required computation if parameters are to be selected via cross-validation.

Finally, we note that in some cases, it is useful to change the $\|L + S - A\|_F$ term to $\|\mathcal{L}(L + S - A)\|_F$ where $\mathcal{L}$ is a linear operator. For example, let Ω be a subset of the indices of an $m \times n$ matrix. We may only observe A restricted to these entries, denoted $\mathcal{P}_\Omega(A)$, in which case we choose $\mathcal{L} = \mathcal{P}_\Omega$. Most existing RPCA/SPCP algorithms adapt to the case $\mathcal{L} = \mathcal{P}_\Omega$ but this is due to the strong properties of the projection operator $\mathcal{P}_\Omega$. The advantage of our approach is that it seamlessly handles arbitrary linear operators $\mathcal{L}$. In fact, it also generalizes to smooth misfit penalties that are more robust than the Frobenius norm, including the Huber loss. Our results also generalize to some other penalties on S besides the 1-norm.

The paper proceeds as follows. In Section 3.1.3, we describe previous work and algorithms for SPCP and RPCA. In Section 3.2, we provide the necessary convex analysis background to understand our algorithms and results. In Section 3.3, we cast the relationships between pairs of problems (flip-SPCP_{sum}), (SPCP_{sum}) and (flip-SPCP_{max}), (SPCP_{max}) into a general variational framework, and highlight the product-space regularization structure that enables us solve the formulations of interest using corresponding flipped problems. In particular, we discuss computationally efficient projections as optimization workhorses in Section 3.3.1, and develop new accelerated projected quasi-Newton methods for the flipped and Lagrangian formulations in Section 3.3.2. We present a view of dual smoothing, describe the TFOCS algorithm, and show how to apply it to RPCA in Section 3.4. In particular, we describe the general class of problems solvable by TFOCS in Section 3.4.1, detail the dual smoothing approach in Section 3.4.2, and present new convergence results in Section 3.4.5. Finally, we demonstrate the efficacy of the new solvers and the overall formulation on synthetic problems and real data problems in Section 3.5.

3.1.3 Prior Art

While problem (SPCP_{sum}) with $\varepsilon = 0$ has several solvers (e.g., it can be solved by applying the widely known Alternating Directions Method of Multipliers (ADMM)/Douglas-Rachford method [38]), the formulation assumes the data are noise free. Unfortunately, the presence of noise we consider in this paper introduces a third term in the ADMM framework, where extra care must be taken to develop a convergent variant of ADMM [34]. Interestingly, there are only a handful of methods that can handle this case. Those using smoothing techniques no longer promote exactly sparse and/or exactly low-rank solutions. Those using dual decomposition techniques require high iteration counts. Because each step requires a partial singular value decomposition (SVD) of a large matrix, it is critical that the methods only take a few iterations.

As a rough comparison, we start with related solvers that solve (SPCP_{sum}) for $\varepsilon = 0$. [88] solves an instance of (SPCP_{sum}) with $\varepsilon = 0$ and an 800×800 system in 8 hours. By switching to the (lag-SPCP) formulation, [50] uses the accelerated proximal gradient method [7] to solve a 1000×1000 matrix in under one hour. This is improved further in [61] which again solves (SPCP_{sum}) with $\varepsilon = 0$ using the augmented Lagrangian and ADMM methods and solves a 1500×1500 system in about a minute. As a prelude to our results, our method can solve some systems of this size in about 10 seconds (c.f., Fig. 3.4).

In the case of (SPCP_{sum}) with $\varepsilon > 0$, [82] propose the alternating splitting augmented Lagrangian method (ASALM), which exploits separability of the objective in the splitting scheme, and can solve a 1500×1500 system in about 5 minutes.

The partially smooth proximal gradient (PSPG) approach of [4] smooths just the nuclear norm term and then applies the well-known FISTA algorithm [7]. [4] show that the proximity step can be solved efficiently in closed-form, and the dominant cost at every iteration is that of the partial SVD. They include some examples on video, solving 1500×1500 formulations in under half a minute.

The nonsmooth adaptive Lagrangian (NSA) algorithm of [5] is a variant of the ADMM for (SPCP_{sum}), and makes use of the insight of [4]. The ADMM variant is interesting in that it splits the variable L, rather than the sum $L + S$ or residual $L + S - A$. Their experiments solve a 1500×1500 synthetic problem in between 16 and 50 seconds (depending on accuracy) of the recovery.

[80] develops a method exploiting a low-rank matrix factorization scheme, maintaining $L = UV^T$. This technique has also been effectively used in practice for matrix completion [3, 57], but lacks a full convergence theory in either context. The method of [80] was an order of magnitude faster than ASALM, but encountered difficulties in some experiments where the sparse component dominated the low-rank component. We note that the factorization technique may potentially speed up some of the methods presented here, but we leave this to future work, and only work with convex formulations.

3.2 Convex Analysis Background

We work in finite dimensional spaces $\mathbb{R}^n$ (with Euclidean inner product) unless otherwise specified; we note however that much of the general theory below generalizes immediately to Hilbert spaces. Standard definitions are not referenced, but can be found in standard convex analysis textbooks [74, 77] or in review papers such as [39].

3.2.1 Key Definitions

In this section, we provide definitions of objects that we use throughout the chapter.

We work with functions that take on values from the extended real line $\overline{\mathbb{R}} := \mathbb{R} \cup \{\pm\infty\}$. For example, we define the indicator function as follows:

DEFINITION 3.1 [Indicator Function of a set C]

$$\chi_C(x) = \begin{cases} 0 & x \in C \\ +\infty & x \notin C \end{cases}$$

and thus for any functional f on $\mathbb{R}^n$,

$$\min_{x \in C} f(x) \;=\; \min_{x \in \mathbb{R}^n} f(x) + \chi_C(x).$$

This allows a unified treatment of constraints and objectives by encoding constraints using indicator functions.

The class $\Gamma_0(\mathbb{R}^n)$ denotes **convex**, **lower semi-continuous** (lsc), **proper** functionals from $\mathbb{R}^n$ to $\overline{\mathbb{R}}$. A function is lsc if and only if its graph is closed, and in particular a continuous function is lsc. A proper function is not identically equal to $+\infty$ and is never $-\infty$. We write $\operatorname{dom} f = \{x \mid f(x) < \infty\}$. Further background is widely available, e.g., [39, 74].

DEFINITION 3.2 [Subdifferential and subgradient] Let $f \in \Gamma_0(\mathbb{R}^n)$, then the **subdifferential** of f at the point $x \in \operatorname{dom} f$ is the set

$$\partial f(x) = \{d \in \mathbb{R}^n \mid \forall y \in \mathbb{R}^n,\ f(y) \geq f(x) + \langle d,\, y - x\rangle\}$$

and elements of the set are known as **subgradients**.

The sub-differential of an indicator function χ_C at x is the normal cone to C at x. Fermat's rule is that $x \in \operatorname{argmin} f(x)$ iff $0 \in \partial f(x)$, which follows by the definition of the subdifferential. If $f, g \in \Gamma_0(\mathbb{R}^n)$, then $\partial(f + g) = \partial f + \partial g$ in many situations (i.e., under constraint qualifications such as f or g having full domain). In finite dimensions, Gâteaux and Fréchet differentiability coincide on $\Gamma_0(\mathbb{R}^n)$ and $\partial f(x) = \{d\}$ iff f is differentiable at x with $\nabla f(x) = d$.

We now introduce a key generalization of projections that will be used widely.

DEFINITION 3.3 [Proximity operator] If $f \in \Gamma_0(\mathbb{R}^n)$ and $\lambda > 0$, define

$$\operatorname{prox}_{\lambda f}(y) = \operatorname*{argmin}_x\ \lambda f(x) + \frac{1}{2}\|x - y\|^2 = (I + \lambda\partial f)^{-1}(y)$$

Note that even though ∂f is potentially multi-valued, the proximity operator is always uniquely defined, since it is the minimizer of a strongly convex function. When we say that a proximity operator for f is easy to compute, we mean that the proximity operator for λf is easy to compute for all $\lambda > 0$. Computational complexity will be explored in more detail in subsequent sections.

The proximity operator generalizes projection, since $\operatorname{prox}_{\chi_C}(y) = \mathcal{P}_C(y)$ where $\mathcal{P}$ denotes orthogonal projection onto a set. Another example is the proximity operator of the ℓ^1 norm, which is equivalent to soft-thresholding. The proximity operator is firmly non-expansive [39], just like orthogonal projections.

DEFINITION 3.4 [Conjugate function] Let $f \in \Gamma_0(\mathbb{R}^n)$, then the conjugate function f^* is defined

$$f^*(y) = \sup_x\ \langle x,\, y\rangle - f(x).$$

Furthermore, $\partial f^* = (\partial f)^{-1}$ and $f^{**} = f$.

DEFINITION 3.5 [Gauge] For a convex set C containing the origin, the gauge $\gamma\,(x \mid C)$ is defined by

$$\gamma\,(x \mid C) = \inf_\lambda\{\lambda : x \in \lambda C\}. \tag{3.6}$$

For any norm $\|\cdot\|$, the set defining it as a gauge is simply the unit ball $\mathbb{B}_{\|\cdot\|} = \{x : \|x\| \leq 1\}$. We introduce gauges for two reasons. First, they are more general (a gauge is a norm only if C is bounded with nonempty interior and symmetric about the origin). For example, gauges trivially allow inclusion of non-negativity constraints.

We make extensive use of the theory of dual functions. For example, if one can compute prox_f, then one can compute $\operatorname{prox}_{f^*}$ and related quantities as well, using

$$\operatorname{prox}_{f^*}(x) = x - \operatorname{prox}_f(x) \tag{3.7}$$

$$\operatorname{prox}_{\check{f}}(x) = -\operatorname{prox}_f(-x) \tag{3.8}$$

where $\check{f}(x) = f(-x)$.

DEFINITION 3.6 [Relative interior] The **relative interior** (ri) of a set $C \subset \mathbb{R}^n$ is the interior of C relative to its affine hull (the smallest affine space containing C).

DEFINITION 3.7 [Lipschitz continuity] A function $F : \mathbb{R}^n \to \mathbb{R}^m$ is Lipschitz continuous with constant ℓ if ℓ is the smallest real number such that for all $x, x' \in \mathbb{R}^n$,

$$\|F(x) - F(x')\|_{\mathbb{R}^m} \leq \ell \|x - x'\|_{\mathbb{R}^n}.$$

3.3 Variational Framework for Residual-Constrained Problems

Recall that both (SPCP_{sum}) and (SPCP_{max}) can be written as follows:

$$\min \phi(L, S) \quad \text{s.t.} \quad \rho(L + S - A) \leq \varepsilon. \tag{3.9}$$

Earlier, we discussed that both ϕ and ρ can be chosen by the modeler; in particular, sum and max formulations come from choosing ϕ_{sum} vs. ϕ_{max}. While classic formulations assume ρ to be the Frobenius norm, this restriction is not necessary, and we consider ρ to be smooth and convex. In particular, ρ can be taken to be the robust Huber penalty [56]. Even more importantly, this formulation allows pre-composition of a smooth convex penalty with an arbitrary linear operator $\mathcal{L}$. In particular, note that the RPCA model is described by a simple linear operator:

$$L + S = \begin{bmatrix} I \; I \end{bmatrix} \begin{bmatrix} L \\ S \end{bmatrix}. \tag{3.10}$$

Projection onto a set of observed indices Ω is also a simple linear operator that can be included in ρ. Operators may include different transforms (e.g., Fourier) applied to either L or S.

The problem class (3.9) falls into the class of problems studied by [83, 84] for $\rho(\cdot) = \|\cdot\|^2$ and by [2] for arbitrary convex ρ. Following these references, we define *value function* $v(\tau)$ as

$$v(\tau) = \min_{L,S} \rho(\mathcal{L}(L, S) - A) \quad \text{s.t. } \phi(L, S) \leq \tau, \tag{3.11}$$

This value function provides the bridge between formulations of type (3.9) and their 'flipped' counterparts. Specifically, one can use Newton's method to find a solution to $v(\tau) = \varepsilon$. The approach is agnostic to the linear operator $\mathcal{L}$ (it can be of the simple form (3.10); include restriction in the missing data case, etc.).

For both formulations of interest, ϕ is a norm defined on a product space $\mathbb{R}^{n \times m} \times \mathbb{R}^{n \times m}$, since we can write

$$\phi_{\text{sum}}(L, S) = \left\| \begin{matrix} |||L|||_* \\ \lambda_{\text{sum}} \|S\|_1 \end{matrix} \right\|_1, \tag{3.12}$$

$$\phi_{\text{max}}(L, S) = \left\| \begin{matrix} |||L|||_* \\ \lambda_{\text{max}} \|S\|_1 \end{matrix} \right\|_\infty. \tag{3.13}$$

In particular, both $\phi_{\text{sum}}(L, S)$ and $\phi_{\text{max}}(L, S)$ are gauges as well as norms, and since we are able to treat this level of generality, we focus our theoretical results on this wider class.

In order to implement Newton's method for (3.11), the optimization problem to evaluate $v(\tau)$ must be solved (fully or approximately) to obtain $(\overline{L}, \overline{S})$. Then the τ parameter for the next (3.11) problem is updated via

$$\tau^{k+1} = \tau^k - \frac{v(\tau) - \epsilon}{v'(\tau)}. \tag{3.14}$$

Given $(\overline{L}, \overline{S})$, $v'(\tau)$ can be written in closed form using [2, Theorem 5.2], which simplifies to

$$v'(\tau) = -\phi^\circ(\mathcal{L}^T \nabla\rho(\mathcal{A}(\overline{L}, \overline{S}) - A)), \tag{3.15}$$

with ϕ° denoting the polar gauge to ϕ. The polar gauge is precisely $\gamma\,(x \mid C^\circ)$, with

$$C^\circ = \{v : \langle v, x\rangle \le 1 \quad \forall x \in C\}. \tag{3.16}$$

In the simplest case, where $\mathcal{L}$ is given by (3.10), and ρ is the least squares penalty, the formula (3.15) becomes

$$v'(\tau) = -\phi^\circ\left(\begin{bmatrix}\overline{L} + \overline{S} - A\\ \overline{L} + \overline{S} - A\end{bmatrix}\right).$$

The main computational challenge in the approach outlined in (3.11)–(3.15) is to design a fast solver to evaluate $v(\tau)$. Section 3.3.2 does just this.

The key to RPCA is that the regularization functional ϕ is a gauge over the product space used to decompose A into summands L and S. This makes it straightforward to compute polar results for both ϕ_{sum} and ϕ_{max}.

THEOREM 3.1 *[Max-Sum Duality for Gauges on Product Spaces] Let γ_1 and γ_2 be gauges on $\mathbb{R}^{n_1}$ and $\mathbb{R}^{n_2}$, and consider the function*

$$g(x, y) = \max\{\gamma_1(x), \gamma_2(y)\}.$$

Then g is a gauge, and its polar is given by

$$g^\circ(z_1, z_2) = \gamma_1^\circ(z_1) + \gamma_2^\circ(z_2).$$

PROOF 3.1 Let C_1 and C_2 denote the canonical sets corresponding to gauges γ_1 and γ_2. It immediately follows that g is a gauge for the set $C = C_1 \times C_2$, since

$$\begin{aligned}\inf\{\lambda \ge 0 | (x, y) \in \lambda C\} &= \inf\{\lambda | x \in \lambda C_1 \text{ and } y \in \lambda C_2\}\\ &= \max\{\gamma_1(x), \gamma_2(y)\}.\end{aligned}$$

By [75, Corollary 15.1.2], the polar of the gauge of C is the support function of C, which is given by

$$\begin{aligned}\sup_{x\in C_1, y\in C_2} \langle (x, y), (z_1, z_2)\rangle &= \sup_{x\in C_1} \langle x, z_1\rangle + \sup_{y\in C_2} \langle y, z_2\rangle\\ &= \gamma_1^\circ(z_1) + \gamma_2^\circ(z_2).\end{aligned}$$

This theorem allows us to easily compute the polars for ϕ_{sum} and ϕ_{max} in terms of the polars of $|||\cdot|||_*$ and $\|\cdot\|_1$, which are the dual norms, the spectral norm and infinity norm, respectively.

COROLLARY 3.1 [Explicit variational formulas for (SPCP$_{\text{sum}}$) and (SPCP$_{\text{max}}$)] We have

$$\begin{aligned}\phi_{\text{sum}}^\circ(Z_1, Z_2) &= \max\left\{|||Z_1|||_2, \frac{1}{\lambda_{\text{sum}}}\|Z_2\|_\infty\right\}\\ \phi_{\text{max}}^\circ(Z_1, Z_2) &= |||Z_1|||_2 + \frac{1}{\lambda_{\text{max}}}\|Z_2\|_\infty,\end{aligned} \tag{3.17}$$

where $\|\|X\|\|_2$ denotes the spectral norm (largest eigenvalue of $X^T X$).

We now have closed form solutions for $v'(\tau)$ in (3.15) for both formulations of interest. The remaining challenge is to design a fast solver for (3.11) for formulations (SPCP_{sum}) and (SPCP_{max}). We focus on this challenge in the remaining sections of the paper. We also discuss the advantage of (SPCP_{max}) from this computational perspective.

3.3.1 Projections

In this section, we consider the computational issues of projecting onto the set defined by $\phi(L, S) \leq \tau$. For $\phi_{\max}(L, S) = \max(\|\|L\|\|_*, \lambda_{\max}\|S\|_1)$ this is straightforward since the set is just the product set of the nuclear norm and ℓ_1 norm balls, and efficient projectors onto these are known. In particular, projecting an $m \times n$ matrix (without loss of generality let $m \leq n$) onto the nuclear norm ball takes $\mathcal{O}(m^2 n)$ operations, and projecting it onto the ℓ_1 ball can be done on $\mathcal{O}(mn)$ operations using fast median-finding algorithms [22, 47].

For $\phi_{\text{sum}}(L, S) = \|\|L\|\|_* + \lambda_{\text{sum}}\|S\|_1$, the projection is no longer straightforward. Nonetheless, the following lemma shows this projection can be efficiently implemented.

PROPOSITION 3.1 Projection onto the scaled ℓ_1 ball, that is, $\{x \in \mathbb{R}^d \mid \sum_{i=1}^d \alpha_i |x_i| \leq 1\}$ for some $\alpha_i > 0$, can be done in $\mathcal{O}(d \log(d))$ time.

We conjecture that fast median-finding ideas could reduce this to $\mathcal{O}(d)$ in theory, the same as the optimal complexity for the ℓ_1 ball. The proof of the proposition follows by noting that the solution can be written in a form depending only on a single scalar parameter, and this scalar can be found by sorting $(|x_i|/\alpha_i)$ followed by appropriate summations. Armed with the above proposition, we state an important lemma below. For our purposes, we may think of S as a vector in $\mathbb{R}^{mn}$ rather than a matrix in $\mathbb{R}^{m \times n}$.

LEMMA 3.1 Let $L = U\Sigma V^T$ and $\Sigma = \text{diag}(\sigma)$, and let $(S_i)_{i=1}^{mn}$ be any ordering of the elements of S. Then the projection of (L, S) onto the ϕ_{sum} ball is $(U \,\text{diag}(\hat{\sigma}) V^T, \hat{S})$, where $(\hat{\sigma}, \hat{S})$ is the projection onto the scaled ℓ_1 ball $\{(\sigma, S) \mid \sum_{j=1}^{\min(m,n)} |\sigma_j| + \sum_{i=1}^{mn} \lambda_{\text{sum}} |S_i| \leq 1\}$.

PROOF 3.2 [Sketch of proof] We need to solve

$$\min_{\{(L',S') \mid \phi_{\text{sum}}(L',S') \leq 1\}} \frac{1}{2}\|L' - L\|_F^2 + \frac{1}{2}\|S' - S\|_F^2.$$

Alternatively, solve

$$\min_{S'} \min_{\{L' \mid \|\|L'\|\|_* \leq 1 - \lambda_{\text{sum}}\|S'\|_1\}} \frac{1}{2}\|L' - L\|_F^2 + \frac{1}{2}\|S' - S\|_F^2.$$

The inner minimization is equivalent to projecting onto the nuclear norm ball, and this is well-known to be soft-thresholding of the singular values. Since it depends only on the singular values, recombining the two minimization terms gives exactly a joint projection onto a scaled ℓ_1 ball.

All the references to the ℓ_1 ball can be replaced by the intersection of the ℓ_1 ball and the non-negative cone, and the projection is still efficient. As noted in Section 3.3, imposing non-negativity constraints is covered by the gauge results of Theorem 3.1 and Corollary 3.1. Therefore, both the variational and efficient computational framework can be applied to this interesting case.

3.3.2 Solving the Max 'Flipped' Sub-Problem via Projected Quasi-Newton Methods

If we adopt $\phi_{\max}$ as the regularizer, then the subproblem (flip-SPCP$_{\max}$) takes the explicit form

$$\begin{aligned} &\underset{L,S}{\text{minimize}}\ \frac{1}{2}\|L+S-A\|_F^2 \\ &\text{s.t.}\quad |||L|||_* \leq \tau_{\max}, \quad \|S\|_1 \leq \tau_{\max}/\lambda_{\text{sum}}. \end{aligned} \tag{3.18}$$

The computational bottle-neck is solving this problem quickly, once at each outer iteration. While first-order methods can be used, we can exploit the structure of the objective by using quasi-Newton methods. The main challenge here is that for the $|||L|||_*$ term, it is tricky to deal with a weighted quadratic term (whereas for $\|S\|_1$, we can obtain a low-rank Hessian and solve it efficiently via coordinate descent).

Let $X = (L, S)$ be the full variable, so we can write the objective function as $f(X) = \frac{1}{2}\|\mathcal{L}(X) - A\|_F^2$. To simplify the exposition, we take $\mathcal{L} = (I, I)$, but the presented approach applies to general linear operators (including terms like $\mathcal{P}_\Omega$). The matrix structure of L and S is not yet important here, so we can think of them as reshaped vectors instead of matrices.

The gradient is $\nabla f(X) = \mathcal{L}^T(\mathcal{L}(X) - A)$. For convenience, we use $r(X) = \mathcal{L}(X) - A$ and

$$\nabla f(X) = \begin{pmatrix} \nabla_L f(X) \\ \nabla_S f(X) \end{pmatrix} = \mathcal{L}^T \begin{pmatrix} r(X) \\ r(X) \end{pmatrix}, \quad r_k \equiv r(X_k).$$

The Hessian is $\mathcal{L}^T\mathcal{L} = \begin{pmatrix} I\, I \\ I\, I \end{pmatrix}$. We cannot simultaneously project (L, S) onto their constraints with this Hessian scaling (doing so would solve the original problem!), since the Hessian removes separability. Instead, we use (L_k, S_k) to approximate the cross-terms.

The true function is a quadratic, so the following quadratic expansion around $X_k = (L_k, S_k)$ is exact:

$$\begin{aligned} f(L,S) &= f(X_k) + \left\langle \begin{pmatrix} \nabla_L f(X_k) \\ \nabla_S f(X_k) \end{pmatrix}, \begin{pmatrix} L - L_k \\ S - S_k \end{pmatrix} \right\rangle \\ &\quad + \frac{1}{2}\left\langle \begin{pmatrix} L - L_k \\ S - S_k \end{pmatrix}, \nabla^2 f \begin{pmatrix} L - L_k \\ S - S_k \end{pmatrix} \right\rangle \\ &= f(X_k) + \left\langle \begin{pmatrix} r_k \\ r_k \end{pmatrix}, \begin{pmatrix} L - L_k \\ S - S_k \end{pmatrix} \right\rangle \\ &\quad + \frac{1}{2}\left\langle \begin{pmatrix} L - L_k \\ S - S_k \end{pmatrix}, \begin{pmatrix} I\, I \\ I\, I \end{pmatrix} \begin{pmatrix} L - L_k \\ S - S_k \end{pmatrix} \right\rangle \\ &= f(X_k) + \left\langle \begin{pmatrix} r_k \\ r_k \end{pmatrix}, \begin{pmatrix} L - L_k \\ S - S_k \end{pmatrix} \right\rangle \\ &\quad + \frac{1}{2}\left\langle \begin{pmatrix} \mathbf{L} - L_k \\ \mathbf{S} - S_k \end{pmatrix}, \begin{pmatrix} L - L_k + \mathbf{S} - S_k \\ \mathbf{L} - L_k + S - S_k \end{pmatrix} \right\rangle \end{aligned}$$

The coupling of the second order terms, shown in bold, prevents direct 1-step minimization of f, subject to the nuclear and 1-norm constraints. The FISTA [7] method replaces the Hessian $\begin{pmatrix} I\, I \\ I\, I \end{pmatrix}$ with the upper bound $2\begin{pmatrix} I\, 0 \\ 0\, I \end{pmatrix}$, which solves the coupling issue, but potentially loses too much second-order information. For (flip-SPCP$_{\text{sum}}$), FISTA is about the best we can do (we actually use SPG [89], which did slightly better in our tests). However, for (flip-SPCP$_{\max}$) (and for (lag-SPCP), which has no constraints but rather non-smooth terms, which can be treated like constraints using proximity operators), the constraints are uncoupled and we can take a "middle road" approach, replacing

$$\left\langle \begin{pmatrix} \mathbf{L} - L_k \\ \mathbf{S} - S_k \end{pmatrix}, \begin{pmatrix} L - L_k + \mathbf{S} - S_k \\ \mathbf{L} - L_k + S - S_k \end{pmatrix} \right\rangle$$

with

$$\left\langle \begin{pmatrix} L - L_k \\ S - S_k \end{pmatrix}, \begin{pmatrix} L - L_k + \mathbf{S_k} - \mathbf{S_{k-1}} \\ \mathbf{L_{k+1}} - \mathbf{L_k} + S - S_k \end{pmatrix} \right\rangle.$$

The first term is decoupled, allowing us to update L_k, and then this is plugged into the second term in a Gauss-Seidel fashion. In practice, we also scale this second-order term with a number slightly greater than 1 but less than 2 (e.g., 1.25) which leads to more robust behavior. We expect this "quasi-Newton" trick to do well when $S_{k+1} - S_k$ is similar to $S_k - S_{k-1}$.

3.4 Dual Smoothing and the Proximal Point Method

This section describes the successful approach of the TFOCS algorithm [12] and software.[8] The method is based on the proximal point algorithm and handles generic convex minimization problems. The original analysis in [12] was in terms of convex *cones*, but we re-analyze the method here in terms of extended valued convex *functions*, and find stronger results.

Secondly, we compare to alternatives in the literature; the basic ingredients involved in TFOCS are well-known in the optimization community, and there are many variants and applications. We discuss in detail the relationship with the family of preconditioned ADMM methods popularized by [31]. The TFOCS algorithm also motivated [45], which promotes an alternative approach that smooths both the primal and the dual. This approach obtains stronger guarantees, but there is a price to pay — smoothing the primal necessarily yields a less sparse solution. We show that with the improved analysis of TFOCS, it enjoys the same strong guarantees, while avoiding smoothing the primal.

Finally, we apply this method to RPCA. The advantage of the TFOCS formulation is that it flexible and can solve all standard variants of RPCA and SPCP, and can easily add non-negativity or other types of additional constraints. We briefly detail how the algorithm can be specialized for the RPCA problem in particular. Even without specializing the algorithm for RPCA, TFOCS has performed well. The results of tests from [20] are that "LSADM [51] and TFOCS [12] solvers seem to be the most adapted ones in the field of video surveillance [90]."

3.4.1 General Form of Our Optimization Problem

In this section, we provide a general notation that captures all optimization problems of interest. Note that even though we do not explicitly write constraints in this formulation, through the use of extended-value functions we capture constraints, and so in particular can express residual-constrained formulations using this notation.

We consider the following generic problem

$$\min_x \; \omega(x) + \psi_0(x) + \sum_{i=1}^{m} \psi_i(\mathcal{L}_i x - b_i) \tag{3.19}$$

[8] `http://cvxr.com/tfocs`

where

- ω and ψ_i for $i = 0, \ldots, m$ are proper convex lsc functions on their respective spaces,
- ω is differentiable everywhere, with Lipschitz continuous gradient; note that we can consider $\omega(\mathcal{L}x - b)$ trivially, since this is also differentiable,
- ψ_i for $i = 0, \ldots, m$ has an easily computable proximity function,
- $\mathcal{L}_i$ for $i = 1, \ldots, m$ is a linear operator, and b_i is a constant offset.

We distinguish ψ_0 from ψ_i, $i \geq 1$, since ψ_0 is not composed with a linear operator. This is significant since being able to easily compute the proximity operator of ψ does not imply one can easily compute the proximity operator of $\psi_i \circ \mathcal{L}_i$ nor of $\psi_i + \psi_j$, so we deal with the $i = 1, \ldots, m$ terms specially.

REMARK 3.1 Unlike the $\mathcal{L}$ terms, the offsets b can be absorbed into ψ, since if $\widetilde{\psi}(x) = \psi(x - b)$ then $\text{prox}_{\widetilde{\psi}}(x) = b + \text{prox}_{\psi}(x - b)$ [39]. Thus we make these offsets explicit or implicit as convenient.

RPCA in the general setting (3.3) can be recovered from the above by setting $x = (L, S)$, $\psi_0(x) = \phi(L, S)$, $m = 1$, and $\psi_1 = \rho$ with $\mathcal{L}_1 x = -L - S$ and $b_1 = -A$, and $\omega = 0$. In fact, many convex problems from science and engineering fit into this framework [39]. The strength of this particular model is that it is often easy to decompose a complicated function f by a finite sum of simple functions ψ_i composed with linear operators. In this case, f may not be differentiable, and prox_f need not be easy to compute, so the model allows us to exploit the structure of the smaller building blocks.

3.4.2 Dual Smoothing Approach

We re-derive the algorithm described in [12], but from a *conjugate-function* viewpoint, whereas [12] used a *dual-conic* viewpoint. The latter viewpoint is subsumed in the former, and is arguably less elegant.

Consider the problem

$$\min_x \; f(x) := \psi_0(x) + \sum_{i=1}^{m} \psi_i(\mathcal{L}_i x - b_i), \tag{3.20}$$

which is similar to (3.19) but without the differentiable term ω. In addition to our previous assumptions on these functions (convex, lsc, proper), we now assume that at least one minimizer exists.

Our main observation is that instead of solving (3.20) directly, we can instead use the proximal point method to minimize f, which exploits the fact that

$$\min_x \; f(x) = \min_y \; \min_x \; f(x) + \frac{\mu}{2}\|x - y\|^2. \tag{3.21}$$

This fact follows, since $f(x) \leq f(x) + \frac{\mu}{2}\|x - y\|^2$, and equality is achieved by setting y to one of the minimizers of f.

Thus we solve a sequence of problems of the form

$$\min_x \; f(x) + \frac{\mu}{2}\|x - y\|^2. \tag{3.22}$$

for a fixed y. The exact proximal point method is

$$y_{k+1} = \operatorname*{argmin}_x \; F_{k(x)} := f(x) + \frac{\mu_k}{2}\|x - y_k\|^2 \tag{3.23}$$

where μ_k is any sequence such that $\limsup \mu_k < \infty$, and y_0 is arbitrary. Note that F_k depends on μ_k and y_k.

The benefit of (3.22) over (3.20) is that F_k is strongly convex, whereas f need not be, and therefore the dual problem of F_k is easy to solve, which we will make precise.

Rewriting the objective, and ignoring offset terms b_i for simplicity (see Remark 3.1), we have

$$\min_x \underbrace{\psi_0(x) + \frac{\mu}{2}\|x-y\|^2}_{\Phi(x)} + \sum_{i=1}^m \psi_i(\mathcal{L}_i x). \tag{3.24}$$

For $i = 1, \ldots, m$, each $\mathcal{L}_i$ is a linear operator from $\mathbb{R}^n$ to $\mathbb{R}^{m_i}$. We can further simplify notation by defining a linear operator $\mathbf{L}$ and a vector $\mathbf{z} \in \mathbb{R}^{\sum_{i=1}^m m_i}$ (e.g., $\mathbf{z} = \mathbf{L}(X)$) such that

$$\mathbf{L}(x) = \begin{pmatrix} L_1(x) \\ L_2(x) \\ \vdots \\ L_m(x) \end{pmatrix}, \quad \mathbf{z} = \begin{pmatrix} z_1 \\ z_2 \\ \vdots \\ z_m \end{pmatrix}$$

Then define

$$\Psi(\mathbf{z}) = \sum_{i=1}^m \psi_i(z_i), \text{ so } \operatorname{prox}_\Psi(\mathbf{z}) = \begin{pmatrix} \operatorname{prox}_{\psi_1}(z_1) \\ \operatorname{prox}_{\psi_2}(z_2) \\ \vdots \\ \operatorname{prox}_{\psi_m}(z_m) \end{pmatrix} \text{ and } \operatorname{prox}_{\Psi^*}(\mathbf{z}) = \begin{pmatrix} \operatorname{prox}_{\psi_1^*}(z_1) \\ \operatorname{prox}_{\psi_2^*}(z_2) \\ \vdots \\ \operatorname{prox}_{\psi_m^*}(z_m) \end{pmatrix}. \tag{3.25}$$

We now rewrite (3.24) in the following compact representation:

$$\min_x \Phi(x) + \Psi(\mathbf{L}x). \tag{3.26}$$

We are now in a position to apply standard Fenchel-Rockafellar duality ([6, 74]), perturbing $\mathbf{L}x$ to arrive at the dual problem

$$\min_{\mathbf{z}} \underbrace{\Phi^*(\mathbf{L}^*\mathbf{z}) + \Psi^*(-\mathbf{z})}_{q(\mathbf{z})}. \tag{3.27}$$

Standard constraint qualifications for finite dimensional problems (e.g., Thm. 15.23 and Prop. 6.19x in [6]) guarantee a zero duality gap if

$$\operatorname{ri}(\operatorname{dom}\Psi) \cap \mathbf{L}(\operatorname{ri}(\operatorname{dom}\Phi)) \neq \emptyset.$$

The primal problem (3.26) is not amenable to computation because even though we can calculate the proximity operators of Ψ and Φ, we cannot easily calculate the proximity operator of $\Psi \circ \mathbf{L}$. The dual formulation (3.27) circumvents this because instead of asking for the proximity operator of $\Phi^* \circ \mathbf{L}^*$, which is not easy, we will use its gradient, and in this case the linear term causes no issue. We can do this because Φ is at least μ strongly convex, so we have the following well-known result (see, e.g., Prop. 12.60 in [77]).

LEMMA 3.2 The function Φ^* is continuously differentiable and the gradient is Lipschitz continuous with constant μ^{-1}, and hence $\Phi^* \circ \mathbf{L}^*$ is also continuously differentiable with Lipschitz constant $\|\mathbf{L}\|^2/\mu$.

Note we are taking the operator norm of $\mathbf{L}$, and $\|\mathbf{L}\|^2 = \|\mathbf{L}\mathbf{L}^*\| = \sum_{i=1}^n \|\mathcal{L}_i\|^2$. The actual gradient can be determined by exploiting the relation

$$\nabla\Phi^*(w) = \partial\Phi^*(w) = \partial\Phi^{-1}(w),$$

which follows from Fenchel's equality (see [74, Theorem 23.5]), i.e., if $x = \nabla\Psi^*(w)$ then $0 \in \partial\Phi(x) - w$, so x minimizes $\Phi(\cdot) - \langle w, \cdot\rangle$. Thus

$$\begin{aligned}
\nabla\Phi^*(w) &= \underset{x}{\operatorname{argmin}}\ \Phi(x) - \langle x, w\rangle \\
&= \underset{x}{\operatorname{argmin}}\ \psi_0(x) + \frac{\mu}{2}\|x-y\|^2 - \langle x, w\rangle \\
&= \underset{x}{\operatorname{argmin}}\ \psi_0(x) + \frac{\mu}{2}\|x-(y+w/\mu)\|^2 \\
&= \operatorname{prox}_{\psi_0/\mu}(y + w/\mu) \qquad (3.28)
\end{aligned}$$

so we can calculate the gradient using $\operatorname{prox}_{\psi_0}$. Furthermore, via the chain rule, we have $\nabla(\Phi^* \circ \mathbf{L}^*)(\mathbf{z}) = \mathbf{L}\nabla\Phi^*(\mathbf{L}^*\mathbf{z})$.

Thus we have shown that the dual problem (3.27) is a sum of two functions, one of which has a Lipschitz continuous gradient and the other admits an easily computable proximity operator. Such problems can be readily solved via proximal gradient methods [41] and accelerated proximal gradient methods [8, 9].

If strong duality holds, then if $\mathbf{z}^\star$ is a solution of the dual problem, $x^\star = \nabla\Psi^*(\mathbf{L}^*\mathbf{z}^\star)$ is the unique solution to the primal problem (cf. Prop. 19.3 [6]), and this was used in [12] to motivate solving the dual. We actually have much stronger results that provide approximate optimality guarantees for approximate dual solutions.

Algorithm 3.1 - FISTA [8] applied to (3.27); enforcing $t_k = 1$ recovers proximal gradient descent

Require: $\ell \geq \|\mathbf{L}\|^2/\mu$ bound on Lipschitz constant; $\mathbf{z}_0$ arbitrary
1: $\mathbf{w}_1 = \mathbf{z}_0$, $t_1 = 1$.
2: **for** $k = 1, 2, \ldots$ **do**
3: Compute $\widetilde{x}_k = \nabla\Phi^*(\mathbf{L}^*\mathbf{w}_k)$ using (3.28)
4: Set $\mathbf{G} = \mathbf{L}\widetilde{x}_k = \mathbf{L}\nabla\Phi^*(\mathbf{L}^*\mathbf{w}_k)$
5: $\mathbf{z}_k = -\operatorname{prox}_{\ell^{-1}\Psi^*}\left(-\mathbf{w}_k + \ell^{-1}\mathbf{G}\right)$ using (3.25) and (3.7)
6: $t_{k+1} = \frac{1+\sqrt{1+4t_k^2}}{2} \geq t_k + 1/2$
7: $\mathbf{w}_{k+1} = \mathbf{z}_k + \frac{t_k - 1}{t_{k+1}}(\mathbf{z}_k - \mathbf{z}_{k-1})$
8: **end for**

We can bound the rate of convergence of the dual objective function q:

THEOREM 3.2 *[Thm. 4.4 in [8]] The sequence $(\mathbf{z}_k)$ generated by Algorithm 3.1 satisfies*

$$q(\mathbf{z}_k) - \min_{\mathbf{z}} q(\mathbf{z}) \leq \frac{2\ell d_0^2}{k^2}$$

where d_0 is the distance from $\mathbf{z}_0$ to the optimal set.

From this, we can recover a remarkable bound on the primal sequence.

THEOREM 3.3 *[Thm. 4.1 in [9]] Let (x_k) be the sequence generated by*

$$x_k = \nabla\Phi^*(\mathbf{L}^*\mathbf{z}_k)$$

(similar to $\widetilde{x}_k$ but evaluated at $\mathbf{z}_k$ not $\mathbf{w}_k$). Let $x^\star$ be the (unique) optimal point to (3.26). *Then*

$$\frac{\mu}{2}\|x_k - x^\star\|^2 \le q(\mathbf{z}_k) - \min_{\mathbf{z}} q(\mathbf{z})$$

for any point $\mathbf{z}_k$, and hence for $(\mathbf{z}_k)$ from Algorithm 3.1,

$$\|x_k - x^\star\| \le \frac{2\sqrt{\ell}d_0}{\sqrt{\mu}k}. \tag{3.29}$$

Note that in practice, one typically uses $\widetilde{x}_k$ for any convergence tests, since it is a by-product of computation, whereas x_k is expensive to compute. Since $0 \le (t_k - 1)/t_{k+1} < 1$ for $t_k \in [1, \infty)$, then if $\mathbf{z}_k$ converges, it follows that $\mathbf{w}_k$ converges to the same limit, so asymptotically $x_k \simeq \widetilde{x}_k$.

The result of the above theorem holds regardless of how the sequence $(\mathbf{z}_k)$ is generated, so if the dual method has better than worst-case convergence (or if, e.g., line search is used), then the primal sequence enjoys the same improvements.

Since $\sqrt{\ell} = \|\mathbf{L}\|/\sqrt{\mu}$, combining with the other factor of $1/\sqrt{\mu}$ shows that $\|x_k - x^\star\| \propto 1/(\mu k)$, so choosing μ large leads to fast convergence (since the dual problem is very smooth). The trade-off is that the outer loop (the proximal point method) will converge more slowly with μ large.

3.4.3 Comparison with literature

Dual methods

The proposed method has been formulated in several contexts; part of the novelty of [12] is the generality of the method and pre-built function routines from which a wide variety of functions could be constructed. The basic concepts of duality and smoothing are widely used, and using duality to avoid difficult affine terms goes back to Uzawa's method (see [35]) and the general concept of *domain decomposition*.

More recent and specific approaches include those of [37, 62, 63, 66], which particularly deal with signal processing problems. The work [37] considers a single smoothed problem, not the full proximal point sequence, and uses proximal gradient descent to solve the dual. They establish convergence of the primal variable but without a bound on the convergence rate. Explicit use of the proximal point algorithm is mentioned in [62], which focuses on the nuclear norm minimization problem, but uses Newton-CG methods, which requires a third level of algorithm hierarchy and good heuristic values for stopping criteria of the conjugate-gradient method. The algorithm SDPNAL [98] is similar to [62] and uses a Newton-CG augmented Lagrangian framework to solve general semi-definite programs (SDP). The work of [63] focuses on a more specific version of the problem but contains the same general ideas, and has inner and outer iterations. The algorithm in [66] focuses on standard ADMM settings that have non-trivial linear terms, and smooths the dual problem; they follow [93] and have specific complexity bounds when the appropriate constraint set is compact.

Primal-dual methods

Another method to remove effects of the linear terms $\mathcal{L}_i$ is to solve a primal-dual formulation. Many of these are based on duplicating variables into a product space and then enforcing *consensus* of the duplicated variables. This can replace many terms, such as $\sum_{i=1}^m \psi_i(\mathcal{L}_i x)$, with two generalized functions, which is inherently easier, since it is often amenable to the Douglas-Rachford algorithm [38]. Specific examples of this approach are [13, 16–19, 36, 40]. The paper [21] is slightly unique in that it reformulates the primal-dual problem into one that can be solved by the obscure forward-backward-forward algorithm of Tseng (the forward-backward algorithm does not apply since in the primal-dual setting, Lipschitz continuity does not imply co-coercivity).

Another main line of primal-dual methods was motivated in [48] as a preconditioned variant of ADMM and then analyzed in [31], and an improved analysis by [54] allowed a generic formulation to be proposed independently by [42, 86]. In more particular settings, it is known as *linearized* ADMM or primal-dual hybrid gradient (PDHG), and has seen a recent surge of interest and analysis [46, 52, 55, 60, 73, 87, 90, 91, 96]. Several recent survey papers [29, 58] review these algorithms in more detail.

The PDHG has not been applied specifically to the RPCA problem, to our knowledge. The next section describes this method in more detail.

Detailed comparison the the PDHG

On certain classes of problems, our approach is quite similar to the PDHG approach. Consider the following simplified version of (3.20):

$$\min_x \psi_0(x) + \psi_1(\mathcal{L}x - b).$$

At each step of the proximal point algorithm, we minimize the above objective perturbed by $\mu/2\|x-y\|^2$, where $\mu > 0$ is arbitrary. Applying FISTA to the dual problem leads to steps of the following form:

$$\begin{aligned} x_{k+1} &= \operatorname*{argmin}_x \ \psi_0(x) - \langle \overline{z}, \mathcal{L}x - b\rangle + \frac{\mu}{2}\|x-y\|^2 \\ z_{k+1} &= \operatorname*{argmin}_z \ \psi_1^*(z) - \langle z, \mathcal{L}x_{k+1}\rangle + \frac{1}{2t}\|z - z_k\|^2 \\ \overline{z} &= z_{k+1} + \theta_k\left(z_{k+1} - z_k\right) \end{aligned}$$

where $\theta_k = (t_k - 1)/t_{k-1}$ as in Algorithm 3.1 (and $\lim_{k\to\infty}\theta_k = 1$). We require the stepsize t to satisfy $t \le \mu/\ell^2$.

For the PDHG method, pick stepsizes $\tau\sigma < 1/\ell^2$. There is no outer loop over y, and the full algorithm is:

$$\begin{aligned} z_{k+1} &= \operatorname*{argmin}_z \ \psi_1^*(z) - \langle z, \mathcal{L}\overline{x}\rangle + \frac{1}{2\sigma}\|z - z_k\|^2 \\ x_{k+1} &= \operatorname*{argmin}_x \ \psi_0(x) - \langle \overline{z}, \mathcal{L}x - b\rangle + \frac{1}{2\tau}\|x - x_k\|^2 \\ \overline{x} &= x_{k+1} + \theta\left(x_{k+1} - x_k\right) \end{aligned}$$

with $\theta = 1$ (see Algorithm 1 in [31]).

The two algorithms are extremely similar, the main differences being that the TFOCS approach updates y occasionally, while PDHG updates $y = x_k$ every iteration of the inner algorithm and thus avoids the outer iteration completely. The lack of the outer iteration is an advantage, mainly since it avoids the issue of a stopping criteria. However, the advantage of an inner iteration is that we can apply an accelerated Nesterov method, which can only be done in the PDHG if one has further assumptions on the objective function.

We present a numerical comparison of the two algorithms applied to RPCA in Figures 3.2 and 3.3.

Detailed comparison with double-smoothing approach

For minimizing smooth but not strongly convex functions f, classical gradient descent generates a sequence of iterates (x_k) such that the objective converges at rate $f(x_k) - \min_x f(x) \le \mathcal{O}(1/k)$, and (x_k) itself converges, with $\|\nabla f(x_k)\| \le \mathcal{O}(1/\sqrt{k})$. The landmark work of Nesterov in 1983 [67] showed that a simple acceleration technique similar to the heavy-ball method generates a sequence (x_k) such that $f(x_k) - \min_x f(x) \le \mathcal{O}(1/k^2)$, although there are no guarantees about convergence of the sequence[9] (x_k) nor strong bounds on $\|\nabla f(x_k)\|$. Note that our dual scheme uses FISTA [7], which is a generalization of Nesterov's scheme; we refer to any such scheme with $\mathcal{O}(1/k^2)$ as "accelerated."

Defining an ϵ-solution to be a point x such that $f(x) - \min_x f(x) \le \epsilon$, we see that it takes $\mathcal{O}(1/\epsilon)$ and $\mathcal{O}(1/\sqrt{\epsilon})$ iterations to reach such a point using the classical and accelerated gradient descent schemes, respectively.

In 2005 Nesterov introduced a smoothing technique [93] that applies to minimizing non-smooth functions over compact sets. The standard algorithm is sub-gradient descent, in which the worst-case convergence of the objective is $\mathcal{O}(1/\sqrt{k})$. By smoothing the primal function by a sufficiently small fixed amount, then due to the compact constraints, the smooth function differs from the original function by less than $\epsilon/2$. One can then apply Nesterov's accelerated method, which generates an $\epsilon/2$-optimal point (to the smoothed problem, and hence ϵ optimal to the original problem) in $\mathcal{O}(\ell/\sqrt{\epsilon})$ iterations. The catch is that ℓ is the Lipschitz constant of the smoothed objective, which is proportional to $\epsilon^{1/2}$, so the overall convergence rate is $\mathcal{O}(1/\epsilon)$, which is still better than the subgradient schemes that would take $\mathcal{O}(1/\epsilon^2)$.

The aforementioned smoothing technique is an alternative to our approach, but the two are not directly comparable, since we do not assume the objective has the same form, nor is our domain necessarily bounded. Furthermore, smoothing the primal objective can have negative consequences. For example, it is common to solve ℓ_1 regularized problems in order to generate sparse solutions. Our dual-smoothing technique keeps the primal non-smooth and therefore still promotes sparsity, whereas a primal-smoothing technique would replace ℓ_1 with the Huber function, and this does not promote sparsity; see Fig. 1 in [12].

Another option is the *double-smoothing* technique proposed by [45]. This is the approach most similar with our own. As with our approach, a strongly convex term is added to the primal problem in order to make the dual problem smooth, and then the dual problem is solved with an accelerated method. Departing from our approach, they additionally smooth the primal as well as in [93], which makes the dual problem strongly convex. The reason for this is subtle. Without the strong convexity in the dual (i.e., our approach), we only have a bound on the dual objective function. To translate this into a bound on the primal variable, measured in terms of objective function or distance to the feasible set, requires using the gradient of the dual variable. As mentioned above, accelerated methods have faster rates of convergence in the objective but not of the gradients. For this reason, one must resort to a classical gradient descent method, which has slower rates of convergence.

Making the dual problem strongly convex allows the use of special variants of Nesterov's accelerated method (see [68]) that converge at a linear rate, and, importantly, so do the iterates and their gradients. The convergence is in terms of the smooth and perturbed problem, so the size of these perturbations is controlled in such a manner (again, the domains are assumed to be bounded) such that one recovers a $\mathcal{O}(1/\epsilon \log(1/\epsilon))$ convergence rate.

The analysis of [45] suggests that our method of single smoothing is flawed, but this seems to be an artifact of their analysis. Using Theorem 3.3, which is a recent result, we have a rate on the convergence of the primal sequence, rather than on its objective value or distance to optimality. In many situations this is a stronger measure of convergence, depending on the purpose of solving the optimization problem. For robust PCA, the distance to the true

---

[9] There is no experimental evidence that the sequence does not converge, and indeed a recent preprint shows that a slight modification of the algorithm can guarantee convergence [30]

solution is indeed a natural metric, whereas sub-optimality of the objective function is rather artificial, and distance to the feasible set depends on the choice of model parameters, which maybe somewhat arbitrary.

Furthermore, the lack of bounds on the iterate *sequence* generated by an accelerated method, which was the issue in the analysis of [45], is mainly a theoretical one, since in most practical situations, the variables and their gradients do appear to converge at a fast rate. Our situation is also different since the constraints need not be compact, and we use the proximal point method to reduce the effect of the smoothing.

3.4.4 Effect of the Smoothing Term

The next section discusses convergence of the proximal point method, but we first discuss the phenomenon that sometimes, the proximal point method converges to the exact solution in a finite number of iterations. This is quite unlike classical exact penalty results [15] which only apply when the perturbation is non-smooth (e.g., $\|x - y\|$) whereas we use a smooth perturbation $\|x - y\|^2$).

Whenever the functions and constraints are polyhedral, such as for linear programs, finite convergence (or the "exact penalty" property) will occur. This was known since the 1970s; see Prop. 8 in [76] and [14, 64, 72]. The special case of noiseless basis pursuit was recently analyzed in [92] using different techniques. More general results, allowing a range of penalty functions, were proved in [49].

For non-polyhedral problems, exact penalty does not occur in general. For example, one can construct an example of nuclear norm minimization that does not have the exact penalty property. However, under additional assumptions that are typical to guarantee exact recovery in the sense of [28], it is possible to obtain exact penalty results. Research in this is motivated by the popularity of the [25] algorithm, which is a special case of the TFOCS framework applied to matrix completion. Results are in [59], as well as [95] (and the correction [94]), which also provides results for the RPCA problem in particular. Some results on generalizations to tensors are also available [81].

3.4.5 Convergence

Certificates of accuracy of the sub-problem

For solving the smoothed sub-problem $\min_x F_k(x)$, we assume the proximity operator of each ψ_i is easy to compute, and given this, it is reasonable to expect that it is easy to compute a point in $\partial\psi_i$ as well. Furthermore, since the algorithm computes the effect of the linear operators on the current iterate, these values are already known and do not have to be re-computed. Thus, computing a point in ∂F_K may be relatively cheap since it is just the sum of the $\partial\psi_i$ composed with the appropriate linear operators. We can now obtain accuracy guarantees via the following proposition.

PROPOSITION 3.2 [76, Prop. 3] Let $y = \operatorname{argmin}_x F_k(x)$, then for all points x and all $d \in \partial F_k(x)$,

$$\|x - y\| \leq \mu_k^{-1}\|d\|. \tag{3.30}$$

Convergence of the proximal point method

The convergence of the proximal point method is well-understood, but we are particularly interested in the case when the update step is computed inaccurately. There has been recent work on this (see, e.g., [44, 78, 85]) but often under the assumption that the computed point is feasible, i.e., it is inside $\operatorname{dom} f$. Using the dual method, this cannot be guaranteed in general (though it certainly applies to many special cases). One can apply the analysis of gradient descent from [44] to the proximal point algorithm (viewed as gradient descent on the Moreau-Yosida envelope), and compute an inexact gradient in the sense that the primal point is the exact gradient of a perturbed point. This perturbed point is based on the sub-optimality of the dual variable (see (3.28)), which, per the discussion of [45] above, does not necessarily convergence when using an accelerated algorithm, and hence we do not pursue this line of analysis.

We start with an extremely broad theorem that guarantees convergence under minimal assumptions, albeit without an explicit convergence rate:

THEOREM 3.4 *[76, Thm. 1] The approximate proximal point method defined by*

$$\begin{aligned} \tilde{y}_{k+1} &= \operatorname{argmin} F_k(x) \\ y_{k+1} &\textit{ any point satisfying } \|y_{k+1} - \tilde{y}_{k+1}\| \le \epsilon_k \end{aligned}$$

with $\sum_{k=1}^{\infty} \epsilon_k < \infty$, y_0 *arbitrary,* F_k *as defined in* (3.23), *and* $\limsup_{k\to\infty} \mu_k < \infty$, *will generate a sequence* $\{y_k\}$ *that converges to a minimizer of* f.

We note that the boundedness of iterates follows by our assumption that a minimizer exists; in infinite dimensions, convergence is in the weak topology. To guarantee $\|y_{k+1} - \tilde{y}_{k+1}\| \le \epsilon_k$, we can either bound this *a priori* using Thm. 3.3, or we can bound it *a posteriori* by explicitly checking using Prop. 3.2.

We state a second theorem that guarantees local linear convergence under standard assumptions. This assumption is that there is a unique solution to $\min f(x)$ and that f has sufficient curvature nearby; it is related to the standard second-order sufficiency condition, but slightly weaker. See [76] for an early use, and [24] for a more recent discussion.

Assumption 3.1 *There is a unique solution* $x^\star$ *to* $\min f(x)$, *i.e.,* $\partial f^{-1}(0) = x^\star$; *and* ∂f^{-1} *is locally Lipschitz continuous at* 0 *with constant* a, *i.e., there is some* r *such that* $\|w\| \le r$ *implies* $\|x - x^\star\| \le a\|w\|$ *whenever* $x \in \partial f^{-1}(w)$.

Recall that via basic convex analysis, $0 \in \partial f(x^\star)$. Finding $\operatorname{argmin} F_k(x)$ is the same as computing the proximity operator $P_k \stackrel{\text{def}}{=} (I + \mu_k^{-1}\partial f)^{-1}$. Define $Q_k = I - P_k$, then we have that P_k (and Q_k) are firmly non-expansive [6, 65], meaning

$$\forall x, x', \ \|P_k(x) - P_k(x')\|^2 + \|Q_k(x) - Q_k(x')\|^2 \le \|x - x'\|^2.$$

Furthermore, $x^\star = P_k(x^\star)$ (this is independent of μ_k), and $Q_k(x^\star) = 0$. Now, we state a novel theorem, where for simplicity we have assumed $\mu_k \equiv \mu \neq 0$:

THEOREM 3.5 *Under Assumption 3.1, the approximate proximal point method defined by*

$$\begin{aligned} \tilde{y}_{k+1} &= \operatorname{argmin} F_k(x) \\ y_{k+1} &\textit{ any point satisfying } \|y_{k+1} - \tilde{y}_{k+1}\| \le (\gamma/2)^k \end{aligned}$$

with

$$\gamma = \frac{a}{\sqrt{a^2 + \mu^{-2}}} < 1$$

generates a sequence (y_k) that converges linearly to $x^\star$ for all k sufficiently large, and with rate γ.

PROOF 3.3 Note that we have defined $\tilde{y}_{k+1} = P_k(y_k)$. Observe that the assumption of Theorem 3.4 holds since the errors are clearly summable, hence (y_k) converges, and $\|y_{k+1} - y_k\| \to 0$, so this is arbitrarily small for k sufficiently large. We also have

$$\|Q_k(y_k)\| = \|y_k - P_k(y_k)\| \leq \|y_k - y_{k+1}\| + \|y_{k+1} - P_k(y_k)\|$$

and both the terms on the right side go to zero. By basic calculation (e.g., Prop. 1a in [76]),

$$P_k(y_k) \in \partial f^{-1}(\mu Q_k(y_k))$$

and so for k large enough (say, $k \geq k_0$), we are in the Lipschitz region of the assumption, so

$$\|P_k(y_k) - x^\star\| \leq a\|\mu Q_k(y_k)\|. \tag{3.31}$$

Now using the firmly non-expansiveness and properties of $x^\star$ mentioned above,

$$\|x^\star - P_k(y_k)\|^2 + \|Q_k(y_k)\|^2 \leq \|x^\star - y_k\|^2 \tag{3.32}$$

so combining (3.31) with (3.32) gives

$$\|x^\star - P_k(y_k)\| \leq \gamma\|x^\star - y_k\|,$$

which in effect proves the eventual linear convergence in the exact case where $y_{k+1} = P_k(y_k)$ (to this point, the proof follows Thm. 2 from [76]).

Now bound

$$\begin{aligned}
\|y_{k+1} - x^\star\| &\leq \|y_{k+1} - P_k(y_k)\| + \|P_k(y_k) - x^\star\| \\
&\leq (\gamma/2)^k + \gamma\|y_k - x^\star\| \\
&\leq (\gamma/2)^k + \gamma\left((\gamma/2)^{k-1} + \gamma\|y_{k-1} - x^\star\|\right) \\
&\vdots \\
&\leq \gamma^k \sum_{i=k_0}^{k} 2^{-i} + \gamma^{k+1-k_0}\|y_{k_0} - x^\star\| \\
&\leq \gamma^k \left(1 + \gamma^{1-k_0}\|y_{k_0} - x^\star\|\right)
\end{aligned}$$

which proves our result.

Again, we can certify that $y_{k+1} \simeq P_k(y_k)$ either use Thm. 3.3, or we can bound it *a posteriori* by explicitly checking using Prop. 3.2. Since the linear convergence only occurs locally, it is not possible to provide an overall iteration-complexity of the inner and outer iterations (it is possible with further assumptions on f, such as f having full domain; see [10]). Without some form of strong convexity near the solution, it is generally not possible to bound the rate on the iterates, but rather only bound the rate of the objective function, and this is not possible with the dual approach since the point may not be feasible.

If we assume that the linear converge occurs globally, then we can combine this with our complexity bound on the sub-problem from (3.29). Converting that rate to our new notation, and using j to index the inner loop, and setting the initial dual variable $\mathbf{z}_0$ to the one corresponding to y_k, we have

$$\|x_j - P_k(y_k)\| \leq 2\frac{\|\mathbf{L}\|d_k}{\mu \cdot j}$$

where d_k is the distance from $\mathbf{z}_0$ (corresponding to $y_k = \nabla\Phi^*(\mathbf{L}^*\mathbf{z}_0)$) to the optimal set of dual solutions. Bounding this explicitly can be done in some cases (see [23]), but for the sake of analysis we will simply assume d_k is upper-bounded by some d.

3.4.6 Solving the Dual Problem Efficiently

For solving (3.2), we set $\psi_0(L,S) = |||L|||_* + \lambda\|S\|_1$ and $\psi_1(\mathcal{L}(L,S) - A) = \iota_{\{0\}}(L+S-A)$ to enforce the constraint exactly. In this case, ψ_1^* is the constant function, and so $\text{prox}_{\psi_1^*}$ is the identity — that is to say, the dual problem is unconstrained.

In that case, instead of using FISTA to solve the dual, one may use techniques from *unconstrained* optimization, such as non-linear conjugate gradient and L-BFGS [69]. These algorithms work extremely well in practice. We do not go into further detail since we find the exact constraint formulation of RPCA to be artificial. With inequality constraints, the dual problem becomes non-smooth so it is necessary to use a proximal gradient method. Due to the cost of the objective function, it may be worthwhile to use quasi-Newton projected gradient methods such as that of [79].

3.5 Numerical Experiments

3.5.1 Numerical Results for TFOCS

To highlight the flexibility of TFOCS, we consider a background subtraction problem of a surveillance video in which we do not wish to enforce $A = L + S$ but instead we wish to separate A into components up to the quantization level. The video A is quantized to integer values between 0 and 255, so we can think of this as being the quantized version of some real-valued video $\widehat{A}$, and thus $\|\widehat{A} - A\|_\infty \leq 0.5$ since the quantized version is rounded to the nearest integer. Hence we solve the following:

$$\min_{L,S} \; |||L|||_* + \lambda\|S\|_1 \quad \text{s.t.} \quad \|A - L - S\|_\infty \leq 0.5. \tag{3.33}$$

In the TFOCS software, we work with the primal variable `X={L,S}`, and one specifies the fucntion ψ_0 in the term `obj` like

```
obj    = { prox_nuclear(1), prox_l1(lambda) };
```

Next we encode ψ_1 and $\mathcal{L}_1$ and b_1. The linear term, applied to `X={L,S}`, is $\mathcal{L}_1 = [I, I]$, and the offset is $b_1 = -A$. This is encoded as

```
affine = { 1, 1, -X };
```

and ψ_1 is represented implicitly by giving its conjugate. For standard quality constrained RPCA, the constraint is $A - L - S = 0$, to ψ_1 is the indicator function of the set $\{0\}$, and the conjugate of this is the function that is constant everywhere, e.g., projection on $\mathbb{R}^n$, since this is the identity. This is written

```
dualProx = proj_Rn
```

If instead we wish to solve (3.33), then ψ_1 is the indicator function of the ℓ_∞ ball of radius 1/2, and so its conjugate is $1/2\|\cdot\|_1$, which is written

```
dualProx = prox_l1(0.5);
```

The code can then be called as follows:

```
X = tfocs_SCD( obj, affine, dualProx, mu, X0 );
```

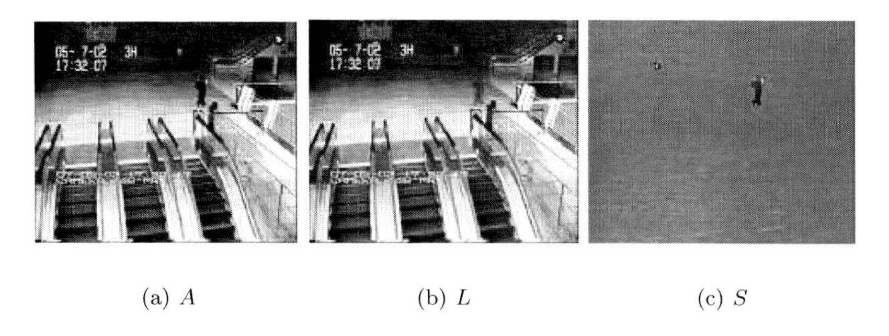

(a) A (b) L (c) S

FIGURE 3.1 Frame 110 from the movie, showing original A, low-rank L and sparse S components.

More detailed code is available as a demo at;[10] the video data is from. [11] Results are shown in Figure 3.1. Although the walking person is correctly identified in the S component, a small amount of the person appears in L. However, it is remarkable that the low-rank component mostly captures the moving escalator, which is a feat that most background subtraction cannot do without a specially targeted algorithm.

Comparison with PDHG

As discussed in §3.4.3, the primal-dual hybrid gradient (PDHG) method is similar to the TFOCS algorithm. In TFOCS, one controls the value of μ and then runs the proximal point algorithm, and the sub-problem is solved by FISTA (or variants) that use linesearch techniques and therefore do not need a stepsize. In PDHG, there is no line search, but there are two stepsizes τ and σ, which are linked in the fashion $\tau\sigma < 1/\ell^2$. Larger stepsizes generally lead to better performance, so by insisting on $\tau\sigma = 0.99/\ell^2$, there is only one effective parameter choice.

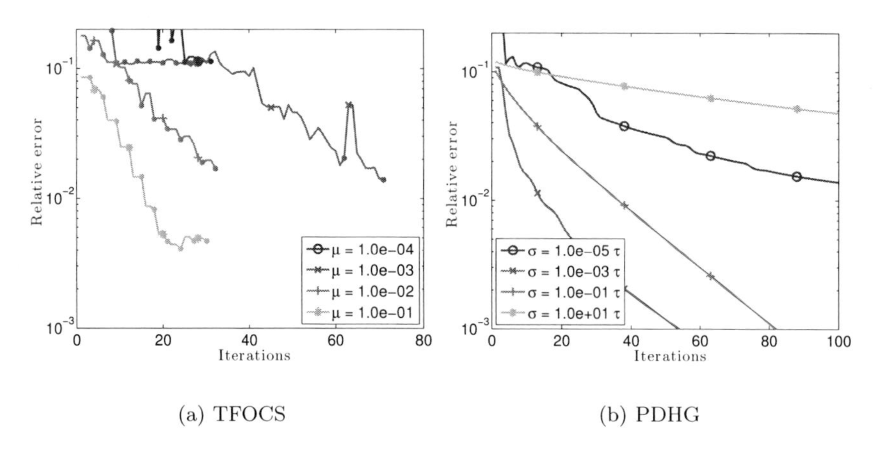

(a) TFOCS (b) PDHG

FIGURE 3.2 Convergence plots on the elevator data, for various parameter values. The small dots in the TFOCS plot show where the proximal point algorithm took another outer step.

[10] `http://cvxr.com/tfocs/demos/rpca/`

[11] `http://perception.i2r.a-star.edu.sg/bk_model/bk_index.html`

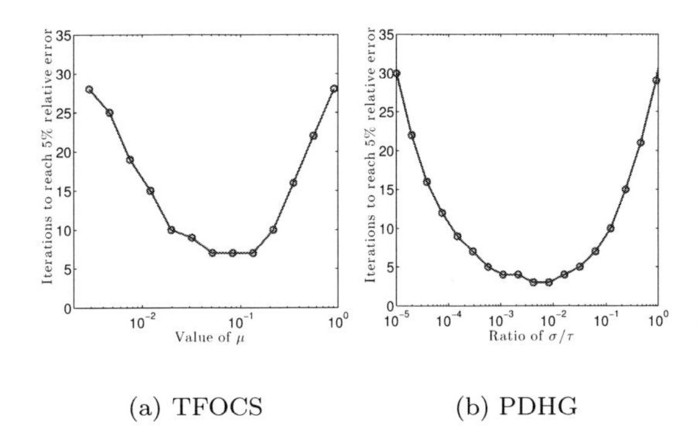

(a) TFOCS (b) PDHG

FIGURE 3.3 Number of iterations to reach a fixed tolerance, as function of parameter value.

Using a version of the same escalator film discussed in the previous section, we compare PDHG and TFOCS on (SPCP_{sum}). In TFOCS, we encode this with

```
dualProx = prox_l2(epsilon);
```

The parameters δ and λ were chosen by running the quasi-Newton solver on (lag-SPCP) and tuning by hand until results looked acceptable. Running the special purpose solver gives a very accurate reference answer. From the solution to (lag-SPCP), one can infer ϵ and λ_{sum}.

Neither method is specific for RPCA problems, so we do not expect cutting-edge performance, but we do see reliable performance, and the ability to adapt to variations in the model. We focus on parameter selection. Both methods perform roughly equally, and both are strongly dependent on the parameter choice. A major weakness of all current methods is lack of guidance for choosing parameters in practice; the effort of [71] to find good values resulted in mixed success. The software TFOCS automatically rescales variables in order to make all $\mathcal{L}_i$ terms have the same spectral norm, which has a small beneficial effect.

Figure 3.2 shows the decay of the relative error $\sqrt{\|L - L_0\|_F^2 + \|S - S_0\|_F^2}/\sqrt{\|L_0\|_F^2 + \|S_0\|_F^2}$ where (L_0, S_0) is the accurate reference solution computed via the quasi-Newton algorithm. TFOCS has the advantage that the sub-problems can use a fast solver with a good linesearch, but the disadvantage that with the two levels of iterations, the inner iteration must be terminated at the right time. If it is stopped too early, the solution is not accurate enough, while if it is stopped too late, the algorithm wastes time on a very precise answer to a useless intermediate problem.

Figure 3.3 shows the number of iterations required to reach a fixed tolerance. Confirming the behavior in the previous figure, it is clear that convergence can be rapid for good parameter choices, and slow for poor parameter choices.

3.5.2 Numerical Results for Quasi-Newton Algorithm

We compare new algorithms and formulations to PSPG [4], NSA [5], and ASALM [82].[12] We modified the other software as needed for testing purposes. PSPG, NSA, and ASALM all solve (SPCP_{sum}), but ASALM has another variant that solves (lag-SPCP) so we test this as well. All three programs also use versions of PROPACK from [11] to compute partial

[12]PSPG, NSA and ASALM available from the experiment package at `http://www2.ie.psu.edu/aybat/codes.html`

SVDs. We measure error as a function of time, since cost of a single iteration can vary among the solvers. To fairly compare all the algorithms in the simulated experiments, we measure the (relative) error of a trial solution (L, S) to a reference solution $(L^\star, S^\star)$ as $\|L - L^\star\|_F/\|L^\star\|_F + \|S - S^\star\|_F/\|S^\star\|_F$. Time to compute this error is accounted for (so does not factor into the comparisons). Finally, since stopping conditions are solver dependent, we show plots of error vs time. All tests are done in MATLAB and the dominant computational time was due to matrix multiplications for all algorithms; all code was run in the same quad-core 1.6 GHz i7 computer.

For our implementations of the (flip-SPCP$_{\text{max}}$), (flip-SPCP$_{\text{sum}}$), and (lag-SPCP), we use a randomized SVD [53]. Since the number of singular values needed is not known in advance, the partial SVD may be called several times (the same is true for PSPG, NSA, and ASALM). Our code limits the number of singular values on the first two iterations in order to speed up calculation without affecting convergence. Since the projection required by (flip-SPCP$_{\text{sum}}$) makes a partial SVD difficult, we use MATLAB's dense `SVD` routine.

Synthetic test with exponential noise

We first provide a test with generated data. The observations $A \in \mathbb{R}^{m\times n}$ with $m = 400$ and $n = 500$ were created by first sampling a rank 20 matrix A_0 with random singular vectors (i.e., from the Haar measure) and singular values drawn from a uniform distribution with mean 0.1, and then adding exponential random noise (with mean equal to one tenth the median absolute value of the entries of A_0). This exponential noise, which has a longer tail than Gaussian noise, is expected to be captured partly by the S term and partly by the $\|L + S - A\|_F$ term.

Given A, the reference solution $(L^\star, S^\star)$ was generated by solving (lag-SPCP) to very high accuracy; the values $\lambda_{\text{L}} = 0.25$ and $\lambda_{\text{S}} = 10^{-2}$ were picked by hand tuning $(\lambda_{\text{L}}, \lambda_{\text{S}})$ to find a value such that both $L^\star$ and $S^\star$ are non-zero. The advantage to solving (lag-SPCP) is that knowledge of $(L^\star, S^\star, \lambda_{\text{L}}, \lambda_{\text{S}})$ allows us to generate the parameters for all the other variants, and hence we can test different problem formulations.

With these parameters, $L^\star$ was rank 17 with nuclear norm 6.754, $S^\star$ had 54 non-zero entries (most of them positive) with ℓ_1 norm 0.045, the normalized residual was $\|L^\star + S^\star - A\|_F/\|A\|_F = 0.385$, and $\varepsilon = 1.1086$, $\lambda_{\text{sum}} = 0.04$, $\lambda_{\max} = 150.0593$, $\tau_{\text{sum}} = 6.7558$ and $\tau_{\max} = 6.7540$.

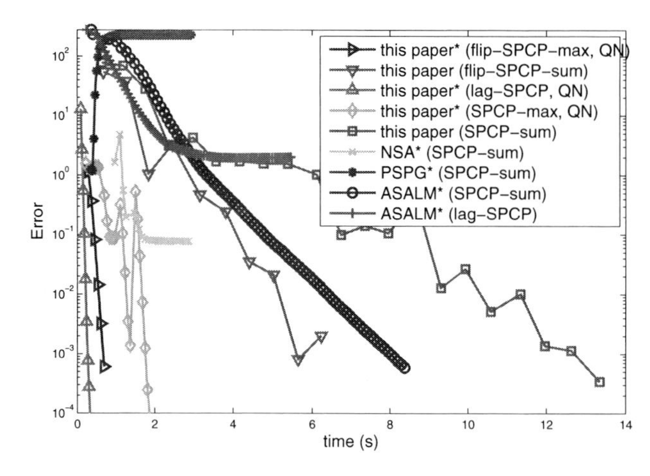

FIGURE 3.4 (See color insert.) The exponential noise test. The asterisk in the legend means the method uses a fast SVD.

Results are shown in Figure 3.4. Our methods for (flip-SPCP$_{\max}$) and (lag-SPCP) are extremely fast, because the simple nature of these formulations allows the quasi-Newton acceleration scheme of Section 3.3.2. In turn, since our method for solving (SPCP$_{\max}$) uses the variational framework of Section 3.3 to solve a sequence of (flip-SPCP$_{\max}$) problems, it is also competitive (shown in cyan in Figure 3.4). The jumps are due to re-starting the sub-problem solver with a new value of τ, generated according to (3.14).

Our proximal gradient method for (flip-SPCP$_{\text{sum}}$), which makes use of the projection in Lemma 3.1, converges more slowly, since it is not easy to accelerate with the quasi-Newton scheme due to variable coupling, and it does not make use of fast SVDs. Our solver for (SPCP$_{\text{sum}}$), which depends on a sequence of problems (flip-SPCP$_{\text{sum}}$), converges slowly.

The ASALM performs reasonably well, which was unexpected since it was shown to be worse than NSA and PSPG in [4, 5]. The PSPG solver converges to the wrong answer, most likely due to a bad choice of the smoothing parameter μ; we tried choosing several different values other than the default but did not see improvement for this test (for other tests, not shown, tweaking μ helped significantly). The NSA solver reaches moderate error quickly but stalls before finding a highly accurate solution.

Synthetic test from [5]

We show some tests from the test setup of [5] in the $m = n = 1500$ case. The default setting of $\lambda_{\text{sum}} = 1/\sqrt{\max(m, n)}$ was used, and then the NSA solver was run to high accuracy to obtain a reference solution $(L^\star, S^\star)$. From the knowledge of $(L^\star, S^\star, \lambda_{\text{sum}})$, one can generate $\lambda_{\max}, \tau_{\text{sum}}, \tau_{\max}, \varepsilon$, but not λ_{S} and λ_{L}, and hence we did not test the solvers for (lag-SPCP) in this experiment. The data was generated as $A = L_0 + S_0 + Z_0$, where L_0 was sampled by multiplication of $m \times r$ and $r \times n$ normal Gaussian matrices, S_0 had p randomly chosen entries uniformly distributed within $[-100, 100]$, and Z_0 was white noise chosen to give a SNR of 45 dB. We show three tests that vary the rank from $\{0.05, 0.1\} \cdot \min(m, n)$ and the sparsity ranging from $p = \{0.05, 0.1\} \cdot mn$. Unlike [5], who report error in terms of a true noiseless signal (L_0, S_0), we report the optimization error relative to $(L^\star, S^\star)$.

For the first test (with $r = 75$ and $p = 0.05 \cdot mn$), $L^\star$ had rank 786 and nuclear norm 111363.9; $S^\star$ had 75.49% of its elements nonzero and ℓ_1 norm 5720399.4, and $\|L^\star + S^\star - A^\star\|_F / \|A\|_F = 1.5 \cdot 10^{-4}$. The other parameters were $\varepsilon = 3.5068$, $\lambda_{\text{sum}} = 0.0258$, $\lambda_{\max} = 0.0195$, $\tau_{\text{sum}} = 2.5906 \cdot 10^5$ and $\tau_{\max} = 1.1136 \cdot 10^5$. An interesting feature of this test is that while L_0 is low-rank, $L^\star$ is nearly low-rank but with a small tail of significant singular values until number 786. We expect methods to converge quickly to low-accuracy where only a low-rank approximation is needed, and then slow down as they try to find a larger rank highly-accurate solution.

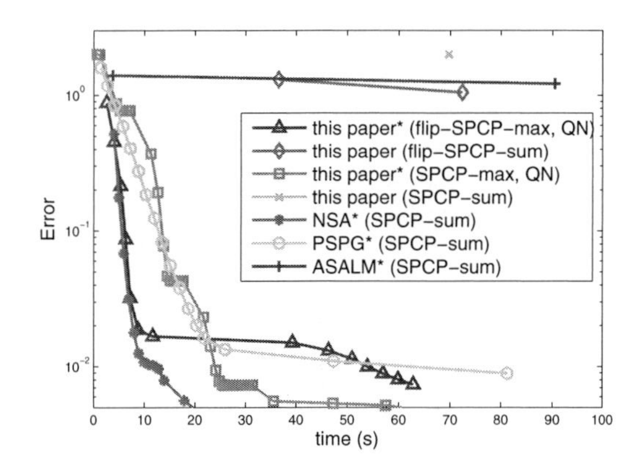

FIGURE 3.5 (See color insert.) The 1500×1500 synthetic noise test.

The results are shown in Figure 3.5. Errors barely dip below 0.01 (for comparison, an error of 2 is achieved by setting $L = S = 0$). The NSA and PSPG solvers do quite well. In contrast to the previous test, ASALM does poorly. Our methods for (flip-SPCP$_{\text{sum}}$), and hence (SPCP$_{\text{sum}}$), are not competitive, since they use dense SVDs. We imposed a time-limit of about one minute, so these methods only manage a single iteration or two. Our quasi-Newton method for (flip-SPCP$_{\text{max}}$) does well initially, then takes a long time due to a long partial SVD computation. Interestingly, (SPCP$_{\text{max}}$) does better than pure (flip-SPCP$_{\text{max}}$). One possible explanation is that it chooses a fortuitous sequence of τ values, for which the corresponding (flip-SPCP$_{\text{max}}$) subproblems become increasingly hard, and therefore benefit from the warm-start of the solution of the easier previous problem. This is consistent with empirical observations regarding continuation techniques; see, e.g., [43, 89].

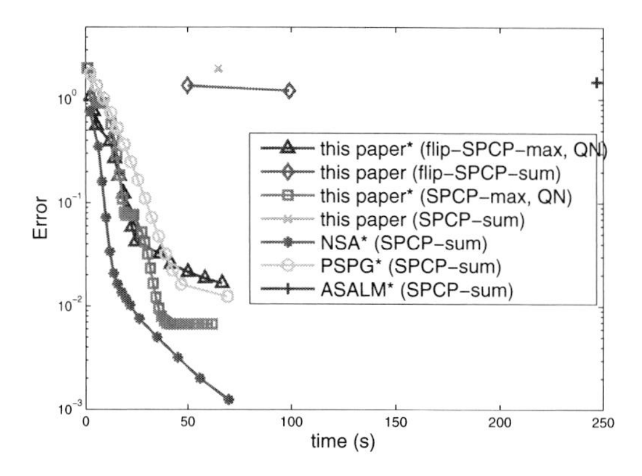

FIGURE 3.6 (See color insert.) Second 1500×1500 synthetic noise test.

Figure 3.6 is the same test but with $r = 150$ and $p = 0.1 \cdot mn$, and the conclusions are largely similar.

Cloud removal

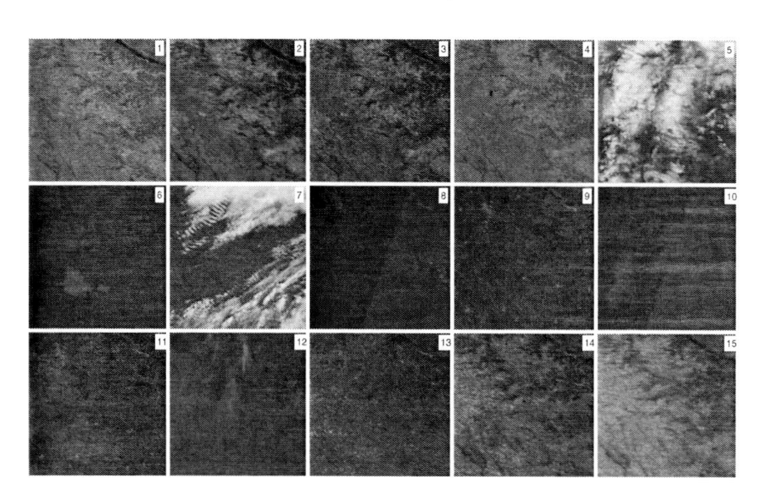

FIGURE 3.7 Satellite photos of the same location on different days.

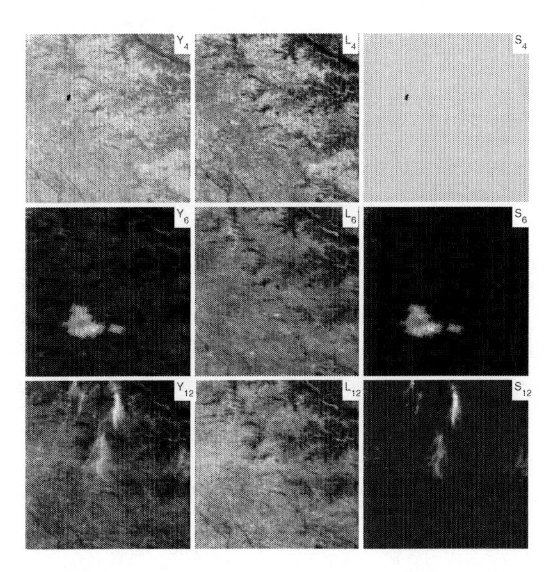

FIGURE 3.8 Showing frames 4, 5 and 12. Leftmost column is original data, middle column is low-rank term of the solution, and right column is sparse term of the solution. Data have been processed slightly to enhance contrast for viewing.

Figure 3.7 shows 15 images of size 300 $\times$ 300 that come from the MODIS satellite,[13] after some transformations to turn images from different spectral bands into one grayscale image. Each image is a photo of the same rural location but at different points in time over the course of a few months. The background changes slowly and the variability is due to changes in vegetation, snow cover, and different reflectance. There are also outlying sources of error, mainly due to clouds (e.g., major clouds in frames 5 and 7, smaller clouds in frames 9, 11, and 12), as well as artifacts of the CCD camera on the satellite (frame 4 and 6) and issues stitching together photos of the same scene (the lines in frames 8 and 10).

There are hundreds of applications for clean satellite imagery, so removing the outlying error is of great practical importance. Because of slow changing background and sparse errors, we can model the problem using the robust PCA approach. We use the (flip-SPCP$_{\text{max}}$) version due to its speed, and pick parameters $(\lambda_{\text{max}}, \tau_{\text{max}})$ by using a Nelder-Mead simplex search. For an error metric to use in the parameter tuning, we remove frame 1 from the data set (call it y_1) and set A to be frames 2–15. From this training data A, the algorithm generates L and S. Since L is a $300^2 \times 14$ matrix, it has far from full column span. Thus our error is the distance of y_1 from the span of L, i.e., $\|y_1 - \mathcal{P}_{\text{span}(L)}(y_1)\|_2$.

Our method takes about 11 iterations and 5 seconds, and uses a dense SVD instead of the randomized method due to the extremely high aspect ratio of the matrix. Some results of the obtained (L, S) outputs are shown in Figure 3.8, where one can see that some of the anomalies in the original data frames A are picked up by the S term and removed from the L term. Frame 4 has what appears to be a camera pixel error; frame 6 has another artificial error (that is, caused by the camera and not the scene); and frame 12 has cloud cover.

---

[13]Publicly available at `http://ladsweb.nascom.nasa.gov/`

Analysis of brain activity in the zebra fish

Recent work [1] has produced video recordings of brain activity, in vivo, of zebra fish. These datasets are used to confirm scientists, theories about the inner workings of the brain as well as to discover unexpected connections. Ultimately the goal is to discover causal, not just correlated, relationships. PCA on these datasets is one of the standard tools used by biologists.

Using a public video of the dataset, we focus on a single 2D slice, sub-sampled spatially (and perhaps with video compression artifacts). We plan to use (SPCP_{max}) as the RPCA technique, and therefore need to estimate ϵ and λ_{max}. To find ϵ, we first take the SVD of the data matrix A. The corresponding singular vales $\sigma(A)$ are plotted in Figure 3.9. This gives us an idea of the compressibility of the data. Keeping about 30 singular values explains over 99% of the data, so if we look for L with approximately rank 30, then, not including the sparse term S (which we expect to be very sparse), we should pick $\epsilon \simeq \sqrt{\sum_{i=31}^{752} \sigma_i(A)}$. This value works well in practice (see Figure 3.10) and did not require cross-validation.

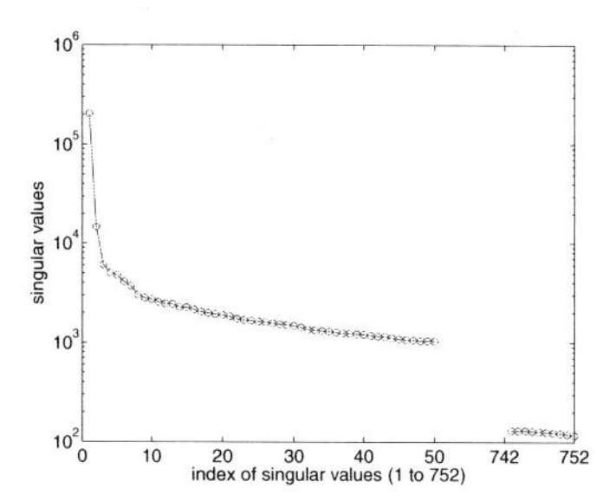

FIGURE 3.9 Decay of singular values of A for the zebra fish dataset. Singular values 51–742 are not shown.

The λ_{max} parameter is tuned by hand, but only takes 3 runs to find a reasonable value. This is much simpler than tuning λ_{max} and ϵ by hand simultaneously. Figure 3.10 shows the resulting top left singular vectors of A and of L, as well as their difference. We see that their difference is sparse, as expected. Since these are singular vectors, not just individual frames from the movie, these sparse differences are persistent over time, and perhaps meaningful. These unpredictable locations could be caused by sensor/microscope error, or they could mean that they come from a part of the brain that is not well correlated with general brain activity. Either way, it is useful to be able to separate out this effect.

3.6 Conclusion

We have discussed both specific algorithms for the RPCA problem, and general algorithm frameworks ("TFOCS," and "flipped" variational value-function approaches) that incorporate RPCA and variants. The custom RPCA algorithm works extremely well in practice, and the process of "flipping" the objective has been studied rigorously, but the inner "quasi-Newton" algorithm lacks a rigorous convergence theory. The general algorithm TFOCS, as well as similar proposals such as the PDHG algorithm, have more established theory but lack practical guidance on setting parameters, and are in practice slower than the special purpose algorithms. An obvious goal of future work is to either improve the analysis of these algorithms (for example, this may give insight into parameter selection), or derive new algorithms that inherent all the advantages.

The running theme of this work has been solving variants of RPCA, in particular ($\text{flip-SPCP}_{\text{sum}}$) and the new variant ($\text{flip-SPCP}_{\text{max}}$). These versions sometimes allow a

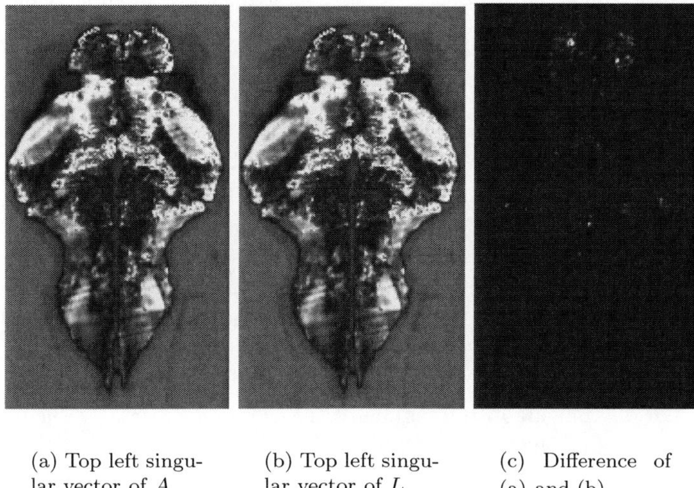

(a) Top left singular vector of A.

(b) Top left singular vector of L.

(c) Difference of (a) and (b).

FIGURE 3.10 (See color insert.) Top left singular vector of A and of L, as well as their difference.

good estimate of τ, thus reducing the parameter selection of the model to the single scalar λ_{sum} or λ_{max}. We also consider different regularizers and data fidelity terms; using these in practice is interesting future work.

References

1. M. Ahrens, M. Orger, D. Robson, J. Li, and P. Keller. Whole-brain functional imaging at cellular resolution using light-sheet microscopy. *Nature Methods*, 10(413-420), 2013.
2. A. Aravkin, J. Burke, and M. Friedlander. Variational properties of value functions. *SIAM Journal on Optimization*, 23(3):1689–1717, 2013.
3. A. Aravkin, R. Kumar, H. Mansour, B. Recht, and F. Herrmann. A robust SVD-free approach to matrix completion, with applications to interpolation of large scale data. *http://arxiv.org/abs/1302.4886*, 2013.
4. N. Aybat, D. Goldfarb, and S. Ma. Efficient algorithms for robust and stable principal component pursuit. *Computational Optimization and Applications*, 58(1), May 2014.
5. N. Aybat and G. Iyengar. A fast first-order method for stable principal component pursuit. *http://arxiv.org/abs/1309.6553*, 2013.
6. H. Bauschke and P. Combettes. *Convex analysis and monotone operators theory in Hilbert spaces*. Springer-Verlag, 2011.
7. A. Beck and M. Teboulle. A Fast Iterative Shrinkage-Thresholding Algorithm for Linear Inverse Problems. *SIAM Journal on Imaging Sciences*, 2(1):183–202, January 2009.
8. A. Beck and M. Teboulle. A fast iterative shrinkage-thresholding algorithm for linear inverse problems. *SIAM Journal on Imaging Sciences*, 2(1):183–202, 2009.
9. A. Beck and M. Teboulle. A fast dual proximal gradient algorithm for convex minimization and applications. *Operations Research Letters*, 42(1):1–6, 2014.
10. S. Becker. *Practical Compressed Sensing: modern data acquisition and signal processing*. PhD thesis, California Institute of Technology, Pasadena, CA, 2011.
11. S. Becker and E. Candès. Singlar value thresholding toolbox, 2008.
12. S. Becker, E. Candès, and M. Grant. Templates for convex cone problems with applications to sparse signal recovery. *Mathematical Programming Computation*, 3(3):165–218, 2011.

13. S. Becker and P. Combettes. An algorithm for splitting parallel sums of linearly composed monotone operators, with applications to signal recovery. *Journal of Nonlinear and Convex Analysis*, 15(1):137–159, 2014.
14. D. Bertsekas. Necessary and sufficient conditions for a penalty method to be exact. *Math. Program.*, 9:87–99, 1975.
15. D. Bertsekas, A. Nedić, and A. Ozdaglar. *Convex Analysis and Optimization.* Athena Scientific, 2003.
16. R. Boţ and E. Csetnek. Forward-backward and Tsengs type penalty schemes for monotone inclusion problems. *Set-Valued and Variational Analysis*, 22(2):313–331, 2014.
17. R. Boţ, E. Csetnek, and A. Heinrich. A primal-dual splitting algorithm for finding zeros of sums of maximal monotone operators. *SIAM Journal on Optimization*, 23(4):2011–2036, 2013.
18. R. Boţ, E. Csetnek, A. Heinrich, and C. Hendrich. On the convergence rate improvement of a primal-dual splitting algorithm for solving monotone inclusion problems. *Mathematical Programming*, 150(2):251–279, 2015.
19. R. Boţ, E. Csetnek, and A. Nagy. Solving systems of monotone inclusions via primal-dual splitting techniques. *Taiwanese Journal of Mathematics*, 17(6):1983, 2013.
20. T. Bouwmans and E. Zahzah. Robust PCA via principal component pursuit: A review for a comparative evaluation in video surveillance. *Computer Vision and Image Understanding*, 122(0):22–34, 2014.
21. L. Briceno-Arias and P. Combettes. A monotone+skew splitting model for composite monotone inclusions in duality. *SIAM Journal on Optimization*, 21(4):1230–1250, 2011.
22. P. Brucker. An O(n) algorithm for quadratic knapsack problems. *Operations Research Letters*, 3(3):163 – 166, 1984.
23. J. Bruer, J. Tropp, V. Cevher, and S. Becker. Time-data tradeoffs by aggressive smoothing. In *Advances in Neural Information Processing Systems*, pages 1664–1672, 2014.
24. J. Burke and M. Qian. A variable metric proximal point algorithm for monotone operators. *SIAM Journal on Control and Optimization*, 37(2):353–375, 1997.
25. J. Cai, E. Candès, and Z. Shen. A singular value thresholding algorithm for matrix completion. *SIAM Journal on Optimization*, 20:1956–1982, 2010.
26. E. Candès, X. Li, Y. Ma, and J. Wright. Robust principal component analysis? *Journal of the Association for Computing Machinery*, 58(3):1–37, May 2011.
27. E. Candès, X. Li, Y. Ma, and J. Wright. Robust principal component analysis? *Journal of ACM*, 58(3):11:1–11:37, May 2011.
28. E. Candés and Y. Plan. Matrix completion with noise. *Proceedings of the IEEE*, 98(6):925–936, 2010.
29. V. Cevher, S. Becker, and M. Schmidt. Convex optimization for big data: Scalable, randomized, and parallel algorithms for big data analytics. *IEEE Signal Processing Magazine*, 31(5):32–43, 2014.
30. A. Chambolle and C. Dossal. On the convergence of the iterates of "FISTA". Technical Report hal-01060130, HAL, 2014. `https://hal.inria.fr/hal-01060130v3`.
31. A. Chambolle and T. Pock. A first-order primal-dual algorithm for convex problems with applications to imaging. *Journal of Mathematical Imaging and Vision*, 40(1):120–145, 2010.
32. V. Chandrasekaran, P. Parrilo, and A. Willsky. Latent variable graphical model selection via convex optimization. *The Annals of Applied Statistics*, 40(4):1935–2357, 2012.
33. V. Chandrasekaran, S. Sanghavi, P. A. Parrilo, and A. S. Willsky. Sparse and low-rank matrix decompositions. In *SYSID 2009*, Saint-Malo, France, July 2009.
34. C. Chen, B. He, Y. Ye, and X. Yuan. The direct extension of ADMM for multi-block convex minimization problems is not necessarily convergent. *Optimization Online*, 2013.
35. P. Ciarlet. *Introduction to numerical linear algebra and optimisation.* Cambridge Univer-

sity Press, 1989.

36. P. Combettes. Systems of structured monotone inclusions: duality, algorithms, and applications. *SIAM Journal on Optimization*, 23(4):2420–2447, 2013.
37. P. Combettes, D. Dũng, and B. Vũ. Dualization of signal recovery problems. *Set-Valued and Variational Analysis*, 18:373–404, 2010.
38. P. Combettes and J. Pesquet. A Douglas-Rachford Splitting Approach to Nonsmooth Convex Variational Signal Recovery. *IEEE Journal of Selected Topics in Signal Processing, J-STSP 2007*, 1(4):564–574, December 2007.
39. P. Combettes and J. Pesquet. Proximal splitting methods in signal processing. In H. Bauschke, R. Burachik, P. Combettes, V. Elser, D. Luke, and H. Wolkowicz, editors, *Fixed-Point Algorithms for Inverse Problems in Science and Engineering*, pages 185–212. Springer-Verlag, New York, 2011.
40. P. Combettes and J. Pesquet. Primal-dual splitting algorithm for solving inclusions with mixtures of composite, Lipschitzian, and parallel-sum type monotone operators. *Set-Valued and Variational Analysis*, 20(2):307–330, 2012.
41. P. Combettes and V. Wajs. Signal recovery by proximal forward-backward splitting. *Multiscale Modeling and Simulation*, 4(4):1168–1200 (electronic), 2005.
42. L. Condat. A primal-dual splitting method for convex optimization involving Lipschitzian, proximable and linear composite terms. *Journal of Optimization Theory and Applications*, pages 460–479, 2013.
43. E. Van den berg and M. Friedlander. Probing the Pareto frontier for basis pursuit solutions. *SIAM Journal on Scientific Computing*, 31(2):890–912, 2008.
44. O. Devolder, F. Glineur, and Y. Nesterov. First-order methods of smooth convex optimization with inexact oracle. *Mathematical Programming*, 116(1), August 2014.
45. O. Devolder, F. Glineur, and Y. Nesterov. Double smoothing technique for large-scale linearly constrained convex optimization. *SIAM Journal on Optimization*, 22(2):702–727, 2012.
46. Q. Dinh and V. Cevher. Constrained convex minimization via model-based excessive gap. In Z. Ghahramani, M. Welling, C. Cortes, N. Lawrence, and K. Weinberger, editors, *Advances in Neural Information Processing Systems 27*, pages 721–729. Curran Associates, Inc., 2014.
47. J. Duchi, S. Shalev-Shwartz, Y. Singer, and T. Chandra. Efficient projections onto the l_1-ball for learning in high dimensions. In *International Conference on Machine Learning, ICML 2008*, pages 272–279, July 2008.
48. E. Esser, X. Zhang, and T. Chan. A general framework for a class of first order primal-dual algorithms for tv minimization. Technical Report 09-67, UCLA, Center for Applied Math, 2009.
49. M. Friedlander and P. Tseng. Exact regularization of convex programs. *SIAM Journal on Optimization*, 18(4):1326–1350, 2007.
50. A. Ganesh, Z. Lin, J. Wright, L. Wu, M. Chen, and Y. Ma. Fast algorithms for recovering a corrupted low-rank matrix. *Computational Advances in Multi-Sensor Adaptive Processing, CAMSAP 2009*, pages 213–215, December 2009.
51. D. Goldfarb, S. Ma, and K. Scheinberg. Fast alternating linearization methods for minimizing the sum of two convex functions. *Mathematical Programming, Series A*, 141(1-2):349–382, 2013.
52. T. Goldstein, E. Esser, and R. Baraniuk. Adaptive primal-dual hybrid gradient methods for saddle-point problems. *Preprint*, 2013.
53. N. Halko, P. Martinsson, and J. Tropp. Finding structure with randomness: Probabilistic algorithms for constructing approximate matrix decompositions. *SIAM review*, 53(2):217–288, 2011.
54. B. He and X. Yuan. Convergence analysis of primal-dual algorithms for a saddle-point problem: from contraction perspective. *SIAM Journal on Imaging Sciences*, pages 119–149,

2012.

55. B. He and X. Yuan. On the $o(1/n)$ convergence rate of the Douglas-Rachford alternating direction method. *SIAM Journal on Numerical Analysis*, 50(700-709), 2012.
56. P. Huber. *Robust Statistics.* John Wiley and Sons, 2nd edition, 2004.
57. J.Lee, B. Recht, R. Salakhutdinov, N. Srebro, and J. Tropp. Practical large-scale optimization for max-norm regularization. *Neural Information Processing Systems, NIPS 2010*, 2010.
58. N. Komodakis and J. Pesquet. Playing with duality: An overview of recent primal-dual approaches for solving large-scale optimization problems. *IEEE Sig. Pro. Magazine,* to appear, May 2015. `http://arxiv.org/abs/1406.5429v2`.
59. M. Lai and W. Yin. Augmented l_1 and nuclear-norm models with a globally linearly convergent algorithm. *SIAM Journal on Imaging Sciences*, 6(2):1059–1091, 2013.
60. X. Li, L. Mo, X. Yuan, and J. Zhang. Linearized alternating direction method of multipliers for sparse group and fused LASSO models. *Comput. Statis. Data Anal.*, 79:203–221, 2014.
61. Z. Lin, M. Chen, and Y. Ma. The augmented Lagrange multiplier method for exact recovery of corrupted low-rank matrices. *arXiv preprint arXiv:1009.5055*, 2010.
62. Y. Liu, D. Sun, and K.Toh. An implementable proximal point algorithmic framework for nuclear norm minimization. *Mathematical Programming*, 2011.
63. F. Malgouyres and T. Zeng. A predual proximal point algorithm solving a non negative basis pursuit denoising model. *Int. J. Comp. Vision*, 83(3):294–311, July 2009.
64. O. Mangasarian and R. Meyer. Nonlinear perturbation of linear programs. *SIAM Journal on Control and Optimization*, 17:745–752, 1979.
65. J. Moreau. Proximité et dualité dans un espace hilbertien. *Bulletin de la Socit Mathmatique de France*, 93:273–299, 1965.
66. I. Necoara and J. Suykens. Application of a smoothing technique to decomposition in convex optimization. *IEEE Transactions on Automatic Control*, 53(11):2674–2679, 2008.
67. Y. Nesterov. A method of solving a convex programming problem with convergence rate $\mathcal{O}(1/k^2)$, in Soviet mathematics doklady. *Soviet Mathematics Doklady*, 27, 1983.
68. Y. Nesterov. *Introductory lectures on convex optimization: a basic course*, volume 87 of *Applied Optimization.* Kluwer Academic Publishers, 2004.
69. J. Nocedal and S. Wright. *Numerical Optimization.* Springer, 2nd edition, 2006.
70. Y. Peng, A. Ganesh, J. Wright, W. Xu, and Y. Ma. RASL: Robust alignment by sparse and low-rank decomposition for linearly correlated images. *IEEE Trans. Pattern Analysis and Machine Intelligence*, 34(11):2233–2246, 2012.
71. T. Pock and A. Chambolle. Diagonal preconditioning for first order primal-dual algorithms in convex optimization. *IEEE International Conference on Computer Vision, ICCV 2011*, pages 1762–1769, November 2011.
72. B. Poljak and N. Tretjakov. An iterative method for linear programming and its economic interpretation. *Matecon*, 10:81–100, 1974.
73. X. Ren and Z. Lin. Linearized alternating direction method with adaptive penalty and warm starts for fast solving transform invariant low-rank textures. *International Journal on Computer Vision*, 104:1–14, 2013.
74. R. Rockafellar. *Convex Analysis.* Priceton Landmarks in Mathematics. Princeton University Press, 1970.
75. R. Rockafellar. *Convex analysis.* Princeton Mathematical Series, No. 28. Princeton University Press, Princeton, N.J., 1970.
76. R. Rockafellar. Monotone operators and the proximal point algorithm. *SIAM Journal on Control and Optimization*, 14:877–898, 1976.
77. R. Rockafellar and R. Wets. *Variational Analysis*, volume 317 of *A Series of Comprehensive Studies in Mathematics.* Springer, 1998.

78. M. Schmidt, N. Le Roux, and F. Bach. Convergence rates of inexact proximal-gradient methods for convex optimization. In *NIPS 2011*, 2011.
79. M. Schmidt, E. van den Berg, M. Friedlander, and K. Murphy. Optimizing costly functions with simple constraints: A limited-memory projected quasi-Newton algorithm. *International Conference on Artificial Intelligence and Statistics, AISTATS 2009*, 2009.
80. Y. Shen, Z. Wen, and Y. Zhang. Augmented Lagrangian alternating direction method for matrix separation based on low-rank factorization. *Optimization Methods and Software*, 29(2):239–263, March 2014.
81. Z. Shi, J. Han, T. Zheng, and J. Li. Guarantees of augmented trace norm models in tensor recovery. *International Joint Conference on Artificial Intelligence*, pages 1670–1676, 2013.
82. M. Tao and X. Yuan. Recovering low-rank and sparse components of matrices from incomplete and noisy observations. *SIAM Journal on Optimization*, 21:57–81, 2011.
83. E. van den Berg and M. Friedlander. Probing the Pareto frontier for basis pursuit solutions. *SIAM Journal on Scientific Computing*, 31(2):890–912, 2008.
84. E. van den Berg and M. Friedlander. Sparse optimization with least-squares constraints. *SIAM Journal on Optimization*, 21(4):1201–1229, 2011.
85. S. Villa, S. Salzo, L. Baldassare, and A. Verri. Accelerated and inexact forward-backward algorithms. *SIAM Journal on Optimization*, 23(3):1607–1633, 2013.
86. B. Vũ. A splitting algorithm for dual monotone inclusions involving cocoercive operators. *Advances in Computational Mathematics*, 38(3):667–681, 2013.
87. X. Wang and X. Yuan. The linearized alternating direction method of multipliers for Dantzig selector. *SIAM Journal on Scientific Computing*, 34:2782–2811, 2012.
88. J. Wright, A. Ganesh, S. Rao, and Y. Ma. Robust principal component analysis: Exact recovery of corrupted low-rank matrices by convex optimization. *Neural Information Processing Systems, NIPS 2009*, 2009.
89. S. Wright, R. Nowak, and M. Figueiredo. Sparse reconstruction by separable approximation. *IEEE Transactions on Signal Processing*, 57(7):2479–2493, July 2009.
90. J. Yang and X. Yuan. Linearized augmented Lagrangian and alternating direction methods for nuclear norm minimization. *Mathematical and Computer Modelling*, 82:301–329, 2013.
91. J. Yang and Y. Zhang. Alternating direction algorithms for l_1-problems in compressive sensing. *SIAM Journal on Scientific Computing*, 33(1-2):250–278, 2011.
92. W. Yin. Analysis and generalizations of the linearized Bregman method. *SIAM Journal on Imaging Sciences*, 3(4):856–877, 2010.
93. Y. Nesterov. Smooth minimization of non-smooth functions. *Mathematical Programming, Series A*, 103:127–152, 2005.
94. Q. You, Q. Wan, and Y. Liu. A short note on strongly convex programming for exact matrix completion and robust principal component analysis. *Inverse Problems and Imaging*, 7(1):305–306, 2013.
95. J. Zhang, J. Cai, L. Cheng, and J. Zhu. Strongly convex programming for exact matrix completion and robust principal component analysis. *Inverse Problems and Imaging*, 6(2):357–372, 2012.
96. X. Q. Zhang, M. Burger, and S. Osher. A unified primal-dual algorithm framework based on Bregman iteration. *SIAM Journal on Scientific Computing*, 46:20–46, 2010.
97. Z. Zhang, X. Liang, A. Ganesh, and Y. Ma. TILT: Transform invariant low-rank textures. In R. Kimmel, R. Klette, and A. Sugimoto, editors, *Computer Vision - ACCV 2010*, volume 6494 of *Lecture Notes in Computer Science*, pages 314–328. Springer, 2011.
98. X. Zhao, D. Sun, and K. Toh. A Newton-CG augmented Lagrangian method for semidefinite programming. *SIAM Journal on Optimization*, 20(4):1737–1765, 2010.

4

Robust Principal Component Analysis Based on Low-Rank and Block-Sparse Matrix Decomposition

Qiuwei Li
Colorado School of Mines, USA

Gongguo Tang
Colorado School of Mines, USA

Arye Nehorai
Washington University, St. Louis, USA

4.1 Introduction

The classical principal component analysis (PCA) [8] is arguably one of the most important tools in high-dimensional data analysis. However, the classical PCA is not robust to gross corruptions on even only a few entries of the underlying low-rank matrix L containing the principal components. The robust principal component analysis (RPCA) [3, 4, 19] models the gross corruptions as a sparse matrix S superpositioned on the low-rank matrix L. The authors of [3, 4, 19] presented various incoherence conditions under which the low-rank matrix L and the sparse matrix B can be accurately decomposed by solving a convex program, the principal component pursuit:

$$\min_{A,E} \|E\|_* + \lambda\|A\|_1 \text{ subject to } D = A + E, \tag{4.1}$$

where D is the observation matrix, λ is a tuning parameter, and $\|\cdot\|_*$ and $\|\cdot\|_1$ denote the nuclear norm and the sum of the absolute values of the matrix entries, respectively. Many algorithms have been designed to efficiently solve the resulting convex program, *e.g.*, the singular value thresholding [2], the alternating direction method [17], the accelerated proximal gradient (APG) method [11], and the augmented Lagrange multiplier method [10]. Some of these algorithms also apply stably when the observation matrix is corrupted by small, entry-wise noise [19]. The applications of RPCA include video surveillance [3], face recognition [12, 14], latent semantic indexing [13], image retrieval and management [20], and computer vision [16, 18], to name a few. We comment that, in certain applications, the sparse matrix S is actually of more interest than the low-rank matrix, which contains the principal components.

However, RPCA is not robust to column/row outliers in the case where E is a column/row block sparse matrix. Many data sets arranged in matrix formats are of low-rank due to the correlations and connections within the data samples. The ubiquity of low-rank matrices is also suggested by the success and popularity of principal component analysis in many applications [8]. Examples of special interest are network traffic matrices [1], and the data matrix formed in face recognition [15], where the columns correspond to vectorized versions of face images. When a few columns of the data matrix are generated by mechanisms different from the rest of the columns, the existence of these outlying columns tends to destroy the low-rank structure of the data matrix. A decomposition that enforces the low-rankness of one part and the block-sparsity of the other part would separate the principal components from the outliers. In this work, we design algorithms for low-rank and block-sparse matrix decomposition as a robust version of principal component analysis insensitive to column/row outliers.

We use a convex program to separate the low-rank part and block-sparse part of the observed matrix. The decomposition involves the following model:

$$D = A + E. \tag{4.2}$$

The observation matrix D is now the sum of the low-rank component A and the block-sparse component E. The block-sparse matrix E contains mostly zero columns, with several non-zero ones corresponding to outliers. In order to eliminate ambiguity, the columns of the low-rank matrix A corresponding to the outlier columns are assumed to be zeros. We demonstrate that the following convex program recovers the low-rank matrix A and the block-sparsity matrix E:

$$\begin{aligned} &\min_{A,E} \quad \|A\|_* + \kappa(1-\lambda)\|A\|_{2,1} + \kappa\lambda\|E\|_{2,1} \\ &\text{subject to } D = A + E, \end{aligned} \tag{4.3}$$

where $\|\cdot\|_*$, $\|\cdot\|_{2,1}$ denote, respectively, the Frobinius norm, and the ℓ_1 norm of the vector formed by taking the ℓ_2 norms of the columns of the underlying matrix. Note that the extra term $\kappa(1-\lambda)\|A\|_{2,1}$ actually ensures the recovered A has exact zero columns corresponding to the outliers.

We design efficient algorithms to solve the convex program (4.3) based on the augmented Lagrange multiplier (ALM) method. The ALM method was employed to successfully perform low-rank and sparse matrix decomposition for large-scale problems [10]. The challenges here are the special structure of the $\|\cdot\|_{2,1}$ norm, and the existence of the extra term $\kappa(1-\lambda)\|A\|_{2,1}$. We demonstrate the validity of (4.3) in decomposition and compare the performance of the ALM algorithm using numerical simulations. In future work, we will test the algorithms by applying them to face recognition and network traffic anomaly detection problems. Due to the ubiquity of low-rank matrices and the importance of outlier detection, we expect low-rank and block-sparse matrix decomposition to have wide applications, especially in computer vision and data mining.

The chapter is organized as follows. In Section 4.2, we introduce notations and the problem setup. Section 4.3 is devoted to the methods of augmented Lagrange multipliers for solving the convex program (4.3). We provide implementation details in Section 4.4. Numerical experiments are used to demonstrate the effectiveness of our method and the results are reported in Section 4.5. Section 4.6 summarizes our conclusions.

4.2 Notations and Problem Setup

In this section, we introduce notations and the problem setup used throughout the paper. For any matrix $A \in \mathbb{R}^{n \times p}$, we use A_{ij} to denote its ijth element, and A_j to denote its jth column

The usual ℓ_p norm is denoted by $\|\cdot\|_p, p \geq 1$ when the argument is a vector. The nuclear norm $\|\cdot\|_*$, the Frobinius norm $\|\cdot\|_{\mathrm{F}}$, and the $\ell_{2,1}$ norm $\|\cdot\|_{2,1}$ of matrix A are defined as follows:

$$\|A\|_* = \sum_i \sigma_i(A), \tag{4.4}$$

$$\|A\|_{\mathrm{F}} = \sqrt{\sum_{i,j} A_{ij}^2}, \tag{4.5}$$

$$\|A\|_{2,1} = \sum_j \|A_j\|_2, \tag{4.6}$$

where $\sigma_i(A)$ is the ith largest singular value of A. We consider the following model:

$$D = A + E. \tag{4.7}$$

Here E has at most s non-zero columns. These non-zero columns are considered as outliers in the entire data matrix D. The corresponding columns of A are assumed to be zeros. In addition, A has low rank, *i.e.*, $\text{rank}(A) \leq r$.

We use the following optimization to recover A and E from the observation D:

$$\begin{aligned} &\min_{A,E} \quad \|A\|_* + \kappa(1-\lambda)\|A\|_{2,1} + \kappa\lambda\|E\|_{2,1} \\ &\text{subject to } D = A + E. \end{aligned} \tag{4.8}$$

This procedure separates the outliers in E and the principal components in A. As we will see in the numerical examples, simply enforcing the sparsity of E and the low-rankness of A would not separates A and E well.

4.3 The Methods of Augmented Lagrange Multipliers

The Augmented Lagrange Multipliers (ALM) method is a general procedure to solve the following equality constrained optimization problem:

$$\min f(S) \text{ subject to } \ h(S) = 0, \tag{4.9}$$

where $f : \mathbb{R}^n \to \mathbb{R}$ and $h : \mathbb{R}^n \to \mathbb{R}^m$. The procedure defines the augmented Lagrangian function as

$$\mathcal{L}(S, \boldsymbol{\lambda}; \mu) = f(S) + \langle \boldsymbol{\lambda}, h(S) \rangle + \frac{\mu}{2}\|h(S)\|_2^2, \tag{4.10}$$

where $\boldsymbol{\lambda} \in \mathbb{R}^m$ is the Lagrange multiplier vector, and μ is a positive scalar. The ALM iteratively solves S and the Lagrangian multiplier vector $\boldsymbol{\lambda}$ as shown in Algorithm 4.1.

To apply the general ALM to problem (4.8), we define

$$f(X) = \|A\|_* + \kappa(1-\lambda)\|A\|_{2,1} + \kappa\lambda\|E\|_{2,1}, \tag{4.11}$$

$$h(X) = D - A - E \tag{4.12}$$

Algorithm 4.1 - General augmented Lagrange multiplier method

1: **initialize:** Given $\rho > 1, \mu_0 > 0$, starting point S_0^{s} and $\boldsymbol{\lambda}_0$; $k \leftarrow 0$
2: **while** not converged **do**
3: Approximately solve $S_{k+1} = \operatorname{argmin}_S \mathcal{L}(S, \boldsymbol{\lambda}_k; \mu_k)$ using starting point S_k^{s}.
4: Set starting point $S_{k+1}^{\mathrm{s}} = S_{k+1}$.
5: Update the Lagrange multiplier vector $\boldsymbol{\lambda}_{k+1} = \boldsymbol{\lambda}_k + \mu_k h(S_{k+1})$.
6: Update $\mu_{k+1} = \rho\mu_k$.
7: **end while**
8: **output:** $S \leftarrow S_k, \boldsymbol{\lambda} \leftarrow \boldsymbol{\lambda}_k$.

with $X = (A, E)$. The corresponding augmented Lagrangian function is

$$\begin{aligned}\mathcal{L}(A, E, Y; \mu) &= \|A\|_* + \kappa(1-\lambda)\|A\|_{2,1} + \kappa\lambda\|E\|_{2,1} \\ &+ \langle Y, D - A - E\rangle + \frac{\mu}{2}\|D - A - E\|_{\mathrm{F}}^2,\end{aligned} \tag{4.13}$$

where we use Y to denote the Lagrange multiplier. Given $Y = Y_k$ and $\mu = \mu_k$, one key step in applying the general ALM method is to solve $\min_{A,E} \mathcal{L}(A, E, Y_k; \mu_k)$. We adopt an alternating procedure: for fixed $E = E_k$, solve

$$A_{k+1} = \operatorname{argmin}_A \mathcal{L}(A, E_k, Y_k; \mu_k), \tag{4.14}$$

and for fixed $A = A_{k+1}$, solve

$$E_{k+1} = \operatorname{argmin}_E \mathcal{L}(A_{k+1}, E, Y_k; \mu_k). \tag{4.15}$$

We first address (4.15), which is equivalent to solving

$$\min_E \kappa\lambda\|E\|_{2,1} + \frac{\mu_k}{2}\left\|D - A_{k+1} + \frac{1}{\mu_k}Y_k - E\right\|_{\mathrm{F}}^2. \tag{4.16}$$

Denote $G^E = D - A_{k+1} + \frac{1}{\mu_k}Y_k$. The following lemma gives the closed-form solution of the minimization with respect to E.

Lemma 1 *The solution to*

$$\min_E \eta\|E\|_{2,1} + \frac{1}{2}\|E - G^E\|_{\mathrm{F}}^2 \tag{4.17}$$

is given by

$$E_j = G_j^E \max\left(0, 1 - \frac{\eta}{\|G_j^E\|_2}\right) \tag{4.18}$$

for $j = 1, \ldots, p$.

Proof: Note the objective function expands as

$$\begin{aligned}&\eta\|E\|_{2,1} + \frac{1}{2}\|E - G^E\|_{\mathrm{F}}^2 \\ &= \sum_{j=1}^{n}\left(\eta\|E_j\|_2 + \frac{1}{2}\|E_j - G_j^E\|_2^2\right).\end{aligned} \tag{4.19}$$

Now we can minimize with respect to E_j separately. Denote

$$\begin{aligned} h(\boldsymbol{e}) &= \eta\|\boldsymbol{e}\|_2 + \frac{1}{2}\|\boldsymbol{e}-\boldsymbol{g}\|_2^2 \\ &= \eta\|\boldsymbol{e}\|_2 + \frac{1}{2}\left[\|\boldsymbol{e}\|_2^2 - 2\langle\boldsymbol{e},\boldsymbol{g}\rangle + \|\boldsymbol{g}\|_2^2\right]. \end{aligned} \tag{4.20}$$

First consider $\|\boldsymbol{g}\|_2 \leq \eta$. Cauchy-Schwarz inequality leads to

$$h(\boldsymbol{e}) \geq \eta\|\boldsymbol{e}\|_2 + \frac{1}{2}\left[\|\boldsymbol{e}\|_2^2 - 2\|\boldsymbol{e}\|_2\|\boldsymbol{g}\|_2 + \|\boldsymbol{g}\|_2^2\right]$$

$$= \frac{1}{2}\|\boldsymbol{e}\|_2^2 + (\eta - \|\boldsymbol{g}\|_2)\|\boldsymbol{e}\|_2 + \frac{1}{2}\|\boldsymbol{g}\|_2^2 \tag{4.21}$$

$$\geq \frac{1}{2}\|\boldsymbol{g}\|_2^2, \tag{4.22}$$

where for the last inequality we used (4.21) is an increasing function on $[0,\infty)$ if $\|\boldsymbol{g}\|_2 \leq \eta$. Apparently, $h(\boldsymbol{e}) = \|\boldsymbol{g}\|_2/2$ is achieved by $\boldsymbol{e} = 0$, which is also unique. Therefore, if $\|\boldsymbol{g}\|_2 \leq \eta$, then the minimum of $h(\boldsymbol{e})$ is achieved by the unique solution $\boldsymbol{e} = 0$

In the second case when $\|\boldsymbol{g}\|_2 > \eta$, setting the derivative of $h(\boldsymbol{e})$ with respect to $\boldsymbol{e}$ to zero yields

$$\boldsymbol{e}\left(\frac{\eta}{\|\boldsymbol{e}\|_2} + 1\right) = \boldsymbol{g}. \tag{4.23}$$

Taking the ℓ_2 norm of both sides yields

$$\|\boldsymbol{e}\|_2 = \|\boldsymbol{g}\|_2 - \eta > 0. \tag{4.24}$$

Plugging $\|\boldsymbol{e}\|_2$ into (4.23) gives

$$\boldsymbol{e} = \boldsymbol{g}\left(1 - \frac{\eta}{\|\boldsymbol{g}\|_2}\right). \tag{4.25}$$

In conclusion, the minimum of (4.17) is achieved by E with

$$E_j = G_j^E \max\left(0, \left(1 - \frac{\eta}{\|G_j^E\|_2}\right)\right) \tag{4.26}$$

for $j = 1, 2, \ldots, p$. ■

We denote the operator solving (4.17) as $T_\eta(\cdot)$. So $E = T_\eta(G)$ sets the columns of E to be zero vectors if the ℓ_2 norms of the corresponding columns of G are less than η, and scales down the columns otherwise by a factor $1 - \frac{\eta}{\|G_j\|_2}$.

Now we turn to solve

$$A_{k+1} = \operatorname{argmin}_A \mathcal{L}(A, E_k, Y_k; \mu_k) \tag{4.27}$$

which can be rewritten as

$$A_{k+1} = \operatorname{argmin}_A \left\{\|A\|_* + \kappa(1-\lambda)\|A\|_{2,1} + \frac{\mu_k}{2}\|G^A - A\|_F^2\right\}, \tag{4.28}$$

where $G^A = D - E_k + \frac{1}{\mu_k} Y_k$. We know that without the extra $\kappa(1-\lambda)\|A\|_{2,1}$ term, a closed form solution is simply given by the soft-thresholding operator [2]. Unfortunately, a closed form solution to (4.28) is not available. We use the Douglas/Peaceman-Rachford (DR) monotone operator splitting method [5–7] to iteratively solve (4.28). Define

$$f_1(A) = \kappa(1-\lambda)\|A\|_{2,1} + \frac{\mu_k}{2}\|G^A - A\|_F^2$$

and

$$f_2(A) = \|A\|_*$$

For any $\beta > 0$ and a sequence

$$\alpha_t \in (0, 2)$$

the DR iteration for (4.28) is expressed as

$$A^{(j+1/2)} = \text{prox}_{\beta f_2}(A^{(j)}), \tag{4.29}$$

$$A^{(j+1)} = A^{(j)} + \alpha_j \left(\text{prox}_{\beta f_1}(2A^{(j+1/2)} - A^{(j)}) - A^{(j+1/2)} \right). \tag{4.30}$$

Here for any proper, lower semi-continuous, convex function f, the proximity operators $\text{prox}_f(\cdot)$ give the unique point $\text{prox}_f(S)$, achieving the infimum of the function

$$z \mapsto \frac{1}{2}\|S - z\|_2^2 + f(z). \tag{4.31}$$

The following lemmas gives the explicit expression for the proximity operator involved in our DR iteration when $f(A) = \|A\|_*$ and $f(A) = \kappa(1-\lambda)\|A\|_{2,1} + \frac{\mu_k}{2}\|G^A - A\|_F^2$.

Lemma 2 *For $f(\cdot) = \|\cdot\|_*$, suppose the singular value decomposition of A is $A = U\Sigma V^T$, then the proximity operator is*

$$\text{prox}_{\beta f}(A) = U \S_\beta(\Sigma) V^T, \tag{4.32}$$

where the non-negative soft-thresholding operator is defined as

$$\S_\nu(x) = \max(0, x - \nu), x \geq 0, \nu > 0, \tag{4.33}$$

for a scalar x and extended entry-wisely to vectors and matrices.

For $f(\cdot) = \eta\|\cdot\|_{2,1} + \frac{\mu}{2}\|G - \cdot\|_F^2$, the proximity operator is

$$\text{prox}_{\beta f}(A) = T_{\frac{\beta\eta}{1+\beta\mu}}\left(\frac{A + \beta\mu G}{1 + \beta\mu}\right). \tag{4.34}$$

With these preparations, we present the algorithm for solving (4.8) using the ALM, called RPCA-LBD, in Algorithm 4.2. Note that instead of alternating between (4.14) and (4.15) many times until convergence, the RPCA-LBD algorithm in Algorithm 4.2 executes them only once. This inexact version greatly reduces the computational burden and yields sufficiently accurate results for appropriately tuned parameters. The strategy was also adopted in [10] to perform the low-rank and sparse matrix decomposition.

Algorithm 4.2 - Block-sparse and low-rank matrix decomposition via ALM

1: RPCA-LBD(D, κ, λ)
2: **input:** Data matrix $D \in \mathbb{R}^{n \times p}$ and the parameters κ, λ.
3: **output:** The low-rank part A and the block-sparse part E.
4: **initialize:** $A_0 \leftarrow D$; $E_0 \leftarrow 0$; $\mu_0 = 30/\|\text{sign}(D)\|_2$; $\rho > 0, \beta > 0, \alpha \in (0,1)$, $k \leftarrow 0$.
5: **while** not converged **do**
6: \\ Lines 6 - 14 solve $A_{k+1} = \text{argmin}_A \mathcal{L}(A, E_k, Y_k; \mu_k)$.
7: $G^A = D - E_k + \frac{1}{\mu_k} Y_k$.
8: $j \leftarrow 0$, $A_{k+1}^{(0)} = G^A$.
9: **while** not converged **do**
10: $A_{k+1}^{(j+1/2)} = U \S_\beta(\Sigma) V^T$ where $A_{k+1}^{(j)} = U \Sigma V^T$ is the SVD of $A_{k+1}^{(j)}$.
11: $A_{k+1}^{(j+1)} = A_{k+1}^{(j)} + \alpha \left(T_{\frac{\beta\kappa(1-\lambda)}{1+\beta\mu_k}} \left(\frac{2A_{k+1}^{(j+1/2)} - A_{k+1}^{(j)} + \beta\mu_k G^A}{1+\beta\mu_k} \right) - A_{k+1}^{(j+1/2)} \right)$.
12: $j \leftarrow j+1$.
13: **end while**
14: $A_{k+1} = A_{k+1}^{(j+1/2)}$.
15: $G^E = D - A_{k+1} + \frac{1}{\mu_k} Y_k$.
16: $E_{k+1} = T_{\frac{\kappa\lambda}{\mu_k}} (G^E)$.
17: $Y_{k+1} = Y_k + \mu_k * (D - A_{k+1} - E_{k+1})$.
18: $\mu_{k+1} = \rho\mu_k$.
19: $k \leftarrow k+1$.
20: **end while**
21: **output:** $A \leftarrow A_k, E \leftarrow E_k$.

4.4 Implementation

Parameter selection and initialization: The tuning parameters in (4.8) are chosen to be $\kappa = 1.1$ and $\lambda = 0.61$. We do not have a systematic way to select these parameters. But these empirical values work well for all tested cases. For the DR iteration (4.29) and (4.30), we have $\beta = 0.2$ and $\alpha_j \equiv 1$. The parameters for the ALM method are $\mu_0 = 30/\|\text{sign}(D)\|_2$ and $\rho = 1.1$. The low-rank matrix A, the block-sparse matrix E, and the Lagrange multiplier Y are initialized, respectively, to D, $\mathbf{0}$, and $\mathbf{0}$. The outer loop in Algorithm 4.2 terminates when it reaches the maximal iteration number 500 or the error tolerance 10^{-7}. The error in the outer loop is computed by $\|D - A_k - E_k\|_F / \|D\|_F$. The inner loop for the DR iteration has maximal iteration number 20 and tolerance error 10^{-6}. The error in the inner loop is the Frobenius norm of the difference between successive $A_k^{(j)}$s.

Performing SVD: The major computational cost in the RPCA-LBD algorithm is the singular value decomposition (SVD) in the DR iteration (Line 10 in Algorithm 4.2). We usually do not need to compute the full SVD because only those that are greater than β are used. We use the PROPACK [9] to compute the first (largest) few singular values. In order to both save computation and ensure accuracy, we need to predict the number of singular values of $A_{k+1}^{(j)}$ that exceed β for each iteration. We adopt the following rule [10]:

$$\text{sv}_{k+1} = \begin{cases} \text{svn}_k + 1, & \text{if } \text{svn}_k < \text{sv}_k; \\ \min(\text{svn}_k + 10, d), & \text{if } \text{svn}_k = \text{sv}_k. \end{cases},$$

where $\text{sv}_0 = 10$ and

$$\text{svn}_k = \begin{cases} \text{svp}_k, & \text{if maxgap}_k \leq 2; \\ \min(\text{svp}_k, \text{maxid}_k), & \text{if maxgap}_k > 2. \end{cases}$$

Here $d = \min(n, p)$, sv_k is the predicted number of singular values that are greater than β, and svp_k is the number of singular values in the sv_k singular values that are larger than β. In addition, maxgap_k and maxid_k are, respectively, the largest ratio between successive singular values of $A_{k+1}^{(j)}$ when arranged in a decreasing order and the corresponding index.

4.5 Numerical Simulations

Throughout this section, we consider only square matrices D with $n = p$, which is generated as follows:

Construction of matrix D

- Construct matrix A as the product of two Gaussian matrices with dimensions $n \times r$ and $r \times n$. This construction guarantees $\text{rank}(A) = r$.
- Normalize the columns of A to have unit norm.
- Generate a support S of size s as a random subset of $\{1, \ldots, p\}$.
- Set the columns of A corresponding to S to zeros.
- Sample the columns of E corresponding to the support S from i.i.d. Gaussian distribution with mean zeros and unit variance.
- Normalize the non-zero columns of E.
- Construct D as the sum of A and E.

4.5.1 Convergence Behavior

In this subsection, we focus on the convergence behavior of the RPCA-LBD and compare it with the classical RPCA solved by the ALM algorithm proposed in [10]. In order to illustrate the convergence behavior of these two algorithms, we apply them to the same matrix D with the same number of iterations $K = 20$. The trajectories of the reconstruction errors $\rho_A^{RPCA-LBD}(k)$, $\rho_A^{RPCA}(k)$, $\rho_E^{RPCA-LBD}(k)$ and $\rho_E^{RPCA}(k)$ are recorded.

DEFINITION 4.1 Assume $A_{RPCA-LBD}(k)$ is the approximation to A and $E_{RPCA-LBD}(k)$ is the approximation to E by running k iterations of the algorithm RPCA-LBD. Assume $A_{RPCA}(k)$ is the approximation to A and $E_{RPCA}(k)$ is the approximation to E by running k_s iteration of the algorithm RPCA. Then,

$$\rho_A^{RPCA-LBD}(k) \triangleq \frac{\|A - A_{RPCA-LBD}(k)\|_F}{\|A\|_F} \tag{4.35}$$

$$\rho_E^{RPCA-LBD}(k) \triangleq \frac{\|E - E_{RPCA-LBD}(k)\|_F}{\|E\|_F} \tag{4.36}$$

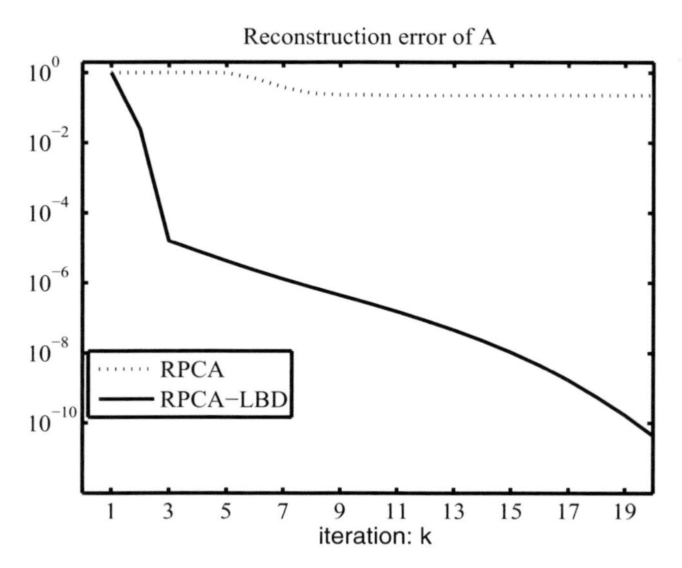

FIGURE 4.1 Reconstruction errors $\rho_A^{RPCA-LBD}(k)$ and $\rho_A^{RPCA}(k)$ versus iteration number k for $n = 100, p = 100, r = 4, s = 4$.

$$\rho_A^{RPCA}(k) \triangleq \frac{\|A - A_{RPCA}(k)\|_F}{\|A\|_F} \tag{4.37}$$

$$\rho_E^{RPCA}(k) \triangleq \frac{\|E - E_{RPCA}(k)\|_F}{\|E\|_F} \tag{4.38}$$

Case I: $n = 100, p = 100, r = 4, s = 4$

Figure 4.1 shows the reconstruction errors $\rho_A^{RPCA-LBD}(k)$ and $\rho_A^{RPCA}(k)$ versus iteration number k for the case $n = 100, p = 100, r = 4, s = 4$ using RPCA-LBD and RPCA, respectively.

Figure 4.2 shows the reconstruction errors $\rho_E^{RPCA-LBD}(k)$ and $\rho_E^{RPCA}(k)$ versus iteration number k for the case $n = 100, p = 100, r = 4, s = 4$ using RPCA-LBD and RPCA, respectively.

Case II: $n = 200, p = 200, r = 8, s = 8$

Figure 4.3 shows the reconstruction errors $\rho_A^{RPCA-LBD}(k)$ and $\rho_A^{RPCA}(k)$ versus iteration number k for the case $n = 200, p = 200, r = 8, s = 8$ using RPCA-LBD and RPCA, respectively.

Figure 4.4 show the reconstruction error $\rho_E^{RPCA-LBD}(k)$ and $\rho_E^{RPCA}(k)$ versus iteration number k for the case $n = 200, p = 200, r = 8, s = 8$ using RPCA-LBD and RPCA, respectively.

Case III: $n = 300, p = 300, r = 12, s = 12$

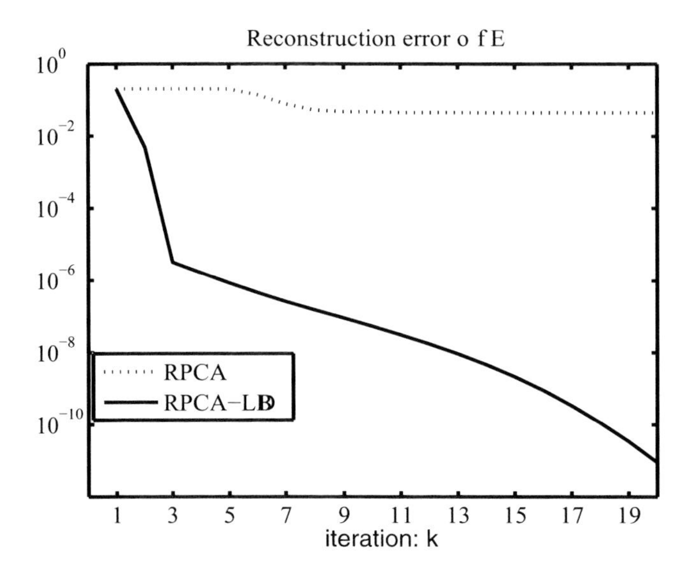

FIGURE 4.2 Reconstruction error $\rho_E^{RPCA-LBD}(k)$ and $\rho_E^{RPCA}(k)$ versus iteration number k for $n = 100, p = 100, r = 4, s = 4$.

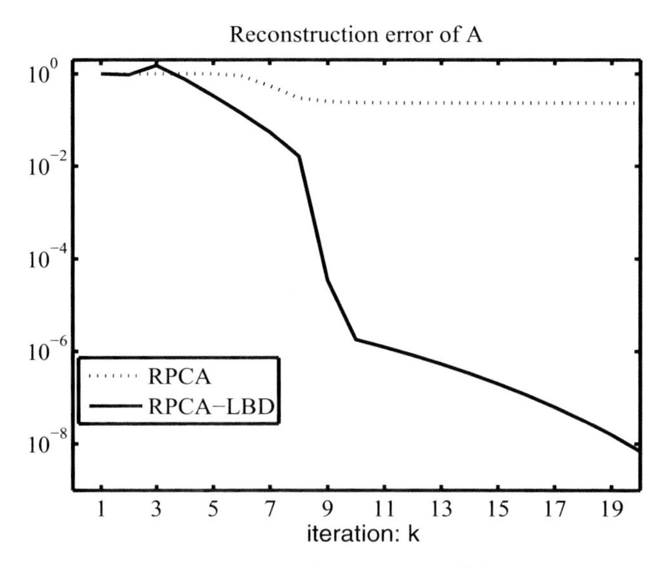

FIGURE 4.3 Reconstruction error $\rho_A^{RPCA-LBD}(k)$ and $\rho_A^{RPCA}(k)$ versus iteration number k for $n = 200, p = 200, r = 8, s = 8$.

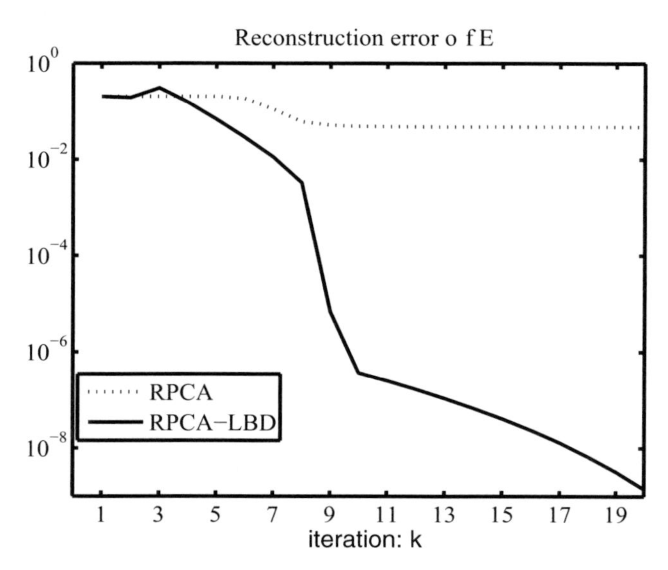

FIGURE 4.4 Reconstruction error $\rho_E^{RPCA-LBD}(k)$ and $\rho_E^{RPCA}(k)$ versus iteration number k for $n = 200, p = 200, r = 8, s = 8$.

Figure 4.5 shows the reconstruction errors $\rho_A^{RPCA-LBD}(k)$ and $\rho_A^{RPCA}(k)$ versus iteration number k for the case $n = 300, p = 300, r = 12, s = 12$ using RPCA-LBD and RPCA, respectively.

Figure 4.6 shows the reconstruction errors $\rho_E^{RPCA-LBD}(k)$ and $\rho_E^{RPCA}(k)$ versus iteration number k for the case $n = 200, p = 300, r = 12, s = 12$ using RPCA-LBD and RPCA, respectively.

Remarks 5.1: As seen from the simulations, in all the cases, i.e., $n = 100, n = 200, n = 300$, the proposed algorithm RPCA-LBD converges to the true solution faster than the RPCA. For example, from the comparison of the evolution of the indicators $\rho_A^{RPCA-LBD}(k)$ and $\rho_A^{RPCA}(k)$, we can see that the RPCA gets trapped into a local minimizer in about 9 iterations for all three cases. Conversely, the proposed RPCA-LBD could recover the true low rank matrix A for all the three cases ($n = 100, n = 200, n = 300$).

4.5.2 Comparison of Recovery Performance

In this subsection, we compare the relative recovery errors $\rho_A^{RPCA-LBD}$, ρ_A^{RPCA}, $\rho_E^{RPCA-LBD}$, and ρ_E^{RPCA} defined as follows.

DEFINITION 4.2 Denote by $A_{RPCA-LBD}$ and $E_{RPCA-LBD}$ the estimates of A and E, respectively, produced by the algorithm RPCA-LBD. Similarly, denote by A_{RPCA} and E_{RPCA} the esimates of A and E produced by the algorithm RPCA. Then, the relative recovery errors are defined as:

$$\rho_A^{RPCA-LBD} \triangleq \frac{\|A - A_{RPCA-LBD}\|_F}{\|A\|_F} \tag{4.39}$$

$$\rho_E^{RPCA-LBD} \triangleq \frac{\|E - E_{RPCA-LBD}\|_F}{\|E\|_F} \tag{4.40}$$

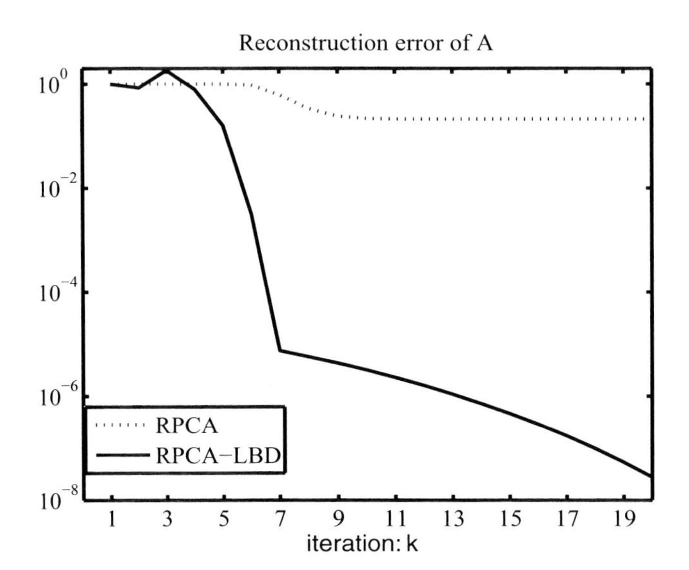

FIGURE 4.5 Reconstruction error $\rho_A^{RPCA-LBD}(k)$ and $\rho_A^{RPCA}(k)$ versus iteration number k for $n = 300, p = 300, r = 12, s = 12$.

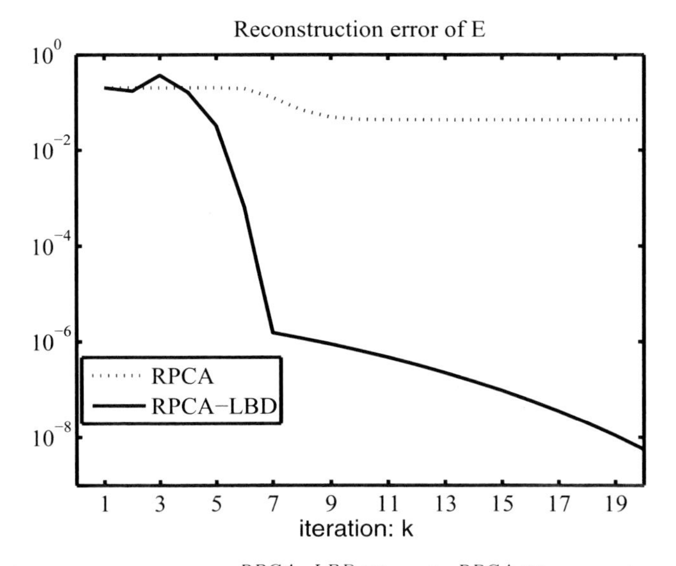

FIGURE 4.6 Reconstruction error $\rho_E^{RPCA-LBD}(k)$ and $\rho_E^{RPCA}(k)$ versus iteration number k for $n = 300, p = 300, r = 12, s = 12$.

$$\rho_A^{RPCA} \triangleq \frac{\|A - A_{RPCA}\|_F}{\|A\|_F} \tag{4.41}$$

$$\rho_E^{RPCA} \triangleq \frac{\|E - E_{RPCA}\|_F}{\|E\|_F} \tag{4.42}$$

In Table 4.1, we compare the four relative recovery errors of the RPCA-LBD and the RPCA: $\rho_A^{RPCA-LBD}$, ρ_A^{RPCA}, $\rho_E^{RPCA-LBD}$ and ρ_E^{RPCA}. We consider $n = 100, 200, 300, 400, 500, 600, 700, 800$ and 900.

Remarks 5.2:

- The rank of the low-rank matrix A is $r = \text{round}(0.04n)$ and the number of the non-zero columns of the block sparse matrix E is $s = \text{round}(0.04n)$.
- We observe from Table 4.1 that the RPCA-LBD can accurately recover the low-rank matrix A and the block sparse matrix E. The relative recovery error of matrix A by RPCA-LBD $\rho_A^{RPCA-LBD}$ is at the order of 10^{-12}. Conversely, the relative recovery error of matrix A by RPCA ρ_A^{RPCA} ρ_A^{RPCA} is at the order of 10^{-2}. Similarly, the relative recovery error of matrix E by RPCA-LBD is at the order of 10^{-8} while the relative recovery error of matrix E by RPCA ρ_E^{RPCA} is at the order of 10^{-1}.

4.5.3 Comparison of the Recovered Matrices $E_{RPCA-LBD}$ and E_{RPCA}

We visually illustrate the reconstructed matrices $E_{RPCA-LBD}$ and E_{RPCA} given by the proposed RPCA-LBD and the classical RPCA, respectively. In the following, we choose $n = p$, $r = \text{round}(0.03n)$ and $s = \text{round}(0.3n)$. In Figure 4.7, we show the original E and those recovered by the RPCA-LBD and the classical RPCA. Here $n = 100, p = 100, r = 3, s = 30$. We can see that our algorithm RPCA-LBD recovers the block sparse matrix perfectly, while the RPCA returns a matrix whose non-zero columns tend to be sparse. The relative error for L recovered by the RPCA-LBD $\rho_E^{RPCA-LBD}$ is 5.6271×10^{-12} while the relative error ρ_E^{RPCA} is 0.3415. The same comments apply to Figures 4.8 and 4.9 for $n = 200, p = 200, r = 6, s = 60$, and $n = 300, p = 300, r = 9, s = 90$.

TABLE 4.1 Comparison of our algorithm (RPCA-LBD) and the RPCA.

n	$\rho_A^{RPCA-LBD}$	ρ_A^{RPCA}	$\rho_E^{RPCA-LBD}$	ρ_E^{RPCA}
100	2.6996e-08	0.0485	1.9262e-12	0.2377
200	5.2486e-08	0.0440	3.9758e-12	0.2157
300	2.9306e-08	0.0438	2.6342e-12	0.2146
400	7.9004e-08	0.0480	1.3780e-12	0.2351
500	7.9182e-08	0.0463	1.2861e-12	0.2269
600	8.0911e-08	0.0460	1.2964e-12	0.2251
700	1.6499e-08	0.0453	2.1736e-12	0.2219
800	1.2771e-08	0.0451	1.9037e-12	0.2211
900	9.2575e-08	0.0464	1.3934e-12	0.2272

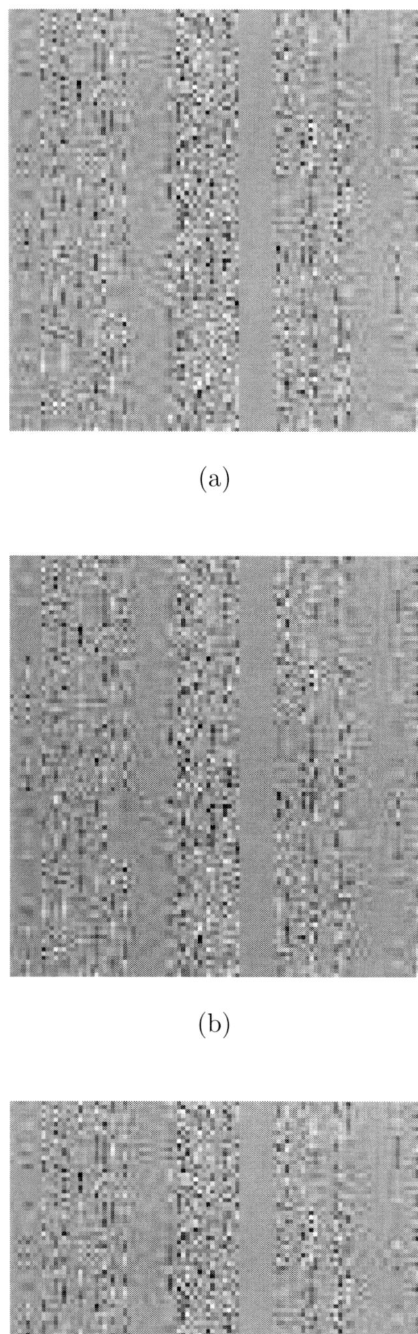

(a)

(b)

(c)

FIGURE 4.7 For the case $n = 100$, $p = 100$, $r = 3$, $s = 30$. (a) The original E; (b) E_{RPCA}; (c) $E_{RPCA-LBD}$.

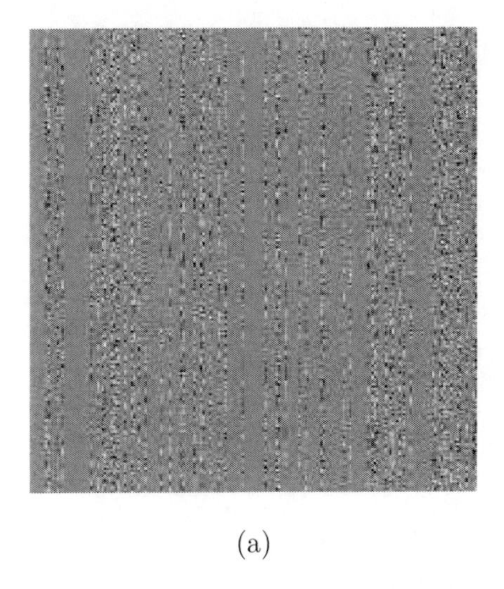

(a)

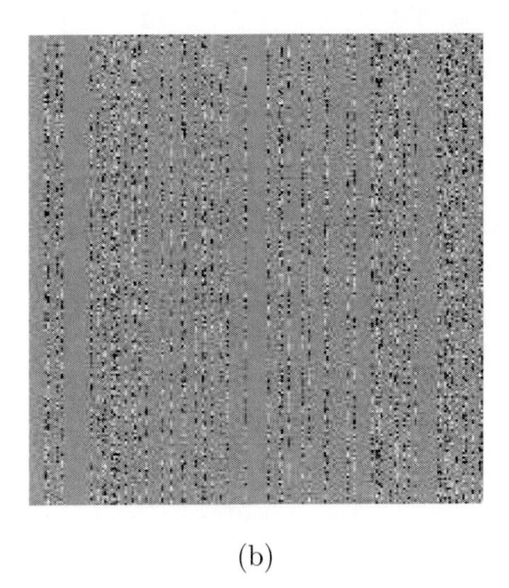

(b)

(c)

FIGURE 4.8 For the case $n = 200$, $p = 200$, $r = 6$, $s = 60$. (a) The original E; (b) E_{RPCA}; (c) $E_{RPCA-LBD}$.

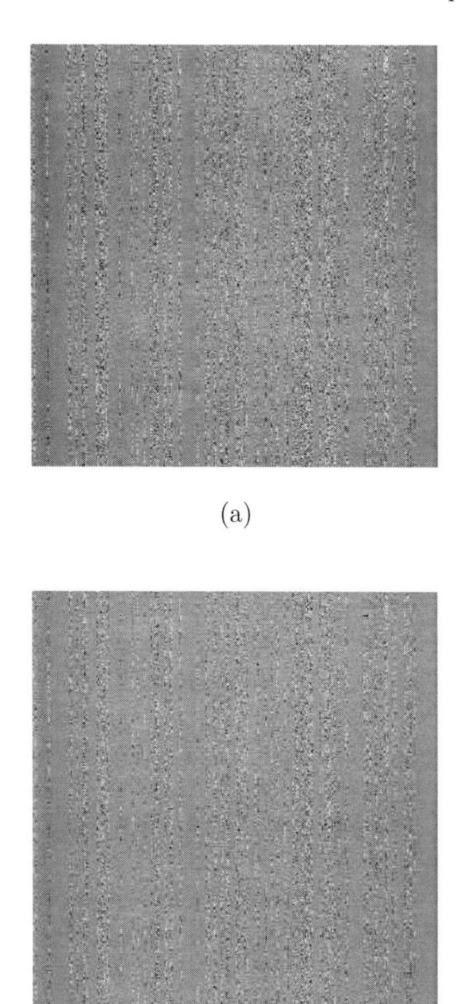

(a)

(b)

(c)

FIGURE 4.9 For the case $n = 300$, $p = 300$, $r = 9$, $s = 90$. (a) The original E; (b) E_{RPCA}; (c) $E_{RPCA-LBD}$.

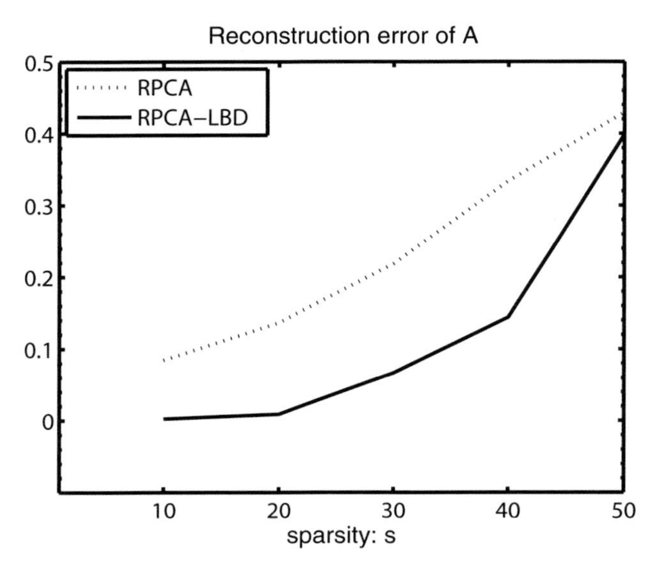

FIGURE 4.10 Reconstruction error $\rho_A^{RPCA-LBD}$ and ρ_A^{RPCA} versus block sparsity s varying from 10 to 50 with $n = 100$, $p = 100$, $r = 4$.

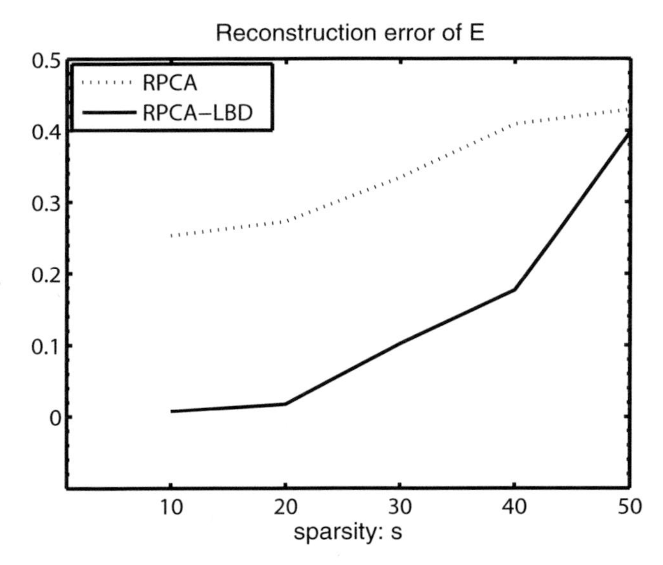

FIGURE 4.11 Reconstruction error $\rho_E^{RPCA-LBD}$ and ρ_E^{RPCA} versus block sparsity s varying from 10 to 50 with $n = 100, p = 100, r = 4$.

4.5.4 Examination of the Effect of the Block Sparsity s

In Figure 4.10, we compare the relative errors ρ_A^{RPCA} and $\rho_A^{RPCA-LBD}$ corresponding to RPCA and RPCA-LBD, respectively, for block sparsity s varying from 10 to 50 with $n = 100, p = 100, r = 4$. The relative errors for the reconstructed matrices E are shown in Figure 4.11. One observes that the classical algorithm RPCA failed to recover the block sparse matrix E and the low-rank matrix A for all the different kinds of block sparsity. However, RPCA-LBD can recovery them exactly when $s = 10, 20$. For $s = 30, 40, 50$, although RPCA-LBD can not exactly recover the matrices A, E, its relative errors are much smaller than those achieved by RPCA.

4.6 Conclusion

In this work we proposed a convex program for accurate low-rank and block-sparse matrix decomposition. We solved the convex program using the augmented Lagrange multiplier method. We used the Douglas/Peaceman-Rachford (DR) monotone operator splitting method to solve a subproblem involved in the augmented Lagrange multiplier method. We demonstrated the accuracy of our program and compared its results with those given by the robust principal component analysis. The proposed algorithm is potentially useful for outlier detection.

References

1. A. Abdelkefi, Y. Jiang, W. Wang, A. Aslebo, and O. Kvittem. Robust traffic anomaly detection with principal component pursuit. In *Proceedings of the ACM CoNEXT Student Workshop*, CoNEXT '10 Student Workshop, pages 10:1–10:2, New York, NY, USA, 2010.
2. J. Cai, E. Candès, and Z. Shen. A singular value thresholding algorithm for matrix completion. *SIAM Journal on Optimization*, 20(4):1956–1982, 2008.
3. E. Candès, X. Li, Y. Ma, and J. Wright. Robust principal component analysis? *J. ACM*, 58(3):1–37, June 2011.
4. V. Chandrasekaran, S. Sanghavi, P. Parrilo, and A. Willsky. Sparse and low-rank matrix decompositions. In *Allerton Conference on Communication, Control, and Computing, 2009. Allerton 2009*, pages 962–967. IEEE, 2009.
5. P. Combettes and J. Pesquet. A Douglas-Rachford splittting approach to nonsmooth convex variational signal recovery. *IEEE Journal of Selected Topics in Signal Processing*, 1(4):564–574, 2007.
6. M. Fadili, J. Starck, and F. Murtagh. Inpainting and zooming using sparse representations. *The Computer Journal*, 52(1):64, 2009.
7. M. J. Fadili, J. L. Starck, and F. Murtagh. Inpainting and zooming using sparse representations. *The Computer Journal*, 52:64–79, 2007.
8. I. Jolliffe. *Principal component analysis*. Springer series in statistics. Springer, 2002.
9. R. Larsen. Lanczos bidiagonalization with partial reorthogonalization. Department of computer science, Aarhus University, Technical report, DAIMI PB-357, code available at http://soi.stanford.edu/ rmunk/PROPACK/, 1998.
10. Z. Lin, M. Chen, L. Wu, and Y. Ma. The augmented Lagrange multiplier method for exact recovery of corrupted low-rank matrices. *UIUC Technical Report*, 1(UILU-ENG-09-2215), November 2009.
11. Z. Lin, A. Ganesh, J. Wright, L. Wu, M. Chen, and Y. Ma. Fast convex optimization algorithms for exact recovery of a corrupted low-rank matrix. *Computational Advances in Multi-Sensor Adaptive Processing, CAMSAP 2009*, 61, 2009.
12. G. Liu, Z. Lin, and Y. Yu. Robust subspace segmentation by low-rank representation. In *International Conference on Machine Learning, ICML 2010*, pages 663–670, 2010.
13. K. Min, Z. Zhang, J. Wright, and Y. Ma. Decomposing background topics from keywords by principal component pursuit. In *ACM International Conference on Information and Knowledge Management*, pages 269–278. ACM, 2010.
14. Y. Peng, A. Ganesh, J. Wright, W. Xu, and Y. Ma. RASL: robust alignment by sparse and low-rank decomposition for linearly correlated images. *IEEE Transactions on Pattern Analysis and Machine Intelligence*, 34(11):2233–2246, 2012.
15. J. Wright, A. Yang, A. Ganesh, S. Sastry, and Y. Ma. Robust face recognition via sparse representation. *IEEE Transactions on Pattern Analysis and Machine Intelligence*,

31(2), February 2009.
16. L. Wu, A. Ganesh, B. Shi, Y. Matsushita, Y. Wang, and Y. Ma. Robust photometric stereo via low-rank matrix completion and recovery. In *Computer Vision, ACCV 2010*, pages 703–717. Springer, 2011.
17. X. Yuan and J. Yang. Sparse and low-rank matrix decomposition via alternating direction methods. *Preprint*, 2009.
18. Z. Zhang, A. Ganesh, X. Liang, and Y. Ma. TILT: transform invariant low-rank textures. *International Journal of Computer Vision*, 99(1):1–24, 2012.
19. Z. Zhou, X. Li, J. Wright, E. Candes, and Y. Ma. Stable principal component pursuit. In *IEEE International Symposium on Information Theory Proceedings, ISIT 2010*, pages 1518–1522. IEEE, 2010.
20. G. Zhu, S. Yan, and Y. Ma. Image tag refinement towards low-rank, content-tag prior and error sparsity. In *International Conference on Multimedia*, pages 461–470. ACM, 2010.

5

Robust PCA by Controlling Sparsity in Model Residuals

Gonzalo Mateos
University of Rochester, USA

Georgios B. Giannakis
University of Minnesota, USA

5.1 Introduction

Principal component analysis (PCA) is the workhorse of high-dimensional data analysis and dimensionality reduction, with numerous applications in statistics, engineering, and the biobehavioral sciences; see, e.g., [26]. Nowadays ubiquitous e-commerce sites, the Web, and urban traffic surveillance systems generate massive volumes of data. As a result, the problem of extracting the most informative, yet low-dimensional structure from high-dimensional datasets is of paramount importance [22, 45]. To this end, PCA provides least-squares (LS) optimal linear approximants in $\mathbb{R}^q$ to a data set in ambient space $\mathbb{R}^p$, for $q \leq p$. The desired linear subspace is obtained from the q-dominant eigenvectors of the sample data covariance matrix, or equivalently from the q-dominant singular vectors of the data matrix [26].

Data obeying postulated low-rank models also include outliers, which are samples not adhering to those nominal models. Unfortunately, LS is known to be very sensitive to outliers [24, 42], and this undesirable property is inherited by PCA as well [26]. Early efforts to robustify PCA have relied on robust estimates of the data covariance matrix; see, e.g., [7]. Related approaches are driven from statistical physics [54], and also from M-estimators [13]. A fast algorithm for computer vision applications was put forth in [49]. Recently, polynomial-time algorithms with remarkable performance guarantees have emerged for low-rank matrix recovery in the presence of sparse – but otherwise arbitrarily large – errors [8, 10]. This pertains to an 'idealized robust' PCA setup, since those entries not affected by outliers are assumed error to be free. Stability in reconstructing the low-rank and sparse matrix components in the presence of 'dense' noise have been reported in [53, 58]. A hierarchical Bayesian model was proposed to tackle the aforementioned low-rank plus sparse matrix decomposition problem in [14].

In the present chapter, a robust PCA approach is pursued requiring minimal assumptions on the outlier model. A natural least-trimmed squares (LTS) PCA estimator is first shown closely related to an estimator obtained from an ℓ_0-(pseudo)norm-regularized criterion, adopted to fit a low-rank bilinear factor analysis model that explicitly incorporates an unknown *sparse* vector of outliers per datum (Section 5.2). As in compressive sampling [51], efficient (approximate) solvers are obtained in Section 5.2.3, by surrogating the ℓ_0-norm of the outlier matrix with its closest convex approximant. This leads naturally to an M-type PCA estimator, which subsumes Huber's optimal choice as a special case [18]. Unlike Huber's formulation, however, results here are not confined to an outlier contamination model. A tunable parameter controls the sparsity of the estimated matrix, and the number of outliers as a byproduct. Hence, effective data-driven methods to select this parameter are of paramount importance, and systematic approaches are pursued by efficiently exploring the entire *robustifaction* (a.k.a. homotopy) path of (group-) Lasso solutions [22, 57]. In this sense, the method here capitalizes on but *is not limited to* sparse settings where outliers are sporadic, since one can examine all sparsity levels along the robustification path. The outlier-aware generative data model and its sparsity-controlling estimator are quite general, since minor modifications discussed in [34, Sec. III-C] enable robustifiying linear regression [19], dictionary learning [30, 50], and K-means clustering as well [17, 22]. Section 5.3.2 deals with further modifications for bias reduction through nonconvex regularization, and automatic determination of the reduced dimension q is explored in Section 5.4 by drawing connections with nuclear-norm minimization [8, 10].

Beyond its ties to robust statistics, the developed outlier-aware PCA framework is versatile to accommodate scalable *robust* algorithms to: i) track the low-rank signal subspace, as new data are acquired in real time (Section 5.4.1); and ii) determine principal components in (possibly) infinite-dimensional feature spaces, thus robustifying kernel PCA [44], and spectral clustering as well [22, p. 544] (Section 5.5). The vast literature on *non-robust* subspace tracking algorithms includes [30, 55], and [2]; see also [23] for a first-order algorithm that is robust to outliers and incomplete data. Relative to [23], the online robust (OR)-PCA algorithm of [32, 33] (described in Section 5.4.1) is a second-order method, which minimizes an outlier-aware exponentially-weighted LS estimator of the low-rank factor analysis model. Since the outlier and subspace estimation tasks decouple nicely in OR-PCA, one can readily devise a first-order counterpart when minimal computational loads are at a premium. In terms of performance, online algorithms are known to be markedly faster than their batch alternatives [2, 23], e.g., in the timely context of low-rank matrix completion [40, 41]. While the focus here is not on incomplete data records, extensions to account for missing data are immediate and have been reported in [33].

Numerical tests with real data are presented throughout to corroborate the effectiveness of the proposed batch and online robust PCA schemes, when used to identify aberrant responses from a questionnaire designed to measure the Big-Five dimensions of personality traits [25], as well as unveil communities in a (social) network of college football teams [20], and intruders from video surveillance data [13]. For additional comprehensive tests and comparisons with competing alternatives (omitted here due to lack of space), the reader is referred to [34, Sec. VII-A]. Concluding remarks are given in Section 5.6.

Notation: Bold uppercase (lowercase) letters will denote matrices (column vectors). Operators $(\cdot)'$ and $\text{tr}(\cdot)$, will denote transposition and matrix trace, respectively. Vector $\text{diag}(\mathbf{M})$ collects the diagonal entries of $\mathbf{M}$, whereas the diagonal matrix $\text{diag}(\mathbf{v})$ has the entries of $\mathbf{v}$ on its diagonal. The ℓ_p-norm of $\mathbf{x} \in \mathbb{R}^n$ is $\|\mathbf{x}\|_p := \left(\sum_{i=1}^{n} |x_i|^p\right)^{1/p}$ for $p \geq 1$; and $\|\mathbf{M}\|_F := \sqrt{\text{tr}\,(\mathbf{M}\mathbf{M}')}$ is the matrix Frobenious norm. The $n \times n$ identity matrix will be represented by $\mathbf{I}_n$, while $\mathbf{0}_n$ will denote the $n \times 1$ vector of all zeros, and $\mathbf{0}_{n\times m} := \mathbf{0}_n\mathbf{0}'_m$. Similar notation will be adopted for vectors (matrices) of all ones. The i-th vector of the canonical basis in $\mathbb{R}^n$ will be denoted by $\mathbf{b}_{n,i}$, $i = 1, \ldots, n$.

5.2 Robustifying PCA

Consider the standard PCA formulation, in which a set of training data $\mathcal{T}_y := \{\mathbf{y}_n\}_{n=1}^N$ in the p-dimensional Euclidean *input* space is given, and the goal is to find the best q-rank $(q \leq p)$ linear approximation to the data in $\mathcal{T}_y$; see e.g., [26]. Unless otherwise stated, it is assumed throughout that the value of q is given. One approach to solving this problem is to adopt a low-rank bilinear (factor analysis) model

$$\mathbf{y}_n = \mathbf{m} + \mathbf{U}\mathbf{s}_n + \mathbf{e}_n, \quad n = 1, \ldots, N \tag{5.1}$$

where $\mathbf{m} \in \mathbb{R}^p$ is a location (mean) vector; matrix $\mathbf{U} \in \mathbb{R}^{p\times q}$ has orthonormal columns spanning the signal subspace; $\{\mathbf{s}_n\}_{n=1}^N$ are the so-termed *principal components*, and $\{\mathbf{e}_n\}_{n=1}^N$ are zero-mean i.i.d. random errors. The unknown variables in (5.1) can be collected in $\mathcal{V} := \{\mathbf{m}, \mathbf{U}, \{\mathbf{s}_n\}_{n=1}^N\}$, and they are estimated using the LS criterion as

$$\min_{\mathcal{V}} \sum_{n=1}^{N} \|\mathbf{y}_n - \mathbf{m} - \mathbf{U}\mathbf{s}_n\|_2^2, \quad \text{s. to} \quad \mathbf{U}'\mathbf{U} = \mathbf{I}_q. \tag{5.2}$$

PCA in (5.2) is a nonconvex optimization problem due to the bilinear terms $\mathbf{U}\mathbf{s}_n$, yet a global optimum $\hat{\mathcal{V}}$ can be shown to exist; see, e.g., [55]. The resulting estimates are $\hat{\mathbf{m}} = \sum_{n=1}^N \mathbf{y}_n/N$ and $\hat{\mathbf{s}}_n = \hat{\mathbf{U}}'(\mathbf{y}_n - \hat{\mathbf{m}})$, $n = 1, \ldots, N$; while $\hat{\mathbf{U}}$ is formed with columns equal to the q-dominant right singular vectors of the $N \times p$ data matrix $\mathbf{Y} := [\mathbf{y}_1, \ldots, \mathbf{y}_N]'$ [22, p. 535]. The principal components (entries of) $\mathbf{s}_n$ are the projections of the centered data points $\{\mathbf{y}_n - \hat{\mathbf{m}}\}_{n=1}^N$ onto the signal subspace. Equivalently, PCA can be formulated based on maximum variance, or, minimum reconstruction error criteria; see, e.g., [26].

5.2.1 Least-Trimmed Squares PCA

Given training data $\mathcal{T}_x := \{\mathbf{x}_n\}_{n=1}^N$ possibly contaminated with outliers, the goal here is to develop a robust estimator of $\mathcal{V}$ that requires minimal assumptions on the outlier model. Note that there is an explicit notational differentiation between: i) the data in $\mathcal{T}_y$ that adhere to the nominal model (5.1); and ii) the given data in $\mathcal{T}_x$ that may also contain outliers, i.e., those $\mathbf{x}_n$ not adhering to (5.1). Building on LTS regression [42], the desired robust estimate $\hat{\mathcal{V}}_{LTS} := \{\hat{\mathbf{m}}, \hat{\mathbf{U}}, \{\hat{\mathbf{s}}_n\}_{n=1}^N\}$ for a prescribed $\nu < N$ can be obtained via the following LTS PCA estimator [cf. (5.2)]

$$\hat{\mathcal{V}}_{LTS} := \arg\min_{\mathcal{V}} \sum_{n=1}^{\nu} r_{[n]}^2(\mathcal{V}), \quad \text{s. to} \quad \mathbf{U}'\mathbf{U} = \mathbf{I}_q \tag{5.3}$$

where $r_{[n]}^2(\mathcal{V})$ is the n-th order statistic among the squared residual norms $r_1^2(\mathcal{V}), \ldots, r_N^2(\mathcal{V})$, and $r_n(\mathcal{V}) := \|\mathbf{x}_n - \mathbf{m} - \mathbf{U}\mathbf{s}_n\|_2$. The so-termed *coverage* ν determines the breakdown point of the LTS PCA estimator [42], since the $N - \nu$ largest residuals are absent from the estimation criterion in (5.3). Beyond this universal outlier-rejection property, the LTS-based estimation offers an attractive alternative to robust linear regression due to its high breakdown point and desirable analytical properties, namely $\sqrt{N}$-consistency and asymptotic normality under mild assumptions [42].

Because (5.3) is a nonconvex optimization problem, a nontrivial issue pertains to the existence of the proposed LTS PCA estimator, i.e., whether or not (5.3) attains a minimum. Fortunately, existence of $\hat{\mathcal{V}}_{LTS}$ can be readily established as follows: i) for each subset of $\mathcal{T}$ with cardinality ν (there are $\binom{N}{\nu}$ such subsets), solve the corresponding PCA problem to obtain a unique candidate estimator per subset; and ii) pick $\hat{\mathcal{V}}_{LTS}$ as the one among all $\binom{N}{\nu}$ candidates with the minimum cost. Albeit conceptually simple, the aforementioned solution procedure is combinatorially complex, and thus intractable except for small sample sizes N. Algorithms to obtain approximate LTS solutions in large-scale linear regression problems are available; see e.g., [42].

REMARK 5.1 In most PCA formulations, data in $\mathcal{T}_y$ are typically assumed zero mean. This is without loss of generality, since nonzero-mean training data can always be rendered zero mean, by subtracting the sample mean $\sum_{n=1}^{N} \mathbf{y}_n / N$ from each $\mathbf{y}_n$. In modeling zero-mean data, the known vector $\mathbf{m}$ in (5.1) can obviously be neglected. When outliers are present, however, data in $\mathcal{T}_x$ are not necessarily zero mean, and it is unwise to center them using the non-robust sample mean estimator, which has a breakdown point equal to zero [42]. Towards robustifying PCA, a more sensible approach is to estimate $\mathbf{m}$ robustly, and jointly with $\mathbf{U}$ and the principal components $\{\mathbf{s}_n\}_{n=1}^{N}$. For this reason $\mathbf{m}$ is kept as a variable in $\mathcal{V}$ and estimated via (5.3).

5.2.2 Robust Statistics Meets Sparse Recovery

Instead of discarding large residuals, the alternative approach here explicitly accounts for outliers in the low-rank data model (5.1). This becomes possible through the vector variables $\{\mathbf{o}_n\}_{n=1}^{N}$, one per training datum $\mathbf{x}_n$, which take the value $\mathbf{o}_n \neq \mathbf{0}_p$ whenever datum n is an outlier, and $\mathbf{o}_n = \mathbf{0}_p$ otherwise. Thus, the outlier-aware factor analysis model is

$$\mathbf{x}_n = \mathbf{y}_n + \mathbf{o}_n = \mathbf{m} + \mathbf{U}\mathbf{s}_n + \mathbf{e}_n + \mathbf{o}_n, \qquad n = 1, \ldots, N \tag{5.4}$$

where $\mathbf{o}_n$ can be deterministic or random with unspecified distribution. In the *under-determined* linear system of equations (5.4), both $\mathcal{V}$ as well as the $N \times p$ matrix $\mathbf{O} := [\mathbf{o}_1, \ldots, \mathbf{o}_N]'$ are unknown. The percentage of outliers dictates the degree of *sparsity* (number of zero rows) in $\mathbf{O}$. Sparsity control will prove instrumental in efficiently estimating $\mathbf{O}$, rejecting outliers as a byproduct, and consequently arriving at a *robust* estimator of $\mathcal{V}$. To this end, a natural criterion for controlling outlier sparsity is to seek the estimator [cf. (5.2)]

$$\{\hat{\mathcal{V}}, \hat{\mathbf{O}}\} = \arg\min_{\mathcal{V}, \mathbf{O}} \|\mathbf{X} - \mathbf{1}_N \mathbf{m}' - \mathbf{S}\mathbf{U}' - \mathbf{O}\|_F^2 + \lambda_0 \|\mathbf{O}\|_0, \quad \text{s. to } \mathbf{U}'\mathbf{U} = \mathbf{I}_q \tag{5.5}$$

where $\mathbf{X} := [\mathbf{x}_1, \ldots, \mathbf{x}_N]' \in \mathbb{R}^{N \times p}$, $\mathbf{S} := [\mathbf{s}_1, \ldots, \mathbf{s}_N]' \in \mathbb{R}^{N \times q}$, and $\|\mathbf{O}\|_0$ denotes the nonconvex ℓ_0-norm that is equal to the number of nonzero rows of $\mathbf{O}$. Vector (group) sparsity in the rows $\hat{\mathbf{o}}_n$ of $\hat{\mathbf{O}}$ can be directly controlled by tuning the parameter $\lambda_0 \geq 0$.

As with compressive sampling and sparse modeling schemes that rely on the ℓ_0-norm [51], the robust PCA problem (5.5) is NP-hard [35]. In addition, the sparsity-controlling estimator (5.5) is intimately related to LTS PCA, as asserted in the following proposition. (A detailed proof is also included since it is instructive towards revealing the link between both estimators.)

PROPOSITION 5.1 If $\{\hat{\mathcal{V}}, \hat{\mathbf{O}}\}$ minimizes (5.5) with λ_0 chosen such that $\|\hat{\mathbf{O}}\|_0 = N - \nu$, then $\hat{\mathcal{V}}_{LTS} = \hat{\mathcal{V}}$.

PROOF 5.1 Given λ_0 such that $\|\hat{\mathbf{O}}\|_0 = N - \nu$, the goal is to characterize $\hat{\mathcal{V}}$ as well as the positions and values of the nonzero rows of $\hat{\mathbf{O}}$. Because $\|\hat{\mathbf{O}}\|_0 = N - \nu$, the last term in the cost of (5.5) is constant, hence inconsequential to the minimization. Upon defining $\hat{\mathbf{r}}_n := \mathbf{x}_n - \hat{\mathbf{m}} - \hat{\mathbf{U}}\hat{\mathbf{s}}_n$, the rows of $\hat{\mathbf{O}}$ satisfy

$$\hat{\mathbf{o}}_n = \begin{cases} \mathbf{0}_p, & \|\hat{\mathbf{r}}_n\|_2 \leq \sqrt{\lambda_0} \\ \hat{\mathbf{r}}_n, & \|\hat{\mathbf{r}}_n\|_2 > \sqrt{\lambda_0} \end{cases}, \quad n = 1, \ldots, N. \tag{5.6}$$

This follows by noting first that (5.5) is separable across the rows of $\mathbf{O}$. For each $n = 1, \ldots, N$, if $\hat{\mathbf{o}}_n = \mathbf{0}_p$ then the optimal cost becomes $\|\hat{\mathbf{r}}_n - \hat{\mathbf{o}}_n\|_2^2 + \lambda_0\|\hat{\mathbf{o}}_n\|_0 = \|\hat{\mathbf{r}}_n\|_2^2$. If on the other hand $\hat{\mathbf{o}}_n \neq \mathbf{0}_p$, the optimality condition for $\mathbf{o}_n$ yields $\hat{\mathbf{o}}_n = \hat{\mathbf{r}}_n$, and thus the cost reduces to λ_0. In conclusion, for the chosen value of λ_0 it holds that $N - \nu$ squared residuals effectively do not contribute to the cost in (5.5).

To determine $\hat{\mathcal{V}}$ and the row support of $\hat{\mathbf{O}}$, one alternative is to exhaustively test all $\binom{N}{N-\nu} = \binom{N}{\nu}$ admissible row-support combinations. For each one of these combinations (indexed by j), let $\mathcal{S}_j \subset \{1, \ldots, N\}$ be the index set describing the row support of $\hat{\mathbf{O}}^{(j)}$, i.e., $\hat{\mathbf{o}}_n^{(j)} \neq \mathbf{0}_p$ if and only if $n \in \mathcal{S}_j$; and $|\mathcal{S}_j| = N - \nu$. By virtue of (5.6), the corresponding candidate $\hat{\mathcal{V}}^{(j)}$ solves $\min_{\mathcal{V}} \sum_{n \in \mathcal{S}_j} r_n^2(\mathcal{V})$ subject to $\mathbf{U}'\mathbf{U} = \mathbf{I}_q$, while $\hat{\mathcal{V}}$ is the one among all $\{\hat{\mathcal{V}}^{(j)}\}$ that yields the least cost. Recognizing the aforementioned solution procedure as the one for LTS PCA outlined in Section 5.2.1, it follows that $\hat{\mathcal{V}}_{LTS} = \hat{\mathcal{V}}$.

The importance of Proposition 5.1 is threefold. First, it formally justifies model (5.4) and its estimator (5.5) for robust PCA, in light of the well-documented merits of LTS [42]. Second, it establishes a connection between the seemingly unrelated fields of robust statistics and sparsity-aware estimation. Third, problem (5.5) lends itself naturally to efficient (approximate) solvers based on convex relaxation, the subject dealt with next.

5.2.3 Sparsity-Controlling Outlier Rejection

Recall that the row-wise ℓ_2-norm sum $\|\mathbf{B}\|_{2,r} := \sum_{n=1}^{N} \|\mathbf{b}_n\|_2$ of matrix $\mathbf{B} := [\mathbf{b}_1, \ldots, \mathbf{b}_N]' \in \mathbb{R}^{N \times p}$ is the closest convex approximation of $\|\mathbf{B}\|_0$ [51]. This property motivates relaxing problem (5.5) to

$$\min_{\mathcal{V},\mathbf{O}} \|\mathbf{X} - \mathbf{1}_N\mathbf{m}' - \mathbf{S}\mathbf{U}' - \mathbf{O}\|_F^2 + \lambda_2\|\mathbf{O}\|_{2,r}, \quad \text{s. to } \mathbf{U}'\mathbf{U} = \mathbf{I}_q. \tag{5.7}$$

The nondifferentiable ℓ_2-norm regularization term encourages row-wise (vector) sparsity on the estimator of $\mathbf{O}$, a property that has been exploited in diverse problems in engineering, statistics, and machine learning [22]. A noteworthy representative is the group Lasso [57], a popular tool for joint estimation and selection of grouped variables in linear regression.

REMARK 5.2 In computer vision applications, for instance where robust PCA schemes are particularly attractive, one may not wish to discard the entire (vectorized) images $\mathbf{x}_n$, but only specific pixels deemed as outliers [13]. This can be accomplished by replacing $\|\mathbf{O}\|_{2,r}$ in (5.7) with $\|\mathbf{O}\|_1 := \sum_{n=1}^{N} \|\mathbf{o}_n\|_1$, a Lasso-type regularization that encourages entry-wise sparsity in $\hat{\mathbf{O}}$.

After the relaxation it is pertinent to ponder on whether problem (5.7) still has the potential of providing robust estimates $\hat{\mathcal{V}}$ in the presence of outliers. The answer is positive, since (5.7) is equivalent to an M-type PCA estimator

$$\min_{\mathcal{V}} \sum_{n=1}^{N} \rho_v(\mathbf{x}_n - \mathbf{m} - \mathbf{U}\mathbf{s}_n), \quad \text{s. to } \mathbf{U}'\mathbf{U} = \mathbf{I}_q \tag{5.8}$$

where $\rho_v : \mathbb{R}^p \to \mathbb{R}$ is a vector extension to Huber's convex loss function [24]; namely

$$\rho_v(\mathbf{r}) := \begin{cases} \|\mathbf{r}\|_2^2, & \|\mathbf{r}\|_2 \leq \lambda_2/2 \\ \lambda_2\|\mathbf{r}\|_2 - \lambda_2^2/4, & \|\mathbf{r}\|_2 > \lambda_2/2 \end{cases}. \tag{5.9}$$

For a detailed proof of the equivalence, see [34].

M-type estimators (including Huber's) adopt a fortiori an ϵ-contaminated probability distribution for the outliers, and rely on minimizing the *asymptotic* variance of the resultant estimator for the least favorable distribution of the ϵ-contaminated class (asymptotic min-max approach) [24]. The assumed degree of contamination specifies the tuning parameter λ_2 in (5.9) (and thus the threshold for deciding the outliers in M-estimators). In contrast, the present approach is universal in the sense that it is not confined to any assumed class of outlier distributions, and can afford a data-driven selection of the tuning parameter. In a nutshell, optimal M-estimators can be viewed as a special case of the present formulation only for a specific choice of λ_2, which is not obtained via a data-driven approach, but from distributional assumptions instead.

All in all, the sparsity-controlling role of the tuning parameter $\lambda_2 \geq 0$ in (5.7) is central, since model (5.4) and the equivalence of (5.7) with (5.8) suggest that λ_2 is a robustness-controlling constant. Data-driven approaches to select λ_2 are described in detail under Section 5.3.1. Before delving into algorithmic issues to solve (5.7), a remark is in order.

REMARK 5.3 The recent upsurge of research toward compressive sampling and parsimonious signal representations hinges on signals being sparse, either naturally, or, after projecting them on a proper basis. Here instead, a neat link is established between sparsity and a fundamental aspect of statistical inference, namely that of robustness against outliers. It is argued that key to robust methods is the control of sparsity in *model residuals*, i.e., those entries in matrix $\mathbf{O}$, even when the signals in $\mathcal{V}$ are not (necessarily) sparse.

5.3 Algorithms and Real Data Tests

To optimize (5.7) iteratively for a given value of λ_2, an alternating minimization (AM) algorithm is adopted which cyclically updates $\mathbf{m}(k) \to \mathbf{S}(k) \to \mathbf{U}(k) \to \mathbf{O}(k)$ per iteration $k = 1, 2, \ldots$. AM algorithms are also known as block-coordinate-descent methods in the optimization parlance; see, e.g., [4, 52]. To update each of the variable groups, (5.7) is minimized while fixing the rest of the variables to their most up-to-date values. While the overall problem (5.7) is not jointly convex with respect to (w.r.t.) $\{\mathbf{S}, \mathbf{U}, \mathbf{O}, \mathbf{m}\}$, fixing all but one of the variable groups yields subproblems that are efficiently solved, and attain a unique solution.

Towards deriving the updates at iteration k and arriving at the desired algorithm, note first that the mean update is $\mathbf{m}(k) = (\mathbf{X}-\mathbf{O}(k))'\mathbf{1}_N/N$. Next, form the centered and outlier-compensated data matrix $\mathbf{X}_o(k) := \mathbf{X} - \mathbf{1}_N\mathbf{m}(k)' - \mathbf{O}(k-1)$. The principal components are readily given by

$$\mathbf{S}(k) = \arg\min_{\mathbf{S}} \|\mathbf{X}_o(k) - \mathbf{S}\mathbf{U}(k-1)'\|_F^2 = \mathbf{X}_o(k)\mathbf{U}(k-1).$$

Continuing the cycle, $\mathbf{U}(k)$ solves

$$\min_{\mathbf{U}} \|\mathbf{X}_o(k) - \mathbf{S}(k)\mathbf{U}'\|_F^2, \quad \text{s. to } \mathbf{U}'\mathbf{U} = \mathbf{I}_q$$

a constrained LS problem also known as reduced-rank *Procrustes rotation* [59]. The minimizer is given in analytical form in terms of the left and right singular vectors of $\mathbf{X}_o'(k)\mathbf{S}(k)$ [59, Thm. 4]. In detail, one computes the SVD of $\mathbf{X}_o'(k)\mathbf{S}(k) = \mathbf{L}(k)\mathbf{D}(k)\mathbf{R}'(k)$ and updates $\mathbf{U}(k) = \mathbf{L}(k)\mathbf{R}'(k)$. Next, the minimization of (5.7) w.r.t. $\mathbf{O}$ is an orthonormal group Lasso problem. As such, it decouples across rows $\mathbf{o}_n$ giving rise to N ℓ_2-norm regularized subproblems, namely

$$\mathbf{o}_n(k) = \arg\min_{\mathbf{o}} \|\mathbf{r}_n(k) - \mathbf{o}\|_2^2 + \lambda_2\|\mathbf{o}\|_2, \qquad n = 1, \ldots, N$$

where $\mathbf{r}_n(k) := \mathbf{x}_n - \mathbf{m}(k) - \mathbf{U}(k)\mathbf{s}_n(k)$. The respective solutions are given by (see, e.g., [36])

$$\mathbf{o}_n(k) = \frac{\mathbf{r}_n(k)(\|\mathbf{r}_n(k)\|_2 - \lambda_2/2)_+}{\|\mathbf{r}_n(k)\|_2}, \qquad n = 1, \ldots, N \tag{5.10}$$

where $(\cdot)_+ := \max(\cdot, 0)$. For notational convenience, these N parallel vector soft-thresholded updates are denoted as $\mathbf{O}(k) = \mathcal{S}\left[\mathbf{X} - \mathbf{1}_N\mathbf{m}'(k-1) - \mathbf{S}(k)\mathbf{U}'(k), (\lambda_2/2)\mathbf{I}_N\right]$ under Algorithm 5.1, where the thresholding operator $\mathcal{S}$ sets the entire outlier vector $\mathbf{o}_n(k)$ to zero whenever $\|\mathbf{r}_n(k)\|_2$ does not exceed $\lambda_2/2$, in par with the group-sparsifying property of group Lasso. Interestingly, this is the same rule used to decide if datum $\mathbf{x}_n$ is deemed an outlier, in the equivalent formulation (5.8) which involves Huber's loss function. Whenever an ℓ_1-norm regularizer is adopted as discussed in Remark 5.2, the only difference is that updates (5.10) boil down to soft-thresholding the scalar entries of $\mathbf{r}_n(k)$.

The entire AM solver is tabulated under Algorithm 5.1, also indicating the recommended initialization. Algorithm 5.1 is conceptually interesting, since it explicitly reveals the intertwining between the outlier identification process and the PCA low-rank model fitting based on the outlier compensated data $\mathbf{X}_o(k)$. The AM solver is also computationally efficient. Computing the $N \times q$ matrix $\mathbf{S}(k) = \mathbf{X}_o(k)\mathbf{U}(k-1)$ requires Npq operations per iteration, and equally costly is to obtain $\mathbf{X}_o'(k)\mathbf{S}(k) \in \mathbb{R}^{p\times q}$. The cost of computing the SVD of $\mathbf{X}_o'(k)\mathbf{S}(k)$ is of order $\mathcal{O}(pq^2)$, while the rest of the operations including the row-wise soft-thresholdings to yield $\mathbf{O}(k)$ are linear in both N and p. In summary, the total cost of Algorithm 5.1 is roughly $k_{\max}\mathcal{O}(Np + pq^2)$, where $k_{\max}$ is the number of iterations required for convergence (typically $k_{\max} = 5$ to 10 iterations suffice). Because $q \leq p$ is typically small, Algorithm 5.1 is attractive computationally both under the classic setting where $N > p$, and p is not large, as well as in high-dimensional data settings where $p \gg N$, a situation typically arising, e.g., in microarray data analysis.

Because each of the optimization problems in the per-iteration cycles has a unique minimizer, and the nondifferentiable regularization only affects one of the variable groups ($\mathbf{O}$), the general results of [52] apply to establish convergence of Algorithm 5.1. Specifically, as $k \to \infty$ the iterates generated by Algorithm 5.1 converge to a stationary point of (5.7).

5.3.1 Selection of λ_2: Robustification Paths

Selecting λ_2 controls the number of outliers rejected. But this choice is challenging because existing techniques such as cross-validation are not effective when outliers are present [42]. To this end, systematic data-driven approaches were devised in [19], which, e.g., require a rough estimate of the percentage of outliers, or, robust estimates $\hat{\sigma}_e^2$ of the nominal noise variance that can be obtained using median absolute deviation (MAD) schemes [24]. These approaches can be adapted to the robust PCA setting considered here, and leverage the

Algorithm 5.1 - Batch robust PCA solver

Set $\mathbf{U}(0) = \mathbf{I}_p(:, 1:q)$ and $\mathbf{O}(0) = \mathbf{0}_{N \times p}$.
for $k = 1, 2, \ldots$ **do**
  Update $\mathbf{m}(k) = (\mathbf{X} - \mathbf{O}(k-1))'\mathbf{1}_N/N$.
  Form $\mathbf{X}_o(k) = \mathbf{X} - \mathbf{1}_N\mathbf{m}'(k) - \mathbf{O}(k-1)$.
  Update $\mathbf{S}(k) = \mathbf{X}_o(k)\mathbf{U}(k-1)$.
  Obtain $\mathbf{L}(k)\mathbf{D}(k)\mathbf{R}(k)' = \text{svd}[\mathbf{X}_o'(k)\mathbf{S}(k)]$ and update $\mathbf{U}(k) = \mathbf{L}(k)\mathbf{R}'(k)$.
  Update $\mathbf{O}(k) = \mathcal{S}\left[\mathbf{X} - \mathbf{1}_N\mathbf{m}'(k) - \mathbf{S}(k)\mathbf{U}'(k), (\lambda_2/2)\mathbf{I}_N\right]$.
end for

robustification paths of (group-)Lasso solutions [cf. (5.7)], which are defined as the solution paths corresponding to $\|\hat{\mathbf{o}}_n\|_2$, $n = 1, \ldots, N$, for all values of λ_2. As λ_2 decreases, more vectors $\hat{\mathbf{o}}_n$ enter the model, signifying that more of the training data are deemed to contain outliers.

Consider then a grid of G_λ values of λ_2 in the interval $[\lambda_{\min}, \lambda_{\max}]$, evenly spaced on a logarithmic scale. Typically, $\lambda_{\max}$ is chosen as the minimum λ_2 value such that $\hat{\mathbf{O}} \neq \mathbf{0}_{N\times p}$, while $\lambda_{\min} = \epsilon\lambda_{\max}$ with $\epsilon = 10^{-4}$, say. Because Algorithm 5.1 converges quite fast, (5.7) can be efficiently solved over the grid of G_λ values for λ_2. On the order of hundreds of grid points can be easily handled by initializing each instance of Algorithm 1 (per value of λ_2) using *warm starts* [22]. This means that multiple instances of (5.7) are solved for a sequence of decreasing λ_2 values, and the initialization of Algorithm 5.1 per grid point corresponds to the solution obtained for the immediately preceding value of λ_2 in the grid. For sufficiently close values of λ_2, one expects that the respective solutions will also be close (the row support of $\hat{\mathbf{O}}$ will most likely not change), and hence Algorithm 1 will converge after few iterations.

Based on the G_λ samples of the robustification paths and the prior knowledge available on the outlier model (5.4), a couple of alternatives described next are possible for selecting the 'best' value of λ_2 in the grid. A comprehensive survey of options can be found in [19].

Number of outliers is known: By direct inspection of the robustification paths one can determine the range of values for λ_2, such that the number of nonzero rows in $\hat{\mathbf{O}}$ equals the known number of outliers sought. Zooming in to the interval of interest, and after discarding the identified outliers, K-fold cross-validation methods can be applied to determine the 'best' λ_2^*.

Nominal noise covariance matrix is known: Given $\mathbf{m}\Sigma_e := E[\mathbf{e}_n\mathbf{e}_n']$, one can proceed as follows. Consider the estimates $\hat{\mathcal{V}}_g$ obtained using (5.7) after sampling the robustification path for each point $\{\lambda_{2,g}\}_{g=1}^G$. Next, pre-whiten those residuals corresponding to training data not deemed as containing outliers; i.e., form $\hat{\mathcal{R}}_g := \{\bar{\mathbf{r}}_{n,g} = \mathbf{m}\Sigma_e^{-1/2}(\mathbf{x}_n - \hat{\mathbf{m}}_g - \hat{\mathbf{U}}_g\hat{\mathbf{s}}_{n,g}) : n \text{ s. to } \hat{\mathbf{o}}_n = \mathbf{0}\}$, and find the sample covariance matrices $\{\hat{\mathbf{m}\Sigma}_{\bar{r},g}\}_{g=1}^G$. The winner $\lambda_2^* := \lambda_{2,g^*}$ corresponds to the grid point minimizing an absolute variance deviation criterion, namely $g^* := \arg\min_g |\text{tr}[\hat{\mathbf{m}\Sigma}_{\bar{r},g}] - p|$.

5.3.2 Bias Reduction through Nonconvex Regularization

Instead of substituting $\|\mathbf{O}\|_0$ in (5.5) by its closest convex approximation, namely $\|\mathbf{O}\|_{2,r}$, letting the surrogate function be nonconvex can yield tighter approximations, and improve the statistical properties of the estimator. In rank minimization problems, for instance, the logarithm of the determinant of the unknown matrix has been proposed as a smooth surrogate to the rank [16]; an alternative to the convex nuclear norm in, e.g., [40]. Nonconvex penalties such as the smoothly clipped absolute deviation (SCAD) have also been adopted to reduce bias [15], present in uniformly weighted ℓ_1-norm regularized estimators such as (5.7) [22, p. 92]. In the context of sparse signal reconstruction, the ℓ_0-norm of a vector was surrogated in [9] by the logarithm of the geometric mean of its elements; see also [39].

Building on this last idea, consider approximating (5.5) by the formulation

$$\min_{\mathcal{V},\mathbf{O}} \|\mathbf{X} - \mathbf{1}_N\mathbf{m}' - \mathbf{S}\mathbf{U}' - \mathbf{O}\|_F^2 + \lambda_0 \sum_{n=1}^{N} \log(\|\mathbf{o}_n\|_2 + \delta), \quad \text{s. to } \mathbf{U}'\mathbf{U} = \mathbf{I}_q \tag{5.11}$$

where the small positive constant δ is introduced to avoid numerical instability. Since the surrogate term in (5.11) is concave, the overall minimization problem is nonconvex and admittedly more complex to solve than (5.7). Local methods based on iterative linearization of $\log(\|\mathbf{o}_n\|_2+\delta)$ around the current iterate $\mathbf{o}_n(k)$, are adopted to minimize (5.11). Skipping details that can be found in [27], application of the majorization-minimization technique to (5.11) leads to an iteratively-reweighted version of (5.7), whereby $\lambda_2 \leftarrow \lambda_0 w_n(k)$ is used for updating $\mathbf{o}_n(k)$ in Algorithm 5.1. Specifically, per $k = 1, 2, \ldots$ one updates

$$\mathbf{O}(k) = \mathcal{S}\left[\mathbf{X} - \mathbf{1}_N\mathbf{m}'(k-1) - \mathbf{S}(k)\mathbf{U}'(k), (\lambda_0/2)\text{diag}(w_1(k), \ldots, w_N(k))\right]$$

where the weights are given by $w_n(k) = \left(\|\mathbf{o}_n(k-1)\|_2 + \delta\right)^{-1}$, $n = 1, \ldots, N$. Note that the thresholds vary both across rows (indexed by n), and across iterations. If the value of $\|\mathbf{o}_n(k-1)\|_2$ is small, then in the next iteration the regularization term $\lambda_0 w_n(k)\|\mathbf{o}_n\|_2$ has a large weight, thus promoting shrinkage of that entire row vector to zero. If $\|\mathbf{o}_n(k-1)\|_2$ is large, the cost in the next iteration downweighs the regularization, and places more importance on the LS component of the fit.

All in all, the idea is to start from the solution of (5.7) for the 'best' λ_2, which is obtained using Algorithm 5.1. This initial estimate is refined after runnning a few iterations of the iteratively-reweighted counterpart to Algorithm 5.1. Extensive numerical tests suggest that even a couple of iterations of this second-stage refinement suffice to yield improved estimates $\hat{\mathcal{V}}$, in comparison to those obtained from (5.7); see also the detailed numerical tests in [34]. The improvements can be leveraged to bias reduction – and its positive effect with regards to outlier support estimation – also achieved by similar *weighted* norm regularizers proposed for linear regression [22, p. 92].

5.3.3 Video Surveillance

To validate the proposed approach to robust PCA, Algorithm 5.1 was tested to perform background modeling from a sequence of video frames, an approach that has found widespread applicability for intrusion detection in video surveillance systems. The experiments were carried out using the dataset studied in [13], which consists of $N = 520$ images ($p = 120 \times 160$) acquired from a static camera during two days. The illumination changes considerably over the two-day span, while approximately 40% of the training images contain people in various locations. For $q = 10$, both standard PCA and the robust PCA (Algorithm 5.1) were applied to build a low-rank background model of the scenery captured by the camera. For robust PCA, ℓ_1-norm regularization on $\mathbf{O}$ was adopted to identify outliers at a pixel level. The outlier sparsity-controlling parameter was chosen as $\lambda_2 = 9.69 \times 10^{-4}$, whereas a single iteration of the reweighted scheme in Section 5.3.2 was run to reduce the bias in $\hat{\mathbf{O}}$.

Results are shown in Figure 5.1, for three representative images. The first column comprises the original frames from the training set, while the second column shows the corresponding PCA image reconstructions. The presence of undesirable 'ghostly' artifacts is

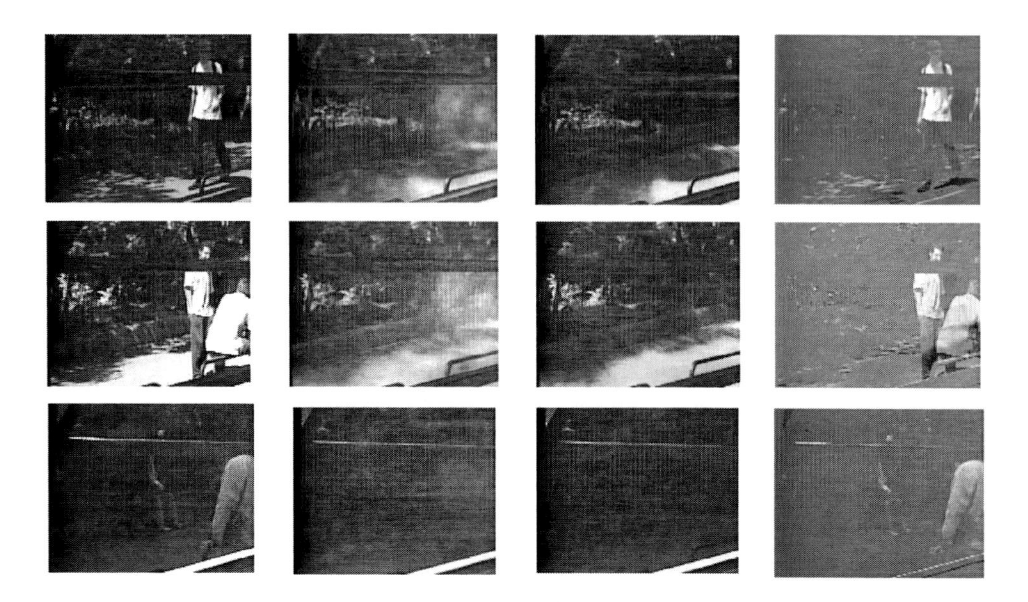

FIGURE 5.1 Background modeling for video surveillance. First column: original frames. Second column: PCA reconstructions, where the presence of undesirable 'ghostly' artifacts is apparent. Third column: robust PCA reconstructions, which recover the illumination changes while successfully subtracting the people. Fourth column: outliers in $\hat{\mathbf{o}}$, which mostly capture the people and abrupt changes in illumination.

apparent, since PCA is unable to completely separate the people from the background. The third column illustrates the robust PCA reconstructions, which recover the illumination changes while successfully subtracting the people. The fourth column shows the reshaped outlier vectors $\hat{\mathbf{o}}_n$, which mostly capture the people and abrupt changes in illumination. See also [34] for additional comparisons with competing methods, including, e.g., the algorithm in [13].

5.3.4 Robust Measurement of the Big Five Personality Factors

The "Big Five" are five factors ($q = 5$) of personality traits, namely extraversion, agreeableness, conscientiousness, neuroticism, and openness; see, e.g., [25]. The Big Five Inventory (BFI) on the other hand, is a brief questionnaire (44 items in total) tailored to measure the Big Five dimensions. Subjects taking the questionnaire are asked to rate in a scale from 1 (disagree strongly) to 5 (agree strongly), items of the form "I see myself as someone who is talkative." Each item consists of a short phrase correlating (positively or negatively) with one factor; see, e.g., [25, pp. 157-58] for a copy of the BFI and scoring instructions.

Robust PCA is used to identify aberrant responses from real BFI data comprising the Eugene-Springfield community sample [21]. The rows of $\mathbf{X}$ contain the $p = 44$ item responses for each one of the $N = 437$ subjects under study. For $q = 5$ and $\lambda_2 = 5.6107$ corresponding to $\|\hat{\mathbf{O}}\|_0 = 100$, Figure 5.2 depicts the norm of the 40 largest outliers. There is an unmistakable break in the scree plot and the 8 largest values are declared as outliers by robust PCA. As a means of validating these results, the following procedure is adopted. Based on the BFI scoring key [25], a list of all pairs of items hypothesized to yield positively correlated responses is formed. For each n, one counts the 'inconsistencies' defined as the number of times that subject n's ratings for these pairs differ in more than four, in absolute

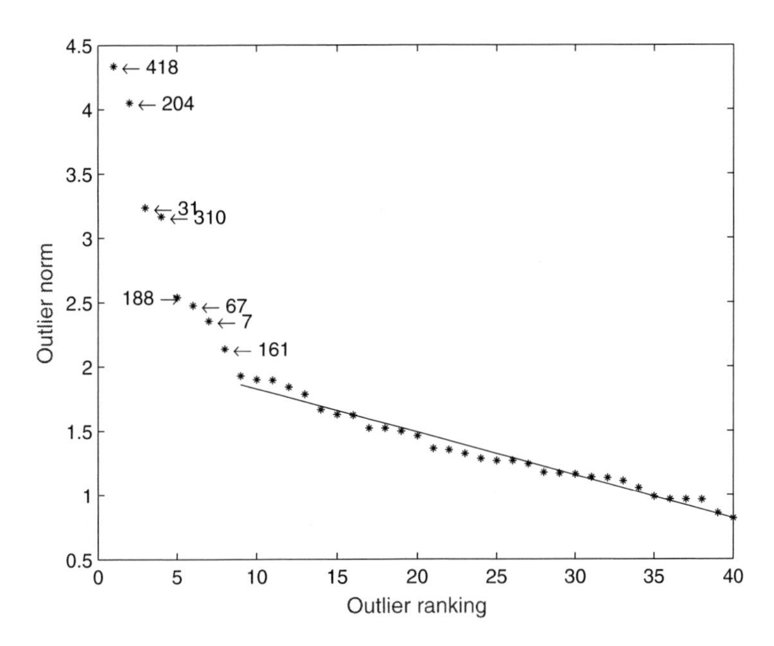

FIGURE 5.2 Pseudo scree plot of outlier size ($\|\hat{\mathbf{o}}_n\|_2$); the fourty largest outliers are shown. Robust PCA declares the largest eight as aberrant responses.

value. Interestingly, after rank-ordering all subjects in terms of this inconsistency score, one finds that $n = 418$ ranks highest with a count of 17, $n = 204$ ranks second (10), and overall the eight outliers found rank in the top twenty.

5.4 Connections with Nuclear-Norm Minimization

Recall that $q \leq p$ is the dimensionality of the subspace where the outlier-free data (5.1) are assumed to live, or equivalently, $q = \text{rank}[\mathbf{Y}]$ in the absence of noise. So far, q was assumed known and fixed. This is reasonable in, e.g., compression/quantization, where a target distortion-rate tradeoff dictates the maximum q. In other cases, the physics of the problem may render q known. This is indeed the case in array processing for direction-of-arrival estimation, where q is the dimensionality of the so-termed *signal subspace*, and is given by the number of plane waves impinging on a uniform linear array; see, e.g., [55].

Other applications, however, call for signal processing tools that can determine the 'best' q, as well as robustly estimate the underlying low-dimensional subspace $\mathbf{U}$ from data $\mathbf{X}$. Noteworthy representatives for this last kind of problems include unveiling traffic volume anomalies in large-scale networks [31, 32], and automatic intrusion detection from video surveillance frames [8, 13], just to name a few. A related approach in this context is (stable) principal components pursuit (PCP) [53, 58], which solves

$$\min_{\mathbf{L},\mathbf{O}} \|\mathbf{X} - \mathbf{L} - \mathbf{O}\|_F^2 + \lambda_* \|\mathbf{L}\|_* + \lambda_2 \|\mathbf{O}\|_{2,r} \tag{5.12}$$

with the objective of reconstructing the low-rank matrix $\mathbf{L} \in \mathbb{R}^{N \times p}$, as well as the sparse matrix of outliers $\mathbf{O}$ in the presence of dense noise with known variance.[14] Note that $\|\mathbf{L}\|_*$ denotes the matrix nuclear norm, a convex surrogate to rank$[\mathbf{L}]$ defined as the sum of the singular values of $\mathbf{L}$. The same way that the ℓ_2-norm regularization promotes sparsity in the rows of $\hat{\mathbf{O}}$, the nuclear norm encourages a low-rank $\hat{\mathbf{L}}$ since it effects sparsity in the vector of singular values of $\mathbf{L}$. Upon solving the convex optimization problem (5.12), it is possible to obtain $\hat{\mathbf{L}} = \hat{\mathbf{S}}\hat{\mathbf{U}}'$ using the SVD. Interestingly, (5.12) does not fix (or require the knowledge of) rank$[\mathbf{L}]$ a fortiori, but controls it through the tuning parameter λ_*. Adopting a Bayesian framework, a similar problem was considered in [14].

Instead of assuming that q is known, suppose that only an upper bound $\bar{q}$ is given. Then, the class of feasible noise-free low-rank matrix components of $\mathbf{Y}$ in (5.1) admit a factorization $\mathbf{L} = \mathbf{SU}'$, where $\mathbf{S}$ and $\mathbf{U}$ are $N \times \bar{q}$ and $p \times \bar{q}$ matrices, respectively. Building on the ideas used in the context of finding minimum rank solutions of linear matrix equations [40], an alternative approach to robustifying PCA is to solve [cf. (5.7)]

$$\min_{\mathbf{U},\mathbf{S},\mathbf{O}} \|\mathbf{X} - \mathbf{SU}' - \mathbf{O}\|_F^2 + \frac{\lambda_*}{2}(\|\mathbf{U}\|_F^2 + \|\mathbf{S}\|_F^2) + \lambda_2\|\mathbf{O}\|_{2,r}. \tag{5.13}$$

Different from (5.12) and (5.7), a Frobenius-norm regularization on both $\mathbf{U}$ and $\mathbf{S}$ is adopted to control the dimensionality of the estimated subspace $\hat{\mathbf{U}}$. Relative to (5.7), $\mathbf{U}$ in (5.13) is not constrained to be orthonormal. It is certainly possible to include the mean vector $\mathbf{m}$ in the cost of (5.13), as well as an ℓ_1-norm regularization for entrywise outliers. The main motivation behind choosing the Frobenius-norm regularization comes from the equivalence of (5.12) with (5.13) provided rank$[\hat{\mathbf{L}}] \leq \bar{q}$, which follows by adapting the results in [40, Lemma 5.1] to the problem formulation considered here; see also the seminal work in [47, 48].

Even though problem (5.13) is nonconvex, the number of optimization variables is reduced from $2Np$ to $Np + (N+p)\bar{q}$, which becomes significant when $\bar{q}$ is small and both N and p are large. Also note that the dominant Np-term in the variable count of (5.13) is due to $\mathbf{O}$, which is sparse and can be efficiently handled. While the factorization $\mathbf{L} = \mathbf{SU}'$ could also have been introduced in (5.12) to reduce the number of unknowns, the cost in (5.13) is separable and much simpler to optimize using, e.g., an AM solver comprising the iterations tabulated in [34, Alg. 2]; see also the discussion on subspace trackers in the ensuing section.

Because (5.13) is a nonconvex optimization problem, most solvers one can think of will at most provide convergence guarantees to a stationary point that may not be globally optimum. Interestingly, the ensuing proposition adapted from [31, Prop. 1] and [6] offers a certificate for stationary points of (5.13), qualifying them as global optima of (5.12).

PROPOSITION 5.2 If $\{\bar{\mathbf{U}}, \bar{\mathbf{S}}, \bar{\mathbf{O}}\}$ is a stationary point of (5.13) and $\|\mathbf{X} - \bar{\mathbf{S}}\bar{\mathbf{U}}' - \bar{\mathbf{O}}\|_2 \leq \lambda_*/2$, then $\{\hat{\mathbf{L}} := \bar{\mathbf{S}}\bar{\mathbf{U}}', \hat{\mathbf{O}} := \bar{\mathbf{O}}\}$ is the optimal solution of (5.12).

The usefulness of the separable Frobenius-norm regularization in (5.13) is further illustrated next, in the context of robust subspace tracking.

5.4.1 Robust Subspace Tracking

E-commerce and Internet-based retailing sites, the World Wide Web, and video surveillance systems generate huge volumes of data, which far outweigh the ability of personal computers to analyze them in real time. Furthermore, observations are oftentimes acquired *sequentially*

[14]Actually, [58] considers entrywise outliers and adopts an ℓ_1-norm regularization on $\mathbf{O}$.

in time, which motivates updating previously obtained 'analytics' rather than re-computing new ones from scratch each time a new datum becomes available [46]. This calls for low-complexity real-time (adaptive) algorithms for robust subspace tracking; see, e.g., [33].

One possible adaptive counterpart to (5.13) is the exponentially-weighted LS (EWLS) estimator found by [32]

$$\min_{\{\mathcal{V},\mathbf{O}\}} \sum_{n=1}^{N} \beta^{N-n} \left[\|\mathbf{x}_n - \mathbf{m} - \mathbf{U}\mathbf{s}_n - \mathbf{o}_n\|_2^2 + \frac{\lambda_*}{2\sum_{u=1}^{N} \beta^{N-u}} \|\mathbf{U}\|_F^2 + \frac{\lambda_*}{2}\|\mathbf{s}_n\|_2^2 + \lambda_2 \|\mathbf{o}_n\|_2 \right] \tag{5.14}$$

where $\beta \in (0, 1]$ is a forgetting factor. In this context, n should be understood as a temporal variable, indexing the instants of data acquisition. Note that in forming the EWLS estimator (5.14) at time N, the entire history of data $\{\mathbf{x}_n\}_{n=1}^{N}$ is incorporated in the real-time estimation process. Whenever $\beta < 1$, past data are exponentially discarded, thus enabling operation in nonstationary environments. For the infinite memory case ($\beta = 1$), on the other hand, the formulation (5.14) coincides with the batch estimator (5.13). This is the reason for the time-varying weight normalizing $\|\mathbf{U}\|_F^2$.

A provably convergent subspace tracker is developed in [32], based on AM of (5.14). In a nutshell, each time a new datum is acquired, outlier estimates are formed via the Lasso [22, p. 68], and the low-rank subspace is refined using recursive LS. For situations where reducing computational complexity is critical, an online stochastic gradient algorithm based on Nesterov's acceleration technique is developed as well [32]. In a stationary setting, the asymptotic subspace estimates obtained offer the well-documented performance guarantees of the batch stable PCP estimator [cf. (5.12) and Proposition 5.2].

Subspace tracking has a long history in signal processing. An early noteworthy representative is the projection approximation subspace tracking (PAST) algorithm [55]; see also [56]. Recently, an algorithm (termed GROUSE) for tracking subspaces from incomplete observations was put forth in [2], based on incremental gradient descent iterations on the Grassmannian manifold of subspaces. Recent analysis has shown that GROUSE can converge locally at an expected linear rate [3], and that it is tightly related to the incremental SVD algorithm [1]. PETRELS is a second-order recursive least-squares (RLS)-type algorithm, that extends the seminal PAST iterations to handle missing data [11]. As noted in [12], the performance of GROUSE is limited by the existence of barriers in the search path on the Grassmanian, which may lead to GROUSE iterations being trapped at local minima; see also [11]. Lack of regularization in PETRELS can also lead to unstable (even divergent) behaviors, especially when the amount of missing data is large. Accordingly, the convergence results for PETRELS are confined to the full-data setting where the algorithm boils down to PAST [11]. When outliers are present, robust counterparts can be found in [23, 37, 38].

REMARK 5.4 Towards addressing the scalability issue outlined at the beginning of this section, the decomposability of the Frobenius-norm regularizer in (5.13) has also been recently exploited for parallel processing across multiple processors when solving large-scale matrix completion problems [41], or to unveil network anomalies [31]. Specifically, [31] puts forth a general framework for *decentralized* sparsity-regularized rank minimization adopting the alternating-direction method of multipliers [5].

5.4.2 Tracking Internet Traffic Flows

Accurate estimation of origin-to-destination (OD) flow traffic in the backbone of large-scale Internet Protocol (IP) networks is of paramount importance for proactive network security

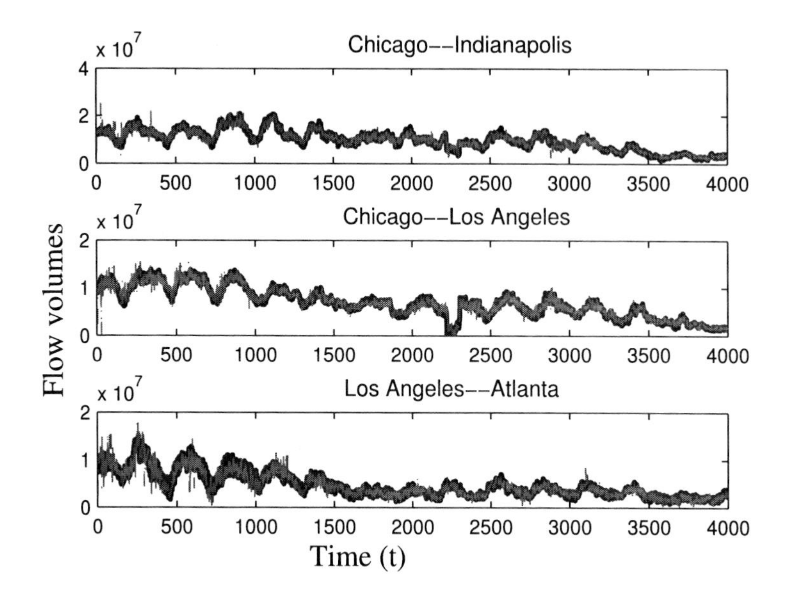

FIGURE 5.3 Online estimated (dashed gray) versus true (solid black) OD flow traffic for 75% missing data, and three representative flows measured from the operation of Internet2.

and management tasks [28]. Several experimental studies have demonstrated that OD flow traffic exhibits a low-intrinsic dimensionality, mainly due to common temporal patterns across OD flows, and periodic trends across time [29]. However, due to the massive number of OD pairs and the high volume of traffic, measuring the traffic of all possible OD flows is impossible for all practical purposes [28, 29]. Only the traffic level for a small fraction of OD flows can be measured via the NetFlow protocol [29].

Here, aggregate OD-flow traffic is collected from the operation of the Internet2 network (Internet backbone across the USA) during December 8–28, 2003, containing 121 OD pairs. The measured OD flows contain spikes (anomalies or outliers), yielding the data stream $\{\mathbf{x}_n\} \in \mathbb{R}^{121}$. The detailed description of the considered dataset can be found in [32]. When only 25% of the total OD flows are sampled by Netflow, Figure 5.3 depicts how the OR-PCA algorithm in [33] accurately tracks three representative OD flows.

5.5 Robustifying Kernel PCA

Kernel (K)PCA is a generalization to (linear) PCA, seeking principal components in a *feature space* nonlinearly related to the *input space* where the data in $\mathcal{T}_x$ live [44]. KPCA has been shown effective in performing nonlinear feature extraction for pattern recognition [44]. In addition, connections between KPCA and spectral clustering [22, p. 548] motivate the KPCA method developed in this section well, to robustly identify cohesive subgroups (communities) from social network data.

Consider a nonlinear function $\mathbf{m}\phi : \mathbb{R}^p \to \mathcal{H}$, that maps elements from the input space $\mathbb{R}^p$ to a feature space $\mathcal{H}$ of arbitrarily large – possibly infinite – dimensionality. Given transformed data $\mathcal{T}_{\mathcal{H}} := \{\mathbf{m}\phi(\mathbf{x}_n)\}_{n=1}^N$, the proposed approach to robust KPCA fits the model

$$\mathbf{m}\phi(\mathbf{x}_n) = \mathbf{m} + \mathbf{U}\mathbf{s}_n + \mathbf{e}_n + \mathbf{o}_n, \quad n = 1, \ldots, N \tag{5.15}$$

Algorithm 5.2 : Robust KPCA solver

Initialize $\mathbf{m}\Omega(0) = \mathbf{0}_{N\times N}$, $\mathbf{S}(0)$ randomly, and form $\mathbf{K} = \mathbf{m}\Phi'\mathbf{m}\Phi$.
for $k = 1, 2, \ldots$ **do**
  Update $\mathbf{m}\mu(k) = [\mathbf{I}_n - \mathbf{m}\Omega(k-1)]\mathbf{1}_N/N$.
  Form $\mathbf{m}\Phi_o(k) = \mathbf{I}_N - \mathbf{m}\mu(k)\mathbf{1}'_N - \mathbf{m}\Omega(k-1)$.
  Update $\mathbf{m}\Upsilon(k) = \mathbf{m}\Phi_o(k)\mathbf{S}(k-1)[\mathbf{S}'(k-1)\mathbf{S}(k-1) + (\lambda_*/2)\mathbf{I}_{\bar{q}}]^{-1}$.
  Update $\mathbf{S}(k) = \mathbf{m}\Phi'_o(k)\mathbf{K}\mathbf{m}\Upsilon(k)[\mathbf{m}\Upsilon(k)'\mathbf{K}\mathbf{m}\Upsilon(k) + (\lambda_*/2)\mathbf{I}_{\bar{q}}]^{-1}$.
  Form $\mathbf{m}\rho_n(k) = \mathbf{b}_{N,n} - \mathbf{m}\mu(k) - \mathbf{m}\Upsilon(k)\mathbf{s}_n(k)$, $n = 1, \ldots, N$.
  Form $\mathbf{m}\Lambda(k) = \text{diag}\left(\frac{(\mathbf{m}\rho'_1(k)\mathbf{K}\mathbf{m}\rho_1(k) - \frac{\lambda_2}{2})_+}{\mathbf{m}\rho'_1(k)\mathbf{K}\mathbf{m}\rho_1(k)}, \ldots, \frac{(\mathbf{m}\rho'_N(k)\mathbf{K}\mathbf{m}\rho_N(k) - \frac{\lambda_2}{2})_+}{\mathbf{m}\rho'_N(k)\mathbf{K}\mathbf{m}\rho_N(k)}\right)$.
  Update $\mathbf{m}\Omega(k) = [\mathbf{I}_N - \mathbf{m}\mu(k)\mathbf{1}'_N - \mathbf{m}\Upsilon(k)\mathbf{S}'(k)]\mathbf{m}\Lambda(k)$.
end for

by solving ($\mathbf{m}\Phi := [\mathbf{m}\phi(\mathbf{x}_1), \ldots, \mathbf{m}\phi(\mathbf{x}_N)]$)

$$\min_{\mathbf{U},\mathbf{S},\mathbf{O}} \|\mathbf{m}\Phi' - \mathbf{1}_N\mathbf{m}' - \mathbf{S}\mathbf{U}' - \mathbf{O}\|_F^2 + \frac{\lambda_*}{2}(\|\mathbf{U}\|_F^2 + \|\mathbf{S}\|_F^2) + \lambda_2\|\mathbf{O}\|_{2,r}. \tag{5.16}$$

It is certainly possible to adopt the criterion (5.7) as well, but (5.16) is chosen here for simplicity in exposition. Except for the principal components' matrix $\mathbf{S} \in \mathbb{R}^{N\times\bar{q}}$, both the data and the unknowns in (5.16) are now vectors/matrices of generally infinite dimension. In principle, this challenges the optimization task since it is impossible to store or perform updates of such quantities directly.

Interestingly, this hurdle can be overcome by endowing $\mathcal{H}$ with the structure of a reproducing kernel Hilbert space (RKHS), where inner products between any two members of $\mathcal{H}$ boil down to evaluations of the reproducing kernel $K_{\mathcal{H}} : \mathbb{R}^p \times \mathbb{R}^p \to \mathbb{R}$, i.e., $\langle \mathbf{m}\phi(\mathbf{x}_i), \mathbf{m}\phi(\mathbf{x}_j)\rangle_{\mathcal{H}} = K_{\mathcal{H}}(\mathbf{x}_i, \mathbf{x}_j)$. Specifically, it is possible to form the kernel matrix $\mathbf{K} := \mathbf{m}\Phi'\mathbf{m}\Phi \in \mathbb{R}^{N\times N}$, without directly working with the vectors in $\mathcal{H}$. This so-termed *kernel trick* is the crux of most kernel methods in machine learning [22], including kernel PCA [44]. The problem of selecting $K_{\mathcal{H}}$ (and $\mathbf{m}\phi$ indirectly) will not be considered here.

Building on these ideas, it is asserted next that Algorithm 5.1 can be *kernelized*, to solve (5.16) at affordable computational complexity and memory storage requirements that do not depend on the dimensionality of $\mathcal{H}$. A proof of Proposition 5.3 is available in [34].

PROPOSITION 5.3 For $k \geq 1$, the sequence of iterates generated by Algorithm 5.1 when applied to solve (5.16) can be written as $\mathbf{m}(k) = \mathbf{m}\Phi\mathbf{m}\mu(k)$, $\mathbf{U}(k) = \mathbf{m}\Phi\mathbf{m}\Upsilon(k)$, and $\mathbf{O}'(k) = \mathbf{m}\Phi\mathbf{m}\Omega(k)$. The quantities $\mathbf{m}\mu(k) \in \mathbb{R}^N$, $\mathbf{m}\Upsilon(k) \in \mathbb{R}^{N\times\bar{q}}$, and $\mathbf{m}\Omega(k) \in \mathbb{R}^{N\times N}$ are recursively updated as in Algorithm 5.2, without the need of operating with vectors in $\mathcal{H}$.

Proposition 5.3 asserts that if the iterates are initialized with outlier estimates in the range space of $\mathbf{m}\Phi$, then all subsequent iterates will admit a similar expansion in terms of feature vectors. This is weaker than claiming that each minimizer of (5.16) admits such an expansion – the latter would require checking whether the regularization term in (5.16) satisfies the conditions of the Representer Theorem [43].

In order to run the robust KPCA algorithm (tabulated as Algorithm 5.2), one does not have to store or process the quantities $\mathbf{m}(k)$, $\mathbf{U}(k)$, and $\mathbf{O}(k)$. As per Proposition 5.3, the iterations of a provably convergent AM solver can be equivalently carried out by cycling through *finite-dimensional* 'sufficient statistics' $\mathbf{m}\mu(k) \to \mathbf{m}\Upsilon(k) \to \mathbf{S}(k) \to \mathbf{m}\Omega(k)$. In other words, the iterations of the robust kernel PCA algorithm are devoid of algebraic operations among vectors in $\mathcal{H}$. Recall that the size of matrix $\mathbf{S}$ is independent of the

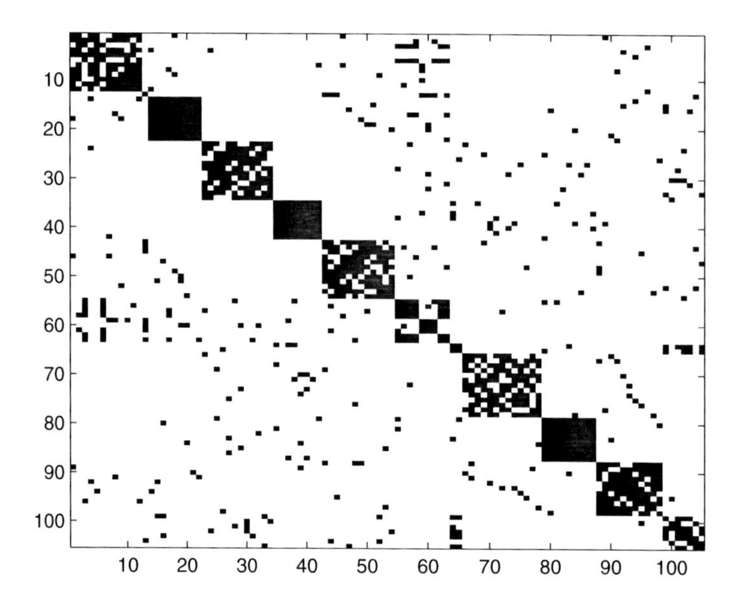

FIGURE 5.4 Entries of $\mathbf{K}$ after removing the outliers, where rows and columns are permuted to reveal the clustering structure found by robust KPCA. The eleven-conference (community) structure is apparent.

dimensionality of $\mathcal{H}$.

Because $\mathbf{O}'(k) = \mathbf{m}\Phi\mathbf{m}\Omega(k)$ and upon convergence of the algorithm, the outlier vector norms are computable in terms of $\mathbf{K}$, i.e., $[\|\mathbf{o}_1(\infty)\|_2^2, \ldots, \|\mathbf{o}_N(\infty)\|_2^2]' = \text{diag}[\mathbf{m}\Omega'(\infty)\mathbf{K}\mathbf{m}\Omega(\infty)]$. These are critical to determine the robustification paths needed to carry out the outlier sparsity control methods in Section 5.3.1. Moreover, the principal component corresponding to any given new data point $\mathbf{x}$ is obtained through the projection $\mathbf{s} = \mathbf{U}(\infty)'[\mathbf{m}\phi(\mathbf{x}) - \mathbf{m}(\infty)] = \mathbf{m}\Upsilon'(\infty)\mathbf{m}\Phi'\mathbf{m}\phi(\mathbf{x}) - \mathbf{m}\Upsilon'(\infty)\mathbf{K}\mathbf{m}\mu(\infty)$, which is again computable after N evaluations the kernel function $K_{\mathcal{H}}$.

5.5.1 Unveiling Communities in Social Networks

Next, robust KPCA is used to identify communities and outliers in a social network of $N = 115$ college football teams, by capitalizing on the connection between KPCA and spectral clustering [22, p. 548]. Nodes in the network graph represent teams belonging to eleven conferences (plus five independent teams), whereas (unweighted) edges joining pairs of nodes indicate that both teams played against each other during the Fall 2000 Division I season [20].

The kernel matrix used to run robust KPCA is $\mathbf{K} = \zeta\mathbf{I}_N + \mathbf{D}^{-1/2}\mathbf{A}\mathbf{D}^{-1/2}$, where $\mathbf{A}$ and $\mathbf{D}$ denote the graph adjacency and degree matrices, respectively; while $\zeta > 0$ is chosen to render $\mathbf{K}$ positive semi-definite. The tuning parameters are chosen as $\lambda_2 = 1.297$ so that $\|\hat{\mathbf{O}}\|_0 = 10$, while $\lambda_* = 1$, and $\bar{q} = 3$. Figure 5.4 shows the entries of $\mathbf{K}$, where rows and columns are permuted to reveal the clustering structure found by robust KPCA (after removing the outliers); see also [34, Figure 6 (top)] for a depiction of the partitioned network.

The quality of the clustering is assessed through the adjusted rand index (ARI) after excluding outliers [17], which yielded the value 0.8967. Four of the teams deemed as outliers are Connecticut, Central Florida, Navy, and Notre Dame, which are indeed teams not belonging to any major conference. The community structure of traditional powerhouse conferences such as Big Ten, Big 12, ACC, Big East, and SEC was identified exactly.

5.6 Conclusion

Outlier-robust PCA methods were developed in this chapter, to obtain low-dimensional representations of (corrupted) data. Bringing together the seemingly unrelated fields of robust statistics and sparse recovery, the surveyed robust PCA framework was found rooted at the crossroads of outlier-resilient estimation, learning via (group-) Lasso and kernel methods, and decentralized as well as real-time adaptive signal processing. Social network analysis, video surveillance, and psychometrics, were highlighted as relevant application domains.

References

1. L. Balzano. On GROUSE and incremental SVD. In *Workshop on Computational Advances in Multi-Sensor Adaptive Processing, CAMSAP 2013*, St. Martin, December 2013.
2. L. Balzano, R. Nowak, and B. Recht. Online identification and tracking of subspaces from highly incomplete information. In *Allerton Conference on Communication, Control, and Computing*, Monticello, IL, USA, September 2010.
3. L. Balzano and S. Wright. Local convergence of an algorithm for subspace identification from partial data. *arXiv preprint arXiv:1306.3391*, 2013.
4. D. Bertsekas. *Nonlinear Programming.* Athena-Scientific, second edition, 1999.
5. S. Boyd, N. Parikh, E. Chu, B. Peleato, and J. Eckstein. Distributed optimization and statistical learning via the alternating direction method of multipliers. *Found. Trends Mach. Learning*, 3:1–122, 2010.
6. S. Burer and R. Monteiro. Local minima and convergence in low-rank semidefinite programming. *Mathematical Programming*, 103(3):427–444, December 2005.
7. N. Campbell. Robust procedures in multivariate analysis i: Robust covariance estimation. *Applied Stat.*, 29:231–237, 1980.
8. E.. Candes, X. Li, Y. Ma, and J. Wright. Robust principal component analysis? *Journal of the ACM*, 58, March 2011.
9. E. Candes, M. Wakin, and S. Boyd. Enhancing sparsity by reweighted ℓ_1 minimzation. *Journal of Fourier Analysis and Applications*, 14:877–905, December 2008.
10. V. Chandrasekaran, S. Sanghavi, P. Parillo, and A. Willsky. Rank-sparsity incoherence for matrix decomposition. *SIAM Journal on Optimization*, 21:572–596, 2011.
11. Y. Chi, Y. Eldar, and R. Calderbank. PETRELS: Parallel subspace estimation and tracking using recursive least squares from partial observations. *IEEE Transactions on Signal Processing*, 61(23):5947–5959, November 2013.
12. W. Dai, O. Milenkovic, and E. Kerman. Subspace evolution and transfer (SET) for low-rank matrix completion. *IEEE Transactions on Signal Processing*, 59(7):3120–3132, July 2011.
13. F. de la Torre and M. Black. A framework for robust subspace learning. *International Journal of Computer Vision*, 54:1183–209, 2003.
14. X. Ding, L. He, and L. Carin. Bayesian robust principal component analysis. *IEEE Transactions on Image Processing*, 20, 2011.
15. J. Fan and R. Li. Variable selection via nonconcave penalized likelihood and its oracle properties. *J. Amer. Stat. Assoc.*, 96:1348–1360, 2001.
16. M. Fazel, H. Hindi, and S. Boyd. Log-det heuristic for matrix rank minimization with appli-

cations to Hankel and Euclidean distance matrices. In *the American Control Conf.*, pages 2156–2162, Denver, CO, June 2003.
17. P. Forero, V. Kekatos, and G. B. Giannakis. Outlier-aware robust clustering. In *International Conference on Acoustics, Speech and Signal Processing, ICASSP 2011*, pages 2244–2247, Prague, Czech Republic, May 2011.
18. J. Fuchs. An inverse problem approach to robust regression. In *International Conference on Acoustics, Speech and Signal Processing, ICASSP 1999*, pages 180–188, Phoenix, AZ, March 1999.
19. G. Giannakis, G. Mateos, S. Farahmand, V. Kekatos, and H. Zhu. USPACOR: Universal sparsity-controlling outlier rejection. In *International Conference on Acoustics, Speech and Signal Processing, ICASSP 2011*, pages 1952–1955, Prague, Czech Republic, May 2011.
20. M. Girvan and M. Newman. Community structure in social and biological networks. *Proc. Natl. Acad. Sci. USA*, 99:7821–7826, 2002.
21. L. R. Goldberg. The Eugene-Springfield community sample: Information available from the research participants. Technical Report vol. 48, no. 1, Oregon Research Institue, 2008.
22. T. Hastie, R. Tibshirani, and J. Friedman. *The Elements of Statistical Learning.* Springer, second edition, 2009.
23. J. He, L. Balzano, and A. Szlam. Incremental gradient on the Grassmannian for online foreground and background separation in subsampled video. In *IEEE Conference on Computer Vision and Pattern Recognition, CVPR 2012*, Providence, Rhode Island, June 2012.
24. P. Huber and E. Ronchetti. *Robust Statistics.* Wiley, New York, 2009.
25. O. John, L. Naumann, and C. Soto. Paradigm shift to the integrative big-five trait taxonomy: History, measurement, and conceptual issues. In *Handbook of personality: Theory and research.* Guilford Press, New York, NY, 2008.
26. I. Jolliffe. *Principal Component Analysis.* Springer, New York, 2002.
27. V. Kekatos and G. Giannakis. From sparse signals to sparse residuals for robust sensing. *IEEE Transactions on Signal Processing*, 59:3355–3368, July 2011.
28. E.. Kolaczyk. *Statistical Analysis of Network Data: Methods and Models.* Springer, 2009.
29. A. Lakhina, K. Papagiannaki, M. Crovella, C. Diot, E. Kolaczyk, and N. Taft. Structural analysis of network traffic flows. In *ACM SIGMETRICS*, New York, NY, July 2004.
30. J. Mairal, J. Bach, J. Ponce, and G. Sapiro. Online learning for matrix factorization and sparse coding. *Journal of Machine Learning Research*, 11:19–60, January 2010.
31. M. Mardani, G. Mateos, and G. Giannakis. Decentralized sparsity regularized rank minimization: Applications and algorithms. *IEEE Transactions on Signal Processing*, 61:5374–5388, November 2013.
32. M. Mardani, G. Mateos, and G. Giannakis. Dynamic anomalography: tracking network anomalies via sparsity and low rank. *IEEE Journal of Selected Topics in Signal Processing*, 7(11):50–66, February 2013.
33. M. Mardani, G. Mateos, and G. Giannakis. Subspace learning and imputation for streaming big data matrices and tensors. *IEEE Transactions on Signal Processing*, 63:2663–2667, March 2015.
34. G. Mateos and G. Giannakis. Robust PCA as bilinear decomposition with outlier-sparsity regularization. *IEEE Transactions on Signal Processing*, 60:5176–5190, 2012.
35. B. Natarajan. Sparse approximate solutions to linear systems. *SIAM Journal on Computing*, 24:227–234, 1995.
36. A. Puig, A. Wiesel, and A. Hero. Multidimensional shrinkage-thresholding operator and group LASSO penalties. *IEEE Signal Process. Letters*, 18:363–366, June 2011.
37. C. Qiu and N. Vaswani. Recursive sparse recovery in large but correlated noise. In *Proc. of Allerton Conf. on Communication, Control, and Computing*, Monticello, IL, 2011.

38. C. Qiu, N. Vaswani, B. Lois, and L. Hogben. Recursive robust PCA or recursive sparse recovery in large but structured noise. *IEEE Transactions on Information Theory*, 60:5007–5039, August 2014.
39. I. Ramirez, F. Lecumberry, and G. Sapiro. Universal priors for sparse modeling. In *International Workshop on Computational Advances in Multi-Sensor Adaptive Processing, CAMSAP 2009*, pages 197–200, Aruba, Dutch Antilles, December 2009.
40. B. Recht, M. Fazel, and P. A. Parrilo. Guaranteed minimum-rank solutions of linear matrix equations via nuclear norm minimization. *SIAM Review*, 52:471–501, 2010.
41. B. Recht and C. Re. Parallel stochastic gradient algorithms for large-scale matrix completion. *Mathematical Programming Computation*, 5:201–226, 2013.
42. P. Rousseeuw and A. Leroy. *Robust regression and outlier detection.* Wiley, New York, 1987.
43. B. Scholkopf, R. Herbrich, and A. J. Smola. A generalized representer theorem. *Computation Learning Theory: Lec. Notes in Computer Science*, 2111:416–426, 2001.
44. B. Scholkopf, A. J. Smola, and K. R. Muller. Nonlinear component analysis as a kernel eigenvalue problem. *Neural Computation*, 10:1299–1319, 1998.
45. K. Slavakis, G. Giannakis, and G. Mateos. Modeling and optimization for big data analytics. *IEEE Signal Processing Magazine*, 31:18–31, September 2014.
46. K. Slavakis, S. Kim, G. Mateos, and G. Giannakis. Stochastic approximation vis-a-vis online learning for big data. *IEEE Signal Processing Magazine*, 31:124–129, November 2014.
47. N. Srebro, J. Rennie, and T. Jaakkola. Maximum-margin matrix factorization. In *Advances in Neural Information Processing Systems*, pages 1329–1336, Vancouver, Canada, December 2004.
48. N. Srebro and A. Shraibman. Rank, trace-norm and max-norm. In *Learning Theory*, pages 545–560. Springer, 2005.
49. M. Storer, P. M. Roth, M. Urschler, and H. Bischof. Fast-robust PCA. *Image Analysis: Lec. Notes in Computer Science*, 5575:430–439, 2009.
50. I. Tosic and P. Frossard. Dictionary learning. *IEEE Signal Processing Magazine*, 28:27–38, March 2010.
51. J. Tropp. Just relax: Convex programming methods for identifying sparse signals. *IEEE Transactions on Information Theory*, 51:1030–1051, March 2006.
52. P. Tseng. Convergence of block coordinate descent method for nondifferentiable maximization. *Journal of Optimization Theory and Applications*, 109:473–492, 2001.
53. H. Xu, C. Caramanis, and S. Sanghavi. Robust PCA via outlier pursuit. *IEEE Transactions on Information Theory*, 58:3047–3064, May 2012.
54. L. Xu and A. Yuille. Robust principal component analysis by self-organizing rules based on statistical physics approach. *IEEE Transactions on Neural Nets*, 6:131–143, January 1995.
55. B. Yang. Projection approximation subspace tracking. *IEEE Transactions on Signal Processing*, 43:95–107, January 1995.
56. J. Yang and M. Kaveh. Adaptive eigensubspace algorithms for direction or frequency estimation and tracking. *IEEE Transactions on Acoustics Speech and Signal Processing*, 36(2):241–251, February 1988.
57. M. Yuan and Y. Lin. Model selection and estimation in regression with grouped variables. *Journal of the Royal Statistical Society: Series B*, 68:49–67, 2006.
58. Z. Zhou, X. Li, J. Wright, E. Candes, and Y. Ma. Stable principal component pursuit. In *International Symposium on Information Theory*, pages 1518–1522, Austin, TX, June 2010.
59. H. Zou, T. Hastie, and R. Tibshirani. Sparse principal component analysis. *Journal of Comp. and Graphical Statistics*, 15(2):265–286, 2006.

II

Robust Matrix Factorization

6

Unifying Nuclear Norm and Bilinear Factorization Methods

Ricardo Cabral
ISR/Instituto Superior Tecnico, Portugal - Carnegie Mellon University, USA

Fernando De la Torre
Carnegie Mellon University, USA

Joao Paulo Costeira
ISR/Instituto Superior Tecnico, Portugal

Alexandre Bernardino
ISR/Instituto Superior Tecnico, Portugal

6.1 Introduction: The Role of Rank

The computer vision research area presently stands in an exciting time, with the ubiquity of imaging sensors in DSLRs, cellphones, and laptops. Together with the advent of large computing power and global internet connectivity, these factors have eased the restrictions that limited amounts of data impose on the statistical learning of visual models, turning the so-called *curse of dimensionality* into a *blessing of dimensionality* [30]. Nonetheless, this new paradigm comes with its unique set of challenges: First, scalable algorithmic solutions are needed to harness this data, which cannot be stored or processed in a single computer; second, models that enforce Occam's razor's notion of simplicity become of utmost importance, so as to preserve model interpretability and avoid the risk of overfitting.

In this chapter, we study the topic of complexity penalization for visual learning tasks through rank minimization models. The use of rank criteria has been pervasive in computer vision applications as a mean of exploiting physical constraints of a model [23, 32, 89] or to minimize its complexity, be it in degrees of freedom [3] or in data redundancy [92, 98]. All these problems are directly or indirectly related to the problem of recovering a rank-k

matrix $\mathbf{Z}$ (see footnote[15] for notation) from a corrupted data matrix $\mathbf{X}$, by minimizing

$$\begin{aligned} \min_{\mathbf{Z}} \quad & f(\mathbf{X}-\mathbf{Z}) \\ \text{subject to} \quad & \operatorname{rank}(\mathbf{Z}) = k, \end{aligned} \tag{6.1}$$

where $f(\cdot)$ denotes a loss function. Due to its intractability, the *hard*-rank constraint in (6.1) has typically been imposed by the inner dimensions of a bilinear factorization $\mathbf{Z} = \mathbf{U}\mathbf{V}^\top$,

$$\min_{\mathbf{U},\mathbf{V}} \quad f(\mathbf{X} - \mathbf{U}\mathbf{V}^\top). \tag{6.2}$$

The factorization approach in (6.2) has been popularized in computer vision by the seminal work on structure from motion of Tomasi and Kanade [89]. Since then, it has been applied to many problems, including non-rigid and articulated structure from motion, as well as photometric stereo [12] and motion segmentation [23], or even classification [75, 87, 96]. It has been shown that when the loss function $f(\cdot)$ is the Least-squares loss, i.e., $f(\mathbf{X}-\mathbf{U}\mathbf{V}^\top) = \|\mathbf{X} - \mathbf{U}\mathbf{V}^\top\|_F^2$, then (6.2) does not have local minima and also that a closed form solution can be obtained via the Singular Value Decomposition (SVD) of $\mathbf{X}$ [7].

Unfortunately, this bilinear factorization approach has several caveats: The Least-squares loss is highly susceptible to outliers; also, the presence of missing data in $\mathbf{X}$ results in local minima. Outliers can be addressed with robust loss functions [48, 92] and optimal algorithms exist when missing data follows a Young diagram pattern [2]. However, missing data in computer vision typically exhibits random or band patterns, and factorization with missing data has been shown to be an NP-Hard problem [36], where many state-of-the-art algorithms fail to even reach good local optima [71]. For this reason, the optimization of (6.2) remains an active research topic, with many works focusing on algorithms that are robust to initialization [11, 32, 48, 70, 93, 104], initialization strategies [43], or incorporating additional problem constraints to achieve better optima [12].

Recently, Candes and Recht [21] have stated that minimizing the rank function – under broad conditions of *incoherence*, i.e., the unalignment of the singular vectors with the canonical axis – can be achieved by its convex surrogate, the nuclear norm. Initially proposed by Fazel [34], the nuclear norm permeated through many of the aforementioned computer vision problems such as structure from motion [4, 25, 26, 104], Robust PCA [98] and motion segmentation [56]. Here, the *soft*-rank regularization provided by the nuclear norm replaces the hard-rank constraints in the factorization approach of (6.2), by minimizing instead

$$\min_{\mathbf{Z}} \quad f(\mathbf{X} - \mathbf{Z}) + \lambda\|\mathbf{Z}\|_*, \tag{6.3}$$

where λ is a trade-off parameter between the error and the low-rank regularization induced by the nuclear norm $\|\mathbf{Z}\|_*$, the sum of singular values of $\mathbf{Z}$. We provide a simple intuition as to why the nuclear norm is in fact the largest possible convex underestimator of the rank function, as proved by [76]: Since the singular values of matrices are always positive, the nuclear norm can be interpreted as an ℓ_1-norm of the singular values. Under this interpretation, one can easily identify it as the convex envelope of the rank function, which is

[15] Bold capital letters denote matrices (e.g., $\mathbf{D}$). All non-bold letters denote scalar variables. d_{ij} denotes the scalar in the row i and column j of $\mathbf{D}$. $\langle \mathbf{d}_1, \mathbf{d}_2 \rangle$ denotes the inner product between two vectors $\mathbf{d}_1$ and $\mathbf{d}_2$. $\|\mathbf{d}\|_2^2 = \langle \mathbf{d}, \mathbf{d} \rangle = \sum_i d_i^2$ denotes the squared Euclidean norm of the vector $\mathbf{d}$. $\operatorname{tr}(\mathbf{A}) = \sum_i a_{ii}$ is the trace of $\mathbf{A}$. $\|\mathbf{A}\|_F^2 = \operatorname{tr}(\mathbf{A}^\top \mathbf{A}) = \sum_{ij} a_{ij}^2$ designates the squared Frobenius norm of $\mathbf{A}$. $\|\mathbf{A}\|_* = \sum_i \sigma_i$ designates the nuclear norm (sum of singular values σ_i) of $\mathbf{A}$. $\odot$ denotes the Hadamard or element-wise product. $\otimes$ denotes the Kronecker product. $\mathbf{I}_K \in \mathbb{R}^{K \times K}$ denotes the identity matrix. $\operatorname{diag}(\mathbf{X})$ is the vector of the diagonal elements of $\mathbf{X}$. $\operatorname{Diag}(\mathbf{X})$ is a matrix containing only the diagonal elements of $\mathbf{X}$.

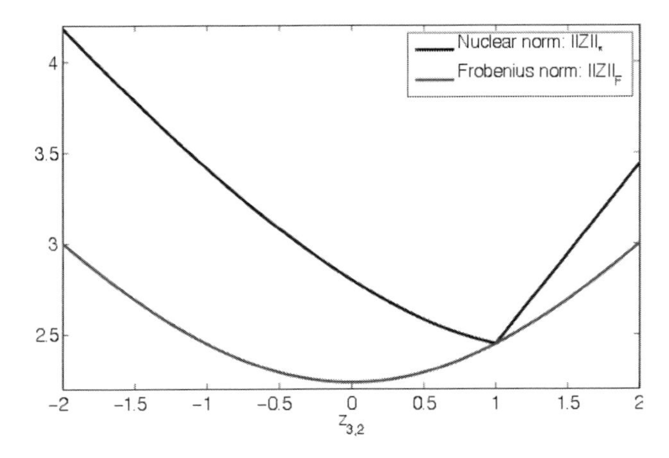

FIGURE 6.1 Comparison of Nuclear and Frobenius norms as function of one single unknown entry $z_{3,2}$ for the matrix in (6.4).

the cardinality (or ℓ_0-norm) of the singular values. To further understand why the singular value sparsity induced by the nuclear norm is important, let us consider completing the matrix

$$\mathbf{Z} = \begin{bmatrix} 1\,1 \\ 1\,1 \\ 1\,? \end{bmatrix}, \tag{6.4}$$

where only one entry $z_{3,2}$ is unknown such that the resulting rank is the smallest possible. The results shown in Figure 6.1 plot the nuclear norm and Frobenius norm of $\mathbf{Z}$ for all possible completions in a range around the value that minimizes its rank $z_{3,2} = 1$. In this case, the sparsity induced by the nuclear norm (ℓ_1-norm on the singular values) yields the optimal solution for $\mathbf{Z}$ with singular values $\sigma = [2.4495\ 0]$, a rank-1 matrix. In opposition, the Frobenius Norm (ℓ_2-norm of singular values) will set the entries to zero, thus leading to a solution with singular values $\sigma = [2.1358\ 0.6622]$, a rank-2 matrix. This key difference can be attributed to the fact that completing a matrix under the rank or nuclear norm favors the interaction between rows and columns to find a global solution, while the Frobenius norm treats each entry in the matrix independently (recall that $\|\mathbf{Z}\|_F^2 = \sum_{ij} z_{ij}^2$).

Contrary to hard-rank models, soft-rank regularization models have further extended the use of low-rank priors to many applications where the rank is not known *a priori*: colorization [95], subspace alignment [74] and clustering [33], segmentation [22], texture unwarping [102], camera calibration [103], tag refinement [65, 106], background modeling [93, 98] and tracking [100]. Soft-rank regularizers such as the nuclear norm or the max norm have also been proposed in machine learning as good regularizers for classification and collaborative filtering [5, 44]. Specifically, they has surfaced as a way to penalize complexity in image classification and regression tasks [42, 58, 75, 81, 86, 97, 101, 106], to reduce model degrees of freedom [6, 29, 69, 90], or to share properties among different classifiers [3, 29, 38].

Despite their convexity and theoretical results for the choice of λ [20], nuclear norm models such as the one in (6.3) also suffer from several drawbacks. On the one hand, it is unclear how to impose a certain rank in $\mathbf{Z}$: we showed in [17] that adjusting λ such that $\mathbf{Z}$ has a predetermined rank typically provides worse results than imposing this rank directly as in (6.2). Also, the inability to access the factorization of $\mathbf{Z}$ in (6.3) hinders the use of the "kernel trick" in classification and component analysis methods, and hence disallows for non-linear kernel extensions [28]. On the other hand, (6.3) is a Semidefinite Program (SDP). Current off-the-shelf SDP optimizers only scale to hundreds of variables, not amenable to the high dimensionality feature inputs typically found in computer vision problems. Several works [19–21, 55, 59, 88] ameliorate this issue by exploiting the fact that the proximal operator of the nuclear norm

$$\arg\min_{\mathbf{Z}} \quad \|\mathbf{X}-\mathbf{Z}\|_F^2 + \frac{1}{2}\|\mathbf{Z}\|_*, \tag{6.5}$$

has a closed form solution based on singular value thresholding. However, they still perform a SVD of $\mathbf{Z}$ in each iteration. Other approaches incrementally optimize (6.3) using gradient methods on the Grassmann manifold [8, 41, 50]. However, they rely on a rank selection heuristic, which fails when data is not missing at random. Zaid et al. [40] decompose the nuclear norm into a surrogate infinite-dimensional optimization, but their coordinate descent only applies to smooth losses $f(\cdot)$. Thus, nuclear norm approaches are currently unsuitable for handling dense, large-scale datasets.

6.1.1 Summary of Contributions and Organization

In summary, the main result of this chapter is that **soft-rank nuclear norm models can be reformulated as hard-rank factorization models** through the variational definition of the nuclear norm. While the variational definition was previously known in the literature [64, 93], we are the first to propose a unification of soft and hard-rank approaches in computer vision under one formulation.

Several implications stem from the unification of soft and hard-rank models: in Section 6.2, we show that nuclear norm (soft-rank) formulations can be reformulated as factorization (hard-rank) formulations and unify them in a single model. We propose an augmented Lagrange multiplier algorithm to solve the unified model. We then split the use of soft and hard-rank models into two regions of our unified model: when rank is known a priori or when rank is known to be low but not precisely known [17].

For soft models, this unification brings advantages such as scalability and kernelization, and explicitly shows their limitations on problems where the output rank is predetermined. In addition to the contents of this chapter, we have proposed two convex soft-rank models for visual weakly supervised learning in images [16, 18] and video [1], as well as a fully supervised robust regression model [42]. These methods apply to a variety of problems in computer vision, since they can easily cope with labeling errors, missing data, background noise, partial occlusions, and outliers that are common in realistic training sets due to occlusion, specular reflections, or noise.

In Section 6.3, we focus on problems where rank is predetermined or known *a priori*. We show the limitations of soft-rank models on these problems, and present "rank continuation," a deterministic strategy that empirically attains good solutions in a significant number of cases when the rank is known *a priori*. This is the case of the NP-Hard problem of factorization with missing data, applicable in structure from motion or photometric stereo. We extend this strategy to the case of Binary Quadratic Problems such as the Quadratic Assignment Problem, which can be reformulated as rank-1 problems.

6.2 Unification of Soft and Hard-Rank Models

As mentioned in Section 6.1, finding models that favor low-rank solutions is an essential tool for solving computer vision and machine learning problems: Low-rank representations allow for reducing degrees of freedom, exploiting redundancy, and enforcing simplicity when representating shape, appearance, or motion. There are two main approaches for imposing low rank, which we will formally define as hard-rank and soft-rank models.

Definition 1 (Hard-rank models) *Optimization models that aim to recover a rank-k matrix* $\mathbf{Z} \in \mathbb{R}^{M \times N}$ *from a data matrix* $\mathbf{X} \in \mathbb{R}^{M \times N}$ *according an error function* $f(\cdot)$, *as*

$$\begin{aligned} \min_{\mathbf{Z}} \quad & f(\mathbf{X} - \mathbf{Z}) \\ \textit{subject to} \quad & \operatorname{rank}(\mathbf{Z}) = k. \end{aligned} \tag{6.6}$$

This constraint is typically directly imposed on the solution by optimizing a bilinear product $\mathbf{Z} = \mathbf{U}\mathbf{V}^\top$ *and specifying the inner dimensions of this product as* k, *as*

$$\min_{\mathbf{U},\mathbf{V}} \; f(\mathbf{X} - \mathbf{U}\mathbf{V}^\top). \tag{6.7}$$

Definition 2 (Soft-rank models) *Optimization models that aim to recover a rank-k matrix* $\mathbf{Z} \in \mathbb{R}^{M \times N}$ *from a data matrix* $\mathbf{X} \in \mathbb{R}^{M \times N}$ *according an error function* $f(\cdot)$, *and a low-rank solution is sought. Thus, the problem is regularized by adding to the cost function a regularizer such as the nuclear norm, as*

$$\min_{\mathbf{Z}} \; f(\mathbf{X} - \mathbf{Z}) + \lambda \|\mathbf{Z}\|_*. \tag{6.8}$$

In this section, we show that nuclear norm (soft-rank) formulations can be reformulated as factorization (hard-rank) formulations and thus unify them in a single model. Let us start by considering the nuclear norm problem in (6.8) with convex $f(\cdot)$: Without loss of generality, we can rewrite (6.8) as the SDP [77]

$$\begin{aligned} \min_{\mathbf{Z},\mathbf{B},\mathbf{C}} \quad & f(\mathbf{X} - \mathbf{Z}) + \frac{\lambda}{2}\left(\operatorname{tr}(\mathbf{B}) + \operatorname{tr}(\mathbf{C})\right) \\ \text{subject to} \quad & \mathbf{Q} = \begin{bmatrix} \mathbf{B} & \mathbf{Z} \\ \mathbf{Z}^\top & \mathbf{C} \end{bmatrix} \succeq 0. \end{aligned} \tag{6.9}$$

For any positive semidefinite matrix $\mathbf{Q}$, we can write $\mathbf{Q} = \mathbf{R}\mathbf{R}^\top$ for some $\mathbf{R}$. Thus, we can replace matrix $\mathbf{Q}$ in (6.9) by

$$\mathbf{Q} = \begin{bmatrix} \mathbf{B} & \mathbf{Z} \\ \mathbf{Z}^\top & \mathbf{C} \end{bmatrix} = \begin{bmatrix} \mathbf{U} \\ \mathbf{V} \end{bmatrix} \begin{bmatrix} \mathbf{U}^\top & \mathbf{V}^\top \end{bmatrix}, \tag{6.10}$$

where $\mathbf{U} \in \mathbb{R}^{M \times r}$, $\mathbf{V} \in \mathbb{R}^{N \times r}$ and $r \leq \min(N, M)$ upper bounds $\operatorname{rank}(\mathbf{Z})$. Merging (6.10) into (6.9) yields

Definition 3 (Unified model)

$$\min_{\mathbf{U},\mathbf{V}} \; f(\mathbf{X} - \mathbf{U}\mathbf{V}^\top) + \frac{\lambda}{2}\left(\|\mathbf{U}\|_F^2 + \|\mathbf{V}\|_F^2\right), \tag{6.11}$$

where the SDP constraint was dropped because it is satisfied by construction. This reformulation seems counterintuitive, as we changed the convex problem in (6.8) into a non-convex one, which may be prone to local minima (e.g., in the case of missing data under the least-squares loss [36]). However, we show that the existence of local minima in (6.11) depends only on the dimension r imposed on matrices $\mathbf{U}$ and $\mathbf{V}$. We extend the analysis of Burer and Monteiro [13] to prove that:

THEOREM 6.1 *Let* $f(\mathbf{X} - \mathbf{Z})$ *be convex in* $\mathbf{Z}$ *and* $\mathbf{Z}^*$ *be an optimal solution of the convex nuclear norm model in* (6.8) *for a given* λ *and let* $\operatorname{rank}(\mathbf{Z}^*) = k^*$. *Then, any solution* $\mathbf{Z} = \mathbf{U}\mathbf{V}^\top$ *of* (6.11) *with* $r \geq k^*$ *is a global minima solution of* (6.8).

To show this, we first note that (6.11) uses the variational formulation of the nuclear norm in (6.13), and that [64] showed the following result:

LEMMA 6.1 For any $\mathbf{Z} \in \mathbb{R}^{M \times N}$, the following holds: If $\text{rank}(\mathbf{Z}) = k^* \leq \min(M, N)$, then the minimum of (6.13) is attained at a factor decomposition $\mathbf{Z} = \mathbf{U}_{M \times k^*} \mathbf{V}_{N \times k^*}^\top$.

This result allows us to prove the desired equivalence:

PROOF 6.1 Applying Lemma 6.1, we can reduce (6.11) to

$$\begin{aligned} &\min_{\mathbf{U},\mathbf{V}} && f(\mathbf{X} - \mathbf{U}\mathbf{V}^\top) + \lambda \|\mathbf{U}\mathbf{V}^\top\|_* \\ &= \min_{\mathbf{Z},\text{rank}(\mathbf{Z})=k^*} && f(\mathbf{X} - \mathbf{Z}) + \lambda \|\mathbf{Z}\|_* \\ &= \min_{\mathbf{Z}} && f(\mathbf{X} - \mathbf{Z}) + \lambda \|\mathbf{Z}\|_*. \end{aligned} \tag{6.12}$$

Theorem 6.1 immediately allows us to draw one conclusion: By application of the variational property of the nuclear norm [77],

$$\|\mathbf{Z}\|_* = \min_{\mathbf{Z}=\mathbf{U}\mathbf{V}^\top} \frac{1}{2} \left(\|\mathbf{U}\|_F^2 + \|\mathbf{V}\|_F^2 \right), \tag{6.13}$$

many soft-rank models can be reformulated into hard-rank models. That is, the factorization and the nuclear norm models in (6.7) and (6.8) are special cases of (6.11). Figure 6.2 illustrates the result of Theorem 6.1 in a synthetic case. We plot the output rank of $\mathbf{Z} = \mathbf{U}\mathbf{V}^\top$ in (6.11) as a function of λ for a random 100×100 matrix $\mathbf{X}$ with all entries sampled i.i.d. from a Gaussian distribution $\mathcal{N}(0, 1)$, no missing data, and $f(\cdot)$ is the least squares loss $\|\mathbf{X} - \mathbf{Z}\|_F^2$: The factorization approach in (6.7) corresponds to the case where $\lambda = 0$ and r is fixed, whilst the nuclear norm in (6.8) outputs an arbitrary rank k^* as a function of λ (the black curve). According to Theorem 6.1, for any $r \geq k^*$ (white area), optimizing (6.11) is equivalent to (6.8). On the other hand, when $r < k^*$ (grey area), the conditions of Theorem 6.1 are no longer valid and thus (6.11) can be prone to local minima.

A special case of Theorem 6.1 has been used to recommend the use of nuclear norm approaches in the machine learning community by Mazumder et al. [64]. However, their analysis is restricted to the least-squares loss and the case where the rank is not known *a priori* (i.e., white area of Figure 6.2).

Our analysis instead extends to other convex loss functions and is motivated by the observation that many computer vision problems live in the grey area of Fig. 6.2. That is, their output rank k is predetermined by a domain-specific constraint (e.g., in Structure from Motion $k = 4$ [89]).

The visual interpretation of Theorem 6.1 in Figure 6.2 shows two clear regions of operation of our unified model. As such, for the remainder of this thesis, we will consider the use of soft and hard-rank models as two separate regions of our unified model: when rank is known *a priori* or when rank is known to be low but not precisely known. We advocate the use of our unified model in (6.11) for both cases over the typical soft and hard-rank models, based on two arguments:

When the output rank is unconstrained (white area of Figure 6.2) soft-rank models should be used, but we can always choose $r \geq k^*$ such that (6.11) provides equivalent results to (6.8). Using the result in Theorem 6.1 and the analysis of Burer and Monteiro [13], we propose an ALM algorithm in Section 6.2.1 using the unified model in (6.11) that has the scalability advantages of factorization approaches, yet it is guaranteed to attain the

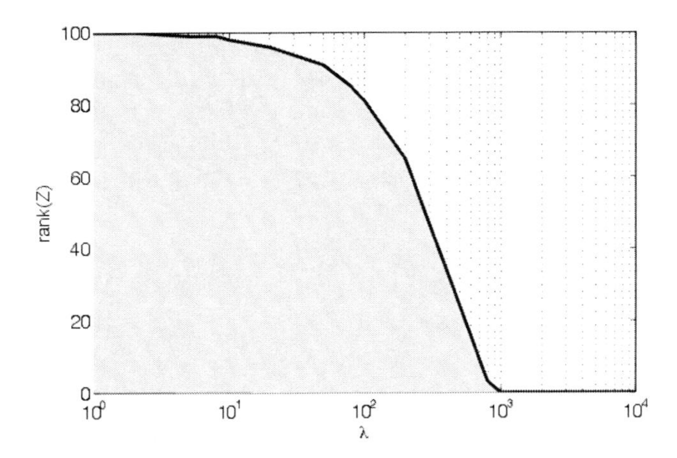

FIGURE 6.2 Region of equivalence between factorization (6.11) and nuclear norm approaches (6.8) for a 100×100 random matrix and least-squares loss. When factorization is initialized in the white area, it is equivalent to the result obtained with the nuclear norm (black line). When the rank is known *a priori*, directly imposing $r = k$ in the factorization approach of (6.11) (gray area) is less prone to local minima than the unregularized problem (6.7) and provides better results than selecting λ in the nuclear norm model (6.8) such that the output rank is k.

global optima of the original nuclear norm model. Also, this equivalence result makes the kernelization of some nuclear norm models trivial. In Section 6.2.2, we show our unified model is faster than state-of-the-art algorithms for optimizing nuclear norm models.

When the output rank is known *a priori* (gray area of Figure 6.2), hard-rank models should be used, but optimizing (6.11) is preferable to (6.7) and (6.8). As we will show in Section 6.3, optimizing (6.11) is less prone to local minima than the unregularized problem (6.7). On the other hand, selecting λ in the nuclear norm model (6.8) such that the output rank k is the desired value typically leads to worse results than directly imposing $r = k$ in (6.11). Based on this analysis, we propose in Section 6.3.1 a "rank continuation" strategy, and empirically show it is able to attain global optimality in several scenarios.

6.2.1 An ALM Algorithm for the Unified Model

In this section, we propose an algorithm for solving (6.11) and show that its complexity is lower than proximal methods [55] for optimizing the nuclear norm model in (6.3). For the remainder of this section, we focus our attention in the LS loss

$$f(\mathbf{W} \odot (\mathbf{X} - \mathbf{Z})) = \|\mathbf{W} \odot (\mathbf{X} - \mathbf{Z})\|_F^2 = \sum_{ij} (w_{ij}(x_{ij} - z_{ij}))^2, \tag{6.14}$$

and the L1 loss

$$f(\mathbf{W} \odot (\mathbf{X} - \mathbf{Z})) = \|\mathbf{W} \odot (\mathbf{X} - \mathbf{Z})\|_1 = \sum_{ij} |w_{ij}(x_{ij} - z_{ij})|, \tag{6.15}$$

where $\mathbf{W} \in \mathbb{R}^{M \times N}$ is a positive weight matrix that can be used to denote missing data (i.e., $w_{ij} = 0$). We note, however, that the results in Theorem 6.1 also apply to many other losses such as the Huber [4, 48] and hinge loss [58, 75].

One important factor to take into account when optimizing (6.11) for the LS and L1 losses is that when either $\mathbf{U}$ or $\mathbf{V}$ is fixed, the remaining part of (6.11) becomes convex, even in the presence of a missing data pattern specified by $\mathbf{W}$. However, it has been reported that pure alternation approaches for this problem are prone to flatlining [11, 32, 71]. For smooth losses such as the LS, this can be circumvented by performing gradient steps jointly in $\mathbf{U}, \mathbf{V}$ [11]. Alternatively, we propose an Augmented Lagrange Multiplier (ALM) method for two reasons: 1) Theorem 6.1 and the analysis in [13] can be used to prove our ALM's convergence to global optima of (6.3) when $r \geq k^*$, and 2) its applicability to the non-smooth L1 norm. Let us rewrite (6.11) as

$$\begin{aligned} \min_{\mathbf{Z},\mathbf{U},\mathbf{V}} \quad & f(\mathbf{W} \odot (\mathbf{X} - \mathbf{Z})) + \frac{\lambda}{2}\left(\|\mathbf{U}\|_F^2 + \|\mathbf{V}\|_F^2\right) \\ \text{subject to} \quad & \mathbf{Z} = \mathbf{U}\mathbf{V}^\top, \end{aligned} \tag{6.16}$$

and its corresponding augmented Lagrangian as

$$\begin{aligned} \min_{\mathbf{Z},\mathbf{U},\mathbf{V},\mathbf{Y}} \quad & f(\mathbf{W} \odot (\mathbf{X} - \mathbf{Z})) + \frac{\lambda}{2}\left(\|\mathbf{U}\|_F^2 + \|\mathbf{V}\|_F^2\right) \\ & + \langle \mathbf{Y}, \mathbf{Z} - \mathbf{U}\mathbf{V}^\top \rangle + \frac{\rho}{2}\|\mathbf{Z} - \mathbf{U}\mathbf{V}^\top\|_F^2, \end{aligned} \tag{6.17}$$

where $\mathbf{Y}$ are Lagrange multipliers and ρ is a penalty parameter to improve convergence [55]. This method exploits the fact that the solution for each subproblem in $\mathbf{U}, \mathbf{V}, \mathbf{Z}$ can be efficiently solved in closed form. For $\mathbf{U}$ and $\mathbf{V}$, the solution is obtained by equating the derivatives of (6.17) in $\mathbf{U}$ and $\mathbf{V}$ to $\mathbf{0}$. For known $\mathbf{U}$ and $\mathbf{V}$, $\mathbf{Z}$ can be updated by solving

$$\min_{\mathbf{Z}} \quad f(\mathbf{W} \odot (\mathbf{X} - \mathbf{Z})) + \frac{\rho}{2}\|\mathbf{Z} - \left(\mathbf{U}\mathbf{V}^\top - \rho^{-1}\mathbf{Y}\right)\|_F^2, \tag{6.18}$$

which can be done in closed form by the element-wise shrinkage operator $\mathcal{S}_\mu(x) = \max(0, x - \mu)$, as

$$\begin{aligned} \mathbf{Z} = {} & \mathbf{W} \odot \left(\mathbf{X} - \mathcal{S}_{\rho^{-1}}(\mathbf{X} - \mathbf{U}\mathbf{V}^\top + \rho^{-1}\mathbf{Y})\right) \\ & + \overline{\mathbf{W}} \odot (\mathbf{U}\mathbf{V}^\top - \rho^{-1}\mathbf{Y}), \end{aligned} \tag{6.19}$$

for the L1 loss, or

$$\begin{aligned} \mathbf{Z} = {} & \mathbf{W} \odot \left(\frac{1}{2+\rho}\left(2\mathbf{X} + \rho(\mathbf{U}\mathbf{V}^\top - \rho^{-1}\mathbf{Y})\right)\right) \\ & + \overline{\mathbf{W}} \odot (\mathbf{U}\mathbf{V}^\top - \rho^{-1}\mathbf{Y}), \end{aligned} \tag{6.20}$$

for the LS loss. Here, $\overline{w}_{ij} = 1, \forall_{ij} w_{ij} \neq 0$ and 0 otherwise. The resulting algorithm is summarized in Algorithm 6.1 and its full derivation is presented in Section 6.2.1. Contrary to pure alternated methods, our numerical experiments show that this method is not prone to flatlining due to the joint optimization being gradually enforced by $\mathbf{Y}$.

Assuming without loss of generality that $\mathbf{X} \in \mathbb{R}^{M \times N}$ and $M > N$, we have that exact state-of-the-art methods for SVD (e.g., Lanczos bidiagonalization algorithm with partial reorthogonalization) take a flop count of $O(MN^2 + N^3)$. The most computationally costly steps in our ALM method are the matrix multiplications in the update of $\mathbf{U}$ and $\mathbf{V}$, which take $O(MNr + Nr^2)$ if done naively. Given that typically $k^* \leq r \ll \min(M, N)$ and k^* can be efficiently estimated [51], Algorithm 6.1 provides significant computational cost savings when compared to proximal methods that use SVDs [55].

We note that there are several recent results in optimization that tackle the complexity issue of SVDs in proximal methods for the nuclear norm. For instance, there has been recent work on online methods for factorization [49], as well as randomized or incremental SVDs [10]. Also, when using a projected sub-gradient method, one can easily avoid the cost of SVD using a polar decomposition of the variable Z, which can be obtained by Halley's method [77]. If the singular values are away from zero, this is much faster than the original

Algorithm 6.1 ALM method for optimizing (6.11)

Require: $\mathbf{X}, \mathbf{W} \in \mathbb{R}^{M \times N}$, params μ, λ, initialization of ρ
  while not converged **do**
    while not converged **do**
      Update $\mathbf{U} = (\rho\mathbf{Z} + \mathbf{Y})\,\mathbf{V}\left(\rho\mathbf{V}^\top\mathbf{V} + \lambda\mathbf{I}_r\right)^{-1}$
      Update $\mathbf{V} = (\rho\mathbf{Z} + \mathbf{Y})^\top\,\mathbf{U}\left(\rho\mathbf{U}^\top\mathbf{U} + \lambda\mathbf{I}_r\right)^{-1}$
      Update $\mathbf{Z}$ via (6.19) for L1 loss or (6.20) for LS loss
    end while
    $\mathbf{Y} = \mathbf{Y} + \rho(\mathbf{Z} - \mathbf{U}\mathbf{V}^\top)$
    $\rho = \min\left(\rho\mu, 10^{20}\right)$
  end while
Ensure: Complete Matrix $\mathbf{Z} = \mathbf{U}\mathbf{V}^\top$

SVD algorithm. Also, there are approaches that minimize models for RPCA in linear time. For instance, [57] solve an initial smaller problem of the dimension of the rank r and then calculate the remainder of the matrix using projections based on the calculated singular vector estimates. However, our result is still relevant in this case for solving the initial problem, as the rank r may still be a large number even if considerably smaller than the matrix dimensions $\min(M, N)$. Moreover, our result allows very scalable solutions recently obtained for factorization methods (e.g., [83]) to be applied to nuclear norm models by resorting to its variational definition.

Full derivation of Algorithm 6.1

We provide the full derivation of Algorithm 6.1 in this section. Let us start by transforming

$$\min_{\mathbf{U},\mathbf{V}} \quad \|\mathbf{W} \odot (\mathbf{X} - \mathbf{U}\mathbf{V}^\top)\|_1 + \frac{\lambda}{2}\left(\|\mathbf{U}\|_F^2 + \|\mathbf{V}\|_F^2\right), \tag{6.21}$$

into the equivalent problem

$$\begin{aligned} \min_{\mathbf{Z},\mathbf{U},\mathbf{V}} \quad & \|\mathbf{W} \odot (\mathbf{X} - \mathbf{Z})\|_1 + \frac{\lambda}{2}\left(\|\mathbf{U}\|_F^2 + \|\mathbf{V}\|_F^2\right) \\ \text{subject to} \quad & \mathbf{Z} = \mathbf{U}\mathbf{V}^\top, \end{aligned} \tag{6.22}$$

and write its augmented Lagrangian function, as

$$\begin{aligned} \min_{\mathbf{Z},\mathbf{U},\mathbf{V}} \mathcal{L} = \; & \|\mathbf{W} \odot (\mathbf{X} - \mathbf{Z})\|_1 + \frac{\lambda}{2}\left(\|\mathbf{U}\|_F^2 + \|\mathbf{V}\|_F^2\right) + \\ & \langle \mathbf{Y}, \mathbf{Z} - \mathbf{U}\mathbf{V}^\top \rangle + \frac{\rho}{2}\|\mathbf{Z} - \mathbf{U}\mathbf{V}^\top\|_F^2, \end{aligned} \tag{6.23}$$

where $\mathbf{Y} \in \mathbb{R}^{d \times n}$ is a Lagrange multiplier matrix, and ρ is a penalty parameter [55]. We solve (6.23) by an iterative Gauss-Siedel method on $\mathbf{Z}, \mathbf{U}, \mathbf{V}$, solved by the subproblems

$$\mathbf{Z}^{(k+1)} = \arg\min_{\mathbf{Z}} \mathcal{L}(\mathbf{Z}^{(k)}, \mathbf{U}, \mathbf{V}, \mathbf{Y}, \rho) \tag{6.24}$$

$$\mathbf{U}^{(k+1)} \arg\min_{\mathbf{U}} \mathcal{L}(\mathbf{Z}, \mathbf{U}^{(k)}, \mathbf{V}, \mathbf{Y}, \rho), \tag{6.25}$$

$$\mathbf{V}^{(k+1)} = \arg\min_{\mathbf{V}} \mathcal{L}(\mathbf{Z}, \mathbf{U}, \mathbf{V}^{(k)}, \mathbf{Y}, \rho), \tag{6.26}$$

where k is the index of iterations. At iteration $k = 0$, the entries of variables $\mathbf{U}, \mathbf{V}, \mathbf{Z}$ are initialized i.i.d. from a standard normal distribution and $\mathbf{Y}, \rho$ are initialized as

$$\mathbf{Y}^{(0)} = \mathbf{0} \tag{6.27}$$
$$\rho^{(0)} = 10^{-5}. \tag{6.28}$$

After initialization,(6.24)–(6.26) are solved sequentially until convergence. In the following subsections, we will derive the solutions of each of these subproblems.

After each Gauss-Siedel convergence, the Lagrange Multiplier matrix $\mathbf{Y}$ is updated by a gradient ascent step

$$\mathbf{Y}^{(k+1)} = \mathbf{Y}^{(k)} + \rho^{(k)}(\mathbf{Z} - \mathbf{U}\mathbf{V}^\top), \tag{6.29}$$

where the penalty variable ρ is updated by the expression

$$\rho^{(k+1)} = \mu\rho^{(k)}, \tag{6.30}$$

and $\mu > 1$ is a constant. A larger μ imposes stronger enforcement of the constraint $\mathbf{Z} = \mathbf{U}\mathbf{V}^\top$, therefore faster convergence of the outer loop, but may result in poor performance of the inner Gauss-Siedel loop and vice versa. In our experiments, we chose $\mu = 1.05$.

***Solving for* U**

Fixing $\mathbf{Z}$ and $\mathbf{V}$, the subproblem (6.25) is reduced to the problem

$$\mathcal{L}(\mathbf{U}) \propto \frac{\lambda}{2}\|\mathbf{U}\|_F^2 + \langle \mathbf{Y}, \mathbf{Z} - \mathbf{U}\mathbf{V}^\top \rangle + \frac{\rho}{2}\|\mathbf{Z} - \mathbf{U}\mathbf{V}^\top\|_F^2, \tag{6.31}$$

whose closed-form solution can be obtained by equating the derivative of (6.31) to $\mathbf{0}$, resulting in

$$\mathbf{U} = (\rho\mathbf{Z} + \mathbf{Y})\,\mathbf{V}\left(\rho\mathbf{V}^\top\mathbf{V} + \lambda\mathbf{I}_r\right)^{-1}. \tag{6.32}$$

***Solving for* V**

Fixing $\mathbf{Z}$ and $\mathbf{U}$, the subproblem (6.26) is reduced to the problem

$$\mathcal{L}(\mathbf{V}) \propto \frac{\lambda}{2}\|\mathbf{V}\|_F^2 + \langle \mathbf{Y}, \mathbf{Z} - \mathbf{U}\mathbf{V}^\top \rangle + \frac{\rho}{2}\|\mathbf{Z} - \mathbf{U}\mathbf{V}^\top\|_F^2, \tag{6.33}$$

whose closed-form solution can be obtained by equating the derivative of (6.33) to $\mathbf{0}$, resulting in

$$\mathbf{V} = (\rho\mathbf{Z} + \mathbf{Y})^\top\,\mathbf{U}\left(\rho\mathbf{U}^\top\mathbf{U} + \lambda\mathbf{I}_r\right)^{-1}. \tag{6.34}$$

***Solving for* Z**

Fixing $\mathbf{U}$, $\mathbf{V}$, the cost function of subproblem (6.24) can be rewritten in an equivalent problem

$$\min_{\mathbf{Z}} \ \|\mathbf{W} \odot (\mathbf{X} - \mathbf{Z})\|_1 + \frac{\rho}{2}\|\mathbf{Z} - \left(\mathbf{U}\mathbf{V}^\top - \rho^{-1}\mathbf{Y}\right)\|_F^2, \tag{6.35}$$

which can be done in closed form from the fact that $\mathbf{0}$ is in the expression of the subdifferential of (6.35). Using the element-wise shrinkage operator $\mathcal{S}_\mu(x) = \max(0, x - \mu)$, this condition can be written as

$$\begin{aligned}\mathbf{Z} = {} & \mathbf{W} \odot \left(\mathbf{X} - \mathcal{S}_{\rho^{-1}}(\mathbf{X} - \mathbf{U}\mathbf{V}^\top + \rho^{-1}\mathbf{Y})\right) \\ & + \overline{\mathbf{W}} \odot (\mathbf{U}\mathbf{V}^\top - \rho^{-1}\mathbf{Y}),\end{aligned} \tag{6.36}$$

where $\overline{w}_{ij} = 1, \forall_{ij} w_{ij} \neq 0$, and 0 otherwise.

Stopping criteria

For the outer loop in Algorithm 6.1 in the main paper, the iteration is not terminated until the equality constraint $\mathbf{Z} = \mathbf{U}\mathbf{V}^\top$ is satisfied up to a given tolerance. In our experiments, we used $\|\mathbf{Z} - \mathbf{U}\mathbf{V}^\top\|_F \leq 10^{-9}\|\mathbf{W} \odot \mathbf{M}\|_F$. For the inner loop, since the global optimum solution is found for (6.24)–(6.26), the objective function monotonically decreases. As such, in our experiments the stopping criteria for the inner Gauss-Siedel loop combines two items:

1. Small decrease of $\mathcal{L}(\cdot)$: $\frac{\|\mathcal{L}(\cdot)^{(k)} - \mathcal{L}(\cdot)^{(k-1)}\|}{\|\mathcal{L}(\cdot)^{(k-1)}\|} \leq 10^{-10}$;

2. Maximum number of iterations is reached: 5000.

6.2.2 Experimental Results on Background Subtraction

In this section, we analyze the application of the unified model proposed in Section 6.2 to Robust PCA, a model typically used for the task of background subtraction. In this problem, one aims to recover a low-rank data matrix $\mathbf{Z}$ from a data matrix $\mathbf{X}$, as

Definition 4 (Robust PCA model)

$$\min_{\mathbf{Z}} \quad \|\mathbf{X} - \mathbf{Z}\|_1 + \lambda\|\mathbf{Z}\|_*. \tag{6.37}$$

As seen in Section 6.2, by using a bilinear factorization of $\mathbf{Z} = \mathbf{U}\mathbf{V}^\top$, the robust PCA model can be equivalently written as (recall (6.11))

$$\min_{\mathbf{U},\mathbf{V}} \quad \|\mathbf{X} - \mathbf{U}\mathbf{V}^\top\|_1 + \frac{\lambda}{2}\left(\|\mathbf{U}\|_F^2 + \|\mathbf{V}\|_F^2\right). \tag{6.38}$$

We show the ALM algorithm proposed in Section 6.2.1 is both faster and more accurate than state-of-the-art nuclear norm algorithms for this problem.

To validate the lower computational complexity of the algorithm proposed in Algorithm 6.1 when the output rank is not known *a priori*, we compared it to state-of-the-art nuclear norm and Grassmann manifold methods: GRASTA [41], PRMF [93], and RPCA-IALM [55] in a synthetic and real-data experiment for background modeling. We use implementations provided in authors' websites for all baselines. For all experiments, we fix $\mu = 1.05$ and initialize $\rho = 10^{-5}$ in Alg. 6.1. All experiments were run on a desktop with a 2.8 GHz Quad-core CPU and 6 GB RAM.

Synthetic data

We mimicked the setup in [55] and generated low-rank matrices $\mathbf{X} = \mathbf{U}\mathbf{V}^\top$. The entries in $\mathbf{U} \in \mathbb{R}^{M \times r}, \mathbf{V} \in \mathbb{R}^{N \times r}$ with $M = N$, and each element was sampled i.i.d. from a Gaussian distribution $\mathcal{N}(0, 1)$. Then, we corrupted 10% of the entries with large errors uniformly distributed in the range $[-50, 50]$. The error support was chosen uniformly at random. Like [55], we set $\lambda = \sqrt{N}$ and used the L1 loss. We varied the dimension N and rank r and measured the algorithm accuracies, defined as $\frac{\|\mathbf{Z}-\mathbf{X}\|_2}{\|\mathbf{X}\|_2}$, and the time they took to run. The results in Table 6.1 corroborate experimentally the complexity analysis of the algorithm performed in Section 6.2.1: As N grows significantly larger than r, the smaller runtime complexity of our method allows for equally accurate reconstructions in a fraction of the time taken by RPCA-IALM. While PRMF and GRASTA are also able to outperform RPCA-IALM in time, these methods achieve less accurate reconstructions due to their alternated nature and sampling techniques, respectively.

TABLE 6.1 Performance comparison of state-of-the-art methods for Robust PCA. Time is in seconds. Error has a factor of 10^{-8}.

Matrix		RPCA-IALM [55]		GRASTA [41]		PRMF [93]		Ours	
N	r	Error	Time	Error	Time	Error	Time	Error	Time
100	3	1.4872	0.3389	226.46	1.7656	3338.7	0.4704	**0.5286**	**0.1734**
200	5	1.5599	2.3575	241.99	2.7282	2687.5	1.0382	**0.7182**	**0.5739**
500	10	3.2595	10.501	263.55	9.5399	1692.4	6.2480	**0.1273**	**3.2373**
1000	15	0.3829	44.111	286.17	23.535	1145.8	30.441	**0.0701**	**14.339**
2000	20	0.6212	196.89	329.11	83.010	808.20	126.95	**0.0308**	**60.658**
5000	25	0.2953	1840.0	379.94	**507.57**	504.08	1307.4	**0.0589**	556.21

Real data

Next, we compared these methods on a real dataset for background modeling. Here, the goal is to obtain the background model of a slowly moving video sequence. Since the background is common across many frames, the matrix concatenating all frames is a low-rank matrix plus a sparse error matrix modeling the dynamic foreground.

We followed the setup of [93] and used the Hall sequence.[16] This dataset consists of 200 frames of video with a resolution of 144×176, and we set the scope of the virtual camera to have the same height, but half the width. We simulated a camera panning by shifting 20 pixels from left to right in frame 100 to simulate a dynamic background. Additionally, we randomly dropped 70% of the pixels. We proceeded as in the previous synthetic experiment. Figure 6.3 shows a visual comparison of the reconstruction of several methods. Results corroborate the experiment in Table 6.1 and show that the lower accuracies of GRASTA and PRMF yield noisier reconstructions than our method.

6.3 Optimizing Hard-Rank Models when Rank Is Known

In Section 6.2, we have presented soft-rank models and showed that these provide a useful technique where a low-rank solution is sought but its rank is not predetermined. However, many problems in computer vision involve the recovery of shape, appearance, or motion representations with a predetermined rank k. Since the recovery of these representations is done from data that is noisy and only partially observed [12, 23, 89], this problem is typically modeled as:

Definition 5 (Rank-k factorization with missing data) *Optimization models that aim to recover a rank-k factorization $\mathbf{U}\mathbf{V}^\top \in \mathbb{R}^{M\times N}$ from a data matrix $\mathbf{X} \in \mathbb{R}^{M\times N}$ with missing entries, as*

$$\min_{\mathbf{Z}} \quad \|\mathbf{W} \odot (\mathbf{X} - \mathbf{U}\mathbf{V}^\top)\|_F^2, \tag{6.39}$$

where $\mathbf{W} \in \mathbb{R}^{M\times N}$ is a positive weight matrix that can be used to denote missing data (i.e., $w_{ij} = 0$), and the rank-k is specified by the inner dimensions of the product $\mathbf{U}\mathbf{V}^\top$.

Unfortunately, (6.39) is an NP-Hard problem where many state-of-the-art algorithms even fail to reach good local minima [36, 71]. For this reason, the optimization of (6.39) remains an active research topic, with many works focusing on algorithms that are robust to initialization [12, 32, 93, 104] or initialization strategies [43]. Buchanan et al. [11] show that alternated minimization algorithms are subject to flatlining and propose a Newton method to jointly optimize $\mathbf{U}$ and $\mathbf{V}$. Okatani et al. [70] show that a Wiberg marginalization strategy on $\mathbf{U}$ or $\mathbf{V}$ is very robust to initialization. However, its high memory usage makes

[16] http://perception.i2r.a-star.edu.sg/bk_model/bk_index.html

it impractical for medium-size datasets. These methods have also been extended to handle outliers [32, 48, 92]. Ke and Kanade [48] suggest replacing the LS error with the L1 norm, minimized by alternated linear programming. Similarly to the LS case, Eriksson et al. [32] show this approach is subject to flatlining and propose a Wiberg extension for L1. Wiberg methods have also been extended to arbitrary loss functions by Strelow [83], but exhibit the same scalability problems as its LS and L1 counterparts. The addition of additional problem specific constraints, e.g., orthogonality of $\mathbf{U}$, has also been shown to help algorithms in attaining better minima in structure from motion [12, 104]. However, these methods are not generalizable to several other computer vision problems that are modeled as low-rank factorization problems [12, 75, 87, 96].

In this section, we show that recent soft rank models (recall (6.8)),

$$\min_{\mathbf{Z}} \quad \|\mathbf{W} \odot (\mathbf{X} - \mathbf{Z})\|_F^2 + \lambda \|\mathbf{Z}\|_*, \tag{6.40}$$

are not ideal to solve problems with a specific predetermined rank-k constraint. To understand why this is the case, let us consider the example of rank-k factorization of a matrix $\mathbf{X}$ under the LS loss with no missing data (i.e., $\mathbf{W} = \mathbf{1}_M \mathbf{1}_N^\top$). For this case, both (6.39) and (6.40) have closed-form solutions in terms of the SVD of $\mathbf{X} = \overline{\mathbf{U}}\Sigma\overline{\mathbf{V}}^\top$, i.e., $\mathbf{U}\mathbf{V}^\top = \overline{\mathbf{U}}\Sigma_{1:k}\overline{\mathbf{V}}^\top$ and $\mathbf{Z} = \overline{\mathbf{U}}\mathcal{S}_{\frac{\lambda}{2}}(\Sigma)\overline{\mathbf{V}}^\top$. In the case of noisy data, while the former yields the optimal rank-k reconstruction, we need to tune λ in the latter such that $\sigma_{k+1} = 0$. If the λ required to satisfy this constraint is high, it may severely distort the non-zero singular values $\sigma_{1:k}$, resulting in poor reconstruction accuracy.

Instead, we argue for using our regularized unified model (recall (6.11))

$$\min_{\mathbf{U},\mathbf{V}} \quad \|\mathbf{W} \odot (\mathbf{X} - \mathbf{Z})\|_F^2 + \frac{\lambda}{2}\left(\|\mathbf{U}\|_F^2 + \|\mathbf{V}\|_F^2\right). \tag{6.41}$$

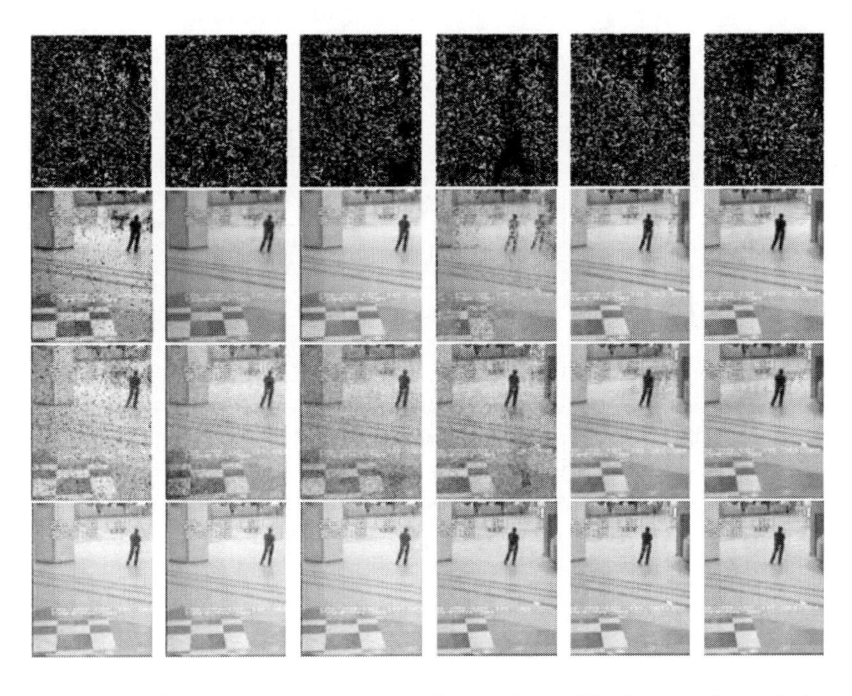

FIGURE 6.3 Results for background modeling with virtual pan. The first row shows the known entries used for training in frames 40, 70, 100, 130, 170, 200. The remaining rows show the results obtained by PRMF, GRASTA and our method, respectively.

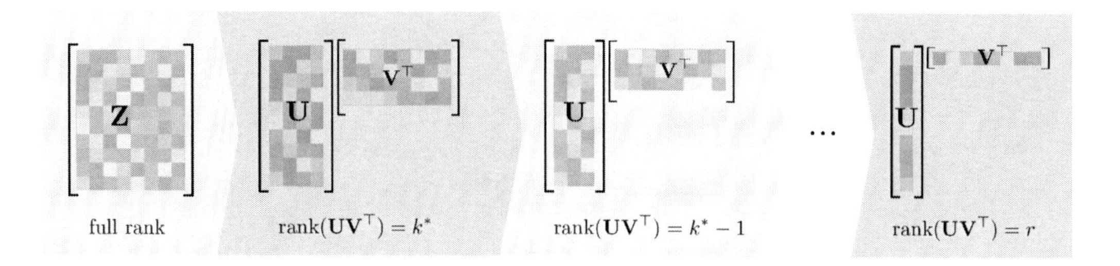

FIGURE 6.4 Illustration of our proposed rank continuation procedure. Starting with a full-rank matrix, we solve the problem and obtain a solution of rank k^*. Under these conditions, Theorem 6.1 guarantees a global solution and equivalence to the convex problem. Then, we solve a sequence of problems of decreasing rank and initialized with the solution of the previous problem, until the desired rank r is attained.

While the analysis in Section 6.2 shows that rank restrictions typically lead to local minima when missing data are present, this problem is exacerbated when regularization is not used (i.e., $\lambda = 0$): in addition, to gauge freedom,[17] it is clear that not all weight matrices $\mathbf{W}$ admit a unique solution [11]. As an extreme example, if $\mathbf{W} = \mathbf{0}$, any choice of $\mathbf{U}$ and $\mathbf{V}$ yields the same (zero) error. Thus, the unregularized factorization in (6.39) will be more prone to local minima than its regularized counterpart (6.41). The two arguments presented against (6.39) and (6.40) provide an argument for choosing our unified model (6.41) and a general guideline for choosing λ: It should be selected as non-zero to ameliorate the local minima problem of (6.39), but small enough such that the first r singular values are not distorted. Moreover, the result in Theorem 6.1 of Section 6.2 that our model is equivalent to the convex nuclear norm model in (6.40) when k is selected to be big enough allows us to provide a deterministic sequence of initializations for this problem. In Section 6.3.1, we propose a "rank continuation" deterministic optimization scheme for the NP-Hard factorization problem that avoids local optima in a significant number of cases. This work has been published in [17]. In Section 6.3.2, we extend the "rank continuation" to the problem of optimizing binary quadratic problems and show an application example for finding correspondences in pairs of images using a max cut formulation.

6.3.1 Rank Continuation for Matrix Factorization

Given that for any fixed λ, as shown in Theorem 6.1 of Section 6.2, (6.41) always has a region with no local minima, we propose the following "rank continuation" strategy: initialize (6.41) with a rank $r \geq k^*$ matrix (i.e., white region of Figure 6.2, where this problem is equivalent to its convex counterpart), to guarantee its convergence to the global solution.[18] Then, use this solution as initialization to a new problem (6.41) where the dimensions r of $\mathbf{U}, \mathbf{V}, \mathbf{\Sigma}$ are decreased by one, until the desired rank is attained. This reduction can be done by using an SVD projection. This approach is summarized in Algorithm 6.2 and illustrated in Figure 6.4. Note this is similar in philosophy to [66] but significantly different in the problem being solved and the continuation path used.

Rank continuation provides a *deterministic* optimization strategy that empirically is

[17] For each solution $\mathbf{U}\mathbf{V}^\top$, any solution $(\mathbf{U}\mathbf{R})(\mathbf{R}^{-1}\mathbf{V}^\top)$ where $\mathbf{R} \in \mathbb{R}^{r \times r}$ is an invertible matrix will provide an equal cost.

[18] Note that in the absence of an estimate for k^*, we can always use $r = \min(M, N)$.

Algorithm 6.2 Rank continuation

Require: $\mathbf{X}, \mathbf{W} \in \mathbb{R}^{M \times N}$, output rank k, parameter λ, an optional estimate of the output rank k^* of (6.3)
  Initialize $\mathbf{U}, \mathbf{V}$ randomly, with $k^* \leq r \leq \min(M, N)$
  Solve for $\mathbf{Z}$ in (6.41) with Alg. 6.1
  for $r = \text{rank}(\mathbf{Z}) - 1, \ldots, k$ **do**
    SVD: $\mathbf{Z} = \overline{\mathbf{U}}\mathbf{\Sigma}\overline{\mathbf{V}}^\top$
    Rank reduce: $\mathbf{U}_r = \overline{\mathbf{U}}\mathbf{\Sigma}_{1:r}^{\frac{1}{2}}$, $\mathbf{V}_r^\top = \mathbf{\Sigma}_{1:r}^{\frac{1}{2}}\overline{\mathbf{V}}^\top$
    Solve $\mathbf{Z}$ in (6.41) with initialization $\mathbf{U}_r, \mathbf{V}_r$ using Alg. 6.1
  end for
Ensure: Complete Matrix $\mathbf{Z}$ with rank k

---

shown to find good optima, compared to other baseline algorithms for this family of problems. In particular, we show in the experimental section that global minima of (6.41) are achieved with this strategy in several cases.

Experimental results

We empirically validated the "rank continuation" strategy proposed in Algorithm 6.2, in several synthetic and real-data problems where the output rank is known *a priori*. We compared our method to state-of-the-art factorization approaches: the damped Newton in [11], the LRSDP formulations in [67], and the LS/L1 Wiberg methods in [32, 70]. Following results reported in the detailed comparisons of [11, 32, 70, 71, 83, 104], we dismissed alternated methods due to their flatlining tendency. To allow direct comparison with published results [11, 67, 70, 104], all methods solved either (6.39) or (6.41) without additional problem-specific constraints, and we fixed $\lambda = 10^{-3}$. For control, we also compared our method to two nuclear norm baselines: NN-SVD, obtained by solving (6.40) with the same λ used for other models and projecting to the desired rank with an SVD; NN-λ, obtained by tuning λ in (6.40) so the desired rank is obtained.

Synthetic data

We assessed the convergence performance of our continuation strategy using synthetic data. We performed synthetic comparisons for two loss choices: LS loss $f(\cdot) = \|\mathbf{W} \odot (\mathbf{X} - \mathbf{Z})\|_F^2$ and the L1 loss $f(\cdot) = \|\mathbf{W} \odot (\mathbf{X} - \mathbf{Z})\|_1$.

For the LS loss, we generated rank-3 matrices $\mathbf{X} = \mathbf{U}\mathbf{V}^\top$. The entries in $\mathbf{U} \in \mathbb{R}^{20 \times 3}$, $\mathbf{V} \in \mathbb{R}^{25 \times 3}$ were sampled i.i.d. from a Gaussian distribution $\mathcal{N}(0, 1)$ and Gaussian noise $\mathcal{N}(0, 0.1)$ was added to every entry of $\mathbf{X}$. For the L1 loss, we proceeded as described for the LS case but additionally corrupted 10% of the entries chosen uniformly at random with outliers uniformly distributed in the range $[-2.5, 2.5]$. We purposely kept the synthetic experiments small, due to the significant memory requirements of the Wiberg algorithms. We varied the percentage of known entries and measured the residual over all *observed* entries, according to the optimized loss function. We chose this measure as it allows for direct comparison between unregularized and regularized models. We ran damped Newton, LRSDP, and Wiberg methods 100 times for each test with random initializations.

Figure 6.5 shows the results for the LS and L1 loss cases. We show two representatives cases for the percentage of known entries (75% and 35%, the breakdown point for L2-Wiberg methods), both for missing data patterns at random (M.A.R.) and with a pattern typical of SfM matrices (Band), generated as in [70]. The theoretical minimum number of entries to reconstruct the matrix is the same as the number of parameters minus factorization ambiguity $Mr + (N - r)(r + 1)$, which for this case is 29.6% [70]. We verified that the

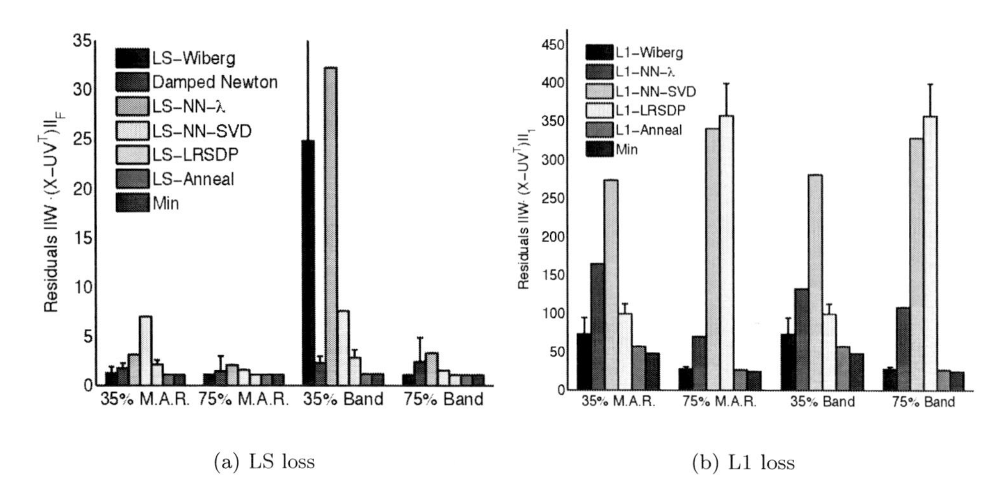

(a) LS loss

(b) L1 loss

FIGURE 6.5 Comparison of convergence to empirical global minima (Min) for the LS and L1 losses in synthetic data. The minima are found as the minimum of all 100 runs of all methods for each test.

behavior of all methods when more than 40% of the entries are known is similar to the result shown for 75%.

For the LS case, results in Figure 6.5(a) show that our deterministic continuation approach always reaches the empirical optima (found as the minimum of all runs of all methods), regardless of the number of known entries or pattern of missing data. Note the minimum error is not zero, due to the variance of the noise. As reported previously [70, 71, 104], we observe that L2-Wiberg is insensitive to initialization for a wide range of missing data entries. However, we note that its breakdown point is not at the theoretical minimum of 35%, due to the lack of regularization. The LRSDP method for optimizing (6.41) outperforms the Wiberg method in this region, suggesting that similar convergence properties of the Wiberg can be obtained without its use of memory. The baseline NN-SVD performed poorly, showing that the estimation of the nuclear norm fits information in its additional degrees of freedom instead of representing it with the true rank.

NN-λ, on the other hand, oversmooths the cost function that might destroy the error function landscape, as is the case for L1 error or when few known entries are available. A visualization of this over-smoothing can be seen in Figure 6.6, where we reproduce the setup of [36] and plot the landscape of the cost function of (6.41) for a factorization of a 3×3 data matrix (i.e., $\mathbf{W} = [1\ 0\ 1; 0\ 1\ 1; 1\ 1\ 1], \mathbf{X} = [1\ 100\ 2; 100\ 1\ 2; 1\ 1\ 1]$) for $r = 1$ and several values of λ spanning the grey and white areas in Figure 6.2. From the figure, it can be seen that while convexifying the landscape is appealing for minimization purposes, some global minima might disappear. For the value of λ chosen, however, it seems we benefit from this smoothing without destroying the global minima landscape. This can be seen in Figure 6.7(a), where we ran Algorithm 6.1 100 times with random initialization for a 20×25 rank-3 matrix with 50% missing entries generated with band pattern as described in the beginning of this section. For each rank, we plotted the minima attained by the algorithm, and compared it with the path obtained by rank continuation. In Figure 6.7, it can be seen that 1) regularized model exhibits a smaller spread in the number of local minima, and 2) the algorithms directly enforcing the solution converge to several local minima with higher cost, whereas rank continuation attains the correct solution in the final stage. One explanation for why rank continuation attains the optima is that the subspace of rank 3

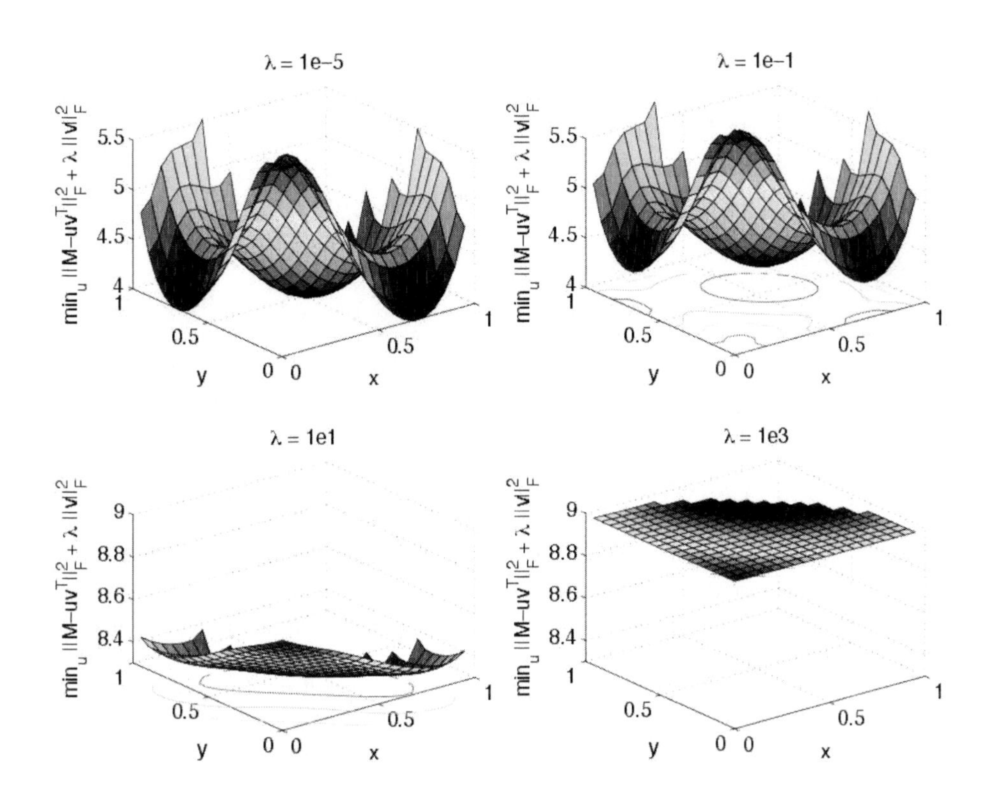

FIGURE 6.6 (See color insert.) Visualization of the cost function in (6.41) for a rank-1 3×3 matrix $\mathbf{X}$ for several values of λ, showing the several local minima existing in the original problem, and that the smoothing induced by the nuclear norm convexifies the problem but its global optima may not necessarily coincide with the original problem's position.

is contained in the one obtained in the convex problem, i.e., the solution obtained when initializing $\mathbf{Z}$ with full rank. This can be seen in Figure 6.7(b), where we measured the Normalized Subspace Inclusion (NSI) [24] between the column subspaces $\mathbf{U}_i$ obtained in each rank step i of the continuation (the row subspaces exhibit the same behavior).

$$NSI(\mathbf{Z}_i, \mathbf{Z}_j) = \frac{\text{trace}(\mathbf{U}_i^\top \mathbf{U}_j \mathbf{U}_j^\top \mathbf{U}_i)}{\min(i, j)} \tag{6.42}$$

For the L1 loss case, results in Figure 6.5(b) show that our continuation strategy no longer attains the empirical optima. We note that this is not surprising since the problem of factorization with missing data is NP-hard. However, its deterministic result is very close to the optima. Our continuation method regained empirical optimality when only 2% of outliers were present in the data, suggesting a dependency on the noise for the L1 case. In this case, our performance is comparable to what is obtained with the L1-Wiberg algorithm [32] on average. Thus, continuation is a viable alternative to this memory-expensive method.

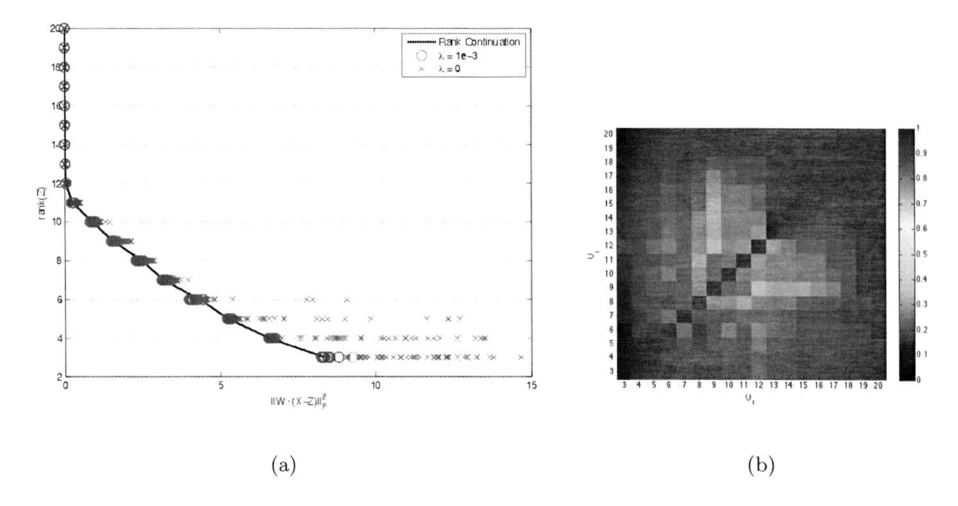

FIGURE 6.7 (See color insert.) Intuition behind Rank Continuation. Figure 6.7(a) Visualization of minima attained for 100 random runs of Algorithm 6.1 for a rank-3 25×20 matrix $\mathbf{X}$ with and without regularization λ shows the several local minima existing in the cost function landscape, and that the smoothing induced by the nuclear norm allows for avoiding some of these. The rank continuation (solid) attains the global optima in the last rank. Figure 6.7(b) Normalized Subspace Inclusion index $NSI(\mathbf{Z}_i, \mathbf{Z}_j)$ measured between the subspaces for each solution step in the continuation for a rank-3 25×20, showing the desired rank-3 subspace is included in the one obtained for the convex region (20).

TABLE 6.2 Real datasets for problems with known output rank-k.

Dataset name	Dataset size	Output rank k	Percentage of known entries
Dino (SFM)	319×72	4	28%
Giraffe (Non-rigid SFM)	240×167	6	70%
Face (Photometric stereo)	2944×20	4	58%
Sculpture (Photometric stereo)	26260×46	3	41%

Real data

Next, we assessed the results of our continuation approach in real data sequences. We used four popular sequences:[19] a) Dino, for affine SfM; b) Giraffe, for non-rigid SfM, and c) Face and d) Sculpture, both photometric stereo sequences. Their details are summarized in Table 6.2. The dimension of these datasets make the usage of the Wiberg algorithms [32] prohibitive in our modest workstation, due to their memory requirements. For the Sculpture dataset, we treated as missing all pixels with intensity greater than 235 or lower than 20 (e.g., in Fig. 6.9(b), the yellow and purple+black masks, resp.). All other datasets provide $\mathbf{W}$.

Table 6.3 shows a comparison of average error over all *observed* entries for the continuation proposed in Algorithm 6.2 and several methods, according to the loss functions L1/LS. "Best" denotes the best-known result in the literature. As explained in Section 6.3.1, we observe that nuclear norm regularized approaches NN-SVD and NN-λ result in bad approximations when a rank restriction is imposed. This can be seen by the high values of λ that have to be used to obtain the desired rank in the variation plots of 6.10. Similar to the results in the synthetic tests, our method always attained or outperformed the state-of-the-art result for the LS loss. The convergence studies in [11, 70] performed optimization on the first three datasets several times with random initializations, so their reported results

[19] http://www.robots.ox.ac.uk/~abm/

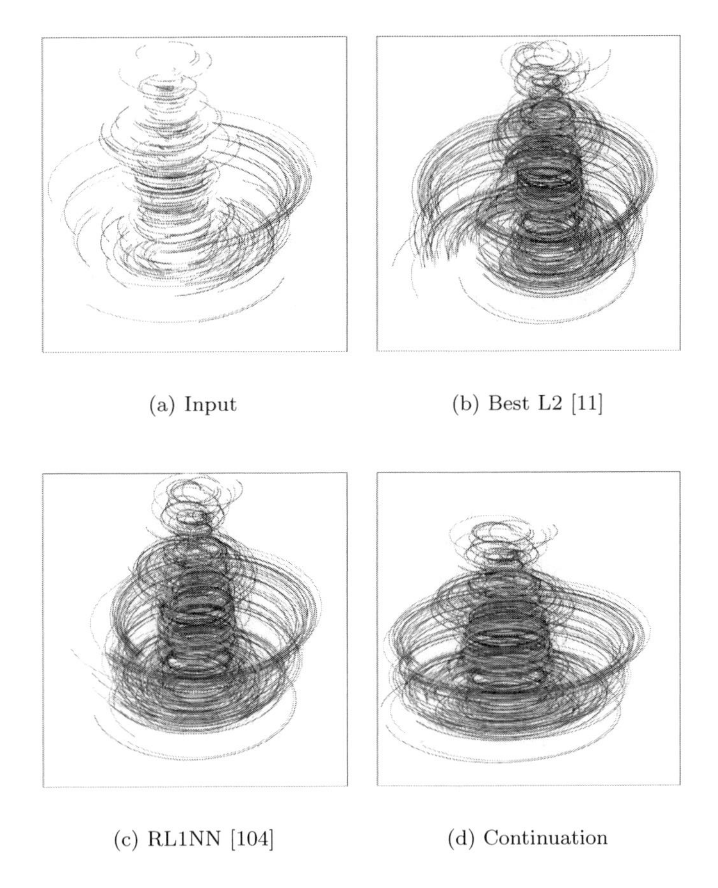

(a) Input (b) Best L2 [11]

(c) RL1NN [104] (d) Continuation

FIGURE 6.8 Structure from motion Dino sequence. Our L1 continuation method 6.8(d) is less prone to local minima and thus can get appealing reconstructions without the use of the additional orthogonality constraints in 6.8(c).

are suspected by the community to be the global optima for these problems. At the cost of solving several rank-constrained problems, our method consistently attains these results in a deterministic fashion, as opposed to state-of-the-art methods that get stuck in local minima several times. As a control experiment, we also ran our continuation strategy for the unregularized case ($\lambda = 0$) on the Dino sequence with LS loss, which resulted in an RMSE of 1.2407 (See Figure 6.8). We attribute this to the fact that this case is more prone to local minima, as mentioned in Section 6.3.1.

For the L1 loss, continuation outperforms the state-of-the art in all datasets. It might be argued that problem-specific constraints are required to obtain clean reconstructions, but we reiterate the importance of escaping local minima. While there are certainly degenerate scenarios that can only be solved with such constraints [63], Algorithm 6.1 (and consequently, Algorithm 6.2) can be trivially extended to handle such cases. For example, the projection step on $\mathbf{U}$ for SfM in [12] can be added to Algorithm 6.1 or the problem can be reformulated as a different SDP [67] with a rank constraint, which can be tackled by our continuation strategy in Algorithm 6.2.

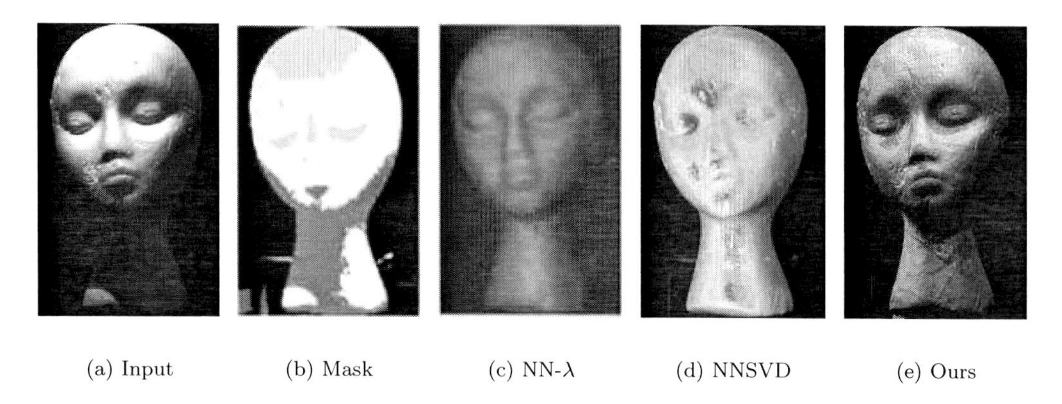

(a) Input (b) Mask (c) NN-λ (d) NNSVD (e) Ours

FIGURE 6.9 Results for frame 17 of the sculpture sequence for photometric stereo. While 6.9(c) smooths out the image and 6.9(d) fails to reconstruct it, our continuation approach 6.9(e) is able to obtain reconstructions that preserve finer details, such as the imperfections on the cheek or chin.

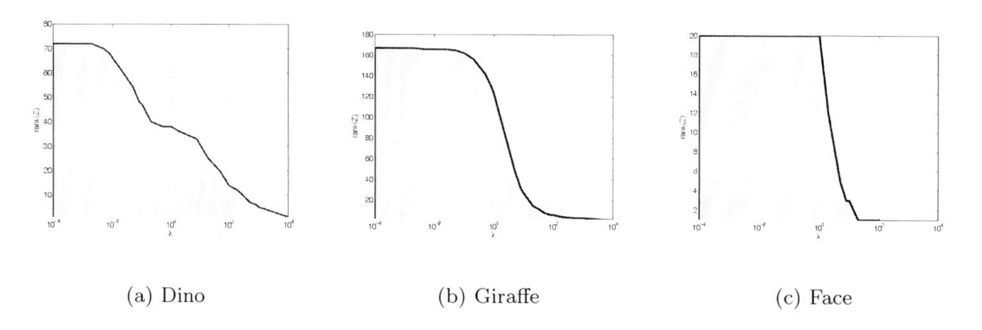

(a) Dino (b) Giraffe (c) Face

FIGURE 6.10 Region of equivalence between factorization (6.41) and nuclear norm approaches (6.40) for several real matrix factorizations with missing data datasets using the least-squares loss.

TABLE 6.3 Comparison of LS/L1 average error over all *observed* entries for structure from motion and photometric stereo datasets. State-of-the-art results were reported in [11, 12, 99, 104]

Dataset name	Error function $f(\cdot)$	Best result reported in the literature	NN-λ	NN-SVD	Ours
Dino	LS	**1.0847**	6.1699	35.8612	**1.0847**
	L1	0.4283	7.6671	80.0544	**0.2570**
Giraffe	LS	**0.3228**	0.4370	0.6519	**0.3228**
	L1	–	1.8974	11.0196	**0.2266**
Face	LS	**0.0223**	0.0301	0.0301	**0.0223**
	L1	–	0.0287	0.6359	**0.0113**
Sculpt	LS	24.6155	44.5859	31.7713	**22.8686**
	L1	17.753	21.828	33.7546	**12.6697**

6.3.2 Rank Continuation for Binary Quadratic Problems

In this Section, we show that the rank continuation strategy devised in Section 6.3.1 can be applied as a black box optimization strategy in problems where a rank constraint exists. One special case of rank-constrained problems are binary quadratic problems (BQPs), where we wish to recover a binary vector $\mathbf{x}$ that maximizes a pairwise cost given by a matrix $\mathbf{C}$, as

$$\begin{aligned} \max_{\mathbf{x}} \quad & \mathbf{x}^\top \mathbf{C}\mathbf{x} \\ \text{subject to} \quad & \mathbf{x} \in \{-1, 1\}. \end{aligned} \tag{6.43}$$

Equation (6.43) is a special case of a rank-constrained model since we can rewrite it without any loss of generality, by lifting the variable product $\mathbf{x}\mathbf{x}^\top$ to a new matrix $\mathbf{X}$, as

$$\begin{aligned} \max_{\mathbf{X}} \quad & \text{trace}(\mathbf{C}^\top \mathbf{X}) \\ \text{subject to} \quad & \text{diag}(\mathbf{X}) = \mathbf{1}, \\ & \mathbf{X} \succeq 0, \\ & \text{rank}(\mathbf{X}) = 1. \end{aligned} \tag{6.44}$$

While important optimization results exist for special cases of these problems — e.g., they become globaly solvable when $\mathbf{C}$ is submodular [9] — the binary constraint in (6.43) makes this general problem NP-hard and thus intractable in high dimensionality settings. These models are very common in computer vision for problems including partitioning and grouping [52]. In fact, graph cuts [9] and Markov Random Fields [47, 85] approaches are the cornerstone to many computer vision algorithms, such as computing depth fields, regularizing image segmentation, or graphical model inference in object classification.

There are two main approaches for solving general large-scale BQPs in computer vision: Semidefinite programming (SDP) approaches, obtained by dropping the rank constraint in (6.44), and spectral approaches.[20] SDP approaches, in particular, work by relaxing intricate constraints of the problem into convex sets in higher-dimensional spaces. Thus, they can be used to obtain upper bounds for combinatorial problems. In fact, these SDP relaxations have been shown to provide better bounds than spectral approaches for many combinatorial problems [27, 37, 52]. Moreover, spectral methods cannot handle inequality constraints, which are necessary in formulations such as segmentation with priors (i.e., biased normalized cuts [62]).

However, three problems remain with SDP approaches. First, they are impaired by the speed of numerical solvers for this problem class. Although off-the-shelf interior point methods can solve SDPs in polynomial time, for many relaxations the exponent in the polynomial complexity bounds is too high for scaling to the large problem sizes typically found in computer vision. Recently, there have been efforts made in the direction of finding scalable and fast approaches for solving this family of SDP problems [94]. However, the effort of obtaining faster algorithmic solutions typically results in bounds that are not as tight as the original SDP formulations. Bie and Cristianini [27] have shown that spectral and SDP relaxations have a continuum of models in between them, and proposed a cascade of relaxations tighter than spectral and looser than SDP. Second, while provable tight bounds have been discovered for specific problems such as the max cut problem, no general result exists on the tightness of bounds when using SDP reformulations for general BQPs. Several efforts have been made in the literature to further tighten the bounds provided by SDP

[20] As can be seen in [72], the "recipe" for obtaining spectral relaxations is to interpret the binary reformulation in (6.43) as $\|\mathbf{x}\|_2^2 = 1$ and reformulate the problem as a generalized eigenvalue problem.

for many problems by adding additional constraints, but often at the cost of exacerbating its scalability problems. Third, the feasible set of the SDP relaxation is convex but not polyhedral, so it is not guaranteed to return a solution in the initial binary domain. Thus, algorithms using these bounds have to rely on postprocessing rounding procedures such as randomized rounding [37], voting schemes, or totally unimodular LP projections [60], whose choice varies according to the problem.

Instead of performing the standard SDP relaxation, we note that the low-rank+SDP formulation in (6.44) has a striking similarity to the formulations of (6.2) and the low-rank SDP models of [13]. The surprising results of [13] allow for the feasibility of large scale SDP problems of this class, by resorting to a factorization model akin to (6.39). Moreover, the deterministic rank continuation strategy we proposed in Sec. 6.3.1 for the NP-hard factorization problem that avoids local optima in a significant number of cases is extendable to this family of problems. That is, we propose to solve a sequence of problems that start in an SDP relaxation and gradually decrease the solution rank until they reach a rank-1 problem, which guarantees a binary solution for (6.43). We show experimentally in Section 6.3.2 that this continuation strategy avoids local optima in a significant number of cases, akin to the results obtained for the factorization problem described in Section 6.3.1. Thus, we believe that rank continuation can be extended to a generic black box optimization strategy for many NP-hard problems of interest in the computer vision domain that can be formulated as rank constrained problems. Contrary to algorithms designed specifically for each problem, our approach covers graph-optimization problems, unsupervised and supervised classification tasks, and inference on Markov random fields without depending on specific assumptions or problem formulations. For instance, image segmentation using normalized cuts [80], matching using the quadratic assignment problem [105], and solving Markov Random Fields have all been formulated as low-rank SDP problems (cf., [27, 78, 91], respectively).

To exemplify how rank continuation problems can be applied to BQPs, let us consider the graph problem below.

Graph Matching

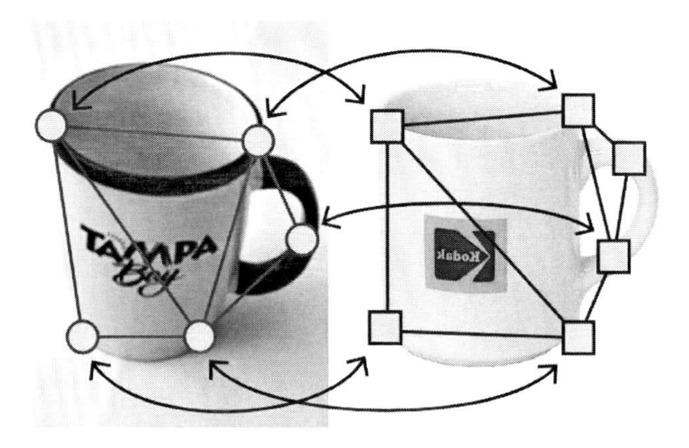

FIGURE 6.11 Example application for a graph matching BQP: finding correspondences between two figures. Adapted from [105].

One BQP problem of interest in computer vision is that of finding correspondences between two images (see Figure 6.11): in this graph-matching problem, each image has a graph of m and n nodes representing interest points, and the goal is to match nodes across images using their similarities and also their shape relationships with neighboring points (modeled as edges on each graph) [105]. This can be given by the formulation

$$\begin{aligned}
\max_{\mathbf{x}\in\{0,1\}^{mn}} \quad & \mathbf{k}^\top \mathbf{x} + \alpha \mathbf{x}^\top \mathbf{K}\mathbf{x} \\
\text{subject to} \quad & \sum_i x_{ij} = 1, \quad \forall 1,\ldots,m \\
& \sum_j x_{ij} \leq 1, \quad \forall 1,\ldots,n,
\end{aligned} \tag{6.45}$$

which maximizes the point similarities k_{ij} of matched pairs and also the edge similarities $K_{ij,kl}$ between the graphs in both images. In this formulation, the element x_{ij} is 1 if the node i on image 1 is to be matched to node j on image 2 and 0 otherwise. By defining $\hat{\mathbf{K}}$ as

$$\hat{\mathbf{K}} = \begin{bmatrix} 0 & 0.5\mathbf{k}^\top \\ 0.5\mathbf{k} & \alpha\mathbf{K} \end{bmatrix} \tag{6.46}$$

and following the "recipe" mentioned in Section 6.3.2, we lift the binary variable $\mathbf{x}$ to a higher dimensional variable

$$\hat{\mathbf{X}} = \begin{bmatrix} 1 & \mathbf{x}^\top \\ \mathbf{x} & \mathbf{x}\mathbf{x}^\top \end{bmatrix} \tag{6.47}$$

and rewrite (6.45) as

$$\begin{aligned}
\max_{\hat{\mathbf{X}}} \quad & \text{trace}(\hat{\mathbf{K}}\hat{\mathbf{X}}) \\
\text{subject to} \quad & \hat{x}_{1,1} = 1, \\
& 2\text{Diag}(\hat{\mathbf{X}}) = \hat{\mathbf{X}}_{1,:} + \hat{\mathbf{X}}_{:,1}^\top, \\
& \mathbf{H}\text{diag}(\mathbf{X}) = \mathbf{1}_m, \\
& \hat{\mathbf{X}} \succeq 0, \\
& \text{rank}(\hat{\mathbf{X}}) = 1,
\end{aligned} \tag{6.48}$$

where $\mathbf{H} = \mathbf{I}_m \otimes \mathbf{1}_n^\top$, and $\hat{\mathbf{X}}_{1,:}$ corresponds to MATLAB notation and denotes the first column of $\hat{\mathbf{X}}$.

The feasible set of the SDP relaxation of (6.48) obtained by dropping the rank constraint is convex but not polyhedral. It contains the set of matrices corresponding to the permutations $\mathbf{x}\mathbf{x}^\top$. But the SDP relaxation solutions discussed above can contain many points not in the affine hull of the constraint set. In particular, it can contain matrices with nonzeros in positions that are zero in the affine hull of the constraint set. So we add additional constraints corresponding to these zeros, which results in

$$\begin{aligned}
\max_{\hat{\mathbf{X}}} \quad & \text{trace}(\hat{\mathbf{K}}\hat{\mathbf{X}}) \\
\text{subject to} \quad & \hat{x}_{1,1} = 1 \\
& 2\text{Diag}(\hat{\mathbf{X}}) = \hat{\mathbf{X}}_{1,:} + \hat{\mathbf{X}}_{:,1}^\top \\
& \mathbf{H}\text{diag}(\mathbf{X}) = \mathbf{1}_m \\
& \mathbf{X} \odot \mathbf{M} = \mathbf{0} \\
& \hat{\mathbf{X}} \succeq 0, \\
& \text{rank}(\hat{\mathbf{X}}) = 1,
\end{aligned} \tag{6.49}$$

where $\mathbf{M} = \mathbf{I}_m \otimes (\mathbf{1}_n^\top \mathbf{1}_n - \mathbf{I}_n) + (\mathbf{1}_m^\top \mathbf{1}_m - \mathbf{I}_m) \otimes \mathbf{I}_n$ is the "gangster operator".[21] The latter rank constraint in (6.49) can be dropped to form an SDP, as

$$\begin{aligned}
\max_{\hat{\mathbf{X}}} \quad & \text{trace}(\hat{\mathbf{K}}\hat{\mathbf{X}}) \\
\text{subject to} \quad & \hat{x}_{1,1} = 1 \\
& 2\text{Diag}(\hat{\mathbf{X}}) = \hat{\mathbf{X}}_{1,:} + \hat{\mathbf{X}}_{:,1}^\top, \\
& \mathbf{H}\text{diag}(\mathbf{X}) = \mathbf{1}_m, \\
& \mathbf{X} \odot \mathbf{M} = \mathbf{0}, \\
& \hat{\mathbf{X}} \succeq 0.
\end{aligned} \tag{6.50}$$

If the optimizer of (6.50) has rank 1, then it is guaranteed to be the optimal result for the original problem (6.45). For the majority of cases, however, the result of (6.50) has higher rank, and thus it is used as an upper bound for (6.45) in the input to a heuristic randomized rounding algorithm [37].

However, the rank of the resulting SDP has been shown to provide useful information when computing bounds [73] or providing strategies for minimization [66]. In fact, if we examine the equivalent reformulation of (6.49) in light of the observations in Section 6.2, it is clear that BQPs are a special case of rank-constrained problems. Thus, the rank continuation proposed in Section 6.3.1 is directly applicable to this problem class. After reformulating the original BQP as an equivalent low-rank formulation, we can solve it by a sequence of problems starting with a convex problem (6.50) and decreasing the rank until a rank-1 problem is achieved. In Section 6.3.2, we show that the rank continuation strategy performs competitvely (and even outperforms in some cases) state-of-the-art algorithms specifically designed for graph matching.

Experimental results

In this section, we compare our rank continuation with SDPCut and several rounding methods, as well as state-of-the-art methods for approximating SDPs in the BQP graph problem of Graph Matching. We report experimental results on two datasets (one synthetic and one real) and compare our method against the state-of-the-art algorithm for graph matching in computer vision [105]. As a baseline, we also compared our method to spectral matching [54] and the use of the minimization algorithm with the rank-1 constraint directly imposed in [13] with a random initialization (sdplr1).

This experiment performed a comparative evaluation of four algorithms on randomly synthesized graphs following the experimental protocol of [105]. For each trial, we constructed two identical graphs, G1 and G2, each of which consists of 10 inlier nodes and later we added outlier nodes to both graphs. For each pair of nodes, the edge is randomly generated according to the edge density parameter $\rho \in [0, 1]$. Each edge in the first graph was assigned a random edge score distributed uniformly, and the corresponding edge in the second graph is perturbed by adding a random Gaussian noise $\mathcal{N}(0, \sigma^2)$. The node-affinity was set to zero. We tested the performance of GM methods under three parameter settings. For each setting, we generated 100 different pairs of graphs and evaluated the average accuracy, obtained by comparing the resulting matrix $\mathbf{x}$ with ground truth, and objective ratio w.r.t. to FGM, by computing the cost function in the original model of (6.45) using the

[21] $\mathbf{M}$ is known in the literature as the "gangster operator" since it shoots holes (zeros) in $\mathbf{X}$. We note that additional constraints can be introduced to tighten the bounds obtained by the SDP, as in [13, 73, 94], but this occurs in even more scalability problems as the number of constraints increase.

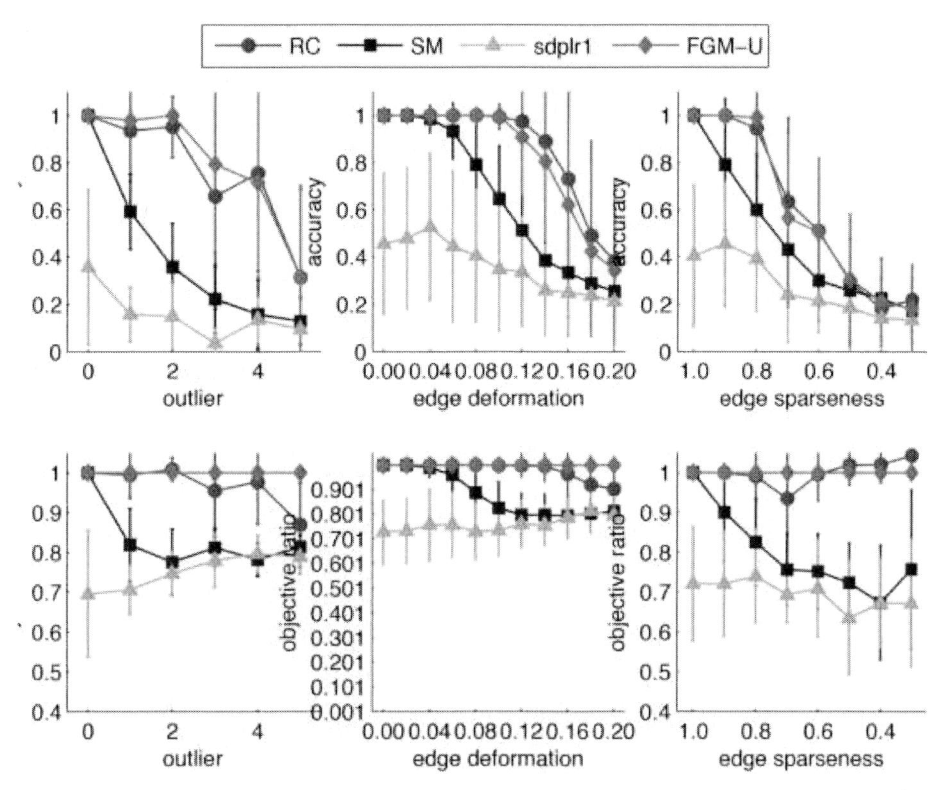

FIGURE 6.12 Accuracy and objective function result for the graph-matching problem of (6.45) in synthetic data for rank continuation (RC), spectral matching (SM), Burer and Monteiro [13] (sdplr1), and Factorized graph matching [105] (FGM). Notice that a ratio bigger than 1 means RC obtains a higher-cost function than that of the baseline (FGM). Left: varying number of outliers with no noise and fully connected graphs. Middle: varying edge deformation by changing the noise parameter σ from 0 to 0.2, with zero outliers and fully connected graphs. Right: varying edge sparsity with zero outliers and no noise.

obtained $\mathbf{x}$ for each method. In the first setting (Figure 6.12 left), we increased the number of outliers from 0 to 10 while fixing the noise to zero and considering only fully connected graphs (i.e., $\rho = 1$). In the second case (Figure 6.12 middle), we perturbed the edge weights by changing the noise parameter σ from 0 to 0.2, while fixing the number of outliers to 0 and $\rho = 1$. In the last case (Figure 6.12 right), we verified the performance of matching sparse graphs by varying ρ from 1 to 0.3.

Under varying parameters, it can be observed that in most cases, our method achieves state-of-the-art performance in terms of both accuracy and objective ratio, being comparable to FGM [105]. We note that there are cases when FGM achieves higher accuracies. This occurs for results that have a smaller cost function than the one obtained by continuation. This can be attributed to the fact that the optimization problem in (6.45) does not model the rigid matching problem entirely, since rigid motion requires higher-order constraints instead of the second-order constraints imposed by the model [31].

Additionally, we compared these methods in a real image sequence. The CMU house image sequence is commonly used to test the performance of graph matching algorithms (see Figure 6.13). This dataset consists of 111 frames of a house, each of which has been manually labeled with 30 landmarks. We used Delaunay triangulation to connect the landmarks. The edge weights are computed as the pairwise distance between the connected nodes, as in [105].

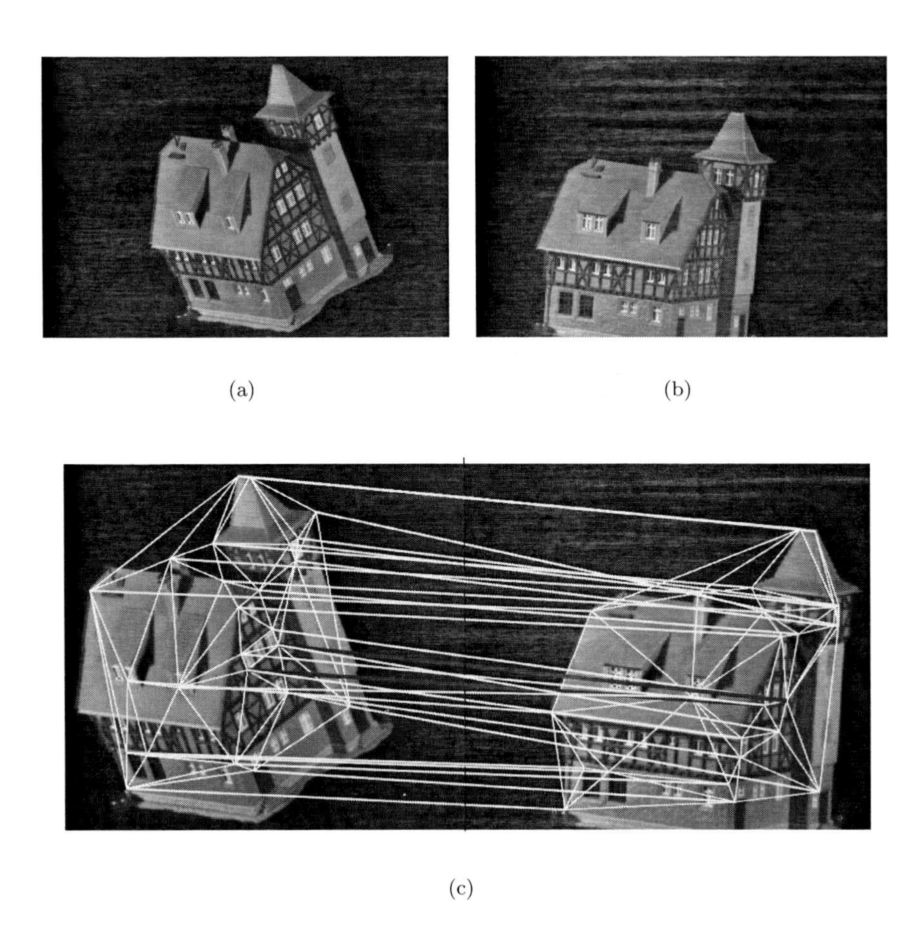

FIGURE 6.13 (See color insert.) An example pair of frames (0 and 99) of the House dataset. An example pair of frames with the correspondence generated by FGM [105], where the blue lines indicate incorrect matches.

We tested the performance of all methods as a function of the separation between frames. We matched all possible image pairs, spaced exactly by 0 : 10 : 90 frames, and computed the average matching accuracy and objective ratio per sequence gap. We tested the performance of graph-matching methods under two scenarios. In the first case (Figure 6.14 left) we used all 30 nodes (i.e., landmarks) and in the second one (Figure 6.14 right) we matched sub-graphs by randomly picking 25 landmarks from each graph. It can be observed that in the first case, FGM, sdplr1, and our method obtained perfect matching of the original graphs. As some nodes became invisible and the graph got corrupted (Figure 6.14 right), the performance of all the methods was degraded. However, our method consistently achieved the best maximum in the objective function.

The results show that rank continuation outperforms all baselines and performs comparably and in some cases better than FGM, which is considered the state-of-the-art for this problem. We note that this algorithm is sophisticated in its use of the specific structure of the graph-matching problem, whereas our strategy is general for rank-constrained problems.

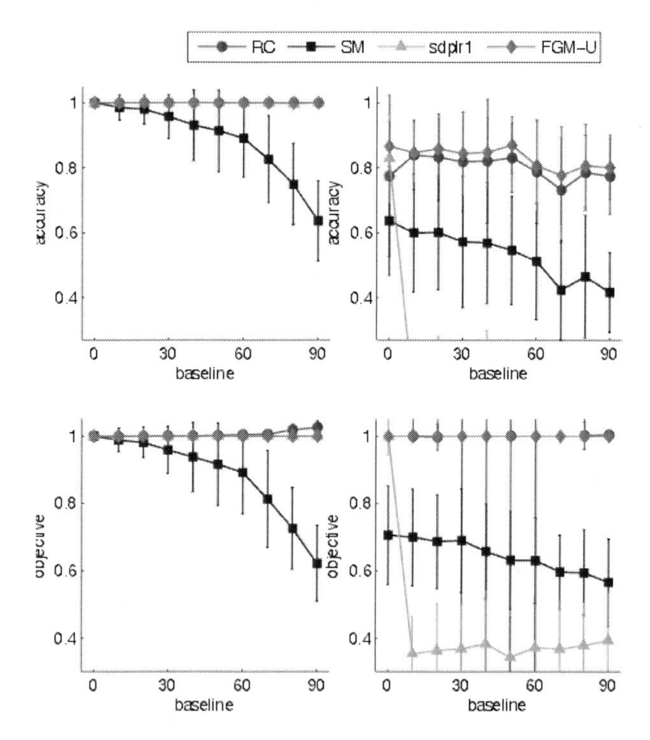

FIGURE 6.14 Accuracy and objective function value results for the graph-matching problem of (6.45) in the CMU House dataset for each baseline (frame distance) using rank continuation (RC), spectral matching (SM), Burer and Monteiro [13] (sdplr1), and Factorized graph matching [105] (FGM). Left: Full data used (30 landmarks). Right: results of randomly picking 25 landmarks on each frame.

6.4 Conclusion

The theoretical results and algorithms presented in Section 6.2 show that future work in factorization algorithms should optimize the presented unified model, since it subsumes and inherits benefits of both traditional factorization and the nuclear norm regularized approaches. In Section 6.2 and [16], we have presented Augmented Lagrange Multiplier and Fixed-Point Continuation methods to optimize nuclear norm problems and have studied their convergence properties. An alternative method for incremental nuclear norm optimization, not included in this thesis, can be found in [14]. Based on this analysis and algorithms, in Section 6.3 we proposed a deterministic "rank continuation" strategy that outperforms state-of-the-art factorization approaches in several computer vision applications with outliers and missing data. Preliminary results show that this strategy is also generalizable for binary quadratic problems such as the quadratic assignment problem.

At the time of writing of this book, some questions still remain open. We provide a summary of these below:

How do distributed alternatives for matrix factorization with constraints compare to Wiberg?

Our work has shown that, for the factorization problem, ALM is a strong contender in terms of attaining global minimum solutions. However, one problem with this framework is its inability to tackle very large-scale datasets, such as the ones in [53, 83]. While there has been a surge of research in distributed algorithms for ALM [68] and parallel implementations for matrix completion using stochastic gradient descent [64], further investigation is required to compare these models to recent work in Wiberg algorithms, which look very promising in terms of its applicability to large-scale problems [83].

Why does Rank Continuation work and are there faster alternatives?

One explanation for why the rank continuation presented in Section 6.3 attains good optima is that the solution subspace of rank-k is contained in the one obtained in the convex problem, i.e., the one obtained when initializing $\mathbf{Z}$ with full rank. The original convex solution containing the desired subspace opens the potential for more efficient ways to select the desired subspace from the former, rather than having to run multiple iterations of the algorithm with decreasing rank. One potential solution would be to tackle the problem as a combinatorial problem, similarly to [46], but formulating it as a selection problem from the basis obtained in $\mathbf{Z}$ and potentially exploiting unimodularity totally in order to obtain a solution from convex programming, as done in [15].

Can Rank problems help explain/improve deep neural nets?

Recently, there has been a surge of impressive results in the area of object classification, provided by features learned by deep neural networks. However, one problem these approaches currently face is the fact that the neural network topology has to be configured manually. While insights exist on how to perform this task [45], it still mainly is done by a process of trial and error, and due to the large size of these networks [79, 84], at the expense of a significant use of computation power.

We notice that in the past, component analysis techniques have been related to neural networks and this connection has been used to explain the inexistence of local minima in the principal component analysis factorization cost function [7], which enabled for least-square based algortihms that enabled the use of this technique in very large datasets [28]. Since early termination has been shown to enforce sparsity in the networks and can be seen as a connection to l1-normalization [82] (early stopping is also a known trick in l1-minimization norma algorithms for obtaining sparse solutions), and since linear auto-encoder networks can be shown to be equivalent to a hard-rank matrix factorization model (PCA), we wonder if modeling auto-encoders as a rank problem and replacing them with soft-rank regularizers could help in automatically discovering good topologies.

Can rank continuation be extended to cardinality problems?

In Section 6.3.2, we showed that the strategy devised for matrix factorization with missing data problems can be extended to Binary Quadratic Problems that can be reformulated as SDP problems with a rank constraint. Since the nuclear norm (the sum of the singular values) can be seen as an ℓ_1-norm in the matrix domain, one could potentially extend our findings to LASSO-like problems, comprised of an error function and an ℓ_1-norm regularization together with a cardinality constraint, as

$$\begin{array}{ll} \min\limits_{\mathbf{x}} & \|\mathbf{y}-\mathbf{x}\|_2^2 + \lambda\|\mathbf{x}\|_1 \\ \text{subject to} & \text{card}(\mathbf{x}) = k. \end{array} \tag{6.51}$$

These models are used in dictionary learning [61] and regression problems. Decimation algorithms with cardinality constraints are also especially important in computer vision for the simplification of noisy meshes of regular structures obtained from multi-view stereo pipelines [35], as is the case of buildings in, e.g., Google street view. One might extend the strategy of rank continuation to cardinality problems of (6.51), by initially solving the convex ℓ_1 problems (dropping the cardinality constraint) and then a sequence of problems with decreasing cardinality constraints. Furthermore, the existing study of parameters in

LASSO problems and the connections of this problem to its matrix counterpart in [39] could yield important insights about the parameter λ in nuclear norm regularized problems.

References

1. E. Adeli Mosabbeb, R. Cabral, F. De la Torre, and M. Fathy. Multi-label discriminative weakly-supervised human activity recognition and localization. In *Asian Conference on Computer Vision, ACCV 2014*, 2014.
2. P. Aguiar, J. Xavier, and M. Stosic. Spectrally optimal factorization of incomplete matrices. In *IEEE Conference on Computer Vision and Pattern, CVPR 2008*, 2008.
3. Y. Amit, M. Fink, N. Srebro, and S. Ullman. Uncovering shared structures in multiclass classification. In *International Conference on Machine Learning, ICML 2007*, New York, New York, USA, 2007.
4. R. Angst, C. Zach, and M. Pollefeys. The generalized trace-norm and its application to structure-from-motion problems. In *ICCV*, 2011.
5. A. Argyriou, C. Micchelli, and M. Pontil. On spectral learning. *JMLR*, 11:935–953, 2010.
6. F. Bach and Inria Willow Project-team. Consistency of Trace Norm Minimization. *Journal of Machine Learning Research*, 8:101–1048, 2008.
7. P. Baldi and K. Hornik. Neural networks and principal component analysis: Learning from examples without local minima. *Neural Networks*, 2(1):53–58, 1989.
8. L. Balzano, R. Nowak, and B. Recht. Online identification and tracking of subspaces from highly incomplete information. In *Annual Allerton Conference*, 2010.
9. Y. Boykov, O. Veksler, and R. Zabih. Fast Approximate Energy Minimization via Graph Cuts. In *International Conference on Computer Vision, ICCV 1999*, 1999.
10. M. Brand. Incremental singular value decomposition of uncertain data with missing values. In *European Conference on Computer Vision, ECCV 2002*, pages 707–720, 2002.
11. A. Buchanan and A.. Fitzgibbon. Damped Newton algorithms for matrix factorization with missing data. In *IEEE Conference on Computer Vision and Pattern, CVPR 2005*, 2005.
12. A. Bue, J. Xavier, L. Agapito, and M. Paladini. Bilinear Modelling via Augmented Lagrange Multipliers (BALM). *PAMI*, 2012.
13. S. Burer and R. Monteiro. Local minima and convergence in low-rank semidefinite programming. *Mathematical Programming*, 103(3):427–444, 2005.
14. R. Cabral, J. Costeira, F. De la Torre, and A. Bernardino. Fast incremental method for matrix completion: an application to trajectory correction. In *International Conference on Image Processing, ICIP 2011*, 2011.
15. R. Cabral, J. Costeira, F. De la Torre, and A. Bernardino. Optimal no-intersection multi-label binary localization for time series using totally unimodular linear programming. In *ICIP*, 2014.
16. R. Cabral, F. De la Torre, J. Costeira, and A. Bernardino. Matrix completion for multi-label image classification. In *Annual Conference on Neural Information Processing Systems, NIPS 2011*, 2011.
17. R. Cabral, F. De la Torre, J. Costeira, and A. Bernardino. Unifying nuclear norm and bilinear factorization approaches for low-rank matrix decomposition. In *International Conference on Computer Vision, ICCV 2013*, 2013.
18. R. Cabral, F. Torre, J. Costeira, and A. Bernardino. Matrix completion for weakly-supervised multi-label image classification. *IEEE Transactions on Pattern Analysis and Machine Intelligence*, 37(1):121–135, Jan 2015.
19. J. Cai, E. Candes, and Z. Shen. A singular value thresholding algorithm for matrix completion. *SIAM Journal on Optimization*, 20(4):1956–1982, 2008.

20. E. Candès, X. Li, Y. Ma, and J. Wright. Robust principal component analysis? *Journal of the ACM*, 58(3), 2011.
21. E. Candès and B. Recht. Exact low-rank matrix completion via convex optimization. In *Allerton*, 2008.
22. B. Cheng, G. Liu, J. Wang, Z. Huang, and S. Yan. Multi-task low-rank affinity pursuit for image segmentation. In *ICCV*, 2011.
23. J. Costeira and T. Kanade. A multibody factorization method for independently moving objects. *International Journal of Computer Vision, IJCV 1998*, 29(3):159–179, 1998.
24. N. da Silva and J. Costeira. The normalized subspace inclusion: Robust clustering of motion subspaces. In *International Conference on Computer Vision, ICCV 2009*, 2009.
25. Y. Dai, H. Li, and M. He. Element-wise factorization for N-view projective reconstruction. In *European Conference on Computer Vision, ECCV 2010*, 2010.
26. Y. Dai, H. Li, and M. He. A simple prior-free method for non-rigid structure-from-motion factorization. In *IEEE Conference on Computer Vision and Pattern Recognition, CVPR 2010*, 2012.
27. T. Bie De and N. Cristianini. Fast SDP relaxations of graph cut clustering, transduction, and other combinatorial problems. *JMLR*, 7:1409–1436, December 2006.
28. F. De la Torre. A least-squares framework for component analysis. *IEEE Transactions on Pattern Analysis and Machine Intelligence, PAMI 2012*, 34(6):1041–1055, 2012.
29. D. DeCoste. Collaborative prediction using ensembles of maximum margin matrix factorizations. In *International Conference on Machine Learning, ICML 2006*, 2006.
30. D. Donoho. Aide-memoire. high-dimensional data analysis: The curses and blessings of dimensionality, 2000.
31. O. Duchenne, F. Bach, In-So Kweon, and J. Ponce. A tensor-based algorithm for high-order graph matching. *IEEE Transactions on Pattern Analysis and Machine Intelligence*, 33(12):2383–2395, December 2011.
32. A. Eriksson and A. van den Hengel. Efficient computation of robust low-rank matrix approximations in the presence of missing data using the L1 norm. In *IEEE Conference on Computer Vision and Pattern, CVPR 2010*, 2010.
33. P. Favaro, R. Vidal, and A. Ravichandran. A closed form solution to robust subspace estimation and clustering. In *CVPR*, 2011.
34. M. Fazel, H. Hindi, and S. Boyd. A rank minimization heuristic with application to minimum order system approximation. In *American Control Conference, ACC 2001*, 2001.
35. Y. Furukawa, B. Curless, S. Seitz, and R. Szeliski. Towards internet-scale multi-view stereo. In *CVPR*, 2010.
36. N. Gillis and F. Glineur. Low-Rank Matrix Approximation with Weights or Missing Data Is NP-Hard. *SIAM Journal on Matrix Analysis and Applications*, 32(4), 2011.
37. M. Goemans and D. Williamson. Improved approximation algorithms for maximum cut and satisfiability problems using semidefinite programming. *Journal of ACM*, 42(6):1115–1145, November 1995.
38. A. Goldberg, X. Zhu, B. Recht, J. Xu, and R. Nowak. Transduction with matrix completion: Three birds with one stone. In *Annual Conference on Neural Information Processing Systems, NIPS 2010*, 2010.
39. E. Grave, G. Obozinski, and F. Bach. Trace Lasso: a trace norm regularization for correlated designs. In *NIPS*, 2011.
40. Z. Harchaoui, M. Douze, M. Paulin, M. Dudik, and J. Malick. Large-scale image classification with trace-norm regularization. In *IEEE Conference on Computer Vision and Pattern, CVPR 2012*, 2012.
41. J. He, L. Balzano, and A. Szlam. Incremental gradient on the Grassmannian for online foreground and background separation in subsampled video. In *IEEE Conference on Computer Vision and Pattern, CVPR 2012*, 2012.

42. D. Huang, R. Cabral, and F. De la Torre. Robust regression. In *ECCV*, 2012.
43. D. Jacobs. Linear fitting with missing data for structure-from-motion. *CVIU*, 82:206–2012, 1997.
44. Martin Jaggi and Marek Sulovský. A Simple Algorithm for Nuclear Norm Regularized Problems. In *ICML*, 2010.
45. K. Jarrett, K. Kavukcuoglu, M. Ranzato, and Y. LeCun. What is the best multi-stage architecture for object recognition? In *International Conference on Computer Vision, ICCV 2009*, 2009.
46. F. Jiang, O. Enqvist, and F. Kahl. A combinatorial approach to l1-matrix factorization. *Journal of Mathematical Imaging and Vision*, 2014.
47. J. Kappes, B. Andres, F. Hamprecht, C. Schnörr, S. Nowozin, D. Batra, S. Kim, B. Kausler, T. Kröger, J. Lellmann, N. Komodakis, B. Savchynskyy, and C. Rother. A comparative study of modern inference techniques for structured discrete energy minimization problems. *CoRR*, abs/1404.0533, 2014.
48. Q. Ke and T. Kanade. Robust l_1 norm factorization in the presence of outliers and missing data by alternative convex programming. In *IEEE Conference on Computer Vision and Pattern, CVPR 2005*, 2005.
49. R. Kennedy, L. Balzano, S. Wright, and C. Taylor. Online algorithms for factorization-based structure from motion. *CoRR*, abs/1309.6964, 2013.
50. R. Keshavan, A. Montanari, and S. Oh. Matrix completion from a few entries. *IEEE Transactions on Information Theory*, 56:2980–2998, June 2010.
51. R. Keshavan and S. Oh. A gradient descent algorithm on the Grassman manifold for matrix completion. *arXiv*, 0910.5260, 2009.
52. J. Keuchel, C. Schnorr, C. Schellewald, and D. Cremers. Binary partitioning, perceptual grouping, and restoration with semidefinite programming. *IEEE Transactions on Pattern Analysis and Machine Intelligence*, 25:1364–1379, 2003.
53. B. Klingner, D. Martin, and J. Roseborough. Street view motion-from-structure-from-motion. In *International Conference on Computer Vision, ICCV 2013*, 2013.
54. M. Leordeanu and M. Hebert. A spectral technique for correspondence problems using pairwise constraints. In *International Conference on Computer Vision, ICCV 2005*, 2005.
55. Z. Lin, M. Chen, and Y. Ma. The Augmented Lagrange Multiplier Method for Exact Recovery of Corrupted Low-Rank Matrices. *Mathematical Programming*, 2010.
56. G. Liu, Z. Lin, and Y. Yu. Robust subspace segmentation by low-rank representation. In *International Conference on Machine Learning, ICML 2010*, 2010.
57. R. Liu, Z. Lin, S. Wei, and Z. Su. Solving principal component pursuit in linear time via l_1 filtering. *CoRR*, abs/1108.5359, 2011.
58. N. Loeff and A. Farhadi. Scene discovery by matrix factorization. In *ECCV*, 2008.
59. S. Ma, D. Goldfarb, and L. Chen. Fixed point and Bregman iterative methods for matrix rank minimization. *Mathematical Programming, to appear*, 2009.
60. J. Maciel and J. Costeira. A global solution to sparse correspondence problems. *IEEE Transactions on Pattern Analysis and Machine Intelligence*, 25(2):187–199, 2003.
61. J. Mairal, F. Bach, J. Ponce, and G. Sapiro. Online dictionary learning for sparse coding. In *International Conference on Machine Learning, ICML 2009*, 2009.
62. S. Maji, N. Vishnoi, and J. Malik. Biased normalized cuts. In *IEEE Conference on Computer Vision and Pattern Recognition, CVPR 2011*, 2011.
63. M. Marques and J. Costeira. Estimating 3D shape from degenerate sequences with missing data. *CVIU*, 113(2):261–272, 2009.
64. R. Mazumder, T. Hastie, and R. Tibshirani. Spectral regularization for learning large incomplete matrices. *Journal of Machine Learning Research, JMLR 2010*, 99:2287–2322, 2010.

65. K. Min, Z. Zhang, J. Wright, and Y. Ma. Decomposing Background Topics from Keywords by Principal Component Pursuit. In *CIKM*, 2010.
66. B. Mishra, G. Meyer, F. Bach, and R. Sepulchre. Low-rank optimization with trace norm penalty. *CoRR*, abs/1112.2318, 2011.
67. K. Mitra, S. Sheorey, and R. Chellappa. Large-scale matrix factorization with missing data under additional constraints. In *NIPS*, 2010.
68. J. Mota, J. Xavier, P. Aguiar, and M. Puschel. D-admm: A communication-efficient distributed algorithm for separable optimization. *IEEE Transactions on Signal Processing*, 61(10):2718–2723, May 2013.
69. F. Ojeda, J. Suykens, and B. De Moor. Low rank updated LS-SVM classifiers for fast variable selection. *Neural networks : the official journal of the International Neural Network Society*, 21(2-3):437–49, 2008.
70. T. Okatani and K. Deguchi. On the Wiberg algorithm for factorization with missing components. *IJCV*, 72(3):329–337, 2007.
71. T. Okatani, T. Yoshida, and K. Deguchi. Efficient algorithm for low-rank matrix factorization with missing components and performance comparison of latest algorithms. In *ICCV*, 2011.
72. C. Olsson, A. Eriksson, and F. Kahl. Solving large scale binary quadratic problems: Spectral methods vs. semidefinite programming. In *IEEE Conference on Computer Vision and Pattern Recognition, CVPR 2007*, 2007.
73. J. Peng, H. Mittelmann, and X. Li. A new relaxation framework for quadratic assignment problems based on matrix splitting. *Mathematical Programming Computation*, 2(1):59–77, 2010.
74. Y. Peng, A. Ganesh, J. Wright, W. Xu, and Y. Ma. Rasl: Robust alignment by sparse and low-rank decomposition for linearly correlated images. In *IEEE Conference on Computer Vision and Pattern, CVPR 2010*, 2010.
75. H. Pirsiavash, D. Ramanan, and C. Fowlkes. Bilinear classifiers for visual recognition. In *Annual Conference on Neural Information Processing Systems, NIPS 2009*, 2009.
76. B. Recht, M. Fazel, and P. Parrilo. Guaranteed minimum-rank solutions of linear matrix equations via nuclear norm minimization. *SIAM Review*, 52(3):471–501, 2010.
77. B. Recht, M. Fazel, and P. Parrilo. Guaranteed minimum-rank solutions of linear matrix equations via nuclear norm minimization. *SIAM Reviews*, 52(3):471–501, August 2010.
78. C. Schellewald and C. Schnörr. Probabilistic Subgraph Matching Based on Convex Relaxation. In *EMMCVPR*, 2005.
79. P. Sermanet, D. Eigen, X. Zhang, M. Mathieu, R. Fergus, and Y. LeCun. Overfeat: Integrated recognition, localization and detection using convolutional networks. In *International Conference on Learning Representations, ICLR 2014*, 2014.
80. J. Shi and J. Malik. Normalized cuts and image segmentation. *IEEE Transactions on Pattern Analysis and Machine Intelligence (PAMI)*, 2000.
81. N. Srebro, J. Rennie, and T. Jaakkola. Maximum-Margin Matrix Factorization. In *Annual Conference on Neural Information Processing Systems, NIPS 2005*, 2005.
82. N. Srivastava, G. Hinton, A. Krizhevsky, I. Sutskever, and R. Salakhutdinov. Dropout: A simple way to prevent neural networks from overfitting. *JMLR*, 15:1929–1958, 2014.
83. D. Strelow. General and nested Wiberg minimization. In *CVPR*, 2012.
84. C. Szegedy, W. Liu, Y. Jia, P. Sermanet, S. Reed, D. Anguelov, D. Erhan, V. Vanhoucke, and A. Rabinovich. Going deeper with convolutions. *CoRR*, abs/1409.4842, 2014.
85. R. Szeliski, R. Zabih, D. Scharstein, O. Veksler, V. Kolmogorov, Aseem Agarwala, M. Tappen, and C. Rother. A comparative study of energy minimization methods for Markov random fields with smoothness-based priors. *IEEE Transactions on Pattern Analysis and Machine Intelligence*, 30(6):1068–1080, June 2008.
86. X. Tan, Y. Li, J. Liu, and L. Jiang. Face Liveness Detection from A Single Image with Sparse

Low Rank Bilinear Discriminative Model. In *ECCV*, 2010.

87. J. Tenenbaum and W. Freeman. Separating Style and Content with Bilinear Models. *Neural Computation*, 1283:1247–1283, 2000.
88. K. Toh and S. Yun. An accelerated proximal gradient algorithm for nuclear norm regularized least squares problems. *Preprint*, 2009.
89. C. Tomasi and T. Kanade. Shape and motion from image streams under orthography: a factorization method. *IJCV*, 9:137–154, 1992.
90. R. Tomioka and K. Aihara. Classifying matrices with a spectral regularization. In *International Conference on Machine Learning, ICML 2007*, 2007.
91. P. Torr. Solving Markov random fields using semi definite programming. In *International Conference on Artificial Intelligence and Statistics, AISTATS 2003*, 2003.
92. F. De La Torre and M. Black. A Framework for Robust Subspace Learning. *International Journal of Computer Vision, IJCV 2003*, 54, 2003.
93. N. Wang, T. Yao, J. Wang, and D. Yeung. A probabilistic approach to robust matrix factorization. In *European Conference on Computer Vision, ECCV 2012*, 2012.
94. P. Wang, C. Shen, and A. van den Hengel. A fast semidefinite approach to solving binary quadratic problems. In *IEEE Conference on Computer Vision and Pattern Recognition,CVPR 2013*, 2013.
95. S. Wang and Z. Zhang. Colorization by matrix completion. In *Association for the Advancement of Artificial Intelligence, AAAI 2012*, 2012.
96. J. Warrell, P. Torr, and S. Prince. StyP-Boost: A Bilinear Boosting Algorithm for Style-Parameterized Classifiers. In *British Machine Vision Conference, BMVC 2010*, 2010.
97. L. Wolf, H. Jhuang, and T. Hazan. Modeling Appearances with Low-Rank SVM. In *IEEE Conference on Computer Vision and Pattern, CVPR 2007*, 2007.
98. J. Wright, A. Ganesh, S. Rao, Y. Peng, and Y. Ma. Robust principal component analysis: Exact recovery of corrupted low-rank matrices via convex optimization. In *Annual Conference on Neural Information Processing Systems, NIPS 2009*, 2009.
99. L. Wu, A. Ganesh, B. Shi, Y. Matsushita, Y. Wang, and Y. Ma. Robust photometric stereo via low-rank matrix completion and recovery. In *Asian Conference on Computer Vision, ACCV 2010*, 2010.
100. F. Xiong, O. Camps, and M. Sznaier. Dynamic context for tracking behind occlusions. In *European Conference on Computer Vision, ECCV 2010*, 2012.
101. O. Yakhnenko and V. Honavar. Multi-Instance Multi-Label Learning for Image Classification with Large Vocabularies. In *BMVC 2011*, 2011.
102. Z. Zhang, X. Liang, and Y. Ma. Unwrapping low-rank textures on generalized cylindrical surfaces. In *ICCV*, 2011.
103. Z. Zhang, Y. Matsushita, and Y. Ma. Camera calibration with lens distortion from low-rank textures. In *IEEE Conference on Computer Vision and Pattern, CVPR 2011*, 2011.
104. Y. Zheng, G. Liu, S. Sugimoto, S. Yan, and M. Okutomi. Practical low-rank matrix approximation under robust l1-norm. In *IEEE Conference on Computer Vision and Pattern Recognition, CVPR 2012*, 2012.
105. F. Zhou and F. De la Torre. Factorized graph matching. In *IEEE Conference on Computer Vision and Pattern Recognition, CVPR 2012*, 2012.
106. G. Zhu and S. Yan. Image tag refinement towards low-rank, content-tag prior and error sparsity. *International Conference on Multimedia, MM 2010*, pages 461–470, 2010.

7

Robust Non-Negative Matrix Factorization under Separability Assumption

Abhishek Kumar
IBM T.J. Watson Research Center, New York, USA

Vikas Sindhwani
Google Research, New York, USA

7.1 Introduction

The problem of non-negative matrix factorization (NMF) is to express a non-negative matrix $\mathbf{X}$ of size $m \times n$, either exactly or approximately, as a product of two non-negative matrices, $\mathbf{W}$ of size $m \times r$ and $\mathbf{H}$ of size $r \times n$. Approximate NMF attempts to minimize a measure of divergence between the matrix $\mathbf{X}$ and the factorization $\mathbf{WH}$. The inner dimension of the factorization r is usually taken to be much smaller than m and n to get interpretable part-based representation of data [22]. NMF is used in a wide range of applications, e.g., topic modeling and text mining, hyper-spectral image analysis, audio source separation, and microarray data analysis [7].

The exact and approximate NMF problem is NP-hard. Hence, traditionally, algorithmic work in NMF has focused on treating it as an instance of non-convex optimization [7, 19, 22, 25] leading to algorithms lacking optimality guarantees beyond convergence to a stationary point. Promising alternative approaches have emerged recently based on a *separability* assumption on the data [2, 4, 12, 15–17, 21], which enables the NMF problem to be solved efficiently and exactly. Under this assumption, the data matrix $\mathbf{X}$ is said to be r-separable if all columns of $\mathbf{X}$ are contained in the conical hull generated by a subset of r columns of $\mathbf{X}$. In other words, if $\mathbf{X}$ admits a factorization $\mathbf{WH}$ then the separability assumption states that the columns of $\mathbf{W}$ are present in $\mathbf{X}$ at positions given by an unknown index set A of size r. Equivalently, the corresponding columns of the right-factor matrix $\mathbf{H}$ constitute the $r \times r$ identity matrix, i.e., $\mathbf{H}_A = \mathbf{I}$. We refer to these columns indexed by A as *anchor columns*.

FIGURE 7.1 RobustXRAY applied to video background-foreground separation problem.

The separability assumption was first investigated by [9] in the context of deriving conditions for uniqueness of NMF. NMF under the separability assumption has been studied for topic modeling in text [1, 21] and hyper-spectral imaging [12, 17], and separability has turned out to be a reasonable assumption in these two applications. In the context of topic modeling where $\mathbf{X}$ is a document-word matrix and $\mathbf{W}$, $\mathbf{H}$ are document-topic and topic-word associations, respectively, it translates to assuming that there is at least one word in every topic that is unique to itself and is not present in other topics.

Our starting point in this chapter is the family of conical hull-finding procedures called XRAY introduced in our earlier work [21] for near-separable NMF problems with Frobenius norm loss. XRAY finds anchor columns one after the other, incrementally expanding the cone and using exterior columns to locate the next anchor. XRAY has several appealing features: (i) it requires no more than r iterations, each of which is parallelizable, (ii) it empirically demonstrates noise-tolerance, (iii) it admits efficient model selection, and (iv) it does not require normalizations or preprocessing needed in other methods. However, in the presence of outliers or different noise characteristics, the use of Frobenius norm approximations is not optimal. In fact, none of the existing near-separable NMF algorithms works with ℓ_1 and Bregman loss functions. On the other hand, there exist local search-based NMF algorithms for ℓ_1 loss and Bregman divergences [27, 29].

In this chapter, we fill this gap and extend XRAY to provide robust factorizations with respect to ℓ_1 loss, and approximations with respect to the family of Bregman divergences. Figure 7.1 shows a motivating application from computer vision. Given a sequence of video frames, the goal is to separate a near-stationary background from the foreground of moving objects that are relatively more dynamic across frames but span only a few pixels. In this setting, it is natural to seek a low-rank background matrix $\mathbf{B}$ that minimizes $\|\mathbf{X}-\mathbf{B}\|_1$ where $\mathbf{X}$ is the frame-by-pixel video matrix, and the ℓ_1 loss imposes a sparsity prior on the residual foreground. Unlike the case of low-rank approximations in Frobenius or spectral norms, this problem does not admit an SVD-like tractable solution. The Robust Principal Component Analysis (RPCA), considered state-of-the-art for this application, uses a nuclear-norm convex relaxation of the low-rank constraints. In this chapter, we instead recover tractability by imposing the separable NMF assumption on the background matrix. This implies that the variability of pixels across the frames can be "explained" in terms of observed variability in a small set of anchor pixels. Under a more restrictive setting, this can be shown to be equivalent to median filtering on the video frames, while a full near-separable NMF model imparts more degrees of freedom to model the background. We show that the proposed near-separable NMF algorithms with ℓ_1 loss are competitive with RPCA in separating foreground from background while outperforming it in terms of computational efficiency.

Our algorithms are empirically shown to be robust to noise (deviations from the pure separability assumption). In addition to the background-foreground problem, we also demonstrate our algorithms on the exemplar selection problem. For identifying exemplars in a data set, the proposed algorithms are evaluated on text documents with classification accuracy as a performance metric and are shown to outperform the recently proposed method of [11].

Related Work: Existing separable NMF methods work either with only a limited num-

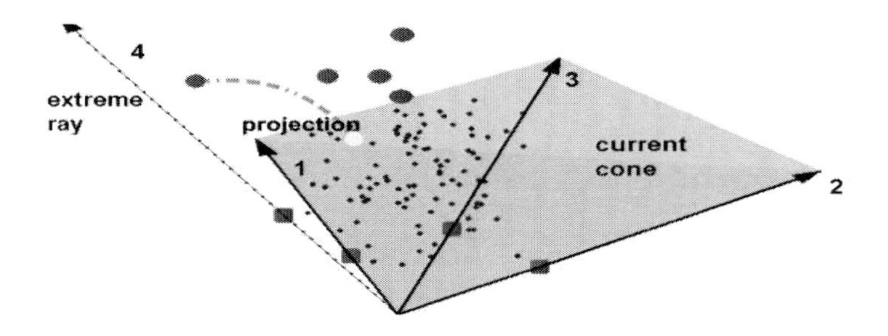

FIGURE 7.2 (See color insert.) Geometrical illustration of the algorithm

ber of loss functions on the factorization error, such as Frobenius norm loss [21], $\ell_{1,\infty}$ norm loss [4], or maximize proxy criteria such as volume of the convex polyhedron with anchor columns as vertices [17] and distance between successive anchors [1] to select the anchor columns. On the other hand, local search-based NMF methods [7] have been proposed for a wide variety of loss functions on the factorization error including ℓ_1 norm loss [20, 27] and instances of Bregman divergence [24, 29]. In this chapter, we close this gap and propose algorithms for near-separable NMF that minimize ℓ_1 loss and Bregman divergence for the factorization.

7.2 Geometric Intuition

The goal in exact NMF is to find a matrix $\mathbf{W}$ such that the cone generated by its columns (i.e., their non-negative linear combinations) contains all columns of $\mathbf{X}$. Under the separability assumption, the columns of matrix $\mathbf{W}$ are to be picked directly from $\mathbf{X}$, also known as anchor columns. The algorithms in this chapter build the cone incrementally by picking a column from $\mathbf{X}$ in every iteration. The algorithms execute r such iterations for constructing a factorization of inner-dimension r. Figure 7.2 shows the cone after three iterations of the algorithm when three anchor columns have been identified. An extreme ray $\{t\mathbf{x} : t > 0\}$ is associated with every anchor point $\mathbf{x}$. The points on an extreme ray cannot be expressed as conic combinations of other points in the cone that do not themselves lie on that extreme ray. To identify the next anchor column, the algorithm picks a point outside the current cone (a green point) and projects it to the current cone so that the distance between the point and the projection is minimized in terms of the desired measure of distance. This projection is then used to set up a specific simple criteria that, when maximized over the data points, identifies a new anchor. This new anchor is then added to the current set of anchors and the cone is expanded iteratively until all anchors have been picked.

These geometric intuitions are inspired by [8, 10], which present linear-programming (LP) based algorithms for general convex and conical hull problems. Their algorithms use ℓ_2 projections of exterior points to the current cone and are also applicable in our NMF setting if the data matrix $\mathbf{X}$ satisfies r-separability exactly. In this case, the ℓ_2 projection and corresponding residual vector of *any* single exterior point can be used to expand the cone and *all* r anchors will be recovered correctly at the end of the algorithm. When $\mathbf{X}$ does not satisfy r-separability exactly, anchor selection criteria derived from *multiple* residuals demonstrate superior noise robustness as empirically shown by [21] which considers the case of Gaussian i.i.d. noise. However, the algorithms of [21] are not suitable for noise distributions other than Gaussian (e.g., other members of the exponential family, sparse noise) as they minimize $\|\mathbf{X} - \mathbf{X}_A\mathbf{H}\|_F^2$. In the following sections, we present algorithms for near-separable NMF that are targeted precisely towards this goal and empirically demonstrate their superiority over existing algorithms under different noise distributions.

7.3 Near-Separable NMF with ℓ_1 loss

This section considers the case when the pure separable structure is perturbed by sparse noise. Hence our aim is to minimize $\|\mathbf{X}-\mathbf{X}_A\mathbf{H}\|_1$ for $\mathbf{H} \geq 0$ where $\|\cdot\|_1$ denotes element-wise ℓ_1 norm of the matrix and $\mathbf{X}_A$ are the columns of $\mathbf{X}$ indexed by set $A \subset \{1, 2, \dots, n\}$. We denote the ith column of $\mathbf{X}$ by $\mathbf{X}_i$. The proposed algorithm proceeds by identifying one anchor column in each iteration and adding it to the current set of anchors, thus expanding the cone generated by anchors. Each iteration consists of two steps: (i) an *anchor selection* step that finds the column of $\mathbf{X}$ to be added as an anchor, and (ii) a *projection* step where all data points (columns of $\mathbf{X}$) are projected to the current cone in terms of minimizing the ℓ_1 norm. Algorithm 7.1 outlines the steps of the proposed algorithm.

Selection Step: In the selection step, we normalize all the points to lie on the hyperplane $\mathbf{p}^T\mathbf{x} = 1$ ($\mathbf{Y}_j = \frac{\mathbf{X}_j}{\mathbf{p}^T\mathbf{X}_j}$) for a strictly positive vector $\mathbf{p}$ and evaluate the selection criterion of Equation 7.3 to select the next anchor column. Note that any exterior point ($i : \|\mathbf{R}_i\|_1 \geq 0$) can be used in the selection criterion – Algorithm 7.1 shows two possibilities for choosing the exterior point. Taking the point with maximum residual ℓ_1 norm to be the exterior point turns out to be far more robust to noise than randomly choosing the exterior point, as observed in our numerical simulations.

Projection Step: The projection step, Equation 7.4, involves solving a multivariate least absolute deviations problem with non-negativity constraints. We use the alternating direction method of multipliers (ADMM) [5] and reformulate the problem as

$$\min_{B \geq 0, \mathbf{Z}} \|\mathbf{Z}\|_1, \text{ such that } \mathbf{X}_A\mathbf{B} + \mathbf{Z} = \mathbf{X}.$$

Thus the non-negativity constraints are decoupled from the ℓ_1 objective and the ADMM optimization proceeds by alternating between two sub-problems – a standard ℓ_1 penalized ℓ_2 proximity problem in variable $\mathbf{Z}$ that has a closed form solution using the soft-thresholding operator, and a non-negative least-squares problem in variable $\mathbf{B}$ that is solved using a cyclic coordinate descent approach (cf. Algorithm 2 in [21]). The standard primal and dual residuals-based criteria are used to declare convergence [5]. The ADMM procedure converges to the global optimum since the problem is convex.

We now show that Algorithm 7.1 correctly identifies all the anchors in a pure separable case.

Lemma 7.3.1 *Let* $\mathbf{R}$ *be the residual matrix obtained after* ℓ_1 *projection of columns of* $\mathbf{X}$ *onto the current cone and* $\mathbf{D}$ *be the set of matrices such that* $\mathbf{D}_{ij} = sign(\mathbf{R}_{ij})$ *if* $\mathbf{R}_{ij} \neq 0$, *else* $\mathbf{D}_{ij} \in [-1, 1]$. *Then, there exists at least one* $\mathbf{D}^\star \in \mathbf{D}$ *such that* ${\mathbf{D}^\star}^T\mathbf{X}_A \leq 0$, *where* $\mathbf{X}_A$ *are anchor columns selected so far by Algorithm 7.1.*

PROOF 7.1 Residuals are given by $\mathbf{R} = \mathbf{X}-\mathbf{X}_A\mathbf{H}$, where $\mathbf{H} = \arg\min_{\mathbf{B}\geq 0}\|\mathbf{X}-\mathbf{X}_A\mathbf{B}\|_1$. Forming the Lagrangian for Equation 7.4, we get $\mathcal{L}(\mathbf{B}, \mathbf{m}\Lambda) = \|\mathbf{X} - \mathbf{X}_A\mathbf{B}\|_1 - tr(\mathbf{m}\Lambda^T\mathbf{B})$, where the matrix $\mathbf{m}\Lambda$ contains the non-negative Lagrange multipliers. The Lagrangian is not smooth everywhere and its sub-differential is given by $\partial\mathcal{L} = -\mathbf{X}_A^T\mathbf{D} - \mathbf{m}\Lambda$ where $\mathbf{D}$ is as defined in the lemma. At the optimum $\mathbf{B} = \mathbf{H}$, we have $0 \in \partial\mathcal{L} \Rightarrow -\mathbf{m}\Lambda \in \mathbf{X}_A^T\mathbf{D}$. Since $\mathbf{m}\Lambda \geq 0$, this means that there exists at least one $\mathbf{D}^\star \in \mathbf{D}$ for which ${\mathbf{D}^\star}^T\mathbf{X}_A \leq 0$.

Algorithm 7.1 - RobustXRAY: Near-separable NMF with ℓ_1 loss

Input: $\mathbf{X} \in \mathbb{R}_+^{m\times n}$, inner dimension r
Output: $\mathbf{W} \in \mathbb{R}_{m\times r}, \mathbf{H} \in \mathbb{R}_{r\times n}$, r indices in A
such that: $\mathbf{X} = \mathbf{W}\mathbf{H}$, $\mathbf{W} = \mathbf{X}_A$
Initialize: $\mathbf{R} \leftarrow \mathbf{X}$, $\mathbf{D}^\star \leftarrow \mathbf{X}$, $A \leftarrow \{\}$
while $|A| < r$ **do**
1. ***Anchor Selection step:***
First, pick any point exterior to the current cone. Two possible criteria are

$$rand: \quad \text{any random } i : \|\mathbf{R}_i\|_1 > 0 \tag{7.1}$$

$$max: \quad i = \arg\max_k \|\mathbf{R}_k\|_1 \tag{7.2}$$

Choose a suitable $\mathbf{D}_i^\star \in \mathbf{D}_i$ where $\mathbf{D}_{ji} = sign(\mathbf{R}_{ji})$ if $\mathbf{R}_{ji} \neq 0$, else $\mathbf{D}_{ji} \in [-1, 1]$ (see Remark (1)).
Select an anchor as follows ($\mathbf{p}$ is a strictly positive vector, not collinear with $\mathbf{D}_i^\star$ (see Remark (3))):

$$j^\star = \arg\max_j \frac{{\mathbf{D}_i^\star}^T \mathbf{X}_j}{\mathbf{p}^T \mathbf{X}_j} \tag{7.3}$$

2. Update: $A \leftarrow A \cup \{j^\star\}$ (see Remark (2))

3. ***Projection step:*** *Project onto current cone.*

$$\mathbf{H} = \arg\min_{\mathbf{B}\geq 0} \|\mathbf{X} - \mathbf{X}_A\mathbf{B}\|_1 \quad \text{(ADMM)} \tag{7.4}$$

4. Update Residuals: $\mathbf{R} = \mathbf{X} - \mathbf{X}_A\mathbf{H}$

end while

Lemma 7.3.2 *For any point $\mathbf{X}_i$ exterior to the current cone, there exists at least one $\mathbf{D}^\star \in \mathbf{D}$ such that it satisfies the previous lemma and ${\mathbf{D}_i^\star}^T \mathbf{X}_i > 0$.*

PROOF 7.2 Let $\mathbf{R} = \mathbf{X} - \mathbf{X}_A\mathbf{H}$, where $\mathbf{H} = \arg\min_{\mathbf{B}\geq 0}\|\mathbf{X} - \mathbf{X}_A\mathbf{B}\|_1$ and $\mathbf{X}_A$ are the current set of anchors. From the proof of the previous lemma, $-\mathbf{m}\Lambda^T \in \mathbf{D}^T\mathbf{X}_A$. Hence, $-\mathbf{m}\Lambda_i^T \in \mathbf{D}_i^T\mathbf{X}_A$ (ith row of both left and right-side matrices). From the complementary slackness condition, we have $\mathbf{m}\Lambda_{ji}\mathbf{H}_{ji} = 0 \; \forall j, i$. Hence, $-\mathbf{m}\Lambda_i^T\mathbf{H}_i = 0 \in \mathbf{D}_i^T\mathbf{X}_A\mathbf{H}_i$.
Since all KKT conditions are met at the optimum, there is at least one $\mathbf{D}^\star \in \mathbf{D}$ that satisfies the previous lemma and for which ${\mathbf{D}_i^\star}^T\mathbf{X}_A\mathbf{H}_i = 0$. For this $\mathbf{D}^\star$, we have ${\mathbf{D}_i^\star}^T\mathbf{X}_i = {\mathbf{D}_i^\star}^T(\mathbf{R}_i + \mathbf{X}_A\mathbf{H}_i) = {\mathbf{D}_i^\star}^T\mathbf{R}_i = \|\mathbf{R}_i\|_1 > 0$ since $\mathbf{R}_i \neq 0$ for an exterior point.

Using the above two lemmas, we prove the following theorem regarding the correctness of Algorithm 7.1 in pure separable case.

Theorem 1 *If the maximizer in Equation 7.3 is unique, the data point $\mathbf{X}_{j^\star}$ added at each iteration in the Selection step of Algorithm 7.1 is an anchor that has not been selected in one of the previous iterations.*

PROOF 7.3 Let the index set A denote all the anchor columns of $\mathbf{X}$. Under the separability assumption, we have $\mathbf{X} = \mathbf{X}_A\mathbf{H}$. Let the index set A^t identify the current set of anchors.

Let $\mathbf{Y}_j = \frac{\mathbf{X}_j}{\mathbf{p}^T\mathbf{X}_j}$ and $\mathbf{Y}_A = \mathbf{X}_A[diag(\mathbf{p}^T\mathbf{X}_A)]^{-1}$ (since $\mathbf{p}$ is strictly positive, the inverse exists). Hence $\mathbf{Y}_j = \mathbf{Y}_A\frac{[diag(\mathbf{p}^T\mathbf{X}_A)]\mathbf{H}_j}{\mathbf{p}^T\mathbf{X}_j}$. Let $\mathbf{C}_j = \frac{[diag(\mathbf{p}^T\mathbf{X}_A)]\mathbf{H}_j}{\mathbf{p}^T\mathbf{X}_j}$. We also have $\mathbf{p}^T\mathbf{Y}_j = 1$ and $\mathbf{p}^T\mathbf{Y}_A = \mathbf{1}^T$. Hence, we have $1 = \mathbf{p}^T\mathbf{Y}_j = \mathbf{p}^T\mathbf{Y}_A\mathbf{C}_j = \mathbf{1}^T\mathbf{C}_j$.

Using Lemma 7.3.1, Lemma 7.3.2, and the fact that $\mathbf{p}$ is strictly positive, we have $\max_{1 \le j \le n} \mathbf{D}_i^{\star T} \mathbf{Y}_j = \max_{j \notin A^t} \mathbf{D}_i^{\star T} \mathbf{Y}_j$. Indeed, for all $j \in A^t$ we have $\mathbf{D}_i^{\star T} \mathbf{Y}_j \le 0$ using Lemma 7.3.1 and there is at least one $j = i \notin A^t$ for which $\mathbf{D}_i^{\star T} \mathbf{Y}_j > 0$ using Lemma 7.3.2. Hence the maximum lies in the set $\{j : j \notin A^t\}$.

Further, we have $\max_{j \notin A^t} \mathbf{D}_i^{\star T} \mathbf{Y}_j = \max_{j \notin A^t} \mathbf{D}_i^{\star T} \mathbf{Y}_A \mathbf{C}_j = \max_{j \in (A \setminus A^t)} \mathbf{D}_i^{\star T} \mathbf{Y}_j$. The second equality is the result of the fact that $\|\mathbf{C}_j\|_1 = 1$ and $\mathbf{C}_j \ge 0$. This implies that if there is a unique maximum at a $j^* = \arg\max_{j \notin A^t} \mathbf{D}_i^{\star T} \mathbf{Y}_j$, then $\mathbf{X}_{j^*}$ is an anchor that has not been selected so far.

Remarks:

(1) For the correctness of Algorithm 7.1, the anchor selection step requires choosing a $\mathbf{D}_i^\star \in \mathbf{D}_i$ for which Lemma 7.3.1 and Lemma 7.3.2 hold true. Here we give a method to find one such $\mathbf{D}_i^\star$ using linear programming. Using KKT conditions, the $\mathbf{D}_i^\star$ satisfying $-\mathbf{X}_A^T \mathbf{D}_i^\star = \mathbf{m}\Lambda_i \in \mathbb{R}_+^{|A|}$ is a candidate. We know $\mathbf{m}\Lambda_{ji} = 0$ if $\mathbf{H}_{ji} > 0$ and $\mathbf{m}\Lambda_{ji} > 0$ if $\mathbf{H}_{ji} = 0$ (complementary slackness). Let $Z = \{j : \mathbf{H}_{ji} > 0\}$ and $\widetilde{Z} = \{j : \mathbf{H}_{ji} = 0\}$. Let $I = \{j : \mathbf{R}_{ji} = 0\}$. Let $\mathbf{u}$ represent the elements of $\mathbf{D}_i^\star$ that we need to find, i.e., $\mathbf{u} = \{\mathbf{D}_{ji}^\star : j \in I\}$. Finding $\mathbf{u}$ is a feasibility problem that can be solved using an LP. Since there can be multiple feasible points, we can choose a dummy cost function $\sum_k \mathbf{u}_k$ (or any other random linear function of $\mathbf{u}$) for the LP. More formally, the LP takes the form:

$$\min_{-1 \le \mathbf{u} \le 1} \mathbf{1}^T \mathbf{u}, \quad \text{such that}$$
$$-\mathbf{X}_A^T \mathbf{D}_i^\star = \mathbf{m}\Lambda_i, \mathbf{m}\Lambda_{ji} = 0 \;\forall\; j \in Z, \mathbf{m}\Lambda_{ji} > 0 \;\forall\; j \in \widetilde{Z}.$$

In principle, the number of variables in this LP is the number of zero entries in residual vector $\mathbf{R}_i$, which can be as large as $m-1$. In practice, we always have the number of zeros in $\mathbf{R}_i$ much less than m since we always pick the exterior point with maximum ℓ_1 norm in the Anchor Selection step of Algorithm 7.1. The number of constraints in the LP is also very small ($= |A| < r$). In our implementation, we simply set $\mathbf{u} = -1$ which, in practice, almost always satisfies Lemma 7.3.1 and Lemma 7.3.2. The LP is called whenever Lemma 7.3.2 is violated, which happens rarely (note that Lemma 7.3.1 will never be violated with this setting of $\mathbf{u}$).

(2) If the maximum of Equation 7.3 occurs at more than one point, it is clear from the proof of Theorem 1 (last paragraph in the proof) that at least two anchors that are not selected so far should attain the maximum, i.e., $\max_{j \notin A^t} \mathbf{D}_i^{\star T} \mathbf{Y}_j = \max_{j \in (A \setminus A^t)} \mathbf{D}_i^{\star T} \mathbf{Y}_j = \mathbf{D}_i^{\star T} \mathbf{Y}_k = \mathbf{D}_i^{\star T} \mathbf{Y}_l$ for $k, l \in (A \setminus A^t), k \ne l$. Hence, if the maximum of Equation 7.3 occurs at exactly two points j_1^* and j_2^*, both these points $\mathbf{X}_{j_1^*}$ and $\mathbf{X}_{j_2^*}$ are anchor points and both are added to anchor set A. If the maximum occurs at more than two points, some of these are the anchors and others are conic combinations of these anchors. We can identify the anchors of this subset of points by calling Algorithm 7.1 recursively.

(3) In Algorithm 7.1, the vector $\mathbf{p}$ needs to satisfy $\mathbf{p}^T \mathbf{x}_i > 0, i = 1 \ldots n$. In our implementation, we simply used $\mathbf{p} = \mathbf{1} + \boldsymbol{\delta} \in \mathbb{R}^m$ where $\boldsymbol{\delta}$ is the small perturbation vector with entries i.i.d. according to a uniform distribution $\mathcal{U}(0, 10^{-5})$. This is done to avoid the possibility of $\mathbf{p}$ being collinear with $\mathbf{D}_i^\star$.

7.4 Near-Separable NMF with Bregman Divergence

Let $\phi : \mathcal{S} \mapsto \mathbb{R}$ be a strictly convex function on domain $\mathcal{S} \subseteq \mathbb{R}$ that is differentiable on its non-empty relative interior $ri(\mathcal{S})$. Bregman divergence is then defined as $D_\phi(x, y) = \phi(x) - \phi(y) - \phi'(y)(x - y)$ where $\phi'(y)$ is the continuous first derivative of $\phi(\cdot)$ at y. Here we will also assume $\phi'(\cdot)$ to be smooth, which is true for most Bregman divergences of interest. A Bregman divergence is always convex in the first argument. Some instances of Bregman divergence are also convex in the second argument (e.g., KL divergence). For two matrices $\mathbf{X}$ and $\mathbf{Y}$, we work with divergence of the form $D_\phi(\mathbf{X}, \mathbf{Y}) := \sum_{ij} D_\phi(\mathbf{X}_{ij}, \mathbf{Y}_{ij})$.

Here we consider the case when the entries of data matrix $\mathbf{X}$ are generated from an exponential family distribution with parameters satisfying the separability assumption, i.e., $\mathbf{X}_{ij} \sim \mathcal{P}_\phi(\mathbf{W}^i\mathbf{H}_j)$, $\mathbf{W} \in \mathbb{R}_+^{m\times r}, \mathbf{H} = [\mathbf{I}\ \mathbf{H}'] \in \mathbb{R}_+^{r\times n}$ ($\mathbf{W}^i$ and $\mathbf{H}_j$ denote the ith row of $\mathbf{W}$ and the jth column of $\mathbf{H}$, respectively). Every member distribution $\mathcal{P}_\phi$ of the exponential family has a unique Bregman divergence $D_\phi(\cdot,\cdot)$ associated with it [3], and solving $\min_{\mathbf{Y}} D_\phi(\mathbf{X}, \mathbf{Y})$ is equivalent to maximum likelihood estimation for parameters $\mathbf{Y}_{ij}$ of the distribution $\mathcal{P}_\phi(\mathbf{Y}_{ij})$. Hence, the projection step in Algorithm 7.1 is changed to $\mathbf{H} = \arg\min_{\mathbf{B}\geq 0} D_\phi(\mathbf{X}, \mathbf{X}_A\mathbf{B})$. We use the coordinate descent-based method of [24] to solve the projection step. To select the anchor columns with Bregman projections $\mathbf{X}_A\mathbf{H}$, we modify the selection criteria as

$$j^\star = \arg\max_j \frac{(\phi''(\mathbf{X}_A\mathbf{H}_i) \odot \mathbf{R}_i)^T\mathbf{X}_j}{\mathbf{p}^T\mathbf{X}_j} \tag{7.5}$$

for any $i : \|\mathbf{R}_i\| > 0$, where $\mathbf{R} = \mathbf{X} - \mathbf{X}_A\mathbf{H}$ and $\phi''(\mathbf{x})$ is the vector of second derivatives of $\phi(\cdot)$ evaluated at individual elements of the vector $\mathbf{x}$ (i.e., $[\phi''(\mathbf{x})]_j = \phi''(\mathbf{x}_j)$), and $\odot$ denotes the element-wise product of vectors. We can show the following result regarding the anchor selection property of this criteria. Recall that an anchor is a column that cannot be expressed as a conic combination of other columns in $\mathbf{X}$.

Theorem 2 *If the maximizer of Equation 7.5 is unique, the data point $\mathbf{X}_{j^\star}$ added at each iteration in the Selection step is an anchor that has not been selected in one of the previous iterations.*

The proof is provided in the Supplementary Material. Again, any exterior point i can be chosen to select the next anchor but our simulations show that taking the exterior point to be $i = \arg\max_k D_\phi(\mathbf{X}_k, \mathbf{X}_A\mathbf{H}_k)$ gives much better performance under noise than randomly choosing the exterior point. Note that for the Bregman divergence induced by function $\phi(x) = x^2$, the selection criteria of Equation 7.5 reduces to the selection criteria of XRAY proposed in [21].

Since Bregman divergence is not generally symmetric, it is also possible to have the projection step as $\mathbf{H} = \arg\min_{\mathbf{B}\geq 0} D_\phi(\mathbf{X}_A\mathbf{B}, \mathbf{X})$. In this case, the selection criteria will change to $j^\star = \arg\max_j \frac{(\phi'(\mathbf{X}_i) - \phi'(\mathbf{X}_A\mathbf{H}_i))^T\mathbf{X}_j}{\mathbf{p}^T\mathbf{X}_j}$ for any point i exterior to the current cone, where $\phi'(\mathbf{x})$ operates element-wise on vector $\mathbf{x}$. However, this variant does not have as meaningful a probabilistic interpretation as the one discussed earlier.

7.5 Empirical Observations

In this section, we present experiments on synthetic and real datasets to demonstrate the effectiveness of the proposed algorithms under noisy conditions. In addition to comparing our algorithms with existing separable NMF methods [4, 17, 21], we also benchmark them against Robust PCA and local-search based low-rank factorization methods, wherever applicable, for the sake of providing a more complete picture.

7.5.1 Anchor Recovery under Noise

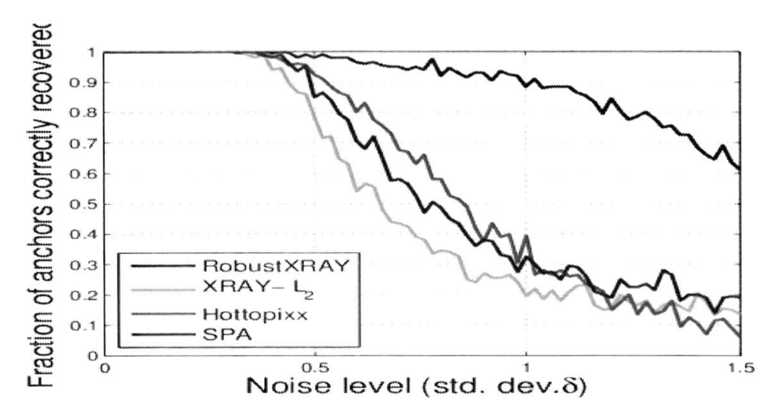

FIGURE 7.3 (See color insert.) Sparse noise case: anchor recovery rate versus noise level (best viewed in color)

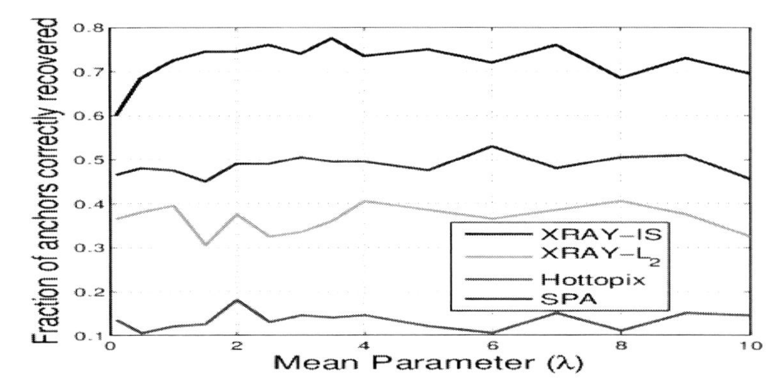

FIGURE 7.4 (See color insert.) Data matrix generated from exponential distribution: anchor recovery rate versus mean parameter (best viewed in color)

Here we test the proposed algorithms for recovery of anchors when the separable structure is perturbed by noise. Since there are no existing separable NMF algorithms that work with ℓ_1 and Bregman loss functions, we compare our method with methods proposed in [17] (abbrv. as SPA for Successive Projection Approximation), [4] (abbrv. as *Hottopixx*), and [21] (abbrv. as XRAY-ℓ_2) to highlight that the choice of a suitable loss function plays a crucial role in column selection. There exist local search methods for ℓ_1- and Bregman NMF [27, 29] but these are not comparable here since they do not recover the anchor columns.

First, we consider the case when the separable structure is perturbed by addition of a sparse noise matrix, i.e., $\mathbf{X} = \mathbf{WH} + \mathbf{N}$, $\mathbf{H} = [\mathbf{I}\ \mathbf{H}']$. Each entry of matrix $\mathbf{W} \in \mathbb{R}_+^{200\times 20}$ is generated i.i.d. from a uniform distribution between 0 and 1. The matrix $\mathbf{H} \in \mathbb{R}^{20\times 210}$ is taken to be $[\mathbf{I}_{20\times 20}\ \mathbf{H}'_{20\times 190}]$ where each column of $\mathbf{H}'$ is sampled i.i.d. from a Dirichlet distribution whose parameters are generated i.i.d. from a uniform distribution between 0 and 1. It is clear from the structure of matrix $\mathbf{H}$ that the first twenty columns are the anchors. The data matrix $\mathbf{X}$ is generated as $\mathbf{WH} + \mathbf{N}$ with $\mathbf{N} = \max(\mathbf{N}_1, 0) \in \mathbb{R}_+^{200\times 210}$, where each entry of $\mathbf{N}_1$ is generated i.i.d. from a Laplace distribution having zero mean and δ standard deviation. Since the Laplace distribution is symmetric around mean, almost half of the entries in matrix $\mathbf{N}$ are 0 due to the max operation. The std. dev. δ is varied from 0 to 1.5 with a step size of 0.02. Figure 7.3 plots the fraction of correctly recovered anchors averaged over 10 runs for each value of δ. The proposed RobustXRAY (Algorithm 7.1) outperforms all other methods including XRAY-ℓ_2 by a huge margin as the noise level

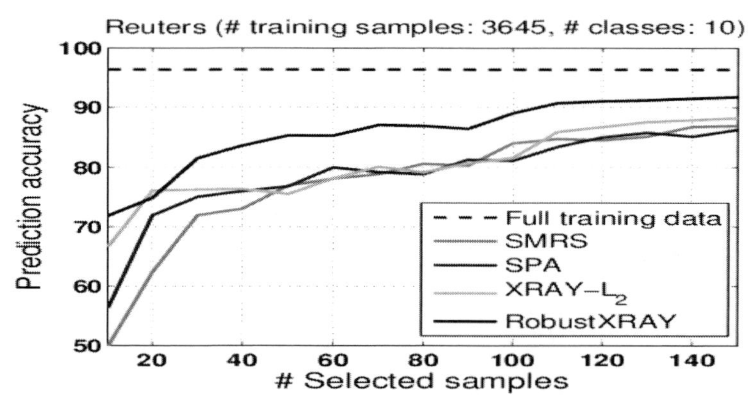

FIGURE 7.5 (See color insert.) Accuracy of SVM trained with selected exemplars on *Reuters data* (best viewed in color)

increases. This highlights the importance of using the right loss function in the projection step that is suitable for the noise model (in this case ℓ_1 loss of Equation 7.4).

Next, we consider the case where the non-negative data matrix is generated from an exponential family distribution other than the Gaussian, i.e., $\mathbf{X}_{ij} \sim \mathcal{P}_\phi(\mathbf{W}^i\mathbf{H}_j)$, $\mathbf{W} \in \mathbb{R}_+^{m\times r}, \mathbf{H} = [\mathbf{I}\ \mathbf{H}'] \in \mathbb{R}_+^{r\times n}$ ($\mathbf{W}^i$ and $\mathbf{H}_j$ denote the ith row of $\mathbf{W}$ and the jth column of $\mathbf{H}$, respectively). As mentioned earlier, every member distribution $\mathcal{P}_\phi$ of the exponential family has a unique Bregman divergence D_ϕ associated with it. Hence we minimize the corresponding Bregman divergence in the projection step of the algorithm as discussed in Section 7.4, to recover the anchor columns. Two most commonly used Bregman divergences are generalized KL-divergence and Itakura-Saito (IS) divergence [3, 14, 29], which correspond to Poisson and Exponential distributions, respectively. We do not report results with generalized KL-divergence here since they were not very informative in highlighting the differences among various algorithms that are considered. The reason is that Poisson distribution with parameter λ has a mean of λ and std. dev. of $\sqrt{\lambda}$, and increasing the noise (std. dev.) actually increases the signal to noise ratio.[22] Hence anchor recovery gets better with increasing λ (perfect recovery after certain value) and almost all algorithms perform as well as XRAY-KL for the full λ range. The anchor recovery results with IS-divergence are shown in Figure 7.4. The entries of data matrix $\mathbf{X}$ are generated as $\mathbf{X}_{ij} \sim \exp(\lambda\mathbf{W}^i\mathbf{H}_j)$ $\mathbf{W} \in \mathbb{R}_+^{200\times 20}, \mathbf{H} = [\mathbf{I}_{20\times 20}\ \mathbf{H}'] \in \mathbb{R}_+^{20\times 210}$. The matrices $\mathbf{W}$ and $\mathbf{H}$ are generated as described in the previous paragraph. The parameter λ is varied from 0 to 10 in the steps of 0.5 and we report the average over 10 runs for each value of λ. The XRAY-IS algorithm significantly outperforms other methods including XRAY-ℓ_2 [21] in correctly recovering the anchor column indices. The recovery rate does not change much with increasing λ since exponential distribution with mean parameter λ has a standard deviation dev. of λ and the signal-to-noise ratio practically stays almost the same with varying λ.

7.5.2 Exemplar Selection

The problem of exemplar selection is concerned with finding a few representatives from a dataset that can summarize the dataset well. Exemplar selection can be used in many applications including summarizing a video sequence, selecting representative images or text

[22] Poisson distribution with parameter λ closely resembles a Gaussian distribution with mean λ and standard deviation dev. $\sqrt{\lambda}$, for large values of λ.

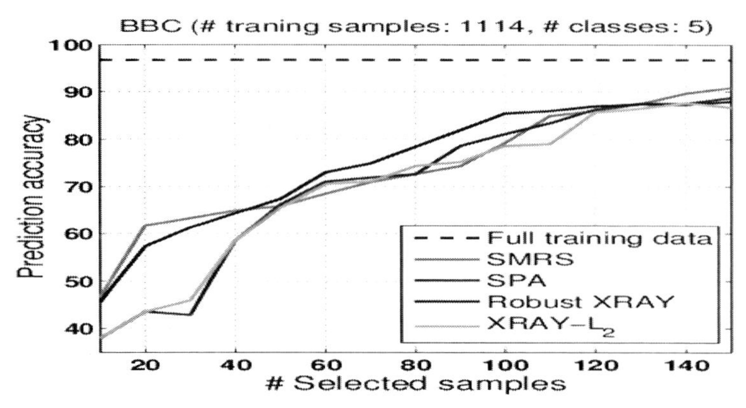

FIGURE 7.6 (See color insert.) Accuracy of SVM trained with selected exemplars on *BBC data* (best viewed in color)

documents (e.g., tweets) from a collection, etc. If $\mathbf{X}$ denotes the data matrix where each *column* is a data point, the exemplar selection problem translates to selecting a few columns from $\mathbf{X}$ that can act as representatives for all the columns. The separable NMF algorithms can be used for this task, working under the assumption that all data points (columns of $\mathbf{X}$) can be expressed as non-negative linear combinations of the exemplars (the anchor columns). To be able to compare the quality of the selected exemplars by different algorithms in an objective manner, we test them on a classification task (assuming that every data point has an associated label). We randomly partition the data in training and test sets, and use only one training set in selecting the exemplars. We train a multiclass SVM classifier [13] with the selected exemplars and look at its accuracy on the held-out test set. The accuracy of the classifier trained with the full training set is taken as a benchmark and is also reported. We also compare with [11] in which a method was recently proposed for exemplar selection, named as Sparse Modeling Representative Selection (SMRS). They assume that the data points can be expressed as a convex linear combination of the exemplars and minimize $\|\mathbf{X}-\mathbf{XC}\|_F^2+\lambda\|\mathbf{C}\|_{1,2}$ s.t. $\mathbf{1}^T\mathbf{C} = \mathbf{1}^T$. The columns of $\mathbf{X}$ corresponding to the non-zero rows of $\mathbf{C}$ are selected as exemplars. We use the code provided by the authors for SMRS. There are multiple possibilities for anchor selection criteria in the proposed RobustXRAY and XRAY-ℓ_2 [21] and we use *max* criterion for both the algorithms.

We report results with two text datasets: Reuters [28] and BBC [18]. We use a subset of Reuters data corresponding to the most frequent 10 classes, which amounts to 7285 documents and 18,221 words ($\mathbf{X} \in \mathbb{R}_+^{18221\times 7285}$). The BBC data consists of 2225 documents and 9635 words with 5 classes ($\mathbf{X} \in \mathbb{R}_+^{9635\times 2225}$). For both datasets, we evenly split the documents into training and test set, and select the exemplars from the training set using various algorithms. Figures 7.5 and 7.6 show the plot of SVM accuracy on the test set against the number of selected exemplars that are used for training the classifier. The number of selected anchors is varied from 10 to 150 in the steps of 10. The accuracy using the full training set is also shown (dotted black line). For Reuters data, the proposed RobustXRAY algorithm outperforms other methods by a significant margin for the whole range of selected anchors. All methods seem to perform comparably on BBC data. An advantage of the SPA and XRAYfamily of methods is that there is no need for a cleaning step to remove near-duplicate exemplars as needed in SMRS [11]. Another advantage is one of computational speed – in all our experiments, SPA and XRAY methods are about 3–10 times faster than SMRS. It is remarkable that even a low number of selected exemplars give reasonable classification accuracy for all methods – SMRS gives 50% accuracy for Reuters data using 10 exemplars (on average 1 training sample per class) while RobustXRAY gives more than 70%.

7.5.3 Foreground-Background Separation

In this section, we consider the problem of foreground-background separation in video. The camera position is assumed to be *almost* fixed throughout the video. In all video frames, the camera captures the background scene superimposed with a *limited* foreground activity (e.g., movement of people or objects). Background is assumed to be stationary or slowly varying across frames (variations in illumination and shadows due to lighting or time of day) while foreground is assumed to be composed of objects that move across frames but span only a few pixels. If we vectorize all video frames and stack them as rows to form the matrix $\mathbf{X}$, the foreground-background separation problem can be modeled as decomposing $\mathbf{X}$ into a low-rank matrix $\mathbf{L}$ (modeling the background) and a sparse matrix $\mathbf{S}$ (modeling the foreground).

Connection to Median Filtering: Median filtering is one of the most commonly used background modeling techniques [6], which simply models the background as the pixel-wise median of the video frames. The assumption is that each pixel location stays in the background for more than half of the video frames. Consider the NMF of inner-dimension 1: $\min_{\boldsymbol{w}\geq 0,\boldsymbol{h}\geq 0} \|\mathbf{X} - \boldsymbol{wh}\|_1$. If we constrain the vector $\boldsymbol{w}$ to be all ones, the solution $\boldsymbol{h}^* = \arg\min_{\boldsymbol{w}=\mathbf{1},\boldsymbol{h}\geq 0} \|\mathbf{X} - \boldsymbol{wh}\|_1$ is nothing but the element-wise median of all rows of $\mathbf{X}$. More generally, if $\boldsymbol{w}$ is constrained to be such that $\boldsymbol{w}_i = c > 0 \;\forall\; i$, the solution $\boldsymbol{h}^*$ is a scaled version of the element-wise median vector. Hence Robust NMF under this very restrictive setting is equivalent to median filtering on the video frames, and we can hope that loosening this assumption and allowing for higher inner-dimension in the factorization can help in modeling more variations in the background.

We use three video sequences for evaluation: *Restaurant*, *Airport Hall*, and *Lobby* [23]. Restaurant and Airport Hall are videos taken at a buffet restaurant and at a hall of an airport, respectively. The lighting is distributed from the ceilings and signicant shadows of moving persons cast on the ground surfaces from different directions can be observed in the videos. The Lobby video sequence was captured from a lobby in an office building and has background changes due to lights being switched on/off. The ground truth (whether a pixel belongs to foreground or background) is also available for these video sequences and we use it to generate the ROC curves. We mainly compare RobustXRAY with Robust PCA, which is widely considered state-of-the-art methodology for this task in the Computer Vision community. In addition, we also compare it with two local-search based approaches: (i) *Robust NMF (local search)*, which solves $\min_{\mathbf{W}\geq 0,\mathbf{H}\geq 0} \|\mathbf{X} - \mathbf{WH}\|_1$ using local search, and (ii) *Robust Low-rank (local search)*, which solves $\min_{\mathbf{W},\mathbf{H}} \|\mathbf{X}-\mathbf{WH}\|_1$ using local search. We use an ADMM, based optimization procedure for both these local-search methods. We also show results with XRAY-ℓ_2 of [21] to highlight the importance of having near-separable NMFs with ℓ_1 loss for this problem. For both XRAY-ℓ_2 and RobustXRAY, we do 1 to 2 refitting steps to refine the solution (i.e., solve $\mathbf{H} = \min_{\mathbf{B}\geq 0} \|\mathbf{X} - \mathbf{X}_{\mathcal{A}}\mathbf{B}\|_1$, then solve $\mathbf{W} = \min_{\mathbf{C}\geq 0} \|\mathbf{X} - \mathbf{CH}\|_1$). For all the methods, we do a grid search on the parameters (inner-dimension or rank parameter for the factorization methods and λ parameter for Robust PCA) and report the best results for each method.

Figure 7.7 shows the ROC plots for the three video datasets. For the Restaurant data, all robust methods (those with ℓ_1 penalty on the foreground) perform almost similarly. For the Airport Hall data, RobustXRAYis tied with local-search based Robust NMF and these two are better than other methods. Surprisingly, XRAY-ℓ_2 performs better than local-search based Robust Low-rank which might be due to bad initialization. For the Lobby data, local-search based Robust low-rank, Robust PCA and RobustXRAY perform almost similarly, and are better than local-search based Robust NMF. The results on these three datasets show that RobustXRAYis a promising method for the problem of foreground-background separation one that has a huge advantage over Robust PCA in terms of speed. Our MAT-

LAB implementation was at least 10 times faster than the inexact Augmented Lagrange Multiplier (i-ALM) implementation of [26].

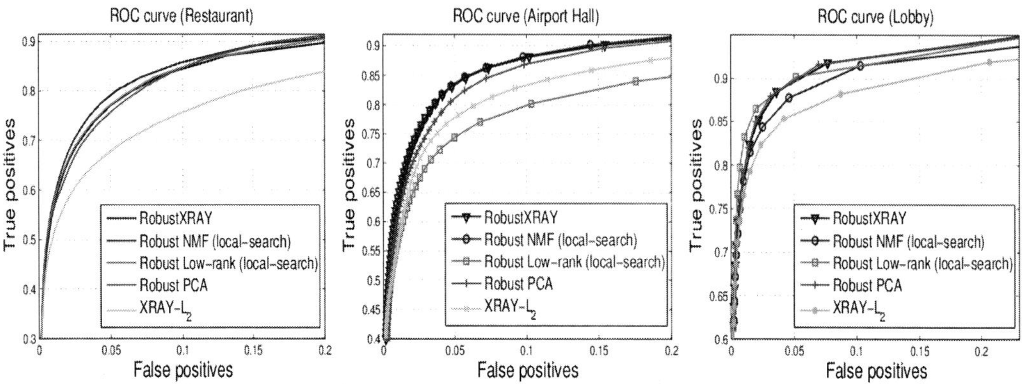

FIGURE 7.7 (See color insert.) Foreground-background separation: ROC curves with various methods for *Restaurant*, *Airport Hall* and *Lobby* video sequences. The ranges for X and Y axes are chosen to better highlight the differences among the ROC curves. (best viewed in color)

7.6 Conclusion

We have proposed generalized conical hull algorithms to extend near-separable NMFs to robust ℓ_1-loss function and Bregman divergences. Empirical results on exemplar selection and video background-foreground modeling problems suggest that this is a promising methodology. Avenues for future work include formal theoretical analysis of noise robustness and applications to online settings.

7.7 Supplementary Material

Let A be the set of anchors selected so far by the algorithm. Let $\phi(\cdot)$ be the strictly convex function that induces the Bregman divergence $D_\phi(x,y) = \phi(x) - \phi(y) - \phi'(y)(x-y)$. For two matrices $\mathbf{X}$ and $\mathbf{Y}$, we consider the Bregman divergence of the form $D_\phi(\mathbf{X},\mathbf{Y}) := \sum_{ij} D_\phi(\mathbf{X}_{ij},\mathbf{Y}_{ij})$. We make the following assumptions for the proofs in this section, which are satisfied by almost all the Bregman divergences of interest:

Assumption 1: The first derivative of ϕ, ϕ' is smooth.

Assumption 2: The second derivative of ϕ, ϕ'' is positive at all nonzero points, i.e., $\phi''(x) > 0$ for $x \neq 0$. Note that strict convexity does not necessarily imply $\phi''(x) > 0$ for all x while the converse is true.

Here we consider the projection step

$$\mathbf{H} = \arg\min_{\mathbf{B}\geq 0} D_\phi(\mathbf{X}, \mathbf{X}_A\mathbf{B}), \tag{7.6}$$

and the following selection criteria to identify the next anchor column:

$$j^\star = \arg\max_j \frac{(\phi''(\mathbf{X}_A\mathbf{H}_i) \odot \mathbf{R}_i)^T\mathbf{X}_j}{\mathbf{p}^T\mathbf{X}_j} \tag{7.7}$$

for any $i : \|\mathbf{R}_i\| > 0$, where $\mathbf{R} = \mathbf{X} - \mathbf{X}_A\mathbf{H}$ and $\phi''(\mathbf{x})$ is the vector of second derivatives of $\phi(\cdot)$ evaluated at individual elements of the vector $\mathbf{x}$ (i.e., $[\phi''(\mathbf{x})]_j = \phi''(\mathbf{x}_j)$), $\mathbf{p}$ is any strictly positive vector not collinear with $(\phi''(\mathbf{X}_A\mathbf{H}_i) \odot \mathbf{R}_i)$, and $\odot$ denotes the element-wise product of vectors. For a matrix $\mathbf{X}$, $\mathbf{X}_{ij}$ denotes the ijth element, $\mathbf{X}_i$ denotes the ith column, and $\mathbf{X}_A$ denotes the columns of $\mathbf{X}$ indexed by set A. $\mathbf{X}_{A_{ij}}$ denotes ijth column of matrix $\mathbf{X}_A$.

Here we show the following result regarding the anchor selection property of Equation 7.7. Recall that an anchor is a column that can not be expressed as a conic combination of other columns in $\mathbf{X}$.

Theorem 3 *If the maximizer of Equation 7.7 is unique, the data point $\mathbf{X}_{j^\star}$ added at each iteration in the Selection step is an anchor that has not been selected in one of the previous iterations.*

The proof of this theorem follows the same style as the proof of Theorem 3.1 in the main paper. We need the following lemmas to prove Theorem 3.

Lemma 7.7.1 *Let $\mathbf{R}$ be the residual matrix obtained after Bregman projection of columns of $\mathbf{X}$ onto the current cone. Then, $(\phi''(\mathbf{X}_A\mathbf{H})\odot\mathbf{R})^T\mathbf{X}_A \leq 0$, where $\mathbf{X}_A$ are anchor columns selected so far by the algorithm.*

PROOF 7.4 Residuals are given by $\mathbf{R} = \mathbf{X} - \mathbf{X}_A\mathbf{H}$, where $\mathbf{H} = \arg\min_{\mathbf{B}\geq 0} D_\phi(\mathbf{X}, \mathbf{X}_A\mathbf{B})$. Forming the Lagrangian for Equation 7.6, we get $\mathcal{L}(\mathbf{B}, \mathbf{m}\Lambda) = D_\phi(\mathbf{X}, \mathbf{X}_A\mathbf{B}) - tr(\mathbf{m}\Lambda^T\mathbf{B})$, where the matrix $\mathbf{m}\Lambda$ contains the non-negative Lagrange multipliers.

At the optimum $\mathbf{B} = \mathbf{H}$, we have $\nabla\mathcal{L} = 0$, which means

$$
\begin{aligned}
&\frac{\partial}{\partial \mathbf{B}_{mn}} D_\phi(\mathbf{X}, \mathbf{X}\mathbf{B})\Big|_{\mathbf{B}=\mathbf{H}} = \mathbf{m}\Lambda_{mn} \\
&\Rightarrow \sum_i \Big[-\mathbf{X}_{A_{im}}\phi'((\mathbf{X}_A\mathbf{H})_{in}) - \mathbf{X}_{A_{im}}\phi''((\mathbf{X}_A\mathbf{H})_{in})(\mathbf{X}_{in} \\
&- (\mathbf{X}_A\mathbf{H})_{in}) + \phi'((\mathbf{X}_A\mathbf{H})_{in})\mathbf{X}_{A_{im}}\Big] = \mathbf{m}\Lambda_{mn} \\
&\Rightarrow \sum_i - \Big[\mathbf{X}_{A_{im}}\phi''((\mathbf{X}_A\mathbf{H})_{in})(\mathbf{X}_{in} - (\mathbf{X}_A\mathbf{H})_{in})\Big] = \mathbf{m}\Lambda_{mn} \\
&\Rightarrow [\phi''(\mathbf{X}_A\mathbf{H}) \odot (\mathbf{X} - \mathbf{X}_A\mathbf{H})]_n^T \mathbf{X}_{A_m} = -\mathbf{m}\Lambda_{mn} \\
&\Rightarrow [\phi''(\mathbf{X}_A\mathbf{H}) \odot (\mathbf{X} - \mathbf{X}_A\mathbf{H})]^T \mathbf{X}_A = -\mathbf{m}\Lambda^T \leq 0,
\end{aligned}
$$

where $\phi''(\cdot)$ is the second derivative of $\phi(\cdot)$ that operates element-wise on the argument (vector or matrix).

Lemma 7.7.2 *For any point $\mathbf{X}_i$ exterior to the current cone, we have $(\phi''(\mathbf{X}_A\mathbf{H}) \odot \mathbf{R})_i^T\mathbf{X}_i > 0$.*

PROOF 7.5 For a vector $\mathbf{v} > 0$ and any vector $\boldsymbol{z}$, we have $(\mathbf{v} \odot \boldsymbol{z})^T\boldsymbol{z} = \sum_i \mathbf{v}_i \boldsymbol{z}_i^2 > 0$. Taking $\mathbf{v}$ to be $\phi''(\mathbf{X}_A\mathbf{H}_i)$ and $\boldsymbol{z}$ to be $\mathbf{R}_i$, we have

$$
\begin{aligned}
&(\phi''(\mathbf{X}_A\mathbf{H}_i) \odot \mathbf{R}_i)^T\mathbf{R}_i > 0 \\
&\Rightarrow (\phi''(\mathbf{X}_A\mathbf{H}_i) \odot \mathbf{R}_i)^T(\mathbf{X}_i - \mathbf{X}_A\mathbf{H}_i) > 0.
\end{aligned}
\tag{7.8}
$$

By complementary slackness condition at the optimum, we have $\mathbf{m}\Lambda_{ji}\mathbf{H}_{ji} = 0$. From the KKT condition in the proof of previous lemma, we have $[\phi''(\mathbf{X}_A\mathbf{H}_i) \odot \mathbf{R}_i]^T \mathbf{X}_A = -\mathbf{m}\Lambda_i^T$. Hence

$[\phi''(\mathbf{X}_A\mathbf{H}_i) \odot \mathbf{R}_i]^T \mathbf{X}_A\mathbf{H}_i = -\mathbf{m}\Lambda_i^T\mathbf{H}_i = 0.$

Hence Eq 7.8 reduces to
$(\phi''(\mathbf{X}_A \mathbf{H}_i) \odot \mathbf{R}_i)^T \mathbf{X}_i > 0.$

Using the above two lemmas, we can prove Theorem 3 as follows.

PROOF 7.6 Denoting $\mathbf{D}_i^\star := (\phi''(\mathbf{X}_A \mathbf{H}_i) \odot \mathbf{R}_i)$ in the proof of Theorem 3.1 of the main paper, all the statements of the proof directly apply in light of Lemma 7.7.1 and Lemma 7.7.2.

References

1. S. Arora, R. Ge, Y. Halpern, D. Mimno, A. Moitra, D. Sontag, Y. Wu, and M. Zhu. A practical algorithm for topic modeling with provable guarantees. *International Conference on Machine Learning, ICML 2013*, 2013.
2. S. Arora, R. Ge, R. Kannan, and A. Moitra. Computing a non-negative matrix factorization - provably. *Annual ACM Symposium on Theory of Computing, STOC 2012*, 2012.
3. A. Banerjee, S. Merugu, I. Dhillon, and J. Ghosh. Clustering with Bregman divergences. *Journal of Machine Learning Research*, 6:1705–1749, 2005.
4. V. Bittorf, B. Recht, C. Re, and J. Tropp. Factoring non-negative matrices with linear programs. *Annual Conference on Neural Information Processing Systems, NIPS 2012*, 2012.
5. S. Boyd, N. Parikh, E. Chu, B. Peleato, and J. Eckstein. Distributed optimization and statistical learning via the alternating direction method of multipliers. *Foundations and Trends® in Machine Learning*, 3(1):1–122, 2011.
6. S. Cheung and C. Kamath. Robust techniques for background subtraction in urban traffic video. In *Electronic Imaging 2004*, pages 881–892, 2004.
7. A. Cichocki, R. Zdunek, A. Phan, and S. Amari. *Non-negative Matrix and Tensor Factorizations*. Wiley, 2009.
8. K. Clarkson. More output-sensitive geometric algorithms. In *Foundations of Computer Science*, 1994.
9. D. Donoho and V. Stodden. When does non-negative matrix factorization give a correct decomposition into parts? In *Annual Conference on Neural Information Processing Systems, NIPS 2003*, 2003.
10. J. Dula, R. Hegalson, and N. Venugopal. An algorithm for identifying the frame of a pointed finite conical hull. *INFORMS Journal on Computing*, 10(3):323–330, 1998.
11. E. Elhamifar, G. Sapiro, and R. Vidal. See all by looking at a few: Sparse modeling for finding representative objects. In *Computer Vision and Pattern Recognition*, 2012.
12. E. Esser, M. Miller, S. Osher, G. Sapiro, and J. Xin. A convex model for non-negative matrix factorization and dimensionality reduction on physical space. *IEEE Transactions on Image Processing*, 21(10):3239–3252, 2012.
13. R. Fan, K. Chang, C. Hsieh, X. Wang, and C. Lin. LIBLINEAR: a library for large linear classification. *Journal of Machine Learning Research, JMLR 2008*, 2008.
14. C. Févotte, N. Bertin, and J. Durrieu. Non-negative matrix factorization with the Itakura-Saito divergence: With application to music analysis. *Neural computation*, 21(3):793–830, 2009.
15. N. Gillis. Successive non-negative projection algorithm for robust non-negative blind source separation. *SIAM Journal on Imaging Sciences*, 7(2):1420–1450, 2014.
16. N. Gillis and R. Luce. Robust near-separable non-negative matrix factorization using linear optimization. *Journal of Machine Learning Research*, 15:1249–1280, April 2014.

17. N. Gillis and S. Vavasis. Fast and robust recursive algorithms for separable non-negative matrix factorization. *arXiv:1208.1237v2*, 2012.
18. D. Greene and P. Cunningham. Practical solutions to the problem of diagonal dominance in kernel document clustering. In *International Conference on Machine Learning, ICML 2006*, 2006.
19. C. Hsieh and I. Dhillon. Fast coordinate descent methods with variable selection for non-negative matrix factorization. *ACM Conference on Knowledge Discovery and Data Mining, KDD 2011*, 2011.
20. H. Kim and H. Park. Sparse non-negative matrix factorizations via alternating non-negativity-constrained least squares for microarray data analysis. *Bioinformatics*, 23:1495–1502, 2007.
21. A. Kumar, V. Sindhwani, and P. Kambadur. Fast conical hull algorithms for near-separable non-negative matrix factorization. In *International Conference on Machine Learning, ICML 2013*, 2013.
22. D. Lee and S. Seung. Learning the parts of objects by non-negative matrix factorization. *Nature*, 401(6755):788–791, 1999.
23. L. Li, W. Huang, I. Gu, and Q. Tian. Statistical modeling of complex backgrounds for foreground object detection. *IEEE Transactions on Image Processing*, 13(11):1459–1472, 2004.
24. L. Li, G. Lebanon, and H. Park. Fast Bregman divergence NMF using Taylor expansion and coordinate descent. *Knowledge Discovery in Databases*, 2012.
25. C. Lin. Projected gradient methods for non-negative matrix factorization. *Neural Computation*, 2007.
26. Z. Lin, M. Cheng, and Y. Ma. The augmented Lagrange multiplier method for exact recovery of corrupted low-rank matrices. *arXiv preprint arXiv:1009.5055*, 2010.
27. G. Naiyang, T. Dacheng, L. Zhigang, and J. Taylor. Mahnmf: Manhattan non-negative matrix factorization. *CoRR abs/1207.3438*, 2012.
28. Reuters. `archive.ics.uci.edu/ml/datasets/Reuters-21578+Text+Categorization+Collection`. (Accessed December, 2015)
29. S. Sra and I. Dhillon. Generalized non-negative matrix approximations with Bregman divergences. *Advances in Neural Information Processing Systems*, pages 283–290, 2005.

8

Robust Matrix Completion through Nonconvex Approaches and Efficient Algorithms

Yuning Yang
KU Leuven, ESAT-STADIUS, Belgium

Yunlong Feng
KU Leuven, ESAT-STADIUS, Belgium

J.A.K. Suykens
KU Leuven, ESAT-STADIUS, Belgium

8.1 Introduction

The goal of matrix completion is to impute missing values of a possibly low-rank matrix with only partial entries observed. This problem arises in online recommendation systems, computer vision, etc. In real-world applications, the matrix to be recovered might be contaminated by noise or outliers, where robust techniques are needed. In this chapter, we introduce a robust matrix completion model, where the robustness benefits from a nonconvex loss function. Efficient algorithms are proposed to solve the introduced robust matrix completion model. Experiments are carried out on synthetic as well as real datasets to validate the efficiency and effectiveness of the proposed models and algorithms.

The problem of matrix completion aims at recovering a matrix from a sampling of its entries, which has arisen from a variety of real-world applications including online recommendation systems [26, 30], image impainting [1, 20], computer vision, and video denoising [17]. The problem itself could be an ill-posed problem without further constraints since we have fewer samples than entries. However, in many applications including those mentioned-above, it is common that the matrix that we are going to recover has some special structures; for example, low-rank or approximately low-rank, which makes it possible to search within all possible completions.

In matrix completion problems, one tries to approximate the observed entries of the matrix as well as possible while also preserving the low-rank property of the recovered matrix. Mathematically, the problem can be formulated as

$$\min_{L\in\mathbb{R}^{m\times n}} \operatorname{rank}(L) \quad \text{s.t. } L_{ij} = B_{ij}, \quad (i,j) \in \Omega,$$

where $L,\ B \in \mathbb{R}^{m\times n}$, and Ω are an index set. Due to the nonconvexity of the rank function rank$(\cdot)$, solving this minimization problem is NP-hard in general. To obtain a tractable convex relaxation, the nuclear norm heuristic is usually employed, which also imposes the low-rank property. To measure the approximation ability of a candidate matrix on the observed entries, the least-squares criterion is usually employed in the data fidelity term. In a seminal work, [3] showed that most low-rank matrices can be recovered exactly from partial sampled entries that may have surprisingly small cardinality by using the convex relaxation introduced in [6], and many algorithms have been introduced to solve the convex optimization problem.

In the noisy setting of the matrix completion problem, the corresponding observed matrix turns out to be

$$B_\Omega = L_\Omega + S,$$

where B_Ω denotes the projection of B onto Ω, and S refers to the noise.

When the observed data are corrupted by gross errors, the resulting matrix could be far away from the ground-truth due to the utilization of the least-squares criterion, which is non-robust. To address this problem, some efforts have been made in the literature. In a seminal work, [2] proposed a robust matrix completion approach, in which the model takes the following form

$$\min_{L,S\in\mathbb{R}^{m\times n}} \quad \|S\|_1 + \lambda\|L\|_* \quad \text{s.t.} \;\; L_\Omega + S = B_\Omega. \tag{8.1}$$

The above model can be further formulated as

$$\min_{L\in\mathbb{R}^{m\times n}} \quad \|L_\Omega - B_\Omega\|_1 + \lambda\|L\|_*,$$

where $\lambda > 0$ is a regularization parameter. The robustness of the model (8.1) results from using the least absolute deviation loss (LAD). This model was later applied to the column-wise robust matrix completion problem in [4].

By further decomposing S into $S = S_1 + S_2$, where S_1 refers to the noise and S_2 stands for the outliers, [10] proposed the following robust reconstruction model

$$\min_{L,S_2\in\mathbb{R}^{m\times n}} \quad \|L_\Omega - B_\Omega - S_2\|_F^2 + \lambda\|L\|_* + \gamma\|S_2\|_1,$$

where $\lambda, \gamma > 0$ are regularization parameters. They further showed that the above estimator is equivalent to the one obtained by using the Huber's criterion when evaluating the data-fitting risk. The Huber's criterion was adopted in [10] to introduce robustness into matrix completion. [25] proposed to use an L_p $(0 < p \leq 1)$ loss to enhance the robustness. However, none of the above approaches can be sufficiently robust to gross errors due to the unboundedness of these loss functions, which cannot remove the impact of the gross errors on the output.

We also note that, to enhance the robustness, several approaches have been proposed in low-rank matrix approximation problems, especially for PCA [2, 35, 36]. However, it is necessary to point out the difference between the PCA and the matrix completion. As suggested in [35], the essential difference lies in that in matrix completion problems the support of missing entries is given, whereas in PCA, corrupted entries are never known. From a statistical learning viewpoint, PCA is a typical unsupervised learning problem while the matrix completion can be interpreted as a supervised learning scenario, e.g., the trace regression problem [19, 28] mentioned above, or a transductive learning scenario [29].

In this chapter, motivated by theoretical investigations presented in [7, 33] and empirical success reported in [11, 21], we propose a nonconvex approach by employing an exponential squared type loss, namely, the Welsch loss, which will be introduced later. The Welsch loss was originally introduced in robust statistics to form robust estimators in linear models [12]. In this chapter, we will show that it can also work efficiently in matrix completion problems and bring us robust output. Moreover, for the proposed nonconvex matrix completion problems, we propose efficient algorithms, where at each iteration, the algorithms first compute a rank-one matrix, and then update the new trial as a linear combination of the current trial and the newly generated rank-one matrix. The rank-one matrix is related to the left and right singular vectors of the leading singular value of a certain matrix, which can be found efficiently by the power method or Lanczos method. Therefore, the whole algorithms are also efficient. We also show the sublinear convergence rate of the proposed algorithms.

We would also like to mention the differences between this chapter and our previous works [37, 38], which are focused on robust and low-rank matrix/tensor completion problems. The models in [37] are similar to those presented in this chapter; however, the computational algorithms proposed here are different from those in [37], along with different convergence analysis. Furthermore, we present more numerical experiments than [37]. The work in [38] is focused on robust tensor completion, while this chapter is restricted to robust matrix completion.

This chapter is organized as follows. In Section 8.2, we formally formulate the proposed robust matrix completion problem and introduce the Welsch loss that will be used in our study. Section 8.3 presents the proposed algorithms. We give convergence analysis of the proposed algorithms in Section 8.4. Numerical experiments are carried out in Section 8.5 to validate the efficiency and effectiveness of the proposed algorithms. We end this chapter with conclusions in Section 8.6.

8.2 Problem Formulation

Formally, the matrix completion problem can be formulated as follows:

$$\min_{L\in\mathbb{R}^{m\times n}} \quad \mathrm{rank}(L) \quad \text{s.t.} \quad L_\Omega = B_\Omega, \tag{8.2}$$

where Ω is the set of indices that indicates the observed entries. When there is noise or outliers, a certain loss function should be introduced to penalize the noise or outliers. Previous work usually employs a least absolute deviation (LAD) loss $|\cdot|$. The advantage of the LAD loss is that its resulting problem is convex, and has a theoretical recovery result [2]. However, it is not as resistant to outliers as some nonconvex robust losses [12]. In view of this, our problem will be formulated as follows: to measure or to penalize the difference between L and B, we adopt the following loss function

$$\ell_\sigma(t) = \sigma^2/2\left(1-\exp(-t^2/\sigma^2)\right),$$

which is known as the Welsch loss in robust statistics [13]. Here $\sigma > 0$ is a parameter. In the following, we would like to mention some properties of this loss from different aspects.

- σ controls the robustness. The smaller the parameter σ, the more robustness it gives the problem.
- The influence function of $\ell_\sigma(t)$ is given by

$$\psi_\sigma(t) = \exp(-t^2/\sigma^2)t.$$

  In fact, the value $\psi_\sigma(t)/t$ can be regarded as a weight of t. One can observe that as t increases, $\psi_\sigma(t)/t$ decreases sharply, which gives a small weight to the value t. On the other hand, the influence function of the LAD loss is only bounded instead of converging to zero.

- Another property of the Welsch loss is that its influence functions are Lipschitz continuous, with Lipschitz constant 1, i.e., for any $t_1, t_2 \in \mathbb{R}$, it holds that
$$|\psi_\sigma(t_1) - \psi_\sigma(t_2)| \leq |t_1 - t_2|.$$
This property is very important and serves as a basis for the convergence of the algorithms presented later.

- As $t \to 0$, ℓ_σ approximates the least-squares loss, which can be seen from their Taylor series. Given a fixed $\sigma > 0$, it holds that $\ell_\sigma(t) = t^2/2 + o(t^2/\sigma^2)$. Therefore, $\ell_\sigma(t) \approx t^2/2$ provided that $t/\sigma \to 0$. This also reminds us that a large σ can lead to the closeness between ℓ_σ and the least-squares loss. Such a property gives more flexibility to the Welsch loss than the LAD loss.

By letting $t_{ij} = L_{ij} - B_{ij}$ where $(i,j) \in \Omega$, and by summing ℓ_σ over all the indices in Ω, we arrive at the following cost function

$$F_\sigma(L) = \sum_{(i,j)\in\Omega} \frac{\sigma^2}{2}\left(1 - \exp\left(-(L_{ij} - B_{ij})^2/\sigma^2\right)\right).$$

Therefore, if the rank information is known a priori, then we can model the problem as

$$\min_{L\in\mathbb{R}^{m\times n}} \ F_\sigma(L) \ \text{ s.t. } \ \text{rank}(L) \leq R. \tag{8.3}$$

Otherwise, we can constrain $F_\sigma(\cdot)$ by nuclear norm constraint, i.e.,

$$\min_{L\in\mathbb{R}^{m\times n}} \ F_\sigma(L) \ \text{ s.t. } \ \|L\|_* \leq \beta, \tag{8.4}$$

where $\beta > 0$ is a parameter to control the complexity of the model.

8.2.1 Extending to the Affine Rank Minimization Problem

It is known that matrix completion is a special case of the following affine rank minimization problem [9, 16, 18, 22, 24, 27]

$$\min_{L\in\mathbb{R}^{m\times n}} \quad \text{rank}(L) \quad \text{s.t.} \quad \mathcal{A}(L) = b, \tag{8.5}$$

where $b \in \mathbb{R}^p$ is given, and $\mathcal{A} : \mathbb{R}^{m\times n} \to \mathbb{R}^p$ is a linear operator defined by

$$\mathcal{A}(\cdot) := \left[\langle A^1, \cdot\rangle, \ \langle A^2, \cdot\rangle, \ldots, \ \langle A^p, \cdot\rangle\right]^T,$$

where $A^i \in \mathbb{R}^{m\times n}$ for each i. (8.5) can be reduced to matrix completion if we set $p = \text{card}(\Omega)$, the cardinality of Ω, and let $A^{(i-1)n+j} = e_i(m)e_j(n)^T$ for each $(i,j) \in \Omega$, where $e_i(m), i = 1, \ldots, m$ and $e_j(n), j = 1, \ldots, n$ are the canonical basis vector of $\mathbb{R}^m$ and $\mathbb{R}^n$, respectively.

(8.3) and (8.4) can be naturally extended to handle cases with noise and outliers of (8.5). Denote the cost function as as follows

$$F_\sigma(L) = \frac{\sigma^2}{2}\sum_{i=1}^{p}\left(1 - \exp\left(-\left(\langle A^i, L\rangle - b_i\right)^2/\sigma^2\right)\right). \tag{8.6}$$

The rank constrained problem can be formulated as

$$\min_{L\in\mathbb{R}^{m\times n}} \ F_\sigma(L) \ \text{ s.t. } \ \text{rank}(L) \leq R, \tag{8.7}$$

and the nuclear norm constrained problem takes the form

$$\min_{L\in\mathbb{R}^{m\times n}} \ F_\sigma(L) \ \text{ s.t. } \ \lambda\|L\|_* \leq \beta. \tag{8.8}$$

8.3 Computational Algorithms

In robust regression, problems associated with a robust loss are usually solved by iterative reweighted least squares approaches [13]. Here, since we want to find low-rank solutions to (8.3) and (8.4), it would be proper to explore the structures of these two problems, and consider different algorithms. The algorithms proposed here are called rank-one matrix updating algorithms. The main idea is that, at each iteration, the algorithms compute a rank-one matrix, which is formed by the left and right singular vectors corresponding to the leading singular value of the matrix $\nabla F_\sigma(L^{(k)})$, by using power method or the Lanczos method. Then the algorithms update the new trial via certain linear combinations of the current trial and the newly generated rank-one matrix. In general, the algorithm framework is presented in Algorithm 8.1.

Algorithm 8.1 - Rank-one matrix updating algorithms for solving (8.3) and (8.4)

Input: Zero matrix $L^{(0)} = 0$.
Output: $L^{(k+1)}$.
for $k = 1$ **to** ... **do**

- Compute a normalized rank-one matrix $W^{(k)} = \mathbf{u}^{(k)}(\mathbf{v}^{(k)})^\top$:

$$(\mathbf{u}^{(k)}, \mathbf{v}^{(k)}) = \arg \max_{\|\mathbf{u}\|_F=1, \|\mathbf{v}\|_F=1} \mathbf{u}^\top \nabla F_\sigma(L^{(k)})\mathbf{v}. \tag{8.9}$$

- Select suitable stepsizes (weights) $(\overline{\alpha}_1, \overline{\alpha}_2)$ and update

$$L^{(k+1)} = \overline{\alpha}_1 L^{(k)} + \overline{\alpha}_2 W^{(k)}.$$

end for

Algorithm 8.1 is related to some recently developed algorithms in the literature. In [15], a simple algorithm for nuclear norm regularized problems was proposed, where at each iteration, the algorithm also computes a rank-one matrix by using the power method or the Lanczos method. Another rank-one matrix updating algorithm was proposed in [34], where the weights are computed as the matching pursuit type methods [23, 32]. Other rank-one updating algorithms have also been developed; see, e.g., [5, 31, 39]. However, a limitation of the above methods is that they are designed for problems with a convex cost function $F(\cdot)$, while in our case, the Welsch loss-based $F_\sigma(\cdot)$ is highly nonconvex.

When solving (8.9), both power method and Lanczos method can be applied and scaled well to large-scale problems. However, to further improve the efficiency of the proposed algorithms, (8.9) may not be solved exactly. In these cases, performing only a few power iterations may be enough to obtain an acceptable output $W^{(k)}$.

The computational complexity of solving (8.9) is at most $O(mn)$. Furthermore, if the matrix $\nabla F_\sigma(L^{(k)})$ is sparse with N nonzero entries, as in our case, then the complexity can be reduced to $O(N)$ [14]. Therefore, the computational complexity of the whole algorithm might be low.

We also note that the proposed algorithms can be easily extended to solving the affine rank minimization problems (8.7) and (8.8), only by replacing the gradient $\nabla F_\sigma(\cdot)$ in (8.9) by the gradient of the cost function defined in (8.6).

In the following, we specify the ways of choosing $(\overline{\alpha}_1, \overline{\alpha}_2)$ in Algorithm 8.1 for the two different problems (8.3) and (8.4). For (8.3), the weights are chosen by the following simple rule:

$$\overline{\alpha}_2 = -\left\langle F_\sigma(L^{(k)}), W^{(k)} \right\rangle / \|W^{(k)}\|_\Omega^2, \quad \overline{\alpha}_1 = 1, \tag{8.10}$$

where $\|W^{(k)}\|_\Omega = \|W_\Omega^{(k)}\|_F$. For (8.4), we first denote

$$D^{(k)} := -\beta \cdot W^{(k)} - L^{(k)}, \tag{8.11}$$

and let

$$L^{(k+1)} = L^{(k)} + \overline{\alpha} D^{(k)},$$

where $\overline{\alpha} \in (0,1)$ is selected by Armijo search rule:

Fixed scalars $l \in (0,1), \mu \in (0,1)$, and we choose the step-size $\overline{\alpha} = l^m$, where m is the first non-negative integer m such that

$$F_\sigma(L^{(k)} + l^m D^{(k)}) - F_\sigma(L^{(k)}) \le \mu l^m \langle \nabla F_\sigma(L^{(k)}), D^{(k)} \rangle. \tag{8.12}$$

The idea behind (8.10) is that we want to minimize a quadratic function that majorizes $F_\sigma(\cdot)$ at $L^{(k+1)}$, which forces $\{F_\sigma(L^{(k)})\}$ to decrease, as will be shown in the next section. The idea behind (8.12) follows the Frank-Wolfe method [8]. First, we notice that

$$\begin{aligned} -\beta \cdot W^{(k)} &= -\beta \cdot \arg \max_{\|\mathbf{u}\|_F = 1, \|\mathbf{v}\|_F = 1} \langle \nabla F_\sigma(L^{(k)}), W \rangle \\ &= -\beta \arg \min_{\|W\|_* \le 1} \langle \nabla F_\sigma(L^{(k)}), W \rangle \\ &= \arg \min_{\|W\|_* \le \beta} \langle \nabla F_\sigma(L^{(k)}), W \rangle, \end{aligned}$$

where the second equality follows from the duality between the matrix spectral norm and the nuclear norm. As a result, $-\beta W^{(k)}$ lies in the nuclear norm ball $\|W\|_* \le \beta$, and $D^{(k)}$ is a descent direction of $F_\sigma(\cdot)$ at $L^{(k)}$. Then, a suitable $\overline{\alpha}$ can be chosen by (8.12) to get a sufficient decrease from $F_\sigma(L^{(k)})$ to $F_\sigma(L^{(k+1)})$. In the next section, we will present their convergence analysis.

8.4 Convergence Analysis

In this section, we will establish the convergence results of (8.1). The convergence analysis is based on the fact that the gradient of $F_\sigma(\cdot)$ is Lipschitz continuous with constant 1, as will be shown in Proposition 8.1. We first present the gradient of $F_\sigma(\cdot)$ at L, which is given by

$$\nabla F_\sigma(L) = \Lambda \circ (L - B),$$

where $\Lambda \in \mathbb{R}^{m \times n}$ is a matrix such that if $(i,j) \in \Omega$, then $\Lambda_{ij} = \exp(-(L_{ij} - B_{ij})^2/\sigma^2)$, and $\Lambda_{ij} = 0$ if $(i,j) \notin \Omega$; $\circ$ denotes the Hadamard operator, i.e., entry-wise product.

PROPOSITION 8.1 [[37], Proposition 1] For any matrices $X, Y \in \mathbb{R}^{m \times n}$, there holds

$$\|\nabla F_\sigma(X) - \nabla F_\sigma(Y)\|_F \le \|X - Y\|_\Omega. \tag{8.13}$$

PROOF 8.1 The proof uses the fact that the influence function $\psi_\sigma(t)$ is Lipschitz continuous, i.e.,

$$|\psi_\sigma(x) - \psi_\sigma(y)| \le |x - y|, \quad \forall\, x, y \in \mathbb{R}.$$

To verify the above inequality, it suffices to verify that the magnitude of

$$\psi_\sigma'(t) = \exp(-t^2/\sigma^2) - 2\exp(-t^2/\sigma^2)t^2/\sigma^2$$

can be upper bounded by 1. We can denote $u = t^2/\sigma^2 \geq 0$, and in fact, the maximum of the function $|\exp(-u)(1-2u)|$ is 1 at $u = 0$. This shows that for any $t \in \mathbb{R}$ and any $\sigma > 0$, there holds $\left|\psi_\sigma'(t)\right| \leq 1$. As a result, it follows

$$\begin{aligned} &\|\nabla F_\sigma(X) - \nabla F_\sigma(Y)\|_F^2 \\ &= \sum_{(i,j)\in\Omega} (\psi_\sigma(X_{ij}) - \psi_\sigma(Y_{ij}))^2 \\ &\leq \|X - Y\|_\Omega^2, \end{aligned}$$

as desired.

Following (8.13) we immediately have

PROPOSITION 8.2 For any matrices $X, Y \in \mathbb{R}^{m\times n}$, there holds

$$F_\sigma(X) \leq F_\sigma(Y) + \langle \nabla F_\sigma(Y), X - Y\rangle + \frac{\|X - Y\|_\Omega^2}{2}. \tag{8.14}$$

Following Proposition 8.2 we have the following convergence result on applying Algorithm 8.1 with strategy (8.10) to solve (8.3).

THEOREM 8.1 *[Convergence result on applying Algorithm 8.1 with strategy* (8.10) *to solve* (8.3)*] Let* $\{L^{(k)}\}$ *be a sequence generated by Algorithm 8.1 with strategy* (8.10). *Then* $\{F_\sigma(L^{(k)})\}$ *is nonincreasing.*

PROOF 8.2 Strategy (8.10) tells us that

$$L^{(k+1)} = L^{(k)} - \frac{\langle \nabla F_\sigma(L^{(k)}), W^{(k)}\rangle}{\|W^{(k)}\|_\Omega^2} W^{(k)},$$

which together with (8.14) implies that

$$\begin{aligned} F_\sigma(L^{(k+1)}) &\leq F_\sigma(L^{(k)}) + \langle \nabla F_\sigma(L^{(k)}), L^{(k+1)} - L^{(k)}\rangle + \frac{\|L^{(k+1)} - L^{(k)}\|_\Omega^2}{2} \\ &= F_\sigma(L^{(k)}) - \frac{\langle \nabla F_\sigma(L^{(k)}), W^{(k)}\rangle^2}{2\|W^{(k)}\|_\Omega^2}, \end{aligned}$$

which shows that $\{F_\sigma(L^{(k)})\}$ is nonincreasing. The proof is completed.

We then consider the convergence result on applying Algorithm 8.1 with strategy (8.12) to solve (8.4).

THEOREM 8.2 *[Convergence result on applying Algorithm 8.1 with strategy* (8.12) *to solve* (8.4)*] Let* $\left\{L^{(k)}\right\}$ *be a sequence generated by Algorithm 8.1 with strategy* (8.12) *to solve* (8.4). *Then every limit point of* $\left\{L^{(k)}\right\}$ *is a critical point of problem* (8.4).

To prove Theorem 8.2, we need some observations and lemmas first.

PROPOSITION 8.3 Let $\overline{W} = \arg\min_{\|W\|_* \le \beta} \langle \nabla F_\sigma(L), W\rangle$. If L is not a critical point of problem (8.4), then

$$\langle \nabla F_\sigma(L), \overline{W} - L\rangle < 0.$$

PROOF 8.3 Suppose $\langle \nabla F_\sigma(L), \overline{W} - L\rangle \ge 0$. Then it follows

$$\langle \nabla F_\sigma(L), W - L\rangle \ge \langle \nabla F_\sigma(L), \overline{W} - L\rangle \ge 0, \quad \forall\ W \text{ satisfying } \|W\|_* \le \beta,$$

which implies that L is a critical point of (8.4), deducing a contradiction.

LEMMA 8.1 Let $\{L^{(k)}\}$ be a sequence generated by Algorithm 8.1 with strategy (8.12) to solve (8.4). Then there holds

$$F_\sigma(L^{(k+1)}) - F_\sigma(L^{(k)}) \le -\frac{2l\mu(1-\mu)\langle \nabla F_\sigma(L^{(k)}), D^{(k)}\rangle^2}{\|D^{(k)}\|_\Omega^2},$$

where $D^{(k)}$ is defined in (8.11).

PROOF 8.4 The Armijo search rule (8.12) implies that

$$F_\sigma(L^{(k)} + \frac{\overline{\alpha}}{l} D^{(k)}) - F_\sigma(L^{(k)}) > \mu \frac{\overline{\alpha}}{l} \langle \nabla F_\sigma(L^{(k)}), D^{(k)}\rangle.$$

Together with Proposition 8.2, it follows

$$\frac{\overline{\alpha}}{l}\langle \nabla F_\sigma(L^{(k)}), D^{(k)}\rangle + \frac{\overline{\alpha}^2}{2l^2}\|D^{(k)}\|_\Omega^2 > \mu \frac{\overline{\alpha}}{l} \langle \nabla F_\sigma(L^{(k)}), D^{(k)}\rangle.$$

Rearranging the terms and noticing Proposition 8.3, we get

$$\overline{\alpha} > \frac{2l(1-\mu)\left|\langle \nabla F_\sigma(L^{(k)}), D^{(k)}\rangle\right|}{\|D^{(k)}\|_\Omega^2}. \tag{8.15}$$

The Armijo search rule again tells us that

$$F_\sigma(L^{(k+1)}) - F_\sigma(L^{(k)}) \le \mu\overline{\alpha}\langle \nabla F_\sigma(L^{(k)}), D^{(k)}\rangle. \tag{8.16}$$

Noticing the non-positivity of $\langle \nabla F_\sigma(L^{(k)}), D^{(k)}\rangle$ and plugging (8.15) into (8.16), we obtain

$$F_\sigma(L^{(k+1)}) - F_\sigma(L^{(k)}) \le -\frac{2l\mu(1-\mu)\langle \nabla F_\sigma(L^{(k)}, D^{(k)}\rangle^2}{\|D^{(k)}\|_\Omega^2},$$

as desired.

PROOF 8.5 [Proof of Theorem 8.2] Denote

$$\mathbf{W} := \{W \mid \|W\|_* \le \beta\},$$

and

$$\mathrm{diam}(\mathbf{W}) := \max_{W_1, W_2 \in \mathbf{W}} \|W_1 - W_2\|$$

as the diameter of $\mathbf{W}$. Since $\mathbf{W}$ is compact, diam($\mathbf{W}$) is finite. From the definition of $D^{(k)}$, it follows

$$\|D^{(k)}\|_\Omega \leq \|D^{(k)}\|_F \leq \text{diam}(\mathbf{W}).$$

This together with Lemma 8.1 shows that

$$F_\sigma(L^{(k+1)}) - F_\sigma(L^{(k)}) < -\frac{2l\mu(1-\mu)\langle \nabla F_\sigma(L^{(k)}, D^{(k)}\rangle^2}{\text{diam}^2(\mathbf{W})}. \tag{8.17}$$

Lemma 8.1 also tells us that $\{F_\sigma(L^{(k)})\}$ is a monotonically decreasing sequence. Since $F_\sigma(\cdot) \geq 0$, we have

$$F_\sigma(L^{(k+1)}) - F_\sigma(L^{(k)}) \to 0,$$

which together with (8.17) implies that

$$|\langle \nabla F_\sigma(L^{(k)}), D^{(k)}\rangle| \to 0.$$

Let L^* be a limit point of $\{L^{(k)}\}$ and let $\{L^{(k)}\}_\mathbf{K}$ be a subsequence of $\{L^{(k)}\}$ such that $\{L^{(k)}\}_\mathbf{K} \to L^*$. Furthermore, let $\overline{\mathbf{K}}$ be a subset of $\mathbf{K}$ such that there exists a subsequence $\{W^{(k)}\}_{\overline{\mathbf{K}}}$ of $\{W^{(k)}\}_\mathbf{K}$ such that $\{W^{(k)}\}_{\overline{\mathbf{K}}} \to W^*$. Without loss of generality we can assume $\overline{\mathbf{K}}$ is $\mathbf{K}$ itself. Then it follows

$$\langle \nabla F_\sigma(L^*), W^* - L^*\rangle = 0.$$

We claim that W^* is a minimizer of $\min_{W\in\mathbf{W}}\langle \nabla F_\sigma(L^*), W\rangle$. Otherwise suppose $\overline{W}$ is a minimizer. Then $\langle \nabla F_\sigma(L^*), W^* - \overline{W}\rangle > 0$. Since $\{\langle \nabla F_\sigma(L^{(k)}), W^{(k)} - \overline{W}\rangle\}_{k\in\mathbf{K}} \to \langle \nabla F_\sigma(L^*), W^* - \overline{W}\rangle$, when k is sufficiently large, it follows

$$\langle \nabla F_\sigma(L^{(k)}), W^{(k)} - \overline{W}\rangle > 0,$$

which shows that $W^{(k)}$ is not a minimizer of $\min_{W\in\mathbf{W}}\langle \nabla F_\sigma(L^{(k)}), W\rangle$, deducing a contradiction. As a result, by the property of W^* we have

$$\langle \nabla F_\sigma(L^*), W - L^*\rangle \geq \langle \nabla F_\sigma(L^*), W^* - L^*\rangle = 0, \quad \forall\, W \in \mathbf{W},$$

which shows that L^* is a critical point of (8.4). The proof is completed.

Next we estimate the rate of convergence in terms of $|\langle \nabla F_\sigma(L^{(k)}), D^{(k)}\rangle|$. We have the following results.

THEOREM 8.3 *Let $\{L^{(k)}\}$ be a sequence generated by Algorithm 8.1 with strategy* (8.12) *to solve* (8.4)*. Then for every $K \geq 1$, we have*

$$\min_{0\leq k\leq K} |\langle \nabla F_\sigma(L^{(k)}), D^{(k)}\rangle|^2 \leq \frac{\text{diam}^2(\mathbf{W})(F_\sigma(L^{(0)}) - F_\sigma^*)}{2l\mu(1-\mu)K},$$

where F_σ^* is the limit of $F_\sigma(L^{(k)})$.

PROOF 8.6 Let $C := \frac{\text{diam}^2(\mathbf{W})}{2l\mu(1-\mu)}$. Then it follows form Lemma 8.1 that

$$\langle \nabla F_\sigma(L^{(k)}), D^{(k)}\rangle^2 \leq C(F_\sigma(L^{(k)}) - F_\sigma(L^{(k+1)})).$$

Summing the above inequality from 0 to K, we get

$$\sum_{k=0}^{K} \langle \nabla F_\sigma(L^{(k)}), D^{(k)} \rangle^2 \leq C(F_\sigma(L^{(0)}) - F_\sigma(L^{(K)}),$$

which implies that

$$\min_{k=0,\ldots,K} \langle \nabla F_\sigma(L^{(k)}), D^{(k)} \rangle^2 \leq \frac{C(F_\sigma(L^{(0)}) - F_\sigma^*)}{K},$$

as desired.

REMARK 8.1 Theorem 8.3 tells us that $|\langle \nabla F_\sigma(L^{(k)}), D^{(k)} \rangle| \to 0$ with rate $O(1/\sqrt{K})$. This together with Theorem 8.2 yields the result that Algorithm 8.1 with the Armijo search rule finds a critical point of (8.4) with convergence rate $O(1/\sqrt{K})$.

8.5 Numerical Experiments

In this section, we present some numerical experiments on synthetic data as well as real data. All the numerical computations are conducted on an Intel i7-3770 CPU desktop computer with 16 GB of RAM. The supporting software is MATLAB R2013a.

8.5.1 Algorithms Setting

We mainly compare the proposed algorithms with RPCA [2], which solves convex optimization problems and employs the LAD loss to penalize the noise or outliers. Algorithm 8.1 with strategy (8.10) for solving the rank constrained problem (8.3) is denoted by RoMu1 for short, while Algorithm 8.1 with strategy (8.12) for solving the nuclear norm constrained problem (8.4) is denoted as RoMu2 for short. The max iteration for all the methods is 400. The stopping criterion for all the methods is that the difference between the current trial and the previous trial is less than a threshold, where the threshold is set to $\epsilon = 10^{-4}$. Parameters are tuned via 5-fold cross validation. All the results are averaged over ten instances.

8.5.2 Synthetic Data

We randomly generate some matrices of size 500×500, and then truncate them to be low rank, where we consider rank $\in \{10, 50\}$. Next, 20% of the entries are contaminated by outliers in $[-10, 10]$. Finally, some entries are randomly missing, where the missing ratio (MR for short) varies between $\{0.2, 0.3, 0.5, 0.7, 0.9\}$. The relative error relerr $= \|X^* - B\|_F / \|B\|_F$ will be used to evaluate performances of the algorithms. Results are reported in Figure 8.1, where the blue curve represents the performance of RoMu1, the red one is that of RoMu2, and the green one stands for RPCA. First we look at Figure 8.1.a, which is the case that rank $= 10$. We observe that when the MR value is less than 0.65, RPCA is better than our methods. However, its performance decreases sharply as the MR value increases, and it achieves 1 when the MR value is 0.9. On the other hand, our method is more stable. Comparing between RoMu1 and RoMu2, we see that RoMu1 is better than RoMu2 when the MR value is less than 0.85. The reason might be that the weights chosen by RoMu1 are more greedy, which leads to a sufficient decrease of the cost function, whereas for RoMu2, choosing the weights has restrictions in $(0, 1)$, as shown in (8.12). We then consider Figure 8.1.b ,

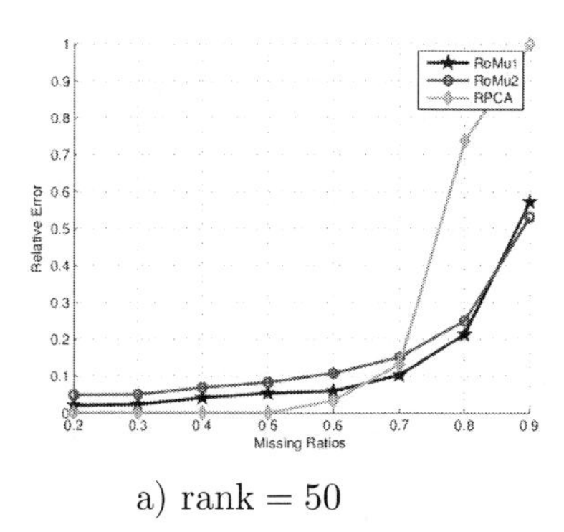

a) rank = 50

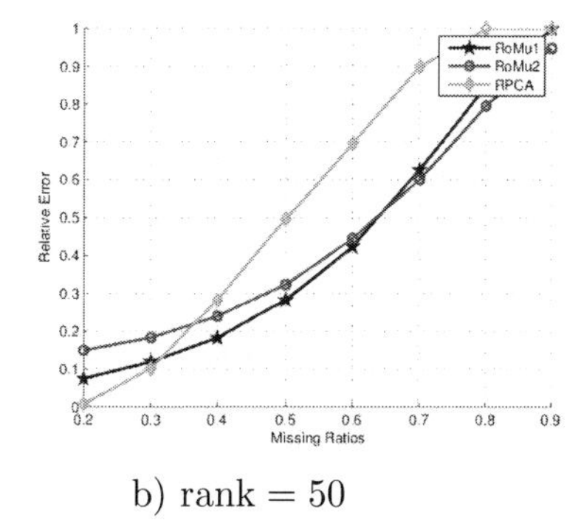

b) rank = 50

FIGURE 8.1 Performance comparisons of RoMu1, RoMu2 and RPCA [2] on synthetic data (500×500), rank $= \{10, 50\}$. The x-axis is the MR value; the y-axis stands for the relative error.

i.e., the case rank = 50. One first notices that all the methods perform worse than the case rank = 10. The reason is evident, since low-rank methods perform better when the rank is not high. One then observes that our methods outperform RPCA when the MR value is larger than 0.4. In summary, our methods are more stable when there are missing values and outliers.

We also report the computational time of the three methods in Table 8.1. From the table, we observe that RoMu1 performs the fastest, followed by RoMu2. This observation confirms the efficiency of our methods. Particularly, an appealing feature of our methods is that as the MR value increases, the computational time decreases, which is due to the fact that the simple structure of Algorithm 8.1 can utilize the sparsity of the matrix: When computing the rank-one matrix, only sparse matrix-vector multiplications are needed; computing the weights is also fast as it can be given by the inner product of sparse matrices, as shown in (8.10).

TABLE 8.1 Efficiency comparisons of RoMu1, RoMu2, and RPCA [2] on synthetic data (500×500), rank $\in \{10, 50\}$.

MR (%)		20	30	40	50	60	70	80	90
	RoMu1	1.49	1.32	1.15	0.89	0.71	0.56	0.38	0.20
rk=10	RoMu2	6.30	5.54	4.77	3.59	2.90	2.26	1.58	0.94
	RPCA [2]	3.77	4.40	5.06	9.26	41.00	40.91	40.83	40.74
	RoMu1	6.04	5.31	4.57	3.47	2.78	2.14	1.51	0.88
rk=50	RoMu2	6.28	5.59	4.82	3.64	2.97	2.26	1.58	0.92
	RPCA [2]	40.93	41.29	41.48	40.64	31.13	36.08	40.59	40.80

8.5.3 Real Data

Image/Video recovery

Gray images can be seen as matrices, while a video can also be treated as a matrix by vectorizing every frame into a vector and arranging them one by one. In real-world applications, due to some reasons, a large fraction of entries of image/video may be missing and may be contaminated by noise or outliers. The goal of this section is to recover such kinds of images/videos. The following datasets are selected: Facade (493×517), Hyperspectral images (50430×96), Brain MRI (39277×181), Incisix (16384×166), and Ocean (17920×32). The first dataset is a gray image, while the last four can be seen as videos. Then 20% of the entries are contaminated by outliers in $[-256, 256]$. Last, some entries are randomly missing, where the missing ratio varies between $\{0.5, 0.6, 0.7, 0.8, 0.9, 0.95\}$. The relative error will also be used to evaluate performances of the algorithms.

The performances are reported in Table 8.2. From the table, we can observe that in most cases, RoMu2 outperforms others, followed by RoMu1. RPCA only performs better than our methods in the image Facade, which has a relatively small size compared to other datasets. This observation implies that our methods may be suitable for large-scale problems. In Figure 8.2 and Figure 8.3, we present part of the recovery results of RoMu1, RoMu2, and RPCA on the datasets Facade and Incisix to intuitively illustrate their performances. On Facade, although Table 8.2 shows that RPCA performs better, from Figure 8.2 it seems that to penalize the outliers, RPCA has to remove more details from the image, while our methods retain more details. Figure 8.3 shows that RPCA cannot correctly recover the dataset. On the other hand, the efficiency is also reported in Table 8.2, where our methods are again much faster than RPCA.

Yale face

As with [2], the goal of this application is to remove shadows from faces, where the datasets are chosen from the extended Yale face database B. We choose two datasets, each of which consists of 64 faces of a person under 64 illumination conditions, and the size of each image is 192×168. We do not add outliers to the datasets, because the shadows in the faces can be regarded as noise or outliers. There do not have to be missing values as well. To show the results, from each dataset we select four images, which are shown in Figure 8.4 and Figure 8.5. From the results, we can observe that all the three methods can remove shadows, while from the second row of Figure 8.4, it seems that our methods perform slightly better than RPCA, as the left eye of the person recovered by RPCA cannot be seen clearly. The first row of Figure 8.5 also indicates that our methods perform better, as the lines in the face have been totally removed by our methods. Finally, we find that empirically we only need the linear combination of around ten rank-one matrices to yield the recovery results, which means that our methods can be stopped within ten iterations, implying that our methods are very efficient.

8.6 Conclusion

In this chapter, we proposed a nonconvex approach for robust matrix completion. Along with the approach, we presented two solution methods, one for solving the rank constrained model, the other one for solving the nuclear norm constrained model. The convergence of the algorithms were verified; particularly, for the second algorithm, we proved that it converges to a stationary point, and showed the iteration complexity, which is $O(1/\sqrt{K})$. Finally, numerical experiments show that the proposed models and algorithms are comparable and better than the state-of-the-art method.

Acknowledgment

EU: The research leading to these results has received funding from the European Research Council under the European Union's Seventh Framework Programme (FP7/2007-2013) / ERC AdG A-DATADRIVE-B (290923). This article reflects only the authors' views; the Union is not liable for any use that may be made of the contained information; Research Council KUL: GOA/10/09 MaNet, CoE PFV/10/002 (OPTEC), BIL12/11T; PhD/Postdoc grants; Flemish Government: FWO: projects: G.0377.12 (Structured systems), G.088114N (Tensor-based data similarity); PhD/Postdoc grants; IWT: projects: SBO POM (100031); PhD/Postdoc grants; iMinds Medical Information Technologies SBO 2014; Belgian Federal

Science Policy Office: IUAP P7/19 (DYSCO, Dynamical systems, control and optimization, 2012–2017). Johan Suykens is a professor at KU Leuven, Belgium.

TABLE 8.2 Performances and efficiency comparisons of RoMu1, RoMu2, and RPCA [2] on some real datasets.

Dataset		RoMu1		RoMu2		RPCA [2]	
		relerr	time	relerr	time	relerr	time
	0.50	9.84E-02	2.57	9.60E-02	3.54	**3.55E-02**	42.02
	0.60	1.14E-01	2.09	1.09E-01	2.96	**4.17E-02**	42.14
Facade	0.70	1.45E-01	1.76	1.30E-01	2.36	**5.22E-02**	42.12
(493 × 517)	0.80	1.73E-01	1.16	1.48E-01	1.56	**8.14E-02**	41.46
	0.90	2.63E-01	0.67	**1.83E-01**	0.93	2.15E-01	40.89
	0.95	3.98E-01	0.43	**2.21E-01**	0.60	1.00E+00	2.55
	0.50	2.18E-02	52.40	1.66E-02	57.19	**1.17E-02**	141.14
	0.60	3.12E-02	43.70	2.34E-02	50.60	**2.03E-02**	143.09
Hyperspectral	0.70	4.76E-02	39.76	**3.85E-02**	34.61	3.91E-02	148.80
(50430 × 96)	0.80	8.42E-02	27.37	**6.22E-02**	26.24	1.18E-01	156.54
	0.90	1.95E-01	15.83	**1.53E-01**	10.41	1.00E+00	14.73
	0.95	5.37E-01	8.14	**3.39E-01**	10.92	1.00E+00	5.72
	0.50	4.44E-02	58.86	**3.77E-02**	86.69	3.03E-01	264.82
	0.60	6.43E-02	49.22	**5.56E-02**	72.10	3.85E-01	265.54
Brain	0.70	9.79E-02	38.58	**8.74E-02**	58.28	5.19E-01	260.55
(39277 × 181)	0.80	1.61E-01	28.59	**1.46E-01**	42.47	8.48E-01	262.08
	0.90	3.12E-01	14.78	**2.79E-01**	23.69	1.00E+00	256.96
	0.95	4.93E-01	7.19	**4.41E-01**	12.88	1.00E+00	237.04
	0.50	2.48E-01	37.94	**2.22E-01**	52.76	2.67E-01	99.42
	0.60	2.80E-01	31.87	**2.58E-01**	44.52	2.99E-01	101.58
Incisix	0.70	3.19E-01	23.60	**3.01E-01**	33.97	3.60E-01	100.22
(16384 × 166)	0.80	3.88E-01	15.84	**3.68E-01**	23.38	4.60E-01	98.81
	0.90	5.00E-01	8.55	**4.76E-01**	12.75	1.00E+00	30.83
	0.95	6.02E-01	4.64	**5.74E-01**	7.42	1.00E+00	6.10
	0.50	1.02E-01	7.39	**9.82E-02**	11.39	1.34E-01	17.14
	0.60	1.21E-01	5.70	**1.18E-01**	7.77	1.87E-01	17.01
Ocean	0.70	1.51E-01	4.51	**1.43E-01**	4.62	3.02E-01	16.74
(17920 × 32)	0.80	2.12E-01	3.03	**1.86E-01**	5.34	1.00E+00	13.55
	0.90	4.06E-01	1.82	**3.54E-01**	3.72	1.00E+00	6.87
	0.95	6.52E-01	1.15	**5.97E-01**	1.84	1.00E+00	3.43

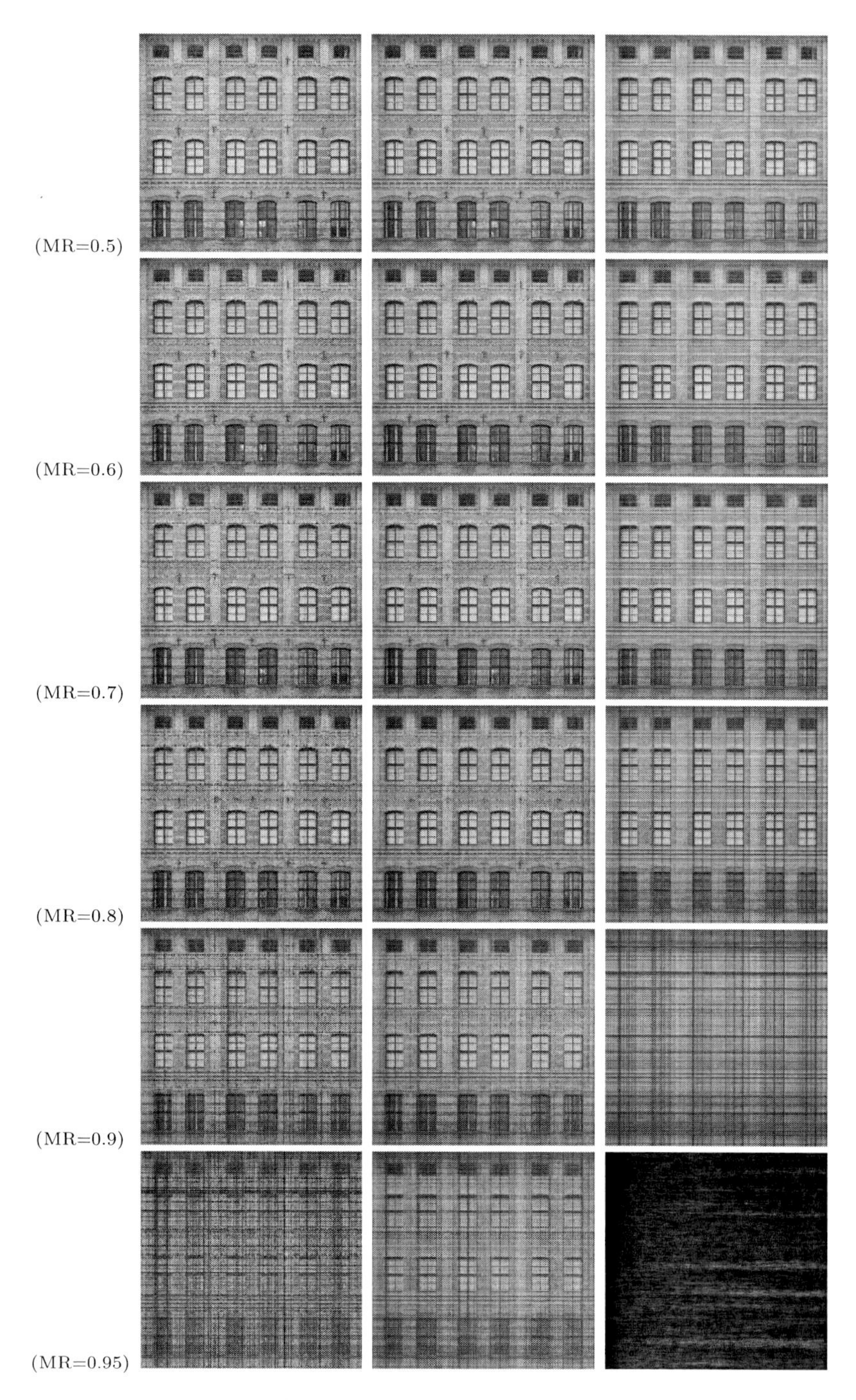

FIGURE 8.2 Comparison of RoMu1 (Column 1), RoMu2 (Column 2), and RPCA [2] (Column 3) on recovering the gray image Facade.

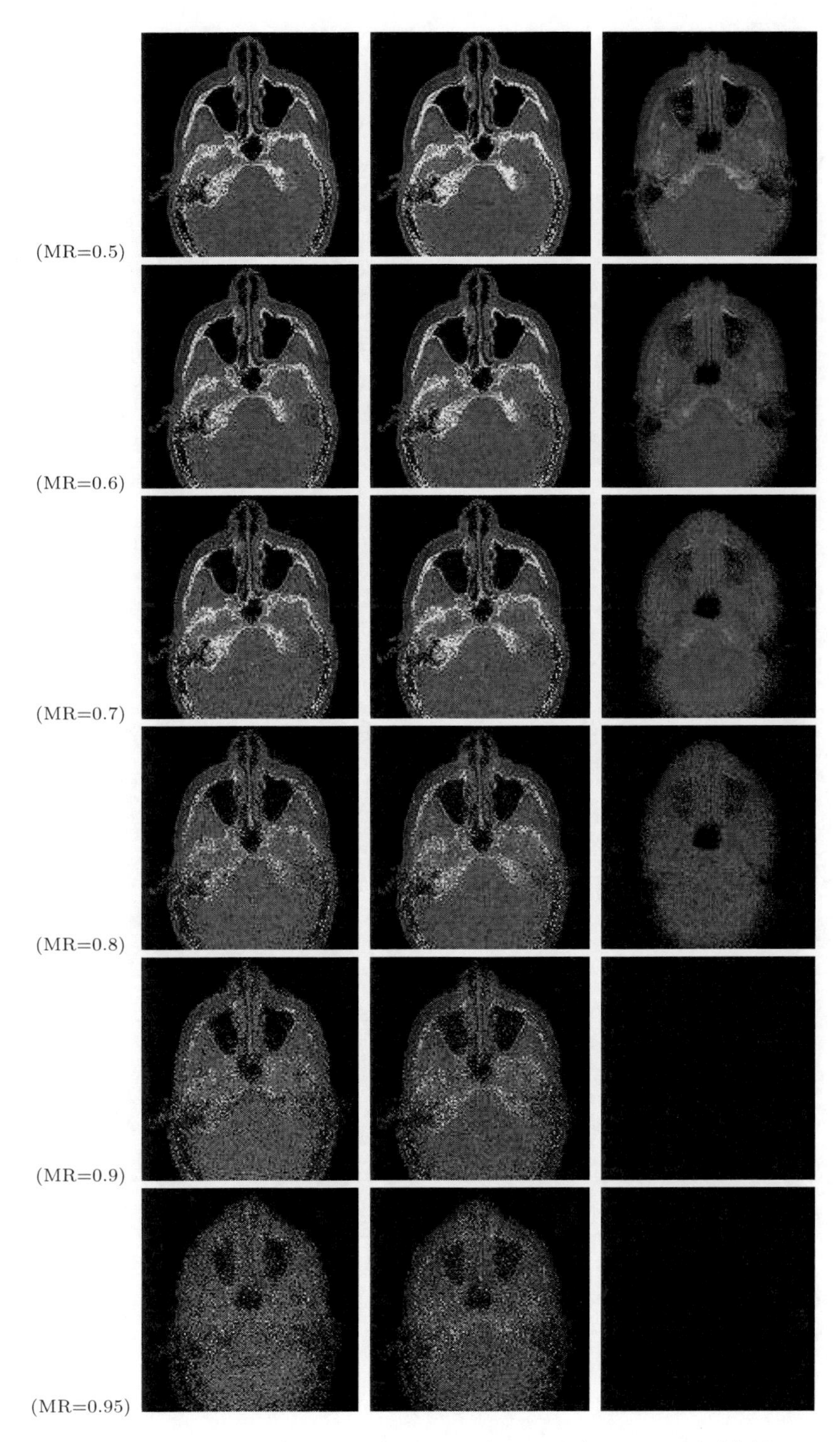

FIGURE 8.3 Comparison of RoMu1 (Column 1), RoMu2 (Column 2), and RPCA [2] (Column 3) on recovering one slide of the Incisix dataset.

FIGURE 8.4 Comparison of RoMu1 (Column 1), RoMu2 (Column 2), and RPCA [2] (Column 3) on removing shadows from faces.

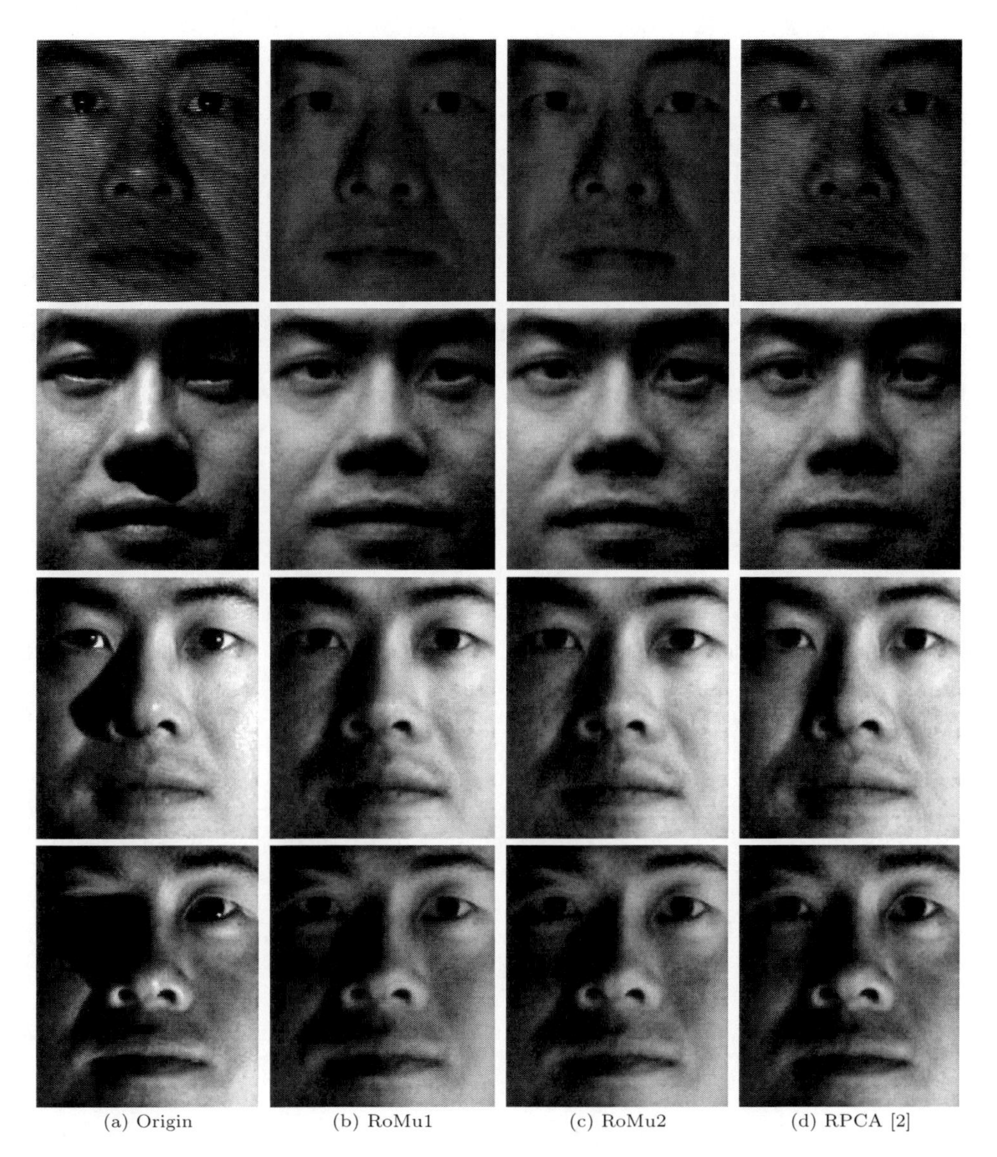

FIGURE 8.5 Comparison of RoMu1 (Column 1), RoMu2 (Column 2), and RPCA [2] (Column 3) on removing shadows from faces.

References

1. M. Bertalmio, G. Sapiro, V. Caselles, and C. Ballester. Image inpainting. In *Annual conference on computer graphics and interactive techniques*, pages 417–424, 2000.
2. E. Candès, X. Li, Y. Ma, and J. Wright. Robust principal component analysis? *Journal of the ACM*, 58(3):11, 2011.
3. E. Candès and B. Recht. Exact matrix completion via convex optimization. *Foundations of Computational Mathematics*, 9(6):717–772, 2009.
4. Y. Chen, H. Xu, C. Caramanis, and S. Sanghavi. Robust matrix completion with corrupted columns. *arXiv preprint arXiv:1102.2254*, 2011.
5. M. Dudik, Z. Harchaoui, and J. Malick. Lifted coordinate descent for learning with trace-norm regularization. In *International Conference on Artificial Intelligence and Statistics, AISTATS 2012*, volume 22, pages 327–336, 2012.
6. M. Fazel, H. Hindi, and S. Boyd. A rank minimization heuristic with application to minimum order system approximation. In *American Control Conference, ACC 2001*, volume 6, pages 4734–4739, 2001.
7. Y. Feng, X. Huang, L. Shi, Y. Yang, and J. Suykens. Learning with the maximum correntropy criterion induced losses for regression. *Journal of Machine Learning Research*, 2015.
8. M. Frank and P. Wolfe. An algorithm for quadratic programming. *Naval research logistics quarterly*, 3(1-2):95–110, 1956.
9. D. Goldfarb and S. Ma. Convergence of fixed-point continuation algorithms for matrix rank minimization. *Foundations of Computational Mathematics*, 11(2):183–210, 2011.
10. T. Hastie. Matrix completion and large-scale SVD computations. *Slides*, May 2012.
11. R. He, W. Zheng, T. Tan, and Z. Sun. Half-quadratic based iterative minimization for robust sparse representation. *IEEE Transactions on Pattern Analysis and Machine Intelligence*, 36(2):261–275, 2013.
12. P. Holland and R. Welsch. Robust regression using iteratively reweighted least-squares. *Communications in Statistics - Theory and Methods*, 6:813–827, 1977.
13. P. Huber. *Robust statistics.* Springer, 2011.
14. M. Jaggi. Revisiting Frank-Wolfe: Projection-free sparse convex optimization. *International Conference on Machine Learning, ICML 2013*, pages 427–435, 2013.
15. M. Jaggi and M. Sulovsk. A simple algorithm for nuclear norm regularized problems. *International Conference on Machine Learning, ICML 2010*, pages 471–478, 2010.
16. P. Jain, R. Meka, and I. Dhillon. Guaranteed rank minimization via singular value projection. In *NIPS*, volume 23, pages 937–945, 2010.
17. H. Ji, C. Liu, Z. Shen, and Y. Xu. Robust video denoising using low rank matrix completion. In *IEEE Conference on Computer Vision and Pattern Recognition, CVPR 2010*, pages 1791–1798. IEEE, 2010.
18. S. Ji and J. Ye. An accelerated gradient method for trace norm minimization. In *Annual International Conference on Machine Learning*, pages 457–464. ACM, 2009.
19. V. Koltchinskii, K. Lounici, and A. Tsybakov. Nuclear-norm penalization and optimal rates for noisy low-rank matrix completion. *The Annals of Statistics*, 39(5):2302–2329, 2011.
20. N. Komodakis. Image completion using global optimization. In *IEEE Conference on Computer Vision and Pattern Recognition, CVPR 2006*, volume 1, pages 442–452. IEEE, 2006.
21. W. Liu, P. Pokharel, and J. Principe. Correntropy: properties and applications in non-Gaussian signal processing. *IEEE Transactions on Signal Processing*, 55(11):5286–5298, 2007.
22. S. Ma, D. Goldfarb, and L. Chen. Fixed point and Bregman iterative methods for matrix rank minimization. *Mathematical Programming*, 128(1-2):321–353, 2011.

23. S. Mallat and Z. Zhang. Matching pursuits with time-frequency dictionaries. *IEEE Transactions on Signal Processing*, 41(12):3397–3415, 1993.
24. K. Mohan and M. Fazel. Iterative reweighted algorithms for matrix rank minimization. *Journal of Machine Learning Research*, 13:3441–3473, 2012.
25. F. Nie, H. Wang, X. Cai, H. Huang, and C. Ding. Robust matrix completion via joint schatten p-norm and l_p-norm minimization. In *IEEE International Conference on Data Mining, ICDM 2012*, pages 566–574. IEEE, 2012.
26. Netflix prize website. http://www.netflixprize.com. *Netflix*, 2009.
27. B. Recht, M. Fazel, and P. Parrilo. Guaranteed minimum-rank solutions of linear matrix equations via nuclear norm minimization. *SIAM Review*, 52(3):471–501, 2010.
28. A. Rohde and A. Tsybakov. Estimation of high-dimensional low-rank matrices. *The Annals of Statistics*, 39(2):887–930, 2011.
29. M. Signoretto, Q. Dinh, L. De Lathauwer, and J.A.K Suykens. Learning with tensors: a framework based on convex optimization and spectral regularization. *Machine Learning*, 94(3):303–351, 2014.
30. N. Srebro and T. Jaakkola. Weighted low-rank approximations. In *International Conference on Machine Learning, ICML 2003*, volume 3, pages 72–727, 2003.
31. A. Tewari, P. Ravikumar, and I. Dhillon. Greedy algorithms for structurally constrained high-dimensional problems. In *Advances in Neural Information Processing Systems*, pages 882–890, 2011.
32. J. Tropp. Greed is good: Algorithmic results for sparse approximation. *IEEE Transactions on Information Theory*, 50(10):2231–2242, 2004.
33. X. Wang, Y. Jiang, M. Huang, and H. Zhang. Robust variable selection with exponential squared loss. *Journal of the American Statistical Association*, 108(502):632–643, 2013.
34. Z. Wang, M. Lai, Z. Lu, and J. Ye. Orthogonal rank-one matrix pursuit for low rank matrix completion. *SIAM Journal on Scientific Computing*, 37:A488A514, 2015.
35. J. Wright, A. Ganesh, S. Raor, Y. Peng, and Y. Ma. Robust principal component analysis: Exact recovery of corrupted low-rank matrices via convex optimization. In *Advances in neural information processing systems*, pages 2080–2088, 2009.
36. H. Xu, C. Caramanis, and S. Sanghavi. Robust PCA via outlier pursuit. *IEEE Transactions on Information Theory*, 58(5):3047–3064, 2012.
37. Y. Yang, Y. Feng, and J. Suykens. A nonconvex relaxation approach to robust matrix completion. *Internal Report 14-61, ESAT-SISTA, KU Leuven*, 2014.
38. Y. Yang, Y. Feng, and J. Suykens. Robust low rank tensor recovery with regularized redescending M-estimator. *Internal Report 14-97, ESAT-SISTA, KU Leuven*, 2014.
39. X. Zhang, D. Schuurmans, and Y. Yu. Accelerated training for matrix-norm regularization: A boosting approach. In *Advances in Neural Information Processing Systems*, pages 2906–2914, 2012.

9

Factorized Robust Matrix Completion

Hassan Mansour
Mitsubishi Electric Research Laboratories, USA

Dong Tian
Mitsubishi Electric Research Laboratories, USA

Anthony Vetro
Mitsubishi Electric Research Laboratories, USA

9.1 Introduction

The problem of reconstructing large-scale matrices from incomplete and noisy observations has attracted a lot of attention in recent years. Of particular interest is the reconstruction or completion of low-rank matrices, which finds many practical applications in recommender systems, collaborative filtering, system identification, and video surveillance.

The robust matrix reconstruction/completion problem can be formulated as the task of determining a low-rank matrix L from observations A that are contaminated by sparse outliers S and noise E. Moreover, the observations A are acquired through a linear operator $\mathcal{P}$ that generates a smaller number of samples than those in L. In the case of matrix completion, the operator $\mathcal{P}$ is a restriction operator that selects a subset Ω of the samples in L. The general observation model is given as follows:

$$A = \mathcal{P}(L) + S + E. \tag{9.1}$$

A natural approach for recovering L from A involves solving the rank minimization problem

$$\min_L \operatorname{rank}(L) \text{ subject to } h(A - \mathcal{P}(L)) \leq \sigma, \tag{9.2}$$

where $h(\cdot)$ is some penalty function suitable for the noise statistics, and σ is a mismatch tolerance. However, Problem (9.2) is noncovex and generally difficult to compute even when the function $h(\cdot)$ is convex. Alternatively, Fazel et al. [12] introduced the nuclear norm heuristic as the convex envelope of the rank to replace the rank objective, resulting in the nuclear norm minimization problem

$$\min_L \|L\|_* \text{ subject to } h(A - \mathcal{P}(L)) \leq \sigma, \tag{9.3}$$

where the nuclear norm $\|X\|_*$ is equal to the sum of the singular values of the matrix X. The attractiveness of the nuclear norm minimization problem lies in the ability to derive efficient algorithms for solving it [2, 6, 18, 24]. Moreover, Recht et al. [24] derived conditions for which the solution to the nuclear norm minimization problem (9.3) coincides with that of the rank minimization problem (9.2) for the case where $\mathcal{P}$ is a random matrix and $h(\cdot)$ is the Frobenius norm. Recovery guarantees for the case where $\mathcal{P}$ is a matrix completion operator was also extensively studied in [8, 23]. Another approach (SpaRCS) was proposed in [32], which follows a greedy approach that iteratively estimates the low-rank subspace of L as well as the support of S followed by truncated SVD and least-squares inversion to compute estimates for L and S.

The choice of the penalty function $h(\cdot)$ determines the sensitivity of the solution of Problem (9.3) to the noise model. For example, letting $h(\cdot)$ be the matrix Frobenius norm assumes that there are no sparse outliers S and the error E is Gaussian distributed. In the case where the observations A are contaminated with large sparse outliers, robust penalty functions such as the ℓ_1 norm, Huber, or the Student's t penalties have been shown to produce more robust reconstructions [1]. When the penalty function is the ℓ_1 norm, an equivalent formulation to (9.3) can be realized through the stable principal component pursuit (SPCP) problem given by

$$\min_{L,S} \|L\|_* + \lambda\|S\|_1 \text{ subject to } \|A - \mathcal{P}(L) + S\|_F \leq \sigma_E, \tag{9.4}$$

where λ is a regularization parameter that corresponds to the tolerance σ in (9.3), and σ_E is a bound on the noise level E. The SPCP problem has the advantage of also determining the sparse component S, which can be the real target in some applications such as video background subtraction. When $\sigma_E = 0$, the problem is also known as robust principal component analysis (RPCA) or sparse matrix separation.

In this chapter, we discuss a factorization-based approach to in the context of solving (9.4). We begin our discussion with a gauge optimization perspective to robust matrix completion in Section 9.2. We then discuss in Section 9.3 how our approach replaces the solution over the matrix L with its low-rank factors U and V, such that $L = UV^T$, similar to the approach adopted by Aravkin et al. [2]. In this context, we develop a gauge minimization algorithm and an alternating direction method of multipliers algorithm that take advantage of the factorized matrix decomposition. We then focus in Section 9.4 on the particular application of video background subtraction, which is the problem of finding moving objects in a video sequence that move independently from the background scene. The segmentation of moving objects helps in analyzing the trajectory of moving targets and in improving the performance of object detection and classification algorithms. In scenes that exhibit camera motion, we first extract the motion vectors from the coded video bitstream and fit the global motion of every frame to a parametric perspective model. The frames are then aligned to match the perspective of the first frame in a group of pictures (GOP) and use our factorized robust matrix completion algorithm to fill in the background pixels that are missing from the individual video frames in the GOP. We also discuss the case where additional depth information is available for the video scene and develop a depth-weighted group-wise PCA algorithm that improves the foreground/background separation by incorporating the depth information into the reconstruction. Finally, we conclude the chapter by discussing the suitability of the proposed method for background subtractions.

9.1.1 Notation

Throughout the chapter we use upper case letters to refer to matrices or their vectorization interchangeably, depending on the context. Consider a matrix $X \in \mathbb{R}^{m\times n}$ of rank $r <$

$\min\{m, n\}$ with singular values σ_i, $i = 1 \dots r$ indexed in decreasing order. We denote by $\|X\|_F = \sqrt{\sum_{i,j} X_{ij}^2} = \sum_{i=1}^{r} \sigma_i$ and $\|X\|_{\text{op}} = \sigma_1$ the Frobenius norm and the operator norm of a matrix X, respectively. Also, we denote by $\|X\|_1 = \sum_{i,j} |X_{ij}|$ and $\|X\|_\infty = \max_{i,j} |X_{ij}|$ the ℓ_1 norm and the ℓ_∞ norm of the vectorization of the matrix X, respectively. For any two matrices $X, Y \in \mathbb{R}^{m \times n}$, we denote the trace of the inner product by $\langle X, Y \rangle = \text{Tr}\left[X^T Y\right]$.

9.2 Robust Matrix Completion

Consider the case where we are given incomplete measurements A of a data matrix $X_0 \in \mathbb{R}^{m \times n}$ that is composed of the superposition of a low-rank matrix L_0 and a sparse outlier matrix S_0, such that, $X_0 = L_0 + S_0$. Let $\mathcal{P}_\Omega : \mathbb{R}^{m \times n} \to \mathbb{R}^p$ be a restriction operator that selects a subset Ω of size p of the mn samples in X_0. We define the robust matrix completion problem as the problem of finding L_0 from the incomplete measurements $A = \mathcal{P}_\Omega(L_0) + S_0$ using the following ℓ_1-norm constrained nuclear norm minimization problem

$$\min_L \|L\|_* \text{ subject to } \|A - \mathcal{P}_\Omega(L)\|_1 \le \sigma, \tag{9.5}$$

where the tolerance σ has to be set equal to an upper bound on the ℓ_1-norm of S_0. We may also write Problem (9.5) by introducing the variable S such that

$$\begin{aligned} \min_{L,S} \|L\|_* \text{ subject to } & A = \mathcal{P}_\Omega(L) + S, \\ & \|S\|_1 \le \sigma. \end{aligned} \tag{9.6}$$

Note that for an appropriate choice of λ, the problem in equation (9.6) is equivalent to the robust PCA problem

$$\min_{L,S} \|L\|_* + \lambda \|S\|_1 \text{ subject to } A = L + S. \tag{9.7}$$

Problems (9.5) and (9.7) belong to the general class of constrained gauge minimization problems [13] for which a solution framework was developed by van den Berg and Friedlander in [30, 31] for the ℓ_2 constrained case and later extended by Aravkin et al. [1] for arbitrary convex constraints. A gauge $\kappa(\cdot)$ is a convex, non-negative, positively homogeneous function that vanishes at the origin. Moreover, we characterize a general gauge optimization problem in L and S as

$$\min_{L,S} \kappa(L, S) \text{ subject to } h(L, S) \le \sigma, \tag{9.8}$$

where $h(\cdot)$ is a convex function. Note that norms are special cases of gauge functions that are finite everywhere, symmetric, and zero only at the origin [13]. Following the framework of [31], we can define the value function

$$\phi(\tau) = \min_{L,S} h(L, S) \text{ subject to } \kappa(L, S) \le \tau, \tag{9.9}$$

and update τ using Newton's method to find τ^* that solves $\phi(\tau) = \sigma$. The solution (L^*, S^*) of $\phi(\tau^*)$ will then be the minimizer of (9.8). The jth Newton update of τ is given by

$$\tau_{j+1} = \tau_j + \frac{\phi(\tau_j) - \sigma}{\phi'(\tau_j)}, \tag{9.10}$$

which requires the evaluation of the derivative $\phi'(\tau)$ with respect to τ. In particular, it was shown in [31] that $\phi'(\tau) = \kappa^o(h'(L, S))$, where κ^o is the polar of κ, and h' is the derivative of h with respect to L and S.

The framework described above is attractive when the optimization problem defining the value function $\phi(\tau)$ is easy to solve. Consider Problem (9.5), the penalty function $h(L,S) = \|A - \mathcal{P}_\Omega(L)\|_1$, which is convex but non-differentiable. On the other hand, the penalty function of Problem (9.7) is the Frobenius norm $h(L,S) = \|A - L - S\|_F$ and the Newton step terminates when $\phi(\tau) = 0$. Consequently, [31] derives the RPCA value function as

$$\phi(\tau) = \min_{L,S} \|A - \mathcal{P}_\Omega L - S\|_F \text{ subject to } \|L\|_* + \lambda\|S\|_1 \leq \tau, \tag{9.11}$$

where τ is updated according to the Newton step

$$\tau_{j+1} = \tau_j + \frac{\|R_j\|_F}{\max\{\|\mathcal{P}_\Omega^T R_j\|_{\text{op}}, \|\mathcal{P}_\Omega^T R_j\|_\infty/\lambda\}}, \tag{9.12}$$

where $R_j = \frac{A - \mathcal{P}_\Omega(L_j + S_j)}{\|A - \mathcal{P}_\Omega(L_j + S_j)\|_F}$ is the normalized residual matrix at the jth Netwon iteration, and $S_j = \mathcal{P}_\Omega S_j$ since $\mathcal{P}_\Omega$ is a mask. Note here that the denominator $\max\{\|\mathcal{P}_\Omega^T R_j\|_{\text{op}}, \|\mathcal{P}_\Omega^T R_j\|_\infty/\lambda\}$ is the expression for the polar of the gauge function $\kappa(X,Y) = \|X\|_* + \lambda\|Y\|_1$, where $X = Y = \mathcal{P}_\Omega^T R_j$ in this case.

Finally, we point out that it was shown in [7] that a choice of $\lambda = \hat{n}^{-1/2}$, $\hat{n} := \max\{m,n\}$, is sufficient to guarantee the recovery of L_0 and S_0 with high probability when the $\text{rank}(L_0) \leq C\hat{n}\,(\log \hat{n})^{-2}$ for some constant C that depends on the coherence of the subspace of L_0.

9.3 Factorized Robust Matrix Completion

One drawback of Problem (9.7) is that it requires the computation of full (or partial) singular value decompositions of L in every iteration of the algorithm, which could become prohibitively expensive when the dimensions are large. To overcome this problem, we adopt a proxy for the nuclear norm of a rank-r matrix L defined by the following factorization from Lemma 8 in [28]

$$\|L\|_* = \inf_{U \in \mathbb{R}^{m,r}, V \in \mathbb{R}^{n,r}} \frac{1}{2}\left(\|U\|_F^2 + \|V\|_F^2\right) \text{ subject to } UV^T = L. \tag{9.13}$$

The nuclear norm proxy has recently been used in standard nuclear norm minimization algorithms [2, 25] that scale to very large matrix completion problems.

In this section, we discuss two algorithms that rely on the factorization in (9.13); the first algorithm follows the gauge minimization technique described in Section 9.2, whereas the second is an alternating direction method (ADM) that minimizes the augmented Lagrangian of Problem (9.7).

9.3.1 Factorized Gauge Minimization Algorithm

The gauge minimization framework discussed in Section 9.2 is summarized in Algorithm 9.1 and requires the following three basic tools:

1. Defining and computing the value function $\phi(\tau)$.

2. Solving the projection onto the gauge constraint $\kappa(\cdot) \leq \tau$.

3. Specifying the polar function κ^o to update τ.

Algorithm 9.1 - Gauge minimization algorithm for factorized robust matrix completion (Gauge-FRMC)

1: **Input** Measurement matrix A, measurement operator $\mathcal{P}_\Omega$, regularization parameter λ, mismatch tolerance σ
2: **Output** Low-rank factors U, V, and sparse component S
3: **Initialize** $\tau_0 = 0$, residual signal $R_0 = A$, $j = 0$
4: **while** $\|R_j\|_F > \sigma$ **do**
5: $j = j + 1$
6: Update the gauge constraint τ_j:

$$\tau_j = \tau_{j-1} + \frac{\|R_{j-1}\|_F}{\max\{\|\mathcal{P}_\Omega^T R_{j-1}\|_{\text{op}}, \|\mathcal{P}_\Omega^T R_{j-1}\|_\infty / \lambda\}}$$

7: Compute the value function $\phi(\tau_j)$ by solving for (U, V, S):

$$\phi(\tau_j) = \min_{U,V,S} \|A - \mathcal{P}_\Omega(UV^T + S)\|_F \text{ subject to } \frac{1}{2}\left(\|U\|_F^2 + \|V\|_F^2\right) + \lambda\|S\|_1 \leq \tau_j$$

8: Update the residual R_j:

$$R_j = A - \mathcal{P}_\Omega(UV^T + S)$$

9: **end while**

The value function $\phi(\tau)$

The factorization-based counterpart of the value function (9.11) is expressed as follows:

$$\phi(\tau) = \min_{U,V,S} \|A - \mathcal{P}_\Omega(UV^T + S)\|_F \text{ subject to } \frac{1}{2}\left(\|U\|_F^2 + \|V\|_F^2\right) + \lambda\|S\|_1 \leq \tau. \quad (9.14)$$

The applicability of the factorization (9.13) follows from Theorem 9.1 and Corollary 9.1 proved in [2] and listed below after specializing to our problem.

THEOREM 9.1 *[2, Theorem 4.1] Consider an optimization problem of the form*

$$\min_{Z \succeq 0} \; h(Z) \quad \textit{subject to} \quad \kappa(Z) \leq 0, \qquad \operatorname{rank}(Z) \leq r, \quad (9.15)$$

where $Z \in \mathbb{R}^{n\times n}$ is positive semidefinite, and h, κ are continuous. Using the change of variable $Z = SS^T$, take $S \in \mathbb{R}^{n\times r}$, and consider the problem

$$\min_{S} \; h(SS^T) \quad \textit{subject to} \quad \kappa(SS^T) \leq 0. \quad (9.16)$$

Let $\bar{Z} = \bar{S}\bar{S}^T$, where $\bar{Z}$ is feasible for (9.15). *Then $\bar{Z}$ is a local minimum of* (9.15) *if and only if $\bar{S}$ is a local minimum of* (9.16).

COROLLARY 9.1 [2, Corollary 4.2] Any optimization problem of the form

$$\min_{X} \; h(X) \quad \text{subject to} \quad \|X\|_* \leq \tau, \quad \operatorname{rank}(X) \leq r, \quad (9.17)$$

where h is continuous, has an equivalent problem in the class of Problems (9.15) characterized by Theorem 9.1.

In particular, Corollary 9.1 implies that when the factors U and V have a rank greater than or equal to the true rank of L_0, then a local minimizer of the value function in (9.9) with a least-squares objective and the factorization in (9.13) as a constraint coincides with the solution of the same value function when the exact nuclear norm is in the constraint. This result follows from the semidefinite programming (SDP) characterization of the nuclear norm established in [24].

In order to solve (9.14), we use a first-order projected gradient algorithm with a simple line search. The approach is similar to that developed in [30] without employing the Barzilai-Borwein line-search method. In every iteration of the algorithm, the gradient updates $(\widehat{U}, \widehat{V}, \widehat{S})$ of the variables (U, V, S) are projected onto the gauge constraint. The following section describes an efficient Newton root-finding algorithm for performing this projection.

Projecting onto the gauge constraint

A critical component for the success of the gauge optimization framework is the efficiency of the projection onto the gauge constraint. Therefore, we developed a fast algorithm that can efficiently solve the following projection problem

$$\min_{U,V,S} \frac{1}{2}\|U - \widehat{U}\|_F^2 + \frac{1}{2}\|V - \widehat{V}\|_F^2 + \frac{1}{2}\|S - \widehat{S}\|_F^2 \text{ subject to } \frac{1}{2}\left(\|U\|_F^2 + \|V\|_F^2\right) + \lambda\|S\|_1 \leq \tau. \tag{9.18}$$

The projection can be realized as shown in Proposition 9.1 by finding a scalar variable γ using the Newton method described in Algorithm 9.2. Clearly, the projection is only performed when the current iterates $(\widehat{U}, \widehat{V}, \widehat{S})$ violate the constraint, i.e., when $\frac{1}{2}\left(\|\widehat{U}\|_F^2 + \|\widehat{V}\|_F^2\right) + \lambda\|\widehat{S}\|_1 > \tau$.

PROPOSITION 9.1 The solution to the projection problem (9.18) is given by

$$\begin{aligned} S(\gamma\lambda) &= \mathcal{T}_{\gamma\lambda}(\widehat{S}) \\ U(\gamma) &= \tfrac{1}{\gamma+1}\widehat{U} \\ V(\gamma) &= \tfrac{1}{\gamma+1}\widehat{V}, \end{aligned} \tag{9.19}$$

where $\mathcal{T}_{\gamma\lambda}(\widehat{S}) = \operatorname{sign}(\widehat{S}) \odot \max\{0, |\widehat{S}| - \gamma\lambda\}$ is the soft-thresholding operator, and γ is the scalar that satisfies the inequality $\frac{1}{2(\gamma+1)^2}\left(\|\widehat{U}\|_F^2 + \|\widehat{V}\|_F^2\right) + \lambda\|S(\gamma\lambda)\|_1 \leq \tau$. Moreover, Algorithm 9.2 specifies a Newton's method for computing γ.

PROOF 9.1 In order to evaluate the projection, we first form the Lagrange dual $\mathcal{L}(U, V, S, \gamma)$ of Problem (9.18) with dual variable $\gamma \geq 0$,

$$\begin{aligned} \mathcal{L}(U, V, S, \gamma) &= \inf_{U,V,S} \tfrac{1}{2}\|U - \widehat{U}\|_F^2 + \tfrac{1}{2}\|V - \widehat{V}\|_F^2 + \tfrac{1}{2}\|S - \widehat{S}\|_F^2 \\ &\qquad + \gamma\left(\tfrac{1}{2}\left(\|U\|_F^2 + \|V\|_F^2\right) + \lambda\|S\|_1 - \tau\right) \\ &= \inf_{S}\left\{\tfrac{1}{2}\|S - \widehat{S}\|_F^2 + \gamma\lambda\|S\|_1\right\} \\ &\quad + \inf_{U,V}\left\{\tfrac{1}{2}\|U - \widehat{U}\|_F^2 + \tfrac{1}{2}\|V - \widehat{V}\|_F^2 + \tfrac{\gamma}{2}\left(\|U\|_F^2 + \|V\|_F^2\right)\right\} - \gamma\tau \end{aligned}$$

The two infimums admit the closed-form solutions shown in (9.19), where the expression for S is a soft-thresholding of the signal $\widehat{S}$ with threshold $\gamma\lambda$, and the symbol $\odot$ is an element-wise Hadamard product. Consequently, the projection is performed by finding γ that satisfies $\frac{1}{2(\gamma+1)^2}\left(\|\widehat{U}\|_F^2 + \|\widehat{V}\|_F^2\right) + \lambda\|S(\gamma\lambda)\|_1 \leq \tau$.

Denote by T the support of the elements in $\widehat{S}$ with magnitude larger then $\gamma\lambda$, and let $k = |T|$ be the cardinality of the set T. The ℓ_1 norm of the soft-thresholded signal $S(\gamma\lambda)$ is then equal to $\|S(\gamma\lambda)\|_1 = \|\widehat{S}_T\|_1 - \gamma\lambda k$, where $\widehat{S}_T$ is a restriction of the signal $\widehat{S}$ to the support set T. The projection is then achieved using Newton's method to find the root of the function

$$f(\gamma) = \lambda(\|\widehat{S}_T\|_1 - \gamma\lambda k) + \frac{1}{2(\gamma+1)^2}\left(\|\widehat{U}\|_F^2 + \|\widehat{V}\|_F^2\right) - \tau = 0. \tag{9.20}$$

Algorithm 9.2 - Newton's method for solving the factorized RPCA projection

1: **Input** Current iterates $\widehat{U}, \widehat{V}, \widehat{S}$, parameter λ, τ
2: **Output** Projected iterates U, V, S
3: **Initialize** $U = \widehat{U}$, $V = \widehat{V}$, $S = \widehat{S}$, $\gamma = 0$, $C = \left(\|U\|_F^2 + \|V\|_F^2\right)$
4: **while** $f(\gamma) > \tau$ **do**
5: Compute the gradient: $g(\gamma) = -\lambda^2\|S\|_0 - (\gamma+1)^{-3}C$
6: Update γ: $\gamma = \gamma - f(\gamma)/g(\gamma)$
7: Soft-threshold the sparse component: $S = \text{sign}(\widehat{S}) \odot \max\{0, |\widehat{S}| - \gamma\lambda\}$
8: Scale the low-rank factors: $U = \frac{1}{\gamma+1}\widehat{U}$, $V = \frac{1}{\gamma+1}\widehat{V}$
9: Update the function value $f(\gamma) = \frac{1}{2}\left(\|U\|_F^2 + \|V\|_F^2\right) + \lambda\|S\|_1 - \tau$.
10: **end while**

The polar function

In determining the polar function, we follow the approach in [2] where the low-rank matrix L is evaluated from the current estimates of the factors U and V, and the Newton update is treated in the nuclear norm of L sense as opposed to the Frobenius norm of the factor U and V. The benefit of this approach is that it allows us to reuse the Newton update shown in (9.12) for the standard RPCA problem.

9.3.2 Factorized ADM Algorithm

The factorization approach can also be applied to the alternating direction method (ADM) for minimizing the augmented Lagrangian of Problem (9.7) resulting in the factorized augmented Lagrangian shown below:

$$\mathcal{L}(U,V,S,Y) = \frac{1}{2}\left(\|U\|_F^2 + \|V\|_F^2\right) + \lambda\|S\|_1 + \langle Y, A - \mathcal{P}_\Omega(UV^T + S)\rangle + \frac{\mu}{2}\|A - \mathcal{P}_\Omega(UV^T + S)\|_F^2, \tag{9.21}$$

where $Y \in \mathbb{R}^{m\times n}$ is the dual multiplier, and μ is the augmented Lagrangian smoothing parameter. The above formulation (9.21) adds outlier robustness to the factorization-based ADM method for matrix completion presented in [24] by introducing the ℓ_1 norm of the sparse component S into the augmented Lagrangian. On the other hand, (9.21) improves the stability of the reconstruction compared to the low-rank matrix fitting (LMaFit) approach [33], which also employs low-rank factors by introducing the nuclear norm proxy into the augmented Lagrangian.

Algorithm 9.3 shows the alternating minimization steps of (ADM-FRMC) for optimizing $\mathcal{L}(U,V,S,Y)$. The non-factorized nuclear norm formulation of ADM algorithms apply a singular value thresholding of L

$$L_{j+1} = \mathcal{D}_{\mu^{-1}}\left(A - S_j - E_j + \mu^{-1}Y_j\right),$$

where E_j is only supported on the complement of the set Ω, and $\mathcal{D}_{\mu^{-1}}$ is the singular value thresholding operator that soft-thresholds the singular values of $A - S_j - E_j + \mu^{-1} Y_j$ by μ^{-1}. On the other hand, Algorithm 9.3 replaces the singular value thresholding operation with steps 5, 6, and 7 that require the inversion of $r \times r$ matrices. Consequently, when the rank of the factors is relatively small compared to the dimensions m and n, the ADM-FRMC algorithm can result in a significant reduction in computational complexity compared to singular value thresholding-based ADM algorithms.

Algorithm 9.3 - Alternating direction method for factorized robust matrix completion (ADM-FRMC)

1: **Input** Measurement matrix A, measurement operator $\mathcal{P}_\Omega$, regularization parameter λ, mismatch tolerance σ, smoothing parameter μ
2: **Output** Low-rank factors U, V, and sparse component S
3: **Initialize** $V_0 =$ random $n \times r$ matrix, $Y_0 = 0$, $S_0 = 0$, $E_0 = 0$, $L_0 = 0$, $j = 0$
4: **while** $\|A - L_{j+1} - S_{j+1} - E_{j+1}\|_F > \sigma$ **do**
5: $\quad U_{j+1} = (Y_j + \mu(A - S_j - E_j))\, V_j (I_r + \mu V_j V_j^T)^{-1}$
6: $\quad V_{j+1} = (Y_j + \mu(A - S_j - E_j))^T\, U_{j+1} (I_r + \mu U_{j+1} U_{j+1}^T)^{-1}$
7: $\quad L_{j+1} = U_{j+1} V_{j+1}^T$
8: $\quad E_{j+1} = \mathcal{P}_{\Omega^c} \left[A - L_{j+1} + \mu^{-1} Y_j\right]$
9: $\quad S_{j+1} = \mathcal{P}_\Omega \left[\mathcal{T}_{\lambda\mu^{-1}} \left(A - L_{j+1} + \mu^{-1} Y_j\right)\right]$
10: $\quad Y_{j+1} = Y_j + \mu(A - L_{j+1} - S_{j+1} - E_{j+1})$
11: $\quad j = j + 1$
12: **end while**

9.3.3 Numerical Evaluation

To evaluate the performance of Gauge-FRMC and ADM-FRMC, we plot the reconstruction error versus runtime of the algorithms and compare the performance with respect to PSPG [3] and ADMIP [4] that use Lanczos SVDs for fast computation of partial singular value decompositions.[23] We generate a synthetic data matrix A of size $m \times n$ that is composed of the sum of a rank r matrix L_0, and a sparse matrix S_0 with $mn/5$ non-zero entries. In all test cases, we set the rank of the factors U and V in FRMC equal to $1.2 \times r$ and choose $\lambda = /\sqrt{\max\{m,n\}}$. The tests were run in MATLAB on a 2.5 GHz Intel Core i5 machine. We define the relative error of variables (L_j, S_j) as follows

$$\mathrm{Err}(L_j, S_j) = \frac{\sqrt{\|L_j - L_0\|_F^2 + \|S_j - S_0\|_F^2}}{\sqrt{\|L_0\|_F^2 + \|S_0\|_F^2}}. \tag{9.22}$$

In the first test, we assume the entries of A are fully observed and set $m = n = 500$, and $r = 20$. The location of the non-zero entries of S_0 are randomly chosen from a Bernoulli distribution with probability $\frac{1}{5}$. The magnitudes of the non-zero entires in S_0 are drawn from a standard normal distribution and scaled such that in one case $\|S_0\|_\infty \leq \sqrt{2\|L_0\|_\infty}$ in order to blend the sparse components with the low-rank signal, and in another case $\|S_0\|_\infty \leq \sqrt{200\|L_0\|_\infty}$ so that the sparse component constitutes large outliers. The performance evaluation for the two cases above are presented in Figure 9.1 (a) and (b),

[23] ADMIP and PSPG codes available from http://www2.ie.psu.edu/aybat/codes.html

respectively. The performance on a larger dataset $m = n = 1500$, $r = 100$ with small outliers is also presented in Figure 9.1 (c). A zoomed-in frame that excludes Gauge-FRMC is shown in Figure 9.1 (d) to highlight the difference between ADM-FRMC, ADMIP, and PSPG.

The figures show that ADM-FRMC and ADMIP have a comparable performance for both small and large outliers and for the small dataset. As the data size increases, Figure 9.1 (d) shows that ADM-FRMC converges quickly to a 10^{-3} relative error point in one third of the time it takes ADMIP or PSPG. After that, the convergence slows down as is typical of ADM-type algorithms. On the other hand, the Gauge-FRMC algorithm is robust to large outliers but converges relatively slowly compared to the other algorithms, primarily due to the slow projected gradient steps used to solve the value function $\phi(\tau)$. Using faster solvers for the value function should improve the overall speed of the algorithm. Finally, we point out that the PSPG algorithm fails to recover the correct signal when the variance of the outliers is too large. This could be the result of a bad choice for the smoothing parameter in the algorithm. We also tested the algorithms on the large dataset with large outliers and found the performance to be similar to that of the small dataset. Therefore, we excluded the corresponding figure from the chapter.

In the second test, we only observe 50% of the entries in A and run the experiment on the $m = n = 500$, $r = 20$ data set. We exclude PSPG and ADMIP from the comparison since the available codes do not support missing data entries. The results for small and large outliers are shown in Figure 9.1 (a) and (b), respectively. The figures demonstrate that both algorithms correctly solve the problem; however, Gauge-FRMC is significantly slower.

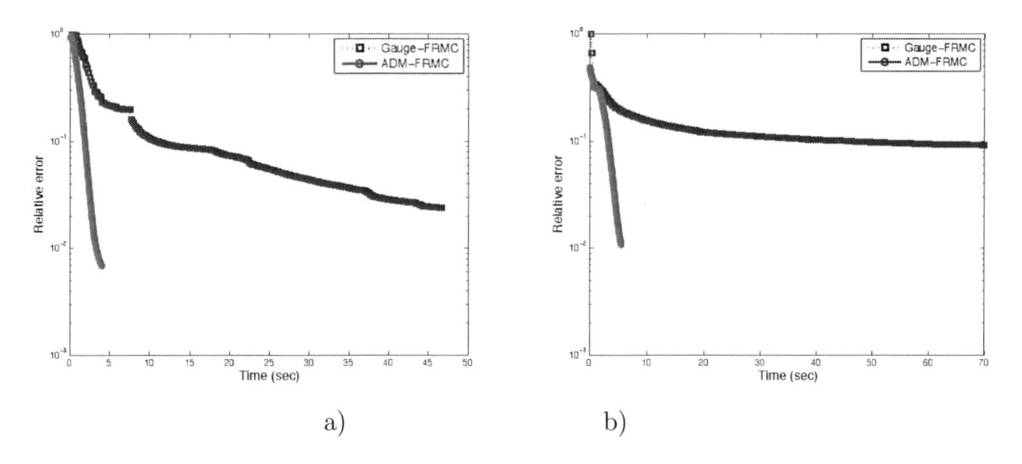

FIGURE 9.1 (See color insert.) Performance evaluation on a synthetic dataset of size $m = n = 500$, $r = 20$ with 50% missing entries for (a) low-variance sparse components, and (b) high-variance sparse components.

9.4 Application to Video Background Subtraction

Video background subtraction algorithms can be classified into algebraic decomposition techniques [7, 14, 15, 34] and statistical motion flow techniques [11, 21, 26, 27]. Algebraic approaches generally model the background scene as occupying a low-dimensional subspace. Moving objects in the video scene can then be modeled as sparse outliers that do not occupy the same subspace as the background scene. When the camera is stationary, the low-dimensional subspace is also stationary and the video signal can be decomposed into a low-rank matrix representing the background pixels as well as a sparse matrix representing the foreground moving objects. In this case, robust PCA has been shown to successfully segment the foreground from the background [7, 32, 34]. When the camera is moving, the low-rank structure no longer holds, thus requiring adaptive subspace estimation techniques [15, 19, 22] when the change in the subspace is smooth. However, this is rarely the case in real-world videos, which exhibit rotation, translation, and zoom among other types of motion. Therefore, a more robust approach performs global motion alignment prior to the matrix decomposition [20] when these significant motion distortions occur.

Once the video images are warped and aligned using global motion compensation, the images are vectorized and stacked into a matrix A of size $m \times n$, where m is the number of pixels in the video frame and n is the number of frames in a group of pictures (GOP). The warped images may contain large areas that have no content. Therefore, the problem can be posed as a robust matrix completion problem where a restriction operator $\mathcal{P}_\Omega$ identifies the set of pixels Ω that contains intensity values. Our objective then is to extract a low-rank component L from A that corresponds to the background pixels, and a sparse component S that captures the foreground's moving objects in the scene.

9.4.1 Stationary Background

For stationary background scenes, we apply the FRMC algorithm directly to the pixel domain, skipping the frame alignment. We test our algorithm on the Shopping Mall video sequence,[24] which is composed of 1000 frames of resolution 320×256. Figure 9.2 compares the qualitative separation performance of FRMC to that of the state-of-the-art algorithm GRASTA [14]. The FRMC algorithm completes the recovery 7 to 8 times faster than GRASTA and results in a comparable separation quality. For a quantitative comparison, we plot the ROC curves of the two algorithms in Figure 9.3. The curves show that GRASTA achieves a slightly better accuracy than FRMC; however, the computational cost is considerably higher.

9.4.2 Depth-Weighted Group-Wise PCA

In practical image sequences, the foreground objects (sparse components) tend to be clustered both spatially and temporally rather than evenly distributed. This observation led to the introduction of group sparsity into RPCA approaches by [9, 16, 17] pushing the sparse component into more structured groups. Our method utilizes the depth map of the video sequence to define the group structures in a depth-weighted group-wise PCA (DG-PCA) method.

In order to deal with structured sparsity, we replace the l_1-norm in the factorized RPCA problem with a mixed $l_{2,1}$-norm defined as

$$\|S\|_{2,1} = \sum_{g=1}^{s} w_g \|S_g\|_2, \tag{9.23}$$

where S_g is the component corresponding to group g, $g = 1, ..., s$, and w_g's are weights

[24] Available from: http://perception.i2r.a-star.edu.sg/bk_model/bk_index.html

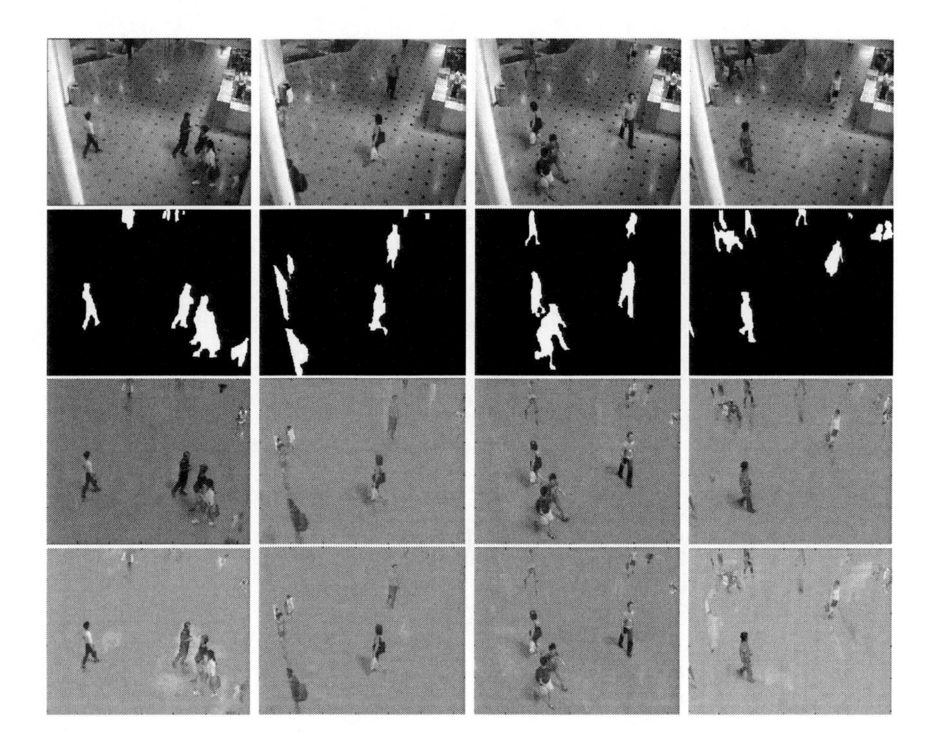

FIGURE 9.2 Background subtraction of four frames from the Shopping Mall sequence. Row one shows the original four frames. Row two shows the ground truth foreground objects. Row three shows the output of the GRASTA algorithm, which required 389.7 seconds to complete. Row four shows the output of our FRMC algorithm running in batch mode and completing in 47.1 seconds. ©(2014) IEEE

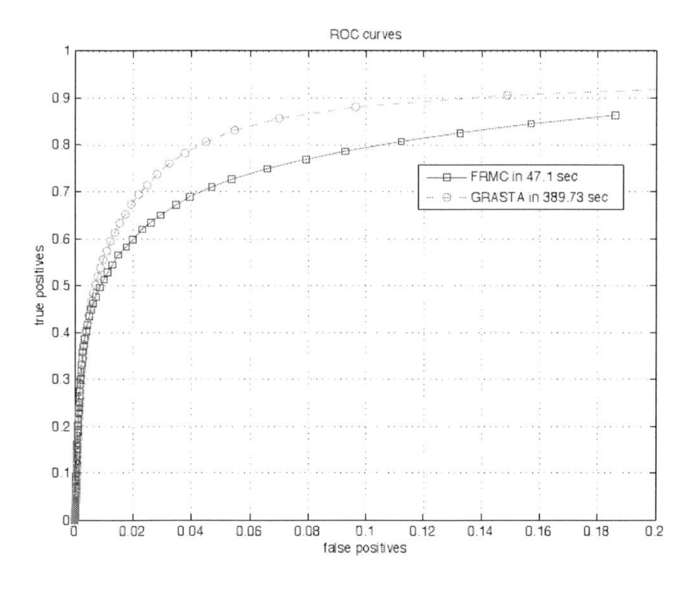

FIGURE 9.3 ROC curves comparing the stationary background subtraction performance between GRASTA and our FRMC algorithm. ©(2014) IEEE

associated to each group. The resulting problem is shown below:

$$\begin{aligned}(U, V, S, Y) = & \arg\min_{U,V,S,Y} \frac{1}{2}\|U\|_F^2 + \frac{1}{2}\|V\|_F^2 + \\ & \lambda\|S\|_{2,1} + \langle Y, A - UV^T - S\rangle + \frac{\mu}{2}\|A - UV^T - S\|_F^2.\end{aligned} \tag{9.24}$$

Algorithm 9.4 - Depth-weighted group-wise PCA (DGPCA) algorithm

Require: Input data A, λ, μ, error tolerance τ, maximum iteration number N, and depth map D

1: Init: $i = 0$, U_i and $V_i \leftarrow$ random matrix, $G \leftarrow \mathcal{G}(D)$
2: **repeat**
3: $\quad U_{i+1} = (\mu(A - S_i) + Y_i)V_i(I_r + \mu V_i^T V_i)^{-1}$
4: $\quad V_{i+1} = (\mu(A - S_i) + Y_i)^T U_{i+1}(I_r + \mu U_{i+1}^T U_{i+1})^{-1}$
5: $\quad S_{i+1,g} = \mathcal{T}_{\lambda/\mu,g}(A_g - U_{i+1,g}V_{i+1,g}^T + \mu^{-1}Y_{i,g})$
6: $\quad E = A - U_{i+1}V_{i+1}^T - S_{i+1}$
7: $\quad Y_{i+1} = Y_i + \mu E$
8: $\quad i = i + 1$
9: **until** $i \geq N$ or $\|E\|_F \leq \tau$
10: **return** U, V, S, i and $\|E\|_F$

Algorithm 9.4 describes the proposed DG-PCA framework. In order to define pixel groups G using the depth map D, an operator $\mathcal{G}(D)$ segments the depth map into s groups using the following procedure. Supposing that the depth level ranges from 0 to 255, a pixel with depth value d will be classified into group $g = \lfloor d/\frac{256}{s} \rfloor + 1$. Consequently, the input data A can be clustered into A_g with $g \in \{1, .., s\}$. Each A_g is composed of elements from A that are marked into segment g. In the same way, U_g, V_g, and Lagrangian multiplier Y_g are also grouped.

Next, the operator $\mathcal{T}_{\lambda/\mu,g}$ in Algorithm 9.4 is a group-wise soft-thresholding, as shown below,

$$\mathcal{T}_{\lambda/\mu,g}(e_g) = \max(\|e_g\|_2 - w_g\lambda/\mu, 0)\frac{e_g}{\|e_g\|_2 + \epsilon}, \tag{9.25}$$

where $e_g = A_g - U_g V_g^T + \frac{1}{\mu}Y_g$, and ϵ is a small constant to avoid division by 0, and w_g defines group weights in (9.23). Since a foreground object has a higher chance of being closer to the camera, i.e., to have a higher depth value than a background object, we propose the following equation to set group weights,

$$w_g = c^{1-\frac{d_g}{255}}, \tag{9.26}$$

where c is some constant, and d_g is the mean depth value of pixels in group g. w_g is equal to 1 for objects nearest to the camera, $d = 255$, and it is equal to c for objects farthest from the camera, $d = 0$. The choice of c controls the value of the threshold that permits foreground pixels to be selected based on their location in the depth field. Finally, after S_g is calculated for each group g, the sparse component S is obtained by summing up all S_g together.

Note that the above setup favors group structures where the foreground objects are closer to the camera. It is also possible within our framework to define the groups as the sets of pixels that are spatially connected and have a constant depth, or connected pixels where the spatial gradient of the depth is constant.

9.4.3 Global Motion Parametrization

In videos where the camera itself is moving, applying the FRMC algorithm directly to the video frames fails in segmenting the correct motion since the background itself is non-stationary. A non-stationary background does not live in a low-rank subspace; therefore, we can only expect the algorithm to fail. Therefore, we first estimate the global motion parameters in the video in order to compensate for the camera motion. We then align the background and apply the FRMC algorithm to segment the moving objects.

Global motion estimation received a lot of attention from the research community during the development of the MPEG-4 Visual standard [10]. One approach relates the coordinates (x_1, y_1) in a reference image I_1 to the coordinates (x_2, y_2) in a target image I_2 using an 8-parameter homography vector h such that

$$\begin{aligned} x_2 &= \tfrac{h_0+h_2x_1+h_3y_1}{1+h_6x_1+h_7y_1} \\ y_2 &= \tfrac{h_1+h_4x_1+h_5y_1}{1+h_6x_1+h_7y_1}. \end{aligned} \tag{9.27}$$

Given the homography vector $h = [h_0\ h_1\ h_2\ h_3\ h_4\ h_5\ h_6\ h_7]^T$ that relates two images, we can warp the perspective of image I_2 to match that of image I_1, thereby aligning the backgrounds of both images. However, estimating h from the raw pixel domain requires finding point-to-point matches between a subset of the pixels of the two images.

In order to compute h, we propose to use the horizontal and vertical motion vectors (m_x, m_y) that are readily available from the compressed video bitstream or during the encoding process. Here we assume that motion estimation is performed using the previous video frame as the only reference picture. The motion vectors provide relatively accurate point matches between the two images. Note, however, that we are only interested in matching pixels from the moving background. Therefore, we first compute a 32-bin histogram of each of the motion vectors m_x and m_y. Next, we extract a subset Λ of the indices of pixels that exclude foreground objects in order to capture the motion of the background and fit the homography parameters to the background pixels alone. The homography parameter vector h is computed by solving the following least-squares problem:

$$h = \arg\min_{\tilde{h}} \left\| \begin{bmatrix} x_{2\Lambda} \\ y_{2\Lambda} \end{bmatrix} - E\tilde{h} \right\|_2, \tag{9.28}$$

where $x_{2\Lambda} = x_{1\Lambda} + m_{x\Lambda}$, $y_{2\Lambda} = y_{1\Lambda} + m_{x\Lambda}$, and the matrix

$$E = \begin{bmatrix} \mathbf{1}\,\mathbf{0}\,x_{1\Lambda}\,y_{1\Lambda} & \mathbf{0} & \mathbf{0} & -x_{2\Lambda}x_{1\Lambda} & -x_{2\Lambda}y_{1\Lambda} \\ \mathbf{0}\,\mathbf{1}\ \ \mathbf{0} \quad \mathbf{0} & x_{1\Lambda} & y_{1\Lambda} & -y_{2\Lambda}x_{1\Lambda} & -y_{2\Lambda}y_{1\Lambda} \end{bmatrix},$$

where the subscript $_\Lambda$ indicates a restriction of the indices to the set Λ.

Next, we align the pictures relative to the perspective of the first frame in a GOP by sequentially warping the pictures using the coordinates of the previously warped frame $\hat{I}_1$ as a reference to warp the coordinates of the next frame I_2 by applying (9.27). Finally, we note that due to the camera motion, the warped frames $\hat{I}_2$ generally occupy a larger viewing area relative to the reference frame I_2. Consequently, applying a forward map $f : (x_1, y_1) \rightarrow (\hat{x}_2, \hat{y}_2)$ often results in holes in the warped frame. To remedy this problem, we compute the reverse mapping $g : (\hat{x}_2, \hat{y}_2) \rightarrow (x_2, y_2)$ as a function of h and warp the frame to obtain $\hat{I}_2(\hat{x}_2, \hat{y}_2) = I_2(g(\hat{x}_2, \hat{y}_2))$. Figure 9.4 illustrates the global motion compensation procedure applied to frame 26 of the Bus sequence.

For non-stationary background sequences, we run our FRMC algorithm with global motion compensation on the reference video sequence Bus composed of 150 CIF resolution

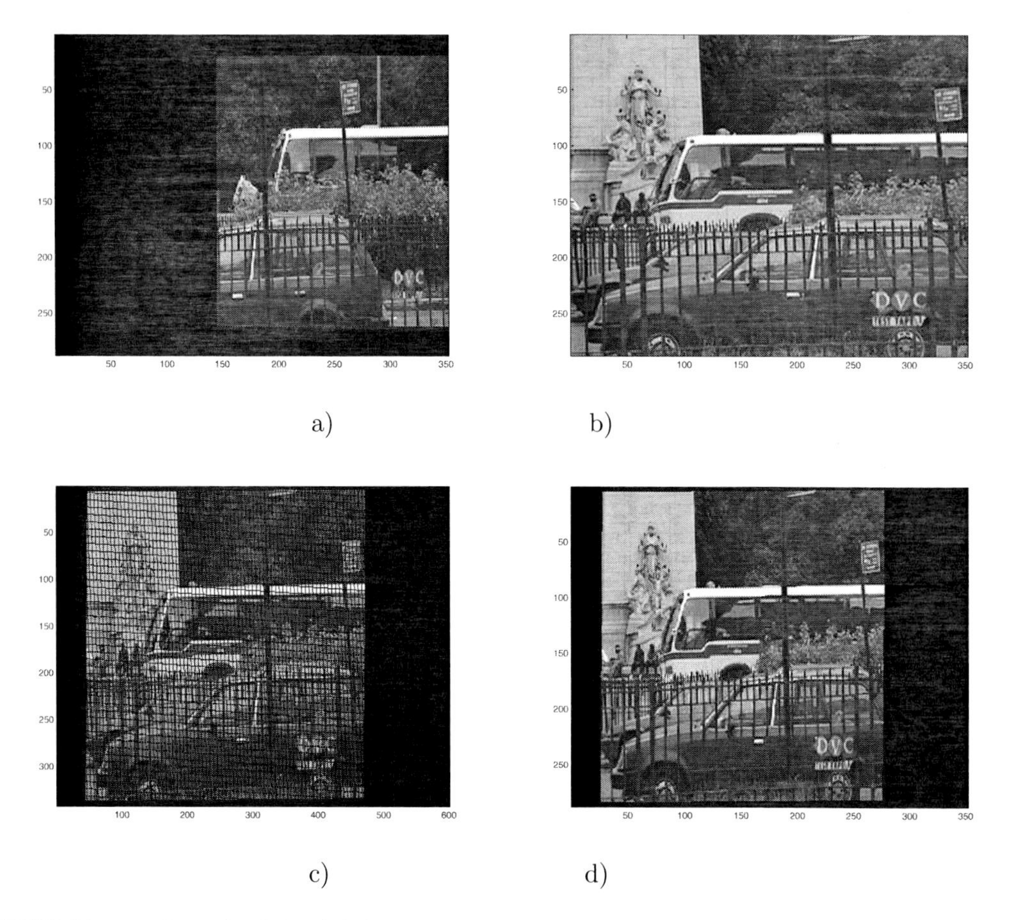

FIGURE 9.4 Example of the global motion compensation procedure used to align the backgrounds of images in a GOP. (a) First frame in the GOP aligned and scaled to its relative location. (b) Original frame 26 as input image I_2. (c) Frame 26 warped and aligned as $\hat{I}_2(\hat{x}_2, \hat{y}_2)$, (d) Warped and reverse mapped frame $\hat{I}_2(g(\hat{x}_2, \hat{y}_2))$. ©(2014) IEEE

(352×288 pixels) frames.[25] The Bus sequence exhibits translation and zooming out. We use the HEVC test model (HM) 11 reference software[26] [5] to encode the sequence and run our FRMC with GME algorithm in batch mode with a batch size of 30 frames. The recovery performance is illustrated in Figure 9.5. Notice how the recovered background expands and stretches relative to the original frames in order to cover the translation and zoom of the 30-frame GOP. Notice also how stationary foreground objects are successfully classified as part of the background subspace and are excluded from the segmented moving objects.

9.4.4 Depth-Enhanced Homography Model

The eight-parameter homography model assumes planar motion. However, motion in a video sequence is generally not planar. Therefore, it is still very common to find large motion es-

[25]Available from: http://trace.eas.asu.edu/yuv/

[26]Available from: https://hevc.hhi.fraunhofer.de/svn/svn_HEVCSoftware/

FIGURE 9.5 Background subtraction of four frames from the Bus sequence. Row one shows the original four frames. Row two shows the motion aligned and FRMC separated background relative to a 30-frame GOP. Row three shows the motion-aligned and FRMC-separated foreground. The total runtime for global motion compensation and background subtraction of 150 frames took 19.8 seconds. ©(2014) IEEE

timation errors in sequences that have a wide depth range. This could dramatically degrade the detection rate in the separation problem. Therefore, we propose a depth-enhanced homography model. Specifically, six new parameters related to depth are added, and we have $h = [h_1, ..., h_8, h_9, ..., h_{14}]^T$. Let z_1 and z_2 stand for the depth of the corresponding pixels, and the proposed depth-enhanced homography model is given as follows:

$$\begin{aligned} x_2 &= \frac{h_1 + h_3 x_1 + h_4 y_1 + h_9 z_1}{1 + h_7 x_1 + h_8 y_1}, \\ y_2 &= \frac{h_2 + h_5 x_1 + h_6 y_1 + h_{10} z_1}{1 + h_7 x_1 + h_8 y_1}, \\ z_2 &= \frac{h_{11} + h_{12} x_1 + h_{13} y_1 + h_{14} z_1}{1 + h_7 x_1 + h_8 y_1}. \end{aligned} \tag{9.29}$$

Note in the above equation, depth value 0 means the object is far from the camera. A larger depth value means that the object is closer to the camera.

To evaluate the performance of the DG-PCA approach, we tested the separation on fr3/walking_rpy sequence from the "dynamic objects" category in the RGB-D benchmark provided by TUM [29]. The dataset contains dynamic objects with a low- to high-level global motion.

The accompanying depth in the dataset is captured by a Microsoft Kinect sensor and denoted by z. The depth map d is computed from z as follows:

$$d = 255 \times \frac{\frac{1}{z} - \frac{1}{z_{\text{far}}}}{\frac{1}{z_{\text{near}}} - \frac{1}{z_{\text{far}}}}. \tag{9.30}$$

where z_{near} and z_{far} denote the nearest and farthest depth extracted from the raw depth data z.

In order to perform FG/BG separation, two consecutive video frames are first aligned using global motion compensation with and without the depth-enhancement. The aligned frames are then processed using FRMC and DG-PCA to separate the background L from the foreground S. The rank of the background is set equal to 2. We used sequence fr3/walking_static with minor camera motion to tune the algorithm parameters and then run tests on the other three sequences with higher motion. We set the parameter $\lambda = 0.05(\|e\|_2/\sqrt{\text{size}(A)}) \times \mu$, where a constant 0.05 is selected empirically to limit the iteration step for a finer background subtraction. When updating the group-wise sparse component in Algorithm 9.4, we use $\lambda_g = \lambda\sqrt{\text{size}(A_i)}$ instead of the image level λ. This scaling in λ_g ensures the dependence on the size of the group since the thresholding operation is applied to the ℓ_2 norm of the group instead of the magnitudes of individual pixels. Moreover, we set $c = 10$ in (9.26). We set the number of groups $s = 32$ since we found no significant difference in the performance of the algorithm when s was varied in the range $[16, 32]$. We also denoised the depth-based grouping map G using a 5×5 median filter.

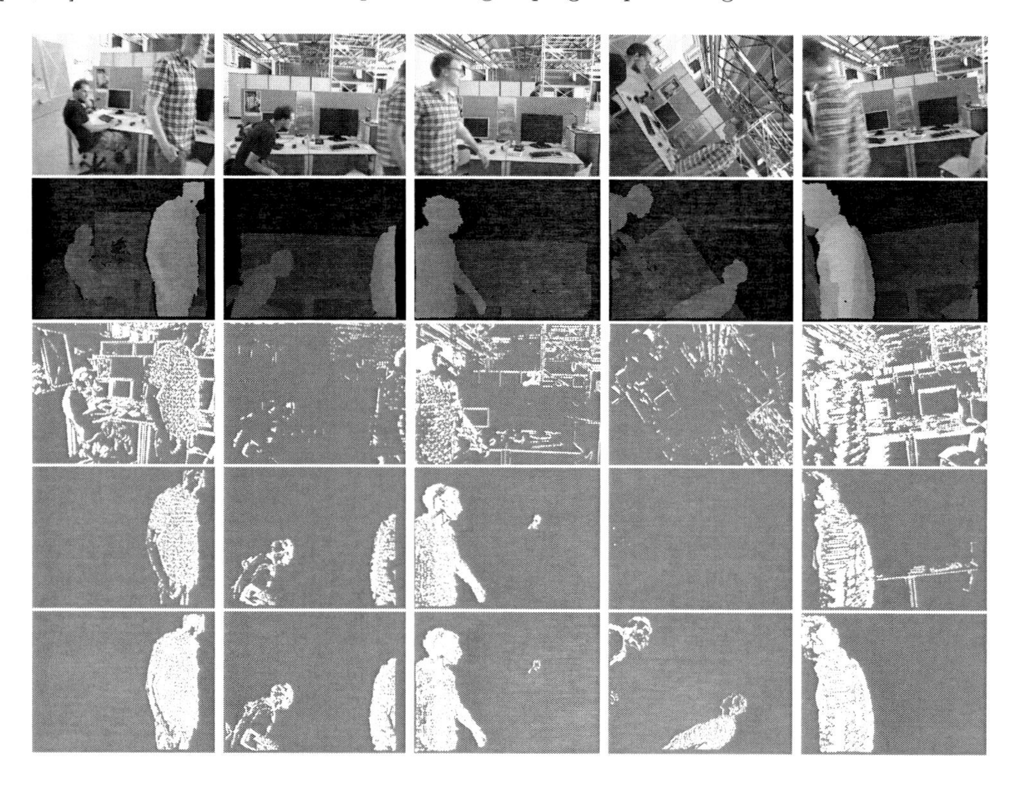

FIGURE 9.6 Performance evaluation. Row 1: color images. Row 2: depth maps. Row 3: Factorized RPCA. Row 4: DG-PCA w/o depth-refined global MC. Row 5: DG-PCA with depth-refined global MC.

Figure 9.6 shows 5 snapshots across fr3/walking_rpy with 910 frames at VGA resolution, which has the greatest global motion in the dataset. The figures show that the two DG-PCA methods (row 4 and 5) produce a much cleaner foreground segmentation compared to FRMC (row 3). For example, in the third snapshot, the person walking at a further distance behind the office partition can also be detected successfully by DG-PCA. Comparing DG-PCA without depth-enhanced global MC (row 4), and with depth-enhanced global MC (row 5), shows that the depth-enhanced homography model helps improve the motion alignment compared to the conventional homography model.

9.5 Conclusion

We developed a factorization-based approach to solving the robust matrix completion that replaces the solution over the low-rank matrix with its low-rank factors. We showed how the factorization approach can be applied in a gauge optimization framework resulting in the Gauge-FRMC algorithm, and in an alternating direction method (ADM) of multiplier framework resulting in the ADM-FRMC algorithm. Performance evaluation of the two algorithms showed that while both algorithms correctly solve the robust matrix completion problem, the ADM approach enjoys faster convergence than the gauge optimization approach. Moreover, the speed of convergence and accuracy of the ADM-FRMC algorithms matches and in some cases exceeds that of state-of-the-art algorithms. The main bottleneck for the gauge minimization approach comes from the use of a first-order projected gradient step to evaluate the value function.

In the second part of the chapter, we focused on video background subtraction as an application to factorized robust matrix completion. With the help of motion vector information available from a coded video bitstream, we showed that our framework is capable of subtracting the background from stationary and moving camera sequences. We also extended our model to incorporate scene depth information by assigning group structures to the sparse data outliers corresponding to foreground objects. Finally, we demonstrated that incorporating depth information into the problem formulation, we were able to improve the foreground/background separation.

References

1. A. Aravkin, J. Burke, and M. Friedlander. Variational properties of value functions. *SIAM Journal*, 23(3):1689–1717, 2013.
2. A. Aravkin, R. Kumar, H. Mansour, B. Recht, and F. Herrmann. A robust SVD free approach to matrix completion, with applications to interpolation of large scale data. *Preprint. http://arxiv.org/abs/1302.4886*, 2013.
3. N. Aybat, D. Goldfarb, and S. Ma. Efficient algorithms for robust and stable principal component pursuit. *Computational Optimization and Applications*, 2013.
4. N. Aybat and G. Iyengar. An alternating direction method with increasing penalty for stable principal component pursuit. *Computational Optimization and Applications*, (61)3: July 2015.
5. B. Bross, W. Han, J. Ohm, G. Sullivan, Y. Wang, and T. Wiegand. *High Efficiency Video Coding (HEVC) text specification draft 10.* JCT-VC of ITU-T SG 16 WP 3 and ISO/IEC JTC 1/SC 29/WG 11, January 2013.
6. J. Cai, E. Candès, and Z. Shen. A singular value thresholding algorithm for matrix completion. *SIAM Journal on Optimization*, 20(4):1956–1982, March 2010.
7. E. Candès, X. Li, Y. Ma, and J. Wright. Robust principal component analysis? *ACM Journal*, 58(3):11:1–11:37, June 2011.
8. E. Candès and T. Tao. The power of convex relaxation: Near-optimal matrix completion. *IEEE Transactions on Information Theory*, 56(5):2053–2080, May 2010.
9. W. Deng, W. Yin, and Y. Zhang. Group sparse optimization by alternating direction method. In *SPIE Optical Engineering+ Applications*, pages 88580R–88580R, 2013.
10. F. Dufaux and J. Konrad. Efficient, robust, and fast global motion estimation for video coding. *IEEE Transactions on Image Processing*, 9(3):497–501, 2000.
11. A. Elqursh and A. Elgammal. Online moving camera background subtraction. In *European Conference on Computer Vision, ECCV 2013*, 2013.
12. M. Fazel, H. Hindi, and S. Boyd. A rank minimization heuristic with application to minimum

order system approximation. In *American Control Conference, ACC 2001*, pages 4734–4739, 2001.

13. M. Friedlander, I. Macêdo, and T. Pong. Gauge optimization and duality. *SIAM Journal on Optimization*, 24(4):1999–2022, 2014.
14. J. He, L. Balzano, and J. Lui. Online robust subspace tracking from partial information. *Preprint, http://arxiv.org/abs/1109.3827*, 2011.
15. J. He, D. Zhang, L. Balzano, and T. Tao. Iterative grassmannian optimization for robust image alignment. *Preprint, http://arxiv.org/abs/1306.0404*, 2013.
16. J. Huang, X. Huang, and D. Metaxas. Learning with dynamic group sparsity. In *IEEE International Conference on Computer Vision, ICCV 2009*, pages 64–71, 2009.
17. Z. Ji, W. Wang, and K. Lv. Foreground detection utilizing structured sparse model via $l_{1,2}$ mixed norms. In *IEEE International Conference on Systems, Man, and Cybernetics, SMC 2013*, pages 2286–2291, 2013.
18. K. Lee and Y. Bresler. ADMiRA: atomic decomposition for minimum rank approximation. *CoRR*, abs/0905.0044, 2009.
19. H. Mansour and X. Jiang. A robust online subspace estimation and tracking algorithm. In *IEEE International Conference on Acoustics, Speech and Signal Processing, ICASSP 2015*, 2015.
20. H. Mansour and A. Vetro. Video background subtraction using semi-supervised robust matrix completion. In *IEEE International Conference on Acoustics, Speech and Signal Processing, ICASSP 2014*, pages 6528–6532, 2014.
21. M. Narayana, A. Hanson, and E. Learned-Miller. Coherent motion segmentation in moving camera videos using optical flow orientations. In *International Conference on Computer Vision, ICCV 2013*, 2013.
22. C. Qiu and N. Vaswani. ReProCS: a missing link between recursive robust PCA and recursive sparse recovery in large but correlated noise. *CoRR*, abs/1106.3286, 2011.
23. B. Recht. A simpler approach to matrix completion. *Journal of Machine Learning Research*, 12:3413–3430, December 2011.
24. B. Recht, M. Fazel, and P. Parrilo. Guaranteed minimum rank solutions to linear matrix equations via nuclear norm minimization. *SIAM Review*, 52(3):471–501, 2010.
25. B. Recht and C. Ré. Parallel stochastic gradient algorithms for large-scale matrix completion. *Mathematical Programming Computation*, 2013.
26. Y. Sheikh, O. Javed, and T. Kanade. Background subtraction for freely moving cameras. In *International Conference on Computer Vision, ICCV 2009*, 2009.
27. J. Shi and J. Malik. Motion segmentation and tracking using normalized cuts. In *International Conference on Computer Vision, ICCV 1998*, 1998.
28. N. Srebro. Learning with matrix factorizations. *Massachusetts Institute of Technology, Cambridge, MA, USA*, 2004.
29. J. Sturm, N. Engelhard, F. Endres, W. Burgard, and D. Cremers. A benchmark for the evaluation of RGB-D SLAM systems. In *International Conference on Intelligent Robot Systems, IROS 2012*, October 2012.
30. E. van den Berg and M. Friedlander. Probing the Pareto frontier for basis pursuit solutions. *SIAM Journal on Scientific Computing*, 31(2):890–912, 2008.
31. E. van den Berg and M. Friedlander. Sparse optimization with least-squares constraints. *SIAM Journal on Optimization*, 21(4):1201–1229, 2011.
32. A. Waters, A. Sankaranarayanan, and R. Baraniuk. SpaRCS: recovering low-rank and sparse matrices from compressive measurements. *Advances in Neural Information Processing Systems 24*, pages 1089–1097, 2011.
33. Z. Wen, W. Yin, and Y. Zhang. Solving a low-rank factorization model for matrix completion by a nonlinear successive over-relaxation algorithm. *Mathematical Programming Computation*, 4(4):333–361, 2012.

34. J. Wright, A. Ganesh, S. Rao, and Y. Ma. Robust principal component analysis: Exact recovery of corrupted low-rank matrices via convex optimization. In *Advances in Neural Information Processing Systems 22*, 2009.

III

Robust Subspace Learning and Tracking

10

Online (Recursive) Robust Principal Components Analysis

Namrata Vaswani
Iowa State University, Ames IA, USA

Chenlu Qiu
Iowa State University, Ames IA, USA

Brian Lois
Iowa State University, Ames IA, USA

Jinchun Zhan
Iowa State University, Ames IA, USA

10.1 Introduction

This work studies the problem of sequentially recovering a sparse vector S_t and a vector from a low-dimensional subspace L_t from knowledge of their sum $M_t := L_t + S_t$. If the primary goal is to recover the low-dimensional subspace in which the L_t's lie, then the problem is one of online or recursive robust principal components analysis (PCA). An example of where such a problem might arise is in separating a sparse foreground and a slowly changing dense background in a surveillance video. In this chapter, we describe our recently proposed algorithm, called Recursive Projected Compressed Sensing (ReProCS), to solve this problem and demonstrate its significant advantage over other robust PCA-based methods for this video layering problem. We also summarize the performance guarantees for ReProCS. Lastly, we briefly describe our work on modified PCP, which is a piecewise batch approach that removes a key limitation of ReProCS; however, it retains a key limitation of existing work. Principal Components Analysis (PCA) is a widely used dimension reduction technique that finds a small number of orthogonal basis vectors, called principal components, along which most of the variability of the dataset lies. It is well known that PCA is sensitive to outliers. Accurately computing the principal subspace in the presence of outliers is called robust PCA [9, 11, 44, 47]. Often, for time series data, the principal subspace changes gradually over time. Updating its estimate on-the-fly (recursively) in the presence of outliers, as more data comes in, is referred to as online or recursive robust PCA [1, 29, 45]. "Outlier" is a loosely defined term that refers to any corruption that is not small compared to the true data vector and that occurs occasionally. As suggested in [9, 51], an outlier can be nicely modeled as a sparse vector whose nonzero values can have any magnitude.

A key application where the robust PCA problem occurs is in video analysis where the goal is to separate a slowly changing background from moving foreground objects [9, 47]. If we stack each frame as a column vector, the background is well modeled as being dense and lying in a low-dimensional subspace that may gradually change over time, while the moving foreground objects constitute the sparse outliers [9, 51]. Other applications include detection of brain activation patterns from functional MRI (fMRI) sequences (the "active" part of the brain can be interpreted as a sparse outlier), detection of anomalous behavior in dynamic social networks, and sensor-networks-based detection and tracking of abnormal events such as forest fires or oil spills. In most of these applications, an online solution is desirable.

The moving objects or the active regions of the brain or the oil spill region may be "outliers" for the PCA problem, but in most cases, these are actually the signals of interest whereas the background image is the noise. Also, all the above signals of interest are sparse vectors. Thus, this problem can also be interpreted as one of recursively recovering a time sequence of sparse signals, S_t, from measurements $M_t := S_t + L_t$ that are corrupted by (potentially) large magnitude but dense and structured noise, L_t. The structure that we require is that L_t be dense and lie in a low-dimensional subspace that is either fixed or changes "slowly enough."

10.1.1 Related Work

There has been a large amount of work on robust PCA, e.g., [9, 11, 35, 44, 47, 52, 56], and recursive or online or incremental robust PCA, e.g., [1, 29, 45]. In most of these works, either the locations of the missing/corruped data points are assumed known [1] (not a practical assumption); or they first detect the corrupted data points and then replace their values using nearby values [45]; or weight each data point in proportion to its reliability (thus soft-detecting and down-weighting the likely outliers) [29, 47]; or just remove the entire outlier vector [35, 52]. Detecting or soft-detecting outliers (S_t) as in [29, 45, 47] is easy when the outlier magnitude is large, but not otherwise. When the signal of interest is S_t, the most difficult situation is when nonzero elements of S_t have small magnitude compared to those of L_t and in this case, these approaches do not work.

In recent works [9, 11], a new and elegant solution to robust PCA called Principal Components' Pursuit (PCP) has been proposed, which does not require a two-step outlier location detection/correction process and also does not throw out the entire vector. It redefines batch robust PCA as a problem of separating a low-rank matrix, $\mathcal{L}_t := [L_1, \dots, L_t]$, from a sparse matrix, $\mathcal{S}_t := [S_1, \dots, S_t]$, using the measurement matrix, $\mathcal{M}_t := [M_1, \dots, M_t] = \mathcal{L}_t + \mathcal{S}_t$.

Let $\|A\|_*$ be the nuclear norm of A (sum of singular values of A) while $\|A\|_1$ is the ℓ_1

norm of A seen as a long vector. It was shown in [9] that, with high probability (w.h.p.), one can recover $\mathcal{L}_t$ and $\mathcal{S}_t$ exactly by solving PCP:

$$\min_{\mathcal{L},\mathcal{S}} \|\mathcal{L}\|_* + \lambda\|\mathcal{S}\|_1 \text{ subject to } \mathcal{L} + \mathcal{S} = \mathcal{M}_t \tag{10.1}$$

provided that (a) the left and right singular vectors of $\mathcal{L}_t$ are dense; (b) any element of the matrix $\mathcal{S}_t$ is nonzero w.p. ϱ, and zero w.p. $1 - \varrho$, independent of all others; and (c) the rank of $\mathcal{L}_t$ and the support size of $\mathcal{S}_t$ are bounded by small-enough values.

10.1.2 Our Work: Provably Correct and Practically Usable Recursive Robust PCA Solutions

As described earlier, many applications where robust PCA is required, such as video surveillance, require an online (recursive) solution. Even for offline applications, a recursive solution is typically faster than a batch one. Moreover, in many of these applications, the support of the sparse part changes in a correlated fashion over time, e.g., where video foreground objects do not move randomly, and may also be static for short periods of time. In recent work [22, 39, 40, 43], we introduced a novel solution approach, called Recursive Projected Compressive Sensing (ReProCS), which recursively recovered S_t and L_t at each time t. Moreover, as we showed later in [31, 32], it can also provably handle correlated support changes significantly better than batch approaches because it uses extra assumptions (accurate initial subspace knowledge and slow subspace change). In simulation and real-data experiments (see [22] and `http://www.ece.iastate.edu/~chenlu/ReProCS/ReProCS_main.htm`), it was faster than batch methods such as PCP and also significantly outperformed them in situations where the support changes were correlated over time (as long as there was *some* support change every few frames). In [43], we studied a simple modification of the original ReProCS idea and obtained a performance guarantee for it. This result needed mild assumptions except one: It needed an assumption on intermediate algorithm estimates and hence it was not a correctness result. In very recent work [31, 32], we also have a complete correctness result for ReProCS that replaces this restrictive assumption by a simple and practical assumption on support change of S_t.

The ReProCS algorithm itself has a key limitation that is removed by our later work on modified-PCP [54, 55] (see Section 10.4); however, as explained in Section 10.5.3, modified PCP does not handle correlated support change as well as ReProCS. In this bookchapter we provide an overview of both ReProCS and modified PCP and a discussion of when which is better and why. We also briefly explain their correctness results and show simulation and real-data experiments comparing them with various existing batch and online robust PCA solutions.

To the best of our knowledge, [43] provided the first performance guarantee for an online robust PCA approach or equivalently for an online sparse + low-rank matrix recovery approach. Of course it was only a partial result (depending on intermediate algorithm estimates satisfying a certain property). Another partial result for online robust PCA was given in more recent work by Feng et al. [17]. In very recent work [31, 32], we provide the first complete correctness result for an online robust PCA/online sparse + low-rank matrix recovery approach. Another online algorithm that addresses the online robust PCA problem is Grassmannian Robust Adaptive Subspace Tracking Algorithm (GRASTA) [24]; however; it does not contain any performance analysis. We should mention here that our results directly apply to the recursive version of the matrix completion problem [5] as well since it is a simpler special case of the current problem (the support set of S_t is the set of indices of the missing entries and is thus known) [9]. We explicitly give the algorithm and result for online MC in [32].

The proof techniques used in our work are very different from those used to analyze other recent batch robust PCA works [9–11, 20, 25, 34–36, 46, 50, 52, 56]. The works of [35, 52] also study a different case: that where an entire vector is either an outlier or an inlier. Our proof utilizes (a) sparse recovery results [3]; (b) results from matrix perturbation theory that bound the estimation error in computing the eigenvectors of a perturbed Hermitian matrix with respect to eigenvectors of the original Hermitian matrix (the famous sin θ theorem of Davis and Kahan [13]) and (c) high probability bounds on eigenvalues of sums of independent random matrices (matrix Hoeffding inequality [48]).

A key difference of our approach to analyzing the subspace estimation step compared with most existing work analyzing finite sample PCA, e.g. [38] and references therein, is that it needs to provably work in the presence of error/noise that is correlated with L_t. Most existing works, including [38] and the references it discusses, assume that the noise is independent of (or at least uncorrelated with) the data. However, in our case, because of how the estimate $\hat{L}_t$ is computed, the error $e_t := L_t - \hat{L}_t$ is correlated with L_t. As a result, the tools developed in these earlier works cannot be used for our problem. This is also the reason why simple PCA cannot be used and we need to develop and analyze a projection-PCA algorithm for subspace tracking (this fact is explained in detail in the Appendix of [43]).

The ReProCS approach is related to that of [7, 26, 37] in that all of these first try to nullify the low-dimensional signal by projecting the measurement vector into a subspace perpendicular to that of the low-dimensional signal, and then solve for the sparse "error" vector (outlier). However, the big difference is that in all of these works the basis for the subspace of the low-dimensional signal is *perfectly known.* Our work studies *the case where the subspace is not known.* We have an initial approximate estimate of the subspace, but over time it can change significantly. In this work, to keep things simple, we mostly use ℓ_1 minimization done separately for each time instant (also referred to as basis pursuit denoising (BPDN)) [3, 12]. However, this can be replaced by any other sparse recovery algorithm, either recursive or batch, as long as the batch algorithm is applied to α frames at a time, e.g., one can replace BPDN by modified-CS or support-predicted modified-CS [41]. We have used a combination of simple ℓ_1 minimization and modified-CS/weighted ℓ_1 in the practical version of the ReProCS algorithm that was used for our video experiments (see Section 10.3.3).

10.1.3 Chapter Organization

This chapter is organized as follows. In Section 10.2, we give the problem definition and the key assumptions required. We explain the main idea of the ReProCS algorithm and develop its practically usable version in Section 10.3. We develop the modified-PCP solution in Section 10.4. The performance guarantees for both ReProCS and modified-PCP are summarized in Section 10.5. Numerical experiments, both for simulated data and for real video sequences, are shown in Section 10.6. Conclusions and future work are discussed in Section 10.7.

10.1.4 Notation

For a set $\mathcal{T} \subset \{1, 2, \ldots, n\}$, we use $|\mathcal{T}|$ to denote its cardinality, i.e., the number of elements in $\mathcal{T}$. We use $\mathcal{T}^c$ to denote its complement w.r.t. $\{1, 2, \ldots n\}$, i.e. $\mathcal{T}^c := \{i \in \{1, 2, \ldots n\} : i \notin \mathcal{T}\}$.

We use the interval notation, $[t_1, t_2]$, to denote the set of all integers between and including t_1 to t_2, i.e., $[t_1, t_2] := \{t_1, t_1 + 1, \ldots, t_2\}$. For a vector v, v_i denotes the ith entry of v and $v_{\mathcal{T}}$ denotes a vector consisting of the entries of v indexed by $\mathcal{T}$. We use $\|v\|_p$ to denote the ℓ_p norm of v. The support of v, supp(v), is the set of indices at which v is nonzero, $\text{supp}(v) := \{i : v_i \neq 0\}$. We say that v is s-sparse if $|\text{supp}(v)| \leq s$.

For a matrix B, B' denotes its transpose, and $B^\dagger$ its pseudo-inverse. For a matrix with linearly independent columns, $B^\dagger = (B'B)^{-1}B'$. We use $\|B\|_2 := \max_{x\neq 0} \|Bx\|_2/\|x\|_2$ to denote the induced 2-norm of the matrix. Also, $\|B\|_*$ is the nuclear norm (sum of singular values) and $\|B\|_{\max}$ denotes the maximum over the absolute values of all its entries. We let $\sigma_i(B)$ denote the ith largest singular value of B. For a Hermitian matrix, B, we use the notation $B \overset{EVD}{=} U\Lambda U'$ to denote the eigenvalue decomposition of B. Here U is an orthonormal matrix and Λ is a diagonal matrix with entries arranged in decreasing order. Also, we use $\lambda_i(B)$ to denote the ith largest eigenvalue of a Hermitian matrix B and we use $\lambda_{\max}(B)$ and $\lambda_{\min}(B)$ to denote its maximum and minimum eigenvalues. If B is Hermitian positive semi-definite (p.s.d.), then $\lambda_i(B) = \sigma_i(B)$. For Hermitian matrices B_1 and B_2, the notation $B_1 \preceq B_2$ means that $B_2 - B_1$ is p.s.d. Similarly, $B_1 \succeq B_2$ means that $B_1 - B_2$ is p.s.d. For a Hermitian matrix B, $\|B\|_2 = \sqrt{\max(\lambda_{\max}^2(B), \lambda_{\min}^2(B))}$ and thus, $\|B\|_2 \leq b$ implies that $-b \leq \lambda_{\min}(B) \leq \lambda_{\max}(B) \leq b$.

We use the notation $B \overset{SVD}{=} U\Sigma V'$ to denote the singular value decomposition (SVD) of B with the diagonal entries of Σ being arranged in non-decreasing order.

We use I to denote an identity matrix of appropriate size. For an index set $\mathcal{T}$ and a matrix B, $B_\mathcal{T}$ is the sub-matrix of B containing columns with indices in the set $\mathcal{T}$. Notice that $B_\mathcal{T} = BI_\mathcal{T}$. Given a matrix B of size $m \times n$ and B_2 of size $m \times n_2$, $[B\ B_2]$ constructs a new matrix by concatenating matrices B and B_2 in the horizontal direction. Let B_{rem} be a matrix formed by some columns of B. Then $B \setminus B_{\text{rem}}$ is the matrix B with columns in B_{rem} removed.

The notation [.] denotes an empty matrix.

For a matrix M,

- span(M) denotes the subspace spanned by the columns of M.

- M is a *basis matrix* if $M'M = I$.

- The b% *left singular values' set* of a matrix M is the smallest set of indices of its singular values that contains at least b% of the total singular values' energy. In other words, if $M \overset{SVD}{=} U\Sigma V'$, it is the smallest set T such that $\sum_{i\in T} (\Sigma)_{i,i}^2 \geq \frac{b}{100} \sum_{i=1}^{n} (\Sigma)_{i,i}^2$.

- The corresponding matrix of left singular vectors, U_T, is referred to as the b% *left singular vectors' matrix*.

DEFINITION 10.1 The *s-restricted isometry constant (RIC)* [7], δ_s for an $n \times m$ matrix Ψ is the smallest real number satisfying $(1 - \delta_s)\|x\|_2^2 \leq \|\Psi_T x\|_2^2 \leq (1 + \delta_s)\|x\|_2^2$ for all sets T with $|T| \leq s$ and all real vectors x of length $|T|$.

10.2 Problem Definition and Model Assumptions

We give the problem definition below followed by the model and then describe the key assumptions needed.

10.2.1 Problem Definition

The measurement vector at time t, M_t, is an n-dimensional vector that can be decomposed as

$$M_t = L_t + S_t \tag{10.2}$$

Here S_t is a sparse vector with support set size at most s and minimum magnitude of nonzero values at least $S_{\min}$. L_t is a dense but low dimensional vector, i.e., $L_t = P_{(t)} a_t$ where $P_{(t)}$ is an $n \times r_{(t)}$ basis matrix with $r_{(t)} < n$, which changes every so often according to the model given below. We are given an accurate estimate of the subspace in which the initial t_{train} L_t's lie, i.e., we are given a basis matrix $\hat{P}_0$ so that $\|(I - \hat{P}_0 \hat{P}_0') P_0\|_2$ is small. Here P_0 is a basis matrix for $\text{span}(\mathcal{L}_{t_{\text{train}}})$, i.e., $\text{span}(P_0) = \text{span}(\mathcal{L}_{t_{\text{train}}})$. Also, for the first t_{train} time instants, S_t is zero. The goal is

1. to estimate both S_t and L_t at each time $t > t_{\text{train}}$, and
2. to estimate $\text{span}(\mathcal{L}_t)$ every so often, i.e. to compute $\hat{P}_{(t)}$ so that the subspace estimation error, $\text{SE}_{(t)} := \|(I - \hat{P}_{(t)} \hat{P}_{(t)}') P_{(t)}\|_2$, is small.

We assume a subspace change model that allows the subspace to change at certain change times t_j rather than continuously at each time. It should be noted that this is only a model for reality. In practice there will typically be some changes at every time t; however this is difficult to model in a simple fashion. Moreover the analysis for such a model will be a lot more complicated. However, we do allow the variance of the projection of L_t along the subspace directions to change continuously. The projection along the new directions is assumed to be small initially and allowed to gradually increase to a large value (see Section 10.2.2).

Model 10.1 (Model on L_t) *We assume that*

1. $L_t = P_{(t)} a_t$ *with* $P_{(t)} = P_j$ *for all* $t_j \le t < t_{j+1}$, $j = 0, 1, 2 \cdots J$. *Here* P_j *is an* $n \times r_j$ *basis matrix with* $r_j < \min(n, (t_{j+1} - t_j))$ *that changes as*

   $$P_j = [P_{j-1} \; P_{j,\text{new}}]$$

   where $P_{j,\text{new}}$ *is an* $n \times c_{j,\text{new}}$ *basis matrix with* $P_{j,\text{new}}' P_{j-1} = 0$. *Thus*

   $$r_j = \text{rank}(P_j) = r_{j-1} + c_{j,\text{new}}.$$

   We let $t_0 = 0$. *Also* $t_{J+1} = t_{\max}$, *which is the length of the sequence. This can be infinite, too. This model is illustrated in Figure 10.1.*
2. *The vector of coefficients,* $a_t := {P_{(t)}}' L_t$, *is a zero-mean random variable (r.v.) with mutually uncorrelated entries, i.e.,* $\mathbb{E}[a_t] = 0$, *and* $\Lambda_t := \text{Cov}[a_t] = \mathbb{E}(a_t a_t')$ *is a diagonal matrix. This assumption follows automatically if we let* $P_{(t)} \Lambda_t {P_{(t)}}'$ *be the eigenvalue decomposition (EVD) of Cov(L_t).*
3. *Assume that* $0 < \lambda^- \le \lambda^+ < \infty$ *where*

   $$\lambda^- := \inf_t \lambda_{\min}(\Lambda_t), \quad \lambda^+ := \sup_t \lambda_{\max}(\Lambda_t).$$

Notice that, for $t_j \le t < t_{j+1}$, a_t is an r_j length vector that can be split as

$$a_t = {P_j}' L_t = \begin{bmatrix} a_{t,*} \\ a_{t,\text{new}} \end{bmatrix}$$

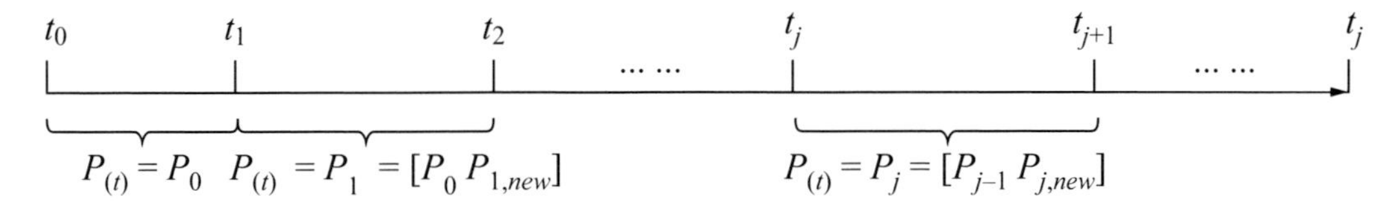

FIGURE 10.1 The subspace change model explained in Sec 10.2.1. Here $t_0 = 0$ and $0 < t_{\text{train}} < t_1$.

where $a_{t,*} := P_{j-1}{}' L_t$ and $a_{t,\text{new}} := P_{j,\text{new}}{}' L_t$. Thus, for this interval, L_t can be rewritten as

$$L_t = [P_{j-1}\ P_{j,\text{new}}] \begin{bmatrix} a_{t,*} \\ a_{t,\text{new}} \end{bmatrix} = P_{j-1} a_{t,*} + P_{j,\text{new}} a_{t,\text{new}}.$$

Also, Λ_t can be split as

$$\Lambda_t = \begin{bmatrix} \Lambda_{t,*} & 0 \\ 0 & \Lambda_{t,\text{new}} \end{bmatrix}$$

where $\Lambda_{t,*} = \text{Cov}[a_{t,*}]$ and $\Lambda_{t,\text{new}} = \text{Cov}[a_{t,\text{new}}]$ are diagonal matrices.

DEFINITION 10.2 Define

$$f := \frac{\lambda^+}{\lambda^-}$$

Define

$$\lambda_{\text{new}}^- := \min_t \lambda_{\min}(\Lambda_{t,\text{new}}), \quad \lambda_{\text{new}}^+ := \max_t \lambda_{\max}(\Lambda_{t,\text{new}}), \quad g := \frac{\lambda_{\text{new}}^+}{\lambda_{\text{new}}^-}.$$

Notice that the set of new directions changes with each subspace change interval, and hence at any time, g is an upper bound on the condition number of $\Lambda_{t,\text{new}}$ for the currrent set of new directions.

The above simple model only allows new additions to the subspace and hence the rank of P_j can only grow over time. The ReProCS algorithm designed for this model can be interpreted as a recursive algorithm for solving the robust PCA problem studied in [9] and other batch robust PCA works. At time t we estimate the subspace spanned by $L_1, L_2, \ldots L_t$. We give this model first for ease of understanding. Both our algorithm and its guarantee will apply without any change even if one or more directions were deleted from the subspace with the following more general model.

Model 10.2 *Assume Model 10.1 with the difference that P_j now changes as*

$$P_j = [(P_{j-1} R_j \setminus P_{j,\text{old}})\ P_{j,\text{new}}]$$

where $P_{j,\text{new}}$ is an $n \times c_{j,\text{new}}$ basis matrix with $P_{j,\text{new}}' P_{j-1} = 0$, R_j is a rotation matrix, and $P_{j,\text{old}}$ is an $n \times c_{j,\text{old}}$ matrix containing columns of $(P_{j-1} R_j)$.

10.2.2 Slow Subspace Change

By slow subspace change we mean all of the following.

First, the delay between consecutive subspace change times is large enough, i.e., for a d large enough,

$$t_{j+1} - t_j \geq d. \tag{10.3}$$

Second, the magnitude of the projection of L_t along the newly added directions, $a_{t,\text{new}}$, is initially small but can increase gradually. We model this as follows. Assume that

$$\lambda_{\text{new}}^{+} \leq g^{+} \lambda_{\text{new}}^{-} \tag{10.4}$$

for a $g^{+} \geq 1$ but not too large; and for an $\alpha > 0$ [27] the following holds:

$$\|a_{t,\text{new}}\|_{\infty} \leq \min\left(v^{\frac{t-t_j}{\alpha}-1} \gamma_{\text{new}}, \gamma_{*}\right) \tag{10.5}$$

when $t \in [t_j, t_{j+1} - 1]$ for a $v \geq 1$ but not too large and with $\gamma_{\text{new}} < \gamma_{*}$ and $\gamma_{\text{new}} < S_{\min}$. Clearly, the above assumption implies that

$$\|a_{t,\text{new}}\|_{\infty} \leq \gamma_{\text{new},k} := \min(v^{k-1} \gamma_{\text{new}}, \gamma_{*})$$

for all $t \in [t_j + (k-1)\alpha, t_j + k\alpha - 1]$.

Third, the number of newly added directions is small, i.e., $c_{j,\text{new}} \leq c_{\max} \ll r_0$.

The last two assumptions are verified for real video data in [43, Section IX-A].

REMARK 10.1 [Large f] Since our problem definition allows large noise, L_t, but assumes slow subspace change, thus $f = \lambda^{+}/\lambda^{-}$ cannot be bounded by a small value. The reason is as follows. Slow subspace change implies that the projection of L_t along the new directions is initially small, i.e., γ_{new} is small. Since $\lambda^{-} \leq \gamma_{\text{new}}$, this means that λ^{-} is small. Since $\mathbb{E}[\|L_t\|^2] \leq r_{\max}\lambda^{+}$ and $r_{\max}$ is small (low-dimensional), thus, large L_t means that λ^{+} needs to be large. As a result, f cannot be upper bounded by a small value.

10.2.3 Measuring Denseness of a Matrix and Its Relation with RIC

Before we can state the denseness assumption, we need to define the denseness coefficient.

DEFINITION 10.3 [denseness coefficient] For a matrix or a vector B, define

$$\kappa_s(B) = \kappa_s(\text{span}(B)) := \max_{|T| \leq s} \|I_T{}' \text{basis}(B)\|_2 \tag{10.6}$$

where $\|.\|_2$ is the vector or matrix ℓ_2-norm.

Clearly, $\kappa_s(B) \leq 1$. First, consider an n-length vector B. Then κ_s measures the denseness (non-compressibility) of B. A small value indicates that the entries in B are spread out, i.e., it is a dense vector. A large value indicates that it is compressible (approximately or exactly sparse). The worst case (largest possible value) is $\kappa_s(B) = 1$, which indicates that B is an s-sparse vector. The best case is $\kappa_s(B) = \sqrt{s/n}$ and this will occur if each entry of B has the same magnitude. Similarly, for an $n \times r$ matrix B, a small κ_s means that most (or all) of its columns are dense vectors.

REMARK 10.2 The following facts should be noted about $\kappa_s(.)$:

1. For a given matrix B, $\kappa_s(B)$ is a non-decreasing function of s.

[27] As we will see in the algorithm, α is the number of previous frames used to get a new estimate of $P_{j,\text{new}}$.

2. $\kappa_s([B_1]) \leq \kappa_s([B_1 \ B_2])$, i.e., adding columns cannot decrease κ_s.
3. A bound on $\kappa_s(B)$ is $\kappa_s(B) \leq \sqrt{s}\kappa_1(B)$. This follows because $\|B\|_2 \leq \left\| \left[\|b_1\|_2 \ldots \|b_r\|_2 \right] \right\|_2$ where b_i is the ith column of B.
4. For a basis matrix $P = [P_1, P_2]$, $\kappa_s([P_1, P_2])^2 \leq \kappa_s(P_1)^2 + \kappa_s(P_2)^2$.

The lemma below relates the denseness coefficient of a basis matrix P to the RIC of $I - PP'$. The proof is in the Appendix.

Lemma 10.2.1 *For an $n \times r$ basis matrix P (i.e., P satisfying $P'P = I$),*

$$\delta_s(I - PP') = \kappa_s^2(P).$$

In other words, if the columns of P are dense enough (small κ_s), then the RIC of $I - PP'$ is small.

In this work, we assume an upper bound on $\kappa_{2s}(P_j)$ for all j, and a tighter upper bound on $\kappa_{2s}(P_{j,\text{new}})$, i.e., there exist $\kappa_{2s,*}^+ < 1$ and a $\kappa_{2s,\text{new}}^+ < \kappa_{2s,*}^+$ such that

$$\max_j \kappa_{2s}(P_{j-1}) \leq \kappa_{2s,*}^+ \text{ and } \max_j \kappa_{2s}(P_{j,\text{new}}) \leq \kappa_{2s,\text{new}}^+. \tag{10.7}$$

The denseness coefficient $\kappa_s(B)$ is related to the denseness assumption required by PCP [9]. That work uses $\kappa_1(B)$ to quantify denseness. To relate to their condition, define the incoherence parameter μ as in [9] to be the smallest real number such that

$$\kappa_1(P_0)^2 \leq \frac{\mu r_0}{n} \quad \text{and} \quad \kappa_1(P_{t_j,\text{new}})^2 \leq \frac{\mu c_{j,\text{new}}}{n} \text{ for all } j = 1, 2, \ldots J. \tag{10.8}$$

Then, using Remark 10.2, it is easy to see that (10.7) holds if

$$\frac{2s(r_0 + Jc_{\max})\mu}{n} \leq \kappa_{2s,*}^{+\,2} \text{ and } \frac{2sc_{\max}\mu}{n} \leq \kappa_{2s,\text{new}}^{+\,2} \tag{10.9}$$

where

$$c_{\max} := \max_j c_{j,\text{new}}.$$

10.2.4 Definitions and Assumptions on the Sparse Outliers

Define the following quantities for the sparse outliers.

DEFINITION 10.4 Let $T_t := \{i : (S_t)_i \neq 0\}$ denote the support set of S_t. Define

$$S_{\min} := \min_{t > t_{\text{train}}} \min_{i \in T_t} |(S_t)_i|, \text{ and } s := \max_t |T_t|.$$

In words, $S_{\min}$ is the magnitude of smallest nonzero entry of S_t for all t, and s is an upper bound on the support size of S_t for all t.

An upper bound on s is imposed by the denseness assumption given in Section 10.2.3. Moreover, we either need a denseness assumption on the basis matrix for the currently unestimated part of span($P_{j,\text{new}}$) (a quantity that depends on intermediate algorithm estimates) or we need an assumption on how slowly the support can change (the support should change at least once every β time instants, and when it changes, the change should be by at least s/ρ indices and be such that the sets of changed indices are mutually disjoint for a period of α frames). The former is needed in the partial result from our recently published work [43]. The latter is needed in our complete correctness result that will appear in the proceedings of ICASSP and ISIT 2015 [31, 32].

10.3 Recursive Projected Compressed Sensing (ReProCS) Algorithm

In Section 10.3.1, we first explain the projection-PCA algorithm that is used for the subspace update step in the ReProCS algorithm. This is followed by a brief explanation of the main idea of the simplest possible ReProCS algorithm in Section 10.3.2. This algorithm is impractical though since it assumes knowledge of model parameters including the subspace change times. In Section 10.3.3, we develop a practical modification of the basic ReProCS idea that does not need model knowledge and can be directly used for experiments with real data. We further improve the sparse recovery and support estimation steps of this algorithm in Section 10.3.4.

10.3.1 Projection-PCA Algorithm for ReProCS

Given a data matrix $\mathcal{D}$, a basis matrix P and an integer r, projection-PCA (proj-PCA) applies PCA on $\mathcal{D}_{\text{proj}} := (I - PP')\mathcal{D}$, i.e., it computes the top r eigenvectors (the eigenvectors with the largest r eigenvalues) of $\frac{1}{\alpha}\mathcal{D}_{\text{proj}}\mathcal{D}_{\text{proj}}{}'$. Here α is the number of column vectors in $\mathcal{D}$. This is summarized in Algorithm 10.1.

If $P = [.]$, then projection-PCA reduces to standard PCA, i.e., it computes the top r eigenvectors of $\frac{1}{\alpha}\mathcal{D}\mathcal{D}'$.

The reason we need projection PCA algorithm in step 3 of Algorithm 10.2 is because the error $e_t = \hat{L}_t - L_t = S_t - \hat{S}_t$ is correlated with L_t; and the maximum condition number of $\text{Cov}(L_t)$, which is bounded by f, cannot be bounded by a small value (see Remark 10.1). This issue is explained in detail in the Appendix of [43]. Most other works that analyze standard PCA, e.g., [38] and references therein, do not face this issue because they assume uncorrelated-ness of the noise/error and the true data vector. With this assumption, one only needs to increase the PCA data length α to deal with the larger condition number.

We should mention that the idea of projecting perpendicular to a partly estimated subspace has been used in other, different contexts in past work [28, 35].

Algorithm 10.1 projection-PCA: $Q \leftarrow \text{proj-PCA}(\mathcal{D}, P, r)$

1. Projection: compute $\mathcal{D}_{\text{proj}} \leftarrow (I - PP')\mathcal{D}$.
2. PCA: compute $\frac{1}{\alpha}\mathcal{D}_{\text{proj}}\mathcal{D}_{\text{proj}}{}' \overset{EVD}{=} \begin{bmatrix} Q \; Q_\perp \end{bmatrix} \begin{bmatrix} \Lambda & 0 \\ 0 & \Lambda_\perp \end{bmatrix} \begin{bmatrix} Q' \\ Q_\perp{}' \end{bmatrix}$ where Q is an $n \times r$ basis matrix and α is the number of columns in $\mathcal{D}$.

10.3.2 Recursive Projected CS (ReProCS)

We summarize the Recursive Projected CS (ReProCS) algorithm in Algorithm 10.2. It uses the following definition.

DEFINITION 10.5 Define the time interval $\mathcal{I}_{j,k} := [t_j + (k-1)\alpha, t_j + k\alpha - 1]$ for $k = 1, \dots K$ and $\mathcal{I}_{j,K+1} := [t_j + K\alpha, t_{j+1} - 1]$.

The key idea of ReProCS is as follows. First, consider a time t when the current basis matrix $P_{(t)} = P_{(t-1)}$ and this has been accurately predicted using past estimates of L_t, i.e., we have $\hat{P}_{(t-1)}$ with $\|(I - \hat{P}_{(t-1)}\hat{P}'_{(t-1)})P_{(t)}\|_2$ small. We project the measurement vector, M_t, into the space perpendicular to $\hat{P}_{(t-1)}$ to get the projected measurement vector $y_t := \Phi_{(t)} M_t$

where $\Phi_{(t)} = I - \hat{P}_{(t-1)}\hat{P}'_{(t-1)}$ (step 1a). The $n \times n$ projection matrix, $\Phi_{(t)}$ has rank $n - r_*$ where $r_* = \text{rank}(\hat{P}_{(t-1)})$, therefore y_t has only $n - r_*$ "effective" measurements,[28] even though its length is n. Notice that y_t can be rewritten as

$$y_t = \Phi_{(t)} S_t + \mathbf{m}b_t, \text{ where } \mathbf{m}b_t := \Phi_{(t)} L_t.$$

Since $\|(I - \hat{P}_{(t-1)}\hat{P}'_{(t-1)})P_{(t-1)}\|_2$ is small, the projection nullifies most of the contribution of L_t and so the projected noise $\mathbf{m}b_t$ is small. Recovering the n-dimensional sparse vector S_t from y_t now becomes a traditional sparse recovery or CS problem in small noise [6, 7, 12, 14, 18, 21]. We use ℓ_1 minimization to recover it (step 1b). If the current basis matrix $P_{(t)}$, and hence its estimate, $\hat{P}_{(t-1)}$, is dense enough, then, by Lemma 10.2.1, the RIC of $\Phi_{(t)}$ is small enough. Using [3, Theorem 1], this ensures that S_t can be accurately recovered from y_t. By thresholding on the recovered S_t, one gets an estimate of its support (step 1c). By computing a least-squares (LS) estimate of S_t on the estimated support and setting it to zero everywhere else (step 1d), we can get a more accurate final estimate, $\hat{S}_t$, as first suggested in [8]. This $\hat{S}_t$ is used to estimate L_t as $\hat{L}_t = M_t - \hat{S}_t$. It is easy to see that if γ_{new} is small enough compared to $S_{\min}$ and the support estimation threshold, ω, is chosen appropriately, we can get exact support recovery, i.e., $\hat{\mathcal{T}}_t = T_t$. In this case, the error $e_t := \hat{S}_t - S_t = L_t - \hat{L}_t$ has the following simple expression:

$$e_t = I_{T_t} (\Phi_{(t)})_{T_t}{}^{\dagger} \mathbf{m}b_t = I_{T_t} [(\Phi_{(t)})'_{T_t} (\Phi_{(t)})_{T_t}]^{-1} I_{T_t}{}' \Phi_{(t)} L_t. \tag{10.10}$$

The second equality follows because $(\Phi_{(t)})_{\mathcal{T}}{}'\Phi_{(t)} = (\Phi_{(t)} I_{\mathcal{T}})'\Phi_{(t)} = I_{\mathcal{T}}{}'\Phi_{(t)}$ for any set $\mathcal{T}$.

Now consider a time t when $P_{(t)} = P_j = [P_{j-1}, P_{j,\text{new}}]$ and P_{j-1} has been accurately estimated but $P_{j,\text{new}}$ has not been estimated, i.e., consider a $t \in \mathcal{I}_{j,1}$. At this time, $\hat{P}_{(t-1)} = \hat{P}_{j-1}$ and so $\Phi_{(t)} = \Phi_{j,0} := I - \hat{P}_{j-1}\hat{P}'_{j-1}$. Let $r_* := r_0 + (j-1)c_{\max}$ (we remove subscript j for ease of notation) and $c := c_{\max}$. Assume that the delay between change times is large enough so that by $t = t_j$, $\hat{P}_{j-1}$ is an accurate enough estimate of P_{j-1}, i.e., $\|\Phi_{j,0}P_{j-1}\|_2 \leq r_*\zeta \ll 1$. Clearly, $\kappa_s(\Phi_0 P_{\text{new}}) \leq \kappa_s(P_{\text{new}}) + r_*\zeta$, i.e., $\Phi_0 P_{\text{new}}$ is dense because P_{new} is dense and because $\hat{P}_{j-1}$ is an accurate estimate of P_{j-1} (which is perpendicular to P_{new}). Moreover, using Lemma 10.2.1, it can be shown that $\phi_0 := \max_{|T| \leq s} \|[(\Phi_0)'_T(\Phi_0)_T]^{-1}\|_2 \leq \frac{1}{1-\delta_s(\Phi_0)} \leq \frac{1}{1-(\kappa_s(P_{j-1})+r_*\zeta)^2}$. The error e_t still satisfies (10.10) although its magnitude is not as small. Using the above facts in (10.10), we get that

$$\|e_t\|_2 \leq \frac{\kappa_s(P_{\text{new}})\sqrt{c}\gamma_{\text{new}} + r_*\zeta(\sqrt{r_*}\gamma_* + \sqrt{c}\gamma_{\text{new}})}{1 - (\kappa_s(P_{j-1}) + r\zeta)^2}.$$

If $\sqrt{\zeta} < 1/\gamma_*$, all terms containing ζ can be ignored and we get that the above is approximately upper bounded by $\frac{\kappa_s(P_{\text{new}})}{1-\kappa_s^2(P_{j-1})}\sqrt{c}\gamma_{\text{new}}$. Using the denseness assumption, this quantity is a small constant times $\sqrt{c}\gamma_{\text{new}}$, e.g., with the numbers assumed in Theorem 10.3 given later, we get a bound of $0.18\sqrt{c}\gamma_{\text{new}}$. Because of slow subspace change, c and γ_{new} are small and hence the error e_t is small, i.e., S_t is recovered accurately. With each projection PCA step, as we explain below, the error e_t becomes even smaller.

Since $\hat{L}_t = M_t - \hat{S}_t$ (step 2), e_t also satisfies $e_t = L_t - \hat{L}_t$. Thus, a small e_t means that L_t is also recovered accurately. The estimated $\hat{L}_t$'s are used to obtain new estimates of $P_{j,\text{new}}$ every α frames for a total of $K\alpha$ frames by projection PCA (step 3). We illustrate the projection PCA algorithm in Figure 10.2. In the first projection PCA step, we get the

[28] i.e., some r_* entries of y_t are linear combinations of the other $n - r_*$ entries.

first estimate of $P_{j,\text{new}}$, $\hat{P}_{j,\text{new},1}$. For the next α frame interval, $\hat{P}_{(t-1)} = [\hat{P}_{j-1}, \hat{P}_{j,\text{new},1}]$ and so $\Phi_{(t)} = \Phi_{j,1} = I - \hat{P}_{j-1}\hat{P}_{j-1}' - \hat{P}_{\text{new},1}\hat{P}_{\text{new},1}'$. Using this in the projected CS step reduces the projection noise, $\mathbf{m}b_t$, and hence the reconstruction error, e_t, for this interval, as long as $\gamma_{\text{new},k}$ increases slowly enough. Smaller e_t makes the perturbation seen by the second projection PCA step even smaller, thus resulting in an improved second estimate $\hat{P}_{j,\text{new},2}$. Within K updates (K chosen as given in Theorem 10.3), it can be shown that both $\|e_t\|_2$ and the subspace error drop down to a constant times $\sqrt{\zeta}$. At this time, we update $\hat{P}_j$ as $\hat{P}_j = [\hat{P}_{j-1}, \hat{P}_{j,\text{new},K}]$.

Algorithm 10.2 - Recursive Projected CS (ReProCS)

Parameters: algorithm parameters: ξ, ω, α, K, model parameters: t_j, $c_{j,\text{new}}$ (set as in Theorem 10.3)

Input: M_t, *Output:* $\hat{S}_t$, $\hat{L}_t$, $\hat{P}_{(t)}$

Initialization: Compute $\hat{P}_0 \leftarrow$ proj-PCA$([L_1, L_2, \cdots, L_{t_{\text{train}}}], [.], r_0)$ where $r_0 =$ rank$([L_1, L_2, \cdots, L_{t_{\text{train}}}])$.

Set $\hat{P}_{(t)} \leftarrow \hat{P}_0$, $j \leftarrow 1$, $k \leftarrow 1$.

For $t > t_{\text{train}}$, do the following:

1. Estimate T_t and S_t via Projected CS:
   (a) Nullify most of L_t: compute $\Phi_{(t)} \leftarrow I - \hat{P}_{(t-1)}\hat{P}_{(t-1)}'$, compute $y_t \leftarrow \Phi_{(t)}M_t$
   (b) Sparse Recovery: compute $\hat{S}_{t,\mathcal{S}}$ as the solution of $\min_x \|x\|_1$ $s.t.$ $\|y_t - \Phi_{(t)}x\|_2 \leq \xi$
   (c) Support Estimate: compute $\hat{T}_t = \{i : \ |(\hat{S}_{t,\mathcal{S}})_i| > \omega\}$
   (d) LS Estimate of S_t: compute $(\hat{S}_t)_{\hat{T}_t} = ((\Phi_{(t)})_{\hat{T}_t})^\dagger y_t$, $(\hat{S}_t)_{\hat{T}_t^c} = 0$
2. Estimate L_t: $\hat{L}_t = M_t - \hat{S}_t$.
3. Update $\hat{P}_{(t)}$: K Projection PCA steps.
   (a) If $t = t_j + k\alpha - 1$,
      i. $\hat{P}_{j,\text{new},k} \leftarrow$ proj-PCA$\left(\left[\hat{L}_{t_j+(k-1)\alpha}, \ldots, \hat{L}_{t_j+k\alpha-1}\right], \hat{P}_{j-1}, c_{j,\text{new}}\right)$.
      ii. set $\hat{P}_{(t)} \leftarrow [\hat{P}_{j-1}\ \hat{P}_{j,\text{new},k}]$; increment $k \leftarrow k+1$.

   Else
      i. set $\hat{P}_{(t)} \leftarrow \hat{P}_{(t-1)}$.

   (b) If $t = t_j + K\alpha - 1$, then set $\hat{P}_j \leftarrow [\hat{P}_{j-1}\ \hat{P}_{j,\text{new},K}]$. Increment $j \leftarrow j+1$. Reset $k \leftarrow 1$.
4. Increment $t \leftarrow t+1$ and go to step 1.

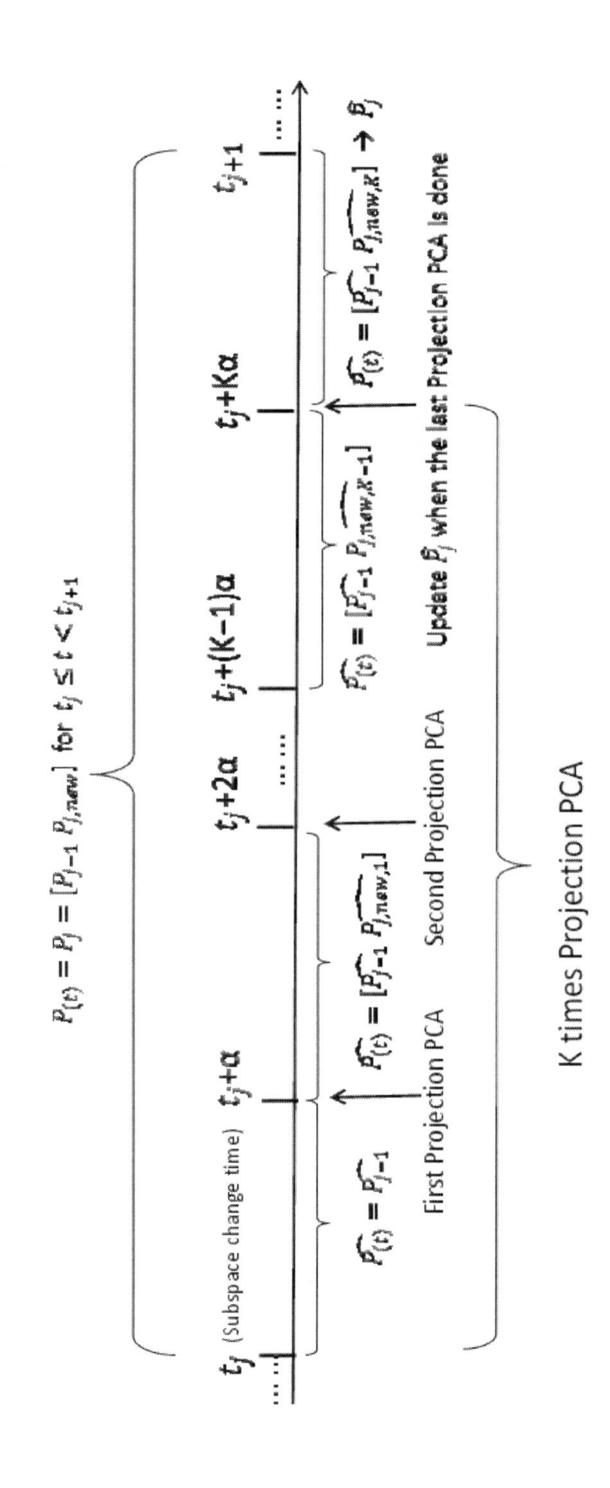

FIGURE 10.2 The K projection PCA steps.

10.3.3 Practical ReProCS

The algorithm described in the previous subsection and in Algorithm 10.2 cannot be used in practice because it assumes knowledge of the subspace change times t_j and of the exact number of new directions added at each change time, $c_{j,\text{new}}$. Moreover, the choices of the algorithm parameters are also impractically large. In this section, we develop a practical modification of the ReProCS algorithm from the previous section that does not assume any model knowledge and that also directly works for various video sequences. The algorithm is summarized in Algorithm 10.3. We use the YALL-1 solver for solving all ℓ_1 norm minimization problems. Code for Algorithm 10.3 is available at `http://www.ece.iastate.edu/~hanguo/PracReProCS.html#Code_`.

Given the initial training sequence, which does not contain the sparse components, $\mathcal{M}_{\text{train}} = [L_1, L_2, \ldots L_{t_{\text{train}}}]$ we compute $\hat{P}_0$ as an approximate basis for $\mathcal{M}_{\text{train}}$, i.e., $\hat{P}_0 = \text{approx-basis}(\mathcal{M}_{\text{train}}, b\%)$. Let $\hat{r} = \text{rank}(\hat{P}_0)$. We need to compute an approximate basis because for real data, the L_t's are only approximately low-dimensional. We use $b\% = 95\%$ or $b\% = 99.99\%$ depending on whether the low-rank part is approximately low-rank or almost exactly low-rank. After this, at each time t, ReProCS proceeds as follows.

Perpendicular Projection. In the first step, at time t, we project the measurement vector, M_t, into the space orthogonal to $\text{span}(\hat{P}_{(t-1)})$ to get the projected measurement vector,

$$y_t := \Phi_{(t)} M_t. \tag{10.11}$$

Sparse Recovery (Recover T_t and S_t). With the above projection, y_t can be rewritten as

$$y_t = \Phi_{(t)} S_t + \mathbf{m}b_t \text{ where } \mathbf{m}b_t := \Phi_{(t)} L_t. \tag{10.12}$$

As explained earlier, because of the slow subspace change assumption, projecting orthogonal to $\text{span}(\hat{P}_{(t-1)})$ nullifies most of the contribution of L_t and hence $\mathbf{m}b_t$ can be interpreted as small "noise." Thus, the problem of recovering S_t from y_t becomes a traditional noisy sparse recovery/CS problem. To recover S_t from y_t, one can use ℓ_1 minimization [3, 12], or any of the greedy or iterative thresholding algorithms from the literature. In this work we use ℓ_1 minimization: we solve

$$\min_x \|x\|_1 \text{ s.t. } \|y_t - \Phi_{(t)} x\|_2 \leq \xi \tag{10.13}$$

and denote its solution by $\hat{S}_{t,cs}$. By the denseness assumption, P_{t-1} is dense. Since $\hat{P}_{(t-1)}$ approximates it, this is true for $\hat{P}_{(t-1)}$ as well [43, Lemma 6.6]. Thus, by Lemma 10.2.1, the RIC of $\Phi_{(t)}$ is small enough. Using [3, Theorem 1], this and the fact that $\mathbf{m}b_t$ is small ensures that S_t can be accurately recovered from y_t. The constraint ξ used in the minimization should equal $\|\mathbf{m}b_t\|_2$ or its upper bound. Since $\mathbf{m}b_t$ is unknown we set $\xi = \|\hat{\beta}_t\|_2$ where $\hat{\beta}_t := \Phi_{(t)} \hat{L}_{t-1}$.

By thresholding on $\hat{S}_{t,cs}$ to get an estimate of its support followed by computing a least squares (LS) estimate of S_t on the estimated support and setting it to zero everywhere else, we can get a more accurate estimate, $\hat{S}_t$, as suggested in [8]. We discuss better support estimation and its parameter setting in Section 10.3.4.

Recover L_t. The estimate $\hat{S}_t$ is used to estimate L_t as $\hat{L}_t = M_t - \hat{S}_t$. Thus, if S_t is recovered accurately, L_t will be as well.

Subspace Update (Update $\hat{P}_{(t)}$). Within a short delay after every subspace change time, one needs to update the subspace estimate, $\hat{P}_{(t)}$. To do this in a provably reliable fashion, we introduced the projection PCA (p-PCA) algorithm [43] in the previous two subsections. In this section, we design a practical version of projection-PCA (p-PCA) that does not need knowledge of t_j or $c_{j,\text{new}}$. This is summarized in Algorithm 10.3. The key idea is as follows. We let $\hat{\sigma}_{\min}$ be the $\hat{r}^{th}$ largest singular value of the training dataset. This serves as the noise threshold for approximately low-rank data. We split projection PCA into two phases: "detect subspace change" and "p-PCA." We are in the detect phase when the previous subspace has been accurately estimated. Denote the basis matrix for this subspace by $\hat{P}_{j-1}$. We detect the subspace change as follows. Every α frames, we project the last α $\hat{L}_t$'s perpendicular to $\hat{P}_{j-1}$ and compute the SVD of the resulting matrix. If there are any singular values above $\hat{\sigma}_{\min}$, this means that the subspace has changed. At this point, we enter the "p-PCA" phase. In this phase, we repeat the K p-PCA steps described earlier with the following change: we estimate $c_{j,\text{new}}$ as the number of singular values above $\hat{\sigma}_{\min}$, but clipped at $\lceil \alpha/3 \rceil$ (i.e., if the number is more than $\lceil \alpha/3 \rceil$ then we clip it to $\lceil \alpha/3 \rceil$). We stop either when the stopping criterion given in step 4(b)iv is achieved ($k \geq K_{\min}$ and the projection of $\hat{L}_t$ along $\hat{P}_{\text{new},k}$ is not too different from that along $\hat{P}_{\text{new},k}$) or when $k \geq K_{\max}$.

REMARK 10.3 The p-PCA algorithm only allows addition of new directions. If the goal is to estimate the span of $[L_1, \ldots L_t]$, then this is what is needed. If the goal is sparse recovery, then one can get a smaller rank estimate of $\hat{P}_{(t)}$ by also including a step to delete the span of the removed directions, $P_{(j),\text{old}}$. This will result in more "effective" measurements available for the sparse recovery step and hence possibly in improved performance. The simplest way to do this is by simple PCA done at the end of the K p-PCA steps. A provably accurate solution is the cluster-PCA approach described in [43, Sec VII].

REMARK 10.4 The p-PCA algorithm works on small batches of α frames. This can be made fully recursive if we compute the SVD needed in projection-PCA using the incremental SVD [1] procedure summarized in Algorithm 10.4 (inc-SVD) for one frame at a time. As explained in [1] and references therein, we can get the left singular vectors and singular values of any matrix $M = [M_1, M_2, \ldots M_\alpha]$ recursively by starting with $\hat{P} = [.], \hat{\Sigma} = [.]$ and calling $[\hat{P}, \hat{\Sigma}] = \text{inc-SVD}(\hat{P}, \hat{\Sigma}, M_i)$ for every column i or for short batches of columns of the size of α/k. Since we use $\alpha = 20$, which is a small value, the use of incremental SVD does not speed up the algorithm in practice.

10.3.4 Exploiting Slow Support Change When Valid and Improved Two-Step Support Estimation

In [42, 43], we always used ℓ_1 minimization followed by thresholding and LS for sparse recovery. However if slow support change holds, one can replace simple ℓ_1 minimization by modified-CS [49], which requires fewer measurements for exact/accurate recovery as long as the previous support estimate, $\hat{\mathcal{T}}_{t-1}$, is an accurate enough predictor of the current support, T_t. In our application, $\hat{\mathcal{T}}_{t-1}$ is likely to contain a significant number of extras and in this case, a better idea is to solve the following weighted ℓ_1 problem [19, 27]

$$\min_x \lambda \|x_T\|_1 + \|x_{T^c}\|_1 \text{ s.t. } \|y_t - \Phi_{(t)} x\|_2 \leq \xi, \quad T := \hat{\mathcal{T}}_{t-1} \tag{10.14}$$

with $\lambda < 1$ (modified-CS solves the above with $\lambda = 0$). Denote its solution by $\hat{S}_{t,cs}$. One way to pick λ is to let it be proportional to the estimate of the percentage of extras in $\hat{\mathcal{T}}_{t-1}$. If slow support change does not hold, the previous support estimate is not a good predictor of the current support. In this case, doing the above is a bad idea and one should instead solve simple ℓ_1, i.e., solve (10.14) with $\lambda = 1$. As explained in [19], if the support estimate contains at least 50% correct entries, then weighted ℓ_1 is better than simple ℓ_1. We use the above criteria with true values replaced by estimates. Thus, if $\frac{|\hat{\mathcal{T}}_{t-2} \cap \hat{\mathcal{T}}_{t-1}|}{|\hat{\mathcal{T}}_{t-2}|} > 0.5$, then we solve (10.14) with $\lambda = \frac{|\hat{\mathcal{T}}_{t-2} \setminus \hat{\mathcal{T}}_{t-1}|}{|\hat{\mathcal{T}}_{t-1}|}$, or else we solve it with $\lambda = 1$.

Improved support estimation

A simple way to estimate the support is by thresholding the solution of (10.14). This can be improved by using the Add-LS-Del procedure for support and signal value estimation [49]. We proceed as follows. First we compute the set $\hat{\mathcal{T}}_{t,\text{add}}$ by thresholding on $\hat{S}_{t,cs}$ in order to retain its k largest magnitude entries. We then compute an LS estimate of S_t on $\hat{\mathcal{T}}_{t,\text{add}}$ while setting it to zero everywhere else. As explained earlier, because of the LS step, $\hat{S}_{\text{t,add}}$ is a less biased estimate of S_t than $\hat{S}_{t,cs}$. We let $k = 1.4|\hat{\mathcal{T}}_{t-1}|$ to allow for a small increase in the support size from $t-1$ to t. A larger value of k also makes it more likely that elements of the set $(T_t \setminus \hat{\mathcal{T}}_{t-1})$ are detected into the support estimate.[29]

The final estimate of the support, $\hat{\mathcal{T}}_t$, is obtained by thresholding on $\hat{S}_{\text{t,add}}$ using a threshold ω. If ω is appropriately chosen, this step helps to delete some of the extra elements from $\hat{\mathcal{T}}_{\text{add}}$ and this ensures that the size of $\hat{\mathcal{T}}_t$ does not keep increasing (unless the object's size is increasing). An LS estimate computed on $\hat{\mathcal{T}}_t$ gives us the final estimate of S_t, i.e., $\hat{S}_t = \text{LS}(y_t, A, \hat{\mathcal{T}}_t)$. We use $\omega = \sqrt{\|M_t\|^2/n}$ except in situations where $\|S_t\| \ll \|L_t\|$ – in this case we use $\omega = 0.25\sqrt{\|M_t\|^2/n}$. An alternate approach is to let ω be proportional to the noise magnitude seen by the ℓ_1 step, i.e., to let $\omega = q\|\beta_t\|_\infty$; however, this approach required different values of q for different experiments (it is not possible to specify one q that works for all experiments).

The complete algorithm with all the above steps is summarized in Algorithm 10.3. Practically, we used $\alpha = 20, K_{\min} = 3, K_{\max} = 10$ in all experiments (α only needs to be large compared to $c_{\max}$). We used $b = 95$ for approximately low-rank data (all real videos and the lake video with simulated foreground) and used $b = 99.99$ for almost exactly low-rank data (simulated data); we used $q = 1$ whenever $\|S_t\|_2$ was of the same order or larger than $\|L_t\|_2$ (all real videos and the lake video) and used $q = 0.25$ when it was much smaller (simulated data with small magnitude S_t).

10.4 Modified PCP: A Piecewise Batch Online Robust PCA Solution

A key limitation of ReProCS is that it does not use the fact that the noise seen by the sparse recovery step of ReProCS, $\mathbf{m}b_t$, approximately lies in a very low-dimensional subspace. To understand this better, consider a time t just after a subspace change time, t_j. Assume that before t_j the previous subspace P_{j-1} has been accurately estimated. For $t \in [t_j, t_j + \alpha)$, $\mathbf{m}b_t$ can be written as

$$\mathbf{m}b_t = P_{j,\text{new}} a_{t,\text{new}} + w_t, \text{ where } w_t := (I - \hat{P}_{j-1}\hat{P}_{j-1}{}')P_{j-1}a_{t,*} + \hat{P}_{j-1}\hat{P}'_{j-1}P_{j,\text{new}}a_{t,\text{new}}$$

Conditioned on accurate recovery so far, w_t is small noise because $\|(I - \hat{P}_{j-1}\hat{P}_{j-1}{}')P_{j-1}\|_2 \le r\zeta$ and $\|\hat{P}_{j-1}\hat{P}'_{j-1}P_{j,\text{new}}\|_2 \le r\zeta$. Thus, $\mathbf{m}b_t$ approximately lies in a $c_{j,\text{new}} \le c_{\max}$ dimen-

[29] Due to the larger weight on the $\|x_{(\hat{\mathcal{T}}^c_{t-1})}\|_1$ term as compared to that on the $\|x_{(\hat{\mathcal{T}}_{t-1})}\|_1$ term, the solution of (10.14) is biased towards zero on $\hat{\mathcal{T}}^c_{t-1}$ and thus the solution values along $(T_t \setminus \hat{\mathcal{T}}_{t-1})$ are smaller than the true ones.

sional subspace. Here $c_{\max}$ is the maximum number of newly added directions, and by assumption, $c_{\max} \ll r \ll n$.

The only way to use the above fact is in a piecewise batch fashion (one cannot impose a low-dimensional subspace assumption on a single vector). The resulting approach can be understood as a modification of the PCP idea that uses an idea similar to our older work on modified CS (for sparse recovery with partial support knowledge) [49] and hence we call it *modified-PCP*. We introduced and studied modified PCP in [54, 55]

10.4.1 Problem Re-Formulation: Robust PCA with Partial Subspace Knowledge

To understand our solution, one can reformulate the online robust PCA problem as a problem of robust PCA with partial subspace knowledge $\hat{P}_{j-1}$. To keep notation simple, we use $\mathbf{G}$ to denote the partial subspace knowledge. With this we have the following problem.

We are given a data matrix $\mathbf{M} \in \mathbb{R}^{n_1 \times n_2}$ that satisfies

$$\mathbf{M} = \mathbf{L} + \mathbf{S} \tag{10.15}$$

where $\mathbf{S}$ is a sparse matrix with support set Ω and $\mathbf{L}$ is a low-rank matrix with reduced singular value decomposition (SVD):

$$\mathbf{L} \overset{\text{SVD}}{=} \mathbf{U\Sigma V}'. \tag{10.16}$$

Let $r := \text{rank}(\mathbf{L})$. We assume that we are given a basis matrix $\mathbf{G}$ so that $(\mathbf{I} - \mathbf{GG}')\mathbf{L}$ has rank smaller than r. The goal is to recover $\mathbf{L}$ and $\mathbf{S}$ from $\mathbf{M}$ using $\mathbf{G}$. Let $r_G := \text{rank}(\mathbf{G})$.

Define $\mathbf{L}_{\text{new}} := (\mathbf{I} - \mathbf{GG}')\mathbf{L}$ with $r_{\text{new}} := \text{rank}(\mathbf{L}_{\text{new}})$ and reduced SVD given by

$$\mathbf{L}_{\text{new}} := (\mathbf{I} - \mathbf{GG}')\mathbf{L} \overset{\text{SVD}}{=} \mathbf{U}_{\text{new}}\mathbf{\Sigma}_{\text{new}}\mathbf{V}'_{\text{new}}. \tag{10.17}$$

We will explain this a little more. With the above, it is easy to show that there exist rotation matrices $\mathbf{R}_U, \mathbf{R}_G$, and basis matrices $\mathbf{G}_{\text{extra}}$ and $\mathbf{U}_{\text{new}}$ with $\mathbf{G}_{\text{extra}}{}'\mathbf{U}_{\text{new}} = 0$, such that

$$\mathbf{U} = [\underbrace{(\mathbf{GR}_G \setminus \mathbf{G}_{\text{extra}})}_{\mathbf{U}_0} \ \mathbf{U}_{\text{new}}]\mathbf{R}'_U. \tag{10.18}$$

Define $r_0 := \text{rank}(\mathbf{U}_0)$ and $r_{\text{extra}} := \text{rank}(\mathbf{G}_{\text{extra}})$. Clearly, $r_G = r_0 + r_{\text{extra}}$ and $r = r_0 + r_{\text{new}} = (r_G - r_{\text{extra}}) + r_{\text{new}}$.

10.4.2 Modified PCP

From the above model, it is clear that

$$\mathbf{L}_{\text{new}} + \mathbf{GX}' + \mathbf{S} = \mathbf{M} \tag{10.19}$$

for $\mathbf{X} = \mathbf{L}'\mathbf{G}$. We recover $\mathbf{L}$ and $\mathbf{S}$ using $\mathbf{G}$ by solving the following **Modified PCP** (mod-PCP) program

$$\begin{aligned} &\text{minimize}_{\tilde{\mathbf{L}}_{\text{new}}, \tilde{\mathbf{S}}, \tilde{\mathbf{X}}} && \|\tilde{\mathbf{L}}_{\text{new}}\|_* + \lambda\|\tilde{\mathbf{S}}\|_1 \\ &\text{subject to} && \tilde{\mathbf{L}}_{\text{new}} + \mathbf{G}\tilde{\mathbf{X}}' + \tilde{\mathbf{S}} = \mathbf{M}. \end{aligned} \tag{10.20}$$

Denote a solution to the above by $\hat{\mathbf{L}}_{\text{new}}, \hat{\mathbf{S}}, \hat{\mathbf{X}}$. Then, $\mathbf{L}$ is recovered as $\hat{\mathbf{L}} = \hat{\mathbf{L}}_{\text{new}} + \mathbf{G}\hat{\mathbf{X}}'$. Modified PCP is inspired by our recent work on sparse recovery using partial support knowledge called modified CS [49]. Notice that modified PCP is equivalent to the following

$$\begin{aligned} &\text{minimize}_{\tilde{\mathbf{L}}, \tilde{\mathbf{S}}} && \|(I - \mathbf{GG}')\tilde{\mathbf{L}}\|_* + \lambda\|\tilde{\mathbf{S}}\|_1 \\ &\text{subject to} && \tilde{\mathbf{L}} + \tilde{\mathbf{S}} = \mathbf{M} \end{aligned} \tag{10.21}$$

and the above is easily understood as being inspired by modified-CS.

10.4.3 Solving the Modified PCP Program

We give below an algorithm based on the Inexact Augmented Lagrange Multiplier (ALM) method [30] to solve the modified-PCP program, i.e., solve (10.20). This algorithm is a direct modification of the algorithm designed to solve PCP in [30]. The only difference is that it uses the idea of [2, 23] for the sparse recovery step.

For the modified-PCP program (10.20), the Augmented Lagrangian function is:

$$\mathbb{L}(\tilde{\mathbf{L}}_{\text{new}}, \tilde{\mathbf{S}}, \mathbf{Y}, \tau) = \|\tilde{\mathbf{L}}_{\text{new}}\|_* + \lambda\|\tilde{\mathbf{S}}\|_1 + \langle \mathbf{Y}, \mathbf{M} - \tilde{\mathbf{L}}_{\text{new}} - \tilde{\mathbf{S}} - \mathbf{G}\tilde{\mathbf{X}}' \rangle + \frac{\tau}{2}\|\mathbf{M} - \tilde{\mathbf{L}}_{\text{new}} - \tilde{\mathbf{S}} - \mathbf{G}\tilde{\mathbf{X}}'\|_F^2,$$

Thus, with similar steps in [30], we have the following algorithm. In Algorithm 10.5, Line 3 solves $\tilde{\mathbf{S}}_{k+1} = \arg\min_{\tilde{\mathbf{S}}} \|\tilde{\mathbf{L}}_{\text{new},k}\|_* + \lambda\|\tilde{\mathbf{S}}\|_1 + \langle \mathbf{Y}_k, \mathbf{M} - \tilde{\mathbf{L}}_{\text{new},k} - \tilde{\mathbf{S}} - \mathbf{G}\tilde{\mathbf{X}}_k' \rangle + \frac{\tau}{2}\|\mathbf{M} - \tilde{\mathbf{L}}_{\text{new},k} - \tilde{\mathbf{S}} - \mathbf{G}\tilde{\mathbf{X}}_k'\|_F^2$; Lines 4–6 solve $[\tilde{\mathbf{L}}_{\text{new},k+1}, \tilde{\mathbf{X}}_{k+1}] = \arg\min_{\tilde{\mathbf{L}}_{\text{new}}, \tilde{\mathbf{X}}} \|\tilde{\mathbf{L}}_{\text{new}}\|_* + \lambda\|\tilde{\mathbf{S}}_{k+1}\|_1 + \langle \mathbf{Y}_k, \mathbf{M} - \tilde{\mathbf{L}}_{\text{new}} - \tilde{\mathbf{S}}_{k+1} - \mathbf{G}\tilde{\mathbf{X}}' \rangle + \frac{\tau}{2}\|\mathbf{M} - \tilde{\mathbf{L}}_{\text{new}} - \tilde{\mathbf{S}}_{k+1} - \mathbf{G}\tilde{\mathbf{X}}_k'\|_F^2$. The soft-thresholding operator is defined as

$$\mathfrak{S}_\epsilon[x] = \begin{cases} x - \epsilon, & \text{if } x > \epsilon; \\ x + \epsilon, & \text{if } x < -\epsilon; \\ 0, & \text{otherwise.} \end{cases} \tag{10.22}$$

Parameters are set as suggested in [30], i.e., $\tau_0 = 1.25/\|\mathbf{M}\|$, $v = 1.5$, $\bar{\tau} = 10^7\tau_0$, and iteration is stopped when $\|\mathbf{M} - \tilde{\mathbf{S}}_{k+1} - \tilde{\mathbf{L}}_{\text{new},k+1} - \mathbf{G}\tilde{\mathbf{X}}_{k+1}\|_F / \|\mathbf{M}\|_F < 10^{-7}$.

10.5 Performance Guarantees for ReProCS and Modified PCP

10.5.1 Results for ReProCS

In recently published work [43], we obtained the following partial result for ReProCS. We call it a partial result because its last assumption depends on intermediate algorithm estimates. In ongoing work that will be presented at ICASSP 2015 [31] and at ISIT 2015 [32], we now have a complete correctness result. We state this later in Theorem 10.4.

THEOREM 10.3 *Consider Algorithm 10.2. Let $c := c_{\max}$ and $r := r_0 + (J-1)c$. Pick a ζ that satisfies*

$$\zeta \leq \min\left(\frac{10^{-4}}{r^2}, \frac{1.5 \times 10^{-4}}{r^2 f}, \frac{1}{r^3\gamma_*^2}\right).$$

Assume that the initial subspace estimate is accurate enough, i.e., $\|(I - \hat{P}_0\hat{P}_0')P_0\| \leq r_0\zeta$. If the following conditions hold:

1. *The algorithm parameters are set as $\xi = \xi_0(\zeta)$, $7\xi \leq \omega \leq S_{\min} - 7\xi$, $K = K(\zeta)$, $\alpha \geq \alpha_{add}(\zeta)$ where these quantities are defined in Definition 4.1 of [43].*
2. *L_t satisfies Model 10.1 or Model 10.2 with*

   (a) *$0 \leq c_{j,\text{new}} \leq c_{\max}$ for all j (thus $r_j \leq r_{\max} := r_0 + Jc_{\max}$),*

   (b) *the a_t's mutually independent over t,*

   (c) *$\|a_t\|_\infty \leq \gamma_*$ for all t (a_t's bounded),*

3. *Slow subspace change holds: (10.3) holds with* $d = K\alpha$; *(10.4) holds with* $g^+ = 1.41$ *and (10.5) holds with* $v = 1.2$; *and* c *and* γ_{new} *are small enough so that* $14\xi_0(\zeta) \le S_{\min}$.

4. *Denseness holds: Equation (10.7) holds with* $\kappa^+_{2s,*} = 0.3$ *and* $\kappa^+_{2s,\text{new}} = 0.15$.

5. *The matrices* $D_{j,\text{new},k} := (I - \hat{P}_{j-1}\hat{P}'_{j-1} - \hat{P}_{j,\text{new},k}\hat{P}'_{j,\text{new},k})P_{j,\text{new}}$ *and* $Q_{j,\text{new},k} := (I - P_{j,\text{new}}P_{j,\text{new}}{}')\hat{P}_{j,\text{new},k}$ *satisfy*

$$\max_j \max_{1\le k\le K} \kappa_s(D_{j,\text{new},k}) \le \kappa^+_s := 0.152$$
$$\max_j \max_{1\le k\le K} \kappa_{2s}(Q_{j,\text{new},k}) \le \tilde{\kappa}^+_{2s} := 0.15;$$

then, with probability at least $(1 - n^{-10})$, *at all times,* t, *all of the following hold:*

1. *at all times,* t,

$$\hat{\mathcal{T}}_t = \mathcal{T}_t \quad \textit{and}$$

$$\|e_t\|_2 = \|L_t - \hat{L}_t\|_2 = \|\hat{S}_t - S_t\|_2 \le 0.18\sqrt{c}\gamma_{\text{new}} + 1.2\sqrt{\zeta}(\sqrt{r} + 0.06\sqrt{c}).$$

2. *the subspace error* $SE_{(t)} := \|(I - \hat{P}_{(t)}\hat{P}'_{(t)})P_{(t)}\|_2$ *satisfies*

$$SE_{(t)} \le \begin{cases} (r_0 + (j-1)c)\zeta + 0.4c\zeta + 0.6^{k-1} \\ \qquad\qquad\qquad \textit{if } t \in \mathcal{I}_{j,k},\ k = 1,2\ldots K \\ (r_0 + jc)\zeta \qquad \textit{if } t \in \mathcal{I}_{j,K+1} \end{cases}$$
$$\le \begin{cases} 10^{-2}\sqrt{\zeta} + 0.6^{k-1} \\ \qquad\qquad\qquad \textit{if } t \in \mathcal{I}_{j,k},\ k = 1,2\ldots K \\ 10^{-2}\sqrt{\zeta} \qquad \textit{if } t \in \mathcal{I}_{j,K+1} \end{cases}$$

3. *the error* $e_t = \hat{S}_t - S_t = L_t - \hat{L}_t$ *satisfies the following at various times*

$$\|e_t\|_2 \le \begin{cases} 0.18\sqrt{c}0.72^{k-1}\gamma_{\text{new}} + \\ \qquad 1.2(\sqrt{r} + 0.06\sqrt{c})(r_0 + (j-1)c)\zeta\gamma_* \\ \qquad\qquad\qquad \textit{if } t \in \mathcal{I}_{j,k},\ k = 1,2\ldots K \\ 1.2(r_0 + jc)\zeta\sqrt{r}\gamma_* \quad \textit{if } t \in \mathcal{I}_{j,K+1} \end{cases}$$
$$\le \begin{cases} 0.18\sqrt{c}0.72^{k-1}\gamma_{\text{new}} + 1.2(\sqrt{r} + 0.06\sqrt{c})\sqrt{\zeta} \\ \qquad\qquad\qquad \textit{if } t \in \mathcal{I}_{j,k},\ k = 1,2\ldots K \\ 1.2\sqrt{r}\sqrt{\zeta} \quad \textit{if } t \in \mathcal{I}_{j,K+1} \end{cases}$$

The above result says the following. Consider Algorithm 10.2. Assume that the initial subspace error is small enough. If the algorithm parameters are appropriately set, if slow subspace change holds, if the subspaces are dense, if the condition number of $\text{Cov}[a_{t,\text{new}}]$ is small enough, and if the currently unestimated part of the newly added subspace is dense enough (this is an assumption on the algorithm estimates), then, w.h.p., we will get exact support recovery at all times. Moreover, the sparse recovery error will always be bounded by $0.18\sqrt{c}\gamma_{\text{new}}$ plus a constant times $\sqrt{\zeta}$. Since ζ is very small, $\gamma_{\text{new}} < S_{\min}$, and c is also small, the normalized reconstruction error for recovering S_t will be small at all times. In the second conclusion, we bound the subspace estimation error, $\text{SE}_{(t)}$. When a subspace change occurs, this error is initially bounded by one. The above result shows that, w.h.p., with each projection PCA step, this error decays exponentially and falls below $0.01\sqrt{\zeta}$ within K projection PCA steps. The third conclusion shows that, with each projection PCA step, w.h.p., the sparse recovery error as well as the error in recovering L_t also decay in a similar fashion.

The most important limitation of the above result is that it requires an assumption on $D_{\text{new},k}$ and $Q_{\text{new},k}$, which depend on intermediate algorithm estimates, $\hat{P}_{j,\text{new},k}$. Moreover, it studies an algorithm that requires knowledge of model parameters. The first limitation was removed in our recent work [31] and both limitations were removed in our very recent work [32]. We briefly summarize the result of [32] using the notation and model of this book chapter (this is the notation and model of our earlier published work [43] and has minor differences from the notation and model used in [32]). The key point is to replace condition 5 (which is an assumption that depends on intermediate algorithm estimates, $\hat{P}_{j,\text{new},k}$) by an assumption on support change of S_t. The new assumption just says that the support should change at least every β frames and when it does, the change should be by at least s/ρ (with $\rho^2\beta$ being small enough) and it should be such that the sets of changed indices are mutually disjoint for a period of α frames.

THEOREM 10.4 *[Complete Correctness Result for ReProCS [32]] Consider Algorithm 10.3 but with sparse recovery done and only simple ℓ_1 minimization and support recovery done by thresholding; and with ξ, ω, α, and K set as given below.*

1. *Replace condition 1) of Theorem 10.3 by the following: set algorithm parameters as $\xi = \sqrt{r_{\text{new}}}\gamma_{\text{new}} + (\sqrt{r} + \sqrt{r_{\text{new}}})\sqrt{\zeta}$; $\omega = 7\xi$; $K = \left\lceil \frac{\log(0.16 r_{\text{new}}\zeta)}{\log(0.83)} \right\rceil$; and $\alpha = C(\log(6(K+1)J) + 11\log(n))$ for a constant $C \geq C_{add}$ with $C_{add} := 32 \cdot 100^2 \frac{\max\{16, 1.2(\sqrt{\zeta}+\sqrt{r_{\text{new}}}\gamma_{\text{new}})^4\}}{(r_{\text{new}}\zeta\lambda^-)^2}$.*

2. *Assume conditions 2), 3), 4) of Theorem 10.3 hold.*

3. *Replace condition 5) of Theorem 10.3 by the following assumption on support change: Let t^k, with $t^k < t^{k+1}$, denote the times at which T_t changes and let $\mathcal{T}^{[k]}$ denote the distinct sets.*

   (a) Assume that $\mathcal{T}_t = \mathcal{T}^{[k]}$ for all times $t \in [t^k, t^{k+1})$ with $(t^{k+1} - t^k) < \beta$ and $|\mathcal{T}^{[k]}| \leq s$.

   (b) Let ρ be a positive integer so that for any k,

   $$\mathcal{T}^{[k]} \cap \mathcal{T}^{[k+\rho]} = \emptyset;$$

   assume that

   $$\rho^2\beta \leq 0.01\alpha.$$

   (c) For any k,

   $$\sum_{i=k+1}^{k+\alpha} \left|\mathcal{T}^{[i]} \setminus \mathcal{T}^{[i+1]}\right| \leq n$$

   and for any $k < i \leq k+\alpha$,

   $$(\mathcal{T}^{[k]} \setminus \mathcal{T}^{[k+1]}) \cap (\mathcal{T}^{[i]} \setminus \mathcal{T}^{[i+1]}) = \emptyset.$$

   (One way to ensure $\sum_{i=k+1}^{k+\alpha} |\mathcal{T}^{[i]} \setminus \mathcal{T}^{[i+1]}| \leq n$ is to require that for all i, $|\mathcal{T}^{[i]} \setminus \mathcal{T}^{[i+1]}| \leq \frac{s}{\rho_2}$ with $\frac{s}{\rho_2}\alpha \leq n$.)

Then all conclusions of Theorem 10.3 hold.

One example application that satisfies the above model on support change is a video application consisting of a foreground with one object of length s or less that can remain static for at most β frames at a time. When it moves, it moves *downwards* (or upwards, but always in one direction) by at least s/ρ pixels, and at most s/ρ_2 pixels. Once it reaches the bottom of the scene, it disappears. The maximum motion is such that, if the object were to move at each frame, it still does not go from the top to the bottom of the scene in a time interval of length α, i.e. $\frac{s}{\rho_2}\alpha \leq n$. Anytime after it has disappeared, another object could appear.

10.5.2 Modified-PCP Correctness Result: Static and Online Robust PCA Cases

Modified-PCP correctness result for the static case

Consider the modified PCP program given in (10.20). As explained in [9], we need that $\mathbf{S}$ is not low rank in order to separate it from $\mathbf{L}_{\text{new}}$ in a batch fashion (recall that modified PCP is a batch program; modified-PCP for online robust PCA is a piecewise batch solution). One way to ensure that $\mathbf{S}$ is full-rank w.h.p. is by selecting the support of $\mathbf{S}$ uniformly at random [9]. We assume this here, too. In addition, we need a denseness assumption on the columns of $\mathbf{G}$ and on the left and right singular vectors of $\mathbf{L}_{\text{new}}$.

Let $n_{(1)} = \max(n_1, n_2)$ and $n_{(2)} = \min(n_1, n_2)$. Assume that the following holds, with a constant ρ_r

$$\max_i \|[\mathbf{G}\ \mathbf{U}_{\text{new}}]'\boldsymbol{e}_i\|^2 \leq \frac{\rho_r n_{(2)}}{n_1 \log^2 n_{(1)}}, \tag{10.23}$$

$$\max_i \|\mathbf{V}'_{\text{new}}\boldsymbol{e}_i\|^2 \leq \frac{\rho_r n_{(2)}}{n_2 \log^2 n_{(1)}}, \tag{10.24}$$

and

$$\|\mathbf{U}_{\text{new}}\mathbf{V}'_{\text{new}}\|_\infty \leq \sqrt{\frac{\rho_r}{n_{(1)} \log^2 n_{(1)}}}. \tag{10.25}$$

THEOREM 10.5 *Consider the problem of recovering* $\mathbf{L}$ *and* $\mathbf{S}$ *from* $\mathbf{M}$ *using partial subspace knowledge* $\mathbf{G}$ *by solving modified-PCP (10.20). Assume that* Ω*, the support set of* $\mathbf{S}$*, is uniformly distributed with size* m *satisfying*

$$m \leq 0.4\rho_s n_1 n_2. \tag{10.26}$$

Assume that $\mathbf{L}$ *satisfies (10.23), (10.24), and (10.25) and* $\rho_s, \rho_r,$ *are small enough and* n_1, n_2 *are large enough. The explicit bounds are available in Assumption 3.2 of [55]. Then, Modified PCP (10.20) with* $\lambda = 1/\sqrt{n_{(1)}}$ *recovers* $\mathbf{S}$ *and* $\mathbf{L}$ *exactly with probability at least* $1 - 23n_{(1)}^{-10}$.

Modified-PCP correctness result for online robust PCA

Consider the online/recursive robust PCA problem where data vectors $M_t := S_t + L_t$ come in sequentially and their subspace can change over time. Starting with an initial knowledge of the subspace, the goal is to estimate the subspace spanned by $L_1, L_2, \ldots L_t$ and to recover the S_t's. For the above model, the following is an easy corollary.

COROLLARY 10.1 [modified PCP for online robust PCA] Assume Model 10.2 holds with $\sum_j (c_{j,\text{new}} - c_{j,\text{old}}) \leq c_{dif}$. Let $\mathbf{M}^j := [M_{t_j}, M_{t_j+1}, \ldots M_{t_{j+1}-1}]$, $\mathbf{L}^j := [L_{t_j}, L_{t_j+1}, \ldots L_{t_{j+1}-1}]$, $\mathbf{S}^j := [S_{t_j}, S_{t_j+1}, \ldots S_{t_{j+1}-1}]$ and let $\mathbf{L}^{\text{full}} := [\mathbf{L}^1, \mathbf{L}^2, \ldots \mathbf{L}^J]$ and $\mathbf{S}^{\text{full}} := [\mathbf{S}^1, \mathbf{S}^2, \ldots \mathbf{S}^J]$. Suppose that the following hold.

1. $\mathbf{S}^{\text{full}}$ satisfies the assumptions of Theorem 10.5.

2. The initial subspace span$(\mathbf{P}_0)$ is exactly known, i.e., we are given $\hat{\mathbf{P}}_0$ with span$(\hat{\mathbf{P}}_0) =$ span$(\mathbf{P}_0)$.

3. For all $j = 1, 2, \ldots J$, (10.23), (10.24), and (10.25) hold with $n_1 = n$, $n_2 = t_{j+1} - t_j$, $\mathbf{G} = \mathbf{P}_{j-1}$, $\mathbf{U}_{\text{new}} = \mathbf{P}_{j,\text{new}}$ and $\mathbf{V}_{\text{new}}$ being the matrix of right singular vectors of $\mathbf{L}_{\text{new}} = (\mathbf{I} - \mathbf{P}_{j-1}\mathbf{P}_{j-1}')\mathbf{L}^j$.

4. We solve modified PCP at every $t = t_{j+1}$, using $\mathbf{M} = \mathbf{M}^j$ and with $\mathbf{G} = \mathbf{G}_j = \hat{\mathbf{P}}_{j-1}$ where $\hat{\mathbf{P}}_{j-1}$ is the matrix of left singular vectors of the reduced SVD of $\hat{\mathbf{L}}_{j-1}$ (the low-rank matrix obtained from modified-PCP on $\mathbf{M}^{j-1}$). At $t = t_1$ we use $\mathbf{G} = \hat{\mathbf{P}}_0$.

Then, modified PCP recovers $\mathbf{S}^{\text{full}}, \mathbf{L}^{\text{full}}$ exactly and in a piecewise batch fashion with probability at least $(1 - 23n^{-10})^J$.

Discussion w.r.t. PCP. Two possible corollaries for PCP can be stated depending on whether we want to compare PCP and mod-PCP when both have the same memory and time complexity or when both are provided the same total data.

The following is the corollary for PCP applied to the entire data matrix $\mathbf{M}^{\text{full}}$ in one go. With doing this, PCP gets all the same data that modified PCP has access to. But PCP needs to store the entire data matrix in memory and also needs to operate on it, i.e. its memory complexity is roughly J times larger than that for modified PCP and its time complexity is $poly(J)$ times larger than that of modified PCP.

COROLLARY 10.2 [PCP for online robust PCA] Assume Model 10.2 holds with $\sum_j (c_{j,\text{new}} - c_{j,\text{old}}) \leq c_{dif}$. If $\mathbf{S}^{\text{full}}$ satisfies the assumptions of Theorem 10.5 and if (10.23), (10.24), and (10.25) hold with $n_1 = n$, $n_2 = t_{J+1} - t_1$, $\mathbf{G}_{PCP} = [\ ]$, $\mathbf{U}_{\text{new},PCP} = \mathbf{U} = [\mathbf{P}_0, \mathbf{P}_{1,\text{new}}, \ldots \mathbf{P}_{J,\text{new}}]$ and $\mathbf{V}_{\text{new},PCP} = \mathbf{V}$ being the right singular vectors of $\mathbf{L}^{\text{full}} := [\mathbf{L}^1, \mathbf{L}^2, \ldots \mathbf{L}^J]$, then, we can recover $\mathbf{L}^{\text{full}}$ and $\mathbf{S}^{\text{full}}$ exactly with probability at least $(1 - 23n^{-10})$ by solving PCP (10.1) with input $\mathbf{M}^{\text{full}}$. Here $\mathbf{M}^{\text{full}} := \mathbf{L}^{\text{full}} + \mathbf{S}^{\text{full}}$.

When we compare this with the result for modified PCP, the second and third condition are clearly significantly weaker than those for PCP. The first conditions cannot be easily compared. The LHS contains at most $r_{\max} + c = r_0 + c_{dif} + c$ columns for modified PCP, while it contains $r_0 + Jc$ columns for PCP. However, the RHS for PCP is also larger. If $t_{j+1} - t_j = d$, then the RHS is also J times larger for PCP than for modified PCP. Thus with the above PCP corollary, the first condition of the above and of modified PCP cannot be compared.

The above advantages for mod-PCP come with two caveats. First, modified PCP assumes knowledge of the subspace change times while PCP does not need this. Secondly, modified PCP succeeds w.p. $(1 - 23n^{-10})^J \geq 1 - 23Jn^{-10}$ while PCP succeeds w.p. $1 - 23n^{-10}$.

Alternatively if PCP is solved at every $t = t_{j+1}$ using $\mathbf{M}^j$, we get the following corollary. This keeps the memory and time complexity of PCP the same as that for modified PCP, but PCP gets access to less data than modified PCP.

COROLLARY 10.3 [PCP for $\mathbf{M}^j$] Assume Model 10.2 holds with $\sum_j (c_{j,\text{new}} - c_{j,\text{old}}) \leq c_{dif}$. Solve PCP, i.e., (10.1), at $t = t_{j+1}$ using $\mathbf{M}^j$. If $\mathbf{S}^{\text{full}}$ satisfies the assumptions of Theorem 10.5 and if (10.23), (10.24), and (10.25) hold with $n_1 = n$, $n_2 = t_{j+1} - t_j$, $\mathbf{G}_{PCP} = [\ ]$, $\mathbf{U}_{\text{new},PCP} = \mathbf{P}_j$ and $\mathbf{V}_{\text{new},PCP} = \mathbf{V}_j$ being the right singular vectors of $\mathbf{L}^j$ for all $j = 1, 2, \ldots, J$, then, we can recover $\mathbf{L}^{\text{full}}$ and $\mathbf{S}^{\text{full}}$ exactly with probability at least $(1 - 23n^{-10})^J$.

When we compare this with modified PCP, the second and third condition are significantly weaker than those for PCP when $c_{j,\text{new}} \ll r_j$. The first condition is exactly the same when $c_{j,\text{old}} = 0$ and is only slightly stronger as long as $c_{j,\text{old}} \ll r_j$.

10.5.3 Discussion

In this discussion we use the correctness result of Theorem 10.4 for ReProCS. This was proved in our very recent work [32]. We should point out first that the matrix completion problem can be interpreted as a special case of the robust PCA problem where the support sets T_t are the set of missing entries and hence are known. Thus an easy corollary of our result is a result for online matrix completion. This can be compared with the corresponding result for nuclear norm minimization (NNM) [15] from [4].

Our result requires accurate initial subspace knowledge. As explained earlier, for video analytics, this corresponds to requiring an initial short sequence of background-only video frames whose subspace can be estimated via SVD (followed by using a singular value threshold to retain a certain number of top-left singular vectors). Alternatively, if an initial short sequence of the video data satisfies the assumptions required by a batch method such as PCP, that can be used to estimate the low-rank part, followed by SVD to get the column subspace.

In Model 10.1 or 10.2 and the slow subspace change assumption, we are placing a slow increase assumption on the eigenvalues along the new directions, $\mathbf{m}P_{t_j,\text{new}}$, only for the interval $[t_j, t_{j+1})$. Thus, after t_{j+1}, the eigenvalues along $\mathbf{m}P_{t_j,\text{new}}$ can increase gradually or suddenly to any large value up to λ^+.

The assumption on T_t is a practical model for moving foreground objects in video. We should point out that this model is one special case of the general set of conditions that we need (see Model 5.1 of [32]). As explained in [32], the model on T_t and the denseness condition of the theorem constrain s and $s, r_0, r_{\text{new}}, J$ respectively. The model on T_t requires $s \leq \rho_2 n/\alpha$ for a constant ρ_2. Using the expression for α, it is easy to see that as long as $J \in \mathcal{O}(n)$, we have $\alpha \in \mathcal{O}(\log n)$ and so this needs $s \in \mathcal{O}(\frac{n}{\log n})$. With $s \in \mathcal{O}(\frac{n}{\log n})$, using (10.9), it is easy to see that the denseness condition will hold if $r_0 \in \mathcal{O}(\log n)$, $J \in \mathcal{O}(\log n)$, and r_{new} is a constant. This is one set of sufficient conditions that we allow on the rank-sparsity product. Let $\mathbf{m}L^{\text{full}} := [L_1, L_2, \ldots, L_{t_{\max}}]$ and $\mathbf{m}S^{\text{full}} := [S_1, S_2, \ldots, S_{t_{\max}}]$. Let $r_{\text{mat}} := \text{rank}(\mathbf{m}L^{\text{full}})$. Clearly $r_{\text{mat}} \leq r_0 + Jr_{\text{new}}$ and the bound is tight. Let $s_{\text{mat}} := t_{\max} s$ be a bound on the support size of the outliers' matrix $\mathbf{m}S^{\text{full}}$. In terms of r_{mat} and s_{mat}, what we need is $r_{\text{mat}} \in \mathcal{O}(\log n)$ and $s_{\text{mat}} \in \mathcal{O}(\frac{nt_{\max}}{\log n})$. This is stronger than what the PCP result from [9] needs ([9] allows $r_{\text{mat}} \in \mathcal{O}\left(\frac{n}{(\log n)^2}\right)$ while allowing $s_{\text{mat}} \in \mathcal{O}(nt_{\max})$), but is similar to what the PCP results from [11, 25] need. Other disadvantages of our result are as follows. (1) Our result needs accurate initial subspace knowledge and slow subspace change of L_t. As explained earlier and in [43, Fig. 6], both of these are often practically valid for video analytics applications. Moreover, we also need the L_t's to be zero mean and mutually independent over time. Zero mean is achieved by letting L_t be the background image at time

t with an empirical 'mean background image,' computed using the training data, subtracted out. The independence assumption then models independent background variations around a common mean. As we explain in Section 10.7, this can be easily relaxed and we can get a result very similar to the current one under a first-order autoregressive model on the L_t's. (2) Moreover, ReProCS needs four algorithm parameters to be appropriately set. The PCP or NNM results need this for non [4, 9] or at most one [11, 25] algorithm parameter. (3) Thirdly, our result for online RPCA also needs a lower bound on $S_{\min}$ while the PCP results do not need this. (4) Moreover, even with this, we can only guarantee accurate recovery of L_t, while PCP or NNM guarantee exact recovery. (1) The advantage of our work is that we analyze an online algorithm (ReProCS) that is faster and needs less storage compared with PCP or NNM. It needs to store only a few $n \times \alpha$ or $n \times r_{\text{mat}}$ matrices, thus the storage complexity is $\mathcal{O}(n \log n)$ while that for PCP or NNM is $\mathcal{O}(nt_{\max})$. In general $t_{\max}$ can be much larger than $\log n$. (2) Moreover, we do not need any assumption on the right singular vectors of $\mathbf{m}L$ while all results for PCP or NNM do. (3) Most importantly, our results allow highly correlated changes of the set of missing entries (or outliers). From the assumption on T_t, it is easy to see that we allow the number of missing entries (or outliers) per row of $\mathbf{m}L$ to be $\mathcal{O}(t_{\max})$ as long as the sets follow the support change assumption. [30] The PCP results from [11, 25] need this number to be $\mathcal{O}(\frac{t_{\max}}{r_{\text{mat}}})$ which is stronger. The PCP result from [9] or the NNM result [4] need an even stronger condition–they need the set $(\cup_{t=1}^{t_{\max}} T_t)$ to be generated uniformly at random.In [17], Feng et. al. propose a method for online RPCA and prove a partial result for their algorithm. The approach is to reformulate the PCP program and use this reformulation to develop a recursive algorithm that converges asymptotically to the solution of PCP as long as the basis estimate $\hat{\mathbf{m}P}_t$ is full rank at each time t. Since this result assumes something about the algorithm estimates, it is also only a *partial* result. Another somewhat related work is that of Feng et. al. [16] on online PCA with contaminated data. This does not model the outlier as a sparse vector but defines anything that is far from the data subspace as an outlier. To compare with the correctness result of modified PCP given earlier, two things can be said. First, like PCP, the result for modified PCP also needs uniformly randomly generated support sets. But its advantage is that its assumption on the rank-sparsity product is weaker than that of PCP, and hence weaker than that needed by ReProCS. Also see the simulations' section.

10.6 Numerical Experiments

10.6.1 Simulation Experiments

Comparing ReProCS and PCP

We first provide some simulations that demonstrate the result we have proven above for ReProCS and how it compares with the result for PCP. The data for Figure 10.3(a) was generated as follows. We chose $n = 256$ and $t_{\max} = 15,000$. Each measurement had $s = 20$ missing or corrupted entries, i.e., $|T_t| = 20$. Each non-zero entry of the sparse vector was drawn uniformly at random between 2 and 6 independent of other entries and other times t. In Fig. 10.3(a) the support of S_t changes as assumed Theorem 10.4 with $\rho = 2$ and $\beta = 18$. So T_t changes by $\frac{s}{2} = 10$ indices every 18 time instants. When it reaches the bottom of the

[30] In a period of length α, the set T_t can occupy index i for at most $\rho\beta$ time instants, and this pattern is allowed to repeat every α time instants. So an index can be in the support for a total of $\rho\beta\frac{t_{\max}}{\alpha}$ time instants and the model assumes $\rho\beta \leq \frac{0.01\alpha}{\rho}$ for a constant ρ.

vector, it starts over again at the top. This pattern can be seen in the bottom half of the figure, which shows the sparsity pattern of the matrix $\mathbf{m}S$.

To form the low-dimensional vectors L_t, we started with an $n \times r$ matrix of i.i.d. Gaussian entries and orthonormalized the columns using Gram-Schmidt. The first $r_0 = 10$ columns of this matrix formed $\mathbf{m}P_{(0)}$, the next 2 columns formed $\mathbf{m}P_{(1),\text{new}}$, and the last 2 columns formed $\mathbf{m}P_{(2),\text{new}}$ We show two subspace changes that occur at $t_1 = 600$ and $t_2 = 8000$. The entries of $\mathbf{m}a_{t,*}$ were drawn uniformly at random between -5 and 5, and the entries of $\mathbf{m}a_{t,\text{new}}$ were drawn uniformly at random between $-\sqrt{3v_i^{t-t_j}\lambda^-}$ and $\sqrt{3v_i^{t-t_j}\lambda^-}$ with $v_i = 1.00017$ and $\lambda^- = 1$. Entries of $\mathbf{m}a_t$ were independent of each other and of the other $\mathbf{m}a_t$'s.

For this simulated data we compare the performance of ReProCS and PCP. The plots show the relative error in recovering L_t, that is $\|L_t - \hat{L}_t\|_2 / \|L_t\|_2$. For the initial subspace estimate $\hat{\mathbf{m}P_0}$, we used $\mathbf{m}P_0$ plus some small Gaussian noise and then obtained orthonormal columns. We set $\alpha = 800$ and $K = 6$. For the PCP algorithm, we perform the optimization every α time instants using all of the data up to that point. So the first time, PCP is performed on $[M_1, \ldots, M_\alpha]$, and the second time it is performed on $[M_1, \ldots, M_{2\alpha}]$, and so on.

Figure 10.3(a) illustrates the result we have proven in Theorem 10.4. That is, ReProCS takes advantage of the initial subspace estimate and slow subspace change (including the bound on γ_{new}) to handle the case when the supports of S_t are correlated in time. Notice how the ReProCS error increases after a subspace change, but decays exponentially with each projection PCA step. For this data, the PCP program fails to give a meaningful estimate for all but a few times. The average time taken by the ReProCS algorithm was 52 seconds, while PCP averaged over 5 minutes. Simulations were coded in MATLAB and run on a desktop computer with a 3.2 GHz processor.

Compare this to Figure 10.3(b) where the only change in the data is that the support of S is chosen uniformly at random from all sets of size $\frac{st_{\max}}{n}$ (as assumed in [9]). Thus the total sparsity of the matrix S is the same for both figures. In Figure 10.3(b), ReProCS performs almost the same as in Fig. 10.3(a), while PCP does substantially better than in the case of correlated supports.

Comparing Modified PCP, ReProCS, PCP and other algorithms

We generated data using Model 10.2 with $n = 256$, $J = 3$, $r_0 = 40$, $t_0 = 200$ and $c_{j,\text{new}} = 4$, $c_{j,\text{old}} = 4$, for each $j = 1, 2, 3$. We used $t_1 = t_0 + 6\alpha + 1$, $t_2 = t_0 + 12\alpha + 1$ and $t_3 = t_0 + 18\alpha + 1$ with $\alpha = 100$, $t_{\max} = 2600$ and $\gamma = 5$. The coefficients, $\mathbf{a}_{t,*} = \mathbf{P}'_{j-1} L_t$, were i.i.d. uniformly distributed in the interval $[-\gamma, \gamma]$; the coefficients along the new directions, $\mathbf{a}_{t,\text{new}} := \mathbf{P}'_{j,\text{new}} L_t$, generated i.i.d. uniformly distributed in the interval $[-\gamma_{\text{new}}, \gamma_{\text{new}}]$ (with a $\gamma_{\text{new}} \leq \gamma$) for the first 1700 columns after the subspace change and i.i.d. uniformly distributed in the interval $[-\gamma, \gamma]$ after that. We vary the value of γ_{new}; small values mean that "slow subspace change" required by ReProCS holds. The sparse matrix $\mathbf{S}$ was generated in two different ways to simulate uncorrelated and correlated support change. For partial knowledge, $\mathbf{G}$, we first did SVD decomposition on $[L_1, L_2, \cdots, L_{t_0}]$ and kept the directions corresponding to singular values larger than $\mathbf{E}(z^2)/9$, where $z \sim \text{Unif}[-\gamma_{\text{new}}, \gamma_{\text{new}}]$.

We solved PCP and the modified PCP every 200 frames by using the observations for the last 200 frames as the matrix $\mathbf{M}$. The ReProCS algorithm was implemented with $\alpha = 100$. The averaged sparse part errors with two different sets of parameters over 20

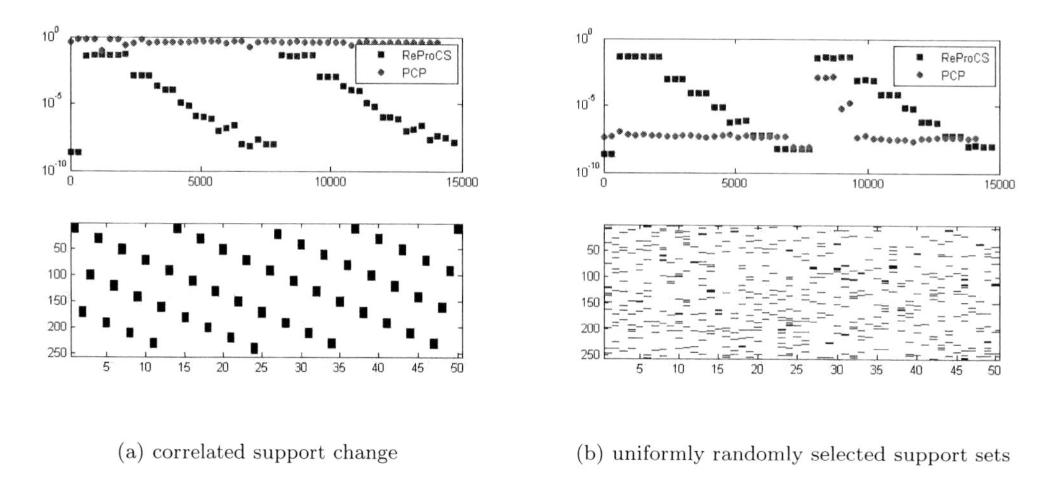

(a) correlated support change

(b) uniformly randomly selected support sets

FIGURE 10.3 Comparison of ReProCS and PCP for the RPCA problem. In each subfigure, the top plot is the relative error $\|L_t - \hat{L}_t\|_2 / \|L_t\|_2$. The bottom plot shows the sparsity pattern of $\mathbf{m}S$ (black represents a non-zero entry). Results are averaged over 100 simulations and plotted every 300 time instants.

Monte Carlo simulations are displayed in Figure 10.4(a) and Figure 10.4(b). In the first case, Figure 10.4(a), we used $\gamma_{\text{new}} = \gamma$ and so "slow subspace change" does not hold. For the sparse vectors S_t, each index is chosen to be in support with probability 0.0781. The nonzero entries are uniformly distributed between [20]. Since "slow subspace change" does not hold, ReProCS does not work well. Since the support is generated independently over time, this is a good case for both PCP and mod-PCP. Mod-PCP has the smallest sparse recovery error. In the second case, Fig. 10.4(b), we used $\gamma_{\text{new}} = 1$ and thus "slow subspace change" holds. For sparse vectors, S_t, the support is generated in a correlated fashion. We used support size $s = 10$ for each S_t; the support remained constant for 25 columns and then moved down by $s/2 = 5$ indices. Once it reached n, it rolled back over to index one. Because of the correlated support change, PCP does not work. In this case, the sparse vectors are highly correlated over time, resulting in sparse matrix $\mathbf{S}$ that is even more low rank, thus neither mod-PCP nor PCP work for this data. In this case, only ReProCS works.

Thus, from simulations, modified PCP is able to handle correlated support change better than PCP but worse than ReProCS. Modified PCP also works when slow subspace change does not hold; this is a situation where ReProCS fails. Of course, modified PCP, GRASTA, and ReProCS are provided the same partial subspace knowledge $\mathbf{G}$.

10.6.2 Video Experiments

We show comparisons on two real video sequences. These are originally taken from `http://perception.i2r.a-star.edu.sg/bk_model/bk_index.html` and `http://research.microsoft.com/en-us/um/people/jckrumm/wallflower/testimages.htm`, respectively.

The first is an indoor video of window curtains moving due to the wind. There was also some lighting variation. The latter part of this sequence also contains a foreground (various persons coming in, writing on the board and leaving). For $t > 1755$, in the foreground, a person with a black shirt walks in, writes on the board, and then walk out, then a sec-

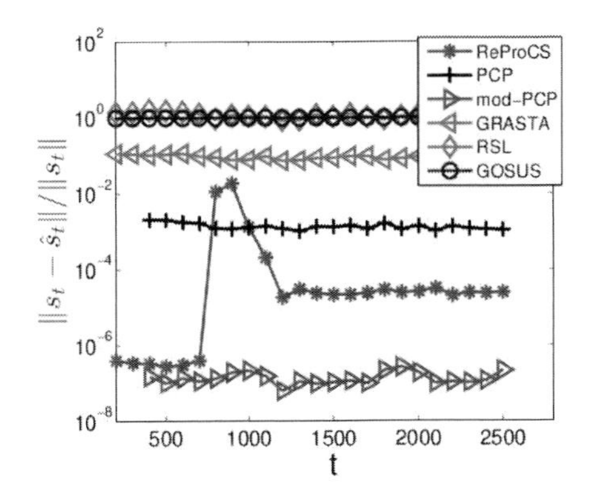

(a) uniformly distributed T_t's, slow subspace change does not hold

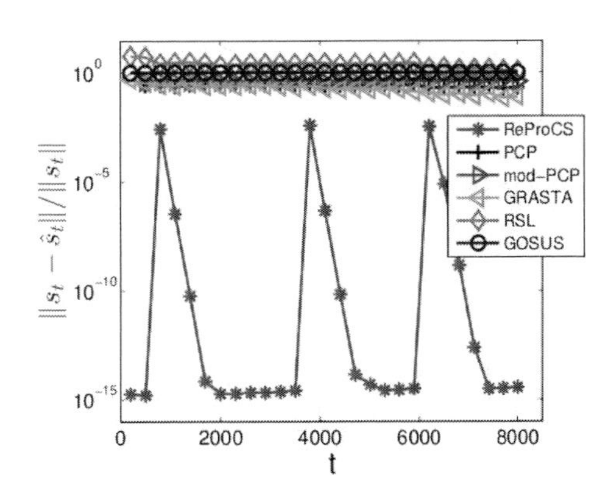

(b) correlated T_t's, slow subspace change holds

FIGURE 10.4 NRMSE of sparse part ($n = 256$, $J = 3$, $r_0 = 40$, $t_0 = 200$, $c_{j,\text{new}} = 4$, $c_{j,\text{old}} = 4$, $j = 1, 2, 3$).

ond person with a white shirt does the same, and then a third person with a white shirt does the same. This video is challenging because (i) the white shirt color and the curtains' color is quite similar, making the corresponding S_t small in magnitude; and (ii) because the background variations are quite large while the foreground person moves slowly. As can be seen from Figure 10.5, ReProCS's performance is significantly better than that of the other algorithms for both foreground and background recovery. This is most easily seen from the recovered background images. One or more frames of the background recovered by PCP, RSL, and GRASTA contains the person, while none of the ReProCS ones do.

The second sequence consists of a person entering a room containing a computer monitor that contains a white moving region. Background changes due to lighting variations and due to the computer monitor. The person moving in the foreground occupies a very large part of the image, so this is an example of a sequence in which the use of weighted ℓ_1 is essential (the support size is too large for simple ℓ_1 to work). As can be seen from Fig. 10.6, for most frames, ReProCS is able to recover the person correctly. However, for the last few frames which consist of the person in a white shirt in front of the white part of the screen, the resulting S_t is too small even for ReProCS to correctly recover. The same is true for the other algorithms.

Videos of all above experiments and of a few others are posted at `http://www.ece.iastate.edu/~hanguo/PracReProCS.html`.

Time Comparisons. The time comparisons are shown in Table 10.1. In terms of speed, GRASTA is the fastest even though its performance is much worse. ReProCS is the second fastest. We expect that ReProCS can be speeded up by using mex files (C/C++ code) for the subspace update step. PCP and RSL are slower because they jointly process the entire image sequence. Moreover, ReProCS and GRASTA have the advantage of being recursive methods, i.e., the foreground/background recovery is available as soon as a new frame appears while PCP or RSL need to wait for the entire image sequence.

Algorithm 10.3 - Practical ReProCS

Input: M_t; **Output:** $\hat{\mathcal{T}}_t$, $\hat{S}_t$, $\hat{L}_t$; **Parameters:** $q, b, \alpha, K_{\min}, K_{\max}$.
Initialization

- $[\hat{P}_0, \hat{\Sigma}_0] \leftarrow$ approx-basis$(\frac{1}{\sqrt{t_{\text{train}}}}[M_1, \dots M_{t_{\text{train}}}], b\%)$.
- Set $\hat{r} \leftarrow \text{rank}(\hat{P}_0)$, $\hat{\sigma}_{\min} \leftarrow ((\hat{\Sigma}_0)_{\hat{r},\hat{r}})$, $\hat{t}_0 = t_{\text{train}}$, flag = detect
- Initialize $\hat{P}_{(t_{\text{train}})} \leftarrow \hat{P}_0$ and $\hat{\mathcal{T}}_t \leftarrow [.]$.

For $t > t_{\text{train}}$ **do**

1. Perpendicular Projection: compute $y_t \leftarrow \Phi_{(t)} M_t$ with $\Phi_{(t)} \leftarrow I - \hat{P}_{(t-1)} \hat{P}'_{(t-1)}$
2. Sparse Recovery (Recover S_t and T_t)
   (a) If $\frac{|\hat{\mathcal{T}}_{t-2} \cap \hat{\mathcal{T}}_{t-1}|}{|\hat{\mathcal{T}}_{t-2}|} < 0.5$
      i. Compute $\hat{S}_{t,\mathcal{S}}$ as the solution of (10.13) with $\xi = \|\Phi_{(t)} \hat{L}_{t-1}\|_2$.
      ii. $\hat{\mathcal{T}}_t \leftarrow \text{Thresh}(\hat{S}_{t,\mathcal{S}}, \omega)$ with $\omega = q\sqrt{\|M_t\|^2/n}$. Here $\mathcal{T} \leftarrow \text{Thresh}(x, \omega)$ means that $\mathcal{T} = \{i : |(x)_i| \geq \omega\}$.

   Else
      i. Compute $\hat{S}_{t,\mathcal{S}}$ as the solution of (10.14) with $\mathcal{T} = \hat{\mathcal{T}}_{t-1}$, $\lambda = \frac{|\hat{\mathcal{T}}_{t-2} \setminus \hat{\mathcal{T}}_{t-1}|}{|\hat{\mathcal{T}}_{t-1}|}$, $\xi = \|\Phi_{(t)} \hat{L}_{t-1}\|_2$.
      ii. $\hat{\mathcal{T}}_{\text{add}} \leftarrow \text{Prune}(\hat{S}_{t,\mathcal{S}}, 1.4|\hat{\mathcal{T}}_{t-1}|)$. Here $\mathcal{T} \leftarrow \text{Prune}(x, k)$ returns indices of the k largest magnitude elements of x.
      iii. $\hat{S}_{t,\text{add}} \leftarrow \text{LS}(y_t, \Phi_{(t)}, \hat{\mathcal{T}}_{\text{add}})$. Here $\hat{x} \leftarrow \text{LS}(y, A, T)$ means that $\hat{x}_{\mathcal{T}} = (A_{\mathcal{T}}{}' A_{\mathcal{T}})^{-1} A_{\mathcal{T}}{}' y$ and $\hat{x}_{\mathcal{T}^c} = 0$.
      iv. $\hat{\mathcal{T}}_t \leftarrow \text{Thresh}(\hat{S}_{t,\text{add}}, \omega)$ with $\omega = q\sqrt{\|M_t\|^2/n}$.

   (b) $\hat{S}_t \leftarrow \text{LS}(y_t, \Phi_{(t)}, \hat{\mathcal{T}}_t)$
3. Estimate L_t: $\hat{L}_t \leftarrow M_t - \hat{S}_t$
4. Update $\hat{P}_{(t)}$: projection PCA
   (a) If flag = detect and $\text{mod}(t - \hat{t}_j + 1, \alpha) = 0$, (here $\text{mod}(t, \alpha)$ is the remainder when t is divided by α)
      i. Compute the SVD of $\frac{1}{\sqrt{\alpha}}(I - \hat{P}_{j-1}\hat{P}'_{j-1})[\hat{L}_{t-\alpha+1}, \dots \hat{L}_t]$ and check if any singular values are above $\hat{\sigma}_{\min}$
      ii. If the above number is more than zero then set flag $\leftarrow$ pPCA, increment $j \leftarrow j + 1$, set $\hat{t}_j \leftarrow t - \alpha + 1$, reset $k \leftarrow 1$

   Else $\hat{P}_{(t)} \leftarrow \hat{P}_{(t-1)}$.

   (b) If flag = pPCA and $\text{mod}(t - \hat{t}_j + 1, \alpha) = 0$,
      i. Compute the SVD of $\frac{1}{\sqrt{\alpha}}(I - \hat{P}_{j-1}\hat{P}'_{j-1})[\hat{L}_{t-\alpha+1}, \dots \hat{L}_t]$,
      ii. Let $\hat{P}_{j,\text{new},k}$ retain all its left-singular vectors with singular values above $\hat{\sigma}_{\min}$ or all $\alpha/3$ top-left singular vectors, whichever is smaller,
      iii. Update $\hat{P}_{(t)} \leftarrow [\hat{P}_{j-1} \ \hat{P}_{j,\text{new},k}]$, increment $k \leftarrow k + 1$
      iv. If $k \geq K_{\min}$ and $\frac{\|\sum_{t-\alpha+1}^{t}(\hat{P}_{j,\text{new},i-1}\hat{P}'_{j,\text{new},i-1} - \hat{P}_{j,\text{new},i}\hat{P}'_{j,\text{new},i})L_t\|_2}{\|\sum_{t-\alpha+1}^{t}\hat{P}_{j,\text{new},i-1}\hat{P}'_{j,\text{new},i-1}L_t\|_2} < 0.01$ for $i = k-2, k-1, k$; or $k = K_{\max}$, then $K \leftarrow k$, $\hat{P}_j \leftarrow [\hat{P}_{j-1} \ \hat{P}_{j,\text{new},K}]$ and reset flag $\leftarrow$ detect.

   Else $\hat{P}_{(t)} \leftarrow \hat{P}_{(t-1)}$.

Algorithm 10.4 - $[\hat{P}, \hat{\Sigma}]$ = inc-SVD($\hat{P}$, $\hat{\Sigma}$, D)

1. Set $D_{\|,proj} \leftarrow \hat{P}'D$ and $D_{\perp} \leftarrow (I - \hat{P}\hat{P}')D$
2. Compute QR decomposition of $D_{\perp}$, i.e., $D_{\perp} \overset{QR}{=} JK$ (here J is a *basis matrix* and K is an upper triangular matrix)
3. Compute the SVD: $\begin{bmatrix} \hat{\Sigma} & D_{\|,proj} \\ 0 & K \end{bmatrix} \overset{SVD}{=} \tilde{P}\tilde{\Sigma}\tilde{V}'$
4. Update $\hat{P} \leftarrow [\hat{P}\ J]\tilde{P}$ and $\hat{\Sigma} \leftarrow \tilde{\Sigma}$

Note: As explained in [1], due to numerical errors, step 4 done too often can eventually result in $\hat{P}$ no longer being a basis matrix. This typically occurs when one tries to use inc-SVD at every time t, i.e., when D is a column vector. This can be addressed using the modified Gram-Schmidt re-orthonormalization procedure whenever loss of orthogonality is detected [1].

Algorithm 10.5 - Algorithm for solving Modified PCP (10.20)

Require: Measurement matrix $\mathbf{M} \in \mathbb{R}^{n_1 \times n_2}$, $\lambda = 1/\sqrt{\max\{n_1, n_2\}}$, $\mathbf{G}$.

1: $\mathbf{Y}_0 = \mathbf{M}/\max\{\|\mathbf{M}\|, \|\mathbf{M}\|_\infty/\lambda\}$; $\tilde{\mathbf{S}}_0 = 0$; $\tau_0 > 0$; $v > 1$; $k = 0$.
2: **while** not converged **do**
3: $\quad \tilde{\mathbf{S}}_{k+1} = \mathfrak{S}_{\lambda\tau_k^{-1}}[\mathbf{M} - \mathbf{G}\tilde{\mathbf{X}}_k - \tilde{\mathbf{L}}_{\text{new},k} + \tau_k^{-1}\mathbf{Y}_k]$.
4: $\quad (\tilde{\mathbf{U}}, \tilde{\mathbf{\Sigma}}, \tilde{\mathbf{V}}) = \text{svd}((I - \mathbf{G}\mathbf{G}')(\mathbf{M} - \tilde{\mathbf{S}}_{k+1} + \tau_k^{-1}\mathbf{Y}_k))$;
5: $\quad \tilde{\mathbf{L}}_{\text{new},k+1} = \tilde{\mathbf{U}}\mathfrak{S}_{\tau_k^{-1}}[\tilde{\mathbf{\Sigma}}]\tilde{\mathbf{V}}^T$.
6: $\quad \tilde{\mathbf{X}}_{k+1} = \mathbf{G}'(\mathbf{M} - \tilde{\mathbf{S}}_{k+1} + \tau_k^{-1}\mathbf{Y}_k)$
7: $\quad \mathbf{Y}_{k+1} = \mathbf{Y}_k + \tau_k(\mathbf{M} - \tilde{\mathbf{S}}_{k+1} - \tilde{\mathbf{L}}_{\text{new},k+1} - \mathbf{G}\tilde{\mathbf{X}}_{k+1})$.
8: $\quad \tau_{k+1} = \min(v\tau_k, \bar{\tau})$.
9: $\quad k \leftarrow k + 1$.
10: **end while**

Ensure: $\hat{\mathbf{L}}_{\text{new}} = \tilde{\mathbf{L}}_{\text{new},k}$, $\hat{\mathbf{S}} = \tilde{\mathbf{S}}_k$, $\hat{L} = \mathbf{M} - \tilde{\mathbf{S}}_k$.

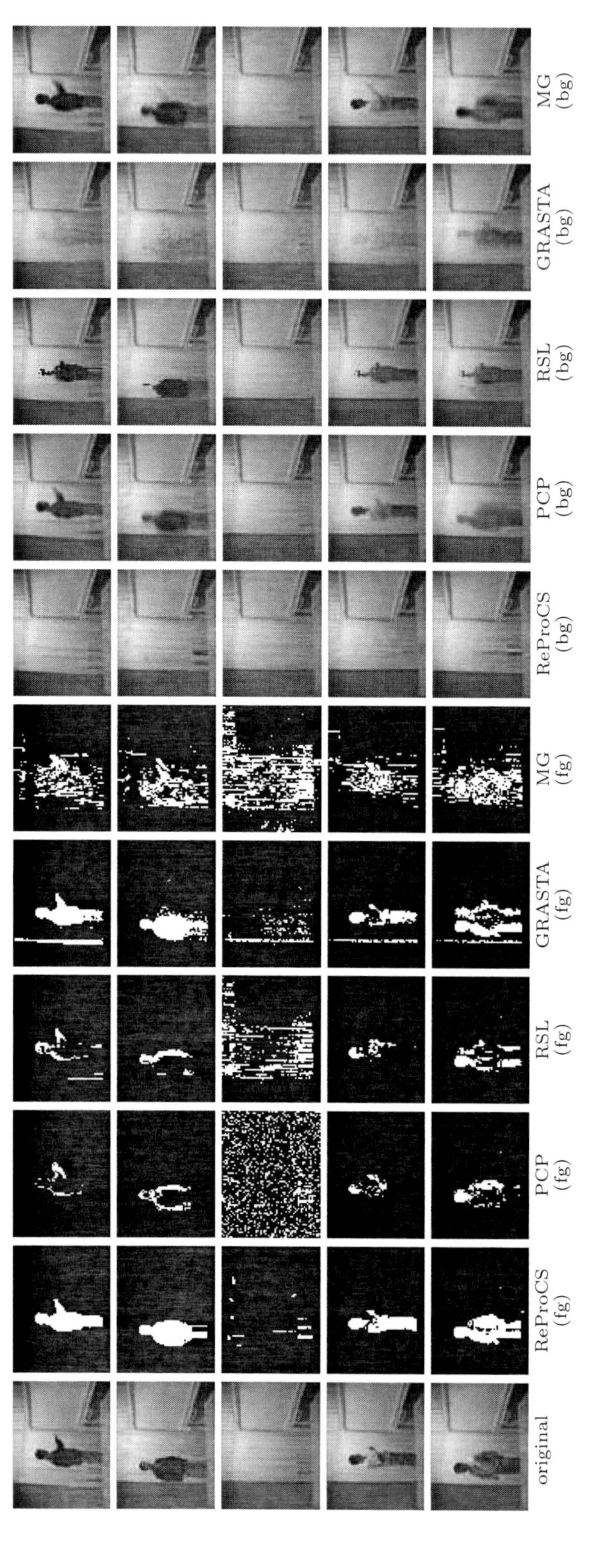

FIGURE 10.5 Original video sequence at $t = t_{\text{train}} + 60, 120, 199, 475, 1148$ and its foreground (fg) and background (bg) layer recovery results using ReProCS (ReProCS-pCA) and other algorithms. For fg, we only show the fg support in white for ease of display.

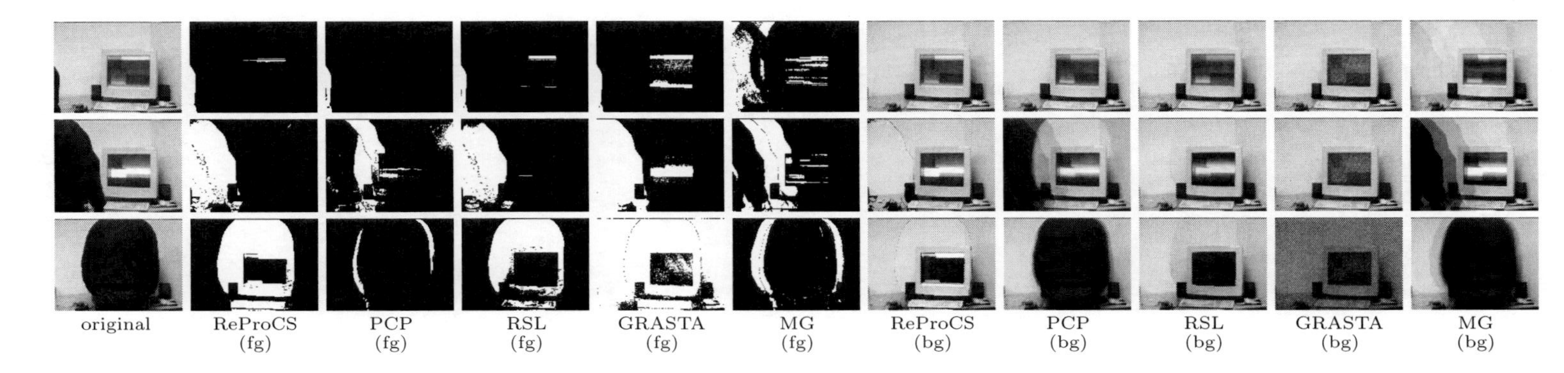

FIGURE 10.6 Original video sequence at $t = t_{\text{train}}+42, 44, 52$ and its foreground (fg) and background (bg) layer recovery results using ReProCS(ReProCS-pCA) and other algorithms. For fg, we only show the fg support in white for ease of display.

TABLE 10.1 Comparison of speed of different algorithms. Experiments were done on a 64-bit Windows 8 laptop with 2.40GHz i7 CPU and 8G RAM. Sequence length refers to the length of sequence for training plus the length of sequence for separation. For ReProCS and GRASTA, the time is shown as training time.

DataSet	Sequence Length	ReProCS-pPCA	ReProCS-Recursive-PCA	PCP	RSL	GRASTA
Lake (72×90)	$1420 + 80$	$2.99 + 19.97$ sec	$2.99 + 19.43$ sec	245.03 sec	213.36 sec	$39.47 + 0.42$ sec
Curtain (64×80)	$1755 + 1209$	$4.37 + 159.02$ sec	$4.37 + 157.21$ sec	1079.59 sec	643.98 sec	$40.01 + 5.13$ sec
Person (120×160)	$200 + 52$	$0.46 + 42.43$ sec	$0.46 + 41.91$ sec	27.72 sec	121.31 sec	$13.80 + 0.64$ sec

10.7 Conclusion

In this book chapter we provided an overview of our recent work on the online or recursive robust PCA problem. We explained both the key idea of the ReProCS algorithm proposed in our recent work to solve this problem, and how to develop its practical modification which can be directly used with real datasets. Extensive numerical experiments — both with simulated data and real videos — were shown demonstrating the advantage of ReProCS over many existing algorithms for robust PCA — both batch and recursive. We also summarized and discussed the performance guarantees for ReProCS in detail. Lastly we briefly described our work on modified PCP, which is a piecewise batch approach that removes a key limitation of ReProCS (it allows a looser bound on the rank-sparsity product as compared with ReProCS); however, it retains a key limitation of PCP (like PCP, it also cannot handle highly correlated support change as well as ReProCS). Its correctness result was also given and discussed.

The tools introduced to prove the correctness result for ReProCS in [31, 32, 43] can also be used to get a correctness result for a practical modification of ReProCS with cluster-PCA (ReProCS-cPCA), which is Algorithm 2 of [43]. This algorithm was introduced to also remove the deleted directions from the subspace estimate. It does this by re-estimating the previous subspace at a time after the newly added subspace has been accurately estimated (i.e., at a time after $\hat{t}_j + K\alpha$). A partial result for this algorithm was proved in [43]. This result will need one extra assumption – it will need the eigenvalues of the covariance matrix of L_t to be clustered for a period of time after the subspace change has stabilized, i.e., for a period of d_2 frames in the interval $[t_j + d + 1, t_{j+1} - 1]$ – but it will have a key advantage. It will need a much weaker denseness assumption and hence a much weaker bound on r or r_{mat}. In particular, with this result we expect to be able to allow $r = r_{\text{mat}} \in \mathcal{O}(n)$ with the same assumptions on s and s_{mat} that we currently allow. This requirement is almost as weak as that of PCP.

The results for ReProCS presented here assume that the L_t's are independent over time and zero mean; this is a valid model when background images have independent random variations about a fixed mean. Using the tools developed in these works, a similar result can also be obtained for the more general case of L_t's following an autoregressive model. This will allow the L_t's to be correlated over time. A partial result for this case was obtained in [53]. The main change in this case will be that we will need to apply the matrix Azuma inequality from [48] instead of matrix Hoeffding. This is will also require algebraic manipulation of sums and some other important modifications, as explained in [53], so that the constant term after conditioning on past values of the matrix is small.

We expect that the tools introduced in [31, 32, 43] can also be used to analyze the noisy case, i.e., the case of $M_t = S_t + L_t + \mathbf{m}w_t$ where $\mathbf{m}w_t$ is small bounded noise. In most practical video applications, while the foreground is truly sparse, the background is only approximately low-rank. The modeling error can be treated as as $\mathbf{m}w_t$. The proposed algorithms already apply without modification to this case (see [22] for results on real videos).

Finally, we expect both the algorithm and the proof techniques to apply with simple changes to the undersampled case $M_t = \mathbf{m}A_t S_t + \mathbf{m}B_t L_t + \mathbf{m}w_t$ as long as $\mathbf{m}B_t$ is *not* time-varying, i.e., $\mathbf{m}B_t = \mathbf{m}B_0$. A partial result for this case was obtained in [33] and experiments were shown in [22].

In other ongoing work, we are working to replace the ell-1 minimization in the projected sparse recovery step by algorithms designed under various other structured sparsity assumptions that may be valid for a given problem. For example, in [41], we introduced the support predicted modified-CS algorithm that uses a motion model to predict a moving object's support in the next frame and uses the predicted support as partial subspace knowledge for modified CS. The motion model parameters are themselves tracked using a Kalman filter.

References

1. M. Brand. Incremental singular value decomposition of uncertain data with missing values. In *Eur. Conf. on Comp. Vis. (ECCV)*, 2002.
2. J. Cai, E. J Candès, and Z. Shen. A singular value thresholding algorithm for matrix completion. *SIAM Journal on Optimization*, 20(4):1956–1982, 2010.
3. E. Candes. The restricted isometry property and its implications for compressed sensing. *Compte Rendus de l'Academie des Sciences, Paris, Serie I*, pages 589–592, 2008.
4. E. Candès and B. Recht. Exact matrix completion via convex optimization. *Foundations of Computational Mathematics*, 6(9):717–772, 2008.
5. E. Candès and B. Recht. Exact matrix completion via convex optimization. *Foundations of Computational Mathematics*, 2009.
6. E. Candes, J. Romberg, and T. Tao. Robust uncertainty principles: Exact signal reconstruction from highly incomplete frequency information. *IEEE Transactions on Information Theory*, 52(2):489–509, February 2006.
7. E. Candes and T. Tao. Decoding by linear programming. *IEEE Trans. Info. Th.*, 51(12):4203–4215, December 2005.
8. E. Candes and T. Tao. The Dantzig selector: statistical estimation when p is much larger than n. *Annals of Statistics*, 2006.
9. E. J. Candès, X. Li, Y. Ma, and J. Wright. Robust principal component analysis? *Journal of ACM*, 58(3), 2011.
10. V. Chandrasekaran, B. Recht, P. Parrilo, and A. Willsky. The convex geometry of linear inverse problems. *Foundations of Computational Mathematics*, 12(6), 2012.
11. V. Chandrasekaran, S. Sanghavi, P. Parrilo, and A. Willsky. Rank-sparsity incoherence for matrix decomposition. *SIAM Journal on Optimization*, 21, 2011.
12. S. Chen, D. Donoho, and M. Saunders. Atomic decomposition by basis pursuit. *SIAM Journal on Scientific Computing*, 20:33–61, 1998.
13. C. Davis and W. Kahan. The rotation of eigenvectors by a perturbation. iii. *SIAM Journal on Numerical Analysis*, 7:1–46, March 1970.
14. D. Donoho. Compressed sensing. *IEEE Transactions on Information Theory*, 52(4):1289–1306, April 2006.
15. M. Fazel. Matrix rank minimization with applications. *PhD thesis, Stanford University*, 2002.
16. J. Feng, H. Xu, S. Mannor, and S. Yan. Online PCA for contaminated data. In *Adv. Neural Info. Proc. Sys. (NIPS)*, 2013.
17. J. Feng, H. Xu, and S. Yan. Online robust PCA via stochastic optimization. In *Adv. Neural Info. Proc. Sys. (NIPS)*, 2013.
18. P. Feng and Y. Bresler. Spectrum-blind minimum-rate sampling and reconstruction of multiband signals. In *IEEE International Conference on Acoustics, Speech and Signal Processing, ICASSP*, volume 3, pages 1688–1691, 1996.
19. M. Friedlander, H. Mansour, R. Saab, and O. Yilmaz. Recovering compressively sampled signals using partial support information. *IEEE Transactions on Information Theory*, 58(2):1122–1134, 2012.
20. A. Ganesh, K. Min, J. Wright, and Y. Ma. Principal component pursuit with reduced linear measurements. *arXiv:1202.6445*, 2012.
21. I. Gorodnitsky and B. Rao. Sparse signal reconstruction from limited data using Focuss: A re-weighted norm minimization algorithm. *IEEE Transactions on Signal Processing*,

pages 600–616, March 1997.

22. H. Guo, C. Qiu, and N. Vaswani. An online algorithm for separating sparse and low-dimensional signal sequences from their sum. *IEEE Transactions on Signal Processing*, August 2014.
23. E. T Hale, W. Yin, and Y. Zhang. Fixed-point continuation for ℓ_1-minimization: Methodology and convergence. *SIAM Journal on Optimization*, 19(3):1107–1130, 2008.
24. J. He, L. Balzano, and A. Szlam. Incremental gradient on the Grassmannian for online foreground and background separation in subsampled video. In *IEEE Conf. on Comp. Vis. Pat. Rec. (CVPR)*, 2012.
25. D. Hsu, S. Kakade, and T. Zhang. Robust matrix decomposition with sparse corruptions. *IEEE Transactions on Information Theory*, November 2011.
26. Y. Jin and B. Rao. Algorithms for robust linear regression by exploiting the connection to sparse signal recovery. In *IEEE Intl. Conf. Acoustics, Speech, Sig. Proc. (ICASSP)*, 2010.
27. A. Khajehnejad, W. Xu, A. Avestimehr, and B. Hassibi. Weighted ℓ_1 minimization for sparse recovery with prior information. *IEEE Transactions on Signal Processing*, 2011.
28. G. Li and Z. Chen. Projection-pursuit approach to robust dispersion matrices and principal components: Primary theory and Monte Carlo. *Journal of the American Statistical Association*, 80(391):759–766, 1985.
29. Y. Li, L. Xu, J. Morphett, and R. Jacobs. An integrated algorithm of incremental and robust PCA. In *IEEE International Conference on Image Procesing (ICIP)*, pages 245–248, 2003.
30. Z. Lin, M. Chen, and Y. Ma. Alternating direction algorithms for l1 problems in compressive sensing. Technical report, University of Illinois at Urbana-Champaign, November 2009.
31. B. Lois and N. Vaswani. A correctness result for online robust PCA. In *IEEE International Conference on Acoustics, Speech and Signal Processing, ICASSP*, 2015.
32. B. Lois and N. Vaswani. Online matrix completion and online robust PCA. In *IEEE International Symposium on Information ISIT*, 2015, longer version at arXiv:1503.03525 [cs.IT].
33. B. Lois, N. Vaswani, and C. Qiu. Performance guarantees for undersampled recursive sparse recovery in large but structured noise. In *GlobalSIP*, 2013.
34. M. Mardani, G. Mateos, and G. Giannakis. Recovery of low-rank plus compressed sparse matrices with application to unveiling traffic anomalies. *IEEE Transactions on Information Theory*, 2013.
35. M. McCoy and J. Tropp. Two proposals for robust PCA using semidefinite programming. *arXiv:1012.1086v3*, 2010.
36. M. McCoy and J. Tropp. Sharp recovery bounds for convex demixing, with applications. *arXiv:1205.1580*, 2012.
37. K. Mitra, A. Veeraraghavan, and R. Chellappa. A robust regression using sparse learning for high dimensional parameter estimation problems. In *IEEE Intl. Conf. Acous. Speech. Sig.Proc.(ICASSP)*, 2010.
38. B. Nadler. Finite sample approximation results for principal component analysis: A matrix perturbation approach. *The Annals of Statistics*, 36(6), 2008.
39. C. Qiu and N. Vaswani. Real-time robust principal components' pursuit. In *Allerton Conference on Communications, Control and Computing*, 2010.
40. C. Qiu and N. Vaswani. Recursive sparse recovery in large but correlated noise. In *Allerton Conference on Communication Control and Computing*, 2011.
41. C. Qiu and N. Vaswani. Support-Predicted Modified-CS for Principal Components' Pursuit. In *IEEE Intl. Symp. on Information Theory (ISIT)*, 2011.
42. C. Qiu and N. Vaswani. Recursive sparse recovery in large but structured noise – part 2. In *IEEE International Symposium on Information ISIT*, 2013.

43. C. Qiu, N. Vaswani, B. Lois, and L. Hogben. Recursive robust PCA or recursive sparse recovery in large but structured noise. *IEEE Trans. Info. Th.*, August 2014, early versions in ICASSP 2013 and ISIT 2013.
44. S. Roweis. EM algorithms for PCA and SPCA. *Advances in Neural Information Processing Systems*, pages 626–632, 1998.
45. D. Skocaj and A. Leonardis. Weighted and robust incremental method for subspace learning. In *IEEE Intl. Conf. on Computer Vision (ICCV)*, volume 2, pages 1494–1501, October 2003.
46. Min Tao and Xiaoming Yuan. Recovering low-rank and sparse components of matrices from incomplete and noisy observations. *SIAM Journal on Optimization*, 21(1):57–81, 2011.
47. F. De La Torre and M. J. Black. A framework for robust subspace learning. *International Journal of Computer Vision*, 54:117–142, 2003.
48. J. A. Tropp. User-friendly tail bounds for sums of random matrices. *Foundations of Computational Mathematics*, 12(4), 2012.
49. N. Vaswani and W. Lu. Modified-cs: Modifying compressive sensing for problems with partially known support. *IEEE Trans. Signal Processing*, September 2010.
50. J. Wright, A. Ganesh, K. Min, and Y. Ma. Compressive principal component pursuit. *arXiv:1202.4596*, 2012.
51. J. Wright and Y. Ma. Dense error correction via l1-minimization. *IEEE Trans. on Info. Th.*, 56(7):3540–3560, 2010.
52. H. Xu, C. Caramanis, and S. Sanghavi. Robust PCA via outlier pursuit. *IEEE Transactions on Information Theory*, 58(5), May 2012.
53. J. Zhan and N. Vaswani. Performance guarantees for reprocs – correlated low-rank matrix entries case. In *IEEE International Symposium on Information ISIT*, 2014.
54. J. Zhan and N. Vaswani. Robust pca with partial subspace knowledge. In *IEEE International Symposium on Information ISIT*, 2014.
55. J. Zhan and N. Vaswani. Robust pca with partial subspace knowledge. *IEEE Transactions on Signal Processing*, 2015, to appear.
56. T. Zhang and G. Lerman. A Novel M-Estimator for Robust PCA. *arXiv:1112.4863v1*, 2011.

11

Incremental Methods for Robust Local Subspace Estimation

Paul Rodriguez
Pontificia Universidad Catolica del Peru, PUCP, Peru

Brendt Wohlberg
Los Alamos National Laboratory, USA

11.1 Introduction

As discussed in detail in the initial chapters of this book, the Robust PCA (RPCA) problem

$$\arg\min_{L,S} \|L\|_* + \lambda \|S\|_1 \ \text{ s.t. } D = L + S \,, \tag{11.1}$$

which decomposes a matrix D into a low-rank, L, and sparse component, S, has been shown to give very good performance for video background modeling, in which context the stationary background is represented by the low-rank component, and the moving foreground is represented by the sparse component. This chapter introduces two distinct types of enhancements to the standard RPCA problem, (i) the development of more computationally efficient algorithms for solving this problem (or a variant thereof), including an incremental algorithm that is able to process a single video one frame at a time, and (ii) modifying the problem form to make it invariant to transformations such as translation and rotation, so that it can be applied to video captured by a non-stationary camera. These two enhancements are combined in the final section of the chapter.

11.2 Alternating Minimization Algorithm for Robust PCA

It is useful to recall that the RPCA problem [14, Eq. 1.1] is derived as the convex relaxation of the original problem [65, Section 2]

$$\arg\min_{L,S} \ \text{rank}(L) + \lambda\|S\|_0 \ \text{s.t.} \ D = L + S \ , \tag{11.2}$$

based on decomposing matrix D such that $D = L+S$, with low-rank L and sparse S. While most RPCA algorithms, including the Augmented Lagrange Multiplier (ALM) and inexact ALM (iALM) algorithms [35, 36] are directly based on (11.1), this is not the only possible tractable problem that can be derived from (11.2). In particular, changing the constraint $D = L + S$ to a penalty and the rank penalty to an inequality constraint leads to the problem

$$\arg\min_{L,S} \frac{1}{2}\|L + S - D\|_F^2 + \lambda\|S\|_1 \quad \text{s.t.} \ \text{rank}(L) \leq r \ . \tag{11.3}$$

A computationally efficient solution can be found via an alternating optimization (AO) [7] procedure, since it seems natural to split (11.3) into a low-rank approximation followed by a shrinkage.

In what follows, it will be shown that an AO method applied to (11.3), i.e., (i) $\arg\min_L \frac{1}{2}\|L + S_0 - D\|_F^2$ s.t. $\text{rank}(L) \leq r$, (ii) $\arg\min_S \frac{1}{2}\|L + S - D\|_F^2 + \lambda\|S\|_1$, with fixed L computed in the previous iteration, not only converges to the global minimum of (11.3) but it is computationally efficient and also provides a smooth transition toward an incremental solution of (11.3). We will also include a brief description of two other works that use very closely related feasible reformulations of (11.2), and use an AO method to compute its solution.

11.2.1 Related Work

To the best of our knowledge, there are three algorithms that use an AO method to solve feasible reformulation of the direct Lagrangian reformulation of the RPCA (11.2): (i) the "Go Decomposition" (GoDec) algorithm [70]; (ii) the "Direct Robust Matrix Factorization" (DRMF) algorithm [66]; and (iii) the "Fast Alternating Minimization PCP" (amFastPCP) algorithm [48], to be described in Section 11.2.2.

GoDec algorithm

The main motivation behind the GoDec algorithm [70] is to estimate the low-rank (L) and sparse (S) approximations of matrix D, s.t. $D = L + S + E$, where E is noise. The original problem GoDec aims to solve is[31]

$$\arg\min_{L,S} \frac{1}{2}\|L + S - D\|_F^2 \quad \text{s.t.} \ \text{rank}(L) \leq r, \ \|S\|_0 \leq c \ . \tag{11.4}$$

This problem can be naively solved via the AO

$$L^{(j+1)} = \arg\min_{L}\|L + S^{(j)} - D\|_F^2 \quad \text{s.t.} \ \text{rank}(L) \leq r \tag{11.5}$$

$$S^{(j+1)} = \arg\min_{S}\|L^{(j+1)} + S - D\|_F^2 \quad \text{s.t.} \ \|S\|_0 \leq c \ . \tag{11.6}$$

[31] In [70] the number of non-zero elements in an S is denoted by $\text{card}(S)$ (cardinality), rather than the more common $\|S\|_0$ (ℓ_0 "norm").

Sub-problem (11.5) is the well-known low-rank approximation problem. Assuming that $\text{SVD}(D - S^{(j)}) = U\Lambda V^T$, then it can be proved that $L^{(j+1)} = U\Lambda_r V^T$, where $\Lambda_r = \{\lambda_1, \lambda_2, \ldots, \lambda_r, 0, \ldots, 0\}$. Sub-problem (11.6) is a cardinality constrained problem, which also has a closed form solution given by entry-wise hard-thresholding the matrix $D - L^{(j+1)}$. In order to reduce the computational cost of computing the full SVD of $D - S^{(j)}$ in the solution of (11.5), the GoDec algorithm makes use of bilateral random projections (BRP) along with a modified power method to solve it. This scheme needs to invert an auxiliary $r \times r$ matrix and then compute its full SVD. Overall in [70] it is reported that the GoDec algorithm needs $O(r^2(m + 3n + 4r) + (4q + 4)mnr)$ flops per loop, where q is a constant greater or equal to zero.

Direct Robust Matrix Factorization algorithm

The Direct Robust Matrix Factorization (DRMF) [66] algorithm also focuses on solving (11.4). As for the GoDec algorithm, DRMF also proposed to solve (11.4) via (11.5)-(11.6). However DRMF uses partial SVD, making use of the PROPACK package [33], sub-problem (11.6) being solved in the same way as for the GoDec. In [66] it is reported that the computational cost per iteration is given by $O(mn(r + \log(c)))$ flops.

11.2.2 Fast Alternating Minimization PCP (amFastPCP) Algorithm

While the GoDec and DRMF algorithms focus on solving the optimization problem described by (11.4), the amFastPCP algorithm [48] aims to solve (11.3), which, as mentioned before, can be easily solved via the AO

$$L^{(j+1)} = \arg\min_{L} \|L + S^{(j)} - D\|_F^2 \quad \text{s.t. } \text{rank}(L) \le r \tag{11.7}$$

$$S^{(j+1)} = \arg\min_{S} \quad \|L^{(j+1)} + S - D\|_F^2 + \lambda\|S\|_1, \tag{11.8}$$

which is summarized in Algorithm 11.1. It is worth noting that (11.7)–(11.8) converge to the global minimum of (11.3) since (i) (11.7) and (11.8) each have a unique global minimizer, and (ii) at each iteration of (11.7)–(11.8) the cost functional of (11.3) is reduced:

- let $f(L, S^{(j)})$ denote $\arg\min_L \|L + S^{(j)} - D\|_F^2$ s.t. $\text{rank}(L) \le r$, then

  $$f(L, S^{(j)}) \ge f(L^{(j+1)}, S^{(j)}) \ \forall L,$$

  where $L^{(j+1)}$ is the solution to (11.7); then, in particular

  $$f(L^{(j)}, S^{(j)}) \ge f(L^{(j+1)}, S^{(j)});$$

- let $f(L^{(j+1)}, S)$ denote $\arg\min_S \|L^{(j+1)} + S - D\|_F^2 + \lambda\|S\|_1$, then

  $$f(L^{(j+1)}, S) \ge f(L^{(j+1)}, S^{(j+1)}) \ \forall S,$$

  where $S^{(j+1)}$ is the solution to (11.8); then, in particular

  $$f(L^{(j+1)}, S^{(j)}) \ge f(L^{(j+1)}, S^{(j+1)});$$

then $f(L^{(j)}, S^{(j)}) \ge f(L^{(j+1)}, S^{(j)}) \ge f(L^{(j+1)}, S^{(j+1)})$. This proves that the AO method based on (11.7)–(11.8) converges to the global minimum of (11.3), since it complies with the necessary conditions for convergence of any given AO method (see [7, Section 6] for more details).

Algorithm 11.1 - Fast alternating minimization PCP [48]

1: **Inputs:** Observed video $D \in \mathbb{R}^{m \times n}$, regularization parameter λ, maximum outer loops *mLoops*, eigen-value tolerance τ.
2: **Initialization:** $S_0 = 0$, initial rank r.
3: **for** $j = 0$: mLoops **do**
4: $[U, \Sigma, V] = \text{partialSVD}(D - S^{(j)}, r)$
5: $L^{(j+1)} = U * \Sigma * V$
6: if $\frac{\sigma_r}{\sum_{l=1}^{r} \sigma_l} > \tau$ then $++r$
7: $S^{(j+1)} = \text{shrink}(D - L^{(j+1)}, \lambda)$
8: **end for**

Clearly, Algorithm 11.1 is a batch method due to the nature of sub-problem (11.7), which is a low-rank approximation that can be solved by computing the partial SVD of $D - S^{(j)}$ with r components. In Algorithm 11.1 this is related to lines 4 and 5, which require $O(m \cdot n \cdot r)$ and $O(2 \cdot m \cdot n \cdot r)$ flops per outer loop respectively. The joint memory requirement for these operations is $O(2 \cdot m \cdot n)$. The solution to (11.8) is simple element-wise shrinkage (soft thresholding): $\text{shrink}(D - L^{(j+1)}, \lambda)$, where

$$\text{shrink}(x, \epsilon) = \text{sign}(x) \max\{0, |x| - \epsilon\} . \tag{11.9}$$

Moreover, based on the observation that the video background modeling application component L typically has very low rank, [48] proposed a simple procedure to estimate an upper bound for r in (11.7), described in line 5 of Algorithm 11.1. Since the singular values of L_{k+1} are a by-product of solving (11.7), if at each outer loop we increase r by one, then we can estimate the contribution of the new singular vector; if such contribution is small enough (less than 1% of the sum all other singular values) then we stop increasing the value of r. Experimental simulations in [48, 50] showed that the rule described in Algorithm 11.1 typically increases r up to a value of 6 or less, which is consistent with the rank estimation performed by other batch PCP algorithms, such the inexact ALM, etc.

The solution obtained via the iterative solution of (11.7)–(11.8) is of comparable quality to the solution of the original PCP problem (see [48] for details), being able to deliver a useful estimate of the sparse component even after a single outer loop, while being approximately an order of magnitude faster than the inexact ALM [35] algorithm to construct a sparse component of similar quality.

11.3 Incremental Algorithm for Robust PCA

11.3.1 Related Work

To the best of our knowledge problem, (i) Recursive Projected Compressive Sensing (ReProCS) [44], (and similar ideas by the same authors [26, 43]), (ii) Grassmannian Robust Adaptive Subspace Tracking Algorithm GRASTA [29], (iii) ℓ_p-norm Robust Online Subspace Tracking pROST [52], and (iv) Grassmannian Online Subspace Updates with Structured-sparsity (GOSUS) [68] are the only RPCA-like methods that are considered to be incremental. All of these methods are only partially incremental in that they have a batch initialization, i.e., they need an initial background subspace estimate (represented by U in (11.10)), which is typically accomplished by analyzing a temporally sub-sampled version of the original dataset via a batch procedure.

Although there are some similarities in the formulation of the ReProCS and the GRASTA methods (see below), they are fundamentally different algorithms: in its original form, ReProCS can take advantage of a known model for the trajectories of the moving objects to improve the performance; however [26] proposed a modification to the original ReProCS algorithm where the optional model assumption is dropped and, more importantly, it makes use of incremental SVD procedures to adapt to slow changes in the background. On the other hand, GRASTA forces the background subspace to be a Grassmannian manifold, and updates it in an incremental fashion.

At a high level, both pROST and GOSUS are very similar to GRASTA; however, the former uses a different norm than GRASTA in order to improve its tracking capabilities and the latter imposes an additional group sparsity constraint on the moving objects.

Let $\mathbf{d}_k$, $\mathbf{l}_k$ and $\mathbf{s}_k$ denote the kth column of matrices D, L, and S (see (11.1)). Starting from $\mathbf{d}_k = \mathbf{l}_k + \mathbf{s}_k$, ReProCS and GRASTA independently proposed (for details see [44, equation after Eq. (1)], and [29, Eq. (1)])

$$\mathbf{l}_k = U\mathbf{x}_k, \qquad \mathbf{d}_k = U\mathbf{x}_k + \mathbf{s}_k \tag{11.10}$$

where noise can be added to (11.10) depending on the method. In both methods U is a $m \times m$ unknown orthonormal matrix, $\mathbf{x}_k$ is a vector of weights, and letting $\mathcal{N}_k$ denote the support of $\mathbf{x}_k$, then $P_k = (U)_{\mathcal{N}_k}$ span the low rank subspace in which the current set of $\mathbf{l}_k$ lies. Assuming P_k is known (or estimated), and $P_{k,\perp}$ is its orthonormal compliment, then ReProCS solves [45, Eq. (6)]

$$\min \|\mathbf{s}_k\|_1 \quad \text{s.t. } 0.5\|P_{k,\perp}\mathbf{s}_k - P_{k,\perp}^T\mathbf{d}_k\|_2^2 \leq \epsilon\ , \tag{11.11}$$

whereas GRASTA solves [29, Eq. (4)]

$$\min \|\mathbf{s}_k\|_1 \quad \text{s.t. } \mathbf{d}_k = U\mathbf{x}_k + \mathbf{s}_k\ . \tag{11.12}$$

ReProCS interprets (11.11) as a compressive sensing problem, whereas GRASTA solves (11.12) via the ADMM method [11]. Due to additional constraints on (11.10), which included a known model for the trajectories of the moving objects, the original ReProCS formulation becomes a non-real-time algorithm since it uses a batch method in its SVD-based initialization step. However, [26], called "practical ReProCS," drops such constraints and improves its performance by means of incremental SVD procedures and it is able to track slow changes in the background but it still needs a batch initialization.

On the other hand, GRASTA [29] can use a reduced number of frames, $q \ll n$, compared to the RPCA problem (11.1), to estimate an initial low-rank sub-space representation of the background U and then processes each frame (which can be spatially sub-sampled) at a time. Alternatively, it can randomly select a few frames (50 frames by default [27, Section 4.5.1]) for its initialization stage. Either initialization procedure is a batch one and can have a relatively high complexity. Once initialized, GRASTA can estimate and track non-stationary backgrounds.

pROST [52] is very similar to the GRASTA algorithm, but it uses a ℓ_p weighted version of (11.12) to track the low-rank sub-space representation of the background. Experimental results in [52] show that pROST can outperform GRASTA in the case of dynamic backgrounds.

Similarly, GOSUS [68] is also closely related to GRASTA; however, GOSUS enforces structured/group sparsity on the sparse component and uses a small number of frames from the initial part of the video to be analyzed for its batch initialization stage, and then proceeds to update the background. Furthermore, GOSUS can also be initialized with a random sub-space (representing the background), although in [68, Section 5.1] it is stated that this initialization takes up to 200 frames to converge to the right background sup-space. Although GOSUS is known to have better tracking properties than GRASTA, its computational cost is higher. Furthermore computational results (presented in Section 11.3.3) suggest that its complexity does not depend linearly on the number of pixels in the analyzed video frame, but it is influenced by the number of moving objects.

11.3.2 Incremental and Rank-1 Modifications for Thin SVD

Given a matrix $D \in \mathbb{R}^{m \times l}$ with thin SVD $D = U_0 \Sigma_0 V_0^T$ where $\Sigma_0 \in \mathbb{R}^{r \times r}$, and column vectors $\mathbf{a} \in \mathbb{R}^m$ and $\mathbf{b} \in \mathbb{R}^l$, note that

$$D + \mathbf{a}\mathbf{b}^T = [U_0\ \mathbf{a}] \begin{bmatrix} \Sigma_0 & \mathbf{0} \\ \mathbf{0}^T & 1 \end{bmatrix} [V_0\ \mathbf{b}]^T , \tag{11.13}$$

where $\mathbf{0}$ is a zero column vector of the appropriate size. Based on [4, 16], as well as on [12], we briefly describe an incremental thin SVD and rank-1 modifications (update, downdate, and replace) for thin SVD. Before proceeding with the description of the rank-1 modifications procedures, note that

- the computational complexity of any of the procedures described below (see [16, Section 3], [12, Section 4]) is upper bounded by $O(10 \cdot m \cdot r) + O(r^3) + O(3 \cdot r \cdot l)$. If $r \ll m, l$ and $l \ll m$ hold, then the complexity is dominated by $O(10 \cdot m \cdot r)$;
- matrices U_0 and V_0 are orthonormal, i.e., $U^T U = I_m$ and $V^T V = I_l$, where I_k is the identity matrix of size $k \times k$, as well as that the diagonal elements Σ_0, are non-negative and non-increasing;
- we assume that the Gram-Schmidt orthonormalization of $\mathbf{a}$ and $\mathbf{b}$ w.r.t. U_0 and V_0 have been computed:

$$\mathbf{x} = U_0^T \mathbf{a}, \quad \mathbf{z}_x = \mathbf{a} - U\mathbf{x}, \quad \rho_x = \|\mathbf{z}_x\|_2, \quad \mathbf{p} = \frac{1}{\rho_x}\mathbf{z}_x \tag{11.14}$$

$$\mathbf{y} = V_0^T \mathbf{b}, \quad \mathbf{z}_y = \mathbf{b} - V\mathbf{y}, \quad \rho_y = \|\mathbf{z}_y\|_2, \quad \mathbf{q} = \frac{1}{\rho_y}\mathbf{z}_y \,. \tag{11.15}$$

Incremental or update thin SVD

Given $D = U_0 \Sigma_0 V_0^T$, with $\Sigma_0 \in \mathbb{R}^{r \times r}$ and $\mathbf{d} \in \mathbb{R}^{m \times 1}$, we want to compute

$$\text{thinSVD}([D\ \mathbf{d}]) = U_1 \Sigma_1 V_1^T , \tag{11.16}$$

with (i) $\Sigma_1 \in \mathbb{R}^{r+1 \times r+1}$ or (ii) $\Sigma_1 \in \mathbb{R}^{r \times r}$. In this case $[D\ \mathbf{0}] = U_0 \Sigma_0 [V_0\ \mathbf{0}]^T$ and $[D\ \mathbf{d}] = [D\ \mathbf{0}] + \mathbf{d}\mathbf{e}^T$, where $\mathbf{e} \in \mathbb{R}^{(l+1) \times 1}$ is a unit vector and $\mathbf{e}(l+1) = 1$, so (11.13) is equivalent to (11.17)–(11.18) with $\hat{\Sigma} \in \mathbb{R}^{(r+1) \times (r+1)}$.

$$[D\ \mathbf{0}] + \mathbf{d}\mathbf{e}^T = [U_0\ \mathbf{p}] \cdot (G \hat{\Sigma} H^T) \cdot \begin{bmatrix} V_0^T & \mathbf{0} \\ \mathbf{0}^T & 1 \end{bmatrix} , \tag{11.17}$$

$$G \hat{\Sigma} H^T = \text{SVD}\left(\begin{bmatrix} \Sigma_0 & \mathbf{x} \\ \mathbf{0}^T & \rho_x \end{bmatrix} \right) . \tag{11.18}$$

Using (11.19) we can compute thinSVD($[D\ \mathbf{d}]$) with (i) $\Sigma_1 \in \mathbb{R}^{(r+1) \times (r+1)}$; similarly, using (11.20), we can compute thinSVD($[D\ \mathbf{d}]$) with (ii) $\Sigma_1 \in \mathbb{R}^{r \times r}$, where MATLAB notation is used to indicate array slicing operations.

$$U_1 = [U_0 \ \mathbf{p}] \cdot G, \quad \Sigma_1 = \hat{\Sigma}, \quad V_1 = \begin{bmatrix} V_0 & \mathbf{0} \\ \mathbf{0}^T & 1 \end{bmatrix} \cdot H \tag{11.19}$$

$$\begin{aligned} &U_1 = U_0 \cdot G(1{:}r, 1{:}r) + \mathbf{p} \cdot G(r{+}1, 1{:}r), \quad \Sigma_1 = \hat{\Sigma}(1{:}r, 1{:}r), \\ &V_1 = [V_0 \cdot H(1{:}r,\ 1{:}r) H(r+1,\ 1{:}r)] \ . \end{aligned} \tag{11.20}$$

Downdate thin SVD

Given $[D \ \mathbf{d}] = U_0 \Sigma_0 V_0^T$, with $\Sigma_0 \in \mathbb{R}^{r \times r}$, we want to compute

$$\text{thinSVD}(D) = U_1 \Sigma_1 V_1^T \tag{11.21}$$

with r singular values. Since $[D \ \mathbf{0}] = [D \ \mathbf{d}] + (-\mathbf{d})\mathbf{e}^T$, the rank-1 modification (11.13) is equivalent to (11.22)–(11.23)

$$[D \ \mathbf{d}] + (-\mathbf{d})\mathbf{e}^T = [U_0 \ \mathbf{0}] \cdot (G\hat{\Sigma}H^T) \cdot [V_0 \ \mathbf{q}]^T, \tag{11.22}$$

$$G\hat{\Sigma}H^T = \text{SVD}\left(\begin{bmatrix} \Sigma_0 \text{-} \Sigma_0 \mathbf{y}\mathbf{y}^T & \text{-}\rho_y \cdot \Sigma_0 \mathbf{y} \\ \mathbf{0}^T & 0 \end{bmatrix} \right), \tag{11.23}$$

from which we can compute thinSVD(D) via (11.24).

$$\begin{aligned} &U_1 = U_0 \cdot G(1{:}r, 1{:}r), \quad \Sigma_1 = \hat{\Sigma}(1{:}r, 1{:}r), \\ &V_1 = V_0 \cdot H(1{:}r, 1{:}r) + \mathbf{q} \cdot H(r+1, 1{:}r) \ . \end{aligned} \tag{11.24}$$

Thin SVD replace

Given $[D \ \mathbf{d}] = U_0 \Sigma_0 V_0^T$, with $\Sigma_0 \in \mathbb{R}^{r \times r}$, we want to compute

$$\text{SVD}([D \ \hat{\mathbf{d}}]) = U_1 \Sigma_1 V_1^T \tag{11.25}$$

with r singular values. Since $[D \ \hat{\mathbf{d}}] = [D \ \mathbf{d}] + \mathbf{c}\mathbf{e}^T$, where $\mathbf{c} = \hat{\mathbf{d}}\text{-}\mathbf{d}$, the rank-1 modification (11.13) is equivalent to (11.26)–(11.27)

$$[D \ \hat{\mathbf{d}}] = [U_0 \ \mathbf{p}] \begin{bmatrix} \Sigma_0 + \mathbf{x}\mathbf{y}^T & \rho_y \cdot \mathbf{x} \\ \rho_x \cdot \mathbf{y}^T & \rho_x \cdot \rho_y \end{bmatrix} [V_0 \ \mathbf{q}]^T \tag{11.26}$$

$$G\hat{\Sigma}H^T = \text{SVD}\left(\begin{bmatrix} \Sigma_0 + \mathbf{x}\mathbf{y}^T & \rho_y \cdot \mathbf{x} \\ \rho_x \cdot \mathbf{y}^T & \rho_x \cdot \rho_y \end{bmatrix} \right) \tag{11.27}$$

from which we can compute thinSVD$([D \ \hat{\mathbf{d}}])$ via (11.28).

$$\begin{aligned} &U_1 = U_0 \cdot G(1:r, 1:r) + \mathbf{p} \cdot G(r+1, 1:r), \ \ \Sigma_1 = \hat{\Sigma}(1:r, 1:r) \\ &V_1 = V_0 \cdot H(1:r, 1:r) + \mathbf{q} \cdot H(r+1, 1:r). \end{aligned} \tag{11.28}$$

11.3.3 Incremental RPCA Based on the amFastPCP Algorithm

In what follows, we assume that we have solved the RPCA up to time $k-1$, i.e.,

$$\underset{L_{k-1}, S_{k-1}}{\arg\min} \ \|L_{k-1}\|_* + \lambda \|S_{k-1}\|_1 \qquad \text{s.t. } D_{k-1} = L_{k-1} + S_{k-1} \tag{11.29}$$

where $D_{k-1} = D(:, 1:k-1)$. Furthermore we also assume that we know the partial (thin) SVD of $L_{k-1} = U_r \Sigma_r V_r^T$, where $\Sigma_r \in \mathbb{R}^{r \times r}$; this result is usually a by-product of any RPCA algorithm. Although (11.29) can be solved in a fully incremental fashion as described at the end of this section, for the sake of simplicity, here we assume that (11.29) has been solved via the amFastPCP algorithm (see Section 11.2.2).

If we were to solve the RPCA problem from scratch when the next frame $\mathbf{d}_k$ is available via the amFastPCP algorithm, then we need to solve the following alternating optimizations:

$$L_k^{(j+1)} = \arg\min_L \|L_k + S_k^{(j)} - D_k\|_F^2 \quad \text{s.t. } \operatorname{rank}(L_k) = r \tag{11.30}$$

$$S_k^{(j+1)} = \arg\min_S \|L_k^{(j+1)} + S_k - D_k\|_F^2 + \lambda\|S_k\|_1, \tag{11.31}$$

where $L_k = [L_{k-1}\,\mathbf{l}_k]$, $S_k = [S_{k-1}\,\mathbf{s}_k]$ and $D_k = [D_{k-1}\,\mathbf{d}_k]$. The minimizer of (11.30), for $j = 0$ is given by

$$L_k^{(1)} = \text{partialSVD}(D_k - S_k^{(0)}) = \text{partialSVD}([D_{k-1}\text{-}S_{k-1}\ \mathbf{d}_k])\ , \tag{11.32}$$

since (i) we know S_{k-1} and (ii) the amFastPCP algorithm is initialized with a zero sparse solution (see Algorithm 11.1). Ideally $[D_{k-1}\text{-}S_{k-1}\ \mathbf{d}_k] = [L_{k-1}\ \mathbf{d}_k]$, and therefore the solution of (11.32) can be computed via the incremental, non-increasing rank, thin SVD procedure previously described in Section 11.3.2. Alternatively, the solution of (11.32) could be computed via the rank-increasing case, which is preferred if the smallest singular value σ_{r+1} has a significant contribution, which could be evaluated in a similar fashions as for the amFastPCP algorithm (see Section 11.2.2 and line 6 in Algorithm 11.1).

In the case of the sparse component, the minimizer of (11.31) for $j = 0$ is given by

$$S_k^{(1)} = \text{shrink}(D_k - L_k^{(1)}, \lambda) = [S_{k-1},\ \text{shrink}(\mathbf{d}_k - \mathbf{l}_k^{(1)}, \lambda)]\ , \tag{11.33}$$

which is only applied to the current frame, since we know S_{k-1}; this is computationally cheap. The solution (11.30) for $j = 1$ (the next inner loop) follows the same logic:

$$L_k^{(2)} = \text{partialSVD}(D_k - S_k^{(1)}) = \text{partialSVD}([D_{k-1}\text{-}S_{k-1}\ \mathbf{d}_k\text{-}\mathbf{s}_k^{(1)}])\ , \tag{11.34}$$

which can be effectively computed using the thin SVD replace procedure, since in the previous step we have computed the partial SVD for $[D_{k-1}\text{-}S_{k-1}\ \mathbf{d}_k]$.

In Algorithm 11.2, lines 5–13, we summarize the method described in the previous paragraphs, where incSVD(·), repSVD(·) and dwnSVD(·) refer respectively to the incremental, replace, and downdate procedures described in Section 11.3.2.

For a video shot from a static camera, it could be assumed that the background does not change, or changes very slowly, but in practice this condition will usually not hold for long. To model these slow changes we could assume that the background is unchanged for at least bL frames. Afterwards we need to "forget" the background frames that are "too old" and always keep a low-rank estimate of a more or less constant background, which can be done via the downdate procedure, as described in Algorithm 11.2, line 14. The resulting algorithm is equivalent to a "sliding window" incremental PCP algorithm for which the background estimate represents the background as observed for the last bL frames.

Finally in Algorithm 11.2, line 17, we check whether the current low-rank approximation is significantly different from the previous one. In real scenarios this would be related to an abrupt background change (e.g., camera is moved, sudden illumination change, etc.), and if this condition is true, then we re-initialize the background estimate using the procedure described next.

Algorithm 11.2 - incPCP: incremental amFastPCP.

1: **Inputs:** Observed video $D \in \mathbb{R}^{m \times n}$, regularization parameter λ, number of inner loops iL, background frames bL, $m = k_0$.
2: **Initialization:** $L + S = D(:, 1 : k_0)$, initial rank r, $[U_r,\ \Sigma_r,\ V_r] = \text{partialSVD}(L,\ r)$
3: **for** $k = k_0 + 1\ :\ n$ **do**
4: $\quad ++m$
5: $\quad [U_k, \Sigma_k, V_k] = \text{incSVD}(D(:, k), U_{k\text{-}1}, \Sigma_{k\text{-}1}, V_{k\text{-}1})$
6: $\quad$ **for** $j = 1\ :\ iL$ **do**
7: $\quad\quad L(:, k) = U_k(:, 1 : r) * \Sigma_k * (V_k(end, :)')$
8: $\quad\quad S(:, k) = \text{shrink}(D(:, k) - L(:, k), \lambda)$
9: $\quad\quad$ **if** $j == iL$ **then**
10: $\quad\quad\quad$ break
11: $\quad\quad$ **end if**
12: $\quad\quad [U_k, \Sigma_k, V_k] = \text{repSVD}(D(:, k), S(:, k),\ U_k, \Sigma_k, V_k)$
13: $\quad$ **end for**
14: $\quad$ **if** $m \geq bL$ **then**
15: $\quad\quad \text{dwnSVD}(\text{"1st column"}, U_k, \Sigma_k, V_k)$
16: $\quad$ **end if**
17: $\quad$ **if** $\|L(:, k) - L(:, k-1)\|_2^2 / \|L(:, k-1)\|_2^2 \geq \tau$ **then**
18: $\quad\quad m = k_0$, use procedure in Section 11.3.3.
19: $\quad$ **end if**
20: **end for**

An incremental initialization for Algorithm 11.2

While Algorithm 11.2 describes an incremental procedure for RPCA, its initialization seems to hint at the need for a batch computation, which will greatly reduce its usefulness to process live streaming videos. In what follows we describe a fully incremental initialization procedure.

In the context of RPCA and video background modeling, the rank of the low-rank matrix $r \in [2,\ 8]$ is adequate when analyzing real videos acquired with a static camera, and thus we assume that r is known. Then we can proceed as follows

- Compute $[U,\ \Sigma] = \text{thinQR}(D(:, 1))$, set $V = I_1$, where $\text{thinQR}(\cdot)$ represents the thin QR decomposition.

- Compute $[U, \Sigma, V] = \text{incSVD}(D(:, k), U, \Sigma, V)$ for $k \in [2,\ r]$, via the rank increasing case.

- Compute $[U, \Sigma, V] = \text{incSVD}(D(:, k), U, \Sigma, V)$ for $k \in [r + 1,\ k - 1]$, via the non-increasing rank case.

to obtain an estimate of the low-rank approximation of the first $k - 1$ frames of the input video.

It was experimentally determined that $r = 1$, $k_0 = 1$ is sufficient, in the long run, to produce very good results. Although this initialization gives a very rough background estimate, this estimate is improved at each iteration (via the incremental SVD procedure, line 5 in Algorithm 11.2) for bL frames, before any downdate operation is performed. Furthermore, at this point the background estimate is equivalent to that which would be computed via a batch PCP algorithm applied to frames $1 : bL$. Once frame $bL+1$ is processed, and after the downdate called, the background estimate is equivalent to that which would be computed via a batch PCP algorithm applied to frames $2 : bL + 1$. This updating process is the key reason that choosing $r = 1$, $k_0 = 1$ for our initialization stage gives good results.

Computational Results

All simulations, related to Algorithm 11.2 and presented in this section, are available at [46]. They have been carried out using single-threaded MATLAB-only code running on an Intel i7-4710HQ quad-core (2.5 GHz, 6MB Cache, 32GB RAM) based laptop with a nvidia GTX980M GPU card. Furthermore, since Algorithm 11.2 behaves in practice as a "sliding window" incremental PCP (we use $bL = 100$ by default), using rank $r = 1$ is adequate for a good background estimate and good tracking properties.

We have used four real and synthetic video sets as test videos:

- V320: 320×256 pixel, 1286-frame color video sequence of 51.4 seconds at 25 fps, from the atrium of a mall, with lots of people walking around [2].
- V640: 640×480 pixel, 400-frame color video sequence of 26.66 seconds at 15 fps, from the Lankershim Boulevard traffic surveillance dataset [3, camera3], where a large number of cars are seen.
- V800: 800×600 pixel, 600-frame color synthetic video sequence of a street with highly variable lighting [13].
- V1920: 1920×1088 pixel, 900-frame color video sequence of 36 seconds at 25 fps, from the Neovison2 public space surveillance dataset with observably variable lighting. [1, Tower 23rd video].

Test Video	Initialization (sec.)		Average per frame (sec.)					
	GRASTA grayscale	GOSUS color	incPCP grayscale		incPCP color		GRASTA grayscale	GOSUS color
	Standard MATLAB	Standard MATLAB	Standard MATLAB	GPU-enabled MATLAB	Standard MATLAB	GPU-enabled MATLAB	Standard MATLAB	Standard MATLAB
V320	23.0	11.0	1.0e-2	1.3e-2	2.0e-2	1.7e-2	3.4e-2	3.0
V640	72.4	39.9	2.9e-2	2.4e-2	6.4e-2	3.7e-2	1.1e-1	26.5
V800	115.5	52.6	4.6e-2	3.7e-2	9.7e-2	5.7e-2	1.7e-1	24.2
V1920	537.5	(*)	1.9e-1	1.2e-1	3.4e-1	2.8e-2	8.1e-1	(*)

TABLE 11.1 Elapsed time to initialize the GRASTA [29] and GOSUS [68] algorithms as well as the average processing time per frame for all algorithms on an Intel-i7 (32 GB RAM) based laptop. Note that "e-k" $= 10^{-k}$. (*) Runs out of memory before completing the processing.

We compare our results with the GRASTA and GOSUS algorithms. ReProCS is not considered since the practical ReProCS [26] implementation, available at [25], has a batch initialization stage that considers all the available frames. We use [28] (a GRASTA implementation, by the authors of [29], that can only process grayscale videos) with its default parameters: 30% and 10% of the information for the batch initialization and sub-space tracking, respectively; we also use [67] (a GOSUS implementation, by the authors of [68], that can only process color videos) with its default parameters, and using the initial 200 frames for the initialization stage. Furthermore, GRASTA and GOSUS implementations are MATLAB based, which takes advantage of MEX interfaces, and while MATLAB implementation of incPCP does not use any MEX interface, it comes in two flavors (i) one that uses standard single-thread MATLAB code and (ii) one that takes advantage of (mainly linear algebra) GPU-enabled MATLAB functions.

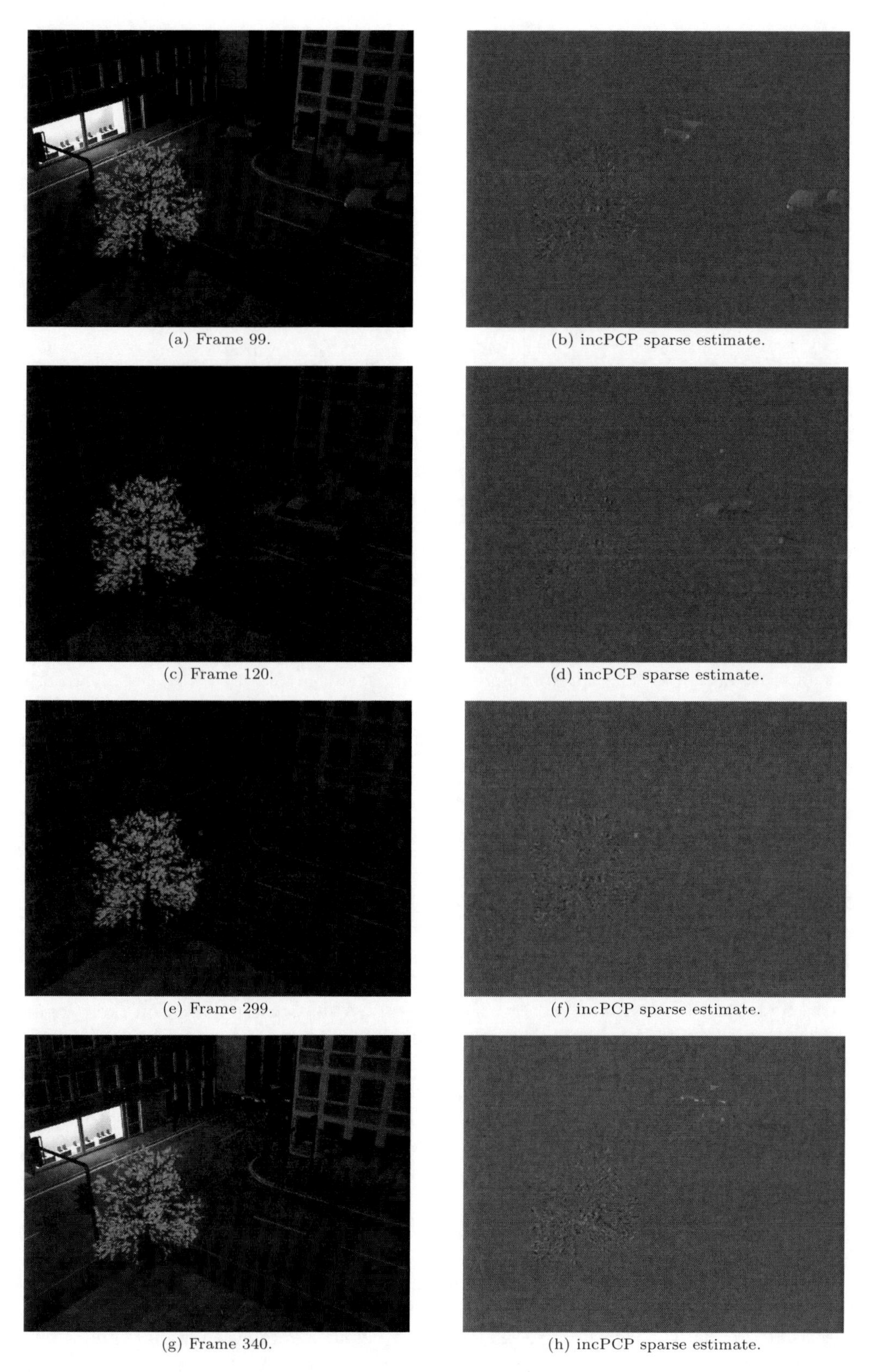

(a) Frame 99.

(b) incPCP sparse estimate.

(c) Frame 120.

(d) incPCP sparse estimate.

(e) Frame 299.

(f) incPCP sparse estimate.

(g) Frame 340.

(h) incPCP sparse estimate.

FIGURE 11.1 Original and sparse estimates via the incPCP algorithm for the V800 (800×600 pixel) test video, where a variable lighting environment can be observed.

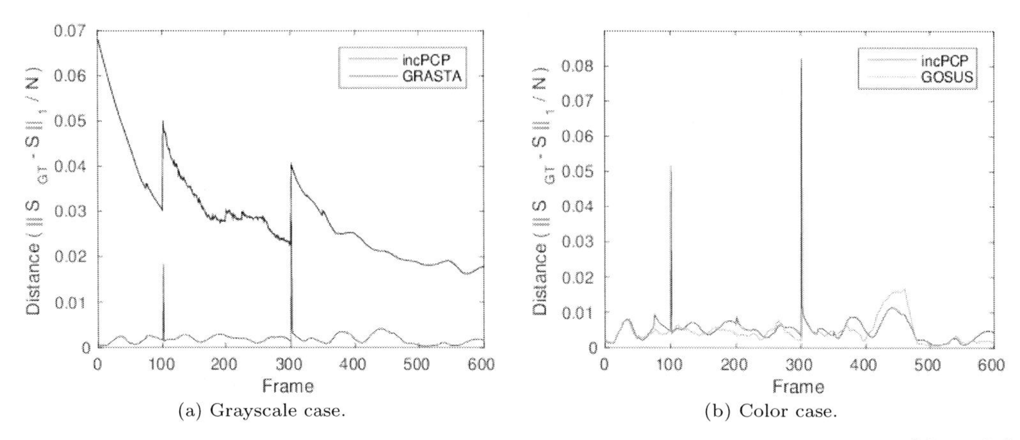

(a) Grayscale case. (b) Color case.

FIGURE 11.2 (See color insert.) Sparse approximation (V800) frame distance measure by $\frac{\|S_{GT}-S_P\|_1}{N}$ where S_{GT} is the ground-truth computed via the (batch) iALM algorithm (20 outer loops) and S_P is the sparse approximation computed via (a) grayscale (blue) GRASTA and (red) incPCP methods, or (b) color (green) GOSUS and (red) incPCP methods, and N is the number of pixels per frame (used as a normalization factor). It is worth noting that the spikes due to background adaptation to light illumination (frames 100 and 300) are less prominent for GOSUS than for incPCP, since GOSUS uses the first 200 frames (by default) for its batch initialization, and thus it has already observed the background change, which is built into its initial subspace approximation.

In order to assess the performance of the incPCP algorithm, we present two kinds of results: (i) computational performance measured as the time to process a given video, summarized in Table 11.1, and (ii) reconstruction quality, depicted in Figures 11.2 and 11.4, measured by $\frac{\|S_{GT}-S_P\|_1}{N}$ where S_{GT} is the "ground truth" sparse video approximation and S_P is the sparse video approximation of either GRASTA, GOSUS, or incPCP algorithms, and N is the number of pixels per frame, which is same value for either the grayscale or color case.

Since one of the main features of the incPCP algorithm is its ability to process large videos (e.g., full-HD or 1920×1080), results for small videos do not illustrate one of its main advantages. Unfortunately we have been unable to find manually segmented ground-truth for large videos. For example, the BMC or background models challenge dataset [57] has videos of size 320×240; some manually segmented ground-truth frames are available for test video V320 (available from [2]), but only for a very small fraction of each test video (V320: 20 ground-truth frames out of 1286 frames), so that results based on these ground truth frames are not representative of the entire test videos, and could be substantially misleading. Therefore for the V320, V640, V800, and V1920 test videos (see note below for the V800 test video case) we use the sparse video approximation computed via the batch iALM algorithm [35] with 20 outer loops (10 for the V1920 case due to memory constraints) as a proxy ground-truth since this result has a high level of confidence, see [10, Tables 6 and 7]. For synthetic test video V800, from the Stuttgart Artificial Background Subtraction Dataset [13], the ground-truth is given in the form of background without any moving object; however, the sudden light change (see Fig. 11.1) is not included for this dataset, limiting its use.

Without taking into account the batch initializations (for GRASTA and GOSUS, since incPCP does not need one), the average processing time per frame listed in Table 11.1 shows that the incPCP algorithm (MATLAB-only version) is between 3 and 4 times faster than GRASTA, when processing grayscale videos, and between 2 and 3 orders of magnitude

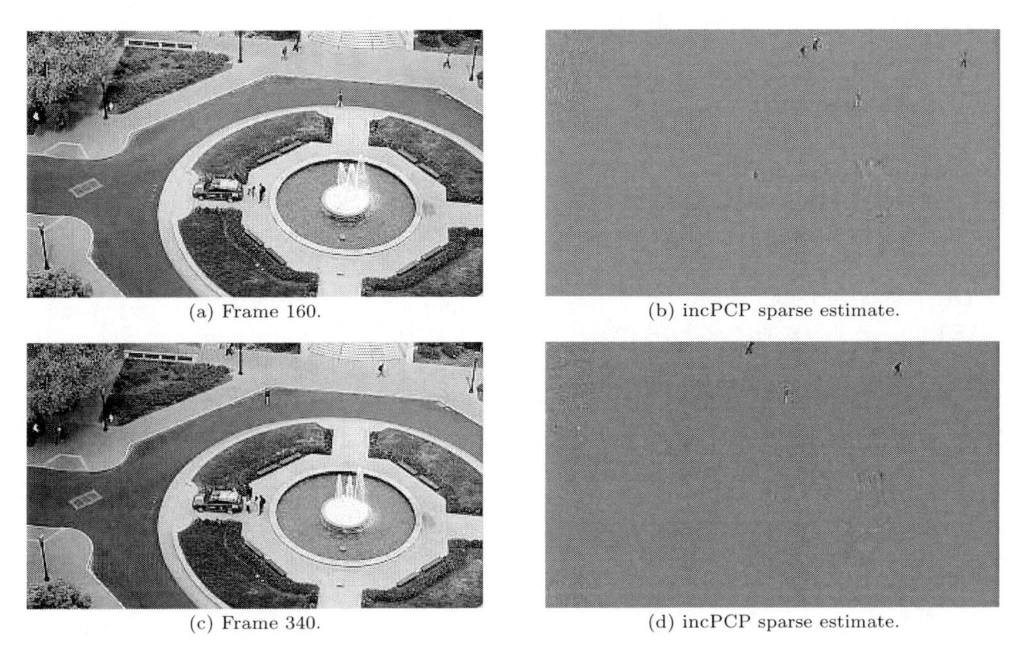

(a) Frame 160. (b) incPCP sparse estimate.

(c) Frame 340. (d) incPCP sparse estimate.

FIGURE 11.3 Original and sparse estimate via the incPCP algorithm for the V1920 (1088×1920 pixel) test video, where a smooth change in the sun-lighting can be observed.

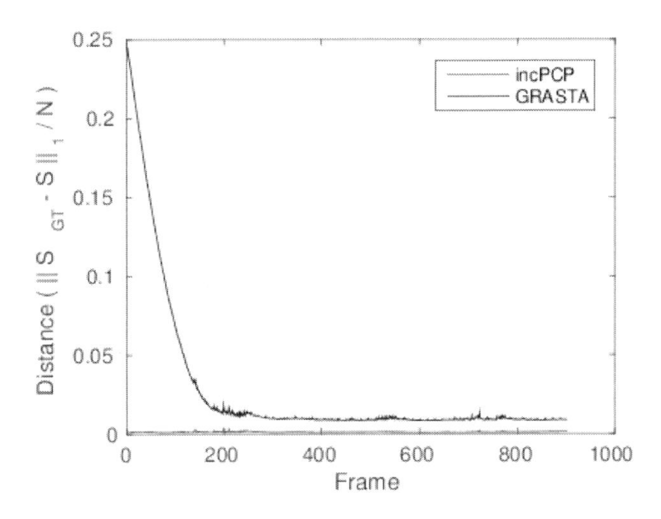

FIGURE 11.4 (See color insert.) Sparse approximation (V1920, grayscale version) frame distance measure by $\frac{\|S_{GT}-S_P\|_1}{N}$ where S_{GT} is the ground-truth computed via the (batch) iALM algorithm (10 nouter loops) and S_P is the sparse approximation computed via the (blue) GRASTA and (red) incPCP method, and N is the number of pixels per frame (used as a normalization factor).

faster than GOSUS when processing color videos; in addition, the MATLAB GPU-enabled incPCP version is about 1.2 $\sim$ 1.5 times faster than the MATLAB-only one. Furthermore, results from Table 11.1 suggest that the computational performance of the GRASTA and incPCP (both versions) algorithms has a more or less linear dependence on the the frame size, whereas this is not true for the GOSUS algorithm: GOSUS takes, on average, 26.5 seconds to process each frame of video V640 (640×480 pixel) and 24.2 seconds to process each frame of video V800 (800×600 pixel). Besides the frame size difference between V640 and V800, they also differ in the number and size of moving objects: V640 shows a street with a large number of (relatively) small moving vehicles, whereas in V800 at most two large moving objects are observed.

Furthermore, the above-mentioned speed-up does not take into consideration the time spent by GRASTA and GOSUS in their batch initializations, respectively: for instance, GRASTA needs between 23 and 537 seconds to complete its initialization for videos of size between 240×320 and 1088×1920, respectively; in the GOSUS case, its batch initialization takes between 11 and 52 seconds for videos of size between 240×320 and 600×800; while GOSUS is much faster than GRASTA, the memory requirements of GOSUS are much larger than those of GRASTA and can thwart any further processing (as for the V1920 case). Furthermore, such initialization procedures hamper the use of either GRASTA or GOSUS as a real-time alternative to the incPCP algorithm, particularly when the background is dynamic.

In Figures 11.2 and 11.4 we present the reconstruction quality measure for videos V800 and V1920; the other considered test videos have similar results and thus are omitted. Test video V800 has an illumination change in frames 100 and frame 300 (see Figure 11.1); since this is a synthetic video, this sudden illumination change spans only one frame; furthermore, this video also includes small changes in the background due to traffic lights, making it a very challenging environment, in which any background modeling algorithm would need fast tracking capabilities as well as adaption capabilities to (more or less) smooth or localized background changes that do not dramatically affect the main background characteristics. In Figure 11.2 we can observed the background tracking properties of the incPCP, GRASTA and GOSUS algorithms. The incPCP algorithm quickly adapts to the above-mentioned sudden illumination changes, while at the same time also adapts to other non-dramatic background changes: this is observed due to the sinusoid-like variation in the incPCP restoration quality plot. A similar behavior is also observed for the GOSUS algorithm.

In Figure 11.4 we present the reconstruction quality measure for the V1920 test video. In this case, there is an observable (more or less) smooth change in the background due to changes in the sun-lighting illumination, as can be observed in Figure 11.3. Although V1920 is a color video, the plots shown in Figure 11.4 correspond to its grayscale version since (i) the GRASTA algorithm can only process grayscale videos, and (ii) the GOSUS algorithm, which can process color videos, was unable to finish any processing due to memory requirements that exceeded that of the computer (RAM 32 GB) used to run the simulations. In Figure 11.4 it is observed that the incPCP restoration quality measure is almost constant with respect to the smooth illumination change; it also is better than the GRASTA sparse approximation, even after GRASTA has adapted to the background (frame 200 onwards).

11.4 Local Subspace Models and Endogenous Sparse Representations

The RPCA decomposition assumes that there is a single, low-dimensional model that describes most components of the elements of the dataset. This provides a good approximation of the stationary background component of a video sequence captured by a stationary cam-

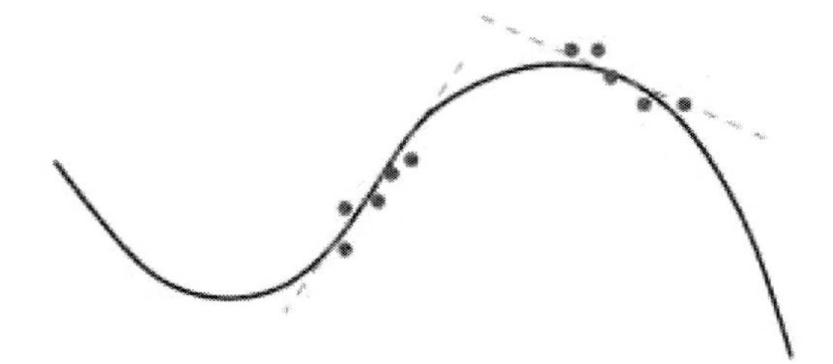

FIGURE 11.5 Illustration of a set of local tangent planes providing location approximations to a manifold.

era, but is not a good model when the camera is non-stationary. In this section we consider generalizing the model from a single low-rank subspace with a sparse set of possibly-large deviations, to a low-dimensional *manifold* [42] with the same type of deviations. As a result, the low-dimensional representation is able to vary across the dataset, modeling, for example, the distinct scenes captured by a moving camera.

Our approach is based on modeling local tangent planes of the manifold, assuming that the local sampling density is adequate to make this possible. The geometric intuition for the representation of these subspaces is that every sample in a local tangent plane may be represented as a sparse linear combination of neighboring samples in the same tangent plane, as illustrated in Figure 11.5.

11.4.1 Endogenous Sparse Representations

This representation corresponds very closely to the union of subspaces model underlying the Sparse Subspace Clustering method, proposed by Elhamifar and Vidal [19–22], for subspace clustering of data belonging to a set of disjoint subspace. The primary difference is that our goal is not clustering, and we do not assume that the subspaces are sufficiently well separated for subspace recovery results [22] to hold, since our goal is merely to estimate the subspace, not to distinguish it from its neighbors.

The core of this method is computing a sparse representation of the data with respect to itself, an unusual form of sparse representation that has been labeled an *endogenous* sparse representation [18]. The basic form of the sparse coding problem is

$$\underset{X,S}{\arg\min} \frac{1}{2}\|DX + S - D\|_F^2 + \lambda_X \|X\|_1 + \lambda_S \|S\|_1 \;\; \text{s.t.} \;\; \text{diag}(X) = 0 \,, \tag{11.35}$$

where D is the data matrix, X is the sparse representation of D with respect to itself, and S represents a sparse set of outliers, as in the RPCA problem. Theoretical aspects of this model, such as conditions under which subspace recovery is possible, have been considered [18, 22, 40, 54], and a number of variants have been proposed, the most significant of which is the Low-Rank Representation [36], constructed by replacing the ℓ_1 norm $\|X\|_1$ with a nuclear norm $\|X\|_*$.

A variant of Equation (??),

$$\underset{X,S}{\arg\min} \frac{1}{2}\|DX + S - D\|^2 + \alpha\|X\|_1 + \beta\|X\|_{2,1} + \gamma\|S\|_1 + \delta \left\| \sqrt{(\nabla_x S)^2 + (\nabla_y S)^2 + (\nabla_z S)^2} \right\|_1 \,. \tag{11.36}$$

was applied to the video background modeling problem in [64]. The main differences are:

- the replacement of the constraint $\text{diag}(X) = 0$ with an $\ell_{2,1}$ norm [59] penalty on X to encourage structured sparsity, which is also effective in avoiding trivial solutions, and
- the inclusion of a Total Variation (TV) [51] penalty on S to encourage spatial contiguity of the sparse deviations, since in this problem they represent contiguous objects in the foreground.

This simple model does not perform very well in practice because the columns of the endogenous dictionary include the moving foreground objects.

An improved, although more computationally expensive, model can be derived by observing that the sparse deviations (or outliers) S of data D are also the deviations of the dictionary D, so that $(D - S)X$ should provide a better locally low-dimensional approximation than DX:

$$\arg\min_{X,S} \frac{1}{2}\|(D - S)X + S - D\|^2 + \alpha\|X\|_1 + \beta\|X\|_{2,1} + \gamma\|S\|_1 + \delta\left\|\sqrt{(\nabla_x S)^2 + (\nabla_y S)^2 + (\nabla_z S)^2}\right\|_1 . \quad (11.37)$$

This method performs well if there is a sufficient number of samples in each local subspace (i.e., the underlying manifold is sufficiently well-sampled to give good estimates of local tangent spaces), as is the case of jitter, very slow panning, or when the camera pans back and forth across the same scene. The performance is demonstrated using a synthetic slowly-panning test video sequence (see Figure 11.6) constructed by taking a moving 240×320 pixel cropping window within the original sequence, a 288-frame traffic video sequence from the Lankershim Boulevard Dataset [3, camera 4, 8:45–9:00 AM]. This window moves slowly to the left, and then back to the original position, at a rate of 1/4 pixel/frame. In this case the background is not very well approximated by any single low-dimensional subspace, but since the background motion is slow with respect to the foreground motion, a locally low-dimensional model provides a substantially better approximation, as illustrated in Figure 11.7.

FIGURE 11.6 Frame 180 of a synthetic slowly-panning test video sequence.

The model of Equation (11.37) does not perform well when the camera motion is less constrained. The next section introduces a modification to the form of representation that allows this method to handle greater translations between frames, corresponding to low sampling density on the manifold of video frames.

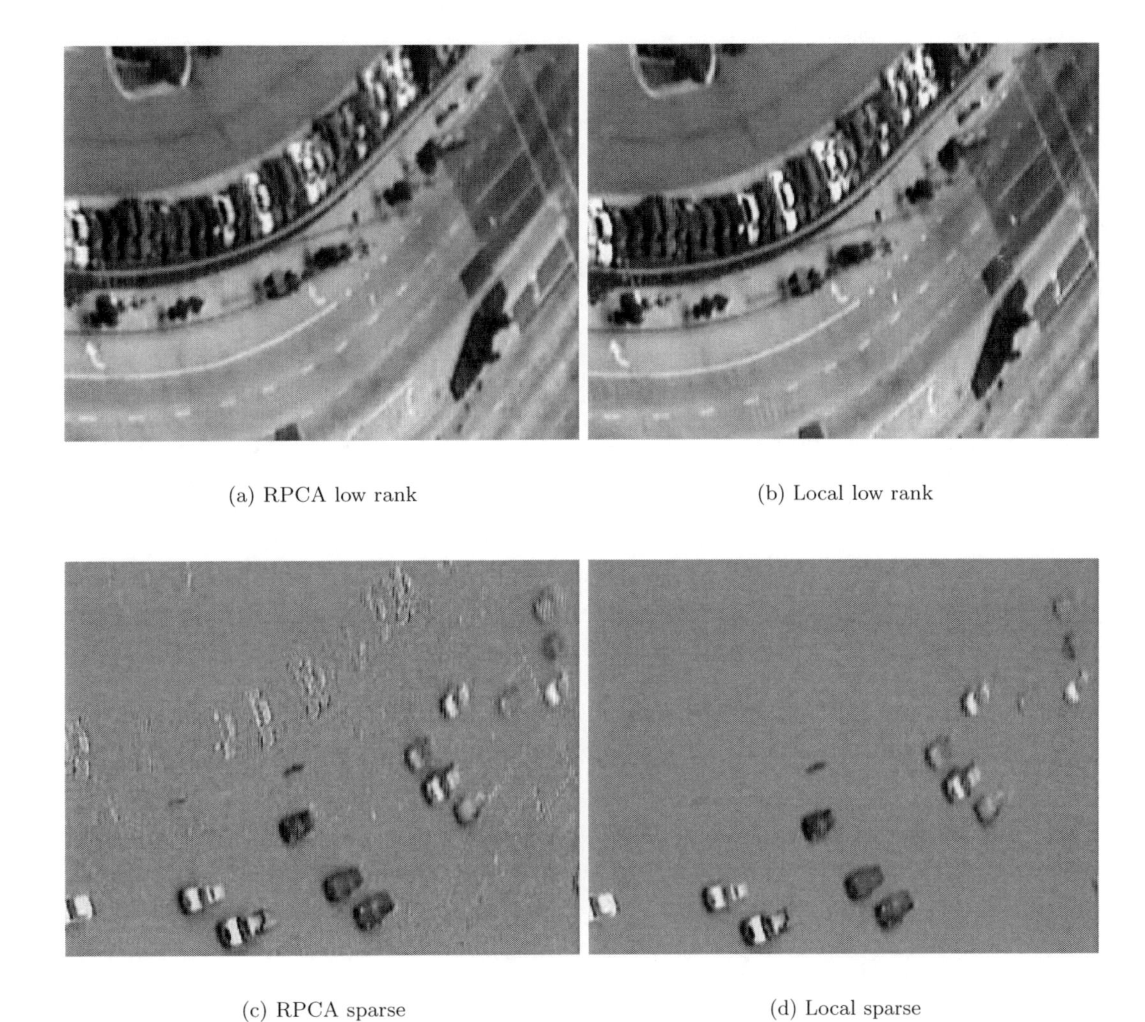

(a) RPCA low rank

(b) Local low rank

(c) RPCA sparse

(d) Local sparse

FIGURE 11.7 Results for frame 180 from the slowly-panning test video sequence. The RPCA results were obtained using RPCA with the standard choice of parameters, and the local subspace representation results were computed using Equation (11.37) with parameters $\alpha = 4.0 \times 10^{-3}, \beta = 8.0 \times 10^{-2}, \gamma = 5.0 \times 10^{-4}$, and $\delta = 3.0 \times 10^{-4}$.

11.4.2 Endogenous Convolutional Sparse Representations

Recall that the standard Basis Pursuit DeNoising (BPDN) sparse coding problem [17] with dictionary matrix A can be expressed as

$$\arg\min_{\mathbf{x}} \frac{1}{2}\|A\mathbf{x} - \mathbf{s}\|_2^2 + \lambda\|\mathbf{x}\|_1 \ . \tag{11.38}$$

The convolutional form [69] is obtained by replacing the linear combination of columns of A with a sum of convolutions by a set of dictionary filters $\{\mathbf{h}_m\}$

$$\arg\min_{\{\mathbf{x}_m\}} \frac{1}{2}\left\|\sum_m \mathbf{h}_m * \mathbf{x}_m - \mathbf{s}\right\|_2^2 + \lambda\sum_m \|\mathbf{x}_m\|_1 \ . \tag{11.39}$$

This is substantially more computationally expensive than standard sparse representations, but considerable progress has recently been made in developing faster algorithms [61].

In an endogenous context, the filters $\{\mathbf{h}_m\}$ are themselves images from the data set of interest, and the corresponding $\{\mathbf{x}_m\}$ are now 2-d coefficient maps, rather than simple scalars in the case of standard sparse representations. The problem to be solved can be expressed as

$$\begin{aligned} \arg\min_{\{\mathbf{x}_{m,k}\},\{\mathbf{s}_k\}} \frac{1}{2}\sum_k \left\|\sum_m \mathbf{h}_m * \mathbf{x}_{m,k} + \mathbf{s}_k - \mathbf{d}_k\right\|_2^2 + \lambda\sum_k\sum_m \|\mathbf{x}_{m,k}\|_1 + \mu\sum_k \|\mathbf{s}_k\|_1 \\ \text{such that } \mathbf{x}_{m,k} = 0 \, \forall m \in \mathcal{E}_k \ , \end{aligned} \tag{11.40}$$

where $\mathcal{E}_k = \{m \mid \mathbf{h}_m = \mathbf{d}_k\}$ as the set of indices m for signal $\mathbf{d}_k$ for which dictionary element $\mathbf{h}_m$ is the same datum as that signal. Expressing this problem in the form

$$\begin{aligned} \arg\min_{\{\mathbf{x}_{m,k}\},\{\mathbf{s}_k\}} \frac{1}{2}\sum_k \left\|\sum_m \mathbf{h}_m * \mathbf{x}_{m,k} + \mathbf{s}_k - \mathbf{d}_k\right\|_2^2 + \lambda\sum_k\sum_m \|\mathbf{x}_{m,k}\|_1 + \mu\sum_k \|\mathbf{s}_k\|_1 \\ + \sum_k \sum_{m\in\mathcal{E}_k} \iota(\mathbf{x}_{m,k}) \text{ such that } \mathbf{x}_{m,k} - \mathbf{y}_{m,k} = 0 \ , \end{aligned} \tag{11.41}$$

where $\iota(\cdot)$ is zero if its argument is a zero vector and infinite otherwise, allows this problem to be solved via relatively minor modification[32] of a recent ADMM algorithm for the standard Convolutional BPDN problem [61].

The advantage that this type of sparse representation brings to video background modeling is that, due to the properties of convolution, integer-pixel translations of the $\{\mathbf{h}_m\}$ can be compensated for by translations within the corresponding coefficient maps $\{\mathbf{x}_{m,k}\}$, allowing translation-invariant subspace modeling [62]. In the simplest case, the set of filters is the same as the set of frames, so that $\mathbf{h}_k = \mathbf{d}_k$, and $\{\mathbf{x}_{m,k}\}$ represents the contribution

[32] In the algorithm described in [62], $\mathbf{s}_k$ is dealt with independently from $\mathbf{x}_{m,k}$. This approach works, but can take many iterations to converge, and convergence is somewhat sensitive to correct choice of the auxiliary parameters λ and μ. More recently, it has become clear that a more effective approach is to absorb the outliers as an additional coefficient map with an impulse filter

$$\sum_m \mathbf{h}_m * \mathbf{x}_{k,m} + \mathbf{s}_k = \sum_m \mathbf{h}_m * \mathbf{x}_{k,m} + \boldsymbol{\delta} * \mathbf{s}_k$$

and then apply the standard Convolutional BPDN algorithm.

of frame m to representing frame k. The goal of the sparsity penalty on $\{\mathbf{x}_{m,k}\}$ is that each coefficient map $\{\mathbf{x}_{m,k}\}$ should be very sparse, with a single very localized concentration of non-zero coefficients, the position of which indicates the relative alignments of frames k and m. Since periodic boundary conditions are not suitable for most video frames, boundary issues arising from the misalignment between frames k and m need to be considered so that the solution is not perturbed by mismatch between the frames in the boundary region. The simplest solution is to mask out a boundary region, the size of which is determined by the largest translation that is desired to be tolerated, in the data fidelity term of Equation (11.40). Unfortunately, such a spatial domain projection is not compatible with the efficient frequency domain solution of the problem [61], but as pointed out in [62], the same effect can instead be achieved by the use of a spatial weighting in the ℓ_1 norm of the $\mathbf{s}_k$ so that there is no penalty at all on $\mathbf{s}_k$ in the boundary region.

	RPCA	ESR	ECSR	ECSR-R	ECSR-M
Background	4.6dB	7.6dB	10.9dB	14.7dB	19.3dB
Foreground	-0.5dB	1.9dB	3.6dB	9.6dB	14.2dB

TABLE 11.2 Comparison of background/foreground separation performance, measured as SNR against ground truth, of RPCA, the endogenous sparse representation method of [64] (ESR), the proposed method without post-processing (ECSR), and the proposed method with RPCA and median filtering post-processing (ECSR-R and ECSR-M, respectively).

This method was tested using a video sequence with synthetic translational panning constructed as in Equation (11.40), but with the 240×320 pixel cropping window moving within the original sequence at a rate of 3 pixels/frame. Due to the increased computational and memory cost of the convolutional form of the problem, the test sequence consisted of only 30 frames. Since the panning view was simulated, approximate ground truth background and foreground could be constructed from the RPCA result for the original uncropped video (see Figures 11.8(a) and 11.8(b)). Direct application of Equation (11.40) to this test sequence (with $\lambda = 50$ and $\mu = 0.06$) gave substantially better performance than the non-convolutional sparse models described in Section 11.4.1 (see Figures 11.8(e) and 11.8(f)), and much better performance than standard RPCA (see Figures 11.8(c) and 11.8(d)), but performance was still somewhat disappointing (see Figures 11.9(a) and 11.9(b)). Inspection of the set of $\mathbf{d}_m * \mathbf{x}_{m,k}$ contributing to the background estimate for frame k revealed that the actual frame alignments produced by the positioning of the non-zero coefficients in the $\mathbf{x}_{m,k}$ were quite accurate, but their sum contained artifacts due to the different positions of the foreground objects in the aligned frames $\mathbf{d}_m$; the same phenomenon that prompted the change from Equation (11.36) to Equation (11.37) in Section 11.4.1.[33]

Since the method was able to produce frame alignment robust to boundary effects and the motion of foreground objects, two alternative methods were considered. Both of these methods use the $\mathbf{x}_{m,k}$ obtained by solving Equation (11.40) to construct, for each frame k, a set of additional frames $\mathbf{d}_m * \mathbf{x}_{m,k}$ with their backgrounds aligned to that of frame k. The first of these methods simply normalizes each member of the set and applies median filtering in the temporal direction, and the second applies RPCA independently to the set of estimates for each frame k. The resulting estimate of the background is then subtracted from the original sequence to obtain an estimate of the foreground that improves on the

[33] The obvious path is to use filters with the sparse component subtracted as the dictionary, in analogy with Equation (11.37). This direction was not pursued, however, since it became apparent that the alternative incremental approach described in Section 11.5 gave good performance at vastly reduced computational cost.

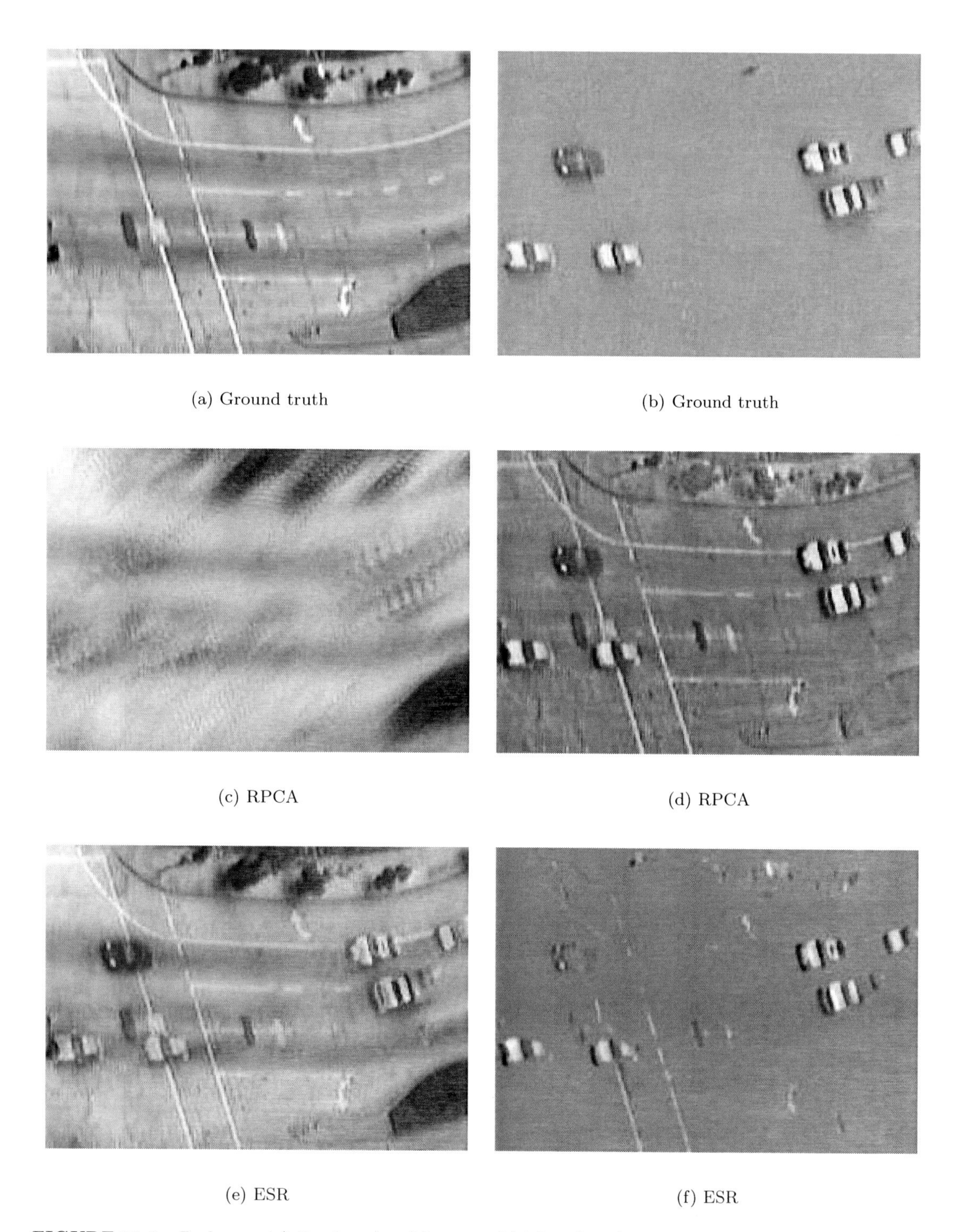

FIGURE 11.8 Background (left column) and foreground (right column) estimates for frame 15 from the fast-panning test video sequence, computed via the endogenous sparse representation method of [64] (ESR) and the proposed method with median filtering post-processing (ECSR-Median).

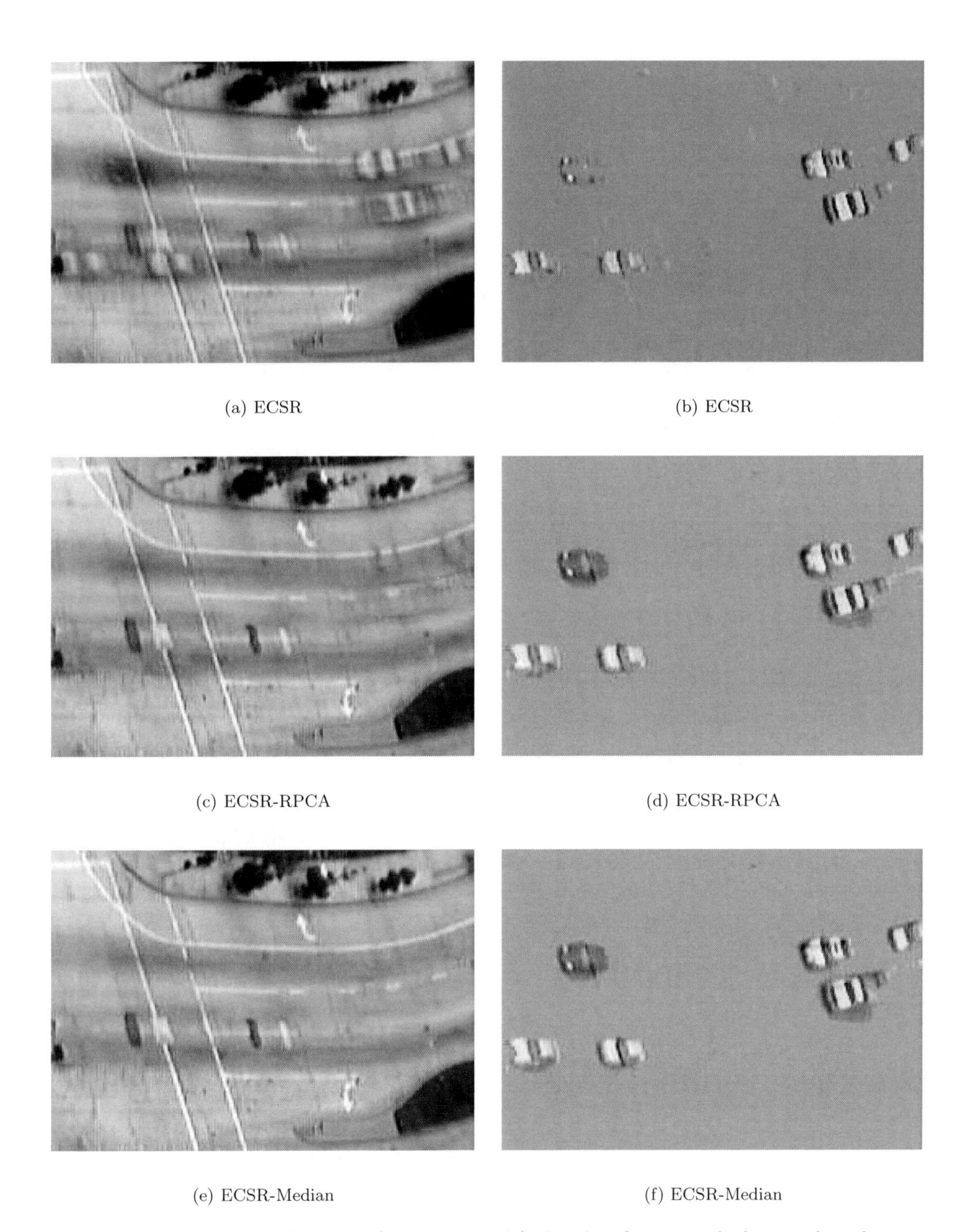

(a) ECSR (b) ECSR

(c) ECSR-RPCA (d) ECSR-RPCA

(e) ECSR-Median (f) ECSR-Median

FIGURE 11.9 Background (left column) and foreground (right column) estimates for frame 15 from the fast-panning test video sequence, computed via the endogenous sparse representation method of [64] (ESR) and the proposed method with median filtering post-processing (ECSR-Median).

$\mathbf{s}_k$ from Equation (11.40) (see Figures 11.9(c) to 11.9(f)). Objective performance as measured by SNR against ground truth of the different methods described here is compared in Table 11.2.

11.5 Transform Invariant Incremental RPCA

The main motivation for the development of an incremental RPCA algorithm, such as the one described in Section 11.3.3, is to have a procedure with an extremely low memory footprint, and a computational complexity that allows real-time processing. Algorithm 11.2 does the job, eliminating the batch processing mode, typically needed for RPCA algorithms whereby a large number of frames have to be observed before starting any processing. However, Algorithm 11.2, as well as most RPCA algorithms, have a high sensitivity to camera jitter, which can affect airborne and space-based sensors [53] as well as fixed ground-based cameras [10] subject to wind.

Here we overcome the jitter sensitivity of Algorithm 11.2 by incrementally solving

$$\min_{L,S,\mathcal{T}} \frac{1}{2}\|D - \mathcal{T}(L^*) - S\| + \lambda\|S\|_1 \quad \text{s.t.} \quad \text{rank}(L^*) \leq r\ , \tag{11.42}$$

where we assume that the observed frames D are misaligned due to camera jitter, the low-rank representation, L^*, is properly aligned, and that $\mathcal{T} = \{\mathcal{T}_k\}$ is a set of invertible and independent transformations such that when applied to each frame $D = \mathcal{T}(L^*) + S$ is satisfied. To make the solution of (11.42) tractable, we focus on the case where $\mathcal{T}_k$ is a rigid transformation, i.e., the camera used to acquire the video frames suffers from translational and rotational jitter. Furthermore, inspired by convolutional sparse representation (see Section 11.4.2) the solution to (11.42) includes a set of filters that describe a translation/rotation invariant sparse representation (see [5, 8, 39, 55] for other applications on this topic).

11.5.1 Related Work

Transform Invariant Sparse Representation

Transform invariant sparse representation is closely related to convolutional sparse representations, although there have been independent streams of research [63]. For this section we specifically focus on [5, 8, 39, 55] because these works are relevant to the transform invariant RPCA algorithm described in Section 11.5.2.

In the traditional sparse representation model, given the overcomplete dictionary Φ we seek to express observed data $\mathbf{s}$ via

$$\mathbf{s} = \Phi\mathbf{x} + \mathbf{m}\epsilon\ , \tag{11.43}$$

where $\mathbf{x}$ is the desired sparse representation and $\mathbf{m}\epsilon$ is noise. There are several algorithms to solve (11.43), among them Basis Pursuit and Matching Pursuit are arguably the most well-known ones.

The sparse representation model can be extended to handle transformation invariance via:

$$\mathbf{s} = \sum_k \tau_k(\Phi)\mathbf{x}^{(k)} + \mathbf{m}\epsilon\ , \tag{11.44}$$

where $\tau_k(\cdot)$ is the operator that accounts for the desired transformation; we note that this model could also be applied to patches in the observed data, giving in this case $\mathbf{s}^{(n)} = \sum_k \tau_k(\Phi)\mathbf{x}^{(n,\,k)} + \mathbf{m}\epsilon$ where index n describes a particular sub-divison of the input data.

The model described in (11.44) has been used in (i) [8], along with a particular sub-divison of the input data, for shift-invariant sparse coding; (ii) [55] adapts (11.44) for the case of digital image sparse representation via

$$\mathbf{s} = \sum_{k,\,n} T_{k,\,n} R_{\theta_{k,\,n}} G_{k,\,n} \mathbf{x}^{(k,\,n)} + \mathbf{m}\epsilon, \tag{11.45}$$

where $T_{k,\,n}$ and $R_{\theta_{k,\,n}}$ represent translation and rotation (angles $\theta_{k,\,n}$ are manually chosen) operators, and $G_{k,\,n}$ represents local structures of the image; (iii) in [39] a transform invariant sparse coding approach is taken, adapting (11.44) to

$$\mathbf{s} = \sum_{k} \sum_{n} \mathbf{m}\alpha^{(k,\,n)} * \tau_n(G_k), \tag{11.46}$$

where $\tau_n(\cdot)$ represents a set of pre-specified transformations, $*$ represents convolution, and G_k is a set of desired feature images; (iv) [5] also focuses on a transform invariant sparse coding, with emphasis on shift and rotation invariance for multivariate signals, where using a representation akin to (11.45) along with a multivariate orthogonal matching pursuit approach is used to propose a shift and rotation invariant new dictionary-learning algorithm able to provide a robust decomposition.

Transform Invariant Algorithms in the RPCA context

- **RASL algorithm**: The Robust Alignment by Sparse and Low-rank decomposition (RASL) algorithm [41] was introduced as a batch PCP method able to handle misaligned video frames by solving

  $$\min_{L,S,\tau} \|L\|_* + \lambda\|S\|_1 \quad \text{s.t.} \quad \tau(D) = L + S \tag{11.47}$$

  where $\tau(\cdot) = \{\tau_k(\cdot)\}$ is a set of independent transformations[34] (one per frame), each having a parametric representation, such that $\tau(D)$ aligns all the observed video frames.

  In [41], the non-linearity in (11.47), is handled via:

  $$\min_{L,S,\Delta\tau} \|L\|_* + \lambda\|S\|_1 \quad \text{s.t.} \quad \tau(D) + \sum_{k=1}^{n} J_k \Delta\tau_k \epsilon_k = L + S \tag{11.48}$$

  where J_k is the Jacobian of frame k with respect to transformation k and ϵ_k denotes the standard basis for $\mathbb{R}^n$.

  Computational results in [41] mainly focus on rigid transformations; it is also stated that as long as the initial misalignment is not too large, (11.48) effectively recovers the correct transformations.

- **t-GRASTA algorithm**: The GRASTA [29] method is an "on-line" algorithm for low-rank subspace tracking. It uses a reduced number of frames, $q \ll n$, compared to the RPCA problem (11.1), to estimate an initial low-rank sub-space representation of the background and then processes one frame at a time. It must be emphasized that this procedure is not fully incremental since it uses a time sub-sampled version of all the available frames for initialization.

[34] Note that $\tau(\cdot)$ is the inverse of $\mathcal{T}(\cdot)$ in (11.42).

The Transformed Grassmannian robust adaptive subspace tracking algorithm (t-GRASTA) [31] is based on the GRASTA [29] and RASL algorithms. It is able to handle misaligned video frames in an online, but not fully incremental, fashion. In particular, the t-GRASTA method handles the misaligned video frames by solving (11.48) via a modified GRASTA algorithm. It uses a batch mode to train an initial low-rank subspace considering the first p frames (the default is $p = 20$ [30]). The initial transformation (τ in (11.48)) is estimated by using a similarity transformation taking a group of points manually chosen from the corresponding original and canonical frames. Computational results in [31] focused on rigid transformations.

11.5.2 Rigid Transformation Invariant Incremental RPCA

We assume that the observed video sequence that suffers from jitter is represented by the matrix $D \in \mathbb{R}^{m\times n}$, where each video frame, labeled as $\mathbf{d}_k$ $k \in \{1, 2, \ldots, n\}$, is a column of D. This notation is also used for L and S, the low-rank and sparse components. Furthermore, we will assume that jitter-free video is represented by D^*, and that the observed video sequence results from applying the set of rigid transformations $\mathcal{T}_k$ (see (11.42)):

$$\mathbf{d}_k = \mathcal{T}_k(\mathbf{d}_k^*) = \mathbf{h}_k * R(\mathbf{d}_k^*, \alpha_k) \tag{11.49}$$

where $R(\mathbf{d}_k^*, \alpha_k)$ denotes rotation by angle α_k and the filter $\mathbf{h}_k$ is (ideally) an "uncentered" Dirac delta function that accounts for the translation; furthermore it is important to notice that the center of rotation could be any given point, since such rotation can always be expressed as the rotation around a fixed point followed by a translation. It is also interesting to note that in this observation model, the transformation $\mathcal{T}_k(\cdot)$ is the inverse of the transformation $\tau_k(\cdot)$, used for RASL and t-GRASTA (see previous section) algorithms.

Since $\mathbf{d}_k = \mathbf{l}_k + \mathbf{s}_k$, then (11.49) implies $\mathbf{l}_k = \mathcal{T}_k(\mathbf{l}_k^*) = \mathbf{h}_k * R(\mathbf{l}_k^*, \alpha_k)$. In general, $\mathbf{h}_k$ and α_k are unknown; however, if we are able to compute the rigid transform-invariant sparse representation given by

$$\mathbf{l}_k = \sum_n \mathbf{g}_{k,n} * R(\mathbf{l}_k^*, \beta_{k,n}), \tag{11.50}$$

where $*$ represents convolution, then (11.42) can be incrementally solved via

$$\begin{aligned}\arg\min_{L,S,G} \frac{1}{2}\sum_k \left\| \sum_n \mathbf{g}_{k,n} * R(\mathbf{l}_k^*, \beta_{k,n}) + \mathbf{s}_k - \mathbf{d}_k \right\|_2^2 + \lambda\|S\|_1 \\ + \sum_k \gamma_k \sum_n \|\mathbf{g}_{k,n}\|_1 \quad \text{s.t. } \operatorname{rank}(L^*) \le r,\end{aligned} \tag{11.51}$$

where G represents the set of filters $\{\mathbf{g}_{k,n}\}$. Before delving into details of (11.51), we first elaborate on the chosen sparse representation (11.50) as well as in its relationship with other transform-invariant sparse representations, such as [5, 8, 39, 55].

While the sparse representation (11.50) seems to consist of both variables $\mathbf{g}_{k,n}$ and $\beta_{k,n}$, we stress that rotation angles $\beta_{k,n}$ are manually sampled, thus making (11.50) an endogenous convolutional sparse representation, akin to that described in Section 11.4.1. Using this direct approach (which has also been used in [5, 39, 55]) we first sample $\beta_{k,n}$, then we compute the rotations $R(\mathbf{l}_k^*, \beta_{k,n})$. Such set of rotated images will be equivalent to the desired feature images, composed of rotated versions of previous low-rank approximations; if we assume that the background has not changed, the result would be an endogenous representation. Since the sought rigid transformation can always be expressed as just one rotation followed by a translation, then the sparse variable in (11.50) will be $\mathbf{g}_{k,n}$ since (ideally) only one filter will be active. Clearly this angle sampling is a very naive approach, which increases the complexity as the sample size increases; in order to reduce such unnecessary increased complexity, several options have been proposed: (i) although a linear sampling of 2π is used in [55], the rotations are applied to small image patches, and the update of the coefficients is reduced to the neighbor search; (ii) [39] used a predefined set of rotation operators (using linear interpolation between the image pixels), which, along the SignSearch algorithm [34], allows us to seek the solution using a gradient descent-like approach; (iii) [5] used complex kernels (filters) and coding coefficient resulting in kernels that are no longer learned through a particular orientation, but in kernels that are shift and rotation invariant.

In its algorithmic form, the incremental solution of (11.51) uses a simple and computationally efficient FFT-based algorithm [56] to rotate a given image around any given point, along with a line search to reduce complexity of the naive approach described above. We will elaborate on this in the following sub-sections; however, we first note that an incremental solution of (11.51) is possible by modifying the incremental amFastPCP algorithm (see Section 11.3.3). If we assume that we have solved problem (11.51) up to frame $k-1$ then we can first solve

$$\hat{\mathbf{h}}_k,\, \hat{\alpha}_k = \arg\min_{G} \frac{1}{2} \| \sum_n \mathbf{g}_{k,n} R(\mathbf{l}_k^*, \beta_{k,n}) + \mathbf{s}_k - \mathbf{d}_k \|_2^2 + \gamma_k \sum_n \|\mathbf{g}_{k,n}\|_1 \,, \tag{11.52}$$

via the algorithm described in Section 11.4.2. In order to simplify the notation we assume that only one filter is activated by the solution of (11.52), thus estimating the pair $\hat{\mathbf{h}}_k$ (translation) and $\hat{\alpha}_k$ (rotation).

With the solution to (11.52), problems (11.53)–(11.54)

$$\arg\min_{\mathbf{l}_k^*} \frac{1}{2} \|\hat{\mathbf{h}}_k * R(\mathbf{l}_k, \hat{\alpha}_k) + \mathbf{s}_k - \mathbf{d}_k\|_2^2 \quad \text{s.t. } \operatorname{rank}(L_k^*) \le r, \quad (L_k^* = [\mathbf{l}_1^*, \mathbf{l}_2^*, \ldots, \mathbf{l}_k^*]) \tag{11.53}$$

$$\arg\min_{\mathbf{s}_k} \frac{1}{2} \|\hat{\mathbf{h}}_k * R(\mathbf{l}_k^*, \hat{\alpha}_k) + \mathbf{s}_k - \mathbf{d}_k\|_2^2 + \lambda \|\mathbf{s}_k\|_1 \tag{11.54}$$

resemble problems (11.30)–(11.31). Furthermore, although sub-problem (11.53) is no longer a low-rank approximation problem, it can also be solved incrementally since it represents an Affinely Constrained Matrix Rank Minimization [24, 38] problem, whereas (11.54) is just element-wise shrinkage.

In the next sub-sections, we delve into the mathematical details of solving (11.51); this encompasses the following topics: (i) image rotation as 1D convolutions [56], (ii) affinely constrained matrix rank minimization [24, 38] and (iii) a practical solution to (11.52). Finally, the Rigid Transformation Invariant Incremental RPCA algorithm is fully listed.

Image rotation as 1D convolutions

In [56] a simple FFT-based algorithm was proposed to rotate an image around any given point. An image $\mathbf{d}^*$ can be rotated by α degrees, denoted by $\mathbf{d} = R(\mathbf{d}^*, \alpha)$, via a collection of independent translations applied to each row and column: (1) translate each row by $\Delta_x = -y \cdot \tan(\alpha/2)$; (2) translate each column by $\Delta_y = x \cdot \sin(\alpha)$; (3) translate each row by $\Delta_x = -y \cdot \tan(\alpha/2)$. The rotation point is defined by the origin of the x and y axes, which easily could be changed to be any point.

The above-described translations can be implemented as 1D convolutions of each row (column) with a filter-based translation operator in the spatial domain, or by 1D FFTs applied to each row (column) and then multiplied with a complex exponential with the proper phase. The computational complexity of such FFT-based implementation for a grayscale image of N_r rows and N_c columns is given by $O(10 \cdot N_c \cdot N_r(2 \cdot \log_2(N_r) + \log_2(N_c))$ (assuming that N_r and N_c are exact powers of 2), which is equivalent to computing three 2D Fourier transforms.

Although there are several methods (e.g., [32, 58]), based on [56], that target rotational alignment, these algorithms assume that either a pseudopolar transformation is applied to the observed (rotated) image, or that the Fourier domain can be sampled with arbitrary trajectories. Next we describe a simple procedure, to the best of our knowledge not previously exploited, to estimate the rotation of two images around a known point.

We start by assuming that two observed images $\mathbf{u}$ and $\mathbf{b}$ are related by $\mathbf{b} = R(\mathbf{u}, \alpha^*)$. Then the solution of

$$\alpha = \arg\min_{\alpha} \frac{1}{2} \| R(\mathbf{u}, \alpha) - \mathbf{b} \|_2^2 \tag{11.55}$$

will estimate the angle α^* that relates $\mathbf{u}$ and $\mathbf{b}$. Unfortunately (11.55) is not convex in α; however, it can be experimentally determined that (11.55) is a concave functional with a unique minimum, within a region close enough to the optimum. Moreover, since the rotation operation is computationally cheap (equivalent to three 2D FFTs), a line search procedure can be used to find its global minimizer (and thus estimate α). In particular, we choose to use a Fibonacci line search method based on [6] to solve (11.55).

Affinely Constrained Matrix Rank Minimization

The affinely constrained matrix rank minimization problem, defined by (11.56)

$$\min_{L} : \frac{1}{2} \| \mathcal{A}(L) - D \|_F^2 \text{ s.t. } \operatorname{rank}(L) \leq r \tag{11.56}$$

arises in many practical applications such as collaborative filtering and matrix completion [15], system identification and optimal control [37], low-dimensional Euclidean embedding [23], etc.

Problem (11.56) is, in general, difficult to solve. However in recent years several iterative hard thresholding (IHT) based algorithms have been proposed [24], as well as iterative reweighted methods [38] that are both simple and computationally appealing. In what follows we briefly describe a simple IHT-based method that suits our needs for an incremental solution (see (11.60) in the next sub-section).

Based on [24], we choose to compute the minimizer to (11.56) via the IHT procedure given by $L^{(j+1)} = \text{partialSVD}(L^{(j)} - \mathcal{A}^*(\mathcal{A}(L^{(j)}) - D), r)$, where $\text{partialSVD}(\cdot, r)$ represents the partial (thin) SVD, which only considers the r largest singular values of the input and $\mathcal{A}^*$ is the adjoint operator of $\mathcal{A}$.

Rigid Transformation Invariant Incremental RPCA Algorithm

As mentioned before, (11.51) includes an endogenous convolutional sparse representation that would allow us to solve the rigid transformation-invariant problem (11.42) incrementally. However the computational cost of such a solution becomes prohibitive as the number of sample angles increases: In our experiments (see Section 11.5.3) a resolution of 0.01 degrees is needed to avoid artifacts in the sparse approximation, and without any a-priori information a linear sampling with such resolution is computationally infeasible.

Instead of solving (11.51), here we directly solve for the ideal case where just one filter is activated; this gives the simplified problem

$$\arg\min_{L,S,H,\mathbf{m}\alpha} \frac{1}{2}\sum_k \|h_k * R(\mathbf{l}_k^*, \alpha_k) + \mathbf{s}_k - \mathbf{d}_k\|_F^2 + \lambda\|S\|_1 + \gamma\sum_k \|h_k\|_1 \quad \text{s.t. } \operatorname{rank}(L^*) \le r \tag{11.57}$$

where H represents the set of filters $\{h_k\}$ (just one per frame), which models the translational jitter, and $\mathbf{m}\alpha$ represents the set of angles $\{\alpha_k\}$ that models the rotational jitter.

Using the ideas mentioned before, (11.57) can be solved incrementally via the alternating minimization, recalling that $L_k^* = [\mathbf{l}_1^*, \mathbf{l}_2^*, \ldots, \mathbf{l}_k^*]$ is the set of the first k unobserved and aligned background (low-rank representation):

$$\mathbf{h}_k^{(j+1)} = \arg\min_{\mathbf{h}_k} \frac{1}{2}\|\mathbf{h}_k * R(\mathbf{l}_k^{*(j)}, \alpha_k^{(j)}) + \mathbf{s}_k^{(j)} - \mathbf{d}_k\|_F^2 + \gamma\|\mathbf{h}_k\|_1 \,, \tag{11.58}$$

$$\alpha_k^{(j+1)} = \arg\min_{\alpha_k} \frac{1}{2}\|\mathbf{h}_k^{(j+1)} * R(\mathbf{l}_k^{*(j)}, \alpha_k) + \mathbf{s}_k^{(j)} - \mathbf{d}_k\|_F^2 \,, \tag{11.59}$$

$$\mathbf{l}_k^{*(j+1)} = \arg\min_{\mathbf{l}_k^*} \frac{1}{2}\|\mathbf{h}_k^{(j+1)} * R(\mathbf{l}_k^*, \alpha_k^{(j+1)}) + \mathbf{s}_k^{(j)} - \mathbf{d}_k\|_F^2 \quad \text{s.t. } \operatorname{rank}(L_k^*) \le r \tag{11.60}$$

$$\mathbf{s}_k^{(j+1)} = \arg\min_{\mathbf{s}_k} \frac{1}{2}\|\mathbf{h}_k^{(j+1)} * R(\mathbf{l}_k^{*(j+1)}, \alpha_k^{(j+1)}) + \mathbf{s}_k^{(j)} - \mathbf{d}_k\|_F^2 + \lambda\|\mathbf{s}_k\|_1 \,. \tag{11.61}$$

Sub-problems (11.58), (11.59), and (11.61) are simple to handle:

- Sub-problem (11.58) can be solved by a variable splitting approach, transforming it into $\arg\min_{\mathbf{h},\mathbf{g}} \frac{1}{2}\|\mathbf{g} * \mathbf{u} - \mathbf{b}\|_F^2 + \gamma\|\mathbf{h}\|_1 + \frac{\mu}{2}\|\mathbf{h} - \mathbf{g}\|_2^2$ and then applying alternating minimization on $\mathbf{h}$ and $\mathbf{g}$. The $\mathbf{h}$ update is solved by soft-thresholding, and the $\mathbf{g}$ update can be efficiently solved in the Fourier domain via a special case of the method described in Section 11.4.2 (see also [61]).

- Sub-problem (11.59) is equivalent to (11.55), and can be effectively solved via the same procedure.

- the solution to sub-problem (11.61) is just element-wise shrinkage.

Sub-problem (11.60) is not as straightforward as all other sub-problems, but by noting that (i) the adjoint operator of rotation $R(., \alpha_k)$ is $R(., -\alpha_k)$ and (ii) the adjoint operator of the translation represented by filter $h_k(x, y)$ is the filter $h_k(-x, -y)$, then (11.60) can be effectively solved via the IHT algorithm used to solve (11.56).

11.5.3 Computational Results for the Rigid Transformation Invariant Incremental RPCA (incPCP_TI) Algorithm

We have used four real video sets for tests:

- V352: a 352 × 224 pixel, 1200-frame color video sequence of 40 seconds at 30 fps, from a sidewalk surveillance camera with real (mainly translational) jitter, from the Change Detection 2014 dataset [60] also used in [9, Ch. 16] and in [30].

- V640: a 640 × 480 pixel, 400-frame color video sequence of 26.66 seconds at 15 fps, from the Lankershim Boulevard traffic surveillance dataset [3, cam. 3].

- V720: a 720 × 480 pixel, 1150-frame color video sequence, from a badminton court, where several players can be seen in action, from the Change Detection 2014 dataset [60].
- V1920: a 1920 × 1088 pixel, 900-frame color video sequence of 36 seconds at 25 fps, from the Neovison2 public space dataset [1, Tower 3rd video].

Test video	Initialize (sec.)	Average per frame (sec.)				
	t-GRASTA grayscale	incPCP_TI grayscale		incPCP_TI color		t-GRASTA grayscale
	Standard MATLAB	Standard MATLAB	GPU-enabled MATLAB	Standard MATLAB	GPU-enabled MATLAB	Standard MATLAB
V352	65.2	0.39	0.27	0.63	0.37	0.88
V640	232.6	1.49	0.60	2.45	0.82	3.40
V720	339.8	1.67	0.55	2.95	0.96	4.28
V1920	1858.7	10.20	2.9	18.04	5.46	25.6

TABLE 11.3 Elapsed time to initialize the t-GRASTA [29] and average processing time per frame for t-GRASTA and incPCP_TI. Video sets V352 and V720 have real jitter, whereas sets V640, V1920 have synthetic jitter: random uniformly distributed translations (±T pixel, T= 5) and rotations ($\pm\alpha$ degrees, $\alpha = 0.5$).

All simulations presented in this section, related to the above-described algorithm, have been carried out using MATLAB-only code[35] running on an Intel i7-4710HQ quad-core (2.5 GHz, 6MB Cache, 32GB RAM) based laptop with a nvidia GTX980M GPU card. Videos "V640" and "V1920" were synthetically jittered with random uniformly distributed translations (±T pixel, T= {5, 10}) and rotations ($\pm\alpha$ degrees, $\alpha = \{0.5, 1\}$), and used as controlled datasets with known alignment, whereas videos "V352" and "V720" do have real jitter. Moreover, video "V720" has manually segmented ground-truth (moving objects) from frame 800 onwards; this is a very challenging video because, besides its mainly translational jitter, blurred frames are observable, generating the side effect of an ever-changing background.

We compare our results with t-GRASTA [31] using a t-GRASTA implementation [30] for grayscale videos by the authors of [31] (using its default parameters). We do not compare it with RASL [41], since it is a batch method. Furthermore, t-GRASTA implementation is MATLAB based and takes advantage of MEX interfaces, and while MATLAB implementation of incPCP_TI does not use any MEX interface, it comes in two flavors: (i) one that uses standard single-thread MATLAB code and (ii) one that takes advantage of (mainly linear algebra, FFT, and convolution) GPU-enabled MATLAB functions.

In order to assess the performance of the incPCP_TI algorithm we present two kinds of results: (i) computational performance measured as the time to process a given video, summarized in Table 11.3, and (ii) reconstruction quality (see Figures 11.11 and 11.14) measured by $\frac{\|S_{GT}-S_P\|_1}{N}$ where S_{GT} is the "ground truth" sparse video approximation and S_P is the sparse video approximation of either t-GRASTA or incPCP_TI algorithms, and N is the number of pixels per frame, which is the same value for either the grayscale or color case.

Without taking into account the batch initializations for t-GRASTA, the average processing time per frame listed in Table 11.3 shows that the incPCP_TI algorithm (MATLAB-

[35] This code is publicly available [47].

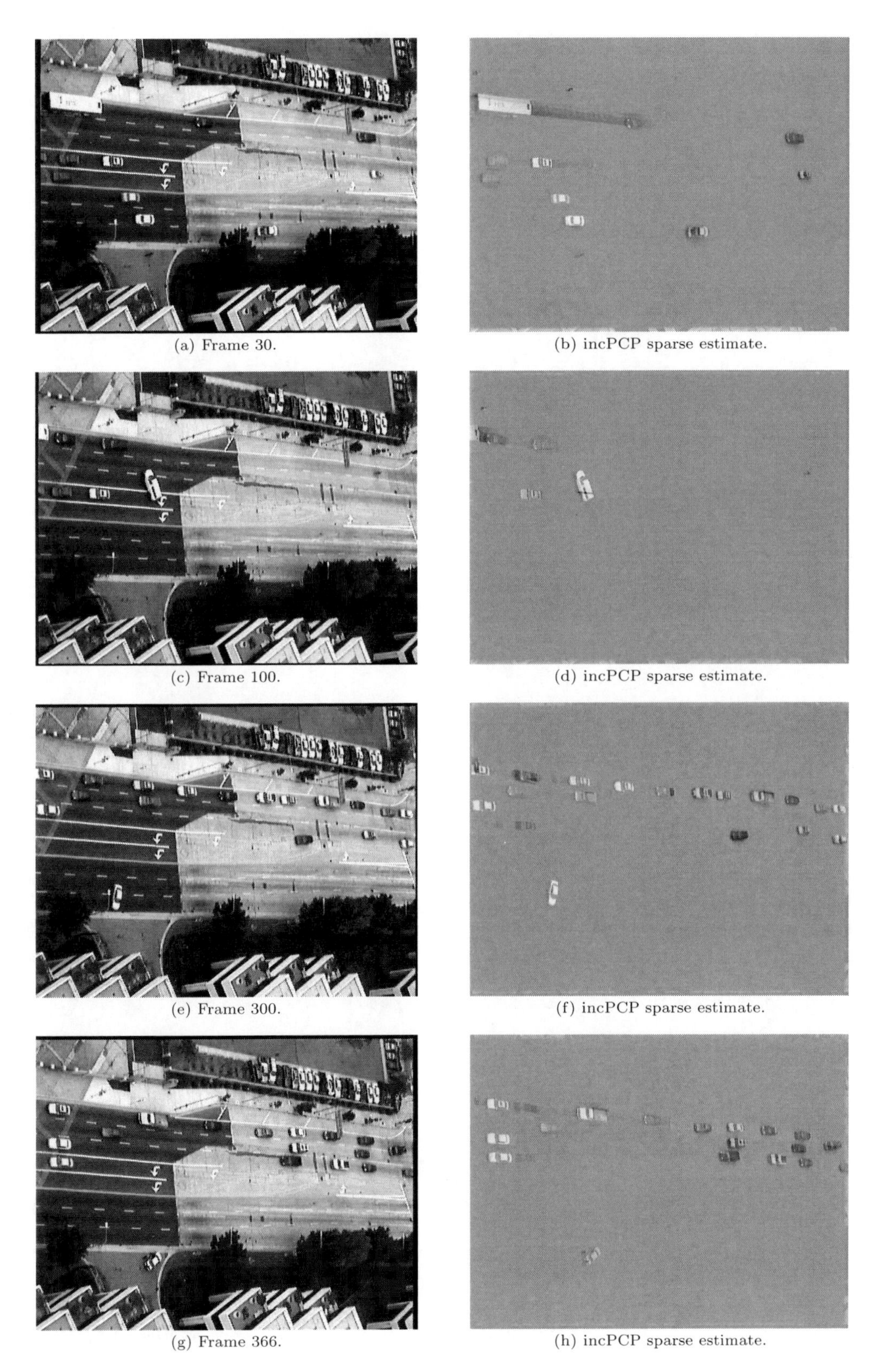

(a) Frame 30.

(b) incPCP sparse estimate.

(c) Frame 100.

(d) incPCP sparse estimate.

(e) Frame 300.

(f) incPCP sparse estimate.

(g) Frame 366.

(h) incPCP sparse estimate.

FIGURE 11.10 Synthetically jittered and sparse estimates via the incPCP_TI algorithm for the V640 (640×480 pixel) test video.

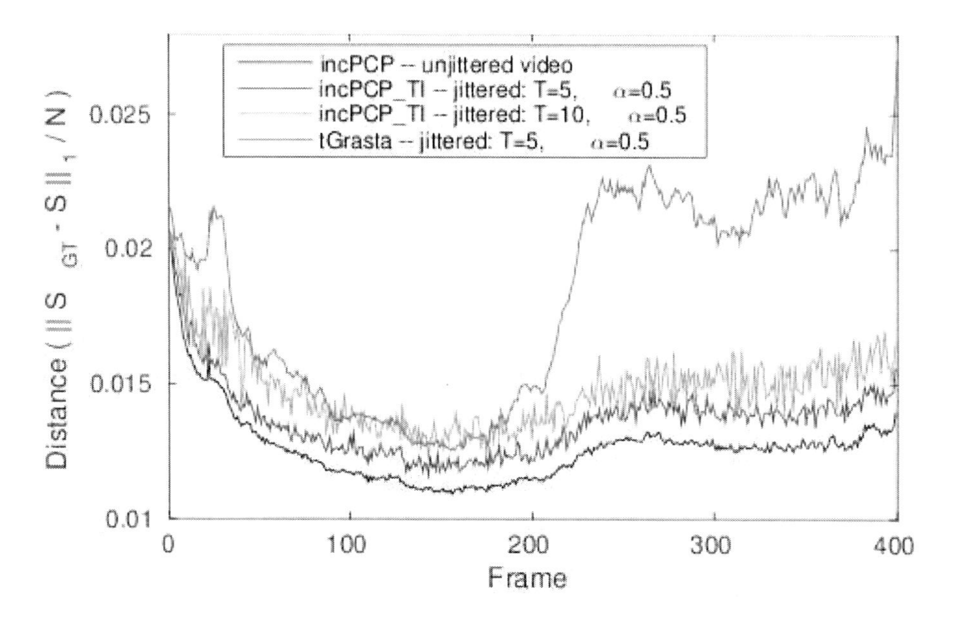

FIGURE 11.11 (See color insert.) Sparse approximation (V640) frame distance measure by $\frac{\|S_{GT}-S_P\|_1}{N}$ where S_{GT} is the ground-truth computed via the (batch) iALM algorithm (20 outer loops) and S_P is the (i) sparse component computed via incPCP_TI method (red, green) and t-Grasta (magenta) for different synthetic jitter conditions or (ii) the sparse component for the unjittered case (blue), computed via [49], provided as a baseline.

only version) is 2.5 times faster than t-GRASTA, when processing grayscale videos. Furthermore, when we consider the initialization time needed by t-GRASTA (about 100 times the time needed to process one frame), the incPCP_TI algorithm looks even more computational appealing. Furthermore, the GPU-enabled incPCP_TI implementation is between $1.3 \sim 3.3$ times faster than the incPCP_TI MATLAB-only version; This performance improvement is specially noticeable for the V640, V720, and V1920 test videos.

Reconstruction quality results are presented for the V640 (synthetically jittered) and V729 (real jitter); for the other two video sets, results are similar and thus omitted. For test video V640, a proxy ground-truth was used: the sparse video approximation computed via the (batch) iALM algorithm [35] with 20 outer loops, whereas for the test video V720, the ground-truth is given (in the form of manually segmented moving objects) from frame 800 onwards.

In Figure 11.11 the V640 reconstruction quality results show that the incPCP_TI algorithm has better properties than t-GRASTA: The red and magenta lines represent the reconstruction quality for the incPCP_TI and t-GRASTA algorithms, respectively, applied to the same (synthetically jittered) video with random uniformly distributed translations (T= 5 pixel) and rotations (α = 0.5 degrees). incPCP_TI has slightly better quantitative results than t-GRASTA and both results follow more or less the same tendency; however, a clear difference is observed from frame 200 onwards; this is due to the characteristics of V640 and t-GRASTA's background adaptation capabilities (which are similar to GRASTA's, see Sections 11.3.1 as well as computational results in Section 11.3.3): (i) There are only a few moving objects in V640 from frame 1 up to (more or less) 200, afterwards a large number of moving objects start to appear and thus this can be interpreted as a change in the background properties, as can be seen in Figure 11.10; (ii) t-GRASTA tries to adapt to this new environment; however, it is known that t-GRASTA has a delay for this type of situation,

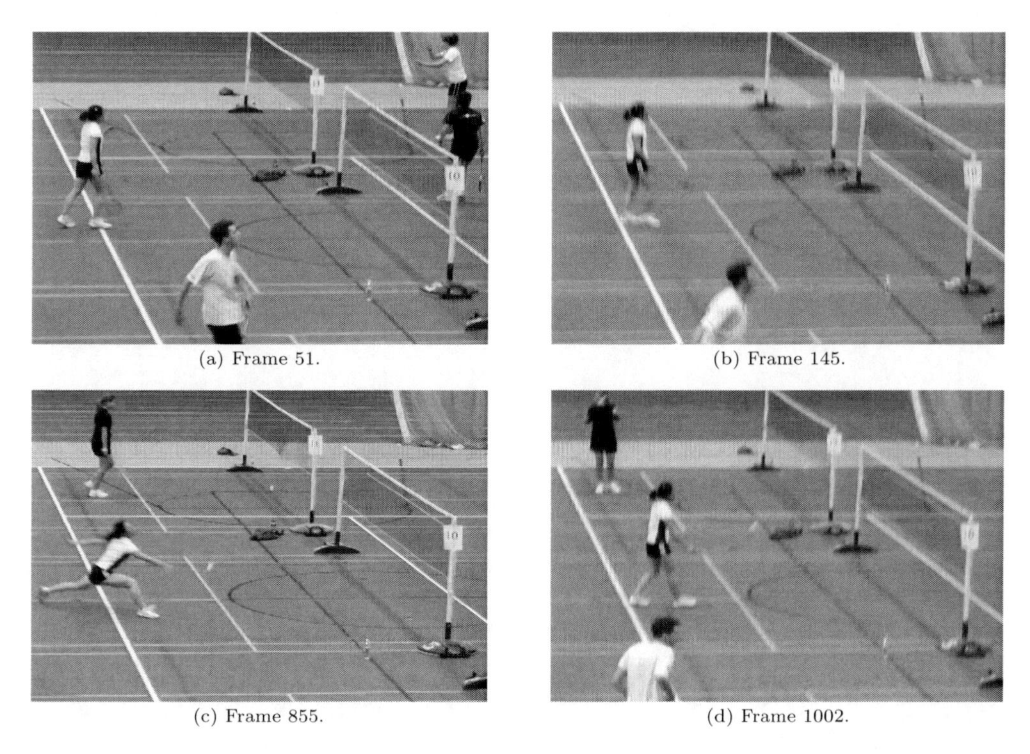

(a) Frame 51. (b) Frame 145.

(c) Frame 855. (d) Frame 1002.

FIGURE 11.12 Selected frames from video V720, where real jitter can be observed. Moreover, several frames are blurred, as shown in sub-figure (b), which severely affects the quality of sparse estimates (see Figure 11.13).

resulting in a large gap between the reconstruction quality for incPCP_TI and t-GRASTA from frame 200 onwards.

The incPCP_TI reconstruction quality for jitter T= 10 pixel, α = 0.5 degrees is also included in Figure 11.11, along with the reconstruction quality of the incPCP algorithm (see Section 11.3.3) applied to the unjittered video, which is provided as baseline.

Similarly, in Figure 11.14 the V720 reconstruction quality results also show that the incPCP_TI has an edge over t-GRASTA. Four selected frames from V720 as well as their corresponding sparse estimates for t-GRASTA and incPCP_TI are displayed in Figures 11.12 and 11.13. It is worth noting that in Figure 11.12 the real jitter that affects V720 is observable and that 11.12(b) is clearly blurred; such blurred frames are not uncommon throughout V720, and do affect the sparse estimates correspondingly, as shown in Figures 11.13(c) and 11.13(d). Finally, for the sake of completeness, in Figure 11.14(b) the reconstruction quality metric is depicted for the color case. Finally, this section is closed by stating that the subjective quality of incPCP_TI's sparse approximation is good at capturing the moving objects (large and small); however, some artifacts are also observable, especially when real jitter is involved.

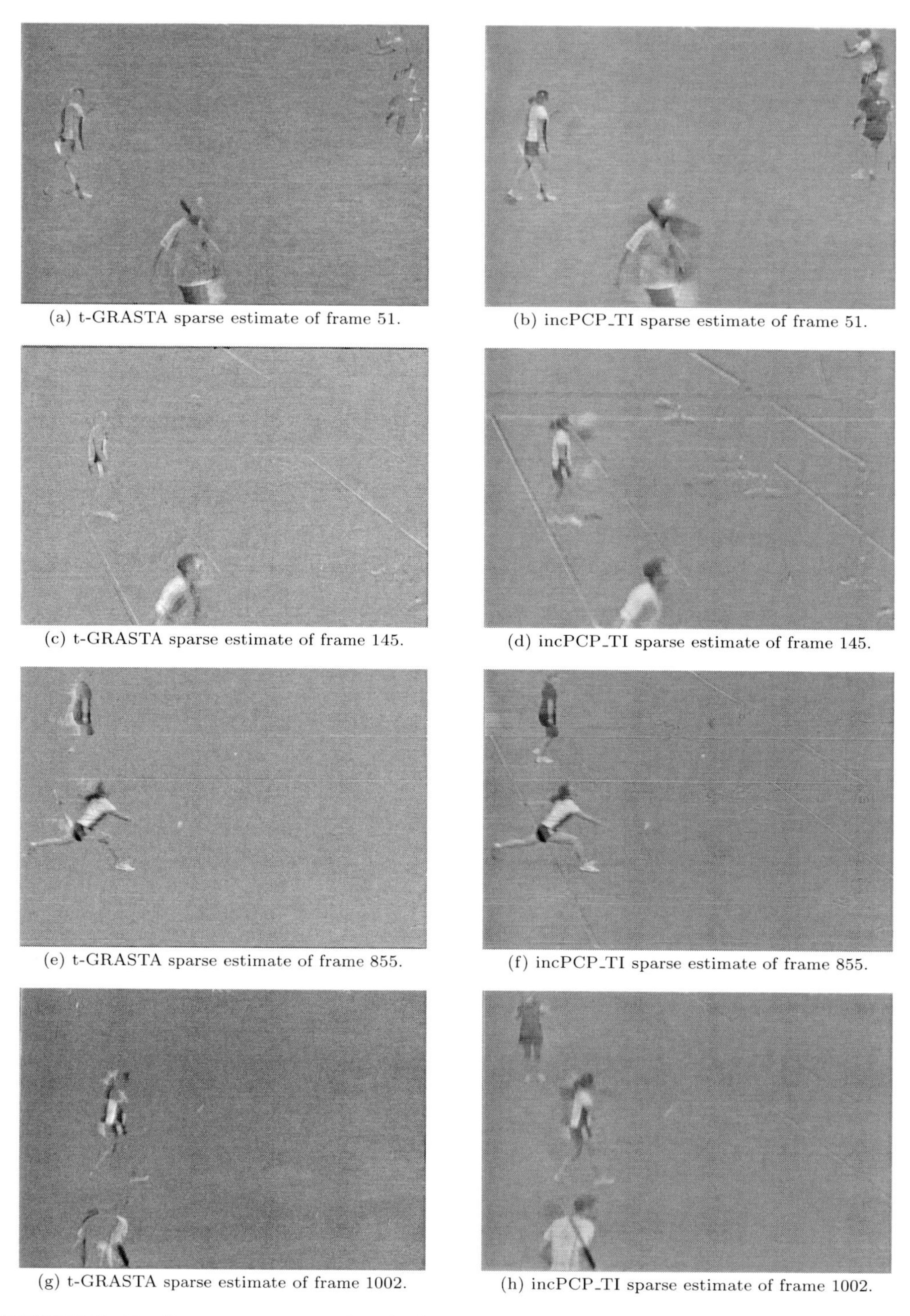

(a) t-GRASTA sparse estimate of frame 51.

(b) incPCP_TI sparse estimate of frame 51.

(c) t-GRASTA sparse estimate of frame 145.

(d) incPCP_TI sparse estimate of frame 145.

(e) t-GRASTA sparse estimate of frame 855.

(f) incPCP_TI sparse estimate of frame 855.

(g) t-GRASTA sparse estimate of frame 1002.

(h) incPCP_TI sparse estimate of frame 1002.

FIGURE 11.13 Sparse estimates via the t-GRASTA and incPCP_TI algorithms for the V720 (720×480 pixel) test video. Note that t-GRASTA can only process grayscale videos.

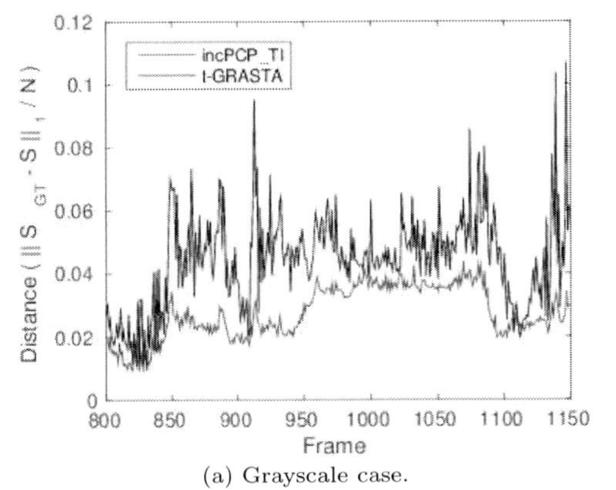

(a) Grayscale case.

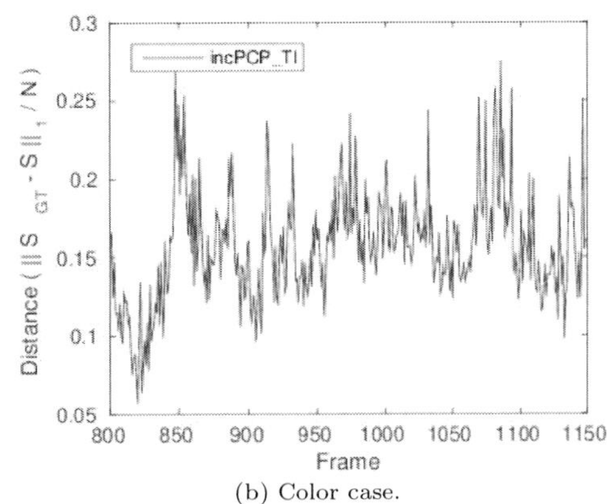

(b) Color case.

FIGURE 11.14 (See color insert.) Sparse approximation (V720) frame distance measure by $\frac{\|S_{GT}-S_P\|_1}{N}$ where S_{GT} is the manually segmented ground-truth and S_P is the sparse approximation computed via (a) grayscale (blue) t-GRASTA and (red) incPCP_TI methods, or (b) color (red) incPCP_TI method, and N is the number of pixels per frame (used as a normalization factor).

11.6 Conclusion

The initial contribution of this chapter is the development of a fully-incremental[36] algorithm for solving a variant of the RPCA problem. It has an extremely low memory footprint, and a computational complexity that allows real-time processing, even when implemented purely in MATLAB. Computational experiments indicate that this algorithm provides similar performance to the partially incremental algorithms while being substantially faster.

The second topic considered is the development of variants of the RPCA problem that are capable of dealing with video sequences with non-stationary backgrounds due to the motion of the camera capturing the scene. A subspace model closely related to that of the Sparse Subspace clustering method is shown to be effective when the background motion is suitably constrained, e.g., moving very slowly, or panning back and forth across the same scene, but does not perform well for more realistic types of motion. An extension of this model using convolutional sparse representations is able to model a much more rapidly translating background, but is computationally expensive and unable to directly model transformations other than translation.

The final contribution is a synthesis of the incremental algorithm with a method that is inspired by the convolutional sparse representation approach, but is vastly less computationally expensive. This algorithm is able to process video with both translating and rotating background while operating at near real-time speed. Computational experiments indicate that it is both faster and delivers a superior sparse representation to t-GRASTA, the only competing algorithm of which we are aware.

Of the possible topics for future work, two stand out as having potential for significant impact. First, the rank-1 update and FFT-based image rotation procedures, which are heavily used by the final form of the presented rigid transformation invariant RPCA algorithm, are easily parallelizable and thus a C/C++ CUDA implementation of the proposed

[36] We describe it as fully incremental to distinguish it from other algorithms that are described as incremental, but that have a batch initialization procedure that requires processing a large block of frames, making them substantially less suitable for real-time application.

algorithm should be explored. Due to the low memory footprint and overall computational complexity properties of our algorithm, such a parallel implementation could be deployed in mobile CUDA-aware processors such the Tegra K1 or X1 for in-situ processing. Second, it is desirable to model more realistic types of motions, such as those observed when a video is acquired by cameras mounted in UAVs or self-driving cars. The framework proposed by (11.42) does allow more general transformations, but substantial work is required to develop computationally efficient methods that would support them.

References

1. USC Neovision2 Project. DARPA Neovision2, data available from `http://ilab.usc.edu/neo2/`.
2. Video sequences for foreground object detection from complex background. available from `http://perception.i2r.a-star.edu.sg/bk_model/bk_index.html`.
3. Lankershim Boulevard dataset, January 2007. U.S. Department of Transportation Publication FHWA-HRT-07-029, data available from `http://ngsim-community.org/`.
4. C. Baker, K. Gallivan, and P. Van Dooren. Low-rank incremental methods for computing dominant singular subspaces. *Linear Algebra and its Applications*, 436(8):2866–2888, 2012.
5. Q. Barthélemy, A. Larue, A. Mayoue, D. Mercier, and J. Mars. Shift & 2d rotation invariant sparse coding for multivariate signals. *IEEE Transactions on Signal Processing*, 60(4):1597–1611, April 2012.
6. H. Benson. An outer approximation algorithm for generating all efficient extreme points in the outcome set of a multiple objective linear programming problem. *Journal of Global Optimization*, 13(1):1–24, 1998.
7. J. Bezdek and R. Hathaway. Some notes on alternating optimization. In *Advances in Soft Computing - AFSS 2002*, volume 2275 of *Lecture Notes in Computer Science*, pages 288–300. Springer Berlin Heidelberg, 2002.
8. T. Blumensath and M. Davies. On shift-invariant sparse coding. *Lecture Notes in Computer Science, Independent Component Analysis and Blind Signal Separation*, 3195:1205–1212, 2004.
9. T. Bouwmans, F. Porikli, B. Hoferlin, and A. Vacavant. *Background Modeling and Foreground Detection for Video Surveillance.* Chapman and Hall/CRC, July 2014.
10. T. Bouwmans and E. Zahzah. Robust PCA via principal component pursuit: A review for a comparative evaluation in video surveillance. *Computer Vision and Image Understanding*, 122:22–34, 2014.
11. S. Boyd, N. Parikh, E. Chu, B. Peleato, and J. Eckstein. Distributed optimization and statistical learning via the alternating direction method of multipliers. *Found. Trends Mach. Learn.*, 3(1):1–122, January 2011.
12. M. Brand. Fast low-rank modifications of the thin singular value decomposition. *Linear Algebra and its Applications*, 415(1):20–30, 2006.
13. S. Brutzer, B. Höferlin, and G. Heidemann. Evaluation of background subtraction techniques for video surveillance. In *Computer Vision and Pattern Recognition, CVPR 2011*, pages 1937–1944. IEEE, 2011.
14. E. Candès, X. Li, Y. Ma, and J. Wright. Robust principal component analysis? *Journal of the ACM*, 58:1–37, June 2011.
15. E. Candes and B. Recht. Exact matrix completion via convex optimization. *Foundations of Computational Mathematics*, 9(6):717–772, 2009.
16. Y. Chahlaoui, K. Gallivan, and P. Van Dooren. Computational information retrieval. In *In Computational Information Retrieval*, chapter, "An Incremental Method for Comput-

ing Dominant Singular Spaces," pages 53–62. SIAM, 2001.
17. S. Chen, D. Donoho, and M. Saunders. Atomic decomposition by basis pursuit. *SIAM Journal on Scientific Computing*, 20(1):33–61, 1998.
18. E. Dyer, A. Sankaranarayanan, and R. Baraniuk. Greedy feature selection for subspace clustering. *Journal of Machine Learning Research*, April 2012. Submitted.
19. E. Elhamifar and R. Vidal. Sparse subspace clustering. In *IEEE Conference on Computer Vision and Pattern Recognition, CVPR 2009*, pages 2790–2797, June 2009.
20. E. Elhamifar and R. Vidal. Sparsity in unions of subspaces for classification and clustering of high-dimensional data. In *Annual Allerton Conference on Communication, Control, and Computing (Allerton)*, pages 1085–1089, 2011.
21. E. Elhamifar and R. Vidal. Structured sparse recovery via convex optimization. Technical Report arXiv:1104.0654v2, arXiv, 2011.
22. E. Elhamifar and R. Vidal. Sparse subspace clustering: Algorithm, theory, and applications. *IEEE Transactions on Pattern Analysis and Machine Intelligence*, 35(11):2765–2781, 2013.
23. M. Fazel, H. Hindi, and S. Boyd. Log-det heuristic for matrix rank minimization with applications to Hankel and Euclidean distance matrices. Proceedings of the 2003, In *American Control Conference*, 2003 volume 3, pages 2156–2162 vol. 3, June 2003.
24. D. Goldfarb and S. Ma. Convergence of fixed-point continuation algorithms for matrix rank minimization. *ArXiv e-prints*, Jun. 2009.
25. H. Guo. Practical ReProCS code. `http://www.ece.iastate.edu/~hanguo/PracReProCS.html`.
26. Han Guo, Chenlu Qiu, and N. Vaswani. An online algorithm for separating sparse and low-dimensional signal sequences from their sum. *IEEE Transactions on Signal Processing*, 62(16):4284–4297, August 2014.
27. J. He, L. Balzano, and J. Lui. Online robust subspace tracking from partial information. *CoRR*, abs/1109.3827, 2011. submitted.
28. J. He, L. Balzano, and A. Szlam. GRASTA code. `https://sites.google.com/site/hejunzz/grasta`.
29. J. He, L. Balzano, and A. Szlam. Incremental gradient on the Grassmannian for online foreground and background separation in subsampled video. In *IEEE Conference on Computer Vision and Pattern Recognition, CVPR 2012*, pages 1568–1575, June 2012.
30. J. He, D. Zhang, L. Balzano, and T. Tao. t-GRASTA code. `https://sites.google.com/site/hejunzz/t-grasta`.
31. J. He, D. Zhang, L. Balzano, and T. Tao. Iterative Grassmannian optimization for robust image alignment. *Image and Vision Computing*, 32(10):800–813, 2014.
32. Y. Keller, A. Averbuch, and M. Israeli. Pseudopolar-based estimation of large translations, rotations, and scalings in images. *IEEE Transactions on Image Processing*, 14(1):12–22, January 2005.
33. R. Larsen. PROPACK. Functions for computing the singular value decomposition of large and sparse or structured matrices, available from `http://sun.stanford.edu/~rmunk/PROPACK`.
34. H. Lee, A. Battle, R. Raina, and A. Ng. Efficient sparse coding algorithms. In *Proceedings of the Neural Information Processing Systems (NIPS)*, pages 801–808, 2007.
35. Z. Lin, M. Chen, and Y. Ma. The augmented Lagrange multiplier method for exact recovery of corrupted low-rank matrices. arXiv:1009.5055v2, 2011.
36. C. Liu, Z. Lin, and Y. Yu. Robust subspace segmentation by low-rank representation. In *International Conference on Machine Learning, ICML 2010*, pages 663–670, 2010.
37. Z. Liu and L. Vandenberghe. Interior-point method for nuclear norm approximation with application to system identification. *SIAM Journal on Matrix Analysis and Applications*, 31(3):1235–1256, November 2009.

38. K. Mohan and M. Fazel. Iterative reweighted algorithms for matrix rank minimization. *Journal of Machine Learning Research*, 13(1):3441–3473, November 2012.
39. M. Mørup and M. Schmidt. Transformation invariant sparse coding. In *IEEE International Workshop on Machine Learning and Signal Processing, MLSP 2011*, pages 1–6, 2011.
40. B. Nasihatkon and R. Hartley. Graph connectivity in sparse subspace clustering. In *IEEE Conference on Computer Vision and Pattern Recognition, CVPR 2011*, pages 2137–2144, June 2011.
41. Y. Peng, A. Ganesh, J. Wright, W. Xu, and Y. Ma. RASL: Robust alignment by sparse and low-rank decomposition for linearly correlated images. *IEEE Transactions on Pattern Analysis and Machine Intelligence*, 34(11):2233–2246, 2012.
42. G. Peyré. Manifold models for signals and images. *Computer Vision and Image Understanding*, 113(2):249–260, 2009.
43. C. Qiu and N. Vaswani. Real-time robust principal components' pursuit. *CoRR*, abs/1010.0608, 2010.
44. C. Qiu and N. Vaswani. Support predicted modified-cs for recursive robust principal components pursuit. In *IEEE International Symposium on Information Theory, ISIT 2011*, 2011.
45. C. Qiu and N. Vaswani. Automated recursive projected CS (ReProCS) for real-time video layering. In *IEEE Conference on Computer Vision and Pattern Recognition, CVPR 2012*, 2012.
46. P. Rodriguez and B. Wohlberg. incremental PCP simulations. http://sites.google.com/a/istec.net/prodrig/Home/en/pubs/incpcp.
47. P. Rodriguez and B. Wohlberg. Rigid transformation invariant incremental PCP simulations. http://sites.google.com/a/istec.net/prodrig/Home/en/pubs/incpcpti.
48. P. Rodriguez and B. Wohlberg. Fast principal component pursuit via alternating minimization. In *IEEE International Conference on Image Processing, ICIP 2013*, pages 69–73, Melbourne, Australia, September 2013.
49. P. Rodriguez and B. Wohlberg. A MATLAB implementation of a fast incremental principal component pursuit algorithm for video background modeling. In *IEEE International Conference on Image Processing, ICIP 2014*, pages 3414–3416, Paris, France, October 2014.
50. P. Rodriguez and B. Wohlberg. Video background modeling under impulse noise. In *IEEE International Conference on Image Processing, ICIP 2014*, pages 1041–1045, Paris, France, October 2014.
51. L. Rudin, S. J. Osher, and E. Fatemi. Nonlinear total variation based noise removal algorithms. *Physica D. Nonlinear phenomena*, 60(1-4):259–268, November 1992.
52. F. Seidel, C. Hage, and M. Kleinsteuber. pROST: a smoothed lp-norm robust online subspace tracking method for background subtraction in video. *Machine Vision and Applications*, 25(5):1227–1240, 2014.
53. K. Simonson and T. Ma. Robust real-time change detection in high jitter. Sandia Report SAND2009-5546, Sandia National Laboratories, 2009.
54. M. Soltanolkotabi and E. Candès. A geometric analysis of subspace clustering with outliers. *CoRR*, abs/1112.4258, 2011.
55. Y. Tomokusa, M. Nakashizuka, and Y. Iiguni. Sparse image representations with shift and rotation invariance constraints. In *International Symposium on Intelligent Signal Processing and Communication Systems, ISPACS 2009*, pages 256–259, January 2009.
56. M. Unser, P. Thevenaz, and L. Yaroslavsky. Convolution-based interpolation for fast high-quality rotation of images. *IEEE Transactions on Image Processing*, 4(10):1371–1381, October 1995.
57. A. Vacavant, T. Chateau, and A. Wilhelmand L. Lequièvre. A benchmark dataset for outdoor foreground/background extraction. In *Computer Vision - ACCV 2012 Workshops*,

volume 7728 of *Lecture Notes in Computer Science*, pages 291–300. Springer Berlin Heidelberg, 2013.
58. G. Vaillant, C. Prieto, C. Kolbitsch, G. Penney, and T. Schaeffter. Retrospective rigid motion correction in k-space for segmented radial MRI. *IEEE Transactions on Medical Imaging*, 33(1):1–10, January 2014.
59. E. van den Berg and M. Friedlander. Theoretical and empirical results for recovery from multiple measurements. *IEEE Transactions on Information Theory*, 56(5):2516–2527, May 2010.
60. Yi Wang, P.-M. Jodoin, F. Porikli, J. Konrad, Y. Benezeth, and P. Ishwar. Cdnet 2014: An expanded change detection benchmark dataset. In *2014 IEEE Conference on Computer Vision and Pattern Recognition Workshops (CVPRW)*, pages 393–400, June 2014.
61. B. Wohlberg. Efficient convolutional sparse coding. In *IEEE International Conference on Acoustics, Speech, and Signal Processing, ICASSP 2014*, pages 7173–7177, Florence, Italy, May 2014.
62. B. Wohlberg. Endogenous convolutional sparse representations for translation invariant image subspace models. In *IEEE International Conference on Image Processing, ICIP 2014*, pages 2859–2863, Paris, France, October 2014.
63. B. Wohlberg. Efficient algorithms for convolutional sparse representations. Manuscript currently under review, 2015.
64. B. Wohlberg, R. Chartrand, and J. Theiler. Local principal component pursuit for nonlinear datasets. In *International Conference on Acoustics, Speech, and Signal Processing, ICASSP 2012*, pages 3925–3928, March 2012.
65. J. Wright, A. Ganesh, S. Rao, Y. Peng, and Y. Ma. Robust principal component analysis: Exact recovery of corrupted low-rank matrices via convex optimization. In *Annual Conference on Neural Information Processing Systems, NIPS 2009*, pages 2080–2088, 2009.
66. L. Xiong, X. Chen, and J. Schneider. Direct robust matrix factorization for anomaly detection. In *IEEE International Conference on Data Mining*, pages 844–853, December 2011.
67. J. Xu, V. Ithapu, L. Mukherjee, J. Rehg, and V. Singh. GOSUS code. `http://pages.cs.wisc.edu/~jiaxu/projects/gosus/`.
68. J. Xu, V. Ithapu, L. Mukherjee, J. Rehg, and V. Singh. GOSUS: Grassmannian online subspace updates with structured-sparsity. *IEEE International Conference on Computer Vision, ICCV 2013*, pages 3376–3383, December 2013.
69. M. Zeiler, D. Krishnan, G. Taylor, and R. Fergus. Deconvolutional networks. *IEEE Conference on Computer Vision and Pattern Recognition, CVPR 2010*, pages 2528–2535, June 2010.
70. T. Zhou and D. Tao. GoDec: randomized low-rank & sparse matrix decomposition in noisy case. In *ACM International Conference on Machine Learning, ICML 2011*, pages 33–40, June 2011.

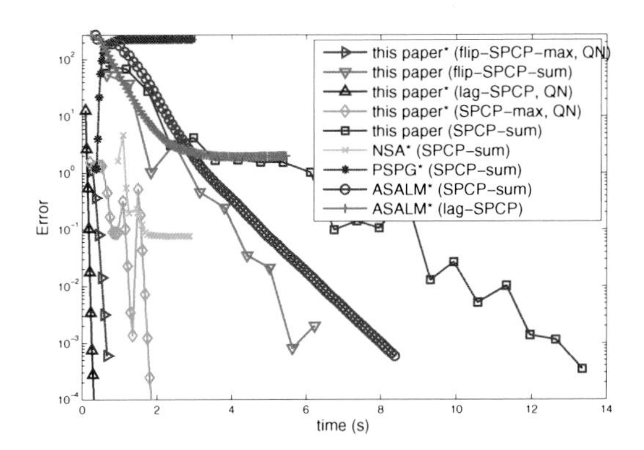

FIGURE 3.4 The exponential noise test. The asterisk in the legend means the method uses a fast SVD.

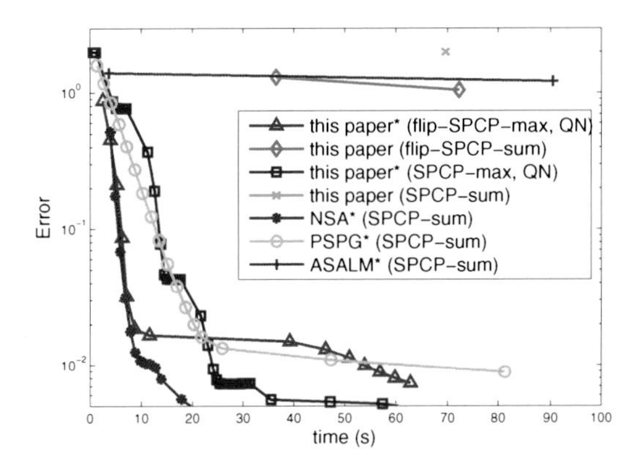

FIGURE 3.5 The 1500×1500 synthetic noise test.

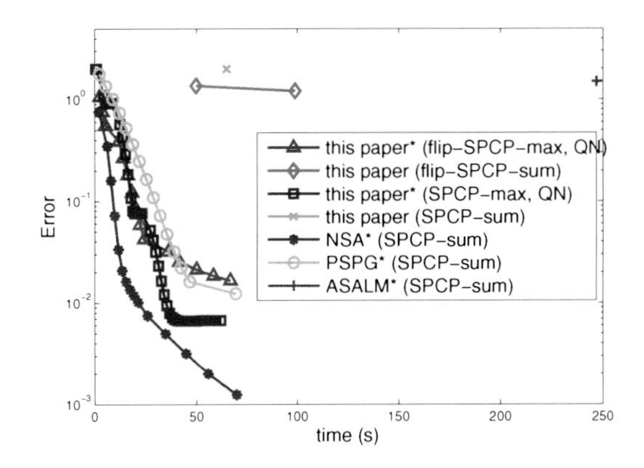

FIGURE 3.6 Second 1500×1500 synthetic noise test.

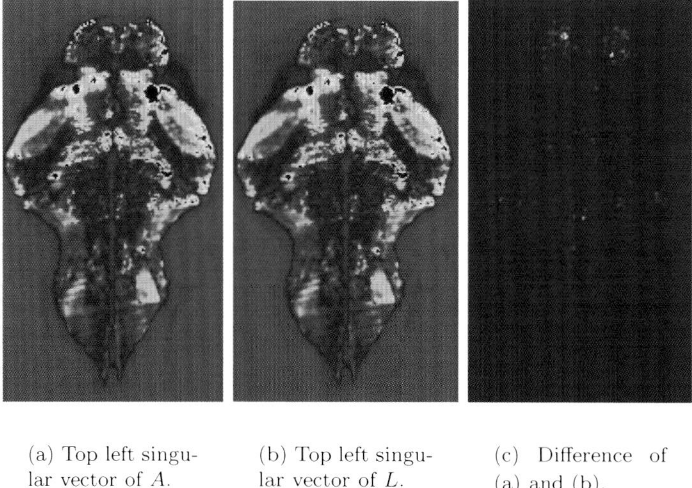

(a) Top left singular vector of A.

(b) Top left singular vector of L.

(c) Difference of (a) and (b).

FIGURE 3.10 Top left singular vector of A and of L, as well as their difference.

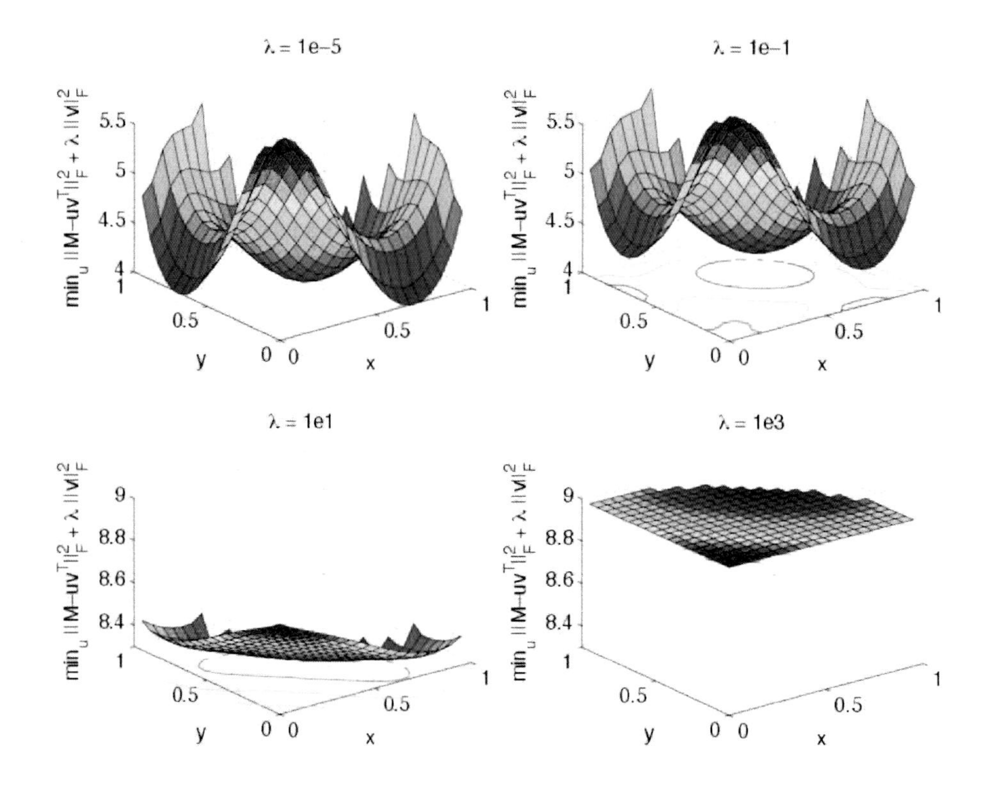

FIGURE 6.6 Visualization of the cost function in (6.41) for a rank-1 3×3 matrix $\mathbf{X}$ for several values of λ, showing the several local minima existing in the original problem, and that the smoothing induced by the nuclear norm convexifies the problem but its global optima may not necessarily coincide with the original problem's position.

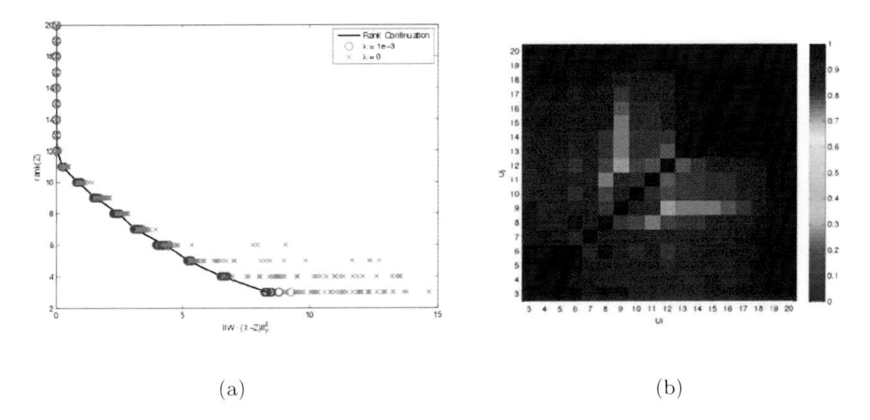

(a) (b)

FIGURE 6.7 Intuition behind Rank Continuation. 6.7(a) Visualization of minima attained for 100 random runs of Algorithm 6.1 for a rank-3 25×20 matrix $\mathbf{X}$ with and without regularization λ shows the several local minima existing in the cost function landscape, and that the smoothing induced by the nuclear norm allows for avoiding some of these. The rank continuation (solid) attains the global optima in the last rank. 6.7(b) Normalized Subspace Inclusion index $NSI(\mathbf{Z}_i, \mathbf{Z}_j)$ measured between the subspaces for each solution step in the continuation for a rank-3 25×20, showing the desired rank-3 subspace is included in the one obtained for the convex region (20).

FIGURE 6.13 An example pair of frames (0 and 99) of the House dataset. An example pair of frames with the correspondence generated by FGM [105], where the blue lines indicate incorrect matches.

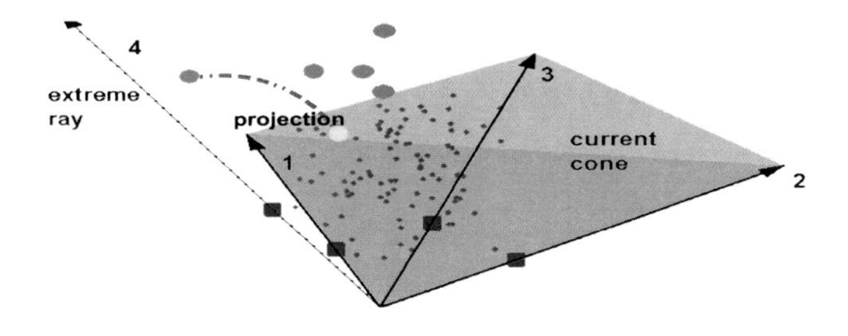

FIGURE 7.2 Geometrical illustration of the algorithm.

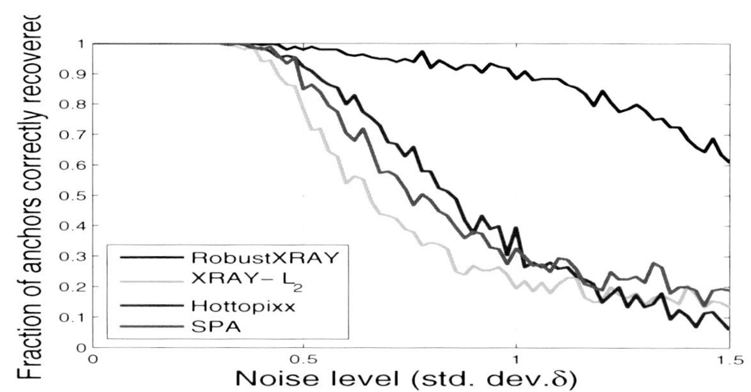

FIGURE 7.3 Sparse noise case: anchor recovery rate versus noise level (best viewed in color)

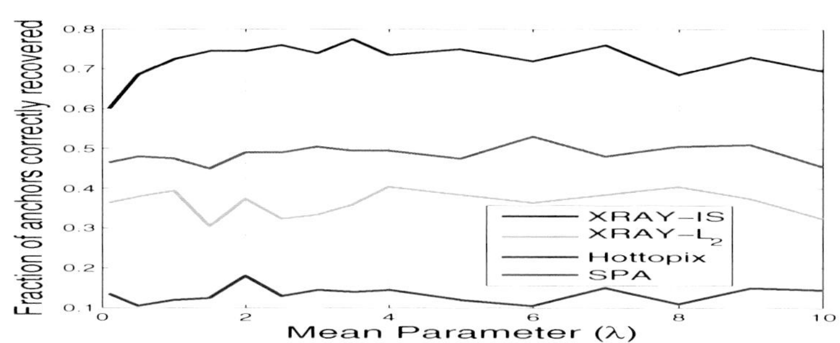

FIGURE 7.4 Data matrix generated from exponential distribution: anchor recovery rate versus mean parameter (best viewed in color)

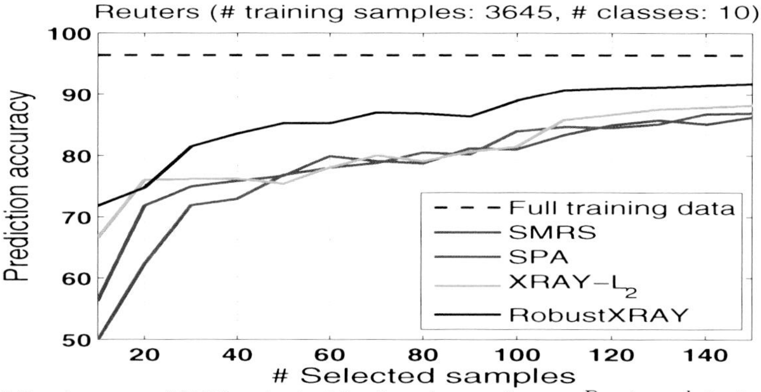

FIGURE 7.5 Accuracy of SVM trained with selected exemplars on *Reuters data* (best viewed in color)

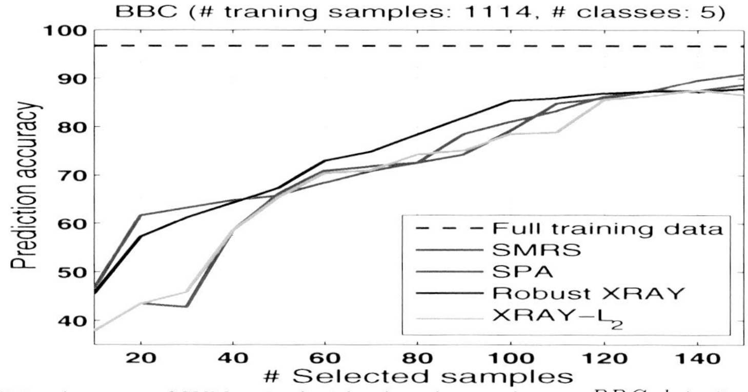

FIGURE 7.6 Accuracy of SVM trained with selected exemplars on *BBC data* (best viewed in color)

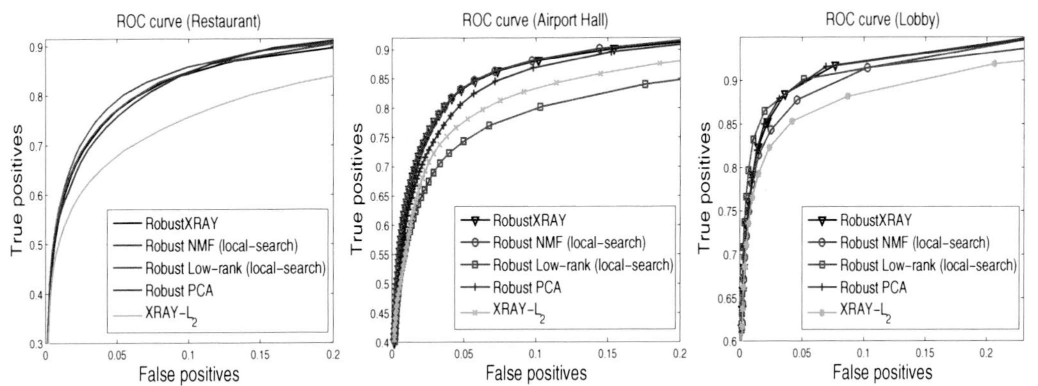

FIGURE 7.7 Foreground-background separation: ROC curves with various methods for *Restaurant*, *Airport Hall* and *Lobby* video sequences. The ranges for X and Y axes are chosen to better highlight the differences among the ROC curves. (best viewed in color)

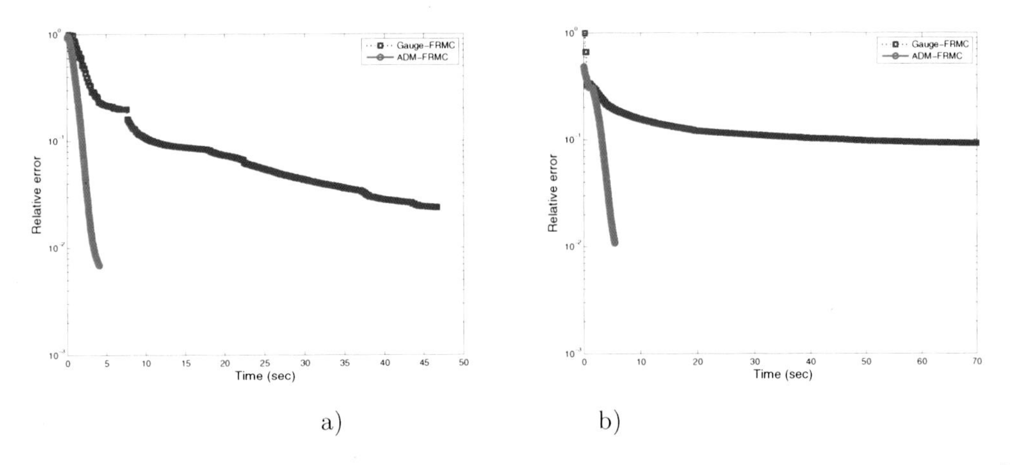

FIGURE 9.1 Performance evaluation on a synthetic dataset of size $m = n = 500$, $r = 20$ with 50% missing entries for (a) low variance sparse components, and (b) high variance sparse components.

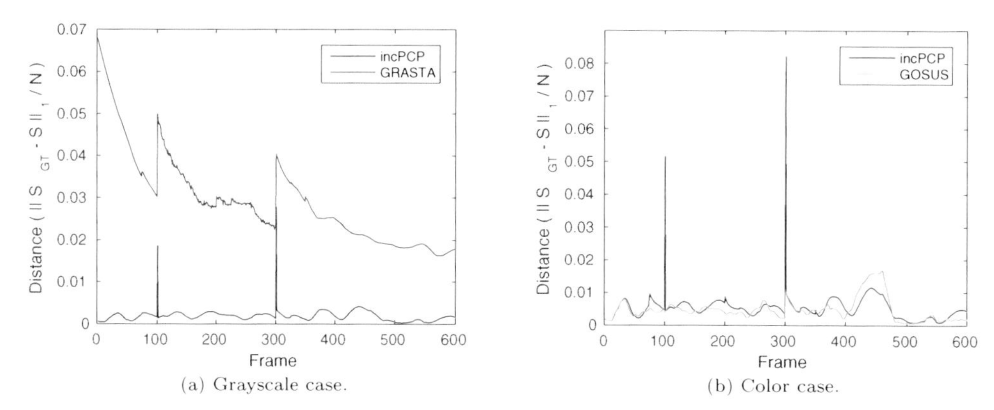

(a) Grayscale case. (b) Color case.

FIGURE 11.2 Sparse approximation (V800) frame distance measure by $\frac{\|S_{GT} - S_P\|_1}{N}$ where S_{GT} is the ground-truth computed via the (batch) iALM algorithm (20 outer loops) and S_P is the sparse approximation computed via (a) grayscale (blue) GRASTA and (red) incPCP methods, or (b) color (green) GOSUS and (red) incPCP methods, and N is the number of pixels per frame (used as a normalization factor). It is worth noting that the spikes due to background adaptation to light illumination (frames 100 and 300) are less prominent for GOSUS than for incPCP, since GOSUS uses the first 200 frames (by default) for its batch initialization, and thus it has already observed the background change which is built into its initial subspace approximation.

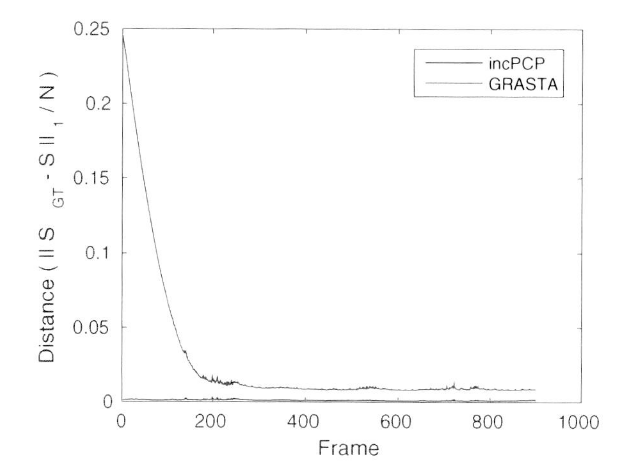

FIGURE 11.4 Sparse approximation (V1920, grayscale version) frame distance measure by $\frac{\|S_{GT} - S_P\|_1}{N}$ where S_{GT} is the ground-truth computed via the (batch) iALM algorithm (10 nouter loops) and S_P is the sparse approximation computed via the (blue) GRASTA and (red) the incPCP method, and N is the number of pixels per frame (used as a normalization factor).

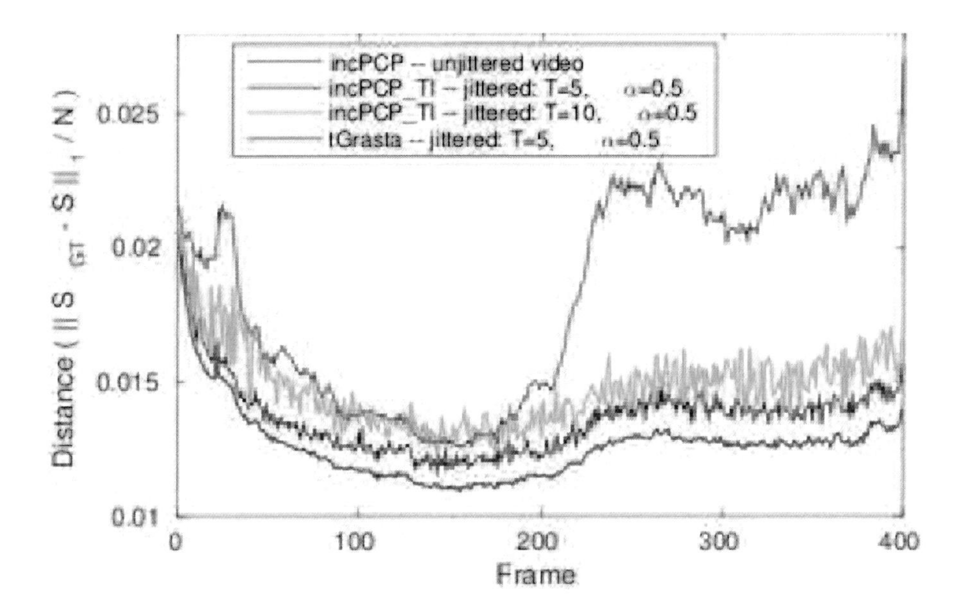

FIGURE 11.11 Sparse approximation (V640) frame distance measure by $\frac{\|S_{GT} - S_P\|_1}{N}$ where S_{GT} is the ground-truth computed via the (batch) iALM algorithm (20 outer loops) and S_P is the (i) sparse component computed via incPCP_TI method (red, green) and t-Grasta (magenta) for different synthetic jitter conditions or (ii) the sparse component for the unjittered case (blue), computed via [49], provided as a baseline.

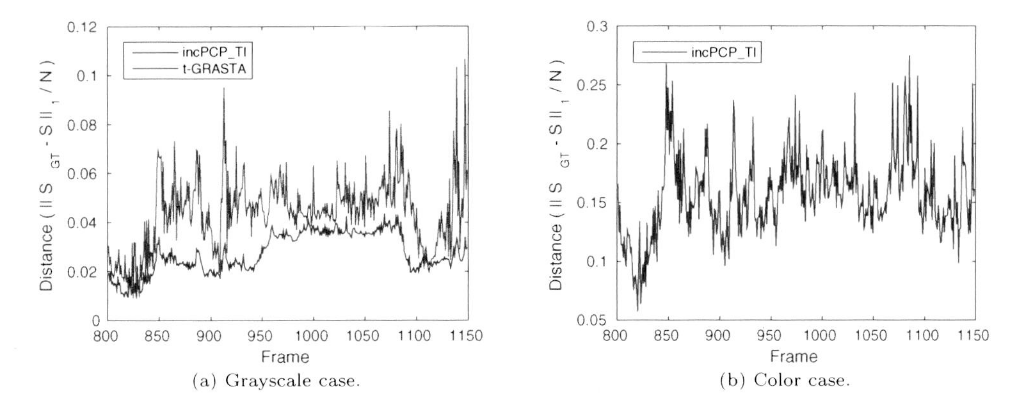

FIGURE 11.14 Sparse approximation (V720) frame distance measure by $\frac{\|S_{GT} - S_P\|_1}{N}$ where S_{GT} is the manually segmented ground-truth and S_P is the sparse approximation computed via (a) grayscale (blue) t-GRASTA and (red) incPCP_TI methods, or (b) color (red) incPCP_TI method, and N is the number of pixels per frame (used as a normalization factor).

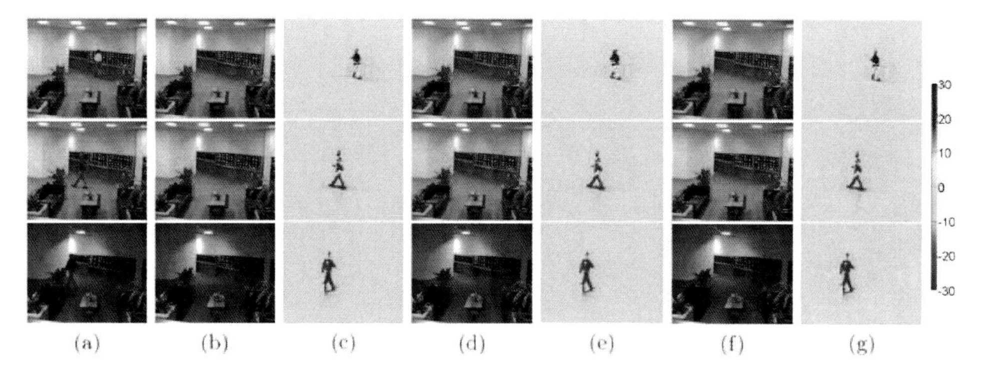

FIGURE 12.5 Comparison of RPCA, ROSL($k = 10$) and ROSL+($l = 50$) in background modeling on the lobby video (size: 160×128, 1060 frames). (a) Original images. Backgrounds recovered by (b) RPCA, (d) ROSL, and (f) ROSL+. Foregrounds recovered by (c) RPCA, (e) ROSL, and (g) ROSL+. ROSL (time: 34.6s) and ROSL+ (time: 3.61s) are significantly ($10\times$, $92\times$) faster than RPCA (time: 334s) while generating almost identical results.

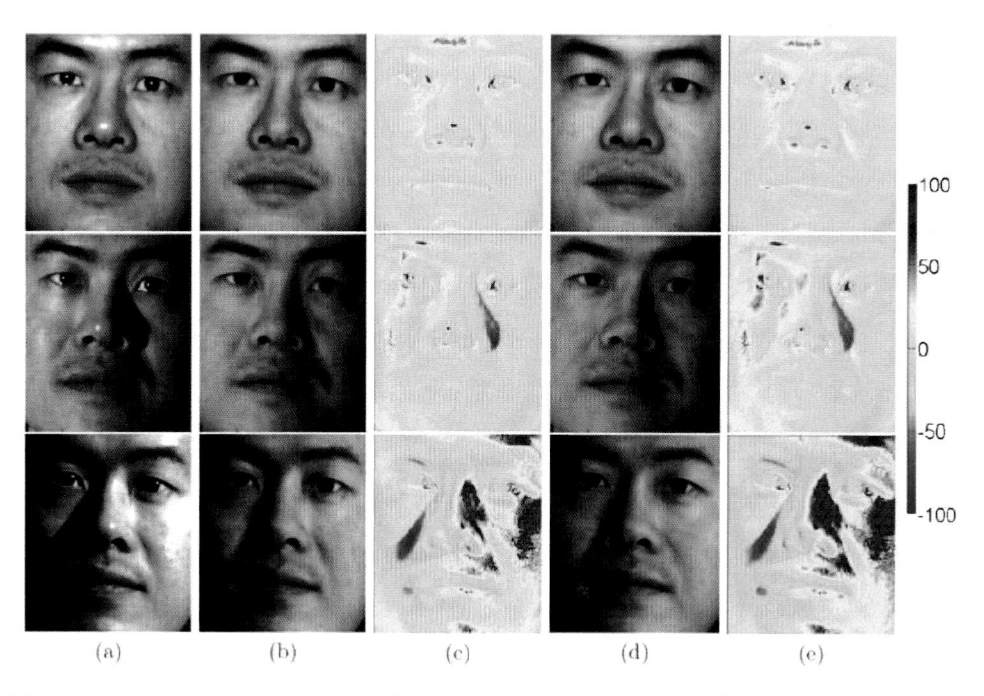

FIGURE 12.6 Visual evaluation of ROSL and RPCA on face images (168×192, 55 frames) under varying illuminations. There is no significant difference between ROSL and RPCA. (a) Original images, diffusive component recovered by (b) RPCA and (d) by ROSL. Non-diffusive component (shadow/specularity) by (c) RPCA (time: 12.16s) and (e) by ROSL (time: 5.85s).

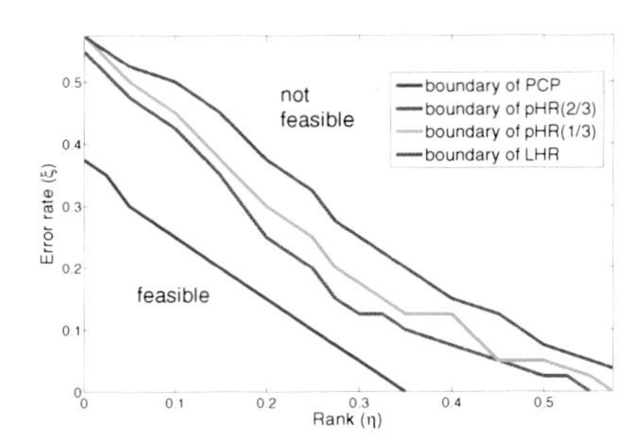

(a) Feasible region verification.

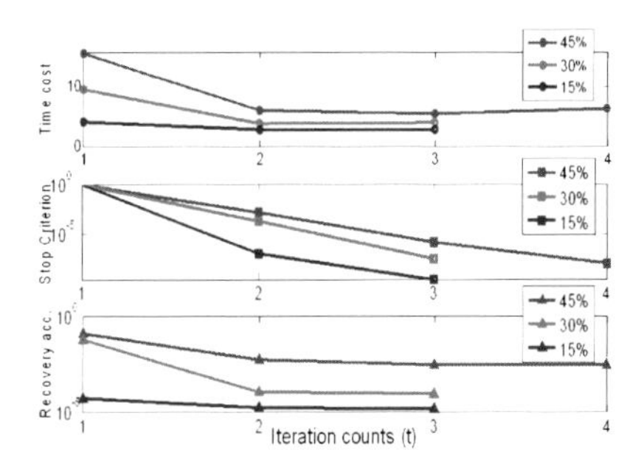

(b) Convergence verification.

FIGURE 13.1 Feasible region and the convergence verifications.

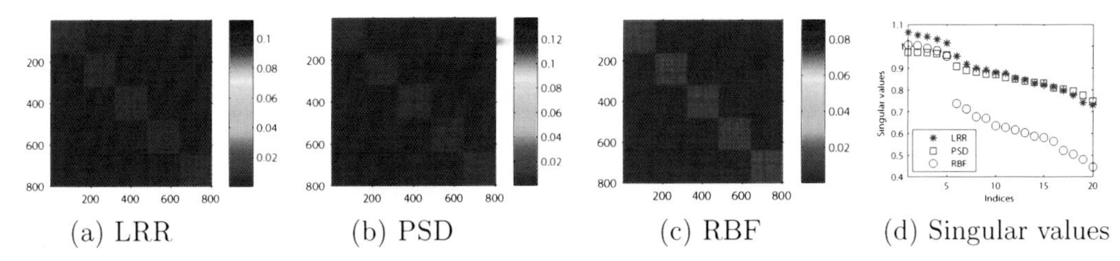

(a) LRR (b) PSD (c) RBF (d) Singular values

FIGURE 15.7 Comparison of affinity matrices produced by LRR, PSD and RBF on the toy data. Note that only the largest 20 singular values of each affinity matrix are shown.

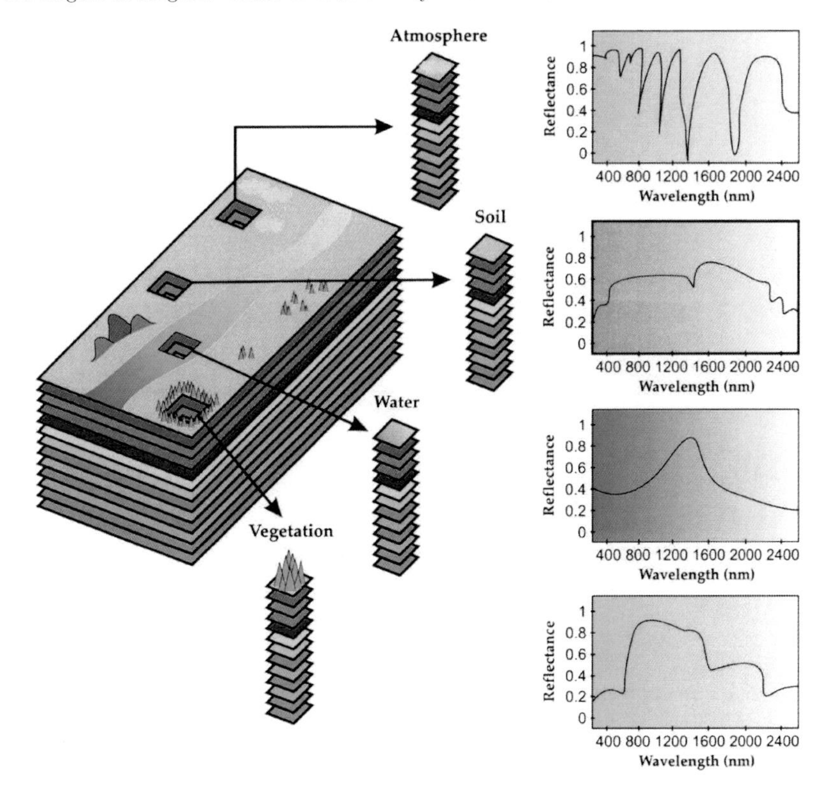

FIGURE 16.1 Example of a hyperspectral image. The image contains many layers that capture different wavelengths of reflected light. Each pixel contains information about the materials present in the scene.

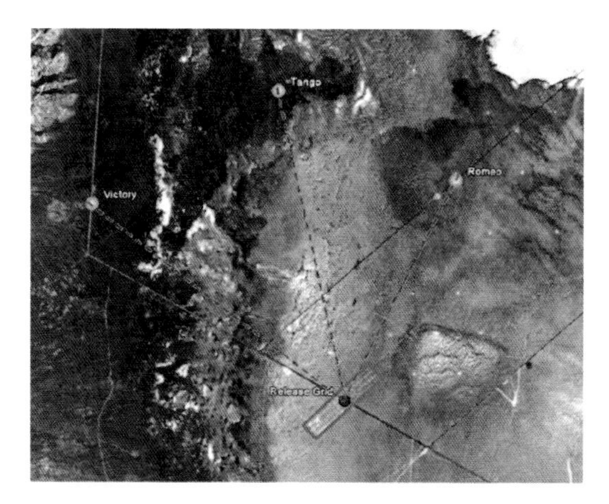

FIGURE 16.3 Placement of the three long wave infrared spectrometers to track the release of the known chemicals.

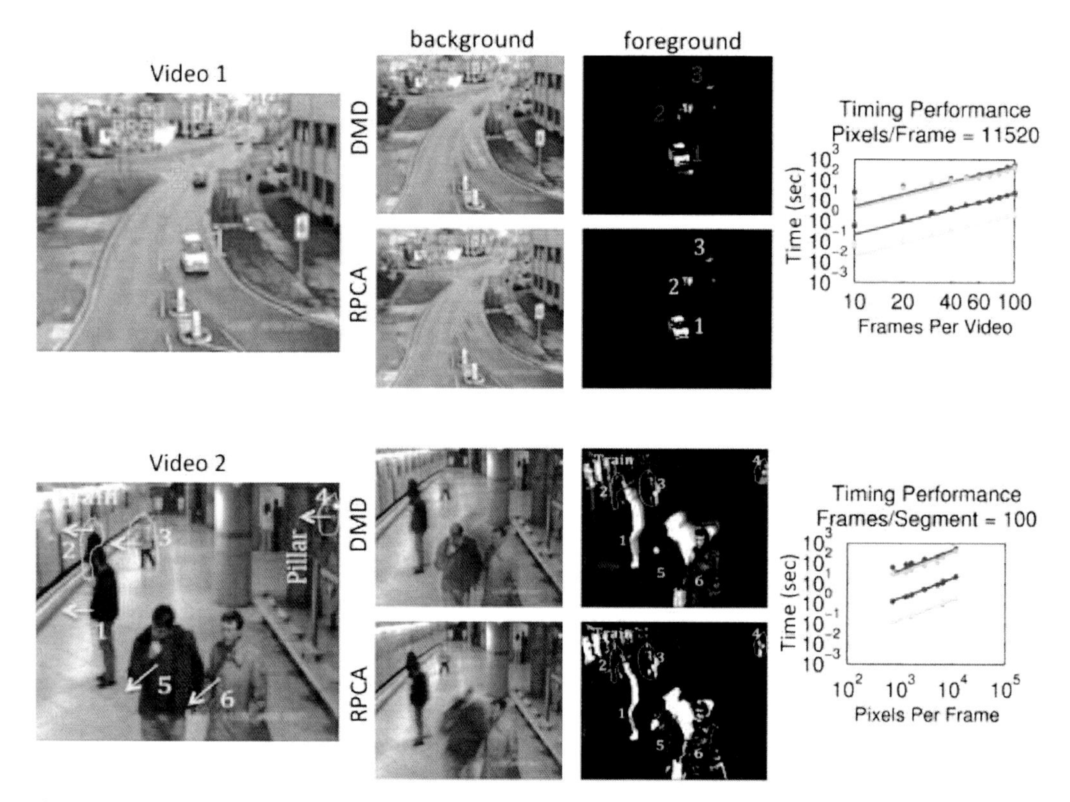

FIGURE 19.2 Demonstration of foreground/background separation using the DMD method and RPCA on "Abandoned Bag" and "Parked Vehicle" videos. The far left panel shows two original clips at a given snapshot in time, or frame number. Both methods provide very similar foreground/background representations. The various items in the foreground have been labeled and numbered. The DMD (green & yellow, respectively), inexact ALM RPCA [14] (red & magenta, respectively), and exact ALM RPCA [14] (blue & cyan, respectively) background/foreground separation methods are graphed on a logarithmic scale. The DMD is an order-of-magnitude faster than competing RPCA methods.

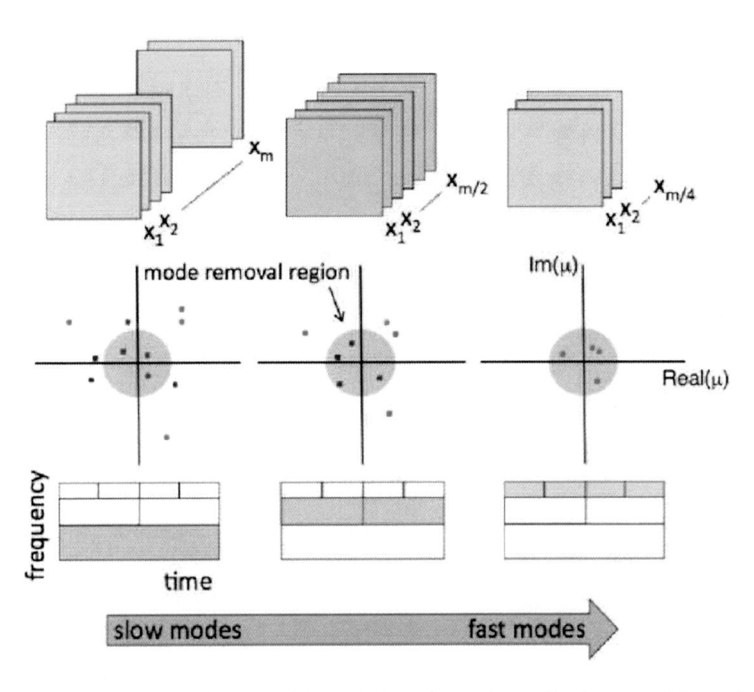

FIGURE 19.3 Representation of the multi-resolution dynamic mode decomposition where successive sampling of the data, initially with M snapshots and decreasing by a factor of two at each resolution level, is shown (top figures). The DMD spectrum is shown in the middle panel where there are m_1 (blue dots) slow-dynamic modes at the slowest level, m_2 (red) modes at the next level and m_3 (green) modes at the fastest time-scale shown. The shaded region represents the modes that are removed at that level. The bottom panels shows the wavelet-like time-frequency decomposition of the data color coded with the snapshots and DMD spectral representations.

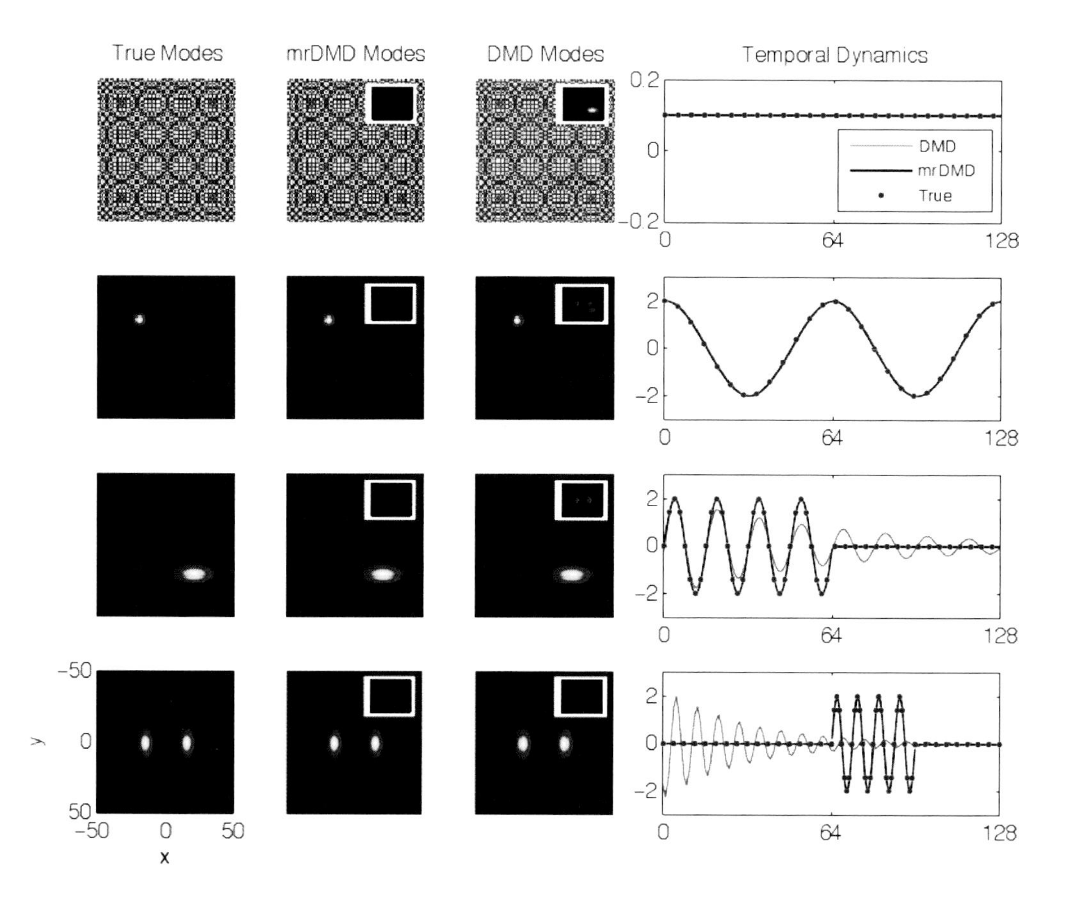

FIGURE 19.4 Example of a multi-resolution analysis using the MRDMD decomposition. The left panel shows the four-modes that are to be mixed. The right panel shows the true time dynamics of each mode, including some modes that turn on and off. The DMD and MRDMD reconstructions of the modes and time dynamics are both illustrated. Without the MRDMD method, the DMD fails to construct the correct dynamics, especially on those modes that turn on and off. The MRDMD works very well in getting both the spatial and temporal dynamics correctly.

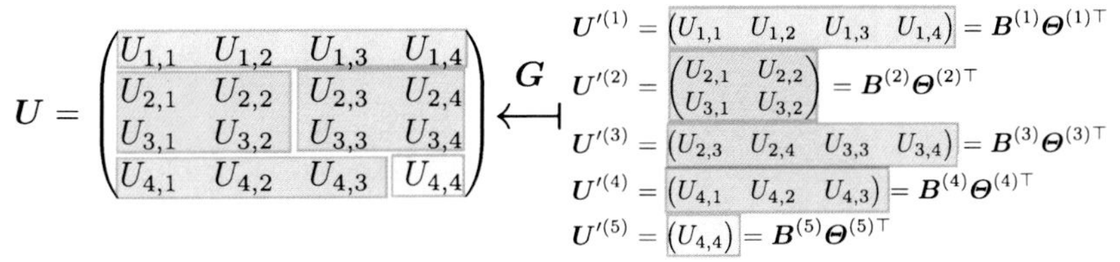

$$
\boldsymbol{U} = \begin{pmatrix} U_{1,1} & U_{1,2} & U_{1,3} & U_{1,4} \\ U_{2,1} & U_{2,2} & U_{2,3} & U_{2,4} \\ U_{3,1} & U_{3,2} & U_{3,3} & U_{3,4} \\ U_{4,1} & U_{4,2} & U_{4,3} & U_{4,4} \end{pmatrix} \overset{\boldsymbol{G}}{\longleftarrow}
\begin{array}{l}
\boldsymbol{U}'^{(1)} = \begin{pmatrix} U_{1,1} & U_{1,2} & U_{1,3} & U_{1,4} \end{pmatrix} = \boldsymbol{B}^{(1)}\boldsymbol{\Theta}^{(1)\top} \\
\boldsymbol{U}'^{(2)} = \begin{pmatrix} U_{2,1} & U_{2,2} \\ U_{3,1} & U_{3,2} \end{pmatrix} = \boldsymbol{B}^{(2)}\boldsymbol{\Theta}^{(2)\top} \\
\boldsymbol{U}'^{(3)} = \begin{pmatrix} U_{2,3} & U_{2,4} & U_{3,3} & U_{3,4} \end{pmatrix} = \boldsymbol{B}^{(3)}\boldsymbol{\Theta}^{(3)\top} \\
\boldsymbol{U}'^{(4)} = \begin{pmatrix} U_{4,1} & U_{4,2} & U_{4,3} \end{pmatrix} = \boldsymbol{B}^{(4)}\boldsymbol{\Theta}^{(4)\top} \\
\boldsymbol{U}'^{(5)} = \begin{pmatrix} U_{4,4} \end{pmatrix} = \boldsymbol{B}^{(5)}\boldsymbol{\Theta}^{(5)\top}
\end{array}
$$

FIGURE 21.1 An example of SMF-term construction. $\boldsymbol{G}(\cdot;\mathcal{X})$ with $\mathcal{X} : (t, d', m') \mapsto (d, m)$ maps the set $\{\boldsymbol{U}'^{(t)}\}_{t=1}^{T}$ of the PR matrices to the target matrix $\boldsymbol{U}$, so that $U'^{(t)}_{d',m'} = U_{\mathcal{X}(t,d',m')} = U_{d,m}$.

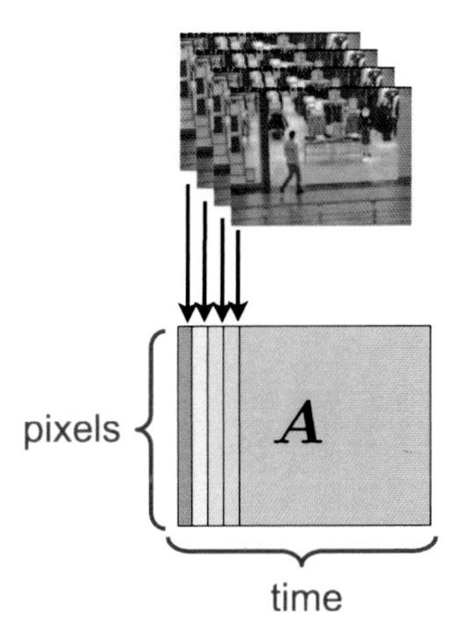

FIGURE 21.3 The observation matrix $\boldsymbol{A}$ is constructed by stacking all pixels in each frame into each column.

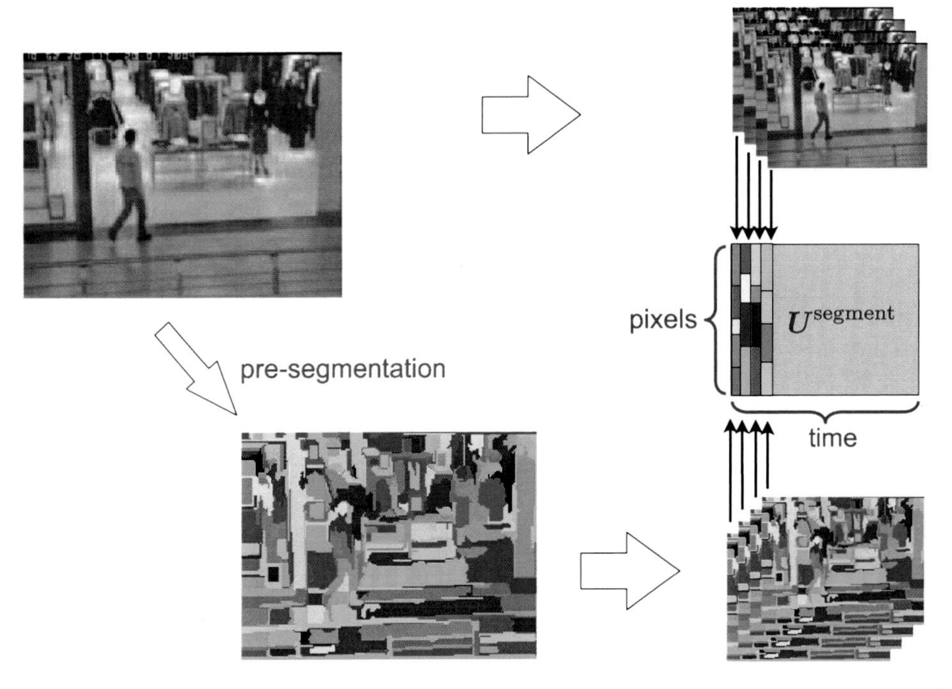

FIGURE 21.4 Construction of a segment-wise sparse term. The original frame is pre-segmented and the sparsity is induced segment-wise..

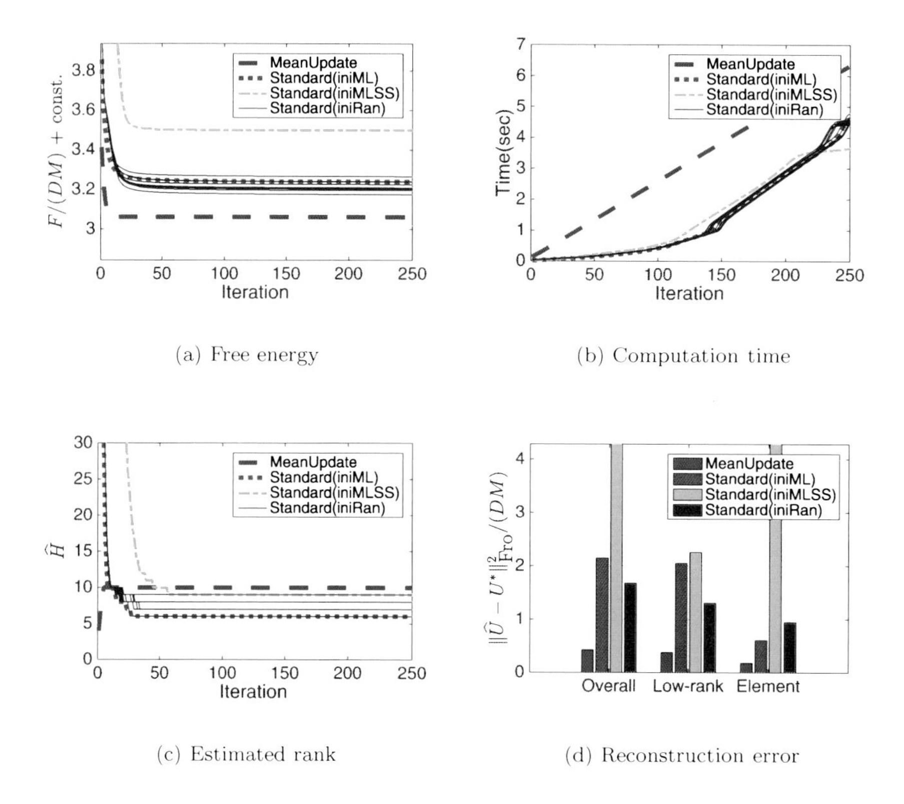

(a) Free energy

(b) Computation time

(c) Estimated rank

(d) Reconstruction error

FIGURE 21.5 Experimental results in robust PCA on an artificial dataset ($D = 40, M = 100, H^* = 10, \rho = 0.05$).

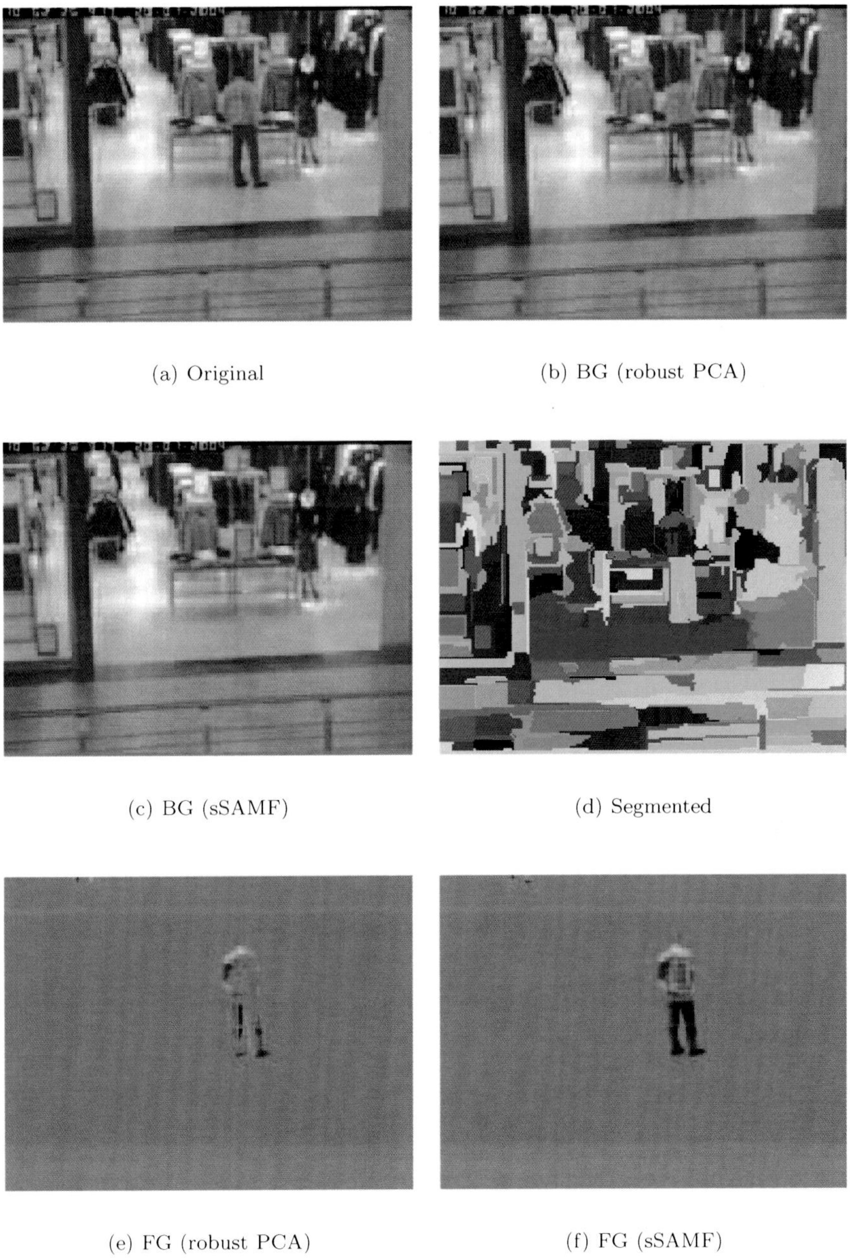

(a) Original

(b) BG (robust PCA)

(c) BG (sSAMF)

(d) Segmented

(e) FG (robust PCA)

(f) FG (sSAMF)

FIGURE 21.6 Robust PCA vs. segmentation-based SAMF (sSAMF) in background(BG)/foreground(FG) video separation.

(a) Segmented ($t = 1$)

(b) Segmented ($t = 10$)

(c) Segmented ($t = 100$)

FIGURE 21.8 Images segmented by the efficient graph-based segmentation (EGS) algorithm with different t values. They are visually different, but with all these segmentations, the BG/FG separation results by sSAMF were almost identical. The original image is shown in Figure 21.6(a).

12

Robust Orthonormal Subspace Learning (ROSL) for Efficient Low-Rank Recovery

Xianbiao Shu
University of Illinois, at Urbana-Champaign, USA

Fatih Porikli
Australian National University/NICTA, Australia

Narendra Ahuja
University of Illinois, at Urbana-Champaign, USA

12.1 Introduction

Learning and exploiting a low-rank structure from the corrupted observation is widely used in machine learning, data mining, and computer vision. This chapter gives a brief overview of two kinds of low-rank recovery methods—convex nuclear-norm based methods and non-convex matrix factorization methods. In general, convex nuclear-norm-based methods are guaranteed to reach the global minima with cubic computational complexity, while non-convex matrix factorization methods are more computationally efficient (quadratic complexity) but suffer from poor convergence. Motivated by these two kinds of low-rank recovery methods, this chapter introduces a computationally efficient low-rank recovery method, called Robust Orthonormal Subspace Learning (ROSL). As is different from convex methods using nuclear norm, ROSL utilizes a novel rank measure on the low-rank matrix that imposes the group sparsity of its coefficients under orthonormal subspace. This new rank measure is proven to be lower bounded by (i.e., has the same global minimum as) nuclear norm, and is experimentally shown to converge to its global minimum with high probability. In addition, this chapter describes a faster version (ROSL+) empowered by random sampling, which further speeds up ROSL from quadratic complexity to linear complexity.

The convex nuclear-norm based method, e.g., Robust PCA (RPCA, also called PCP in [4]) and Sparse Low-Rank Matrix Decomposition (SLRMD) [23], employ the nuclear norm as a surrogate for the highly non-convex rank minimization [15]. RPCA has been shown to be a convex problem with performance guarantee [4]. It assumes the observation matrix $X \in \mathbb{R}^{m \times n}$ is generated by the addition of a low-rank matrix A (rank: $r \ll \min\{m, n\}$) and a sparse matrix E. Supposing that Singular Value Decomposition (SVD) of A is denoted as $A = USV^T$, where S is a diagonal matrix with singular values $S_i, 1 \leq i \leq \min\{m, n\}$) on the diagonal, RPCA recovers the low-rank matrix A from the corrupted observation X as follows:

$$\min_{A,E} \|A\|_* + \lambda \|E\|_1 \quad \text{s.t.} \quad A + E = X \tag{12.1}$$

where nuclear norm $\|A\|_* = \sum_{i=1}^{n} S_i$.

Despite its excellent results, RPCA is computationally expensive with $\mathcal{O}(\min(m^2 n, mn^2))$ complexity due to multiple iterations of SVD. Reducing the number of the required SVD operations is a possible remedy [19], yet the computational load is dominated by SVD itself. Instead of full SVD, partial RPCA [10] computes κ ($r < \kappa$) major singular values, thus it has $\mathcal{O}(\kappa mn)$ complexity. Nevertheless, partial RPCA requires a proper way to preset the optimal value of κ. GoDec [24] uses bilateral random projection to accelerate the low-rank approximation in RPCA. Similarly, RP-RPCA [14] applies random projection P on A (i.e., $A' = PA$) and then minimizes the rank of A'. However, rank minimization using randomized SVD is unstable and might be even slower than RPCA, for it requires conducting SVD on many different projected matrices A' at each iteration.

Non-convex matrix factorization approaches including RMF [8] and LMaFit [16] have also been proposed for fast low-rank recovery. Instead of minimizing the rank of A, these approaches represent A under some preset-rank subspaces (spanned by $D \in \mathbb{R}^{m \times k}$) as $A = D\alpha$, where coefficients $\alpha \in \mathbb{R}^{k \times n}$ and $r < k \ll \min(m, n)$. Due to its SVD-free property, these non-convex matrix factorization approaches are computationally preferable to RPCA. Still, their quadratic complexity $\mathcal{O}(kmn)$ is prohibitive for large-scale low-rank recovery. Besides, they require an accurate initial rank estimate, which is not easy to obtain by itself.

This chapter presents a computationally efficient low-rank recovery method, called Robust Orthonormal Subspace Learning (ROSL). Motivated by the group sparsity (structure) in sparse coding [7, 13, 20, 22], ROSL speeds the rank-minimization of a matrix A by imposing the group sparsity of its coefficients α under orthonormal subspace (spanned by orthonormal bases D). Its underlying idea is that, given the subspace representation $A = D\alpha$, the rank of A is upper bounded by the number of non-zero rows of α. ROSL can be regarded as a non-convex relaxation of RPCA in that it replaces nuclear norm with this rank heuristic. First, this relaxation enables the employment of efficient sparse coding algorithms in low-rank recovery, therefore ROSL has only $\mathcal{O}(rmn)$ ($r < \kappa, k$) complexity, much faster than RPCA. Second, by imposing this rank heuristic, ROSL is able to seek the most compact orthonormal subspace that represents the low-rank matrix A without requiring accurate rank estimate (unlike RMF and LMaFit). Third, this rank heuristic is proven to be lower bounded by the nuclear norm, which means that ROSL has the same global minimum as RPCA.

An efficient ROSL solver is also presented. This solver incorporates a block coordinate descent (BCD) algorithm into an inexact alternating decision method (ADM). Despite its non-convexity, this solver is shown to exhibit strong convergence behavior, given random initialization. Experimental results validate that the solution obtained by this solver is identical or very close to the global optimum of RPCA.

As another contribution, a random sampling algorithm is introduced to further speed up ROSL such that ROSL+ has linear complexity $\mathcal{O}(r^2(m + n))$. Similar sampling-based frameworks for RPCA can be found in DFC [12] and L1 filtering [11]. Although these methods follow the same idea—Nystrom method [9, 18, 21], ROSL+ addresses a different

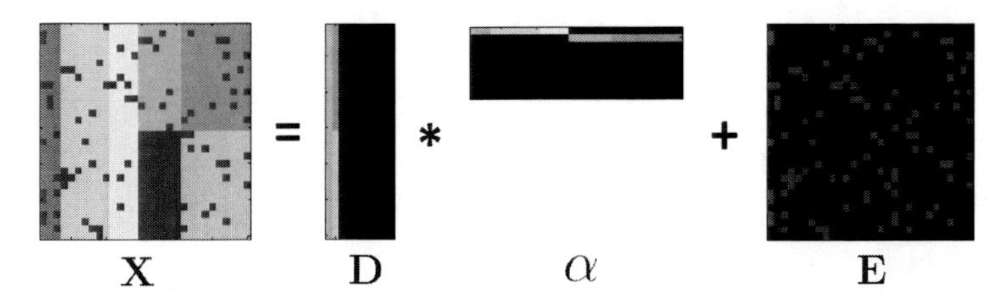

FIGURE 12.1 Illustration of the observation model $X = A + E = D\alpha + E$ in ROSL.

problem, i.e., accelerating orthogonal subspace learning. In addition, ROSL+ elucidates a key point in the Nystrom method—how to estimate multiple sub-matrices, which is missed by DFC.

This chapter is organized as follows. Section 12.2 presents the proposed method (ROSL). Section 12.3 develops its efficient solver. Section 12.4 provides its accelerated version (ROSL+). Section 12.5 presents experimental results. Section 12.7 adds the acknowledgement and Section 12.6 gives the concluding remarks.

12.2 Robust Orthonormal Subspace Learning

As shown in Figure 12.1, similar to RPCA, ROSL assumes that the observation $X \in \mathbb{R}^{m\times n}$ is generated by the addition of a low-rank matrix A (rank: $r \ll \min\{m,n\}$) and a sparse outlier matrix E. As being different from RPCA, which uses the principal subspace, ROSL represents the low-rank matrix A under an ordinary orthonormal subspace (spanned by $D = [D_1, D_2 ..., D_k] \in \mathbb{R}^{m\times k}$), denoted as $A = D\alpha$, where coefficients $\alpha = [\alpha_1; \alpha_2; ...; \alpha_k] \in \mathbb{R}^{k\times N}$ and α_i specify the contribution of D_i to each column of A. The dimension k of the subspace is set as $k = \beta_1 r(\beta_1 > 1$ is a constant).

12.2.1 Group Sparsity under Orthonormal Subspace

ROSL introduces a new formulation of rank minimization to replace the nuclear norm used in RPCA. Although the Frobenius-norm regularization is a valid substitute for the nuclear norm, as shown in Lemma 12.1, it fails to recover the low-rank matrix without a rank estimate.

THEOREM 12.1 *[5, 17]* $\|A\|_* = \min_{D,\alpha} \frac{1}{2}(\|D\|_F^2 + \|\alpha\|_F^2)$ *s.t.* $A = D\alpha$.

Motivated by the group sparsity [7, 13, 20, 22], ROSL represents A under some vector subspace D and constrains the rank of A by imposing the group sparsity of its coefficients α. Its main idea is that, given $A = D\alpha$, the rank of A, or exactly α, is upper bounded by the number of non-zero rows of α, i.e., $\|\alpha\|_{\text{row-0}}$. In order to avoid the vanishing of coefficients α, the subspace bases are constrained to be on the unit sphere, i.e., $D_i^T D_i = 1, \forall i$. To further enable the group sparsity of α as a valid measure of rank (A), we should eliminate the correlation of columns of D by constraining it to be orthonormal, i.e., $D^T D = I_k$, where I_k is an identity matrix. Thus, ROSL recovers the low-rank matrix A from X by minimizing the number of non-zero rows of α, and the sparsity of E as follows:

$$\min_{E,D,\alpha} \|\alpha\|_{\text{row-0}} + \lambda\|E\|_0 \quad \text{s.t.} D\alpha + E = X, D^T D = I_k, \forall i. \tag{12.2}$$

THEOREM 12.2 $\|A\|_* = \|\alpha\|_{\text{row-1}}$, *when* $A = D\alpha, D^T D = I_k$ *and* α *consists of orthogonal rows.*

It is well known that sparsity-inducing ℓ_1-norm is an acceptable substitute for the sparsity measure (i.e., ℓ_0-norm). Similarly, the row-1 norm, which is defined as $\|\alpha\|_{\text{row-1}} = \sum_{i=1}^{k} \|\alpha_i\|_2$, is a good heuristic for the row sparsity (i.e., row-0 norm). Actually, it is easy to reach the conclusion that the nuclear norm $\|A\|_*$ is equal to the group sparsity $\|\alpha\|_{\text{row-1}}$ under orthonormal subspace D, where $A = D\alpha$, if rows of α are orthogonal, as stated in Lemma 12.2. In this case, the subspace bases $D = U$ and coefficients $\alpha = SV^T$, where $A = USV^T$ by SVD. For the computational efficiency, ROSL removes this orthogonal constraint on α and recovers the low-rank matrix A from X by minimizing the row-1 norm of α, and the ℓ_1-norm of E.

$$\min_{E,D,\alpha} \|\alpha\|_{\text{row-1}} + \lambda \|E\|_1 \quad \text{s.t.} D\alpha + E = X, D^T D = I_k, \forall i. \tag{12.3}$$

12.2.2 Bound of Group Sparsity under Orthonormal Subspace

To show ROSL is a valid non-convex relaxation of the performance-guaranteed RPCA, we investigate the relationship between the group-sparsity-based rank formulation with matrix rank/nuclear norm.

PROPOSITION 12.1 Consider a thin matrix $A \in \mathbb{R}^{m\times n}$ $(m \geq n)$; its SVD and orthonormal subspace decomposition are respectively denoted as $A = USV^T$ and $A = D\alpha$, where $D \in \mathbb{R}^{m\times n}, \alpha \in \mathbb{R}^{n\times n}$, and $D^T D = I_n$ without loss of generality. The minima of row-0 group sparsity and row-1 group sparsity of A under orthonormal subspace are respectively rank(A) and nuclear norm $\|A\|_*$:

$$\text{(P12.1)} \qquad \min_{D\alpha=A, D^T D=I_n} \|\alpha\|_{\text{row-0}} = \text{rank}(A), \tag{12.4}$$

$$\text{(P12.2)} \qquad \min_{D\alpha=A, D^T D=I_n} \|\alpha\|_{\text{row-1}} = \|A\|_*. \tag{12.5}$$

Proof of (P12.1) It is straightforward that the rank of A, where $A = D\alpha$, should not be larger than the dimension of α, resulting in that $\|\alpha\|_{\text{row-0}} \geq \text{rank}(\alpha) \geq \text{rank}(A)$. Thus, the row-0 norm of α under orthonormal subspace D is lower bounded by the rank of A.

Proof of (P12.2) This part can be restated as: $\|\alpha\|_{\text{row-1}} = \sum_{i=1}^{n} \|\alpha_i\|_2$, which will reach its minimum $\|A\|_*$, when the orthonormal bases are equal to the principal components, i.e., $D = U$, where $A = USV^T$ by SVD. For simplicity of proof, we ignore other trivial solutions—the variations (column-wise permutation or $\pm$ column vectors) of U. Since both D and U are orthonormal bases, we reach the relationship, $D = U\Omega$ and $\alpha = \Omega^T SV^T$, where Ω is a rotation matrix $(\Omega^T\Omega = I_n, \det(\Omega) = 1)$. Here, we introduce a decreasing sequence of non-negative numbers $\sigma_i, 1 \leq i \leq n$ such that $S_i = \sigma_i, 1 \leq i \leq n$. To validate (P12.2), we need to prove that the following relation holds for any Ω (the equality holds when Ω is the identity matrix).

$$\|\alpha\|_{\text{row-1}} = \|\Omega^T SV^T\|_{\text{row-1}} \geq \sum_{i=1}^{n} S_i = \|A\|_*. \tag{12.6}$$

1. We begin with the special case that all the singular values are identical. Specifically, we decrease the singular values such that $\forall i \in \{1, ..., n\}$, $S_i = \sigma_n$, where σ_n is the

last number in the decreasing sequence $\sigma_i, 1 \le i \le n$. Since each row of the rotation matrix Ω is a unit vector, we reach the following relationship:

$$\|\alpha\|_{\text{row-1}} = \sum_{j=1}^{n} \sqrt{\sum_{i=1}^{n} \Omega_{ij}^2 S_i^2} = n\sigma_n = \sum_{i=1}^{n} S_i = \|A\|_*. \tag{12.7}$$

2. Then, we try to prove that $\|\alpha\|_{\text{row-1}} \ge \|A\|_*$ still holds in the general case, i.e., $S_i = \sigma_i, 1 \le i \le n$. We can transform the special case above into the general case by $n-1$ steps, among which the t-th step is increasing the top $n-t$ singular values $(S_i, 1 \le i \le n-t)$ from σ_{n-t+1} to σ_{n-t}. When increasing $S_i, 1 \le i \le n-1$ from σ_n to σ_{n-1} in the first step, the partial derivative of $\|\alpha\|_{\text{row-1}}$ with respect to S_i is calculated as follows:

$$\frac{\partial \|\alpha\|_{\text{row-1}}}{\partial S_i} = \sum_{j=1}^{n} \frac{\Omega_{ij}^2}{\sqrt{\sum_{t=1}^{n-1} \Omega_{tj}^2 + \Omega_{nj}^2 (S_n^2/S_i^2)}}. \tag{12.8}$$

Since $S_n \le S_i, 1 \le i \le n-1$ and $\sum_{t=1}^{n} \Omega_{tj}^2 = 1$, we reach the following relationship:

$$\frac{\partial \|\alpha\|_{\text{row-1}}}{\partial S_i} \ge \sum_{j=1}^{n} \Omega_{ij}^2 = 1 = \frac{\partial \|A\|_*}{\partial S_i}. \tag{12.9}$$

Thus, $\|\alpha\|_{\text{row-1}} \ge \|A\|_*$ holds when increasing $S_i, 1 \le i \le n-1$ in the first step. In the same way, we can prove that $\|\alpha\|_{\text{row-1}} \ge \|A\|_*$ holds in the following $n-2$ steps.

3. In sum, $\|\alpha\|_{\text{row-1}} \ge \|A\|_*$ in the general case where singular values S_i are not identical, i.e., $S_i = \sigma_i, \forall i \in \{1, ..., n\}$.

According to Proposition 12.1, the minimum of row-1 group sparsity under orthonormal subspace is the nuclear norm, i.e., $\|\alpha\|_{\text{row-1}} \ge \|A\|_*$, where $A = D\alpha$ and $D^T D = I_k$. Suppose, at weight λ, RPCA recovers the low-rank matrix as its ground truth A^*, i.e., $\widehat{A} = A^*$, then, $\|\widehat{\alpha}\|_{\text{row-1}} + \lambda\|X - \widehat{A}\|_1 \ge \|\widehat{A}\|_* + \lambda\|X - \widehat{A}\|_1 \ge \|A^*\|_* + \lambda\|X - A^*\|_1$ holds for any $(\widehat{A}, \widehat{D}, \widehat{\alpha})_{\widehat{A}=\widehat{D}\widehat{\alpha}, \widehat{D}^T\widehat{D}=I_k}$. In sum, at the weight λ, ROSL has the same global minimum $(\widehat{A} = A^*, \widehat{D} = U, \widehat{\alpha} = SV^T)$ as RPCA, where $A^* = USV^T$ by SVD.

12.2.3 A General Framework of Robust Low-Rank Recovery Approaches

To better compare our ROSL with other existing approaches, we present a general framework of robust low-rank recovery approaches, as shown in Table 12.1. All the low-rank recovery methods listed in the table utilize the ℓ_1-norm to the sparsity measure and different low-rank measures. RPCA and its variant RP-RPCA use nuclear norm, which is equivalent to the groups' sparsity under orthonormal subspace, with the constraint orthogonal coefficients. RMF uses the Frobenius-norm regularization as low-rank measure. LMaFit has no low-rank measure. To recover the low-rank structure, our ROSL seeks the groups' sparsity under orthonormal subspace, without the constraint orthogonal coefficients.

ROSL can be considered a compromise between RPCA and ordinary matrix factorization methods (e.g., RMF and LMaFit). On the one hand, ROSL improves upon RMF and LMaFit

TABLE 12.1 A general framework of robust low-rank recovery approaches. Given a corrupted low-rank matrix $X = A + E$, $A \in \mathbb{R}^{m\times n}$ $(m > n)$ and its projected version A' can be represented as $A = D\alpha$ and $A' = D'\alpha'$. All approaches follow the same framework; minimizing the sparsity and rank measures under some constraints, where I_n and Δ_n respectively denote identity and diagonal matrices.

Approaches	RPCA/SLRMD	RP-RPCA	RMF	LMaFit	ROSL
Sparsity Measure	$\|E\|_1$	$\|E\|_1$	$\|E\|_1$	$\|E\|_1$	$\|E\|_1$
Rank Measure	$\|\alpha\|_{\text{row-1}}$	$\|\alpha'\|_{\text{row-1}}$	$\|D\|_F^2 + \|\alpha\|_F^2$	N/A	$\|\alpha\|_{\text{row-1}}$
Constraints	$D^TD = I_n$ $\alpha^T\alpha = \Delta_n$	$D'^TD' = I_n$ $\alpha'^T\alpha' = \Delta_n$	N/A	N/A	$D^TD = I_n$

by seeking the group sparsity of A under orthonormal subspace D. This helps it to recover the low-rank structure of X without requiring accurate rank estimate. On the other hand, ROSL becomes a non-convex relaxation of RPCA by replacing nuclear norm $\|A\|_*$ with the group sparsity $\|\alpha\|_{\text{row-1}}$ under orthonormal subspace. As stated in Lemma 12.2, the nuclear norm $\|A\|_*$ is equal to the group sparsity $\|\alpha\|_{\text{row-1}}$ under orthonormal subspace D, where $A = D\alpha$, if rows of α are orthogonal. By removing the orthogonality constraint on α, ROSL can efficiently solve the low-rank recovery problem by sparse coding algorithms without requiring multiple iterations of SVD. To better compare our ROSL with other existing approaches, we present a general framework of robust low-rank recovery approaches, as shown in Table 12.1.

12.3 Fast Algorithm for ROSL

In this section an efficient algorithm is presented to solve the ROSL problem in Equation (12.3).

Algorithm 12.1 - ROSL Solver by inexact ADM/BCD

Require: $X \in \mathbb{R}^{m\times n}$, k, λ.
Ensure: D, α, E
1: $E^0 = Y^0 = \text{zeros}(m,n); D^0 = \text{zeros}(m,k); \alpha^0 = \text{rand}(k,n); \mu^0 > 0; \rho > 1; i = 0;$
2: **while** E not converged **do**
3: **for** $t = 1 \to k$ **do**
4: Compute the t-th residual: $R_t^i = X - E^i + Y^i/\mu^i - \sum_{j<t} D_j^{i+1}\alpha_j^{i+1} - \sum_{j>t} D_j^i\alpha_j^i$;
5: Orthogonalization:
$R_t^i = R_t^i - \sum_{j=1}^{t-1} D_j^{i+1}(D_j^{i+1})^T R_t^i$;
6: Update: $D_t^{i+1} = R_t^i {\alpha_t^i}^T$;
$D_t^{i+1} = D_t^{i+1}/(\|D_t^{i+1}\|_2)$;
7: Update:$\alpha_t^{i+1} = \mathbb{S}_{1/\mu^i}({D_t^{i+1}}^T R_t^i)$;
8: **end for**
9: Prune: for $t = 1 \to k$, delete $(D_t^{i+1}, \alpha_t^{i+1})$ and set $k = k - 1$, if $\|\alpha_t^{i+1}\|_2^2 = 0$;
10: Update: $E^{i+1} = \mathbb{S}_{\lambda/\mu^i}(X - D^{i+1}\alpha^{i+1} + Y^i/\mu^i)$;
11: Update: $Y^{i+1} = Y^i + \mu^i(X - D^{i+1}\alpha^{i+1} - E^{i+1}); \mu^{i+1} = \rho\mu^i; i = i + 1;$
12: **end while**

12.3.1 Alternating Direction Method

Similar to [10], we apply the augmented Lagrange multiplier (ALM) [3] to remove the equality constraint $X = D\alpha + E$ in Equation (12.3). Its augmented Lagrangian function is written as:

$$\mathcal{L}(D, \alpha, E, Y, \mu) = \|\alpha\|_{\text{row-1}} + \lambda\|E\|_1 + Y(X - D\alpha - E) \\ + \frac{\mu}{2}\|X - D\alpha - E\|_F^2 \quad \text{s.t.} \quad D^T D = I_k. \tag{12.10}$$

where μ is the over-regularization parameter and Y is the Lagrange multiplier. We solve the above Lagrange function by inexact alternating direction method (ADM), which iterates through the following three steps:

1. Solve $(D^{i+1}, \alpha^{i+1}) = \arg\min \mathcal{L}(D, \alpha, E^i, Y^i, \mu^i)$.
2. Solve $E^{i+1} = \arg\min \mathcal{L}(D^{i+1}, \alpha^{i+1}, E, Y^i, \mu^i)$.
3. Update $Y^{i+1} = Y^i + \mu^i(X - D^{i+1}\alpha^{i+1} - E^{i+1}), \mu^{i+1} = \rho\mu^i$, where $\rho > 1$ is a constant.

In the first step, solving D and α simultaneously with constraint $D\alpha + E = X + \frac{Y}{\mu}$ is a non-convex problem. Fortunately, the sub-problem—updating one matrix when fixing the other one—is convex. This indicates solving D and α using the coordinate descent method. In the second step, we can easily update $E^{i+1} = \mathbb{S}_{\lambda/\mu^i}(X - D^{i+1}\alpha^{i+1} + \frac{Y^i}{\mu^i})$, where shrinkage function $\mathbb{S}_a(X) = \max\{\text{abs}(X) - a, 0\} \cdot \text{sign}(X)$ and "$\cdot$" denotes element-wise multiplication.

12.3.2 Block Coordinate Descent

Motivated by group sparse coding [2], we apply block coordinate descent (BCD) to solve D and α in the first step of ADM. Supposing the subspace bases $D = [D_1, ..., D_t, ..., D_k]$ and $\alpha = [\alpha_1; ...; \alpha_t; ...; \alpha_k]$, the BCD scheme sequentially updates the pair (D_t, α_t), by leaving all the other indices intact. In this way, it allows shrinking the group sparsity $\|\alpha\|_{\text{row-1}}$ under the orthonormal subspace D, while sequentially updating (D_t, α_t). In addition, it obtains new subspace bases and coefficients that best fit the constraint $A = D\alpha$ and thus achieves higher convergence rate, as explained in [1, 6]. The BCD scheme sequentially updates each pair $(D_t, \alpha_t), 1 \le t \le k$ such that $D_t\alpha_t$ is a good rank-1 approximation to R_t^i, where the residual is defined as $R_t^i = X + \frac{Y^i}{\mu^i} - E^i - \sum_{j<t} D_j^{i+1}\alpha_j^{i+1} - \sum_{j>t} D_j^i\alpha_j^i$. Thus, if removing the orthonormal constraint on D, the pair (D_t, α_t) can be efficiently updated as follows:

$$D_t^{i+1} = R_t^i {\alpha^i}^T \tag{12.11}$$

$$\alpha_t^{i+1} = \frac{1}{\|D_t^{i+1}\|_2^2}\overline{\mathbb{S}}_{1/\mu^i}({D_t^{i+1}}^T R_t^i) \tag{12.12}$$

where $\overline{\mathbb{S}}_a(X)$ is the magnitude shrinkage function defined as $\overline{\mathbb{S}}_a(X) = \max\{\|X\|_2 - a, 0\}X/\|X\|_2$. Due to the space limit, we refer the readers to [2] for the detailed induction of Equation (12.12).

When taking into account the orthonormal subspace, we need to orthonormalize D_t^{i+1} by the Gram-Schmidt process. As shown in Algorithm 12.1, the new D_t^{i+1} is obtained via three steps: (1) project R_t^i onto the null space of $[D_1, ..., D_{t-1}]$, (2) update D_t^{i+1} as Equation (12.11), and (3) then project it onto the unit sphere by normalization.

The above BCD scheme attempts to keep sequentially fitting the rank-1 subspaces $(D_t^{i+1}\alpha_t^{i+1})$ to the objective $X + \frac{Y^i}{\mu^i} = D^{i+1}\alpha^{i+1} + E^i$, until the fitted subspace is canceled by magnitude shrinkage, i.e., $\|\alpha_t^{i+1}\|_2 = 0$. To improve the computational efficiency,

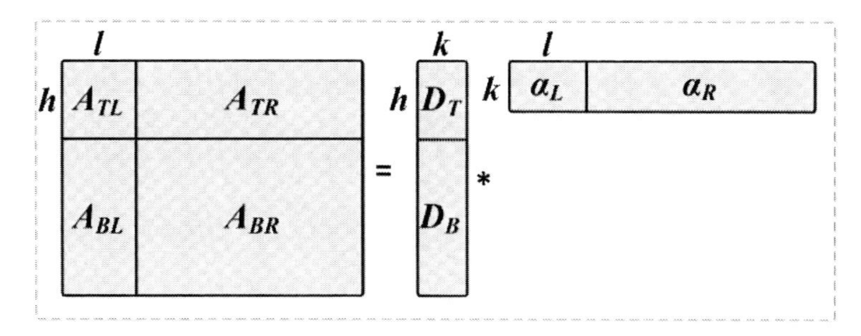

FIGURE 12.2 Decomposition of the low-rank matrix $A \in \mathbb{R}^{m \times n}$.

we shrink the subspace dimension k by pruning the zero pairs, for they will stay zero in the next iteration.

It is possible to run many rounds of BCD to solve D^{i+1} and α^{i+1} exactly in the first step of ADM. In practice, updating $(D_t^{i+1}, \alpha_t^{i+1})$, $1 \leq t \leq k$ once at each round of ADM is shown to be sufficient for the inexact ADM algorithm to converge to a valid solution (D^{i+1}, α^{i+1} and E^{i+1}) to Equation (12.3).

As shown in Algorithm 12.1, ROSL can be solved using inexact ADM at the higher scale and inexact BCD at the lower scale. To the best of our knowledge, there is no established convergence theory, either for ADM algorithms applied to non-convex problems with more than two groups of variables [16], or for BCD algorithms applied to sparse coding [1, 2]. Like all non-convex problems, ROSL has no theoretical guarantee of convergence. However, empirical evidence suggests that the ROSL solver has strong convergence behavior and provides a valid solution: $A^{i+1} = D^{i+1}\alpha^{i+1}$ and E^{i+1}, when it initializes E^0, Y^0 and D^0 as zero matrices, as well as α^0 as a random matrix.

12.3.3 Computational Complexity

Compared with RPCA, which has cubic complexity of $\mathcal{O}(\min(m^2n, mn^2))$, ROSL is much more efficient, when the matrix rank $r \ll \min(m, n)$. Its dominant computational processes are (1) left multiplying the residual matrix $R \in \mathbb{R}^{m \times n}$ by D, and (2) right multiplying it by α. Thus, the complexity of ROSL depends on the subspace dimension k. If we set the initial value of k as several times larger than r (i.e., r and k are of the same order, being much smaller than m and n), ROSL has the quadratic complexity of matrix size, i.e., $\mathcal{O}(mnk)$ or $\mathcal{O}(mnr)$.

12.4 Acceleration by Random Sampling

Motivated by the Nystrom method [9, 18, 21], we present a random sampling algorithm to further speed up ROSL such that its accelerated version (ROSL+) has linear complexity with respect to the matrix size.

12.4.1 Random Sampling in ROSL+

As shown in Figure 12.2, the low-rank matrix $A \in \mathbb{R}^{m \times n}$ is first permuted column-wisely and row-wisely, and then divided into four sub-matrices ($A_{TL} \in \mathbb{R}^{h \times l}$, A_{TR}, A_{BL} and A_{BR}). Accordingly, top sub-matrix A_T and left sub-matrix A_L are respectively defined as $A_T = [A_{TL}, A_{TR}]$ and $A_L = [A_{TL}; A_{BL}]$. The same permutation and division are done on X and E. As shown in Figure 12.2, subspace base D is divided into $D_T \in \mathbb{R}^{h \times k}$ and D_B, as well as coefficient α being divided into $\alpha_L \in \mathbb{R}^{k \times l}$ and α_R, such that

$$A = \begin{bmatrix} A_{TL} \; A_{TR} \\ A_{BL} \; A_{BR} \end{bmatrix} = \begin{bmatrix} D_T \\ D_B \end{bmatrix} [\alpha_L \quad \alpha_R]. \tag{12.13}$$

The Nystrom method is initially used for large dense matrix approximation [9], and extended to speed up RPCA in DFC [12]. Supposing $\text{rank}(A_{TL}) = \text{rank}(A) = r$, instead of recovering the full low-rank matrix A, DFC first recovers its sub-matrices and then approximates $\widehat{A}$ as:

$$\widehat{A} = \widehat{A_L}(\widehat{A_{TL}})^{+}\widehat{A_T} \tag{12.14}$$

where "+" denotes pseudo-inverse. However, DFC does not describe how to estimate the top-left submatrix.

Here, we investigate this specific issue and further simplify the Nystrom method in the framework of robust subspace learning. An intuitive solution would be independently recovering all three sub-matrices. But this requires exhaustively tuning different parameters λ, which eventually prevents us from achieving high accuracy. The feasible way is that ROSL+ directly recovers the left sub-matrix and the top submatrix, i.e., $\widehat{A_L} = \widehat{D}\widehat{\alpha_L}$ and $\widehat{A_T} = \widehat{D_T}\widehat{\alpha}$, and then approximates $\widehat{A_{TL}}$ by the left sub-matrix of $\widehat{A_T}$. Thus, the low-rank matrix A can be reconstructed as follows:

$$\widehat{A} = \widehat{A_L}((\widehat{A_T})_L)^{+}\widehat{A_T} = \widehat{D}\widehat{\alpha_L}((\widehat{\alpha})_L)^{+}\widehat{\alpha}, \tag{12.15}$$

where $(X)_L$ denotes the left sub-matrix of X. Actually, when $\text{rank}(A_{TL}) = \text{rank}(A)$ holds, $\widehat{\alpha_L}$ recovered from the left observation matrix X_L is a good approximation to, or exactly equal to, $(\widehat{\alpha})_L$ recovered from the top observation matrix X_T. The same relationship exists between $(\widehat{D})_T$ and $\widehat{D_T}$, where $(\widehat{D})_T$ denotes the top sub-matrix of $\widehat{D}$. Thus, we can further simplify ROSL+ as

$$\widehat{A} = \widehat{D}\widehat{\alpha} \tag{12.16}$$

where $\widehat{D}$ and $\widehat{\alpha}$ are respectively recovered from X_L and X_T in the following two simple steps.

1. Solve $\widehat{D}$ and $\widehat{\alpha_L}$ by applying ROSL on X_L:

$$\min_{D,\alpha_L,E_L} \|\alpha_L\|_{\text{row-1}} + \lambda\|E_L\|_1 \;\; \text{s.t.} \begin{matrix} X_L = D\alpha_L + E_L \\ D^T D = I_k. \end{matrix} \tag{12.17}$$

2. Solve $\widehat{\alpha}$ by minimizing $\|X_T - \widehat{D_T}\alpha\|_1$ by fixing $\widehat{D_T}$ as $(\widehat{D})_T$.

In other words, ROSL+ first recovers $\widehat{D}$ from the left sub-matrix X_L (complexity: $\mathcal{O}(mlr)$), and then solves $\widehat{\alpha}$ by minimizing the ℓ_1-norm of $X_T - \widehat{D_T}\alpha$ (complexity: $\mathcal{O}(nhr)$). Thus, the complexity of ROSL+ is $\mathcal{O}(r(ml + nh))$. When the matrix rank r is much smaller that its size, i.e., $r \ll \min(m, n)$, the sample number can be set as $l = \beta_2 r$ and $h = \beta_3 r$, where β_2 and β_3 are constants larger than 1. In this case, ROSL+ has the linear complexity of the matrix size, i.e., $\mathcal{O}(r^2(m + n))$.

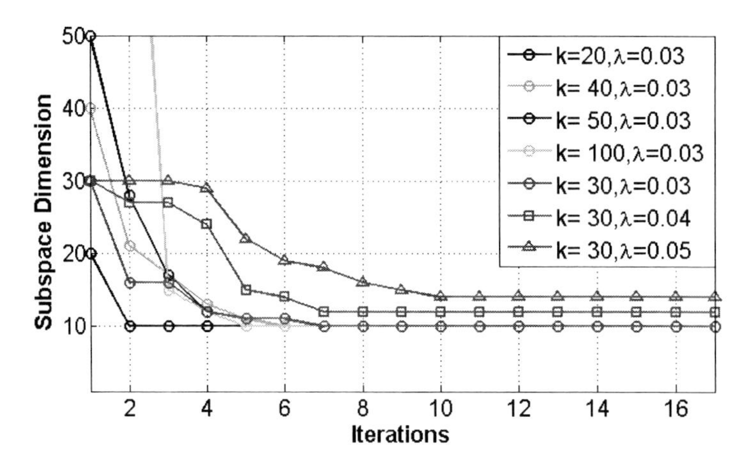

FIGURE 12.3 Convergence rate of ROSL. At the fixed $\lambda = 0.03$, the recovered subspace dimension always converges to $r = 10$ in less than 7 iterations **regardless** of the initial value of k, which indicates the ROSL solver is robust and very stable. The recovered subspace dimension increases as λ increases from 0.03 to 0.05. MAE $\approx 10^{-6}$ at all cases above.

12.5 Experimental Results

We present several experiments to evaluate the performance of ROSL and ROSL+, including (1) simulation on a corrupted synthetic low-rank matrix of varying dimension, (2) visual low-rank recovery on real data for background subtraction. Note that the ROSL algorithm is implemented in MATLAB without using any advanced tools unlike some other methods we compare. All the experimental results are executed on an Intel W3530 CPU and 6GB memory. For simplicity, we set the sample number $h = l$ for ROSL+ and other sampling-based methods we tested.

TABLE 12.2 Evaluation of ROSL, ROSL+, and the existing low-rank recovery approaches on synthetic low-rank matrices (size: $m \times m$ and rank $r = 10$). The experimental parameters are set up as: (1) λ is best tuned for each method, (2) the dimension of D is initialized as $k = 30$, (3) the stop criterion is $\|X - A^{i+1} - E^{i+1}\|_F / \|X\|_F \leq 10^{-6}$, (4) max iteration number (iter) is set to be 300, and (5) the sample number $l = h = 100$. The Mean of Absolute Error (MAE) between A and $\widehat{A}$ is used to gauge the recovery accuracy. The iterations (rounds of ADM) and the total running time (seconds) are reported. Note: aEb denotes $a \times 10^b$.

	RPCA		Partial RPCA		RP-RPCA		LMaFit		ROSL		ROSL-Nys1		ROSL-Nys2		ROSL+	
m	MAE	**Time**	MAE	**Time**	MAE	**Time**	MAE	**Time**	MAE	**Time**	MAE	**Time**	MAE	**Time**	MAE	**Time**
500	2.8E-6	**2.51**	2.2E-6	**1.44**	0.03	**5.9**	0.53	**6.9**	6.3E-6	**0.78**	2.4	**0.42**	4.8E-5	**0.42**	2.9E-5	**0.31**
1000	1.0E-6	**12.7**	1.1E-6	**5.60**	0.37	**23.7**	0.38	**28.7**	6.1E-6	**2.83**	2.6	**0.89**	5.4E-5	**0.89**	3.1E-5	**0.65**
2000	5.7E-7	**112**	7.6E-7	**24.4**	0.42	**110**	0.18	**116**	2.2E-6	**12.8**	2.3	**1.56**	5.0E-5	**1.56**	3.3E-5	**1.1**
4000	1.2E-6	**981**	5.3E-7	**161**	0.77	**669**	0.034	**442**	9.8E-6	**41.8**	3.0	**3.78**	4.3E-5	**3.77**	2.7E-5	**2.5**
8000	N/A	**N/A**	6.7E-7	**802**	1.62	**3951**	0.005	**1750**	2.2E-6	**214**	2.8	**9.0**	4.6E-5	**8.9**	2.2E-5	**5.6**
Iter	18∼20		18∼20		300		300		16∼17		18∼20		18∼20		18∼20	

Similar to [14], a square low-rank matrix $A \in \mathbb{R}^{m \times m}$ is synthesized as a product of an $m \times r$ matrix and an $r \times m$ matrix (r is set to be 10), whose entries obey the normal distribution. Then, the corrupted data X is generated by the addition of A and a sparse matrix $E \in \mathbb{R}^{m \times m}$ (10% of its entries are non-zero and drawn from the uniform distribution on [-50, 50]).

On this synthetic data, we evaluate the recovery accuracy and efficiency of ROSL, compared with those of RPCA, RP-RPCA, and LMaFit (advanced version of RMF). As shown in Table 12.2, ROSL is much faster than these methods without compromising the recov-

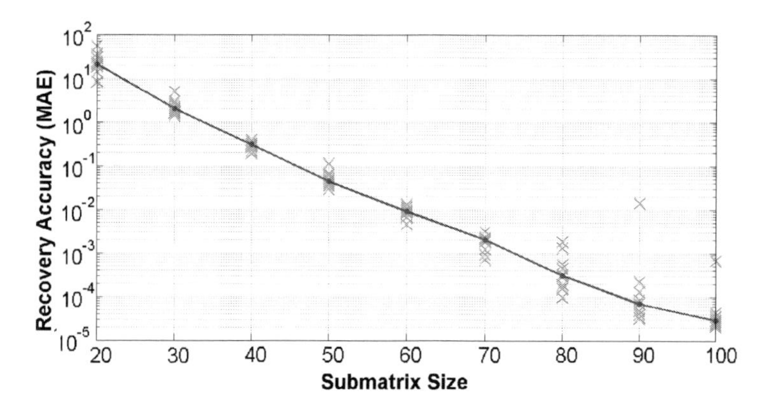

FIGURE 12.4 Recovery accuracy (MAE) of ROSL+ on synthetic data ($m = 1000, r = 10, k = 30$). For each l, the recovery errors (MAE) of ROSL+ in 10 different random-sampling trials are shown in green (their median in red). The recovery error (MAE) of ROSL+ decreases exponentially with the increase of l. These tests also indicate that ROSL+ gets the same global solution as RPCA in almost all cases.

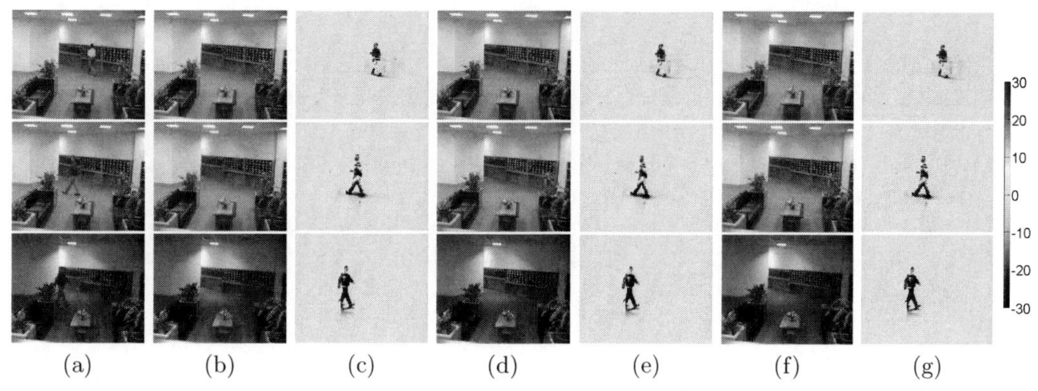

FIGURE 12.5 (See color insert.) Comparison of RPCA, ROSL($k = 10$), and ROSL+($l = 50$) in background modeling on the lobby video (size: 160×128, 1060 frames). (a) Original images. Backgrounds recovered by (b) RPCA, (d) ROSL, and (f) ROSL+. Foregrounds recovered by (c) RPCA, (e) ROSL, and (g) ROSL+. ROSL (time: 34.6s) and ROSL+ (time: 3.61s) are significantly ($10\times, 92\times$) faster than RPCA (time: 334s) while generating almost identical results.

ery accuracy. The original RPCA using full SVD is computationally costly and is almost infeasible when the matrix size $m = 8000$. Even partial RPCA [10] is consistently 4 times slower than ROSL and also requires a proper way to update κ. Although random projection helps reduce the computation of a single SVD, many iterations of SVD are needed to be conducted on different projected matrices. Thus, the total computation of RP-RPCA is costly and its recovery accuracy is low (Table 12.2). In the ideal case that the matrix rank is known, LMaFit has the same accuracy and complexity as ROSL. However, since it is unable to minimize the matrix rank, it fails to obtain accurate low-rank matrix recovery without exactly setting $k = r$. On this synthetic data (rank $r = 10$) in Table 12.2, LMaFit converges very slowly and fails to obtain accurate recovery at $k = 30$, which is true even at $k = 14$.

To evaluate the performance of ROSL+, we apply the generalized Nystrom method (employed in DFC) to ROSL, called ROSL-Nys. Since the performance of ROSL-Nys highly depends on how to recover A_{TL}, we present two different variants of ROSL-Nys, i.e., ROSL-Nys1 recovering sub-matrices (A_{TL}, A_T and A_L) independently, and ROSL-Nys2 recovering A_{TL} by the left sub-matrix of A_T. Actually, DFC also employed another column-sampling

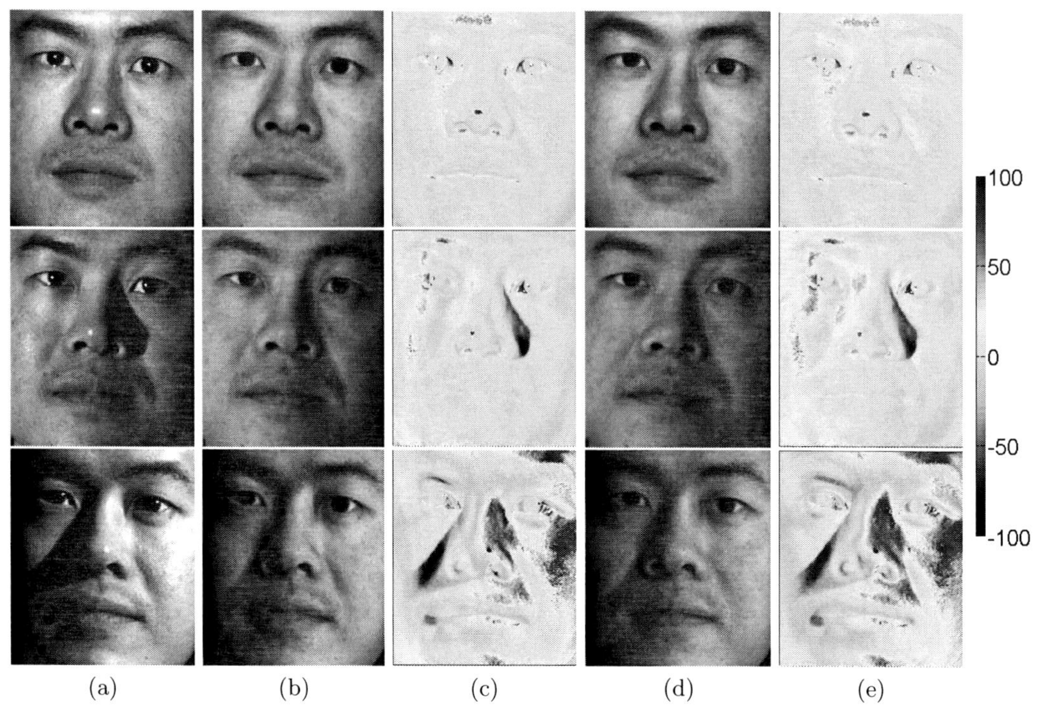

FIGURE 12.6 (See color insert.) Visual evaluation of ROSL and RPCA on face images (168 × 192, 55 frames) under varying illuminations. There is no significant difference between ROSL and RPCA. (a) Original images, diffusive component recovered by (b) RPCA and (d) by ROSL. Non-diffusive component (shadow/specularity) by (c) RPCA (time: 12.16s) and (e) by ROSL (time: 5.85s).

method. But it requires recovering multiple (i.e., $\frac{n}{l}$) sub-matrices (size:$m \times l$) and thus has quadratic complexity, much slower than ROSL+ (linear complexity). As shown in Table 12.2, RPCA-Nys1 fails to obtain accurate recovery. The reason is that tuning a common weight λ cannot guarantee the optimality of three subproblems—estimating A_L, A_T, and A_{TL}. Both the computational complexity and recovery accuracy of ROSL+ are on the same order of that of ROSL-Nys2, and are slightly ($1.5 \sim 2$ times) better that the latter. This better performance is due to the fact that ROSL+ consists of only one time ROSL and one time linear regression.

In addition, we evaluate the stability and convergence rate of ROSL/ROSL+ on the same synthetic matrix by varying the initial rank k, weight λ, or submatrix size l.

First, we observed that the recovery accuracy and convergence rate of ROSL are not sensitive to selection of k, as long as $k > r$. As shown in Figure 12.3, $\forall k \in [20, 100]$, the subspace dimension recovered by ROSL at $\lambda = 0.03$ fast converges to the rank $r = 10$ and the high accuracy (MAE $\approx 10^{-6}$) is achieved.

Second, ROSL produces accurate low-rank recovery at any weight $\lambda \in [0.03, 0.05]$ and the recovered subspace dimension consistently increases with λ. ROSL recovers the 14-dimension orthonormal subspace when $\lambda = 0.05$ and obtains accurate recovery (MAE$\approx 10^{-6}$).

Third, at the fixed sub-matrix size l, the recovery accuracy of ROSL+ is relatively stable in different random sampling trials. As the submatrix size l increases, the recovery error (MAE) of ROSL+ decreases exponentially and reaches as low as 3×10^{-5} when $l = 10r = 100$ (Figure 12.4). This result is in line with the failure probability δ of rank(A_{TL})=rank(A) that exponentially decreases with the increase of l.

To compare the recovery accuracy of ROSL/ROSL+ with that of RPCA, we evaluate them on two standard visual data sets, Yale-B face images and the lobby background subtraction video, similar to [4]. From each video, we build an observation matrix X by vectorizing each frame as one column, and respectively recover the low-rank component A from X by ROSL and RPCA.

In the lobby video, both ROSL and ROSL+ exactly recover the same (accurate) foreground objects and background components as RPCA at much faster speeds (ROSL: **10**×, ROSL+: **92**×) as shown in Figure 12.5.

In the face image experiments, the non-diffusive component E detected by ROSL is almost the same as that detected by RPCA (Figure 12.6). The results of ROSL+ are very close to those of ROSL and thus not included in Figure 12.6, due to the space limit. Note that, the lobby video is a thin matrix (20480× 1060) and the efficiency improvement of ROSL/ROSL+ is expected to be even higher for large-scale square matrices. Such matrices are common in typical applications, e.g., in video summarization (10^5 images of 10^6 pixels) and in face recognition (10^6 images of 10^6 pixels).

12.6 Conclusion

In this chapter, a Robust Orthonormal Subspace Learning (ROSL) approach is proposed for efficient robust low-rank recovery. This approach accelerates the state-of-the-art method, i.e., RPCA, by replacing the nuclear norm on the low-rank matrix by a light-weight measure—the group sparsity of its coefficients under orthonormal subspace. This enables using fast sparse-coding algorithms to solve the robust low-rank recovery problem at the quadratic complexity of matrix size. This novel rank measure is proven to be lower-bounded by the nuclear norm and thus ROSL has the same global optima as RPCA. In addition, a random sampling algorithm is introduced to further speed up ROSL such that ROSL+ has linear complexity of the matrix size. Experimental results on the synthetic and real data show that ROSL and ROSL+ achieve state-of-the-art efficiency at the same level of recovery accuracy.

12.7 Acknowledgement

The support of Mitsubishi Electric Research Lab (MERL) and the support of the National Science Foundation under grant IIS 11-44227 are gratefully acknowledged.

References

1. M. Aharon, M. Elad, and A. Bruckstein. K-svd: An algorithm for designing overcomplete dictionaries for sparse representation. *IEEE Transactions on Signal Processing*, 54(11):4311–4322, 2006.
2. S. Bengio, F. Pereira, Y. Singer, and D. Strelow. Group sparse coding. In *Annual Conference on Neural Information Processing Systems, NIPS 2009*, volume 22, pages 82–89, 2009.
3. D. Bertsekas, editor. *Constrained optimization and Lagrange multiplier method.* Academic Press, 1982.
4. E. Candes, X. Li, Y. Ma, and J. Wright. Robust principal component analysis? *Journal of the ACM*, 58(3):article 11, 2011.
5. M. Fazel, H. Hindi, and S. Boyd. A rank minimization heuristic with application to minimum order system approximation. *American Control Conference, ACC 2001*, 42(165):115–142, 2001.

6. K. Gregor and Y. LeCun. Learning fast approximations of sparse coding. In *International Conference on Machine Learning, ICML 2010*, 2010.
7. J. Huang, T. Zhang, and D. Metaxas. Learning with structured sparsity. In *International Conference on Machine Learning, ICML 2009*, 2009.
8. Q. Ke and T. Kanade. Robust l_1 norm factorization in the presence of outliers and missing data by alternative convex programming. In *IEEE Conference on Computer Vision and Pattern Recognition, CVPR 2005*, volume 1, pages 739–746, 2005.
9. S. Kumar, M. Mohri, and A. Talwalkar. On sampling-based approximate spectral decomposition. In *International Conference on Machine Learning, ICML 2009*, 2009.
10. Z. Lin, M. Chen, and Y. Ma. The augmented Lagrange multiplier method for exact recovery of corrupted low-rank matrices. Technical Report UILU-ENG-09-2214, University of Illinois Urbana-Champaign (UIUC), 2010.
11. R. Liu, Z. Lin, Z. Su, and J. Gao. Solving principal component pursuit in linear time via l1 filtering. *Neurocomputing, to appear*, 2014.
12. L. Mackey, A. Talwakar, and M. Jordan. Divide-and-conquer matrix factorization. In *Annual Conference on Neural Information Processing Systems, NIPS 2011*, 2011.
13. J. Mairal, F. Bach, J. Ponce, G. Sapiro, and A. Zisserman. Non-local sparse models for iamge restoration. In *ICCV*, 2009.
14. Y. Mu, J. Dong, X. Yuan, and S. Yan. Accelerated low-rank visual recovery by random projection. In *IEEE Conference on Computer Vision and Pattern Recognition, CVPR 2011*, 2011.
15. B. Recht, M. Fazel, and P. Parrilo. Guaranteed minimum-rank solutions of linear matrix equations via nuclear norm minimization. *Arxiv preprint:0706.4138*, 2007.
16. Y. Shen, Z. Wen, and Y. Zhang. Augmented Lagrangian alternating direction method for matrix separation based on low-rank factorization. Technical Report TR11-02, Rice University, 2011.
17. N. Srebro, J. Rennie, and T. Jaakkola. Maximum-margin matrix factorization. *Annual Conference on Neural Information Processing Systems, NIPS 2005*, 17:1329–1336, 2005.
18. A. Talwalkar and A. Rostamizadeh. Matrix coherence and the Nystrom method. In *Proceedings of the Twenty-Sixth Annual Conference on Uncertainty in Artificial Intelligence (UAI-10)*, 2010.
19. R. Tomioka, T. Suzuki, M. Sugiyama, and H. Kashima. A fast augmented Lagrangian algorithm for learning low-rank matrices. In *International Conference on Machine Learning, ICML 2010*, 2010.
20. B. Turlach, W. Venables, and S. Wright. Simultaneous variable selction. *Technometrics*, 47:349–363, 2005.
21. C. Williams and M. Seeger. Using the Nystrom method to speed up the kernel machines. In *Annual Conference on Neural Information Processing Systems, NIPS 2000*, 2000.
22. M. Yuan and Y. Lin. Model selection and estimation in regresion with grouped variables. *Journal of the Royal Statistical Society. Series B*, 68:49–67, 2006.
23. X. Yuan and J. Yang. Sparse and low-rank matrix decomposition via alternating direction methods. Technical Report, Hong Kong Baptist University, 2009.
24. T. Zhou and D. Tao. GoDec: randomized low-rank and sparse matrix decomposition in noisy case. In *International Conference on Machine Learning, ICML 2011*, 2011.

13

A Unified View of Nonconvex Heuristic Approach for Low-Rank and Sparse Structure Learning

Yue Deng
University of California, San Francisco, USA

Feng Bao
Tsinghua University, China

Qionghai Dai
Tsinghua University, China

13.1 Introduction

Learning the intrinsic data structures via matrix analysis has received wide attention in many fields, e.g., neural networks [23], machine learning [21] [29], financial engineering [12], computer vision [3] [36] [13] and pattern recognition [11] [31]. There are quite a number of efficient mathematical tools for rank analysis, e.g., Principal Component Analysis (PCA) and Singular Value Decomposition (SVD). However, these typical approaches could only handle some preliminary and simple problems. With the recent progress of compressive sensing [15], a new concept on nuclear norm optimization has emerged into the field of rank minimization [35] and has led to a number of interesting applications, e.g., low-rank structure learning (LRSL) from corruptions and background modeling [10].[37]

[37]Parts of this chapter are reproduced from [10] with the licence No. ♯ 3600341135333 @ IEEE.

LRSL is a general model for many practical problems in the communities of machine learning and signal processing, which considers learning a data of the low-rank structure from sparse errors [6] [8] [22] [30]. Such problem can be formulated as: $\mathbf{P} = f(\mathbf{A}) + g(\mathbf{E})$, where $\mathbf{P}$ is the corrupted matrix observed in the practical world; $\mathbf{A}$ and $\mathbf{E}$ are *low-rank* matrix and *sparse* corruption, respectively and the functions $f(\cdot)$ and $g(\cdot)$ are both linear mappings. Recovering two variables (i.e., $\mathbf{A}$ and $\mathbf{E}$) just from one equation is an ill-posed problem but one that is still possible to be addressed by optimizing:

$$\begin{aligned} (\text{P0}) \quad & \min_{(\mathbf{A},\mathbf{E})} \quad rank(\mathbf{A}) + \lambda \left\|\mathbf{E}\right\|_{\ell_0} \\ & s.t. \quad \mathbf{P} = f(\mathbf{A}) + g(\mathbf{E}). \end{aligned} \tag{13.1}$$

In (P0), $rank(\mathbf{A})$ is adopted to describe the low-rank structure of matrix $\mathbf{A}$, and the sparse errors are penalized via $\|\mathbf{E}\|_{\ell_0}$, where ℓ_0 norm counts the number of all the non-zero entries in a matrix. (P0) is always referred to as sparse optimization since rank term and ℓ_0 norm are sparse measurements for matrices and vectors, respectively. However, such sparse optimization is of little use due to the discrete nature of (P0) and the exact solution to it requires an intractable combinatorial search.

A common approach that makes (P0) trackable tries to minimize its convex envelope, where the rank of a matrix is replaced by the nuclear norm and the sparse errors are penalized via ℓ_1 norm, which are convex envelopes for $rank(\cdot)$ and $\|\cdot\|_{\ell_0}$, respectively. In practical applications, LRSL via ℓ_1 heuristic is powerful enough for many learning tasks with relative low-rank structure and sparse corruptions. However, when the desired matrix becomes complicated, e.g., it has high intrinsic rank structure or the corrupted errors become dense, the convex approach may not achieve promising performances. In order to handle those tough tasks via LRSL, in this chapter we take advantage of non-convex approximation to better enhance the sparseness of signals. The prominent reason that we adopt non-convex terms is mainly due to their enhanced sparsity. Geometrically, non-convex terms generally lie much closer to the essential ℓ_0 norm than the convex ℓ_1 norm [7].

To fully exploit the merits of non-convex terms for LRSL, we formulate (P0) as a semi-definite programming (SDP) problem so that LRSL with two different norms will eventually be converted into an optimization only with ℓ_1 norm. Thanks to the SDP formulation, non-convex terms can be explicitly combined into the paradigm of low-rank structure learning and we will investigate two widely used non-convex terms in this chapter, i.e., ℓ_p norm ($0 < p < 1$) and log-sum term. Accordingly, two non-convex models, i.e., ℓ_p-norm heuristic recovery (pHR) and log-sum heuristic recovery (LHR) will be proposed for corrupted matrix learning. Theoretically, we will analyze the relationship of these two models and reveal that the proposed LHR exhibits the same objective of pHR when p infinitely approaches to 0^+. Therefore, LHR owns more powerful sparseness enhancement capabilities than pHR.

For the sake of accurate solutions, the Majorization-Minimization (MM) algorithm will be applied to solve the non-convex heuristic model. MM algorithm is implemented in an iterative way such that it first replaces the non-convex component of the objective with a convex upper-bound, and then to minimize the constructed surrogate, which makes the non-convex problem fall exactly into the general paradigm of the reweighted schemes. Accordingly, it is possible to solve the non-convex optimization following a sequence of convex optimizations, and we will prove that with the MM framework, non-convex models finally converge to a stationary point after successive iterations. The advantages of LRSL are verified on a number of tasks including simulations and practical applications. Specifically, for practical application, we adopt our LRSL model to a low-rank matrix recovery (LRMR) problem where non-convex heuristic models are used to recover a low-rank matrix from sparse corruptions [14].

The remainder of this chapter is organized as follows. We review previous works in Section 13.2. The background of the typical convex LRSL model and its specific form of semi-definite programming will be provided in Section (13.3). Section 13.4 introduces the general non-convex heuristic models and discusses how to solve the non-convex problem by MM algorithm. We address the low-rank matrix recovery (LRMR) problem and compare the proposed LHR and pHR model with PCP from simulations in Section 13.5. Finally, we show how to integrate our model to the application of depth-maps matrix fusion.

13.2 Previous Works

In this part, we review some related works from the following perspectives. First, we discuss two famous models in LRSL, i.e., Low-Rank Matrix Recovery (LRMR) from corruptions and low-rank representation (LRR). Then, some previous works about the Majorization-Minimization algorithm and reweighted approaches are presented.

13.2.1 Low-Rank Structure Learning

Low-rank matrix recovery
Sparse learning and rank analysis are now drawing more and more attention in both the fields of machine learning and signal processing. In [42], sparse learning is incorporated into a non-negative matrix factorization framework for blind sources separation. Besides, it has also been introduced to the typical subspace-based learning framework for face recognition [24]. In addition to the widely used ℓ_1 norm based sparse learning, some other surrogates have been proposed for signal and matrix learning. In [41], the $\ell_{1/2}$ norm and its theoretical properties are discussed for sparse signal analysis. In [26], elastic-net-based matrix completion algorithms extend the extensively used elastic-net penalty to matrix cases. A fast algorithm for matrix completion by using a ℓ_1 filter has been introduced in [32]. Corrupted matrix recovery considers decomposing a low-rank matrix from sparse corruptions that can be formulated as $\mathbf{P} = \mathbf{A} + \mathbf{E}$, where $\mathbf{A}$ is a low-rank matrix, $\mathbf{E}$ is the sparse error, and $\mathbf{P}$ is the observed data from real-world devices, e.g., cameras, sensors, and other equipment. The rank of $\mathbf{P}$ is not low, in most scenarios, due to the disturbances of $\mathbf{E}$. How can we recover the low-rank structure of the matrix from gross errors? This interesting topic has been discussed in a number of works, e.g., [6], [8], and [22]. Wright et al. proposed the PCP (a.k.a. RPCA) to minimize the nuclear norm of a matrix by penalizing the ℓ_1 norm of errors [8]. PCP could exactly recover the low-rank matrix from sparse corruptions. In some recent works, Ganesh et al. investigated the parameter-choosing strategy for PCP from both the theoretical justifications and simulations [20]. In this chapter, we will introduce the reweighted schemes to further improve the performances of PCP. Our algorithm could exactly recover a corrupted matrix from much denser errors and higher rank.

Low-rank representation
Low-rank representation [29] is a robust tool for subspace clustering [38], the desired task of which is to classify the mixed data in their corresponding subspaces/clusters. The general model of LRR can be formulated as $\mathbf{P} = \mathbf{PA} + \mathbf{E}$, where $\mathbf{P}$ is the original mixed data, $\mathbf{A}$ is the affine matrix that reveals the correlations between different pairs of data, and $\mathbf{E}$ is the residual of such a representation. In LRR, the affine matrix $\mathbf{A}$ is assumed to be low-rank and $\mathbf{E}$ is regarded as sparse corruptions. Compared with existing SC algorithms, LRR is much more robust to noises and archives promising clustering results on public datasets. Fast implementations for LRR solutions are introduced in [28] by iteratively linearizing approaches.

13.2.2 MM Algorithm and Reweighted Approaches

The Majorization-Minimization (MM) algorithm is widely used in machine learning and signal processing. It is an effective strategy for non-convex problems in which the hard problem is solved by optimizing a series of easy surrogates. Therefore, most optimizations via the MM algorithm fall into the framework of reweighted approaches.

In the field of machine learning, the MM algorithm has been applied to parameters selection for bayesian classification [19]. In the area of signal processing, the MM algorithm leads to a number of interesting applications, including wavelet-based processing [18] and total variation (TV) minimization [2]. For compressive sensing, the reweighted method was used in the ℓ_1 heuristic and led to a number of practical applications including portfolio management [16] and image processing [7]. The reweighted nuclear norm was first discussed in [17] and the convergence of such an approach has been proven in [34].

Although there are some previous works on reweighted approaches for rank-minimization, our approach is quite different. First, this chapter tries to consider a new problem of low-rank structure learning from corruptions, while not on the single task of sparse signal or nuclear norm minimization. Besides, existing works on reweighted nuclear norm minimization in [17] [34] are solved by semi-definite programming that could only handle the matrix of relatively small size. In this chapter, we will use the first-order numerical algorithm (e.g., alternating direction method (ADM)) to solve the reweighed problem, which can significantly improve the numerical performance. Due to the distributed optimization strategy, it is possible generalize the learning capabilities to large-scale matrices.

13.3 ℓ_1-Based Low-Rank Structure Learning

In this part, we formulate the LRSL as a semidefinite programming (SDP). With the SDP formulation, it will become apparent that typical LRSL is a kind of general ℓ_1 heuristic optimization. This section serves as the background material for the discussions in Section 13.4.

As stated previously, the basic optimization (P0) is non-convex and generally impossible to solve as its solution usually requires an intractable combinatorial search. In order to make (13.1) trackable, convex alternatives are widely used in a number of works, e.g., [6] [8]. Among these approaches, one prevalent method tries to replace the rank of a matrix by its convex envelope, i.e., the nuclear norm, and the ℓ_0 sparsity is penalized via ℓ_1 norm. Accordingly, by convex relaxation, the problem in (13.2) can actually be recast as a semi-definite programming,

$$\begin{aligned} \min_{(\mathbf{A},\mathbf{E})} & \ \|\mathbf{A}\|_* + \lambda \|\mathbf{E}\|_{\ell_1} \\ s.t. & \ \mathbf{P} = f(\mathbf{A}) + g(\mathbf{E}), \end{aligned} \tag{13.2}$$

where $\|\mathbf{A}\|_* = \sum_{i=1}^{r} \sigma_i(A)$, is the nuclear norm of the matrix that is defined as the summation of the singular values of $\mathbf{A}$; and $\|\mathbf{E}\|_{\ell_1} = \sum_{ij} |E_{ij}|$ is the ℓ_1 norm of a matrix. Although the objective in (13.2) involves two norms: nuclear norm and ℓ_1 norm, its essence is based on the ℓ_1 heuristic. We will verify this point with the following lemma.

LEMMA 13.1 For a matrix $\mathbf{X} \in \mathbb{R}^{m \times n}$, its nuclear norm is equivalent to the following optimization:

$$\|\mathbf{X}\|_* = \left\{ \begin{array}{cc} \min\limits_{(\mathbf{Y},\mathbf{Z},\mathbf{X})} & \frac{1}{2}[tr(\mathbf{Y}) + tr(\mathbf{Z})] \\ s.t. & \begin{bmatrix} \mathbf{Y} & \mathbf{X} \\ \mathbf{X}^T & \mathbf{Z} \end{bmatrix} \succeq 0, \end{array} \right\}, \tag{13.3}$$

where $\mathbf{Y} \in \mathbb{R}^{m \times m}$, $\mathbf{Z} \in \mathbb{R}^{n \times n}$ are both symmetric and positive definite. The operator $tr(\cdot)$ means the trace of a matrix and $\succeq$ represents semi-positive definite.

The proof of Lemma 13.1 may refer to [16] [35]. According to this lemma, we can replace the nuclear norm in (13.2) and formulate it in the form of:

$$\begin{array}{cl} \min\limits_{(\mathbf{Y},\mathbf{Z},\mathbf{A},\mathbf{E})} & \frac{1}{2}[tr(\mathbf{Y}) + tr(\mathbf{Z})] + \lambda \|\mathbf{E}\|_{\ell_1} \\ s.t. & \begin{bmatrix} \mathbf{Y} & \mathbf{A} \\ \mathbf{A}^T & \mathbf{Z} \end{bmatrix} \succeq 0 \\ & \mathbf{P} = f(\mathbf{A}) + g(\mathbf{E}). \end{array} \tag{13.4}$$

From Lemma 13.1, we know that both $\mathbf{Y}$ and $\mathbf{Z}$ are symmetric and positive definite. Therefore, the trace of $\mathbf{Y}$ and $\mathbf{Z}$ can be expressed as a specific form of ℓ_1 norm, i.e., $tr(\mathbf{Y}) = \|diag(\mathbf{Y})\|_{\ell_1}$. $diag(\mathbf{Y})$ is an operator that only keeps the entries on the diagonal position of $\mathbf{Y}$ in a vector. Therefore, the optimization in (13.4) can be expressed as:

$$\min_{\hat{X} \in \hat{D}} \frac{1}{2}(\|diag(\mathbf{Y})\|_{\ell_1} + \|diag(\mathbf{Z})\|_{\ell_1}) + \lambda \|\mathbf{E}\|_{\ell_1}, \tag{13.5}$$

where $\hat{X} = \{\mathbf{Y}, \mathbf{Z}, \mathbf{A}, \mathbf{E}\}$ and

$$\hat{D} = \{(\mathbf{Y}, \mathbf{Z}, \mathbf{A}, \mathbf{E}) : \begin{bmatrix} \mathbf{Y} & \mathbf{A} \\ \mathbf{A}^T & \mathbf{Z} \end{bmatrix} \succeq 0, (\mathbf{A}, \mathbf{E}) \in C\}.$$

$(\mathbf{A}, \mathbf{E}) \in C$ stands for convex constraint.

13.4 Corrupted Matrix Recovery via Non-Convex Heuristic

By Lemma 13.1, the convex problem with two norms in (13.2) has been successfully converted to an optimization only with ℓ_1 norm and therefore it is called ℓ_1-heuristic. ℓ_1 norm is the convex envelope of the concave ℓ_0 norm but a number of previous research works have indicated the limitation of approximating ℓ_0 sparsity with ℓ_1 norm, e.g., [7] [43] [9]. It is natural to ask, for example, whether a different alternative might not only find a correct solution, but also outperform the performance of ℓ_1 norm. One natural inspiration is to use some non-convex terms lying much closer to the ℓ_0 norm than the convex ℓ_1 norm. However, by using the non-convex heuristic terms, two problems come out inevitably: 1) which non-convex functionality is ideal and 2) how to efficiently solve the non-convex optimization. In the following two subsections, we will respectively address these two problems by introducing the log-sum heuristic recovery and its reweighted solution.

13.4.1 Non-Convex Heuristic Recovery

In this section, we will introduce two non-convex terms to enhance the sparsity of model in (13.5). The first one is the widely used ℓ_p norm with $0 < p < 1$. Intuitively, it lies in the scope between the ℓ_0 norm and the ℓ_1 norm. Therefore, it is believed to have a better sparse representation ability than the convex ℓ_1 norm. We define the general concave ℓ_p norm by $f_p(X) = \sum_{ij} |X_{ij}|^p, 0 < p < 1$. Therefore, by taking it into (13.5), the following ℓ_p-norm Heuristic Recovery (pHR) optimization is obtained.

$$\begin{aligned}(p\text{HR})\ H_p(\hat{X}) = \min_{\hat{X}\in\hat{D}} \tfrac{1}{2}[f_p(diag(\mathbf{Y})) + f_p(diag(\mathbf{Z}))] \\ + \lambda f_p(\mathbf{E}).\end{aligned} \tag{13.6}$$

In the formulation of pHR, obviously, it differs from (13.5) only on the selection of the sparse norm, where the latter uses concave ℓ_p norm instead of the typical ℓ_1 norm. Starting from pHR, another non-convex heuristic model with much sparser penalization can be derived. Obviously, $\forall p > 0$, minimizing the above pHR is equivalent to

$$\begin{aligned}\min_{\hat{X}\in\hat{D}} F(\hat{X}) &= \tfrac{1}{p}[H_p(\hat{X}) - (\tfrac{1}{2}m + \tfrac{1}{2}n + \lambda mn)] \\ &= \tfrac{1}{2}\sum_{i=1}^{m} \frac{|Y_{ii}|^p - 1}{p} + \tfrac{1}{2}\sum_{i=1}^{n} \frac{|Z_{ii}|^p - 1}{p}. \\ &+ \lambda \sum_{i=1}^{n}\sum_{j=1}^{m} \frac{|E_{ij}|^p - 1}{p}\end{aligned} \tag{13.7}$$

The optimization in (13.6) is the same as the problem in (13.7) because the multiplied scaler $\frac{1}{p}$ is a positive constant and $\frac{1}{2}m + \frac{1}{2}n + \lambda mn$ is a constant. According to L'Hôspital's rule, we know that $\lim_{p\to 0} \frac{x^p - 1}{p} = \frac{\partial_p(x^p - 1)}{\partial_p(p)} = \log x$, where $\partial_p(f(p))$ stands for the derivative of $f(p)$ with respect to p. Accordingly, by taking the limit $\lim_{p\to 0^+} F(X)$ in (13.7), we get the Log-sum Heuristic Recovery (LHR) model $H_L(\hat{X})$:

$$\begin{aligned}(\text{LHR})\ H_L(\hat{X}) = \min_{\hat{X}\in\hat{D}} \tfrac{1}{2}[f_L(diag(\mathbf{Y})) + f_L(diag(\mathbf{Z}))] \\ + \lambda f_L(\mathbf{E}).\end{aligned} \tag{13.8}$$

For any matrix $\mathbf{X} \in \mathbb{R}^{m\times n}$, the log-sum term is defined as $f_L(\mathbf{X}) = \sum_{ij} \log(|X_{ij}| + \delta)$, where $\delta > 0$ is a small regularization constant.

From (13.6) and (13.8), we know that LHR is a particular case of pHR by taking the limit of p at 0^+. It is known that, when $0 < p < 1$, the closer p approaches to zero, the stronger sparse enhancement the ℓ_p-based optimization exhibits. We also comment here that when p equals zero, the pHR exactly corresponds to the intractable discrete problem in (13.1). When $p = 0$ and $p \to 0^+$, pHR gives two different objectives. This finding does not deny our basic derivation since when $p = 0$ or $p < 0$, the equivalence from (13.6) to (13.7) does not hold any longer. Therefore, LHR only uses a limit to approximate the intractable objective of ℓ_0-based recovery. This is meanwhile the very reason why we denote a "plus" on the superscript of zero in limit $p \to 0^+$. LHR exploits the limit of the ℓ_0 norm in the objective and is regarded as having much stronger sparsity-enhancement capability than general pHR.

Due to much more powerful sparseness of LHR, in this chapter, we advocate the usage of LHR for non-convex-based LRSL. Therefore, in the remainder of this chapter, we will discuss the formulations of LHR for low-rank optimization in detail. However, fortunately, thanks to the natural connections between pHR and LHR, the optimization rule of LHR is also applied to pHR.

13.4.2 Solving LHR via Reweighed Approaches

Although we have placed a powerful term to enhance the sparsity, unfortunately, it also causes non-convexity in the objective function. For example, the LHR model is not convex since the log-function over $\mathbb{R}_{++} = (\delta, \infty)$ is concave. In most cases, non-convex problems can be extremely hard to solve. Fortunately, the convex upper bound of $f_L(\cdot)$ can be easily found and defined by its first order Taylor expansion. Therefore, we will introduce the Majorization-Minimization algorithm to solve the LHR optimization.

The Majorization-Minimization (MM) algorithm replaces the hard problem by a sequence of easier ones. It proceeds in an Expectation Maximization (EM)-like fashion by repeating two steps of **M**ajorization and **M**inimization in an iterative way. During the *Majorization* step, it constructs the convex upper bound of the non-convex objective. In the *Minimization* step, it minimizes the upper bound.

According to Appendix 13.8.2, the first-order Taylor expansion of each component in (13.8) is well defined. Therefore, we can construct the upper bound of LHR and instead to optimize the following problem

$$\begin{aligned}\min_{\hat{X}\in\hat{D}} T(\hat{X}|\hat{\Gamma}) = \tfrac{1}{2}tr[(\mathbf{\Gamma}_Y + \delta\mathbf{I}_m)^{-1}\mathbf{Y}] + \tfrac{1}{2}tr[(\mathbf{\Gamma}_z + \delta\mathbf{I}_n)^{-1}\mathbf{Z}] \\ + \lambda\textstyle\sum_{ij}(\Gamma_{E_{ij}} + \delta)^{-1}E_{ij} + const.\end{aligned} \tag{13.9}$$

In (13.9), set $\hat{X} = \{\mathbf{Y}, \mathbf{Z}, \mathbf{A}, \mathbf{E}\}$ contains all the variables to be optimized and set $\hat{\Gamma} = \{\mathbf{\Gamma}_Y, \mathbf{\Gamma}_Z, \mathbf{\Gamma}_E\}$ contains all the parameter matrices. The parameter matrices define the points at which the concave function is linearized via Taylor expansion. See Appendix 13.8.2 for details. At the end of (13.9), *const* stands for the constants that are irrelative to $\{\mathbf{Y}, \mathbf{Z}, \mathbf{A}, \mathbf{E}\}$. In some previous works of MM algorithms [19] [7] [34], they denote the parameter $\hat{\Gamma}$ in t^{th} iteration with the optimal value of $\hat{X}$ of the last iteration, i.e. $\hat{\Gamma} = \hat{X^t}^*$.

To numerically solve the LHR optimization, we remove the constants that are irrelative to $\mathbf{Y}, \mathbf{Z}$, and $\mathbf{E}$ in $T(\hat{X}|\hat{\Gamma})$ and get the new convex objective

$$\min \frac{1}{2}[tr(\mathbf{W}_Y^2\mathbf{Y}) + tr(\mathbf{W}_Z^2\mathbf{Z})] + \lambda\sum_{ij}(W_E)_{ij}E_{ij}$$

where $\mathbf{W_{Y(Z)}} = (\mathbf{\Gamma_{Y(Z)}} + \delta\mathbf{I_{m(n)}})^{-1/2}$ and $(W_E)_{ij} = (E_{ij} + \delta)^{-1}, \forall ij$. It is worth noting that $tr(\mathbf{W}_Y^2\mathbf{Y}) = tr(\mathbf{W}_Y\mathbf{Y}\mathbf{W}_\mathbf{Y})$. Besides, since both $\mathbf{W}_Y$ and $\mathbf{W}_Z$ are positive definite, the first constraint in (13.8) is equivalent to

$$\begin{bmatrix}\mathbf{W}_Y & \mathbf{0}\\ \mathbf{0} & \mathbf{W}_Z\end{bmatrix}\begin{bmatrix}\mathbf{Y} & \mathbf{A}\\ \mathbf{A}^T & \mathbf{Z}\end{bmatrix}\begin{bmatrix}\mathbf{W}_Y & \mathbf{0}\\ \mathbf{0} & \mathbf{W}_Z\end{bmatrix} \succeq \mathbf{0}.$$

Therefore, after convex relaxation, the optimization in (13.8) is now subjected to

$$\begin{aligned}\min \quad & \tfrac{1}{2}[tr(\mathbf{W}_Y\mathbf{Y}\mathbf{W}_Y) + tr(\mathbf{W}_Z\mathbf{Z}\mathbf{W}_Z)] + \lambda\|\mathbf{W}_E \odot \mathbf{E}\|_{\ell_1}\\ s.t. \quad & \begin{bmatrix}\mathbf{W}_Y\mathbf{Y}\mathbf{W}_Y & \mathbf{W}_Y\mathbf{A}\mathbf{W}_Z\\ (\mathbf{W}_Y\mathbf{A}\mathbf{W}_Z)^T & \mathbf{W}_Z\mathbf{Z}\mathbf{W}_Z\end{bmatrix} \succeq \mathbf{0}\\ & \mathbf{P} = f(\mathbf{A}) + g(\mathbf{E}).\end{aligned} \tag{13.10}$$

Here, we apply Lemma 13.1 to (13.10) once again and rewrite the optimization in (13.10) in the form of the summation of the nuclear norm and ℓ_1 norm,

$$\begin{aligned}\min_{(\mathbf{A},\mathbf{E})} . \|\mathbf{W}_Y\mathbf{A}\mathbf{W}_Z\|_* + \lambda\|\mathbf{W}_E \odot \mathbf{E}\|_{\ell_1}\\ s.t. \ \mathbf{P} = f(\mathbf{A}) + g(\mathbf{E}).\end{aligned} \tag{13.11}$$

In (13.11), the operator $\odot$ in the error term denotes the component-wise product of two variables, i.e., for $\mathbf{W}_E$ and $\mathbf{E}$: $(\mathbf{W}_E \odot \mathbf{E})_{ij} = (W_E)_{ij}E_{ij}$. According to [16], we know that $\mathbf{Y}^* = \mathbf{U}\Sigma\mathbf{U}^T$ and $\mathbf{Z}^* = \mathbf{V}\Sigma\mathbf{V}^T$, if we do singular value decomposition for $\mathbf{A}^* = \mathbf{U}\Sigma\mathbf{V}^T$. Accordingly, the weight matrix $\mathbf{W}_Y = (\mathbf{U}\Sigma\mathbf{U}^T + \delta\mathbf{I}_m)^{-1/2}$ and matrix $\mathbf{W}_Z =$

$(\mathbf{V}\Sigma\mathbf{V}^T + \delta\mathbf{I}_n)^{-1/2}$.[38] We should comment here that Equation 13.11 is also applied to solve the pHR problem. It just uses different weighting matrices, $\mathbf{W}_Y = diag((\mathbf{U}\Sigma\mathbf{U}^T + \delta\mathbf{I}_m))^{(p-1)/2}$, $\mathbf{W}_Z = diag((\mathbf{V}\Sigma\mathbf{V}^T + \delta\mathbf{I}_n))^{(p-1)/2}$ and $\mathbf{W}_E = [(E_{ij} + \delta)^{(p-1)}]$. The derivations of these parameter matrices are so similar to the formulations of LHR in Appendix 13.4 that we omit them here.

Here, based on the MM algorithm, we have converted the non-convex LHR optimization to a sequence of convex reweighted problems. We call this the *reweighted method* (13.11) since in each iteration we should re-denote the weight matrix set $\hat{W}$ and use the updated weights to construct the surrogate convex function. Besides, the objective in (13.11) is convex with a summation of a nuclear norm and an ℓ_1 norm and can be solved by convex optimization. In the next two sections, the general LHR model will be adapted to a specific model and we will provide the optimization strategy. But before that, we first stop here and extend some theoretic discussions of the LHR model.

13.4.3 Theoretical Justifications

In this part, we investigate some theoretical properties of the proposed non-convex heuristic algorithm with the MM optimization and prove its convergence. For simplicity, we define the objective function in (13.8) as $H(\hat{X})$ and the surrogate function in (13.9) is defined as $T(\hat{X}|\hat{\Gamma})$. $\hat{X}$ is a set containing all the variables and set $\hat{\Gamma}$ records the parameter matrices. Before discussing the convergence property of LHR, we will first provide two lemmas.

LEMMA 13.2 If set $\hat{\Gamma}^t := \hat{X}^t$, the MM algorithm could monotonically decrease the non-convex objective function $H(\hat{X})$, i.e., $H(\hat{X}^{t+1}) \leq H(\hat{X}^t)$.

The proof of this lemma may refer to Appendix.13.8.3. According to Lemma 13.2, we can give Lemma 13.3 to prove the convergence of the LHR iterations.

LEMMA 13.3 Let $\hat{X} = \{\hat{X}^0, \hat{X}^1 ... \hat{X}^t ...\}$ be a sequence generated by the MM framework. After successive iterations, such a sequence converges to the same limit point.

The proof of Lemma 13.3 includes two steps. First, we should prove there exists a convergent subsequence $\hat{X}^{t_k} \in \hat{X}$. See the discussions in Appendix 13.8.3 for details. Then, we prove the whole Lemma 13.3 by the contradiction in Appendix 13.8.3. Based on the two lemmas proved previously, we can now provide the convergence theorem of the proposed LHR model.

THEOREM 13.1 *With the MM framework, the LHR model finally converges to a stationary point.*

See Appendix 13.8.3 for the proof. In this part, we have shown that with the MM algorithm, the LHR model could converge to a stationary point. However, it is impossible to claim that the converged point is the global minimum since the objective function of LHR is not convex. Fortunately, with a good starting point, we can always find a desirable

[38] In such cases, the weighting matrices may cause complex numbers due to the inverse operation. Under such conditions, we use the approximating matrices $\mathbf{W}_Y = \mathbf{U}(\Sigma + \delta\mathbf{I}_m)^{-1/2}\mathbf{U}^T$ and $\mathbf{W}_Z = \mathbf{V}(\Sigma + \delta\mathbf{I}_m)^{-1/2}\mathbf{V}^T$ in LHR.

solution by iterative approaches. In this chapter, the solution of the ℓ_1 heuristic model was used as a starting point and it could always lead to a satisfactory result.

13.5 LHR for Low-Rank Matrix Recovery from Corruptions

In this part, we first apply the LHR model to recover a low-rank matrix from corruption, and its performance is compared with the widely used Principal Component Pursuit (PCP).

13.5.1 Proposed Algorithm

Based on the LHR derivations, the corrupted low-rank matrix recovery problem can be formulated as a reweighted problem:

$$\begin{aligned} &\min_{(\mathbf{A},\mathbf{E})} . \|\mathbf{W}_Y \mathbf{A} \mathbf{W}_Z\|_* + \lambda \|\mathbf{W}_E \odot \mathbf{E}\|_{\ell_1} \\ &s.t.\ \ \mathbf{P} = \mathbf{A} + \mathbf{E}. \end{aligned} \tag{13.12}$$

Due to the reweighted weights that are placed in the nuclear norm, it is impossible to directly get the closed-form solution of the nuclear norm minimization. Therefore, inspired by the work in [29], we introduce another variable $\mathbf{J}$ to (13.12) by adding another equality constraint, and to solve

$$\begin{aligned} &\min . \|\mathbf{J}\|_* + \lambda \|\mathbf{W}_E \odot \mathbf{E}\|_{\ell_1} \\ &s.t.\ \ \mathbf{h}_1 = \mathbf{P} - \mathbf{A} - \mathbf{E} = \mathbf{0} \\ &\qquad \mathbf{h}_2 = \mathbf{J} - \mathbf{W}_Y \mathbf{A} \mathbf{W}_Z = \mathbf{0}. \end{aligned} \tag{13.13}$$

Based on the transformation, there is only one single $\mathbf{J}$ in the nuclear norm of the objective such that we can directly get its closed-form update rule by [5]. There are quite a number of methods that can be used to solve it, e.g., with the Proximal Gradient (PG) algorithm [1] or Alternating Direction Methods (ADM) [4]. In this chapter, we will introduce the ADM method since it is more effective and efficient. Using the ALM method [27], it is computationally expedient to relax the equality in (13.13) and instead solve:

$$\begin{aligned} L = &\|\mathbf{J}\|_* + \lambda \|\mathbf{W}_E \odot \mathbf{E}\|_{\ell_1} + < \mathbf{C}_1, \mathbf{h}_1 > \\ &+ < \mathbf{C}_2, \mathbf{h}_2 > + \tfrac{\mu}{2}(\|\mathbf{h}_1\|_F^2 + \|\mathbf{h}_2\|_F^2) \end{aligned} \tag{13.14}$$

where $<,>$ is an inner product and $\mathbf{C}_1$ and $\mathbf{C}_2$ are the Lagrange multipliers, which can be updated via the dual ascending method. (13.14) contains three variables, i.e., $\mathbf{J}, \mathbf{E}$, and $\mathbf{A}$. Accordingly, it is possible to solve the problem via the distributed optimization strategy called the Alternating Direction Method (ADM). The convergence of the ADM for convex problems has been widely discussed in a number of works [4] [25]. By ADM, the joint optimization can be minimized by four steps, as $\mathbf{E}$-minimization, $\mathbf{J}$-minimization, $\mathbf{A}$-minimization, and dual ascending.

The detailed derivations are similar to the previous works in [27] [14] and we omit them here. The whole framework to solve the LHR model for LRMR via reweighted schemes is given in Algorithm 13.1.[39] In lines 8 and 9 of the algorithm, $s_\alpha(\cdot)$ and $d_\alpha(\cdot)$ are defined as signal shrinkage operator and matrix shrinkage operator, respectively [27].

[39] The optimization for pHR is made very similar by changing the weight matrices.

Algorithm 13.1 - Optimization strategy of LHR for corrupted matrix recovery

1: **Input:** Corrupted matrix P and parameter λ
2: **Initialization:** $t := 1, E_{ij}^0 := 1, \forall i,j.\ \mathbf{W}_{Y(Z)}^{(1)} = \mathbf{I}_{m(n)}$.
3: **repeat**
4: Update the weighting matrices $\mathbf{W}_E^{(t)}, \mathbf{W}_Y^{(t)}$, and $\mathbf{W}_Z^{(t)}$ according to current estimation of $\mathbf{A}^{(t)}$ and $\mathbf{E}^{(t)}$
5: Reset $C_0 > 0; \mu_0 > 0; \rho > 1; k = 1; \mathbf{A}^0 = \mathbf{E}^0 = \mathbf{0}$
6: **while** not converged **do**
7: Variables updating.
8: $E_{ij}^k = s_{\lambda\mu^{-1}\left|(W_E^{(t)})_{ij}\right|}(P - A^{k-1} - \mu^{-1}C_1^k)_{ij}, \forall ij$
9: $\mathbf{J}^k = d_{\mu^{-1}}(\mathbf{W}_Y^{(t)}\mathbf{A}^{\mathbf{k-1}}\mathbf{W}_Z^{(t)} + \mu^{-1}\mathbf{C}_2^k)$
10: $\mathbf{A}^{\mathbf{k}} = \mathbf{A}^{k-1} + \gamma[-\mathbf{W}_Y^{(t)}(\mathbf{h}_1^k + \mu^{-1}\mathbf{C}_2^k)\mathbf{W}_Z^{(t)} + (\mathbf{h}_2^k + \mu^{-1}\mathbf{C}_1^k)]$
11: $\mathbf{C}_1^k = \mathbf{C}_1^{k-1} + \mu_k\mathbf{h}_1^k$
12: $\mathbf{C}_2^k = \mathbf{C}_2^{k-1} + \mu_k\mathbf{h}_2^k$
13: $k := k+1, \mu_{k+1} = \rho\mu_k;$
14: **end while**
15: $(\mathbf{A}^{(t)}, \mathbf{E}^{(t)}) = (\mathbf{A}_k, \mathbf{E}_k)$
16: $t := t+1$
17: **until** convergence
18: Output: $(\mathbf{A}^{(t)}, \mathbf{E}^{(t)})$.

13.5.2 Simulation Results and Validation

We have explained how to recover a low-rank matrix via LHR in the preceding sections. In this section, we will conduct some experiments to test its performances with the comparisons to robust PCP from numerical simulations.

Numerical simulations

We will demonstrate the accuracy of the proposed non-convex algorithms on randomly generated matrices. For an equivalent comparison, we adopted the same data-generating method in [6] where all the algorithms are performed on the squared matrices, and the ground-truth low-rank matrix (rank r) with $m \times n$ entries, denoted as $\mathbf{A}^*$, is generated by the independent random orthogonal model [6]; the sparse error $\mathbf{E}^*$ is generated via uniformly sampling the matrix and the error values are randomly generated in the range [-100,100]. The corrupted matrix is generated by $\mathbf{P} = \mathbf{A}^* + \mathbf{E}^*$, where $\mathbf{A}^*$ and $\mathbf{E}^*$ are the ground truth. For simplicity, we denote the rank rate as $\eta = \frac{rank(\mathbf{A}^*)}{max\{m,n\}}$ and the error rate as $\xi = \frac{\|\mathbf{E}\|_{\ell_0}}{m \times n}$.

For an equivalent comparison, we use the code in [33] to solve the PCP problem.[40] [6] indicated that the PCP method could exactly recover a low-rank matrix from corruptions within the region of $\eta + \xi < 0.35$. Here, in order to highlight the effectiveness of our LHR model, we directly consider much more difficult tasks where we set $\eta + \xi = 0.5$. We compare the PCP (p=1) model with the proposed pHR (with p=1/3 and p=2/3) and the LHR (which can be regarded as $p \to 0^+$). Each experiment is repeated ten times and the mean values and their standard deviations (std)are tabulated in Table 13.1. In the table, $\frac{\|\mathbf{A}-\mathbf{A}^*\|_F}{\|\mathbf{A}^*\|_F}$ denotes

[40]In [27], Lin et al. provided two solvers, i.e., exact and inexact solvers, to solve the PCP problem. In this chapter, we use the exact solver for PCP because it performs better than the inexact solver.

TABLE 13.1 Evaluations of low-rank matrix recovery of Robust PCA and Non-convex Heuristic Recovery (Mean ± std).

		$rank(\mathbf{A}^*) = 0.4m$ $\|\|\mathbf{E}^*\|\|_{\ell_0} = 0.1m^2$			$rank(\mathbf{A}^*) = 0.1m$ $\|\|\mathbf{E}^*\|\|_{\ell_0} = 0.4m^2$		
m	methods	$\frac{\|\mathbf{A}-\mathbf{A}^*\|_F}{\|\mathbf{A}^*\|_F}$	$rank(\mathbf{A})$	$time(s)$	$\frac{\|\mathbf{A}-\mathbf{A}^*\|_F}{\|\mathbf{A}^*\|_F}$	$rank(\mathbf{A})$	$time(s)$
200	PCP	$(4.6 \pm 1.1)e^{-1}$	102 ± 12	5.9 ± 1.3	$(1.2 \pm 0.5)e^{-1}$	107 ± 11	7.4 ± 1.5
	$p\mathrm{HR}^{2/3}$	$(3.7 \pm 0.3)e^{-2}$	88 ± 4	16.4 ± 3.1	$(9.3 \pm 0.9)e^{-3}$	20 ± 0	16.3 ± 2.6
	$p\mathrm{HR}^{1/3}$	$(1.8 \pm 0.3)e^{-2}$	83 ± 2	13.1 ± 3.7	$(3.6 \pm 0.6)e^{-3}$	20 ± 0	13.4 ± 3.5
	LHR	$(8.1 \pm 1.2)e^{-4}$	80 ± 0	12.7 ± 2.5	$(1.3 \pm 0.7)e^{-3}$	20 ± 0	14.1 ± 2.7
400	PCP	$(4.5 \pm 1.2)e^{-1}$	205 ± 37	27.4 ± 3.3	$(6.4 \pm 2.3)e^{-1}$	217 ± 52	33.2 ± 4.6
	$p\mathrm{HR}^{2/3}$	$(2.3 \pm 0.9)e^{-2}$	193 ± 23	73.8 ± 12.3	$(5.0 \pm 1.1)e^{-3}$	71 ± 9	63.2 ± 15.7
	$p\mathrm{HR}^{1/3}$	$(1.2 \pm 0.5)e^{-2}$	160 ± 0	64.2 ± 13.4	$(4.0 \pm 0.7)e^{-4}$	41 ± 3	63.2 ± 12.4
	LHR	$(2.3 \pm 0.7)e^{-3}$	160 ± 2	53.4 ± 11.8	$(1.7 \pm 0.5)e^{-4}$	40 ± 0	54.3 ± 10.7
800	PCP	$(4.7 \pm 1.1)e^{-1}$	435 ± 97	36.2 ± 8.4	$(9.1 \pm 1.7)e^{-2}$	348 ± 42	50.1 ± 10.6
	$p\mathrm{HR}^{2/3}$	$(2.3 \pm 0.6)e^{-2}$	361 ± 32	103.6 ± 21.5	$(6.2 \pm 0.9)e^{-3}$	80 ± 3	129.2 ± 25.2
	$p\mathrm{HR}^{1/3}$	(8.7 ± 2.3)e-3	320 ± 0	96.2 ± 27.3	$(5.3 \pm 1.1)e^{-3}$	80 ± 0	119.2 ± 23.6
	LHR	$(1.7 \pm 0.3)e^{-3}$	320 ± 0	89.3 ± 21.2	$(4.1 \pm 0.5)e^{-3}$	80 ± 0	107.6 ± 25.3

the recovery accuracy, *rank* denotes the rank of the recovered matrix **A**, and *time* records the computational costs (in seconds). We do not report the std of the PCP method on the recovered errors because PCP definitely diverges on the tasks and its "std" does not exhibit significant statistic meanings.

From the results, obviously, compared with PCP, the LHR model could exactly recover the matrix from higher ranks and denser errors. The pHR model could correctly recover the matrix in most cases but the recovery accuracy is a bit lower than that of LHR. We also report the processing time in Table 13.1. The computer to implement the experiments is equipped with a 2.3 GHZ CPU processor and a 4GB RAM.

Feasible region

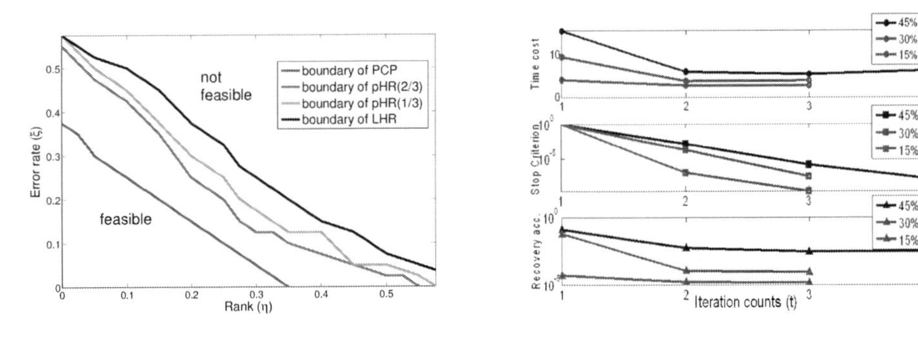

(a) Feasible region verification.

(b) Convergence verification.

FIGURE 13.1 (See color insert.) Feasible region and the convergence verifications.

Since the basic optimization involves two terms, i.e., low-rank matrix and sparse error, in this part we will vary these two variables to test the feasible boundary of PCP pHR and LHR, respectively. The experiments are conducted on the 400×400 matrices with sparse errors uniformly distributed in $[-100, 100]$. In the feasible region verification, when the recovery accuracy is larger than 1% (i.e., $\frac{\|A-A^*\|_F}{\|A^*\|_F} > 0.01$), it is believed that the algorithm diverges. The two rates η and ξ are varied from zero to one with the step of 0.025. On each test point, all the algorithms are repeated ten times. If the median recovery accuracy is less than 1%, the point is regarded as the feasible point. The feasible regions of these two algorithms are shown in Figure 13.1(a).

From Figure 13.1(a), the feasible region of LHR is much larger than the region of PCP. We get the same conclusion as made in [6] that the feasible boundary of PCP roughly fits the curve where $\eta^{PCP} + \xi^{PCP} = 0.35$. The boundary of LHR is around the curve that $\eta^{LHR} + \xi^{LHR} = 0.575$. Moreover, on the two sides of the red curve in Figure 13.1(a), the boundary equation can even be extended to $\eta^{LHR} + \rho^{LHR} = 0.6$. Although the performance of pHR is not as good as that of LHR, it still greatly outperforms that of PCP. When $p = 1/3$ and $p = 2/3$, the boundary equations are subjected to $\eta^{pHR} + \rho^{pHR} = 0.52$ and $\eta^{pHR} + \rho^{pHR} = 0.48$, respectively. These improvements are reasonable since pHR and LHR use the functionalities that are much closer to the ℓ_0 norm. Accordingly, the proposed non-convex heuristic method covers a larger feasible region. From this test, it is apparent that the proposed LHR algorithm covers largest area of the feasible region, which implies that LHR could handle more difficult tasks that robust PCA fails to do.

Convergence verification

Finally, we will experimentally verify the convergence of the LHR. The experiments are conducted on 400×400 matrices with the rank equivalent to 40, and the portion of gross errors are set as 15%, 30%, and 45%, respectively. The experimental results are reported in Figure 13.5.2 where the axes' coordinate denotes the iteration sequences.

The top sub-figure in Figure 13.5.2 reports the time cost of each iteration. It is interesting to note that the denser the error is, the more time cost is required for one iteration. Besides, the most time-consuming part occurs in the first iteration. During the first iteration, (13.11) subjects to the typical PCP problem. However, in the second and the third iteration, the weight matrix is assigned with different values and thus it could make (13.11) converge with fewer iterations. Therefore, the time cost for each iteration is different in LHR. The first iteration needs many computational resources while the later ones can be further accelerated owing to the penalty of the weight matrix.

The middle sub-figure records the stopping criterion, which is denoted as $\frac{\|W^{(t+1)} - W^{(t)}\|_F}{\|W^{(t)}\|_F}$. It is believed that the LHR converges when the stopping criterion is less than $1e-5$. It is apparent from Figure 13.5.2 that the LHR could converge in just three iterations with 15% and 30% gross errors, while for the complicated case with 45% errors, LHR can converge in four steps. The bottom sub-figure shows the recovery accuracy after each iteration. It is obvious that the recovery accuracy increases significantly from the first iteration to the second one. Such an increase phenomenon verifies the advantage of the reweighted approach derived from LHR.

Robustness on initialization

We will discuss the performances of the proposed algorithm with different initialization strategies. In the previous experiments, following the steps in Algorithm 13.1, both the matrices **A** and **E** are initialized to be zero. In this part, to further verify the effectiveness of LHR, we adopt a random initializing strategy. All the entries in **A** and **E** are, respectively, randomly initialized. With such initialization, LHR is implemented on a 400×400 matrix with $rank = 10$ and with $15\%, 30\%$, and 45% corruptions, respectively. Since the algorithm is randomly initialized, we repeat LHR ten times with different initializations on the same matrix. The final recovery accuracy and the time costs are reported in Table 13.2 with standard deviation.

From the table, it is apparent that different initialization methods do not make significant differences in the recovery accuracy. The random initialization method returns much more consistent accuracy with tiny standard deviation. Meanwhile, the accuracy of random

TABLE 13.2 Performance comparisons with different initialization methods.

	Random Initialization		Zero Initialization	
Corruptions	$\frac{\|\mathbf{A}-\mathbf{A}^*\|_F}{\|\mathbf{A}^*\|_F}$	$time(s)$	$\frac{\|\mathbf{A}-\mathbf{A}^*\|_F}{\|\mathbf{A}^*\|_F}$	$time(s)$
15%	$(1.63 \pm 0.07)e^{-5}$	25.7 ± 3.2	$1.58e^{-5}$	19.3
30%	$(5.33 \pm 0.06)e^{-5}$	41.2 ± 4.1	$5.17e^{-5}$	39.6
45%	$(2.61 \pm 0.01)e^{-4}$	64.7 ± 7.2	$2.37e^{-4}$	51.2

initialization-based recovery is very similar to the accuracy obtained by zero initialization, which is advocated in Algorithm 13.1. It is not surprising to see the robustness of LHR with different initializations, since the inner loops of LHR depend on a convex programming. A convex problem has a unique global minimum and thus it is very robust to the initial points. However, with different initialization, the time costs are different. Generally, zero initialization requires less computational costs than random methods. This is because with a bad random initialization, each inner programming requires many loops to get to the optimum.

For the weighting matrices, their initial value cannot be arbitrarily set because there isn't any prior about these two sparse components. An arbitrary random initialization will possibly make the whole programming diverge. Therefore, a good choice is to initialize all the weighting matrices by identity matrices. With such a setting, LHR exactly solves a PCP-like optimization in the first inner programming. After getting the initial guess for the two matrices, in the second and the following inner loops, the weighting matrices can be initialized according to the current estimation of **A** and **E**.

13.6 Practical Application: Nosiy Depth Maps Fusion for Muti-View Stereo

In this section, we demonstrate how to apply our LHR model to a classical vision problem: to fuse a noisy map from multi-view images of a same object [14]. It considers generating the whole geometry of the 3D object from computing the depth maps, a.k.a. the point clouds, generated by multiple views, and then fusing these point clouds to get the entire model of the object. Unfortunately, the rough depth map generated via stereo matching has two prominent drawbacks, i.e., it is redundant and noisy. In this section, we propose to reduce the noise and the redundancy among the multi-view depth maps using our LHR matrix recovery structure.

13.6.1 Matrix Recovery for Depth Maps

The ground truth depth map we used here is the mesh set of the foutain-P11 [37], which is acquired by 3D scanner. Since the fountain model is extremely large, we only select the central part of the fountain model. We directly add the noises to the entries in the fusion matrix obtained from a ground-truth depth map to see whether the LHR model is robust to remove such kinds of noises in the matrix. The two sides of the fountain model are flat walls and the central parts contain all the details about the fountain.

In practical data capture, there are eleven cameras placed around the object. Each view provides an incomplete (due to occlusion) and noisy vector that describes the partial information of the 3D fountain. We have introduced a fusion matrix construction algorithm in [14] to compose the vectors as an incomplete matrix. It is interesting to note that since all the respective views describe the same object, its intrinsic rank will be one. However, due to the noises in the point-cloud generation step, the matrix is not low rank per se. Accordingly, we propose to use the LHR method introduced previously to recover the rank-1 matrix from the noisy and incomplete matrix.

The original noisy point cloud and the rank-one fusion result are shown in Figure 13.2. From the visual comparisons, it is observed that the LHR significantly removes the noises on the 3D model and produced a much cleaner result.

FIGURE 13.2 The matrix completion result for noisy depth-maps fusion. In the figure, the left subfigure shows the original depth maps extracted by typical computer vision algorithms; the right panel provides the fused result with matrix completion.

13.7 Conclusion

This chapter presents a log-sum heuristic recovery algorithm to learn the essential low-rank structures from corrupted matrices. We introduced an MM algorithm to convert the non-convex objective function to a series of convex optimizations via reweighed approaches and proved that the solution may converge to a stationary point. Then, the general model was applied to a practical task: LRMR. We gave the solution/update rules to each variable in the joint optimizations via ADM. For the general PCP problem, LHR extended the feasible region to the boundary of $\eta + \xi = 0.58$. We applied our model to fuse a noisy depth map and achieved a good result.

However, a limitation of the proposed LHR model is for the reweighted phenomenon that requires us to solve convex optimizations multiple times. The implementations of LHR are a bit more time consuming than those for PCP and LRR. Therefore, the LHR model is especially recommend to learn the low-rank structure from data with denser corruptions and higher ranks.

13.8 Appendix

13.8.1 ABBREVIATIONS

ADM: Alternating Direction Method, **MM**: Majorization Minimization, **GPCA**: Generalized Principal Component Analysis, **PCP**: Principal Component Pursuit, **LHR**: Log-sum Heuristic Recovery, **LLMC**: Locally Linear Manifold Clustering, **LRMR**: Low-Rank Matrix Recovery, **LRR**: Low-Rank Representation, **LRSL**: Low-Rank Structure Learning, **LSA**: Local Subspace Affinity, **MoG**: Mixture of Gaussian, **RANSAC**: Random Sample Consensus, **RPCA**: Robust Principal Component Analysis, **SC**: Subspace Clustering, **SSC**: Sparse Subspace Clustering.

13.8.2 Constructing the Upper Bound of LHR

To see how the MM works for LHR, let's recall the objective function in (13.8) and make some simple algebra operations:

$$\begin{aligned} &\tfrac{1}{2}[\|diag(\mathbf{Y})\|_L + \|diag(\mathbf{Z})\|_L] + \lambda \|\mathbf{E}\|_L \\ &= \tfrac{1}{2}[\textstyle\sum_i \log(Y_{ii}+\delta) + \sum_k \log(Z_{kk}+\delta)] \\ &\quad + \lambda \textstyle\sum_{ij} \log(|E_{ij}|+\delta) \\ &= \tfrac{1}{2}[\log\det(\mathbf{Y}+\delta\mathbf{I_m}) + \log\det(\mathbf{Z}+\delta\mathbf{I_n})] \\ &\quad + \lambda \textstyle\sum_{ij} \log(|E_{ij}|+\delta) \end{aligned} \tag{13.15}$$

where $\mathbf{I}_m \in \mathbb{R}^{m\times m}$ is an identity matrix. It is well known that the concave function is bounded by its first-order Taylor expansion. Therefore, we calculate the convex upper bounds of all the terms in (13.15). For the term $\log\det(\mathbf{Y}+\delta\mathbf{I}_m)$,

$$\begin{aligned} \log\det(\mathbf{Y}+\delta\mathbf{I_m}) \le \log\det(&\mathbf{\Gamma}_Y + \delta\mathbf{I}_m) \\ &+ tr[(\mathbf{\Gamma}_Y+\delta\mathbf{I}_m)^{-1}(\mathbf{Y}-\mathbf{\Gamma_Y})]. \end{aligned} \tag{13.16}$$

The inequality in (13.16) holds for any $\mathbf{\Gamma}_Y \succ 0$. Similarly, for any $(\Gamma_E)_{ij} > 0$,

$$\textstyle\sum_{ij} \log(|E_{ij}|+\delta) \le \sum_{ij} [\log[(\Gamma_E)_{ij}+\delta] + \frac{E_{ij}-(\Gamma_E)_{ij}}{(\Gamma_E)_{ij}+\delta}]. \tag{13.17}$$

We replace each term in (13.15) with the convex upper bound and define $T(\hat{X}|\hat{\Gamma})$ as the surrogate function after convex relaxation.

13.8.3 Theoretic Proof of the Convergence of LHR

Proof of Lemma 13.2

In order to prove the monotonically decreased property, we can instead prove:

$$H(\hat{X}^{t+1}) \le T(\hat{X}^{t+1}|\hat{\Gamma}^t) \le T(\hat{X}^t|\hat{\Gamma}^t) = H(\hat{X}^t). \tag{13.18}$$

We prove (13.18) by the following three steps:

(**i**) The first inequality follows from the argument that $T(\hat{X}|\hat{\Gamma})$ is the upper bound of $H(\hat{X})$.

(**ii**) The second inequality holds since the MM algorithm computes $\hat{X}^{t+1} = \arg\min_{\hat{X}} T(\hat{X}|\hat{\Gamma}^t)$. The function $T(\cdot)$ is convex, therefore, $\hat{X}^{t+1}$ is the unique global minimum. This property guarantees that $T(\hat{X}^{t+1}|\hat{\Gamma}^{t+1}) < T(\cdot|\hat{\Gamma}^t)$ with any $\hat{X} \neq \hat{X}^{t+1}$ and $T(\hat{X}^{t+1}|\hat{\Gamma}^{t+1}) = T(\cdot|\hat{\Gamma}^t)$ if and only if $\hat{X} = \hat{X}^{t+1}$.

(**iii**) The last equality can be easily verified by expanding $T(\hat{X}^t|\hat{\Gamma}^t)$ and making some simple algebra. The transformation is straightforward and omitted here.

Proof of Lemma 13.3

We give a proof by contradiction. We assume that sequence $\hat{X}$ diverges, which means that $\lim_{t\to\infty} \|\hat{X}^{t+1} - \hat{X}^t\|_F \neq 0$. According to the discussions in Appendix 13.8.3, we know that there exists a convergent subsequence $\hat{X}^{t_k}$ converging to ϕ, i.e., $\lim_{k\to\infty} \hat{X}^{t_k} = \phi$, and meanwhile, we can construct another convergent subsequence $\hat{X}^{t_k+1}$ that $\lim_{k\to\infty} \hat{X}^{t_k+1} = \varphi$. We assume that $\phi \neq \varphi$. Since the convex upper-bound $T(\cdot|\hat{\Gamma})$ is continuous, we get $\lim_{k\to\infty} T(\hat{X}^{t_k+1}|\hat{\Gamma}^{t_k}) =$

$\underbrace{T(\lim_{k\to\infty} \hat{X}^{t_k+1}}_{\varphi}|\hat{\Gamma}^{t_k}) < \underbrace{T(\lim_{k\to\infty} \hat{X}^{t_k}}_{\phi}|\hat{\Gamma}^{t_k}) = \lim_{k\to\infty} T(\hat{X}^{t_k}|\hat{\Gamma}^{t_k})$. The *strict less-than operator* "$<$" holds because $\varphi \neq \phi$. See **(ii)** in the proof of Lemma 13.2 for details. Therefore, it is straightforward to get the following inequalities: $\lim_{k\to\infty} H(\hat{X}^{t_k+1}) \leq \lim_{k\to\infty} T(\hat{X}^{t_k+1}|\hat{\Gamma}^{t_k}) < \lim_{k\to\infty} T(\hat{X}^{t_k}|\hat{\Gamma}^{t_k}) = \lim_{k\to\infty} H(\hat{X}^{t_k})$. Accordingly,

$$\lim_{k\to\infty} H(\hat{X}^{t_k+1}) < \lim_{k\to\infty} H(\hat{X}^{t_k}). \tag{13.19}$$

Besides, it is obvious that the function of $H(\cdot)$ in (13.8) is bounded below, i.e., $H(\hat{X}) > (mn+m+n)\log\delta$. Moreover, as proved in Lemma 13.2, $H(\hat{X})$ is monotonically decreasing, which guarantee that $\lim_{t\to\infty} H(\hat{X}^t)$ exists, i.e.,

$$\begin{aligned}\lim_{k\to\infty} H(\hat{X}^{t_k}) &= \lim_{t\to\infty} H(\hat{X}^t) = \lim_{t\to\infty} H(\hat{X}^{t+1}) \\ &= \lim_{k\to\infty} H(\hat{X}^{t_k+1}).\end{aligned} \tag{13.20}$$

Obviously, (13.20) contradicts (13.19). Therefore, $\phi = \varphi$ and we get the conclusion that $\lim_{t\to\infty} \|\hat{X}^{t+1} - \hat{X}^t\|_F = 0$.

Convergence of subsequences in the proof of Lemma 13.3

In this part, we provide the discussions about the properties of the convergent subsequences that are used in the proof of Lemma 13.3.

Since sequence $\hat{X}^t = \{\mathbf{Y}^t, \mathbf{Z}^t, \mathbf{A}^t, \mathbf{E}^t\}$ is generated via Equation 13.8, we know that $\hat{X} \in \hat{D}$ strictly holds. Therefore, all the variables (i.e., $\mathbf{Y}^t, \mathbf{Z}^t, \mathbf{A}^t, \mathbf{E}^t$) in set $\hat{X}$ should be bounded. This claim can be easily verified because if any variable in the set $\hat{X}$ goes to infinity, the constraints in domain $\hat{D}$ will not be satisfied. Accordingly, we know that sequence $\hat{X}^t$ is bounded. According to the *Bolzano-Welestrass Theorem* [39], we know that every bounded sequence has a convergent subsequence. Since $\hat{X}^t$ is bounded, it is apparent that there exists a convergent subsequence $\hat{X}^{t_k}$. Without the loss of generality, we can construct another subsequence $\hat{X}^{t_k+1}$, which is also convergent. The proof of the convergence of $\hat{X}^{t_k+1}$ relies on the monotonically decreasing property proved in Lemma 13.2. Since $H(\cdot)$ is monotonically decreasing, it is easy to check that $H(\hat{X}^{t_k}) \geq H(\hat{X}^{t_k+1}) \geq H(\hat{X}^{t_{k+1}}) \geq H(\hat{X}^{t_{k+1}+1}) \geq H(\hat{X}^{t_{k+1+1}})$. According to the above inequalities, we get that

$$\lim_{k\to\infty} H(\hat{X}^{t_k}) \geq \lim_{k\to\infty} H(\hat{X}^{t_{k+1}}) \geq \lim_{k\to\infty} H(\hat{X}^{t_{k+2}}). \tag{13.21}$$

Since subsequence $\hat{X}^{t_k}$ converges, it is obvious that $\lim_{k\to\infty} H(\hat{X}^{t_k}) = \lim_{k\to\infty} H(\hat{X}^{t_{k+2}}) = \beta$. According to the famous *Squeeze Theorem* [40], from (13.21), we get the $\lim_{k\to\infty} H(\hat{X}^{t_k+1}) = H(\lim_{k\to\infty} \hat{X}^{t_k+1}) = \beta$. Since the function $H(\cdot)$ is monotonically decreasing and $\hat{X}$ is bounded, the convergence of $H(\hat{X}^{t_k+1})$ can be obtained if and only if the subsequence $\hat{X}^{t_k+1}$ is convergent.

Proof of Theorem.13.1

As stated in Lemma 13.3, the sequences generated by the MM algorithm converges to a limitation and here we will first prove that the convergence is a *fixed point*. We define the mapping from $\hat{X}^k$ to $\hat{X}^{k+1}$ as $M(\cdot)$, and it is straightforward to get, $\lim_{t\to\infty} \hat{X}^t = \lim_{t\to\infty} \hat{X}^{t+1} =$

$\lim_{t\to\infty} M(\hat{X}^t)$, which implies that $\lim_{t\to\infty} \hat{X}^t = \phi$ is a fixed point. In the MM algorithm, when constructing the upper bound, we use the first-order Taylor expansion. It is well known that the convex surrogate $T(\hat{X}|\hat{\Gamma})$ is tangent to $H(\hat{X})$ at $\hat{X}$ by the property of Taylor expansion. Accordingly, the gradient vector of $T(\hat{X}|\hat{\Gamma})$ and $H(\hat{X})$ are equal when evaluating at $\hat{X}$. Besides, we know that $\mathbf{0} \in \nabla_{\hat{X}=\phi} T(\hat{X}|\hat{\Gamma})$ and because it is tangent to $H(\hat{X})$, we can directly get that $\mathbf{0} \in \nabla_{\hat{X}=\phi} H(\hat{X})$, which proves that the convergent fixed point ϕ is also a *stationary point* of $H(\cdot)$.

References

1. A. Beck and M. Teboulle. A fast iterative shrinkage-thresholding algorithm for linear inverse problems. *SIAM Journal on Image Science*, 2(1):183–202, 2009.
2. J. Bioucas-Dias, M. Figueiredo, and J. Oliveira. Total variation-based image deconvolution: a majorization-minimization approach. *IEEE International Conference on Acoustics, Speech and Signal Processing, ICASSP 2006*, 2:2, May 2006.
3. T. Bouwmans. Traditional and recent approaches in background modeling for foreground detection: An overview. *Computer Science Review*, 11:31–66, 2014.
4. S. Boyd, N. Parikh, E. Chu, B. Peleato, and J. Eckstein. Distributed optimization and statistical learning via the alternating direction method of multipliers. *Foundations and Trends in Machine Learning*, 3(1):1–123, 2010.
5. J. Cai, E. Candes, and Z. Shen. A singular value thresholding algorithm for matrix completion. *Preprint*, 2008.
6. E. Candes, X. Li, Y. Ma, and J. Wright. Robust principal component analysis? *Journal of the ACM*, 59(3):1–37, May 2011.
7. E. Candes, M. Wakin, and S. Boyd. Enhancing sparsity by reweighted ℓ_1 minimization. *Journal of Fourier Analysis and Applications*, pages 877–905, 2007.
8. V. Chandrasekaran, S. Sanghavi, P. Parrilo, and A. Willsky. Rank-sparsity incoherence for matrix decomposition. *arXiv:0906.2220*, June 2009.
9. R. Chartrand and W. Yin. Iteratively reweighted algorithms for compressive sensing. In *IEEE International Conference on Acoustics, Speech and Signal Processing, ICASSP 2008*, pages 3869–3872. IEEE, 2008.
10. Y. Deng, Q. Dai, R. Liu, Z. Zhang, and S. Hu. Low-rank structure learning via nonconvex heuristic recovery. *IEEE Transactions on Neural Networks and Learning Systems*, 24(3):383–396, March 2013.
11. Y. Deng, Q. Dai, and Z. Zhang. Graph Laplace for occluded face completion and recognition. *IEEE Transactions on Image Processing*, 20(8):2329–2338, August 2011.
12. Y. Deng, Y. Kong, F. Bao, and Q. Dai. Sparse coding-inspired optimal trading system for hft industry. *IEEE Transactions on Industrial Informatics*, 11(2):467–475, April 2015.
13. Y. Deng, Y. Li, Y. Qian, X. Ji, and Q. Dai. Visual words assignment via information-theoretic manifold embedding. *IEEE Transactions on Cybernetics*, 44(10):1924–1937, October 2014.
14. Y. Deng, Y. Liu, Q. Dai, Z. Zhang, and Y. Wang. Noisy depth maps fusion for multiview stereo via matrix completion. *IEEE Journal of Selected Topics in Signal Processing*, 6(5):566–582, September 2012.
15. D. Donoho. Compressed sensing. *IEEE Transactions on Information Theory*, 52(4):1289–1306, April 2006.
16. M. Fazel. Matrix rank minimization with applications. *PhD thesis, Stanford University*, 2002.
17. M. Fazel, H. Hindi, and S. Boyd. Log-det heuristic for matrix rank minimization with applications to Hankel and Euclidean distance matrices. In *American Control Conference,*

ACC 2003, volume 3, pages 2156–2162, June 2003.
18. M. Figueiredo, J. Bioucas-Dias, and R. Nowak. Majorization minimization algorithms for wavelet-based image restoration. *IEEE Transactions on Image Processing*, 16(12):2980 –2991, December 2007.
19. C. Foo, C. Do, and A. Ng. A majorization-minimization algorithm for (multiple) hyperparameter learning. In *International Conference on Machine Learning, ICML 2009*, pages 321–328, 2009.
20. A. Ganesh, J. Wright, X. Li, E. Candes, and Y. Ma. Dense error correction for low-rank matrices via principal component pursuit. *International Symposium on Information Theory, ISIT 2010*, June 2010.
21. A. Goldberg, X. Zhu, B. Recht, J. Sui, and R. Nowak. Transduction with Matrix Completion: Three Birds with One Stone. *Annual Conference on Neural Information Processing Systems, NIPS 2010*, 2010.
22. D. Hsu, S. Kakade, and T. Zhang. Robust matrix decomposition with outliers. *arXiv:1011.1518v3*, 2010.
23. S. Hu and J. Wang. Absolute exponential stability of a class of continuous-time recurrent neural networks. *IEEE Transactions on Neural Networks*, 14(1):35–45, January 2003.
24. Z. Lai, W. Wong, Z. Jin, J. Yang, and Y. Xu. Sparse approximation to the eigensubspace for discrimination. *IEEE Transactions on neural networks and learning systems*, 23(12):1948–1960, 2012.
25. M. Lees. A note on the convergence of alternating direction methods. *Mathematics of Computation*, 16(77):70–75, 1963.
26. H. Li, N. Chen, and L. Li. Error analysis for matrix elastic-net regularization algorithms. *IEEE Transactions on Neural Networks and Learning Systems*, 23(5):737–748, 2012.
27. Z. Lin, M. Chen, and Y. Ma. The augmented Lagrange multiplier method for exact recovery of corrupted low-rank matrices. *arXiv:1009.5055v2*, March 2011.
28. Z. Lin, R. Liu, and Z. Su. Linearized alternating direction method with adaptive penalty for low-rank representation. *Neural Information Processing Systems, NIPS 2011*, 2011.
29. G. Liu, Z. Lin, S. Yan, J. Sun, Y. Yu, and Y. Ma. Robust recovery of subspace structures by low-rank representation. *IEEE Transactions on Pattern Analysis and Machine Intelligence*, 35(1):171–184, January 2013.
30. G. Liu, Z. Lin, and Y. Yu. Robust subspace segmentation by low-rank representation. In *International Conference on Machine Learning*, pages 663–670, 2010.
31. R. Liu, Z. Lin, S. Wei, and Z. Su. Feature extraction by learning Lorentzian metric tensor and its extensions. *Pattern Recognition*, 43(10):3298–3306, 2010.
32. R. Liu, Z. Lin, S. Wei, and Z. Su. Solving principal component pursuit in linear time via l_1 filtering. *Technical Report*, 2011.
33. Y. Ma. http://perception.csl.uiuc.edu/matrix-rank/sample_code.html. *University of Illinois, USA*, 2015.
34. K. Mohan and M. Fazel. Reweighted nuclear norm minimization with application to system identification. *American Control Conference*, 2010.
35. B. Recht, M. Fazel, and P. Parrilo. Guaranteed minimum-rank solutions of linear matrix equations via nuclear norm minimization. *SIAM Review*, 52(3):471–501, 2010.
36. A. Sobral, T. Bouwmans, and E. Zahzah. Incremental and multi-feature tensor subspace learning applied for background modeling and subtraction. In *International Conference on Image Analysis and Recognition, ICIAR 2014*, volume 8814, page 94, October 2014.
37. E. Tola, C. Strecha, and P. Fua. http://cvlab.epfl.ch/research/surface/emvs. *CV Laboratory, Switzerland*, 2015.
38. R. Vidal. Subspace clustering. *IEEE Signal Processing Magazine*, 28(2):52–68, 2011.
39. Wikipedia. Bolzano-Welestrass theorem. *http://en.wikipedia.org/wiki/Bolzano*, 2015.
40. Wikipedia. Sequezz theorem. *http://en.wikipedia.org/wiki/Squeezetheorem*, 2015.

41. Z. Xu, X. Chang, F. Xu, and H. Zhang. l_1/l_2-regularization: A thresholding representation theory and a fast solver. *IEEE Transactions on neural networks and learning systems*, 23(7):1013–1027, 2012.
42. Z. Yang, Y. Xiang, S. Xie, S. Ding, and Y. Rong. Non-negative blind source separation by sparse component analysis based on determinant measure. *IEEE Transactions on Neural Networks and Learning Systems*, 23(10):1601–1610, October 2012.
43. H. Zou and T. Hastie. Regularization and variable selection via the elastic net. *Journal of the Royal Statistical Society B*, 67:301–320, 2005.

IV

Applications in Image and Video Processing

14

A Variational Approach for Sparse Component Estimation and Low-Rank Matrix Recovery

Zhaofu Chen
Northwestern University, USA

Rafael Molina
Universidad de Granada, Spain

Aggelos K. Katsaggelos
Northwestern University, USA

14.1 Introduction

The problem of estimating low-rank matrices in the presence of sparse outliers has drawn significant attention recently. A typical example is robust principal component analysis (RPCA), where the high-dimensional data lying in a low-dimensional subspace are subject to the perturbation of a few outliers. Recently, theoretical performance guarantees for RPCA have been provided in [8], where it is shown that the RPCA problem can be solved using convex optimization. In addition, RPCA has been applied to solve a wide range of problems [12, 15, 17, 21] and its advantage has been demonstrated [28, 30, 38, 39].

In this chapter, we consider a generalization of the original RPCA problem, where a linear transformation, through the use of a known measurement matrix, is applied to the outlier-corrupted data. The goal is to estimate the outlier amplitudes given the transformed observation. This problem stems from several practical scenarios, which we will discuss in detail shortly. A regularization-based algorithm, which requires the manual tuning of its parameters, was proposed to solve this problem [24]. In this work, we propose a variational Bayesian-based approach that provides approximate posterior distributions of all the model unknowns. Experiments using real-life datasets as well as computer simulations show performance improvement of the proposed algorithm over its regularization-based counterpart.

This chapter is organized as follows. In Section 14.2 we present the general data model and several areas of applications. A brief overview of the related work in each of these areas is also provided. In Section 14.3 we introduce the proposed hierarchical Bayesian model. Details of the variational inference procedure are provided in Section 14.4. Numerical examples are presented in Section 14.5. Finally we draw conclusion remarks in Section 14.6.

Notation: Matrices and vectors are denoted by uppercase and lowercase boldface letters, respectively. $\text{vec}(\cdot)$, $\text{diag}(\cdot)$ and $\text{Tr}(\cdot)$ are vectorization, diagonalization, and trace operators, respectively. Given a matrix $\mathbf{X}$, we denote as $\mathbf{x}_{i\cdot}$, $\mathbf{x}_{\cdot j}$, and X_{ij} its ith row, j^{th} column, and $(i,j)^{\text{th}}$ element, respectively.

14.2 Problem Statement and Data Model

In this section we formulate the sparse component detection and low-rank matrix estimation problem. We first present a general data model that covers a wide range of applications, and then consider specific scenarios as its special cases.

Let $\mathbf{X} \in \mathbb{R}^{L\times T}$ be a low-rank matrix with $\text{rank}(\mathbf{X}) \ll \min(L,T)$, and $\mathbf{E} \in \mathbb{R}^{F\times T}$ be a sparse matrix with entries of arbitrarily large magnitudes. Matrix $\mathbf{R} \in \mathbb{R}^{L\times F}$ models the linear transformation performed on the data, that is, $\mathbf{R}$ is a known measurement matrix, where $L \leq F$. $\mathbf{R}$ bears specific physical meanings in the scenarios discussed below. Consider also dense measurement noise $\mathbf{N} \in \mathbb{R}^{L\times T}$ added to the observations. $\mathbf{N}$ is assumed to have small amplitudes compared with $\mathbf{X}$ and $\mathbf{E}$. With the quantities defined above, we have the general data model:

$$\mathbf{Y} = \mathbf{X} + \mathbf{R}\mathbf{E} + \mathbf{N}\,. \tag{14.1}$$

The goal of the problem is to obtain accurate estimates for the sparse term $\mathbf{E}$ and the low-rank term $\mathbf{X}$, given the noise-corrupted observation $\mathbf{Y}$. Although $\mathbf{E}$ is sparse, the multiplication with a wide matrix $\mathbf{R}$ has an effect of compression, and hence the product $\mathbf{RE}$ is not necessarily sparse. Therefore conventional RPCA approaches cannot be applied directly to solve this problem.

Given (14.1), we present below examples of $\mathbf{R}$, which stem from different application scenarios.

14.2.1 Robust PCA

With $L = F$ and $\mathbf{R} = \mathbf{I}_F$, (14.1) reduces to

$$\mathbf{Y} = \mathbf{X} + \mathbf{E} + \mathbf{N}, \tag{14.2}$$

where all matrices involved are of dimensions $F \times T$. This is the classical RPCA problem, where the goal is to recover the low-rank component $\mathbf{X}$ and the sparse component $\mathbf{E}$. The RPCA problem has recently drawn significant attention from the research community, and a wealth of literature has been devoted to related studies. Algorithms for solving the RPCA problem can be broadly classified into two categories: regularization-based approaches or statistical inference-based approaches .

For the former category, the problem is formulated as regularized fitting, where the regularizers are convex surrogates for rank and sparsity. Analysis on the exact recovery in the RPCA problem is given in [8]. Examples of algorithms in this category include the singular value thresholding (SVT) algorithm [7], the accelerated proximal gradient (APG) method [23], and the augmented Lagrange multiplier (ALM) method in [22]. In addition, extensive efforts have been made to analyze the RPCA problem from various perspectives [2, 4, 34, 35, 37].

For the latter category, hierarchical statistical models are introduced to formulate the data generation process, and prior distributions are selected to capture the low-rank and sparse properties of the respective terms. The joint distribution involving the observations, unknown variables, and hyperparameters can be determined from the priors and conditional distributions. Posterior distributions of the unknowns are approximated using Bayesian-based approaches. Representative algorithms in this category include [3, 14, 16]. As an example, [3] employs a hierarchical model to capture the properties of the data and applies a variational approach for inference. It is therefore a special case of the algorithm proposed herein when $\mathbf{R} = \mathbf{I}_F$.

14.2.2 Foreground Detection in Blurred and Noisy Video Sequence

Foreground (FG) detection is an important computer vision problem [25, 31]. Denote a video sequence as an $F \times T$ matrix $\mathbf{D}$, where F is the number of pixels per frame and T is the number of frames. Each frame, represented as a column in $\mathbf{D}$, is the superposition of moving FG objects and relatively static background (BG) scene. Since the FG objects are usually small relative to the entire frame and do not persist across the entire sequence, they can be represented as an $F \times T$ sparse matrix $\mathbf{E}$ [3, 13, 14]. On the other hand, the background is more static, and its time invariance can be captured by a low-rank matrix $\mathbf{X}_0$. Let $\mathbf{R}$ model a linear transformation on a frame, such as blurring and resolution scaling. With the above specifications, the transformed and noisy video sequence can be represented as

$$\begin{aligned} \mathbf{Y} &= \mathbf{RD} + \mathbf{N} = \mathbf{R}(\mathbf{X}_0 + \mathbf{E}) + \mathbf{N} \\ &= \mathbf{RX}_0 + \mathbf{RE} + \mathbf{N} = \mathbf{X} + \mathbf{RE} + \mathbf{N}, \end{aligned} \tag{14.3}$$

where we have combined $\mathbf{RX}_0$ as $\mathbf{X}$ since the primary objective is to detect the foreground $\mathbf{E}$. Note $\mathbf{X}$ inherits the low-rank property from $\mathbf{X}_0$. It is clear that (14.3) is identical to (14.1), with $\mathbf{R}$ being an $L \times F$ real-valued matrix modeling the linear transformation performed on video pixels.

14.2.3 Network Anomaly Detection from Link Observations

Another application emerging from (14.1) is network security and anomaly detection [10, 29]. Consider a network consisting of N nodes (e.g., routers) that can send and forward data packets. A link is the physical connection between a pair of nodes, and L denotes the number of links. An origin-destination (OD) flow is defined as a stream of packets sent from one node and received by another node, and F denotes the number of flows. Since the network in general is not strongly connected (i.e., there is not a link between every pair of nodes), we have $L \ll F$. In the network scenario, $\mathbf{R}$ is an $L \times F$ wide routing matrix consisting of binary entries r_{ij}, with $r_{ij} = 1$ denoting flow j passes through link i and $r_{ij} = 0$ otherwise [26].

Let the $F \times T$ matrix $\mathbf{X}_0$ denote the OD flow traffic during normal network operation, where its $(i, j)^{\text{th}}$ component is the sampled traffic of flow i at time j. $\mathbf{X}_0$ is low-rank because the normal network flows roughly follow a temporal pattern, rendering the columns of $\mathbf{X}_0$ approximately linearly dependent. On the other hand, network anomalies can occur sporadically in the OD flows and at different times, possibly due, for instance, to network failures, external attacks, etc. The anomaly-induced traffic can be of large magnitude compared with the normal traffic, and is represented by a sparse $F \times T$ matrix $\mathbf{E}$. The multiplication of the total OD flow $\mathbf{X}_0 + \mathbf{E}$ by $\mathbf{R}$ maps the composite flow traffic into the link measurements. It can be seen that the data model is again the same as (14.1), with $\mathbf{R}$ being an $L \times F$ wide binary routing matrix.

Researchers have performed extensive and in-depth studies into OD flow anomaly detection. For example, [20] presents analysis of OD flow time series using PCA and the decomposition of flows into the normal and abnormal constituents. [1] applies the principal component pursuit (PCP) algorithm [8] to analyze the OD flows. Along the same line of research as [20], authors of [19] apply subspace methods to decompose the link measurements and discuss the subsequent identification of anomalies in the OD flows. From a slightly different perspective, [33] examines the possible types of attacks specifically targeted at PCA-based detectors, which provides valuable insights as to the design of more effective detection technologies. In addition, the authors of [24] extend the APG algorithm [23] to analyze the link measurements for flow anomaly detection. For an overview of anomaly detection techniques the reader is referred to [29].

In the next section we propose a variational Bayesian-based approach to solve the recovery problem (14.1) for general $\mathbf{R}$. Within this framework, the variational Bayesian robust PCA algorithm introduced in [3] can be regarded as a special case of our proposed algorithm when $\mathbf{R} = \mathbf{I}_F$.

14.3 Bayesian Modeling

In this section we present the Bayesian modeling for the sparse component estimation and low-rank matrix recovery problem. Both unknowns and observed quantities are treated as random variables. A hierarchical Bayesian framework is employed to model the data generation process, where the joint distribution of all quantities is factorized into the product of priors and conditional probabilities.

14.3.1 Low-Rank Term Modeling

Given the observation model in (14.1) our goal is to estimate a low-rank matrix $\mathbf{X}$ and a sparse matrix $\mathbf{E}$ from the noisy $\mathbf{Y}$. A natural estimator is to fit $\mathbf{Y}$ in the least-squares (LS) sense as well as to minimize the rank of $\mathbf{X}$ and the number of non-zero entries in $\mathbf{E}$ measured by its l_0-(pseudo) norm. Unfortunately, both l_0-norm and rank minimizations are in general NP-hard problems [11, 27]. Therefore, we start by replacing the l_0-norm with its convex surrogate $||\mathbf{E}||_1$, and the rank of $\mathbf{X}$ with $||\mathbf{X}||_*$, which is equal to the sum of the singular values of $\mathbf{X}$. The nuclear norm $||\mathbf{X}||_*$ can be neatly characterized as [32]

$$\begin{aligned} ||\mathbf{X}||_* = \min_{\{\mathbf{A},\mathbf{B}\}} \frac{1}{2}\{||\mathbf{A}||_{\mathrm{F}}^2 + ||\mathbf{B}||_{\mathrm{F}}^2\} \\ \text{subject to} \quad \mathbf{X} = \mathbf{A}\mathbf{B}^{\mathrm{T}}. \end{aligned} \tag{14.4}$$

Adopting these as relaxation and parametrization, one obtains the following optimization problem

$$\min_{\{\mathbf{A},\mathbf{B},\mathbf{E}\}} \frac{1}{2}||\mathbf{Y}-\mathbf{A}\mathbf{B}^{\mathrm{T}}-\mathbf{R}\mathbf{E}||_{\mathrm{F}}^2+\lambda_* \left(||\mathbf{A}||_{\mathrm{F}}^2+||\mathbf{B}||_{\mathrm{F}}^2\right)+\lambda_1||\mathbf{E}||_1, \tag{14.5}$$

where λ_* and λ_1 are regularization parameters.

We denote an estimated upper bound for the rank of $\mathbf{X}$ as k and let $\mathbf{A}$ and $\mathbf{B}$ be $L \times k$ and $T \times k$ matrices, respectively. In this way $\mathbf{X}$ can be expressed as the sum of k outer-products, i.e.,

$$\mathbf{X} = \mathbf{A}\mathbf{B}^{\mathrm{T}} = \sum_{i=1}^{k} \mathbf{a}_{\cdot i}\mathbf{b}_{\cdot i}^{\mathrm{T}}. \tag{14.6}$$

Note that with the factorization in (14.6) we are able to characterize the rank of $\mathbf{X}$ while accommodating $\mathbf{X}$ of arbitrary dimensions.

To embody the low-rank property of $\mathbf{X}$, we aim at promoting column sparsity in $\mathbf{A}$ and $\mathbf{B}$ such that the sum in (14.6) has only a small number of non-zero terms. To enforce this constraint, we associate the columns of $\mathbf{A}$ and $\mathbf{B}$ with Gaussian priors of precisions $\{\gamma_i\}_{i=1}^k$, i.e.,

$$p(\mathbf{A}|\boldsymbol{\gamma}) = \prod_{i=1}^{k} p(\mathbf{a}_{\cdot i}|\gamma_i) = \prod_{i=1}^{k} \mathcal{N}(\mathbf{a}_{\cdot i}|\mathbf{0}, \gamma_i^{-1}\mathbf{I}_L), \tag{14.7a}$$

$$p(\mathbf{B}|\boldsymbol{\gamma}) = \prod_{i=1}^{k} p(\mathbf{b}_{\cdot i}|\gamma_i) = \prod_{i=1}^{k} \mathcal{N}(\mathbf{b}_{\cdot i}|\mathbf{0}, \gamma_i^{-1}\mathbf{I}_T), \tag{14.7b}$$

where $\boldsymbol{\gamma} = [\gamma_1, \cdots, \gamma_k]$. Since the columns of $\mathbf{A}$ and $\mathbf{B}$ have the same sparsity profile enforced by the common precisions $\{\gamma_i\}_{i=1}^k$, they become small simultaneously when the corresponding γ_i assumes large value. During the inference process, many of the precision parameters $\boldsymbol{\gamma}_i$ will take large values, effectively eliminating the respective outer products from $\mathbf{X}$. In this fashion, we achieve the goal of promoting the low-rank property of $\mathbf{X}$.

In addition to (14.7), we incorporate the i.i.d. conjugate Gamma hyperprior on the precisions $\{\gamma_i\}_{i=1}^k$, i.e.,

$$p(\boldsymbol{\gamma}) = \prod_{i=1}^{k} p(\gamma_i) = \prod_{i=1}^{k} \gamma_i^{a-1} \exp(-b\gamma_i), \tag{14.8}$$

where a and b are treated as deterministic and they are assigned small values to yield broad hyperpriors.

14.3.2 Sparse Term Modeling

To model the sparse term $\mathbf{E}$, we let the entries be independent of each other, and their amplitudes be governed by zero-mean Gaussian distributions with independent precisions. Specifically, we have

$$p(\mathbf{E}|\boldsymbol{\alpha}) = \prod_{i=1}^{F}\prod_{j=1}^{T} p(e_{ij}|\alpha_{ij}) = \prod_{i=1}^{F}\prod_{j=1}^{T} \mathcal{N}(e_{ij}|0, \alpha_{ij}^{-1}), \tag{14.9}$$

where $\boldsymbol{\alpha}$ is the matrix containing all the FT values of the precisions α_{ij}. For large α_{ij}, the corresponding e_{ij} is close to zero with high probability. During inference, a large number of α_{ij} will be set to high values, and consequently the corresponding e_{ij} will be literally zeros. The precisions α_{ij} are assigned i.i.d. non-informative Jeffrey's prior

$$p(\boldsymbol{\alpha}) = \prod_{i=1}^{F}\prod_{j=1}^{T} p(\alpha_{ij}) = \prod_{i=1}^{F}\prod_{j=1}^{T} \alpha_{ij}^{-1}. \tag{14.10}$$

14.3.3 Noisy Observation Modeling

We employ the common i.i.d. Gaussian priors with zero mean and precision β to model the dense observation noise $\mathbf{N}$:

$$p(\mathbf{N}|\beta) = \prod_{i=1}^{L}\prod_{j=1}^{T} p(n_{ij}|\beta) = \prod_{i=1}^{L}\prod_{j=1}^{T} \mathcal{N}(n_{ij}|0, \beta^{-1}), \tag{14.11}$$

Assigning Jeffrey's prior on β, we have

$$p(\beta) = \beta^{-1}\,. \tag{14.12}$$

Given the components defined above, the observation model is given by

$$\begin{aligned} p(\mathbf{Y}|\mathbf{A},\mathbf{B},\mathbf{E},\beta) &= \mathcal{N}(\text{vec}(\mathbf{Y})|\text{vec}(\mathbf{A}\mathbf{B}^{\mathrm{T}}+\mathbf{R}\mathbf{E}),\beta^{-1}\mathbf{I}_{LT}) \\ &\propto \beta^{\frac{LT}{2}}\exp\left\{-\frac{\beta}{2}||\mathbf{Y}-\mathbf{A}\mathbf{B}^{\mathrm{T}}-\mathbf{R}\mathbf{E}||_{\mathrm{F}}^2\right\}. \end{aligned} \tag{14.13}$$

14.3.4 Joint Distribution

By combining the stages of the hierarchical Bayesian model, the joint distribution of the observation and all the unknowns is expressed as

$$\begin{aligned} p(\mathbf{Y},\mathbf{A},\mathbf{B},\mathbf{E},\boldsymbol{\gamma},\boldsymbol{\alpha},\beta) &= p(\mathbf{Y}|\mathbf{A},\mathbf{B},\mathbf{E},\beta)p(\mathbf{A}|\boldsymbol{\gamma})p(\mathbf{B}|\boldsymbol{\gamma}) \\ &\times p(\boldsymbol{\gamma})p(\mathbf{E}|\boldsymbol{\alpha})p(\boldsymbol{\alpha})p(\beta), \end{aligned} \tag{14.14}$$

where $p(\mathbf{Y}|\mathbf{A},\mathbf{B},\mathbf{E},\beta)$, $p(\mathbf{A}|\boldsymbol{\gamma})$, $p(\mathbf{B}|\boldsymbol{\gamma})$, $p(\boldsymbol{\gamma})$, $p(\mathbf{E}|\boldsymbol{\alpha})$, $p(\boldsymbol{\alpha})$, and $p(\beta)$ are given in (14.13), (14.7a), (14.7b), (14.8), (14.9), (14.10), and (14.12), respectively. The dependencies in this joint probability model are shown in graphical form in Figure 14.1, where the arrows are used to denote the generative model.

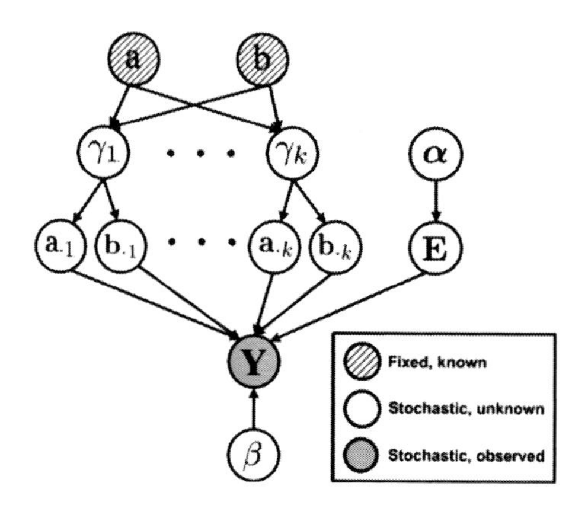

FIGURE 14.1 Directed acyclic graph representation of the hierarchical Bayesian model.

14.4 Approximate Bayesian Inference

As is widely known, Bayesian inference is based on the posterior distribution of unknowns given the observations. Let $\mathbf{z}$ be the vector of all the unknowns such that

$$\mathbf{z} = (\mathbf{A},\mathbf{B},\boldsymbol{\gamma},\mathbf{E},\boldsymbol{\alpha},\beta)\,. \tag{14.15}$$

The posterior is expressed using Bayes rule as

$$p(\mathbf{z}|\mathbf{Y}) = \frac{p(\mathbf{Y},\mathbf{z})}{p(\mathbf{Y})}\,. \tag{14.16}$$

However, the posterior $p(\mathbf{z}|\mathbf{Y})$ is computationally intractable because the denominator

$$p(\mathbf{Y}) = \int p(\mathbf{Y},\mathbf{z})\mathrm{d}\mathbf{z} \tag{14.17}$$

cannot be calculated analytically by marginalizing all unknowns. Therefore, approximation methods must be utilized. Common approaches for approximation include maximum *a posteriori* (MAP) estimation, evidenced-based analysis, and variational Bayesian. Among these options, Bayesian inference is generally more effective in avoiding local minima than deterministic approaches such as MAP, due to the fact that Bayesian methods approximate the full posterior distributions instead of merely providing point estimates of the modes.

In this section we present an inference procedure based on mean field variational Bayes [5, 6] for approximating the posterior distributions. Let $q(\mathbf{z})$ denote the approximate posterior distribution. The goal is to minimize the Kullback-Leibler (KL) divergence between $q(\mathbf{z})$ and the true posterior distribution $p(\mathbf{z}|\mathbf{Y})$, given by

$$\begin{aligned}\mathrm{KL}(q(\mathbf{z})||p(\mathbf{z}|\mathbf{Y})) &= \int q(\mathbf{z}) \log \frac{q(\mathbf{z})}{p(\mathbf{z}|\mathbf{Y})} \mathrm{d}\mathbf{z} \\ &= \int q(\mathbf{z}) \log \frac{q(\mathbf{z})}{p(\mathbf{Y},\mathbf{z})} \mathrm{d}\mathbf{z} + \mathrm{C},\end{aligned} \tag{14.18}$$

where $\mathrm{C} = \log p(\mathbf{Y})$ does not involve $\mathbf{z}$. To simplify notation, we use C to denote constants that do not depend on the variables currently under consideration, and C in different equations may have different meanings. The non-negative $\mathrm{KL}(q(\mathbf{z})||p(\mathbf{z}|\mathbf{Y}))$ measures the difference between $q(\mathbf{z})$ and $p(\mathbf{z}|\mathbf{Y})$ and equals 0 if and only if $q(\mathbf{z}) = p(\mathbf{z}|\mathbf{Y})$. In this sense, minimizing the KL divergence with respect to $q(\mathbf{z})$ is equivalent to finding an approximation to the unknown $p(\mathbf{z}|\mathbf{Y})$.

Another motivation for minimizing the KL divergence is to derive a lower bound for the evidence of the observed data $p(\mathbf{Y})$, with marginalization performed over the unknown variables. This can be seen by decomposing $\log p(\mathbf{Y})$ as follows

$$\begin{aligned}\log p(\mathbf{Y}) &= \log \frac{p(\mathbf{z}|\mathbf{Y})p(\mathbf{Y})}{p(\mathbf{z}|\mathbf{Y})} = \log \frac{p(\mathbf{Y},\mathbf{z})q(\mathbf{z})}{p(\mathbf{z}|\mathbf{Y})q(\mathbf{z})} \\ &= \int q(\mathbf{z}) \log \frac{p(\mathbf{Y},\mathbf{z})q(\mathbf{z})}{p(\mathbf{z}|\mathbf{Y})q(\mathbf{z})} \mathrm{d}\mathbf{z} \\ &= Q(q(\mathbf{z})) + \mathrm{KL}(q(\mathbf{z})||p(\mathbf{z}|\mathbf{Y})),\end{aligned} \tag{14.19}$$

where

$$Q(q(\mathbf{z})) = \int q(\mathbf{z}) \log \frac{p(\mathbf{Y},\mathbf{z})}{q(\mathbf{z})} \mathrm{d}\mathbf{z} \tag{14.20}$$

is a lower bound of $\log p(\mathbf{Y})$ because $\mathrm{KL}(q(\mathbf{z})||p(\mathbf{z}|\mathbf{Y})) \geq 0$. Since $\log p(\mathbf{Y})$ does not depend on $q(\mathbf{z})$, we have

$$\begin{aligned}\underset{q(\mathbf{z})}{\operatorname{argmax}}\, Q(q(\mathbf{z})) &= \underset{q(\mathbf{z})}{\operatorname{argmin}}\, \mathrm{KL}(q(\mathbf{z})||p(\mathbf{z}|\mathbf{Y})) \\ &= \underset{q(\mathbf{z})}{\operatorname{argmin}} \int q(\mathbf{z}) \log \frac{q(\mathbf{z})}{p(\mathbf{Y},\mathbf{z})} \mathrm{d}\mathbf{z},\end{aligned} \tag{14.21}$$

where the second equality follows from (14.18).

In the mean field approach, we make the assumption on the factorization of the approximate posterior distributions

$$q(\mathbf{z}) = \prod_m q(\mathbf{z}_m), \tag{14.22}$$

where $\mathbf{z}_m$ denote the components of $\mathbf{z}$, as is given in (14.15).

With this factorization we have

$$\log q(\mathbf{z}_m) = \mathrm{E}_{\mathbf{z}\backslash\mathbf{z}_m} [\log p(\mathbf{Y},\mathbf{z})] + \mathrm{C}, \tag{14.23}$$

from which we can determine the functional form of the approximate posterior of each unknown. The expectation $\mathrm{E}_{\mathbf{z}\backslash\mathbf{z}_m}[\cdot]$ in (14.23) is taken with respect to $q(\mathbf{z}\backslash\mathbf{z}_m)$. In the following we present the update rules resulting from this inference scheme (14.23) for each unknown.

14.4.1 Inference for A and B

Letting $\mathbf{z}_m = \mathbf{A}$ and invoking (14.23), we have

$$\begin{aligned}\log q(\mathbf{A}) &= \mathrm{E}_{\mathbf{z}\backslash\mathbf{A}}[\log p(\mathbf{Y},\mathbf{z})] + \mathrm{C}\\ &= \mathrm{E}_{\mathbf{z}\backslash\mathbf{A}}[\log p(\mathbf{Y}|\mathbf{A},\mathbf{B},\mathbf{E},\beta) + \log p(\mathbf{A}|\gamma)] + \mathrm{C}\\ &= \sum_{i=1}^{L} \mathrm{E}_{\mathbf{z}\backslash\mathbf{A}}\left[-\frac{1}{2}(\beta\|\mathbf{y}_{i\cdot} - \mathbf{a}_{i\cdot}\mathbf{B}^{\mathrm{T}} - \mathbf{r}_{i\cdot}\mathbf{E}\|_2^2 + \mathbf{a}_{i\cdot}\mathbf{\Gamma}\mathbf{a}_{i\cdot}^{\mathrm{T}})\right] + \mathrm{C}\\ &= \sum_{i=1}^{L} \log q(\mathbf{a}_{i\cdot}) + \mathrm{C},\end{aligned} \tag{14.24}$$

where $\mathbf{\Gamma} = \mathrm{diag}(\boldsymbol{\gamma})$. From (14.24) we see that the approximate posterior distribution of $\mathbf{A}$ decomposes as independent distributions of its rows. Moreover, by multiplying out the terms and taking the appropriate expectations, it follows that

$$\begin{aligned}\log q(\mathbf{a}_{i\cdot}) &= \mathrm{E}_{\mathbf{z}\backslash\mathbf{A}}\left[-\frac{1}{2}(\beta\|\mathbf{y}_{i\cdot} - \mathbf{a}_{i\cdot}\mathbf{B}^{\mathrm{T}} - \mathbf{r}_{i\cdot}\mathbf{E}\|_2^2 + \mathbf{a}_{i\cdot}\mathbf{\Gamma}\mathbf{a}_{i\cdot}^{\mathrm{T}})\right] + \mathrm{C}\\ &= -\frac{1}{2}\left[(\mathbf{a}_{i\cdot} - \langle\mathbf{a}_{i\cdot}\rangle)(\mathbf{\Sigma}^A)^{-1}(\mathbf{a}_{i\cdot} - \langle\mathbf{a}_{i\cdot}\rangle)^{\mathrm{T}}\right] + \mathrm{C}\,.\end{aligned} \tag{14.25}$$

Recognizing the right-hand side of (14.25) as the energy of a Gaussian distribution, we have

$$q(\mathbf{a}_{i\cdot}) = \mathcal{N}(\mathbf{a}_{i\cdot}|\langle\mathbf{a}_{i\cdot}\rangle, \mathbf{\Sigma}^A), \tag{14.26}$$

where

$$\langle\mathbf{a}_{i\cdot}\rangle^{\mathrm{T}} = \langle\beta\rangle\mathbf{\Sigma}^A\langle\mathbf{B}\rangle^{\mathrm{T}}(\mathbf{y}_{i\cdot} - \mathbf{r}_{i\cdot}\langle\mathbf{E}\rangle)^{\mathrm{T}} \tag{14.27}$$

and

$$\mathbf{\Sigma}^A = \left(\langle\beta\rangle\langle\mathbf{B}^{\mathrm{T}}\mathbf{B}\rangle + \langle\mathbf{\Gamma}\rangle\right)^{-1} \tag{14.28}$$

are the approximate posterior mean and covariance matrix, respectively. In (14.27) and (14.28) the expectations on the right-hand side of the equalities are taken with respect to the most recent approximate posterior distributions of the respective terms.

Following steps similar to (14.24), we can see that $q(\mathbf{B})$ also decomposes into the product of Gaussian distributions $q(\mathbf{b}_{i\cdot})$, i.e.,

$$q(\mathbf{B}) = \prod_{i=1}^{T} q(\mathbf{b}_{i\cdot}) = \prod_{i=1}^{T} \mathcal{N}(\mathbf{b}_{i\cdot}|\langle\mathbf{b}_{i\cdot}\rangle, \mathbf{\Sigma}^B), \tag{14.29}$$

where

$$\langle\mathbf{b}_{i\cdot}\rangle^{\mathrm{T}} = \langle\beta\rangle\mathbf{\Sigma}^B\langle\mathbf{A}\rangle^{\mathrm{T}}(\mathbf{y}_{\cdot i} - \mathbf{R}\langle\mathbf{e}_{\cdot i}\rangle) \tag{14.30}$$

and

$$\mathbf{\Sigma}^B = \left(\langle\beta\rangle\langle\mathbf{A}^{\mathrm{T}}\mathbf{A}\rangle + \langle\mathbf{\Gamma}\rangle\right)^{-1}. \tag{14.31}$$

The required expectations in (14.28) and (14.31) can be found as

$$\begin{aligned}\langle\mathbf{A}^{\mathrm{T}}\mathbf{A}\rangle &= \langle\mathbf{A}\rangle^{\mathrm{T}}\langle\mathbf{A}\rangle + L\mathbf{\Sigma}^A\\ \langle\mathbf{B}^{\mathrm{T}}\mathbf{B}\rangle &= \langle\mathbf{B}\rangle^{\mathrm{T}}\langle\mathbf{B}\rangle + T\mathbf{\Sigma}^B\,.\end{aligned} \tag{14.32}$$

14.4.2 Inference for E

Substituting $\mathbf{z}_m = \mathbf{E}$ into (14.23), we have

$$\begin{aligned}\log q(\mathbf{E}) =& \mathrm{E}_{\mathbf{z}\backslash\mathbf{E}}[\log p(\mathbf{Y},\mathbf{z})] + \mathrm{C} \\ =& \mathrm{E}_{\mathbf{z}\backslash\mathbf{E}}[\log p(\mathbf{Y}|\mathbf{A},\mathbf{B},\mathbf{E},\beta) + \log p(\mathbf{E}|\boldsymbol{\alpha})] + \mathrm{C} \\ =& \sum_{i=1}^{T} \mathrm{E}_{\mathbf{z}\backslash\mathbf{E}}\left[-\frac{1}{2}\{\beta\|\mathbf{y}_{\cdot i} - \mathbf{A}\mathbf{b}_{i\cdot}^{\mathrm{T}} - \mathbf{R}\mathbf{e}_{\cdot i}\|_2^2 + \mathbf{e}_{\cdot i}^{\mathrm{T}}\boldsymbol{\Omega}^i\mathbf{e}_{\cdot i}\}\right] + \mathrm{C},\end{aligned} \tag{14.33}$$

where $\boldsymbol{\Omega}^i = \mathrm{diag}([\alpha_{1i}, \alpha_{2i}, \cdots, \alpha_{Fi}])$ contains the prior precisions for the ith column of $\mathbf{E}$.

From (14.33) we see that $q(\mathbf{E})$ factors into the product of the approximate posterior distributions $q(\mathbf{e}_{\cdot i})$, where

$$q(\mathbf{e}_{\cdot i}) = \mathcal{N}(\mathbf{e}_{\cdot i}|\langle\mathbf{e}_{\cdot i}\rangle, \boldsymbol{\Sigma}^{E_i}) \tag{14.34}$$

is a Gaussian distribution. With some algebra, we obtain that the approximate posterior mean and covariance matrix of (14.34) are given by

$$\langle\mathbf{e}_{\cdot i}\rangle = \langle\beta\rangle\boldsymbol{\Sigma}^{E_i}\mathbf{R}^{\mathrm{T}}\left(\mathbf{y}_{\cdot i} - \langle\mathbf{A}\rangle\langle\mathbf{b}_{i\cdot}\rangle^{\mathrm{T}}\right) \tag{14.35}$$

and

$$\boldsymbol{\Sigma}^{E_i} = \left(\langle\beta\rangle\mathbf{R}^{\mathrm{T}}\mathbf{R} + \langle\boldsymbol{\Omega}^i\rangle\right)^{-1}. \tag{14.36}$$

From (14.36) we see that although the elements of $\mathbf{e}_{\cdot i}$ are independent of each other in the prior distribution, they are correlated with each other in the posterior distribution. This is due to the coupling of $\mathbf{R}$, which in effect takes a weighted sum of a column in $\mathbf{E}$ and maps it into a single entry in the observation.

Note that when $\mathbf{R} = \mathbf{I}_F$, as is the case in robust principal component analysis, the elements in $\mathbf{E}$ are independent of each other and their distributions are single-variable Gaussians.

14.4.3 Estimation of Hyperparameters γ and α

By keeping $p(\mathbf{A}|\boldsymbol{\gamma})$, $p(\mathbf{B}|\boldsymbol{\gamma})$, and $p(\boldsymbol{\gamma})$ in (14.23), we have

$$\begin{aligned}\log q(\boldsymbol{\gamma}) =& \mathrm{E}_{\mathbf{z}\backslash\boldsymbol{\gamma}}[\log p(\mathbf{Y},\mathbf{z})] + \mathrm{C} \\ =& \mathrm{E}_{\mathbf{z}\backslash\boldsymbol{\gamma}}[\log p(\mathbf{A}|\boldsymbol{\gamma}) + \log p(\mathbf{B}|\boldsymbol{\gamma}) + \log p(\boldsymbol{\gamma})] + \mathrm{C} \\ =& \sum_{i=1}^{k} \mathrm{E}_{\mathbf{z}\backslash\boldsymbol{\gamma}}[\log p(\mathbf{a}_{\cdot i}|\gamma_i) + \log p(\mathbf{b}_{\cdot i}|\gamma_i) + \log p(\gamma_i)] + \mathrm{C}.\end{aligned} \tag{14.37}$$

Taking the logarithms of (14.7) and (14.8), we have

$$\begin{aligned}\log p(\mathbf{a}_{\cdot i}|\gamma_i) &= \frac{L}{2}\log\gamma_i - \frac{\gamma_i}{2}\|\mathbf{a}_{\cdot i}\|_2^2, \\ \log p(\mathbf{b}_{\cdot i}|\gamma_i) &= \frac{T}{2}\log\gamma_i - \frac{\gamma_i}{2}\|\mathbf{b}_{\cdot i}\|_2^2, \\ \log p(\gamma_i) &= (a-1)\log\gamma_i - b\gamma_i.\end{aligned} \tag{14.38}$$

Substituting (14.38) into (14.37), it follows that

$$\begin{aligned}\log q(\boldsymbol{\gamma}) &= \sum_{i=1}^{k} \mathrm{E}_{\mathbf{z}\backslash\boldsymbol{\gamma}} \left[\left(\frac{L+T+2a}{2} - 1 \right) \log \gamma_i \right. \\ &\quad \left. - \frac{||\mathbf{a}_{\cdot i}||_2^2 + ||\mathbf{b}_{\cdot i}||_2^2 + 2b}{2} \gamma_i \right] + \mathrm{C} \\ &= \sum_{i=1}^{k} \left[\left(\frac{L+T+2a}{2} - 1 \right) \log \gamma_i \right. \\ &\quad \left. - \frac{\langle ||\mathbf{a}_{\cdot i}||_2^2 \rangle + \langle ||\mathbf{b}_{\cdot i}||_2^2 \rangle + 2b}{2} \gamma_i \right] + \mathrm{C} \\ &= \sum_{i=1}^{k} \log q(\gamma_i) + \mathrm{C},\end{aligned} \tag{14.39}$$

where

$$q(\gamma_i) \propto \gamma_i^{\frac{L+T+2a}{2} - 1} \exp\left(-\gamma_i \frac{\langle ||\mathbf{a}_{\cdot i}||_2^2 \rangle + \langle ||\mathbf{b}_{\cdot i}||_2^2 \rangle + 2b}{2} \right). \tag{14.40}$$

is recognized as a Gamma distribution. After resolving terms $\langle ||\mathbf{a}_{\cdot i}||_2^2 \rangle$ and $\langle ||\mathbf{b}_{\cdot i}||_2^2 \rangle$, we have

$$\langle \gamma_i \rangle = \frac{L+T+2a}{||\langle \mathbf{a}_{\cdot i} \rangle||_2^2 + ||\langle \mathbf{b}_{\cdot i} \rangle||_2^2 + L\sigma_{ii}^A + T\sigma_{ii}^B + 2b}, \tag{14.41}$$

where σ_{ii}^A and σ_{ii}^B denote the $(i,i)^{\text{th}}$ element of $\boldsymbol{\Sigma}^A$ and $\boldsymbol{\Sigma}^B$, respectively.

Similarly, setting $\mathbf{z}_m = \boldsymbol{\alpha}$ in (14.23), it follows that

$$\begin{aligned}\log q(\boldsymbol{\alpha}) &= \mathrm{E}_{\mathbf{z}\backslash\boldsymbol{\alpha}} \left[\log p(\mathbf{Y}, \mathbf{z}) \right] + \mathrm{C} \\ &= \mathrm{E}_{\mathbf{z}\backslash\boldsymbol{\alpha}} \left[\log p(\mathbf{E}|\boldsymbol{\alpha}) + \log p(\boldsymbol{\alpha}) \right] + \mathrm{C} \\ &= \sum_{i=1}^{F} \sum_{j=1}^{T} \mathrm{E}_{\mathbf{z}\backslash\boldsymbol{\alpha}} \left[\log p(e_{ij}|\alpha_{ij}) + \log p(\alpha_{ij}) \right] + \mathrm{C} \\ &= \sum_{i=1}^{F} \sum_{j=1}^{T} \mathrm{E}_{\mathbf{z}\backslash\boldsymbol{\alpha}} \left[\frac{1}{2} \log \alpha_{ij} - \frac{1}{2} e_{ij}^2 \alpha_{ij} - \log \alpha_{ij} \right] + \mathrm{C} \\ &= \sum_{i=1}^{F} \sum_{j=1}^{T} \left[-\frac{1}{2} \log \alpha_{ij} - \frac{1}{2} \langle e_{ij}^2 \rangle \alpha_{ij} \right] + \mathrm{C} \\ &= \sum_{i=1}^{F} \sum_{j=1}^{T} \log q(\alpha_{ij}) + \mathrm{C}.\end{aligned} \tag{14.42}$$

From (14.42) $q(\alpha_{ij})$ is recognized as a Gamma distribution with mean

$$\langle \alpha_{ij} \rangle = \frac{1}{\langle e_{ij}^2 \rangle} = \frac{1}{\langle e_{ij} \rangle^2 + (\sigma_{ii}^{E_j})^2}, \tag{14.43}$$

where $(\sigma_{ii}^{E_j})^2$ denotes the $(i,i)^{\text{th}}$ element in $\boldsymbol{\Sigma}^{E_j}$, the approximate posterior covariance matrix for the jth column of $\mathbf{E}$.

14.4.4 Estimation of Noise Precision β

By including the components of $p(\mathbf{Y}, \mathbf{z})$ that depend on β in (14.23), it follows that

$$\begin{aligned}\log q(\beta) &= \mathrm{E}_{\mathbf{z}\setminus\beta}\left[\log p(\mathbf{Y}|\mathbf{A},\mathbf{B},\mathbf{E},\beta) + \log p(\beta)\right] + \mathrm{C} \\ &= \mathrm{E}_{\mathbf{z}\setminus\beta}\left[\frac{LT}{2}\log\beta - \frac{\beta}{2}||\mathbf{Y}-\mathbf{A}\mathbf{B}^{\mathrm{T}}-\mathbf{R}\mathbf{E}||_{\mathrm{F}}^2 - \log\beta\right] + \mathrm{C} \\ &= \left(\frac{LT}{2}-1\right)\log\beta - \frac{1}{2}\langle||\mathbf{Y}-\mathbf{A}\mathbf{B}^{\mathrm{T}}-\mathbf{R}\mathbf{E}||_{\mathrm{F}}^2\rangle\beta + \mathrm{C}.\end{aligned} \tag{14.44}$$

Therefore, $q(\beta)$ is a Gamma distribution

$$q(\beta) \propto \beta^{\frac{LT}{2}-1}\exp\left\{-\frac{\langle||\mathbf{Y}-\mathbf{A}\mathbf{B}^{\mathrm{T}}-\mathbf{R}\mathbf{E}||_{\mathrm{F}}^2\rangle}{2}\beta\right\} \tag{14.45}$$

with mean given by

$$\langle\beta\rangle = \frac{LT}{\langle||\mathbf{Y}-\mathbf{A}\mathbf{B}^{\mathrm{T}}-\mathbf{R}\mathbf{E}||_{\mathrm{F}}^2\rangle}. \tag{14.46}$$

Equation (14.46) intuitively makes sense because the denominator measures the energy of the dense noise and the entire expression is the reciprocal of the noise power (variance), which is exactly the definition of precision.

To evaluate the expectation in the denominator of (14.46), additional steps are needed. We present the result here while leaving the derivation to Appendix A, in order to make the text more readable. With some algebra, the denominator can be computed in closed form using the posterior means of other quantities as

$$\begin{aligned}\langle||\mathbf{Y}-\mathbf{A}\mathbf{B}^{\mathrm{T}}-\mathbf{R}\mathbf{E}||_{\mathrm{F}}^2\rangle &= ||\mathbf{Y}-\langle\mathbf{A}\rangle\langle\mathbf{B}\rangle^{\mathrm{T}}-\mathbf{R}\langle\mathbf{E}\rangle||_{\mathrm{F}}^2 \\ &+ T\mathrm{Tr}\left(\langle\mathbf{A}\rangle^{\mathrm{T}}\langle\mathbf{A}\rangle\boldsymbol{\Sigma}^B\right) + L\mathrm{Tr}\left(\langle\mathbf{B}\rangle^{\mathrm{T}}\langle\mathbf{B}\rangle\boldsymbol{\Sigma}^A\right) \\ &+ LT\mathrm{Tr}\left(\boldsymbol{\Sigma}^A\boldsymbol{\Sigma}^B\right) + \sum_{i=1}^{T}\mathrm{Tr}\left(\mathbf{R}\boldsymbol{\Sigma}^{E_i}\mathbf{R}^{\mathrm{T}}\right).\end{aligned} \tag{14.47}$$

14.4.5 Summary

The proposed variational Bayesian inference procedure estimates the posterior distribution of the unknowns iteratively, where in each iteration the algorithm first computes $q(\mathbf{A})$, $q(\mathbf{B})$, and $q(\mathbf{E})$ using (14.26), (14.29), and (14.34), respectively. Then the hyperparameters are estimated by taking their approximate posterior means according to (14.41), (14.43), and (14.46), respectively. A stopping criterion is that the relative change in the estimated $\langle\mathbf{E}\rangle$ falls below a pre-specified threshold.

14.5 Numerical Examples

In this section we test the proposed algorithm with numerical experiments. We consider both real-life datasets to demonstrate the effectiveness of the proposed approach in solving practical problems, and simulated experiments to evaluate its performance under various conditions.

To make the discussion clear, we name the proposed algorithm that solves (14.1) as variational Bayesian-based sparse estimator (VBSE). The algorithm presented in [3] is a special case of VBSE when $\mathbf{R} = \mathbf{I}_F$. In comparison, we consider two popular regularization-based alternatives to validate the performance of VBSE and furthermore to demonstrate their respective merits and limitations. The algorithm presented in [24] is named as accelerated proximal gradient sparse estimator (APGSE). We have also extended the ALM algorithm, which is the state-of-the-art solver for the RPCA problem, to solve (14.1), and name it augmented Lagrange multiplier sparse estimator (ALMSE). The algorithmic details are left to Appendix B. Note that both APGSE and ALMSE have user parameters controlling the achieved sparsity in $\mathbf{E}$ and the estimated rank of $\mathbf{X}$, respectively. The selection of these parameters has direct impact on the performance of the regularization-based approaches. In each of the experiments, we have determined the empirical optimal values of these parameters. The VBSE algorithm is free of user parameters.

We consider two applications, namely the foreground detection in blurred and noisy video sequences and the detection of network anomalies, whose details have been presented in Sections 14.2.2 and 14.2.3, respectively. The experimental description and performance evaluation of the proposed and alternative algorithms are given below.

14.5.1 Foreground Detection in Blurred and Noisy Video

In this experiment, the *CAVIAR* test video sequence [9] was used. A sample video frame is presented in Figure 14.3(a), showing the hallway in a shopping mall with people moving as the foreground. The original 180×140 frame was divided into 30 non-overlapping blocks of size 30×28, and each block was treated as the basic unit in the experiment. The pixels in each block were vectorized into a column of size $F = 30 \times 28 = 840$, and $T = 100$, such that columns were stacked into an $F \times T$ matrix $\mathbf{D}$. The $F \times F$ matrix $\mathbf{R}$ models pixel-wise blurring, where each output was computed as the average of the 13 inputs within its radius-2 neighborhood, as is illustrated in Figure 14.2. Dense Gaussian noise $\mathbf{N}$ was added to the blurred video $\mathbf{RD}$, resulting in the observed $\mathbf{Y}$ with an SNR value at 23.5 dB. A blurred and noisy sample frame is shown in Figure 14.3(b). Note that due to the presence of noise, the direct application of inverse filtering will magnify the noise and render poor results. Therefore, approaches such as the proposed VBSE have to be employed to deal with the blurred and noisy data.

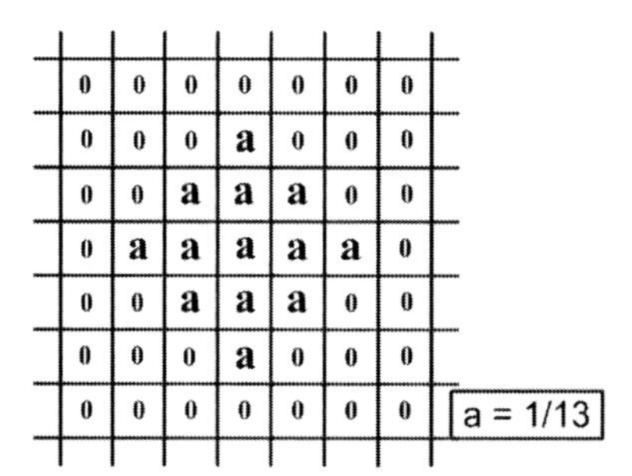

FIGURE 14.2 Illustration of blurring kernel.

In order to obtain a reference for performance evaluation, we estimate the ground truth foreground $\mathbf{E}$ by applying VBSE to the original video $\mathbf{D}$ with $\mathbf{R} = \mathbf{I}_F$. This estimated ground truth, denoted as $\mathbf{E}_{\mathrm{GT}}$ and shown in Figure 14.3(c), provides a reasonable representation of the moving FG objects. From the corresponding binarized FG map in Figure 14.3(d) we see that the FG contains two moving shoppers and their reflected images cast by the floor and the glass. Figures 14.3(c) and 14.3(d) serve as the reference for visually evaluating the performance of the algorithms.

VBSE, APGSE, and ALMSE were then applied on the blurred and noisy $\mathbf{Y}$, and the results are shown in Figures 14.4(a)–14.4(f). From the figures we see that both VBSE and its regularization-based counterparts produce reasonable results that highlight the moving objects in the foreground. The VBSE algorithm gives cleaner FG maps, while the results of APGSE and ALMSE contain more isolated pixels falsely classified as foreground. The

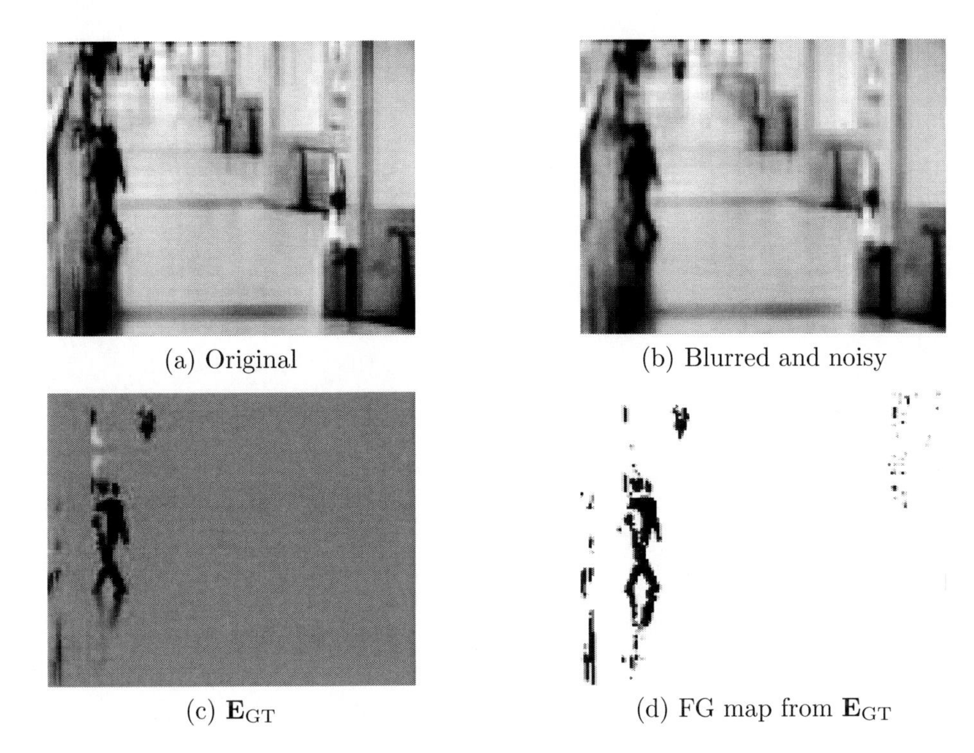

(a) Original (b) Blurred and noisy

(c) $\mathbf{E}_{\text{GT}}$ (d) FG map from $\mathbf{E}_{\text{GT}}$

FIGURE 14.3 Sample frame of *CAVIAR* video sequence and estimated ground truth.

presence of such pixels may result in an increased false detection probability, if the FG maps are to be used subsequently for applications such as surveillance and intruder detection. Also note that the performance of APGSE and ALMSE has been optimized via the manual tuning of input parameters, while the VBSE approach is free of input parameters and is hence more amenable to automated deployment.

14.5.2 Detection of Network Anomalies

In this subsection we apply the VBSE algorithm to the network anomaly detection problem and examine its performance under various experimental conditions. First we consider the real life *Internet2* dataset [18], which records the composite OD flow traffic across the Abilene backbone network.

Figure 14.5 gives an illustration of the network, which consists of 11 nodes located at various cities across the United States, represented by red dots in the figure. The number of OD flows is given by $F = N^2 = 121$, including flows to and from the same nodes. A solid line denotes a pair of bi-directional links between two nodes, and there are a total of 15 such link pairs. In addition, there is a link from every node to itself. Therefore, the total number of links present in the network is $L = 2 \times 15 + 11 = 41$. A binary routing matrix $\mathbf{R}$ maps the $F = 121$ flows onto the $L = 41$ network links. In the experiment we took $T = 210$ temporal snapshots from the data. Given the $F \times T$ OD flow traffic $\mathbf{D}$, the link load $\mathbf{Y}$ is obtained through multiplication with $\mathbf{R}$. Note that besides obtaining $\mathbf{Y}$ from flow measurements, link loads can also be determined using the simple network management protocol (SNMP) traces [36].

In order to quantitatively evaluate the performance of the algorithms, we need to estimate the ground truth from the data. This was done by applying the VBSE algorithm to $\mathbf{D}$ by setting $\mathbf{R} = \mathbf{I}_F$. As a validation, we also applied the APG [23] and ALM [22] algorithms

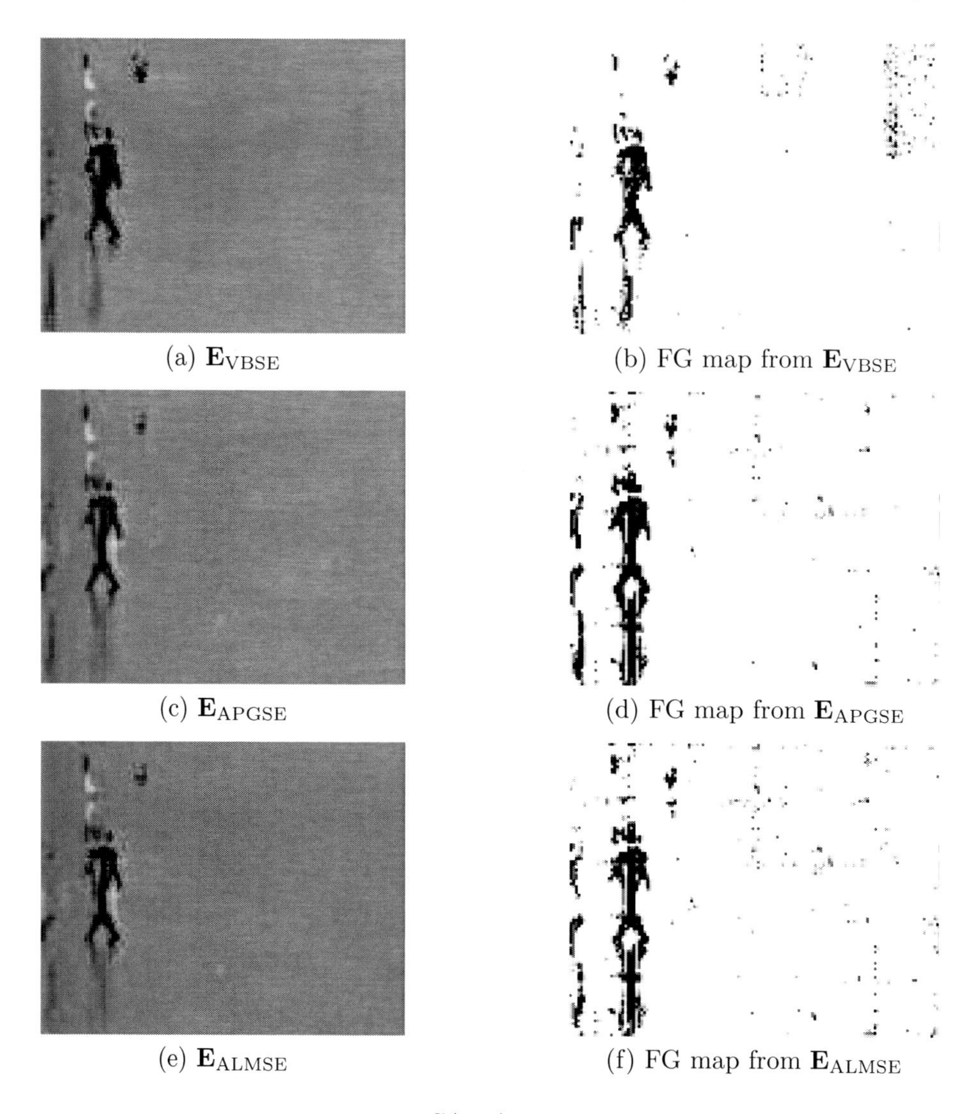

FIGURE 14.4 Estimated foreground from *CAVIAR* video sequence.

to $\mathbf{D}$. These three algorithms produced comparable results, with the estimated anomalies comprising nine numerically non-zero entries at the same spatio-temporal locations and the estimated clean traffic being of rank 4. We define as the estimated ground truth anomalies $\mathbf{E}_{\text{GT}}$ the average of the results produced by VBSE (with $\mathbf{R} = \mathbf{I}_F$), APG, and ALM in what follows.

The link load data $\mathbf{Y}$ were fed into the algorithms for anomaly detection and amplitude estimation. Figure 14.6 shows the algorithmic results superimposed on the ground truth. As can be seen in the figures, VBSE detects all the anomalies present in the ground truth and accurately estimates their amplitudes. In contrast, APGSE tends to produce biased estimates and the amplitudes differ from the ground truth by a margin. ALMSE, although accurately estimating the amplitudes of anomalies present in the ground truth, has additionally detected many spurious anomalies. This will result in an increased false alarm probability, if ALMSE is used for network monitoring.

To further investigate the performance of the various algorithms, we considered additional numerical experiments. Specifically, we artificially added dense Gaussian noise to the

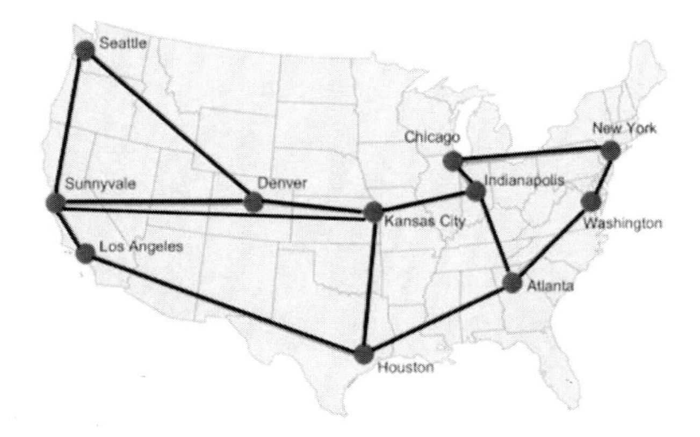

FIGURE 14.5 *Internet2* backbone network map.

link measurements. The performance of VBSE, APGSE, and ALMSE at various SNR levels are recorded in Figures 14.7(a) and 14.7(b). As can be seen, VBSE is more robust to noise and gives uniformly lower estimation error than APGSE and ALMSE. VBSE is also able to precisely identify the number of anomalies, while APGSE yields a noticeably higher number of false detections.

Finally, we carried out a separate simulated experiment to examine the performances at different anomaly densities (i.e., various degrees of sparsity of $\mathbf{E}$). The data generation process is described as follows. A low-rank $\mathbf{X}_0$ was simulated as the product of $F \times r$ and $r \times T$ matrices with i.i.d. entries drawn from $\mathcal{N}(0, 100/F)$, and $\mathcal{N}(0, 100/T)$, respectively. Synthesized $\mathbf{E}$ with amplitudes drawn from i.i.d. $\mathcal{U}(-10, 10)$ was added to $\mathbf{X}_0$. $\mathbf{R}$ is simulated as a random binary matrix with 95% of zeros. The anomaly density was varied across a wide range. As can be seen from Figures 14.8(a) and 14.8(b), VBSE gives uniformly lower estimation error and more accurate estimate of anomaly density than APGSE and ALMSE.

The numerical examples presented above demonstrate the effectiveness of VBSE in solving real-life problems. Compared with its regularization-based counterparts, VBSE has satisfactory performance under a broad range of experimental conditions. Moreover, VBSE is free of user parameters and is hence especially amenable for automated deployment. Last but not least, VBSE provides the posterior distributions of the unknowns rather than only point estimates.

14.6 Conclusion

In this chapter we proposed a variational Bayesian-based algorithm for estimation of the sparse component from an outlier-corrupted low-rank matrix. A general data model was formulated and several specific application scenarios were discussed. The proposed algorithm is based on a hierarchical Bayesian model, and employs a variational approach for inference. The proposed algorithm is free of user parameters and its effectiveness was demonstrated using both real-life and simulated experiments.

References

1. A. Abdelkefi, Y. Jiang, W. Wang, A. Aslebo, and O. Kvittem. Robust traffic anomaly detection with principal component pursuit. In *ACM CoNEXT Student Workshop*, pages 10:1–10:2, 2010.
2. M. Ayazoglu, M. Sznaier, and O. Camps. Fast algorithms for structured robust principal com-

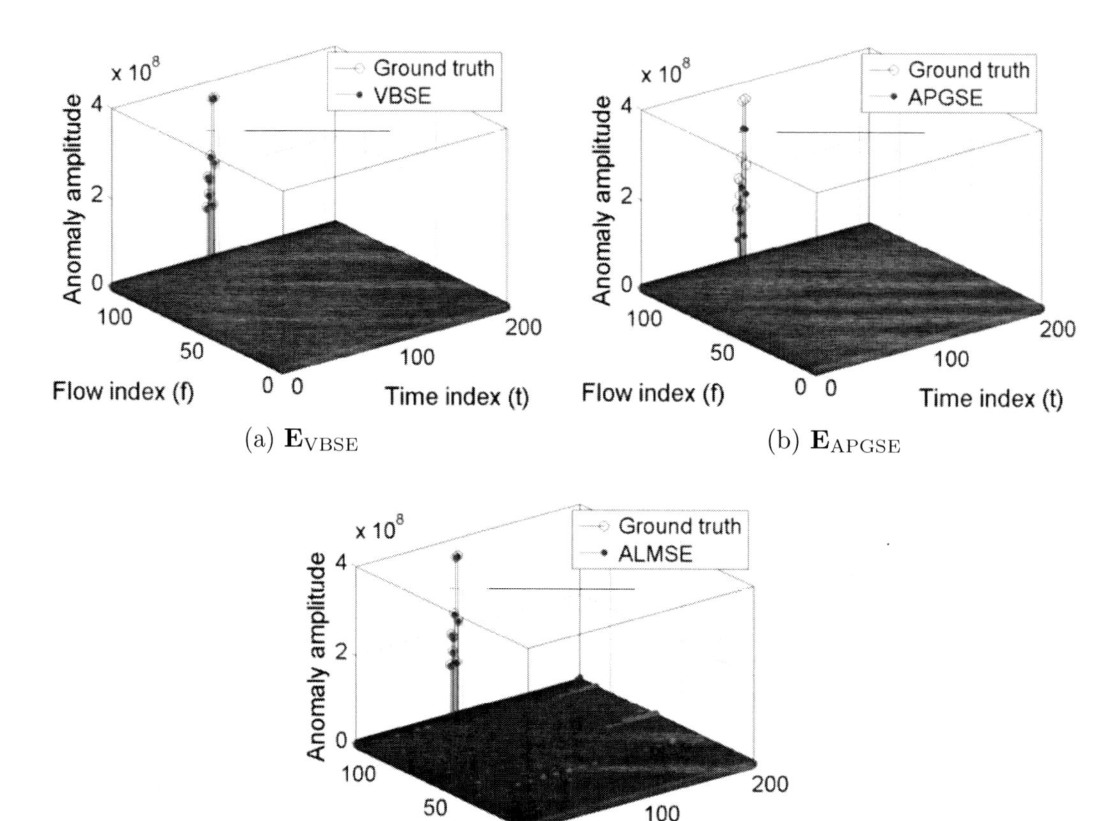

(a) $\mathbf{E}_{\text{VBSE}}$ (b) $\mathbf{E}_{\text{APGSE}}$

(c) $\mathbf{E}_{\text{ALMSE}}$

FIGURE 14.6 Estimated flow anomalies from *Internet2* data.

ponent analysis. In *IEEE Conference on Computer Vision and Pattern Recognition, CVPR 2012*, pages 1704–1711, 2012.

3. S. Babacan, M. Luessi, R. Molina, and A. Katsaggelos. Sparse Bayesian methods for low-rank matrix estimation. *IEEE Transactions on Signal Processing*, 60(8):3964–3977, August 2012.
4. B. Bao. Inductive robust principal component analysis. *IEEE Transactions on Image Processing*, 21(8):3794–3800, 2012.
5. M. Beal. *Variational algorithms for approximate Bayesian inference*. PhD thesis, Gatsby Computational Neuroscience Unit, University College London, 2003.
6. C. Bishop. *Pattern recognition and machine learning*. Springer-Verlag, 2006.
7. J. Cai, E. Candès, and Z. Shen. A singular value thresholding algorithm for matrix completion. *SIAM Journal on Optimization*, 20(4):1956–1982, January 2010.
8. E. Candès, X. Li, Y. Ma, and J. Wright. Robust principal component analysis? *Journal of the ACM*, 58(3), May 2011.
9. CAVIAR. CAVIAR test case scenarios. *http://homepages.inf.ed.ac.uk/rbf/CAVIARDATA1/*, January 2004.
10. V. Chandola, A. Banerjee, and V. Kumar. Anomaly detection: a survey. *ACM Computing Surveys*, 41(3):15:1–15:58, 2009.
11. A. Chistov and D. Grigorév. Complexity of quantifier elimination in the theory of algebraically

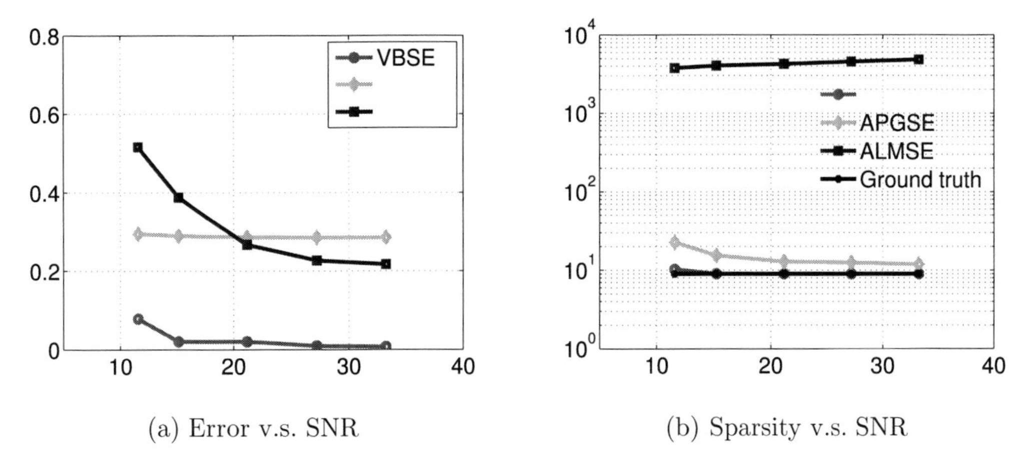

(a) Error v.s. SNR (b) Sparsity v.s. SNR

FIGURE 14.7 Performance of VBSE, APGSE, and ALMSE at various SNR levels.

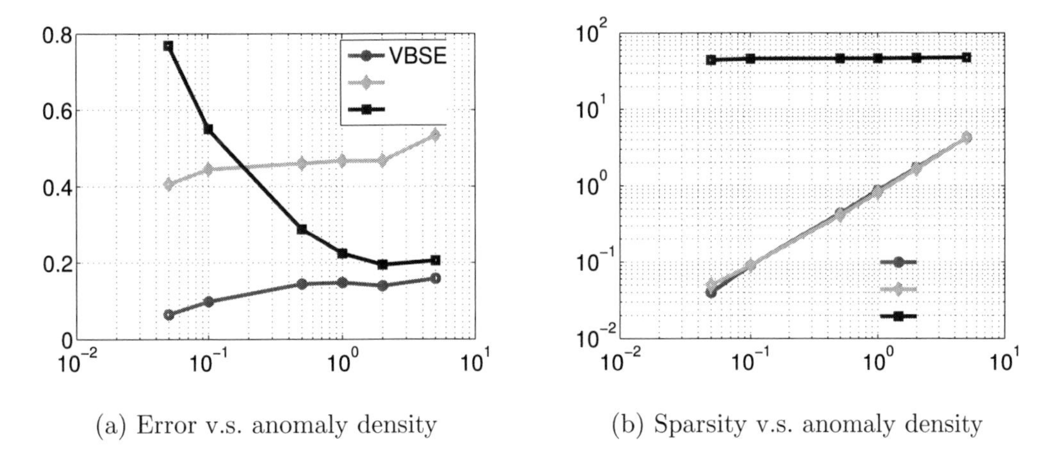

(a) Error v.s. anomaly density (b) Sparsity v.s. anomaly density

FIGURE 14.8 Performance of VBSE, APGSE and ALMSE at various degrees of sparsity of **E**.

closed fields. *Mathematical Foundations of Computer Science, Lecture Notes in Computer Science*, 176:17–31, 1984.

12. N. Chitradevi, V. Palanisamy, K. Baskaran, and U. Nisha. Outlier aware data aggregation in distributed wireless sensor network using robust principal component analysis. In *International Conference on Computing Communication and Networking Technologies*, pages 1–9, 2010.
13. M. Dikmen and T. Huang. Robust estimation of foreground in surveillance videos by sparse error estimation. In *International Conference on Pattern Recognition, ICPR 2008*, pages 1–4, 2008.
14. X. Ding, L. He, and L. Carin. Bayesian robust principal component analysis. *IEEE Transactions on Image Processing*, 20(12):3419–3430, 2011.
15. L. Duarte, E. Nadalin, K. Filho, R. Zanetti, J. Romano, and M. Tygel. Seismic wave separation by means of robust principal component analysis. In *European Signal Processing Conference*, pages 1494–1498, 2012.
16. J. Gai, Y. Li, and R. Stevenson. Robust Bayesian PCA with Student's t-distribution: the variational inference approach. In *IEEE International Conference on Image Processing, ICIP 2008*, pages 1340–1343, 2008.

17. P. Huang, S. Chen, P. Smaragdis, and M. Hasegawa-Johnson. Singing-voice separation from monaural recordings using robust principal component analysis. In *IEEE International Conference on Acoustics, Speech and Signal Processing, ICASSP 2012*, pages 57–60, 2012.
18. Internet2. Internet2 datasets. *http://internet2.edu/observatory/archive/data-collections.html*, January 2013.
19. A. Lakhina, M. Crovella, and C. Diot. Diagnosing network-wide traffic anomalies. In *Conference on Applications, Technologies, Architectures, and Protocols for Computer Communications*, pages 219–230, 2004.
20. A. Lakhina, K. Papagiannaki, M. Crovella, C. Diot, E. Kolaczyk, and N. Taft. Structural analysis of network traffic flows. In *Joint International Conference on Measurement and Modeling of Computer Systems*, pages 61–72, 2004.
21. S. Lang, B. Zhao, S. Wang, X. Liu, and G. Fang. Ionospheric ionogram denoising based on robust principal component analysis. In *International Conference on Systems and Informatics*, pages 1956–1960, 2012.
22. Z. Lin, M. Chen, and Y. Ma. The augmented Lagrange multiplier method for exact recovery of corrupted low-rank matrices. *University of Illinois at Urbana-Champaign technical report #UILU-ENG-09-2215*, 2009.
23. Z. Lin, A. Ganesh, J. Wright, L. Wu, M. Chen, and Y. Ma. Fast convex optimization algorithms for exact recovery of a corrupted low-rank matrix. In *International Workshop on Computational Advances in Multi-Sensor Adaptive Processing*, December 2009.
24. M. Mardani, G. Mateos, and G. Giannakis. Unveiling anomalies in large-scale networks via sparsity and low rank. In *Asilomar Conference on Signals, Systems, and Computers*, Pacific Grove, CA, USA, November 2011.
25. A. McIvor. Background subtraction techniques. In *Image and Vision Computing New Zealand, IVCNZ 2000*, Auckland, New Zealand, 2000.
26. A. Medina, N. Taft, K. Salamatian, S. Bhattacharyya, and C. Diot. Traffic matrix estimation: existing techniques and new directions. *Conference on Applications, Technologies, Architectures, and Protocols for Computer Communications*, pages 161–174, 2002.
27. B. Natarajan. Sparse approximate solutions to linear systems. *SIAM Journal of Computing*, 24(2):227–234, 1995.
28. C. Papadimitriou, P. Raghavan, H. Tamaki, and S. Vempala. Latent semantic indexing: a probabilistic analysis. *Journal of Computer and System Sciences*, 61(2):217–235, 2000.
29. A. Patcha and J. Park. An overview of anomaly detection techniques: existing solutions and latest technological trends. *Elsevier Computer Networks*, 51(12):3448–3470, 2007.
30. Y. Peng. RASL: robust alignment by sparse and low-rank decomposition for linearly correlated images. In *IEEE Conference on Computer Vision and Pattern Recognition, CVPR 2010*, pages 763–770, 2010.
31. M. Piccardi. Background subtraction techniques: a review. In *IEEE International Conference on Systems, Man and Cybernetics, SMC 2004*, volume 4, pages 3099–3104, 2004.
32. B. Recht, M. Fazel, and P. A. Parrilo. Guaranteed minimum-rank solutions of linear matrix equations via nuclear norm minimization. *SIAM Review*, 52(3):471–501, 2010.
33. B. Rubinstein, B. Nelson, L. Huang, A. Joseph, S. Lau, N. Taft, and J. Tygar. Compromising PCA-based anomaly detectors for network-wide traffic. *University of California at Berkeley Technical Report No. UCB/EECS-2008-73*, 2008.
34. G. Tang and A. Nehorai. Constrained Cramér-Rao bound on robust principal component analysis. *IEEE Transactions on Signal Processing*, 59(10):5070–5076, 2011.
35. G. Tang and A. Nehorai. Robust principal component analysis based on low-rank and block-sparse matrix decomposition. In *Annual Conference on Information Sciences and*

Systems, pages 1–5, 2011.

36. M. Thottan and C. Ji. Anomaly detection in IP networks. *IEEE Transactions on Signal Processing*, 51(8):2191–2204, 2003.
37. L. Wang. An iterative algorithm for robust kernel principal component analysis. In *International Conference on Machine Learning and Cybernetics, ICMLC 2007*, pages 3484–3489, 2007.
38. J. Wright, A. Ganesh, S. Rao, Y. Peng, and Y. Ma. Robust principal component analysis: exact recovery of corrupted low-rank matrices via convex optimization. *Advances in Neural Information Processing Systems 22*, pages 2080–2088, 2009.
39. J. Wright, A. Ganesh, Z. Zhou, and Y. Ma. Towards a practical face recognition system: Robust registration and illumination by sparse representation. *IEEE Conference on Computer Vision and Pattern Recognition, CVPR 2009*, pages 597–604, 2009.

15

Recovering Low-Rank and Sparse Matrices with Missing and Grossly Corrupted Observations

Fanhua Shang
The Chinese University of Hong Kong, Hong Kong, China

Yuanyuan Liu
The Chinese University of Hong Kong, Hong Kong, China

James Cheng
The Chinese University of Hong Kong, Hong Kong, China

Hong Cheng
The Chinese University of Hong Kong, Hong Kong, China

15.1 Introduction

Recovering low-rank and sparse matrices from incomplete *or* even corrupted observations is a common problem in many application areas, including statistics [1, 9, 51], bioinformatics [37], machine learning [28, 47, 49, 52], computer vision [5, 7, 42, 43, 58], and signal and image processing [27, 30, 38]. In these areas, data often have high dimensionality, such as digital photographs and surveillance videos, which makes inference, learning, and recognition infeasible due to the "curse of dimensionality."

To process high-dimensional data, traditional Matrix Factorization (MF) methods such as Principal Component Analysis (PCA) and Non-negative Matrix Factorization (NMF) are commonly used, mainly because they are simple to implement, can be solved efficiently, and are often effective in real-world applications such as latent semantic indexing and face recognition. However, one of the main challenges faced by these methods is that the observed data is often contaminated by outliers or missing values [13], or is a small set of linear measurements [51]. To address these issues, a large number of methods based on compressive sensing and rank minimization have been proposed, such as robust PCA [43, 52, 53] (RPCA, also called Principal Component Pursuit (PCP) in [7] and Low-Rank and Sparse matrix Decomposition (LRSD) in [47, 56]) and Low-Rank Matrix Completion (LRMC) [8, 9].

In many applications, we have to recover a matrix from only a small number of observed entries, for example collaborative filtering for recommender systems. This problem is often called LRMC, where missing entries or outliers are presented at arbitrary locations in the measurement matrix. In addition, we would like to recover low-rank and sparse matrices from corrupted data, i.e., RPCA problems. For example, the face images of a person may be corrupted by glasses or shadows [18]. The classical PCA cannot address the issue as its least-squares fitting is sensitive to these gross outliers. Recovering a low-rank matrix in the presence of outliers has been extensively studied and successfully applied in many important applications, such as video surveillance [7, 52] and image alignment [38]. In some more general applications, we also have to simultaneously recover both low-rank and sparse matrices from small sets of linear measurements, which is called Compressive Principal Component Pursuit (CPCP) [51].

In theory, those problems mentioned above can be exactly solved with high probability under mild assumptions via a hybrid convex program involving the trace norm (also called the nuclear norm [14]) or together with the l_1-norm, such as low-rank matrix recovery and completion algorithms [7–9, 51]. However, those algorithms all exploit a closed-form expression for the proximal operator of the trace norm, which involves the Singular Value Decomposition (SVD). Thus, they all have high computational cost and are even not applicable for solving large-scale problems.

To address this issue, we propose a scalable Robust Bilinear Factorization (RBF) framework [42, 44] to recover low-rank and sparse matrices from incomplete, corrupted data or a small set of linear measurements. The proposed RBF framework not only takes into account the fact that the observations are contaminated by additive outliers and missing data, i.e., Robust Matrix Completion problems [9, 19] (RMC, i.e., RPCA plus LRMC), but can also identify both low-rank and sparse noisy components from incomplete and grossly corrupted measurements, i.e., CPCP problems. In the unified RBF framework for both RMC and CPCP problems, repetitively calculating SVD of a large matrix in [43, 51] is replaced by updating two much smaller factor matrices. By imposing the orthogonality constraint, the original RMC and CPCP problems are transformed into two smaller-scale matrix trace norm minimization problems. Moreover, we develop an efficient Alternating Direction Method of Multipliers (ADMM) to solve the RMC problem, and then extend it to solve CPCP problems. Finally, we theoretically analyze the equivalent relationship between the QR scheme and the SVD scheme, and the convergence behavior of the proposed algorithms.

This chapter is structured as follows. In Section 15.2, we introduce the challenges faced by existing RMC and CPCP methods. In Section 15.3, we present two scalable trace norm-regularized RBF models for RMC and CPCP problems, respectively. In Section 15.4, we develop an ADMM algorithm to solve the non-convex RMC problem and then extend it to solve the convex RMC and CPCP problems. In Section 15.5, we provide the theoretical analysis of the proposed algorithms, such as equivalent relationship and convergence analysis. The proofs of the theorems can be found in the original papers [43, 44]. In Section 15.6 we discuss various extensions and related work. We present empirical results in Section 15.7 and conclude this chapter in Section 15.8.

15.2 Background and Challenges

In this chapter, we mainly consider the following trace norm minimization problem,

$$\min_{L,S} f(S) + \lambda\|L\|_*, \quad \text{s.t.}, \psi(L+S) = y$$

where $\lambda \geq 0$ is a regularization parameter, $\|L\|_*$ is the trace norm of a low-rank matrix $L \in \mathbb{R}^{m\times n}$ (sum of singular values), $S \in \mathbb{R}^{m\times n}$ is sparse, $y \in \mathbb{R}^p$ is the linear measurements, $\psi(\cdot)$ is an underdetermined linear operator such as the linear projection operator $\mathcal{P}_\Omega$, and $f(\cdot)$ denotes the loss function associated with the l_1-norm, l_2-norm, or $l_{2,1}$-norm. The above model generalizes several existing models below.

If $f(\cdot)$ is the l_1-norm loss function, and $\psi(\cdot)$ is the operator $\mathcal{P}_Q$, i.e., the projection operator onto a linear subspace $Q \subseteq \mathbb{R}^{m\times n}$, both a low-rank matrix $L \in \mathbb{R}^{m\times n}$ and a sparse one $S \in \mathbb{R}^{m\times n}$ can be recovered from highly corrupted measurements $y = \mathcal{P}_Q(A) \in \mathbb{R}^p$ via the following CPCP model [51],

$$\min_{L,S} \|S\|_1 + \lambda\|L\|_*, \quad \text{s.t.}, \mathcal{P}_Q(L+S) = \mathcal{P}_Q(A) \tag{15.1}$$

where $\|\cdot\|_1$ denotes the l_1-norm (sum of magnitudes). When $\mathcal{P}_Q = \mathcal{P}_\Omega$ and Ω is the index set of observed entries, the model (15.1) is the RMC problem,

$$\min_{L,S} \|S\|_1 + \lambda\|L\|_*, \quad \text{s.t.}, \mathcal{P}_\Omega(L+S) = \mathcal{P}_\Omega(A). \tag{15.2}$$

Theorem 1.2 in [7] and Theorem 2.1 in [51] state that a commensurately small number of measurements are sufficient to accurately recover the low-rank and sparse matrices with high probability. Indeed, if Q is the entire space or all entries of the corrupted matrix are directly observed, the model (15.1) or (15.2) degenerates to the following RPCA problem [7, 52],

$$\min_{L,S} \|S\|_1 + \lambda\|L\|_*, \quad \text{s.t.}, L+S = A \tag{15.3}$$

where A denotes the grossly corrupted observations. In addition, Xu et al. [53] used the $l_{1,2}$-norm to model corrupted columns of S.

If $f(\cdot)$ is the l_2-norm loss function and $\psi(\cdot)$ is the operator $\mathcal{P}_\Omega$, the model (15.1) or (15.2) degenerates to the following trace norm-regularized linear least-squares problem,

$$\min_{L} \frac{1}{2}\|\mathcal{P}_\Omega(L) - \mathcal{P}_\Omega(A)\|_F^2 + \lambda\|L\|_*. \tag{15.4}$$

Since the RPCA problem (15.3) and the LRMC problem (15.4) have been extensively studied, we are particularly interested in the more general RMC and CPCP problems. Some efficient algorithms have been developed to solve the convex optimization problems (15.1) and (15.2), such as CPCP [51] and RMC [43]. Although both models (15.1) and (15.2) are convex, and their algorithms converge to the globally optimal solution, they involve SVD at each iteration and then suffer from a high computational cost of $O(mn^2)$. While there have been a lot of efforts towards fast SVD computation such as partial SVD [20] and approximate SVD [31], the performance of those methods is still unsatisfactory for many real applications [25–27].

15.3 RBF Framework

The dominant cost of existing RMC and CPCP algorithms in each iteration is computing an SVD of the same size as the input data. As a remedy, Robust Bilinear Factorization (RBF) aims to find two smaller low-rank matrices $U \in \mathbb{R}^{m\times d}$ ($U^TU = I$) and $V \in \mathbb{R}^{n\times d}$ whose product is equal to the desired low-rank matrix $L \in \mathbb{R}^{m\times n}$, i.e., $L = UV^T$, where d is an upper bound on the rank of L, i.e., $d \geq r = \text{rank}(L)$. Fortunately, d can be easily computed by several matrix rank estimation strategies in [16, 50].

15.3.1 Convex RMC Model

From the optimization problem (15.2), we easily find the optimal solution $S_{\Omega^C} = \mathbf{0}$ [42, 43], where Ω^C is the complement of Ω, i.e., the index set of unobserved entries. Consequently, we have the following lemma [42, 43].

LEMMA 15.1 The RMC model (15.2) with the linear projection operator $\mathcal{P}_\Omega$ is equivalent to the following convex optimization problem

$$\min_{L,\, S} \|\mathcal{P}_\Omega(S)\|_1 + \lambda\|L\|_*, \quad \text{s.t.}, \mathcal{P}_\Omega(L+S) = \mathcal{P}_\Omega(A),\, \mathcal{P}_{\Omega^C}(S) = \mathbf{0}. \tag{15.5}$$

15.3.2 Non-Convex RMC Model

For the incomplete and corrupted matrix A, the non-convex RBF model for RMC is to find two smaller matrices, whose product approximates L, which can be formulated as follows:

$$\min_{U,\, V,\, S} \|\mathcal{P}_\Omega(S)\|_1 + \lambda\|UV^T\|_*, \ \text{s.t.}, \mathcal{P}_\Omega(UV^T + S) = \mathcal{P}_\Omega(A). \tag{15.6}$$

LEMMA 15.2 [43] Let U and V be two matrices of compatible dimensions, where U has orthogonal columns, i.e., $U^TU = I$, then we have $\|UV^T\|_* = \|V\|_*$.

By imposing $U^TU = I$ and substituting $\|UV^T\|_* = \|V\|_*$ into (15.6), we arrive at a much smaller-scale matrix trace norm minimization problem,

$$\min_{U,\, V,\, S} \|\mathcal{P}_\Omega(S)\|_1 + \lambda\|V\|_*, \ \text{s.t.}, \mathcal{P}_\Omega(UV^T + S) = \mathcal{P}_\Omega(A),\, U^TU = I. \tag{15.7}$$

THEOREM 15.1 *[44] Suppose (L^*, S^*) is a solution to (15.5) with rank$(L^*) = r$, then there exists the solution $U_k \in \mathbb{R}^{m\times d}$, $V_k \in \mathbb{R}^{n\times d}$, and $S_k \in \mathbb{R}^{m\times n}$ to (15.7) with $d \geq r$ and $(S_k)_{\Omega^C} = \mathbf{0}$, such that $(U_kV_k^T, S_k)$ is also a solution to (15.5).*

15.3.3 CPCP Model

For a small set of linear measurements $y \in \mathbb{R}^p$, the non-convex RBF model for CPCP problems is to recover low-rank and sparse matrices as follows,

$$\min_{U,\, V,\, S} \|S\|_1 + \lambda\|V\|_*, \ \text{s.t.}, \mathcal{P}_Q(UV^T + S) = \mathcal{P}_Q(A),\, U^TU = I. \tag{15.8}$$

Algorithm 15.1 - Solving the generic problem (15.9) via ADMM

1: Input: γ_0.
2: Initialization: $x_0 = \mathbf{0}$, $z_0 = \mathbf{0}$ and $\gamma_0 = \mathbf{0}$.
3: **for** $k = 0, 1, \cdots, T$ **do**
4: $\quad x_{k+1} = \arg\min_x \mathcal{L}_\alpha(x, z_k, \gamma_k)$.
5: $\quad z_{k+1} = \arg\min_z \mathcal{L}_\alpha(x_{k+1}, z, \gamma_k)$.
6: $\quad \gamma_{k+1} = \gamma_k + \alpha(Ex_{k+1} + Fz_{k+1} - c)$.
7: **end for**
8: Output: x_k and z_k.

15.4 Optimization

In this section, we propose an efficient algorithm based on Alternating Direction Method of Multipliers (ADMM) to solve the non-convex RMC problem (15.7), and extend it to solve the convex RMC problem (15.5) and the CPCP problem (15.8). We provide the convergence analysis of the proposed algorithms in Section 15.5.

15.4.1 Generic Formulation

The ADMM was introduced for optimization in the 1970s, and its origins can be traced back to techniques for solving partial difference equations in the 1950s. It has received renewed interest due to the fact that it is efficient to tackle large-scale problems and solve optimization problems with multiple non-smooth terms in the objective function [21]. The ADMM can be considered as an approximation of the method of multipliers. It decomposes a large global problem into a series of smaller subproblems, and coordinates the solutions of subproblems to compute the globally optimal solution. The problem solved by ADMM takes the following generic form

$$\min_{x\in\mathbb{R}^n, z\in\mathbb{R}^m} f(x) + g(z), \qquad \text{s.t., } Ex + Fz = c \tag{15.9}$$

where both $f(\cdot)$ and $g(\cdot)$ are convex functions. ADMM reformulates the problem using a variant of the augmented Lagrangian method as follows:

$$\mathcal{L}_\alpha(x, z, \gamma) = f(x) + g(z) + \gamma^T(Ex + Fz - c) + \frac{\alpha}{2}\|Ex + Fz - c\|_2^2$$

where γ is the Lagrangian multiplier and α is a penalty parameter. ADMM solves the problem (15.9) by iteratively minimizing $\mathcal{L}_\alpha(x, z, \gamma)$ over x, z, and then updating γ, as outlined in **Algorithm 15.1** [4].

15.4.2 ADMM for Non-Convex RMC

For efficiently solving the non-convex RMC problem (15.7), we can assume without loss of generality that the unknown entries of A are simply set as zeros, i.e., $A_{\Omega^C} = \mathbf{0}$, and S_{Ω^C} may be any values such that $\mathcal{P}_{\Omega^C}(A) = \mathcal{P}_{\Omega^C}(UV^T) + \mathcal{P}_{\Omega^C}(S)$. Therefore, the constraint with the operator $\mathcal{P}_\Omega$ in (15.7) is simplified into $A = UV^T + S$. Hence, we introduce the constraint $A = UV^T + S$ into (15.7), and obtain the following equivalent form:

$$\min_{U, V, S} \|\mathcal{P}_\Omega(S)\|_1 + \lambda\|V\|_*, \ \text{s.t., } UV^T + S = A, \ U^TU = I. \tag{15.10}$$

The partial augmented Lagrangian function of (15.10) is

$$\mathcal{L}_\alpha(U, V, S, Y) = \lambda\|V\|_* + \|\mathcal{P}_\Omega(S)\|_1 + \langle Y, A - S - UV^T\rangle + \frac{\alpha}{2}\|A - S - UV^T\|_F^2 \tag{15.11}$$

where $Y \in \mathbb{R}^{m \times n}$ is a matrix of Lagrange multipliers, $\alpha > 0$ is a penalty parameter, and $\langle M, N \rangle$ denotes the inner product between matrices M and N of equal sizes.

Updating U:

By fixing V and S at their latest values, removing the terms that do not depend on U, and adding some proper terms that do not depend on U, the problem with respect to U is formulated as follows:

$$\min_{U} \|UV_k^T - P_k\|_F^2, \quad \text{s.t.}, \; U^T U = I \tag{15.12}$$

where $P_k = A - S_k + Y_k/\alpha_k$. In fact, the optimal solution to (15.12) can be given by the SVD of $P_k V_k$ as in [36, 43]. To further speed up the calculation, we introduce the idea in [50] that uses a QR decomposition instead of SVD. The resulting iteration step is

$$U_{k+1} = Q, \quad \text{QR}(P_k V_k) = QR \tag{15.13}$$

where U_{k+1} is an orthogonal basis for the range space $\mathcal{R}(P_k V_k)$, i.e., $\mathcal{R}(U_{k+1}) = \mathcal{R}(P_k V_k)$.

REMARK 15.1 Although U_{k+1} in (15.13) is not an optimal solution to (15.12), the iterative scheme based on QR factorizations and the SVD scheme in [43] are equivalent to solve (15.12) and (15.14), and their equivalent analysis is provided in Section 15.5. Moreover, the use of QR factorizations also makes the proposed iterative scheme highly scalable on modern parallel architectures [2].

Updating V:

Fixing U and S, the optimization problem with respect to V can be rewritten as:

$$\min_{V} \frac{\alpha_k}{2} \|U_{k+1} V^T - P_k\|_F^2 + \lambda \|V\|_*. \tag{15.14}$$

To solve (15.14), the spectral soft-thresholding operation [6] is defined as follows.

DEFINITION 15.1 For any given matrix $M \in \mathbb{R}^{n \times d}$ whose rank is r, and $\mu \geq 0$, the Singular Value Thresholding (SVT) operator is defined as follows:

$$\text{SVT}_{\mu}(M) = \overline{U} \text{diag}(\max\{\overline{\sigma} - \mu, 0\}) \overline{V}^T$$

where $\max\{\cdot, \cdot\}$ should be understood element-wise, and $\overline{U}\,\text{diag}(\overline{\sigma})\,\overline{V}^T$ is the SVD of M.

THEOREM 15.2 *[43] The trace norm minimization problem (15.14) has a closed-form solution given by*

$$V_{k+1} = \text{SVT}_{\lambda/\alpha_k}(P_k^T U_{k+1}). \tag{15.15}$$

REMARK 15.2 Here, only one much smaller matrix $P_k^T U_{k+1} \in \mathbb{R}^{n \times d}$ in (15.15) needs to perform SVD. Thus, this step of the proposed algorithm has a significantly lower computational complexity $O(d^2 n)$ while the computational complexity of existing convex algorithms is $O(n^2 m)$ in each iteration ($m \geq n \gg d$).

Algorithm 15.2 - Solving the non-convex RMC problem (15.7) via ADMM

1: Input: $\mathcal{P}_\Omega(A)$, λ and ε.
2: Initialization: $U_0 = \text{eye}(m, d)$, $V_0 = \mathbf{0}$, $Y_0 = \mathbf{0}$, $\alpha_0 = \frac{1}{\|\mathcal{P}_\Omega(A)\|_F}$, $\alpha_{\max} = 10^{10}$, and $\rho = 1.1$.
3: **while** not converge **do**
4: Update U_{k+1}, V_{k+1} and S_{k+1} by (15.13), (15.15), (15.19), and (15.20), respectively.
5: Update the multiplier Y_{k+1} by $Y_{k+1} = Y_k + \alpha_k(A - U_{k+1}V_{k+1}^T - S_{k+1})$.
6: Update α_{k+1} by $\alpha_{k+1} = \min(\rho\alpha_k, \alpha_{\max})$.
7: Check the convergence condition, $\|A - U_{k+1}V_{k+1}^T - S_{k+1}\|_F < \varepsilon$.
8: **end while**
9: Output: U, V and S, where S_{Ω^C} is set to $\mathbf{0}$.

Updating S:

Fixing U and V, we can update S by solving

$$\min_{S} \|\mathcal{P}_\Omega(S)\|_1 + \frac{\alpha_k}{2}\|S + U_{k+1}V_{k+1}^T - A - Y_k/\alpha_k\|_F^2. \tag{15.16}$$

For solving (15.16), we introduce the following element-wise shrinkage operator $\mathcal{S}_\tau$ [10]:

$$\mathcal{S}_\tau(M_{ij}) := \begin{cases} M_{ij} - \tau, & M_{ij} > \tau, \\ M_{ij} + \tau, & M_{ij} < -\tau, \\ 0, & \text{otherwise}. \end{cases}$$

The optimal solution S_{k+1} can be obtained by solving the following two subproblems with respect to S_Ω and S_{Ω^C}, respectively,

$$\min_{S_\Omega} \frac{\alpha_k}{2}\|\mathcal{P}_\Omega(S + U_{k+1}V_{k+1}^T - A - Y_k/\alpha_k)\|_F^2 + \|\mathcal{P}_\Omega(S)\|_1 \tag{15.17}$$

$$\min_{S_{\Omega^C}} \|\mathcal{P}_{\Omega^C}(S + U_{k+1}V_{k+1}^T - A - Y_k/\alpha_k)\|_F^2. \tag{15.18}$$

By the operator $\mathcal{S}_\tau$ and letting $\tau = 1/\alpha_k$, the closed-form solution to (15.17) is given by

$$(S_{k+1})_\Omega = \mathcal{S}_\tau((A - U_{k+1}V_{k+1}^T + Y_k/\alpha_k)_\Omega). \tag{15.19}$$

We can easily obtain the closed-form solution by zeroing the gradient of (15.18),

$$(S_{k+1})_{\Omega^C} = (A - U_{k+1}V_{k+1}^T + Y_k/\alpha_k)_{\Omega^C}. \tag{15.20}$$

REMARK 15.3 We can replace the l_1-norm in the sparse component recovery problem (15.16) with the $l_{2,1}$-norm for a sparse solution, such as outlier pursuit [53] or Low-Rank Representation (LRR) [22] problems. The optimal solution to the problems with $l_{2,1}$-norm regularization can be obtained by the shrinkage operator in [54].

Summarizing the analysis above, we can obtain an ADMM algorithm to solve the non-convex RMC problem (15.7), as outlined in **Algorithm 15.2**. Our algorithm is essentially a Gauss-Seidel-type scheme of ADMM, and the update strategy of the Jacobi version of ADMM is easily implemented, well suited for parallel and distributed computing, and hence is particularly attractive for solving large-scale problems [41]. In addition, S_{Ω^C} should be set to $\mathbf{0}$ for the expected output S. Algorithm 15.2 is easily used to solve the RPCA problem (15.3), where all entries of the corrupted matrix are directly observed.

Algorithm 15.3 - Solving the convex RMC problem (15.5) via ADMM

1: Input: $\mathcal{P}_\Omega(A)$, λ and ε.
2: Initialization: $L_0 = S_0 = Y_0 = \mathbf{0}$, $\alpha_0 = \frac{1}{\|\mathcal{P}_\Omega(A)\|_F}$, $\alpha_{\max} = 10^{10}$, and $\rho = 1.1$.
3: **while** not converge **do**
4: Update L_{k+1} and S_{k+1} by (15.22) and (15.23), respectively.
5: Update the multiplier Y_{k+1} by $Y_{k+1} = Y_k + \alpha_k(A - L_{k+1} - S_{k+1})$.
6: Update α_{k+1} by $\alpha_{k+1} = \min(\rho\alpha_k, \alpha_{\max})$.
7: Check the convergence condition, $\|A - L_{k+1} - S_{k+1}\|_F < \varepsilon$.
8: **end while**
9: Output: L and S, where S_{Ω^C} is set to $\mathbf{0}$.

15.4.3 ADMM for Convex RMC

Similar to (15.11), the convex RMC problem (15.5) is reformulated as follows:

$$\min_{L,\,S} \|\mathcal{P}_\Omega(S)\|_1 + \lambda\|L\|_*, \quad \text{s.t.}, \; L + S = A. \tag{15.21}$$

The augmented Lagrangian of (15.21) is given by

$$\mathcal{L}_\alpha(L, S, Y) = \|\mathcal{P}_\Omega(S)\|_1 + \lambda\|L\|_* + \langle Y,\; A - L - S\rangle + \frac{\alpha}{2}\|A - L - S\|_F^2.$$

Updating L:

With all other variables fixed, the optimal L is the solution to the following problem:

$$\min_{L} \lambda\|L\|_* + \frac{\alpha_k}{2}\|L - A + S_k - Y_k/\alpha_k\|_F^2. \tag{15.22}$$

Updating S:

The optimal S with all other variables fixed is the solution to the following problem,

$$\min_{S} \|\mathcal{P}_\Omega(S)\|_1 + \frac{\alpha_k}{2}\|S - A + L_k - Y_k/\alpha_k\|_F^2. \tag{15.23}$$

Similar to Algorithm 15.2, we can develop an ADMM algorithm for solving the convex RMC problem (15.5), as outlined in **Algorithm 15.3**.

15.4.4 Extension for CPCP

Algorithm 15.2 can be extended to solve the CPCP problem (15.8) with a general linear operator $\mathcal{P}_Q$, whose augmented Lagrange function is

$$\mathcal{L}_\alpha(U, V, S, Y) = \lambda\|V\|_* + \|S\|_1 + \langle Y,\, y - \mathcal{P}_Q(S + UV^T)\rangle + \frac{\alpha}{2}\|y - \mathcal{P}_Q(S + UV^T)\|_2^2.$$

Considering that $\mathcal{P}_Q$ is a non-identity operator, there are no closed-form solutions to the resulting subproblems. To overcome this difficulty, a common strategy is to introduce some auxiliary variables. However, with more variables and more constraints, more memory is required and the convergence of ADMM also becomes slower [21].

To avoid introducing auxiliary variables, we minimize $\mathcal{L}_\alpha$ with respect to (U, V, S) by using a recently proposed linearized alternating direction method (LADM) [55], which can address such problems with some non-identity operators such as $\mathcal{P}_Q$. Specifically, for updating U and V, let $T = UV^T$ and $\mathcal{H}(T) = \frac{\alpha_k}{2}\|y - \mathcal{P}_Q(S_k + T) + Y_k/\alpha_k\|_2^2$, then $\mathcal{H}(T)$ is approximated by

$$\mathcal{H}(T) \approx \mathcal{H}(T_k) + \langle \nabla\mathcal{H}(T_k), T - T_k\rangle + \mu\|T - T_k\|_F^2 \tag{15.24}$$

where $\nabla\mathcal{H}(T_k) = \alpha_k\mathcal{P}_Q^\star(\mathcal{P}_Q(T_k + S_k) - y - Y_k/\alpha_k)$, $\mathcal{P}_Q^\star$ is the adjoint operator of $\mathcal{P}_Q$, and μ is chosen as $\mu = 1/\|\mathcal{P}_Q^\star\mathcal{P}_Q\|_2$ as in [55], and $\|\cdot\|_2$ the spectral norm of a matrix, i.e., the largest singular value of a matrix.

Similarly, for updating S, let $T_{k+1} = U_{k+1}V_{k+1}^T$ and $\mathcal{J}(S) = \frac{\alpha_k}{2}\|y - \mathcal{P}_Q(S + T_{k+1}) + Y_k/\alpha_k\|_2^2$, then $\mathcal{J}(S)$ is approximated by

$$\mathcal{J}(S) \approx \mathcal{J}(S_k) + \langle \nabla\mathcal{J}(S_k), S - S_k\rangle + \mu\|S - S_k\|_F^2 \tag{15.25}$$

where $\nabla\mathcal{J}(S_k) = \alpha_k\mathcal{P}_Q^\star(\mathcal{P}_Q(S_k + T_{k+1}) - y - Y_k/\alpha_k)$.

15.5 Theoretical Analysis

In this section, we mainly analyze the convergence properties of Algorithm 15.2 and Algorithm 15.3 as in [43, 44]. First, we analyze the equivalent relationship between the iterative scheme based on QR decompositions and the SVD scheme in [43], as shown by the following theorem [44].

15.5.1 Equivalent Relation

THEOREM 15.3 *Let (U_k^*, V_k^*, S_k^*) be the solution of the subproblems (15.12), (15.14), and (15.16) at the k-th iteration, $Y_k^* = Y_{k-1}^* + \alpha_{k-1}(A - U_k^*(V_k^*)^T - S_k^*)$, and (U_k, V_k, S_k, Y_k) be generated by Algorithm 15.2 at the k-th iteration. Then, $\exists O_k \in \mathcal{O} = \{M \in \mathbb{R}^{d\times d} | M^T M = I\}$ such that $U_k^* = U_k O_k$ and $V_k^* = V_k O_k$, and we have $U_k^*(V_k^*)^T = U_k V_k^T$, $\|V_k^*\|_* = \|V_k\|_*$, $S_k^* = S_k$, and $Y_k^* = Y_k$.*

Since the Lagrange function (15.11) is determined by the product UV^T, V, S, and Y, the different values of U and V are essentially equivalent as long as we get the same product UV^T and $\|V\|_* = \|V^*\|_*$. Meanwhile, the proposed iterative scheme replaces SVD by the QR decomposition, which can avoid the SVD computation for solving (15.12) with an orthogonal constraint.

15.5.2 Convergence Analysis

The convergence of ADMM to solve the standard form (15.9) was studied in [4, 11]. We establish the convergence of Algorithm 15.3 by transforming (15.21) into a standard form (15.9), and show that the transformed problem satisfies the condition needed to establish the convergence. In Algorithm 15.3, we state that our algorithm alternates between two blocks of variables, L and S. Let l denote the vectorization of L, i.e., $l = \text{vec}(L) \in \mathbb{R}^{mn\times 1}$, $s = \text{vec}(S) \in \mathbb{R}^{mn\times 1}$, and $a = \text{vec}(A) \in \mathbb{R}^{mn\times 1}$, and $f(L) := \lambda\|L\|_*$ and $g(S) := \|\mathcal{P}_\Omega(S)\|_1$. We can write the equivalence constraint in (15.21) as the form: $El - Fs = a$, where both $E \in \mathbb{R}^{mn\times mn}$ and $F \in \mathbb{R}^{mn\times mn}$ are the identity matrices. By the definition $f(L)$ and $g(S)$, it is easy to verify that (15.21) and Algorithm 15.3 satisfy the conditions in Algorithm 15.1. Thus, the convergence of Algorithm 15.3 is given as follows.

THEOREM 15.4 *Consider the convex RMC problem (15.21), where both $f(L)$ and $g(S)$ are convex functions, and E and F are both identity matrices and have full column rank. The sequence $\{L_k, S_k\}$ generated by Algorithm 15.3 converges to an optimal solution of (15.21).*

Hence, the sequence $\{L_k, S_k\}$ converges to an optimal solution to (15.5), where $(S_k)_{\Omega^C}=\mathbf{0}$. Although the resulting problem (15.7) is non-convex, the global convergence of our derived algorithm is guaranteed, as shown in the following lemmas and theorems [44].

LEMMA 15.3 Let (U_k, V_k, S_k) be a sequence generated by Algorithm 15.2, then we have that (U_k, V_k, S_k) approaches to a feasible solution, i.e., $\lim_{k\to\infty} \|A - U_k V_k^T - S_k\|_F = 0$, and both sequences $U_k V_k^T$ and S_k are Cauchy sequences.

Lemma 15.3 ensures only that the feasibility of each solution has been assessed. In this chapter, we want to show that it is possible to prove the local optimality of the solution produced by Algorithm 15.2. Let k^* be the number of iterations needed by Algorithm 15.2 to stop, and (U^*, V^*, S^*) be defined by $U^* = U_{k^*+1}$, $V^* = V_{k^*+1}$, $S^* = S_{k^*+1}$. In addition, let Y^* (resp. $\widehat{Y}^*$) denote the Lagrange multiplier Y_{k^*+1} (resp. $\widehat{Y}_{k^*+1}$) associated with (U^*, V^*, S^*), i.e., $Y^* = Y_{k^*+1}$, $\widehat{Y}^* = \widehat{Y}_{k^*+1}$, where $\widehat{Y}_{k^*+1} = Y_{k^*} + \alpha_{k^*}(A - U_{k^*+1}V_{k^*+1}^T - S_{k^*})$.

LEMMA 15.4 For the solution (U^*, V^*, S^*) generated by Algorithm 15.2, the following

$$\|\mathcal{P}_\Omega(S)\|_1 + \lambda\|V\|_* \geq \|\mathcal{P}_\Omega(S^*)\|_1 + \lambda\|V^*\|_* + \langle \widehat{Y}^* - Y^*,\ UV^T - U^*(V^*)^T \rangle - mn\varepsilon$$

holds for any feasible solution $(U,\ V,\ S)$ to (15.10), where ε is the stopping tolerance.

To reach the global optimality of (15.10) based on the above lemma, it is required to show that the term $\langle \widehat{Y}^* - Y^*,\ UV^T - U^*(V^*)^T \rangle$ diminishes. Since

$$\|Y^* - \widehat{Y}^*\|_2 \leq \sqrt{mn}\|Y^* - \widehat{Y}^*\|_\infty$$

and by Lemma 13 in [44], we have

$$\|Y^* - \widehat{Y}^*\|_\infty = \|\mathcal{P}_\Omega(Y^*) - \widehat{Y}^*\|_\infty \leq \|\mathcal{P}_\Omega(Y^*)\|_\infty + \|\widehat{Y}^*\|_\infty \leq 1 + \lambda,$$

which means that $\|Y^* - \widehat{Y}^*\|_\infty$ is bounded. By setting the parameter ρ to be relatively small as in [24], $\|Y^* - \widehat{Y}^*\|_\infty$ is small, which means that $\|Y^* - \widehat{Y}^*\|_2$ is also small. Let $\varepsilon_1 = \|Y^* - \widehat{Y}^*\|_2$, then we have the following theorems [44].

THEOREM 15.5 *Let f^g be the globally optimal objective function value of (15.10), and $f^* = \|\mathcal{P}_\Omega(S^*)\|_1 + \lambda\|V^*\|_*$ be the objective function value generated by Algorithm 15.2. We have*

$$f^* \leq f^g + c_1\varepsilon_1 + mn\varepsilon$$

where c_1 is a constant defined by

$$c_1 = \frac{mn}{\lambda}\|\mathcal{P}_\Omega(A)\|_F\left(\frac{\rho(1+\rho)}{\rho - 1} + \frac{1}{2\rho^{k^*}}\right) + \frac{\|\mathcal{P}_\Omega(A)\|_1}{\lambda}.$$

THEOREM 15.6 *Suppose that* (L^0, S^0) *is an optimal solution to the RMC problem (15.5),* $\text{rank}(L^0) = r$ *and* $f^0 = \|\mathcal{P}_\Omega(S^0)\|_1 + \lambda\|L^0\|_*$*. Let* $f^* = \|\mathcal{P}_\Omega(S^*)\|_1 + \lambda\|U^*(V^*)^T\|_*$ *be the objective function value generated by Algorithm 15.2 with* $d > 0$*. We have*

$$f^0 \leq f^* \leq f^0 + c_1\varepsilon_1 + mn\varepsilon + (\sqrt{mn} - \lambda)\sigma_{d+1} \max(r - d,\, 0)$$

where $\sigma_1 \geq \sigma_2 \geq \ldots$ *are the singular values of* L^0*.*

Since the rank parameter d is set to be higher than the rank of the optimal solution to (15.5), i.e., $d \geq r$, Theorem 15.6 directly concludes that

$$f^0 \leq f^* \leq f^0 + c_1\varepsilon_1 + mn\varepsilon.$$

Moreover, the value of ε can be set to be arbitrarily small, and the second term involving ε_1 diminishes. Hence, for the solution (U^*, V^*, S^*) generated by Algorithm 15.2, a solution to the RMC problem (15.5) can be achieved by computing $L^* = U^*(V^*)^T$.

15.5.3 Complexity Analysis

We also discuss the time complexity of the proposed algorithms . For solving (15.7), the main running time of Algorithm 15.2 is consumed by performing SVD on the small matrix of size $n \times d$, the QR decomposition of $P_k V_k$, and some matrix multiplications. The total time complexity of Algorithm 15.2 and Algorithm 15.3 for solving (15.7) and (15.5) is $O(t(d^2m + mnd))$ and $O(tmn^2)$, respectively, where $d \ll n \leq m$ and t is the number of iterations.

15.6 Extensions and Related Work

As the RBF framework introduced for RMC and CPCP is general, there are many possible extensions of our methodology. In this section, we outline novel results and methodology for two extensions we consider most important: LRMC and LRR. Our intent here is not to fully describe each development, but instead to give the reader enough details to understand and use each of these extensions.

15.6.1 Matrix Completion

By introducing an auxiliary variable L, the LRMC problem (15.4) can be written into the following form,

$$\min_{U,V,L} \frac{1}{2}\|\mathcal{P}_\Omega(A) - \mathcal{P}_\Omega(L)\|_F^2 + \lambda\|V\|_*, \text{ s.t., } L = UV^T,\, U^TU = I. \tag{15.26}$$

Similar to Algorithm 15.2, we can develop an efficient ADMM algorithm to solve (15.26). This algorithm can also be easily used to solve the low-rank MF problem, where all entries of the given matrix are observed.

15.6.2 Low-Rank Representation

For the observed matrix $A \in \mathbb{R}^{m\times n}$ drawn from a union of multiple subspaces, the LRR model for multiple subspace clustering such as motive segmentation [23] is given by

$$\min_{L,S} \|S\|_{2,1} + \lambda\|L\|_*, \text{ s.t., } A = AL + S \tag{15.27}$$

where $\|S\|_{2,1} = \sum_{j=1}^{n}\sqrt{\sum_{i=1}^{m} s_{ij}^2}$ is the $l_{2,1}$-norm of S. We also extend the proposed RBF framework to solve the LRR problem (15.27), and formulate it as follows:

$$\min_{U,V,S} \|S\|_{2,1} + \lambda\|V\|_*, \;\; \text{s.t.}, \; A = A(UV^T) + S, \; U^TU = I. \tag{15.28}$$

Similar to Algorithm 15.2, we can propose an efficient ADMM algorithm for solving the LRR problem (15.28). Furthermore, the main difference with Algorithm 15.2 is that the l_1-norm penalty of S is replaced by the $l_{2,1}$-norm. Thus, the updating for S can be efficiently computed via the following lemma [54].

LEMMA 15.5 For any given matrix $M \in \mathbb{R}^{m\times n}$ and $\tau > 0$, the unique closed-form solution to the following problem

$$\min_{S} \tau\|S\|_{2,1} + \frac{1}{2}\|S - M\|_F^2$$

is S^*, then the i-th column of S^* is

$$[S^*]_{:,i} = \begin{cases} \frac{\|[M]_{:,i}\|_2 - \tau}{\|[M]_{:,i}\|_2}[M]_{:,i}, & \text{if } \|[M]_{:,i}\|_2 > \tau; \\ 0, & \text{otherwise.} \end{cases}$$

After solving the problem (15.28), we use the low-rank representation $L^* = U^*(V^*)^T$ to define the affinity matrix of an undirected graph as follows: $W = |L^*| + |(L^*)^T|$. Following [23], we could use the spectral clustering algorithms such as Normalized Cuts [46] to produce the final clustering results.

15.6.3 Connections to Existing Approaches

According to the discussion in Section 15.5, it is clear that the RBF method is a scalable method for both RMC and CPCP problems. Compared with those convex optimization algorithms such as RMC [43] and CPCP [51] methods, which have a computational complexity of $O(mn^2)$ and are impractical for solving relatively large-scale problems, the RBF method has a linear complexity and scales well to handle large-scale problems.

To better understand the superiority of the RBF method, we compare and relate RBF with popular robust MF methods. The model in [33, 45, 58] is formulated as

$$\min_{U,V} \|W \odot (A - UV^T)\|_1 \tag{15.29}$$

where $\odot$ denotes the Hadamard product, and $W \in \mathbb{R}^{m\times n}$ is an indicator matrix that can be used to denote missing data (i.e., $w_{ij} = 0$). It is clear that (15.29) is a special case of (15.7) when $\lambda = 0$. Moreover, the models in [5, 57] focus only on the desired low-rank matrix. In this sense, they can be viewed as the special cases of (15.7). The other major difference is that SVD is used in [57], while QR factorizations are used in this chapter. The use of QR factorizations also makes the update operation highly scalable on modern parallel architectures [2]. Regarding the complexity, it is clear that both schemes have similar computational complexity. However, from the experimental results in Section 15.6, we can see that the proposed algorithm usually runs much faster, but is more accurate than the methods in [5, 57]. The following bilinear spectral regularized MF formulation in [5] is one of the most similar models to (15.7),

$$\min_{L,U,V} \|W \odot (A - L)\|_1 + \frac{\lambda}{2}(\|U\|_F^2 + \|V\|_F^2), \quad \text{s.t.}, \; L = UV^T.$$

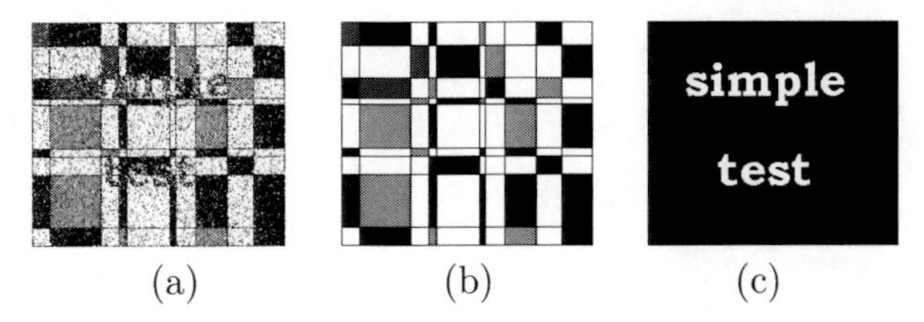

(a) (b) (c)

FIGURE 15.1 Images used in text removal: (a) Input image; (b) Original image; (c) Outlier mask (©(2015) Elsevier).

15.7 Experimental Results

In this section, we apply the proposed algorithms to solve RMC and CPCP problems such as text removal, background modeling, face reconstruction, collaborative filtering, and subspace clustering. We ran experiments on an Intel(R) Core (TM) i5-4570 (3.20 GHz) PC running Windows 7 with 8GB main memory.

15.7.1 Text Removal

We first conduct an experiment by considering a simulated task on artificially generated data, whose goal is to remove some generated text from an image. The ground-truth image is of size 256×222 with rank equal to 10 for the data matrix. We then add to the image a short phrase in text form, which plays the role of outliers. Figure 15.1 shows the image together with the clean image and outliers mask. For fairness, we set the rank of all the algorithms to 20, which is twice the true rank of the underlying matrix. The input data are generated by setting 30% of the randomly selected pixels of the image as missing entries. We compare the RBF method with the state-of-the-art methods, including PCP [7], SpaRCS[41] [49], RegL1[42] [57], BF-ALM [5], and RMC [43] (i.e., Algorithm 15.3). We set the regularization parameter $\lambda = \sqrt{\max(m, n)}$ for RegL1, RMC, and RBF, and the stopping tolerance $\varepsilon = 10^{-4}$ for all algorithms in this section.

The results obtained by different methods are visually shown in Figure 15.2, where the outlier detection accuracy (the score Area Under the receiver operating characteristic Curve, AUC) and the Relative Standard Error (RSE) of low-rank component recovery (i.e., RSE $:= \|A - L\|_F / \|A\|_F$, where A and L denote the ground-truth image matrix and the recovered image matrix, respectively) are also presented. As far as low-rank matrix recovery is concerned, these RMC methods outperform PCP, not only visually but also quantitatively. For outlier detection, it can be seen that the RBF method significantly performs better than the other methods. In short, RBF significantly outperforms the other methods in terms of both low-rank matrix recovery and spare outlier identification. Moreover, all the compared methods are about three to fourty times slower than RBF.

15.7.2 Background Modeling

In this part, we test the RBF method on real surveillance videos for object detection and background subtraction. A video sequence satisfies the low-rank and sparse structures, because the background of all the frames is controlled by few factors and hence exhibits

[41] http://www.ece.rice.edu/ aew2/sparcs.html
[42] https://sites.google.com/site/yinqiangzheng/

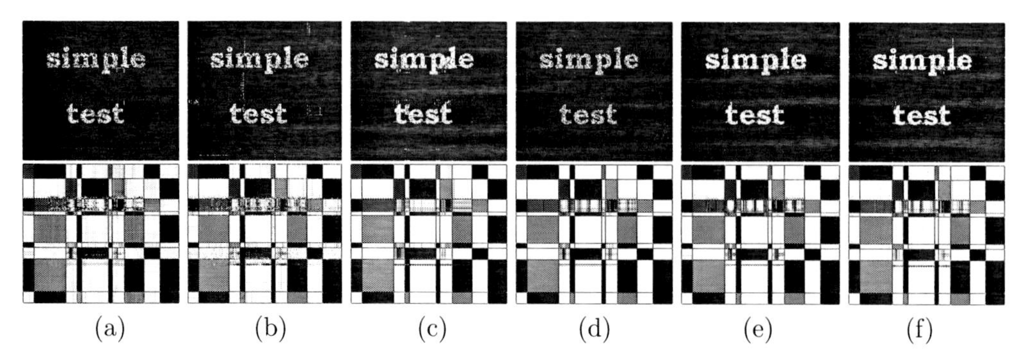

FIGURE 15.2 Text removal results. The first row shows the foreground masks and the second row shows the recovered background images: (a) PCP (AUC: 0.8558; RSE: 0.2516); (b) SpaRCS (AUC: 0.8665; RSE: 0.2416); (c) RegL1 (AUC: 0.8792; RSE: 0.2291); (d) BF-ALM (AUC: 0.8568; RSE: 0.2435); (e) RMC (AUC: 0.9206; RSE: 0.1987); (f) RBF (AUC: 0.9227; RSE: 0.1844).

TABLE 15.1 Comparison of time costs in CPU seconds of GRASTA, RegL1, RMC and RBF on background modeling data sets.

Datasets	Sizes	GRASTA	RegL1	RMC	RBF
Bootstrap	$57,600 \times 400$	153.65	93.17	344.13	36.79
Lobby	$61,440 \times 400$	187.43	139.83	390.78	49.62
Hall	$76,032 \times 400$	315.11	153.45	461.48	65.18
Mall	$245,760 \times 200$	493.92	–	–	93.35

low-rank property, and the foreground is detected by identifying spatially localized sparse residuals [7, 52]. We test the RBF method on four color surveillance videos: Bootstrap, Lobby, Hall, and Mall databases.[43] The data matrix A consists of the first 400 frames of size 144×176. Since all the original videos have colors, we first reshape every frame of the video into a long column vector and then collect all the columns into a data matrix A with size of 76032×400. Moreover, the input data is generated by setting 10% of the randomly selected pixels of each frame as missing entries.

Figure 15.3 illustrates the background extraction results on the Bootstrap data set, where the first and fourth columns represent the input images with missing data, the second and fifth columns show the low-rank recoveries, and the third and sixth columns show the sparse components. It is clear that the background can be effectively extracted by RBF, RMC [43], RegL1, and GRASTA[44] [15]. Moreover, we can see that the decomposition results of RBF, especially the recovered low-rank components, are slightly better than that of the other methods. We also report the running time of these methods in Table 15.1, from which we can see that RBF is more than two to seven times faster than the other methods. This further shows that the RBF method has very good scalability and can address large-scale problems.

15.7.3 Face Reconstruction

We also test the RBF method for the face reconstruction problems with the incomplete and corrupted face data or a small set of linear measurements y as in [51], respectively. The face database used here is a part of Extended Yale Face Database B [18] with the large corruptions. The face images can often be decomposed as a low-rank part, capturing the face

[43]http://perception.i2r.a-star.edu.sg/bkmodel/bkindex

[44]https://sites.google.com/site/hejunzz/grasta

FIGURE 15.3 Background extraction results of different algorithms on the Bootstrap data set, where the first, second, third, and fourth rows show the recovered low-rank and sparse images by GRASTA, RegL1, RMC, and RBF, respectively.

appearances under different illuminations, and a sparse component, representing varying illumination conditions and heavy "shadows." The resolution of all images is 192×168 and the pixel values are normalized to [0, 1], then the pixel values are used to form data vectors of dimension 32,256. The input data are generated by setting 40% of the randomly selected pixels of each image as missing entries.

Figure 15.4 shows some original and reconstructed images by RBF, PCP, RegL1, RMC, and CWM[45] [33], where the average computational time of all these algorithms on each person's face is presented. It can be observed that RBF performs better than the other methods not only visually but also in terms of running time, and effectively eliminates the heavy noise and "shadows" and simultaneously completes the missing entries. In other words, RBF can achieve the latent features underlying the original images regardless of the observed data corrupted by outliers or missing values.

Moreover, we implement a challenging problem to recover face images from incomplete linear measurements. Considering the computational burden of the projection operator $\mathcal{P}_Q$, we resize the original images into 42×48 and normalize the raw pixel values to form data vectors of dimension 2016. Following [51], the input data is $\mathcal{P}_Q(A)$, where Q is a subspace generated randomly with the dimension $0.75mn$.

Figure 15.5 illustrates some reconstructed images by CPCP [51] and RBF, respectively. It is clear that both CPCP and RBF effectively remove "shadows" from face images and simultaneously successfully recover both low-rank and sparse components from the reduced measurements.

15.7.4 Collaborative Filtering

Some LRMC experiments are conducted on three widely used recommendation system data sets: MovieLens100K with 100K ratings, MovieLens1M with 1M ratings, and MovieLens10M

[45]http://www4.comp.polyu.edu.hk/ cslzhang/papers.htm

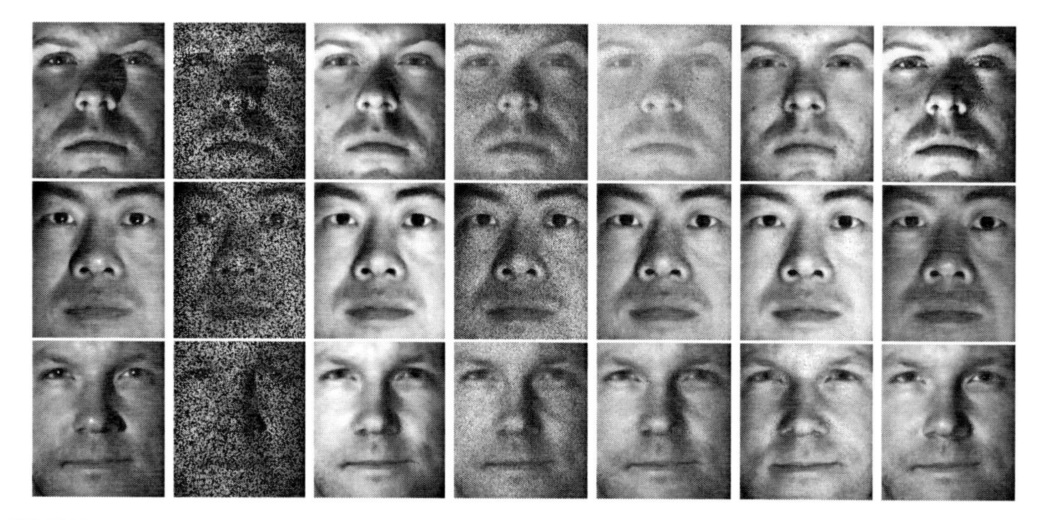

FIGURE 15.4 Face recovery results by these algorithms. From left column to right column: Original images, input corrupted images (black pixels denote missing entries), reconstruction results by PCP (982.56s), CWM (1545.38s), RegL1 (2450.94s), RMC (56.21s), and RBF (23.07s), respectively.

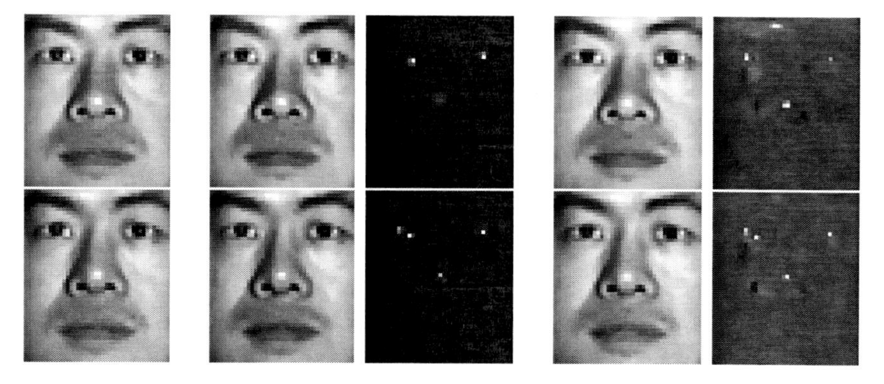

FIGURE 15.5 Face reconstruction results by CPCP and RBF, where the first column shows the original images, the second and third columns show the low-rank and sparse components obtained by CPCP, while the last two columns show the low-rank and sparse components obtained by RBF (©(2015) Elsevier).

with 10M ratings. We randomly split these data sets into training and testing sets such that the ratio of the training set to testing set is 9:1. We compare the RBF method with RMC, APG[46] [48], Soft-Impute[47] [32], LMaFit[48] [50], and three state-of-the-art manifold optimization methods: OptSpace[49] [16], ScGrass[50] [34], and RTRMC[51] [3]. We use the Root

[46] http://www.math.nus.edu.sg/ mattohkc/NNLS.html
[47] http://www.stat.columbia.edu/ rahulm/software.html
[48] http://lmafit.blogs.rice.edu/
[49] http://web.engr.illinois.edu/ swoh/software/optspace/
[50] http://www-users.cs.umn.edu/ thango/
[51] http://perso.uclouvain.be/nicolas.boumal/RTRMC/

TABLE 15.2 RMSE of different methods on three data sets: MovieLens100K, MovieLens1M and MovieLens10M.

Methods	MovieLens100K			MovieLens1M			MovieLens10M		
APG	1.2142			1.1528			0.8583		
Soft-Impute	1.0489			0.9058			0.8615		
RMC	1.2361			1.0149			0.8755		
Ranks	5	6	7	5	6	7	5	6	7
OptSpace	0.9490	0.9398	0.9354	0.9182	0.9061	0.9081	1.1352	1.1265	1.1194
ScGrass	0.9647	0.9809	0.9945	0.8847	0.8852	0.8936	0.8359	0.8290	0.8247
RTRMC	0.9837	1.0617	1.1642	0.8875	0.8893	0.8960	0.8463	0.8442	0.8386
LMaFit	0.9468	0.9540	0.9568	0.8918	0.8920	0.8853	0.8576	0.8530	0.8423
RBF	**0.9393**	**0.9513**	**0.9485**	**0.8672**	**0.8624**	**0.8591**	**0.8193**	**0.8159**	**0.8110**

Mean Squared Error (RMSE) as the evaluation measure, which is defined as

$$\text{RMSE} = \sqrt{\frac{1}{|\Omega|}\Sigma_{(i,j)\in\Omega}(A_{ij} - L_{ij})^2}$$

where $|\Omega|$ is the total number of ratings in the testing set, A_{ij} denotes the ground-truth rating of user i for item j, and L_{ij} denotes the corresponding predicted rating.

The average RMSE on these three data sets is reported over ten independent runs and is shown in Table 15.2, from which we can see that, for some fixed ranks, most MF methods including ScGrass, RTRMC, LMaFit, and RBF, but not OptSpace, usually perform better than the three convex trace norm-minimization methods, APG, Soft-Impute, and RMC. Moreover, the RBF method with trace norm regularization consistently outperforms the other MF methods and the three convex trace norm minimization methods. This confirms that the RBF model with trace norm regularization is reasonable.

15.7.5 Subspace Clustering

Following the setting in [23], we first construct 5 independent subspaces $\{S_i\}_{i=1}^5$ whose bases $\{B_i\}_{i=1}^5$ are generated by $B_{i+1} = TB_i$, $1 \leq i \leq 4$, where T represents a random rotation and B_1 is a random column orthogonal matrix of size 100×4. Therefore, each subspace is a 4-dimensional subspace. We create a data matrix $A = [A_1, \cdots, A_5]$ by sampling 40 data vectors (or more) from each subspace by $A_i = B_iC_i$, $1 \leq i \leq 5$ with C_i being a 4×40 i.i.d. standard Gaussian matrix. Then the size of the low-rank representation matrix L is 200×200. Next, we randomly select some data vectors to be corrupted by using relatively large Gaussian noise with zeros mean and deviation $0.2\|A\|_F$. In the implementation, we set $\lambda = 0.12$ for the RBF method, PSD [35], LRR[52] [23], and LADM[53] [21] (the two latter methods utilize the MATLAB version of PROPACK [17]), and $\lambda = 0.81$ for SSC[54] [12]. The stopping tolerance for all these algorithms is set to $\varepsilon = 10^{-6}$.

We illustrate the average accuracy and average running time of SSC, LRR, PSD, LADM, and the RBF method under different settings of data points with 50 percent of corruption, as shown in Figure 15.6. It is clear that RBF is much faster than the other methods, and the running time of LRR, PSD, and LADM dramatically grows with the increasing number of data points. Moreover, the clustering accuracy of RBF is very stable, that is, it achieves consistently good performance with increasing the number of data points in each subspace, whereas the clustering accuracies of the other methods fall dramatically as the number of data points increases, especially when they are greater than 800.

[52] https://sites.google.com/site/guangcanliu/

[53] http://www.cis.pku.edu.cn/faculty/vision/zlin/zlin.htm

[54] http://www.cis.jhu.edu/ ehsan/code.htm

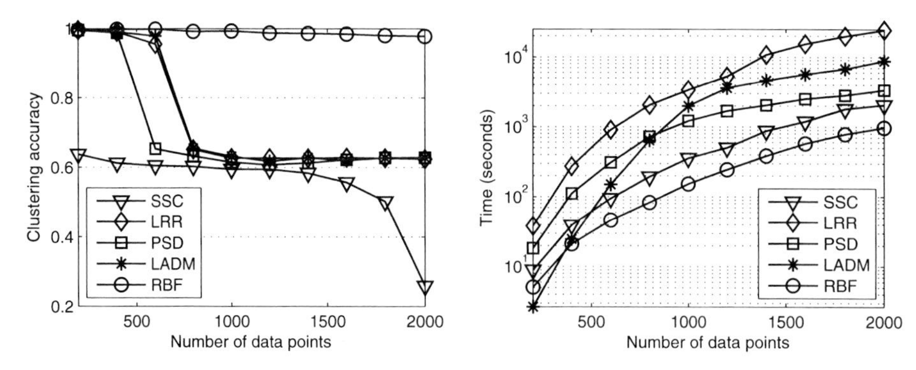

FIGURE 15.6 Comparison of clustering accuracy (left) and running time (right) for various approaches at 50 percent of corruption.

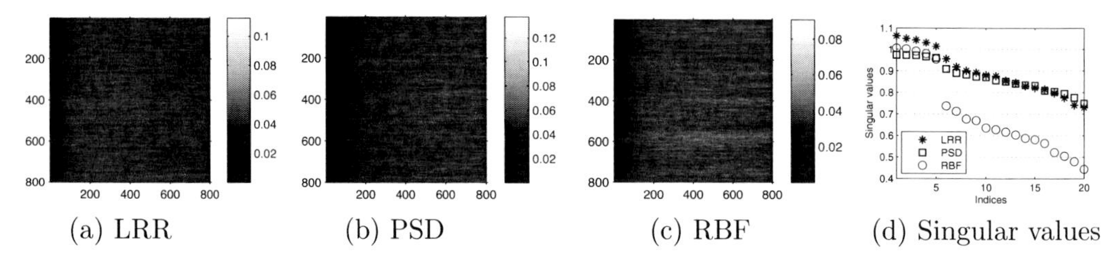

FIGURE 15.7 (See color insert.) Comparison of affinity matrices produced by LRR, PSD, and RBF on the toy data. Note that only the largest 20 singular values of each affinity matrix are shown.

To make it clearer how the RBF method exploits the low-rank structure of multiple subspace data and why it outperforms two state-of-the-art methods, LRR and PSD, we also demonstrate the comparison of affinity matrices produced by LRR, PSD, and RBF on the toy data set of size 100×800 with 50 percent of corruption, as shown in Figure 15.7, from which it is clear that the affinity matrix learned by RBF exhibits clear block structures, meaning that each cluster becomes highly compact and different clusters become far apart, whereas the affinity matrices produced by LRR and PSD show no clear block-diagonal structure. In addition, we illustrate the distribution of the singular values of affinity matrices, as shown in Figure 15.7(d). It is clear that using the large gap in the singular values of the matrix produced by RBF, the correct number of subspaces can be determined.

15.8 Conclusion

In this chapter, a scalable RBF framework is proposed for RMC and CPCP problems. We first presented two smaller-scale matrix trace norm-minimization models for RMC and CPCP problems, respectively. Then, we developed an efficient ADMM algorithm to solve the non-convex RMC problems, and analyzed the suboptimality of the solution produced by the proposed algorithm. Finally, we extended the algorithm to solve CPCP and convex RMC problems. Experimental results demonstrated the superior performance of RBF in comparison with the state-of-the-art methods in terms of both efficiency and effectiveness.

RBF can address various large-scale low-rank matrix recovery problems. Moreover, it can also be extended to various higher-order tensor recovery problems, such as low-rank tensor decomposition and completion [29]. For future work, we plan to explore ways to regularize the RBF model with auxiliary information as in [28, 39, 40], such as graph Laplacian, as well as with general penalties such as the non-negative constraint.

Acknowledgments

This work was supported by SHIAE Grant No. 8115048, GRF No. 411211, and CUHK direct grant Nos. 4055015, 4055043, and 4055048.

References

1. A. Agarwal, S. Negahban, and M. Wainwright. Noisy matrix decomposition via convex relaxation: Optimal rates in high dimensions. *Annals of Statistics*, 40(2):1171–1197, 2012.
2. H. Avron, S. Kale, S. Kasiviswanathan, and V. Sindhwani. Efficient and practical stochastic subgradient descent for nuclear norm regularization. In *International Conference on Machine Learning, ICML 2012*, 2012.
3. N. Boumal and P. Absil. RTRMC: A Riemannian trust-region method for low-rank matrix completion. In *Neural Information Processing Systems, NIPS 2011*, pages 406–414, 2011.
4. S. Boyd, N. Parikh, E. Chu, B. Peleato, and J. Eckstein. Distributed optimization and statistical learning via the alternating direction method of multipliers. *Foundations and Trends in Machine Learning*, 3(1):1–122, 2011.
5. R. Cabral, F. Torre, J. Costeira, and A. Bernardino. Unifying nuclear norm and bilinear factorization approaches for low-rank matrix decomposition. In *International Conference on Computer Vision, ICCV 2013*, pages 2488–2495, 2013.
6. J. Cai, E. Candès, and Z. Shen. A singular value thresholding algorithm for matrix completion. *SIAM Journal on Optimization*, 20(4):1956–1982, 2010.
7. E. Candès, X. Li, Y. Ma, and J. Wright. Robust principal component analysis? *Journal of ACM*, 58(3):1–37, 2011.
8. E. Candès and B. Recht. Exact matrix completion via convex optimization. *Foundations of Computational Mathematics*, 9(6):717–772, 2009.
9. Y. Chen, A. Jalali, S. Sanghavi, and C. Caramanis. Low-rank matrix recovery from errors and erasures. *IEEE Transactions on Information Theory*, 59(7):4324–4337, 2013.
10. I. Daubechies, M. Defrise, and C. De Mol. An iterative thresholding algorithm for linear inverse problems with a sparsity constraint. *Communications on Pure and Applied Mathematics*, 57(11):1413–1457, 2004.
11. J. Eckstein and D. Bertsekas. On the Douglas-Rachford splitting method and the proximal point algorithm for maximal monotone operators. *Mathematical Programming*, 55(1):293–318, 1992.
12. E. Elhamifar and R. Vidal. Sparse subspace clustering. In *IEEE International Conference on Computer Vision and Pattern Recognition, CVPR 2009*, pages 2790–2797, 2009.
13. P. Favaro, R. Vidal, and A. Ravichandran. A closed form solution to robust subspace estimation and clustering. In *IEEE International Conference on Computer Vision and Pattern Recognition, CVPR 2011*, pages 1801–1807, 2011.
14. M. Fazel. *Matrix Rank Minimization with Applications.* PhD thesis, Stanford University, 2002.
15. J. He, L. Balzano, and A. Szlam. Incremental gradient on the Grassmannian for online foreground and background separation in subsampled video. In *IEEE International Conference on Computer Vision and Pattern Recognition, CVPR 2012*, pages 1568–1575, 2012.
16. R. Keshavan, A. Montanari, and S. Oh. Matrix completion from a few entries. *IEEE Transactions on Information Theory*, 56(6):2980–2998, 2010.

17. R. Larsen. PROPACK-software for large and sparse SVD calculations. *Available from http://sun.stanford.edu/srmunk/PROPACK/*, 2005.
18. K. Lee, J. Ho, and D. Kriegman. Acquiring linear subspaces for face recognition under variable lighting. *IEEE Transactions on Pattern Analysis and Machine Intelligence*, 27(5):684–698, 2005.
19. X. Li. Compressed sensing and matrix completion with constant proportion of corruptions. *Journal on Constructive Approximation*, 37:73–99, 2013.
20. Z. Lin, M. Chen, L. Wu, and Y. Ma. The augmented Lagrange multiplier method for exact recovery of corrupted low-rank matrices. *University of Illinois, Urbana-Champaign*, March 2009.
21. Z. Lin, R. Liu, and Z. Su. Linearized alternating direction method with adaptive penalty for low-rank representation. In *Neural Information Processing Systems, NIPS 2011*, pages 612–620, 2011.
22. C. Liu, Z. Lin, S. Yan, J. Sun, Y. Yu, and Y. Ma. Robust recovery of subspace structures by low-rank representation. *IEEE Transactions on Pattern Analysis and Machine Intelligence*, 35(1):171–184, 2013.
23. C. Liu, Z. Lin, and Y. Yu. Robust subspace segmentation by low-rank representation. In *International Conference on Machine Learning, ICML 2010*, pages 663–670, 2010.
24. C. Liu and S. Yan. Active subspace: toward scalable low-rank learning. *Journal of Computational Neuroscience*, 24(12):3371–3394, 2012.
25. Y. Liu, L. Jiao, and F. Shang. An efficient matrix factorization based low-rank representation for subspace clustering. *Pattern Recognition*, 46(1):284–292, 2013.
26. Y. Liu, L. Jiao, and F. Shang. A fast tri-factorization method for low-rank matrix recovery and completion. *Pattern Recognition*, 46(1):163–173, 2013.
27. Y. Liu, L. Jiao, F. Shang, F. Yin, and F. Liu. An efficient matrix bi-factorization alternative optimization method for low-rank matrix recovery and completion. *Neural Networks*, 48:8–18, 2013.
28. Y. Liu, F. Shang, H. Cheng, and J. Cheng. A Grassmannian manifold algorithm for nuclear norm regularized least squares problems. In *Uncertainty in Artificial Intelligence, UAI 2014*, 2014.
29. Y. Liu, F. Shang, W. Fan, J. Cheng, and H. Cheng. Generalized higher-order orthogonal iteration for tensor decomposition and completion. In *Neural Information Processing Systems, NIPS 2014*, pages 1763–1771, 2014.
30. R. Ma, N. Barzigar, A. Roozgard, and S. Cheng. Decomposition approach for low-rank matrix completion and its applications. *IEEE Transactions on Signal Processing*, 62(7):1671–1683, 2014.
31. S. Ma, D. Goldfarb, and L. Chen. Fixed point and Bregman iterative methods for matrix rank minimization. *Mathematical Programming*, 128(1):321–353, 2011.
32. R. Mazumder, T. Hastie, and R. Tibshirani. Spectral regularization algorithms for learning large incomplete matrices. *J. Mach. Learn. Res.*, 11:2287–2322, 2010.
33. D. Meng, Z. Xu, L. Zhang, and J. Zhao. A cyclic weighted median method for l_1 low-rank matrix factorization with missing entries. In *AAAI*, pages 704–710, 2013.
34. T. Ngo and Y. Saad. Scaled gradients on Grassmann manifolds for matrix completion. In *Neural Information Processing Systems, NIPS 2012*, pages 1421–1429, 2012.
35. Y. Ni, J. Sun, X. Yuan, S. Yan, and L. Cheong. Robust low-rank subspace segmentation with semidefinite guarantees. In *IEEE International Conference on Data Mining, ICDM 2010*, pages 1179–1188, 2010.
36. H. Nick. Matrix procrustes problems. *Talk*, 1995.
37. R. Otazo, E. Candès, and D. Sodickson. Low-rank and sparse matrix decomposition for accelerated dynamic MRI with separation of background and dynamic components. *Journal of Magnetic Resonance Imaging*, 73(3):1125–1136, 2015.

38. Y. Peng, A. Ganesh, J. Wright, W. Xu, and Y. Ma. RASL: Robust alignment by sparse and low-rank decomposition for linearly correlated images. *IEEE Transactions on Pattern Analysis and Machine Intelligence*, 34(11):2233–2246, 2012.
39. F. Shang, L. Jiao, and F. Wang. Graph dual regularization non-negative matrix factorization for co-clustering. *Pattern Recognition*, 45(6):2237–2250, 2012.
40. F. Shang, L. Jiao, and F. Wang. Semi-supervised learning with mixed knowledge information. In Knowledge Discovery and Data Mining (*KDD*), pages 732–740, 2012.
41. F. Shang, Y. Liu, and J. Cheng. Generalized higher-order tensor decomposition via parallel ADMM. In *Association for the Advancement of Artificial Intelligence, AAAI 2014*, pages 1279–1285, 2014.
42. F. Shang, Y. Liu, J. Cheng, and H. Cheng. Recovering low-rank and sparse matrices via robust bilateral factorization. In *IEEE International Conference on Data Mining, ICDM 2014*, pages 965–970, 2014.
43. F. Shang, Y. Liu, J. Cheng, and H. Cheng. Robust principal component analysis with missing data. In *ACM International Conference on Information and Knowledge Management, CIKM 2014*, pages 1149–1158, 2014.
44. F. Shang, Y. Liu, H. Tong, J. Cheng, and H. Cheng. Robust bilinear factorization with missing and grossly corrupted observations. *Inform. Sciences*, 307:53–72, 2015.
45. Y. Shen, Z. Wen, and Y. Zhang. Augmented Lagrangian alternating direction method for matrix separation based on low-rank factorization. *Optimization Methods and Software*, 29(2):239–263, 2014.
46. J. Shi and J. Malik. Normalized cuts and image segmentation. *IEEE Transactions on Pattern Analysis and Machine Intelligence*, 22(8):888–905, 2000.
47. M. Tao and X. Yuan. Recovering low-rank and sparse components of matrices from incomplete and noisy observations. *SIAM Journal on Optimization*, 21(1):57–81, 2011.
48. K. Toh and S. Yun. An accelerated proximal gradient algorithm for nuclear norm regularized least squares problems. *Pacific Journal of Optimization*, 6:615–640, 2010.
49. A. Waters, A. Sankaranarayanan, and R. Baraniuk. SpaRCS: Recovering low-rank and sparse matrices from compressive measurements. In *Neural Information Processing Systems, NIPS 2011*, pages 1089–1097, 2011.
50. Z. Wen, W. Yin, and Y. Zhang. Solving a low-rank factorization model for matrix completion by a nonlinear successive over-relaxation algorithm. *Mathematical Programming Computation*, 4(4):333–361, 2012.
51. J. Wright, A. Ganesh, K. Min, and Y. Ma. Compressive principal component pursuit. *Journal on Information Inference*, 2:32–68, 2013.
52. J. Wright, A. Ganesh, S. Rao, Y. Peng, and Y. Ma. Robust principal component analysis: exact recovery of corrupted low-rank matrices by convex optimization. In *Neural Information Processing Systems, NIPS 2009*, pages 2080–2088, 2009.
53. H. Xu, C. Caramanis, and S. Sanghavi. Robust PCA via outlier pursuit. In *Neural Information Processing Systems, NIPS 2010*, pages 2496–2504, 2010.
54. J. Yang, W. Yin, Y. Zhang, and Y. Wang. A fast algorithm for edge-preserving variational multichannel image restoration. *SIAM Journal on Imaging Sciences*, 2(2):569–592, 2009.
55. J. Yang and X. Yuan. Linearized augmented Lagrangian and alternating direction methods for nuclear norm minimization. *Mathematics of Computation*, 82:301–329, 2013.
56. X. Yuan and J. Yang. Sparse and low-rank matrix decomposition via alternating direction methods. *Pacific Journal of Optimization*, 9(1):167–180, 2013.
57. Y. Zheng, C. Liu, S. Sugimoto, S. Yan, and M. Okutomi. Practical low-rank matrix approximation under robust l_1-norm. In *IEEE International Conference on Computer Vision and Pattern Recognition, CVPR 2012*, pages 1410–1417, 2012.

58. T. Zhou and D. Tao. Greedy bilateral sketch, completion & smoothing. In *International Conference on Artificial Intelligence and Statistics, AISTATS 2013*, pages 650–658, 2013.

16

Applications of Low-Rank and Sparse Matrix Decompositions in Hyperspectral Video Processing

Jen-Mei Chang
California State University, Long Beach, USA

Torin Gerhart
Western Digital Corp., USA

16.1 Introduction

This chapter provides a background to those unfamiliar with hyperspectral images and details the application of low-rank and sparse decompositions in hyperspectral video analysis. Traditional RGB images capture light in the red, green, and blue portions of the visible light spectrum. This is due to the way images are displayed on computer monitors and televisions using red, green, and blue pixels. Each layer represents the amount of radiated energy being emitted at a particular wave length. Images using more than 3 layers are referred to as multispectral or hyperspectral images. These images can involve light that is outside the visible spectrum, such as infra-red (IR) and UV (ultra-violet) light. Hyperspectral images have a higher spectral resolution compared to multispectral images while being limited to a narrow spectral bandwidth. By imaging the light that is absorbed and reflected in high detail within a certain region of the electromagnetic spectrum, it is possible to identify particular materials present in the image. One major application of hyperspectral imaging is the detection of invisible gaseous chemical agents and other anomalous harmful particles. This problem comes up in many practical applications such as defense, security, and environmental safety [1].

Herein, we will focus on the detection and identification of gaseous chemical plumes in hyperspectral video data. Hyperspectral video adds a time component to the processing of hyperspectral images. Processes such as gaseous diffusion can be captured, even for gases not visible to the human eye. These video sequences are typically large in size due to the fact that the images themselves are of high resolution. Dimension reduction is often needed to make further processing easier. An algorithm to decompose a hyperspectral video sequence into a low-rank and sparse representation will be derived and applied to the detection of chemical plumes. The particular data set used in this work was provided by the Applied Physics Laboratory at Johns Hopkins University as part of a Defense Threat Reduction Agency (DTRA) research grant (BE-22504).

16.2 Hyperspectral Imaging

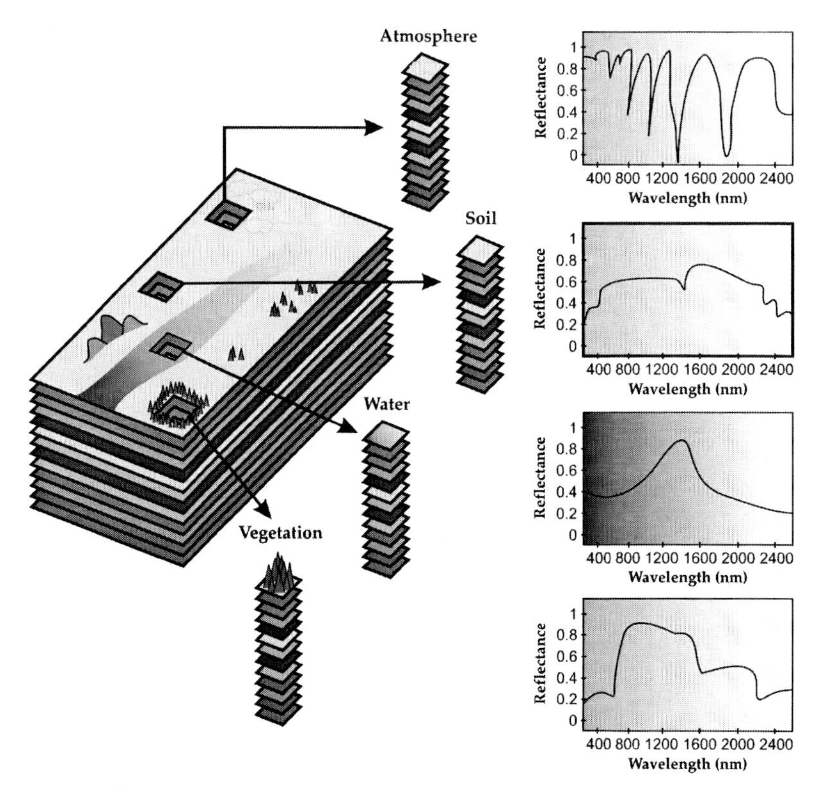

FIGURE 16.1 (See color insert.) Example of a hyperspectral image. The image contains many layers that capture different wavelengths of reflected light. Each pixel contains information about the materials present in the scene.

Traditional RGB images consist of three layers, each corresponding to intensity values of the red, green, and blue components of the image. Hyperspectral images extend this idea by capturing hundreds of different frequencies of light. Hyperspectral imaging aims to capture the absorbed and reflected radiation of a given scene such that a region of the electromagnetic spectrum may be viewed with high spectral resolution. This spectral resolution is what differentiates hyperspectral imaging from other types of spectral imaging, such as multispectral imaging. In order to produce an accurate representation of a region of the electromagnetic spectrum, a fine discretization is required. Hyperspectral images have a frequency gap between spectral bands that is small relative to the measured spectrum.

Hyperspectral imaging finds a lot of application in the geosciences. Measuring the absorbed and reflected radiation allows for the characterization of surface and atmospheric features in geophysical images. Figure 16.1©(2012) SPIE shows how pixels in a hyperspectral image contain spectral information that may be used to distinguish between vegetation, water, soil, and atmosphere. The most widespread application of hyperspectral imaging is the identification of particular materials present in a scene.

Hyperspectral video sequences serve as an excellent example of a problem from modern imaging science that is closely tied to the analysis of large data sets. Each hyperspectral video frame is an individual data cube, which means these video sequences require a large amount of memory. For example, a data cube with a spatial resolution of 150×300 pixels and a spectral resolution of 150 would require 54 megabytes of physical memory (assuming it is stored as 8-byte double precision floating point values). Twenty frames alone would require over a gigabyte in the memory space. The size of these videos may prohibit loading the entire video into memory without sufficient preprocessing or dimension reduction.

Hyperspectral imaging was originally developed to be used for geology and mining applications. The imaging of core samples would detect the presence of particular mineral deposits and could aid in finding oil [5]. Today, hyperspectral imaging is used in many different areas from surveillance and defense to ecology and agriculture.

16.2.1 Mixing Model

A *three-layer mixing model* is a simple method to describe the different components that comprise the spectral radiance measurement for each pixel in a long-wave infrared hyperspectral image. Figure 16.2 ©(2011) JHU illustrates the different objects, or layers, that contribute to the spectral radiance measurement of the long-wave infrared (LWIR) spectrometer. The three layers in this data set are the background, chemical plume, and atmosphere. Each pixel has its own radiance $L(\nu)$, and transmittance $\tau(\nu)$. The transmittance is the ratio of light leaving a surface relative to the amount of light entering the medium. Both the background and plume spectral radiances must pass through other mediums before reaching the long wave infrared spectrometer. Therefore, the spectral radiance measurement of the sensor can be represented as:

$$L(\nu) = \tau_{atm}(\nu)L_p(\nu) + \tau_p(\nu)\tau_{atm}(\nu)L_b(\nu) + L_{atm}(\nu). \qquad (16.1)$$

The subscripts *atm*, *p*, and *b* in Equation (16.1) refer to the atmosphere, plume, and background, respectively. This model may be modified slightly for our application. In the data set we are using, the long-wave infrared spectrometers are placed within two kilometers of the release site. It is reasonable to assume that the contribution to spectral radiance from the atmosphere would be negligible, therefore the L_{atm} term from Equation (16.1) can be dropped for simplicity. In addition, it is assumed that the atmospheric transmittance does not significantly affect the spectral radiance because of the short path length, allowing most of the signal to pass through. These assumptions reduce Equation (16.1) into a simpler *two layer model* equation

$$L(\nu) = \tau_p(\nu)L_b(\nu) + L_p(\nu). \qquad (16.2)$$

According to this model the spectral radiance of the scene measured by the LWIR sensors is a sum of the light emitted by the chemical plume and background mediums at the measured wavelengths in the electromagnetic spectrum [1].

16.2.2 Dugway Proving Ground Data Set

The data set analyzed for this work was provided by the Applied Physics Laboratory at Johns Hopkins University as part of a Defense Threat Reduction Agency (DTRA) research grant. It consists of a series of video sequences recording the release of non-toxic chemical plumes into the atmosphere at the US Army's Dugway Proving Ground facility in Utah. Figure 16.3 ©(2011) JHU shows the three long-wave infrared spectrometers (named Romeo, Victory, and Tango) placed at different locations to track the release of known chemicals. The sensors capture one frame every five seconds, consisting of two spatial dimensions and one spectral dimension. The spatial dimension of each of these data cubes is 128 ×

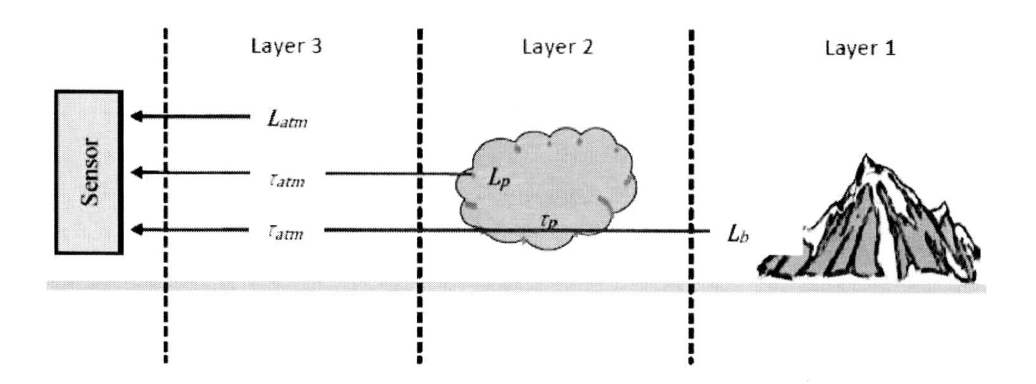

FIGURE 16.2 A three-layer mixing model depicting the composition of spectral radiance in hyperspectral images.

320 pixels, while the spectral dimension measures 129 different wavelengths in the long-wave infrared (LWIR) portion of the electromagnetic spectrum. Each layer in the spectral dimension depicts a particular frequency starting at 7,830 nm and ending with 11,700 nm, with a channel spacing of 30 nm. Low-rank and sparse decompositions of this data set will provide an estimation of the background radiation, i.e., the low-rank component, and the motion captured between frames, i.e., the sparse component.

16.3 Principal Component Pursuit

This section outlines the problem of decomposing a hyperspectral video sequence into low rank and sparse matrices. A data matrix X is constructed through concatenation of video frames and decomposed into $X = L + S$, where L is a low-rank matrix and S is a sparse matrix. The algorithm used to arrive at this decomposition is iterative and involves minimizing a functional consisting of the nuclear norm and the ℓ_1 norm defined on matrices. The version of the ℓ_1 norm we are using is not the typical 1-norm for matrices. It is a slightly modified version of the ℓ_1 vector norm, extended to be defined on matrices, where a matrix is seen as a long vector. This norm is used here as a sparsity-inducing norm, intended to give sparse solutions. The nuclear norm is used as an approximation to the *rank* function. A minimal nuclear norm implies low rank. By minimizing this functional, we are guaranteed solutions that are low rank and sparse. Formally stated, the problem is

$$\begin{aligned} &\underset{L,S}{\text{minimize}} \ \|L\|_\star + \lambda \|S\|_1 \\ &\text{subject to } X = L + S \end{aligned} \tag{16.3}$$

where

$$\begin{aligned} \|X\|_1 &= \sum_{i,j} |X_{i,j}| \ \ (\ell_1 \text{ norm}) \\ \|X\|_\star &= \sum_{i=1}^{r} \sigma_i(X) \text{ (nuclear norm).} \end{aligned} \tag{16.4}$$

Here, $\sigma_i(X)$'s are the singular values of X and r is the rank of X. The objective function of (16.3) has a separable structure associated with it, which allows the problem to be split

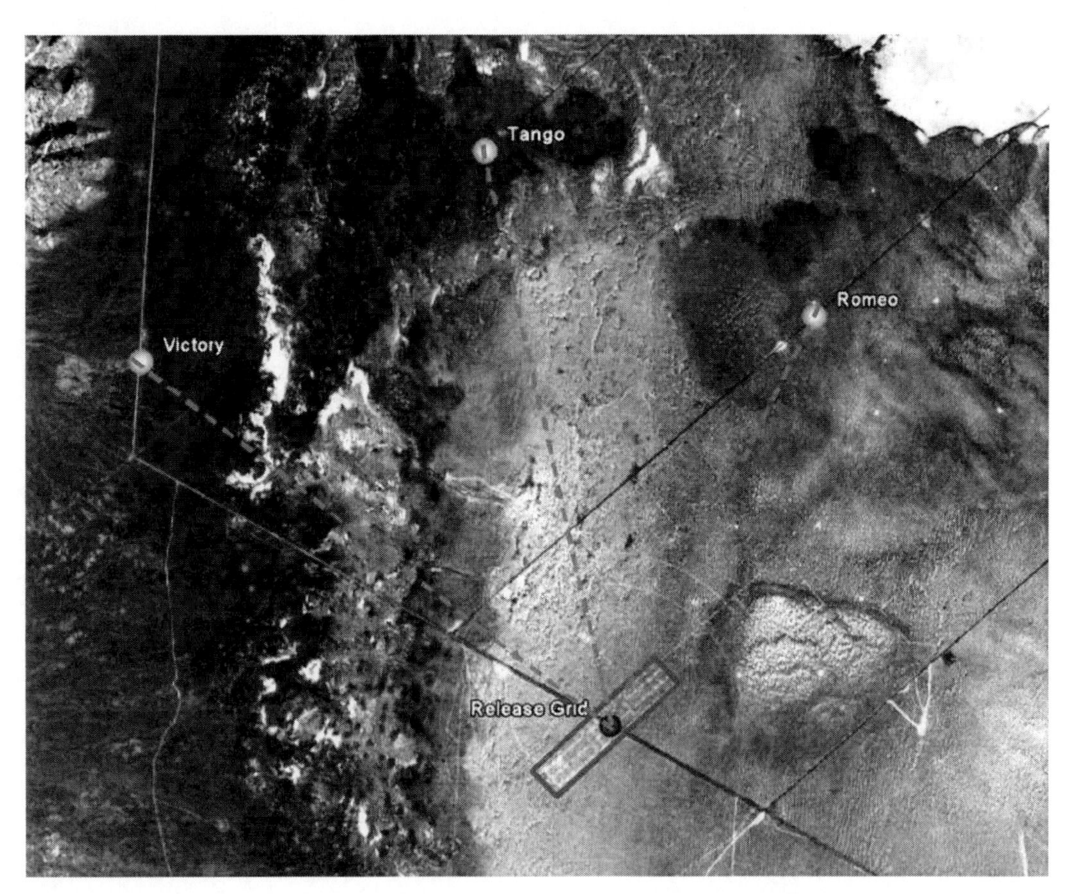

FIGURE 16.3 (See color insert.) Placement of the three long-wave infrared spectrometers to track the release of the known chemicals.

into separate subproblems. The augmented Lagrangian [2] of (16.3) gives

$$\mathcal{L}_A(L, S, Y) = \|L\|_\star + \lambda\|S\|_1 + \langle Y, X - L - S\rangle + \frac{\mu}{2}\|X - L - S\|_F^2. \tag{16.5}$$

Solutions are derived from (16.5) by using the alternating direction method of multipliers (ADMM). One subproblem holds S fixed and minimizes $\mathcal{L}_A(L, S, Y)$ with respect to L; the next subproblem holds the new L fixed and minimizes $\mathcal{L}_A(L, S, Y)$ with respect to S. The subproblem involving the minimization of the nuclear norm is related to the minimization of the *rank* function. The rank function is non-convex, and finding minimizers is NP-hard. Replacing the non-convex *rank* function with a convex approximation makes the problem much easier to solve. The nuclear norm is the "best" convex approximation to the *rank* function [4]. The solution to the subproblem involving minimization of the nuclear norm is given by the singular value thresholding operator, $\mathcal{D}_\tau$. A more detailed derivation may be found in [3]. The other subproblem involving the $\|S\|_1$ term is essentially minimizing the ℓ_1 vector norm and is therefore solved by a thresholding operator, S_τ. The iterations obtained from (16.5) are

$$\begin{aligned}
L_{k+1} &= \underset{L}{\operatorname{argmin}}\ \mathcal{L}_A(L_k, S_k, Y_k) = \mathcal{D}_\tau(X - S_k + \mu^{-1}Y_k) \\
S_{k+1} &= \underset{S}{\operatorname{argmin}}\ \mathcal{L}_A(L_{k+1}, S_k, Y_k) = \mathcal{S}_{\frac{\lambda}{\mu}}(X - L_{k+1} + \frac{1}{\mu}Y_k) \\
Y_{k+1} &= Y_k + \mu(X - L_{k+1} - S_{k+1})
\end{aligned} \tag{16.6}$$

where

$$\mathcal{D}_\tau(X) = U\mathcal{S}_\tau(\Sigma)V^T = U\text{diag}(\max(\sigma_i - \tau, 0))V^T$$
$$\mathcal{S}_\tau(X) = \max(0, x_{ij} - \tau).$$

When applied to video analysis, the low-rank matrix of the final decomposition represents the background, and the sparse matrix represents motion between frames. The algorithm requires a number of other matrices with the same size as X to be held in memory concurrently. The memory requirement of this problem is typically where challenges arise in practice. For example, each frame of the Dugway Proving Ground data set is a 128×320×129 data cube. Concatenating along the spectral dimension produces a vector of length 5,283,840. A data matrix with 100 frames will be of size 5,283,840 × 100. There are many preprocessing techniques that can be utilized to make the task computationally feasible. For example, one can select a subset of the spectral channels based on noise or performing dimension reduction on each frame of the video sequence.

16.4 Results

The Low-Rank Sparse algorithm outlined in (16.6) was applied to a false color RGB video, created by projecting each frame onto the first three principle components. The movie generated for these results was done in RGB to demonstrate different aspects of the decomposition. Movies with more layers were processed and the results are similar to what is presented here. In early frames, the low-rank approximation is able to capture the background very well. After the plume is released, the sparse component captures the movement of the plume through each channel of the video sequence. Applying this method to the original (non-reduced) video sequence results in the background matrix approximating stationary signals and the sparse component showing moving signals and noise. Each frame contained 128 × 320 pixels, with 3 layers. By concatenating the frames into long vectors in $\mathbb{R}^{128\times320\times3}$, a movie of 40 frames was made into the $122,880 \times 40$ matrix M. The original matrix had rank 40, while the low-rank approximation L had rank 12.

Motion was captured on each RGB layer in the sparse component S. The red component of the resulting decomposition may be seen in Figure 16.4. Notice that the frames of the low-rank decomposition are virtually identical and the noise may be seen in the sparse component in the first 20 frames, as illustrated in Figure 16.4(a). On the other hand, Figure 16.4(b) shows the release of the plume in frames 21 through 40. Its motion is captured throughout the sparse component with some anomalies appearing in the low-rank component towards the end of the sequence in areas where the plume has traveled. Similar results in each of the green and blue components of the resulting decomposition may be seen in Figures 16.5 and 16.6, respectively. The final resulting RGB movies may be seen in Figure 16.7. The lighting fluctuations and noise can be seen in the sparse component of 16.7(a). As the gas plume is released, the diffusion process can be observed in the sparse component, as evidenced in 16.7(b). There, a noticeable presence of the plume can also be observed in the low-rank component.

16.5 Conclusion

The results presented here showed that it is possible to isolate the motion of chemical plumes from hyperspectral video data. This was achieved by using a reformulation of the principal components analysis to incorporate sparsity. The estimation of the background was done using a low-rank approximation. There are other means of background estimation such as a non-negative matrix factorization technique, which takes into account other physical

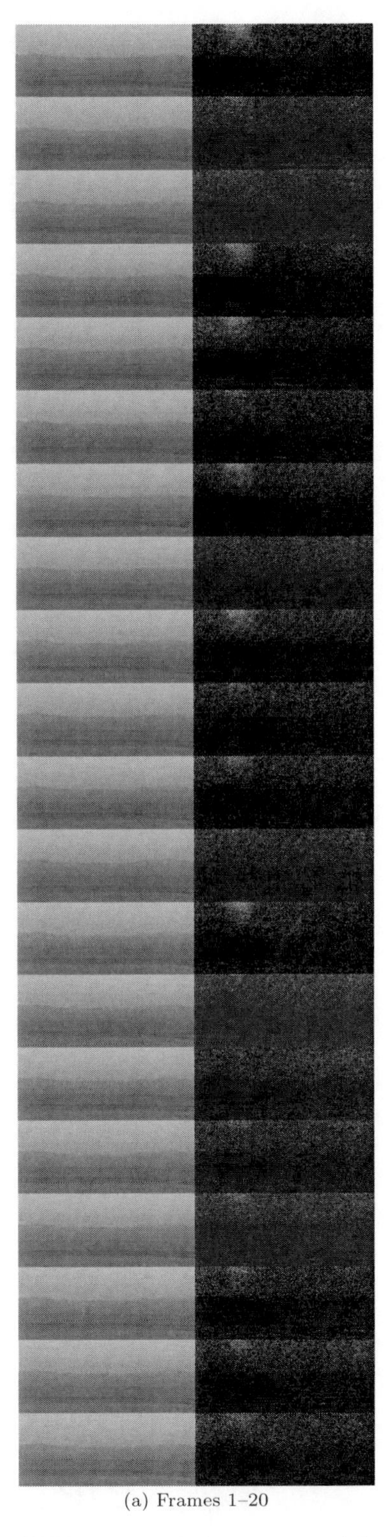
(a) Frames 1–20

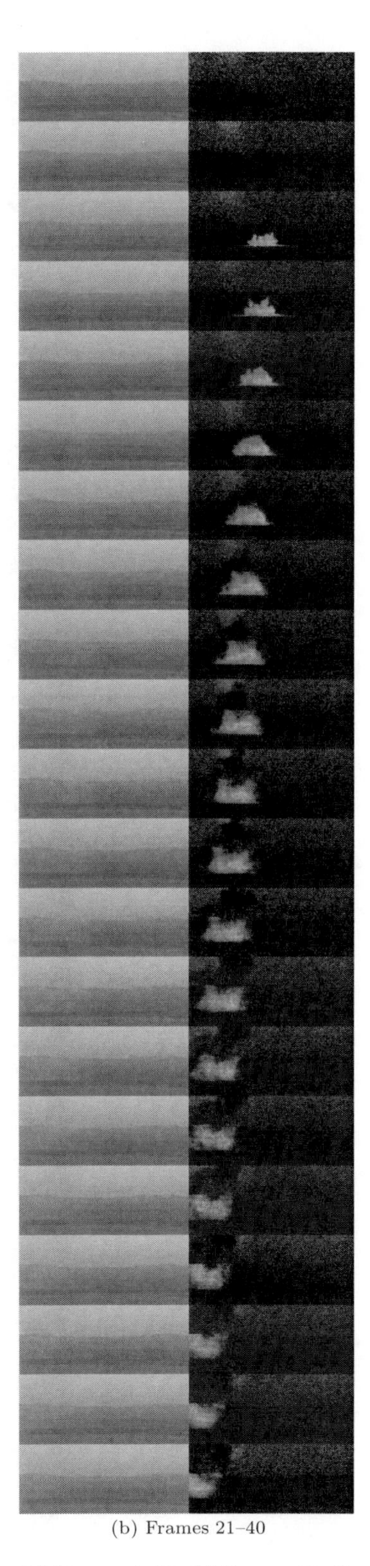
(b) Frames 21–40

FIGURE 16.4 Red component of the decomposed video sequence.(a) Frames 1–20; (b) Frames 21–40 of the red component of the decomposed video sequence. The left panel displays the low-rank approximation and the right panel shows the sparse component within each strip.

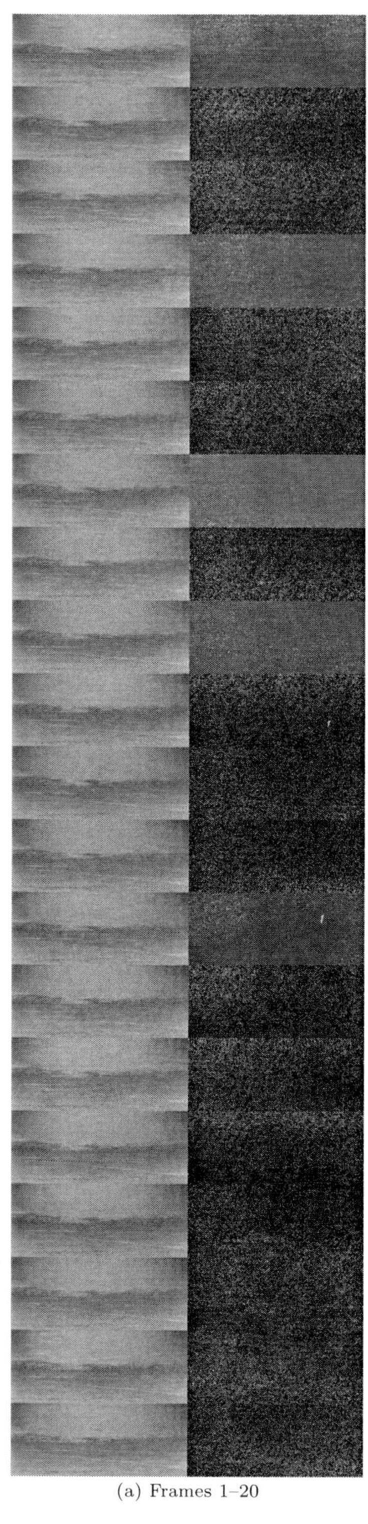

(a) Frames 1–20

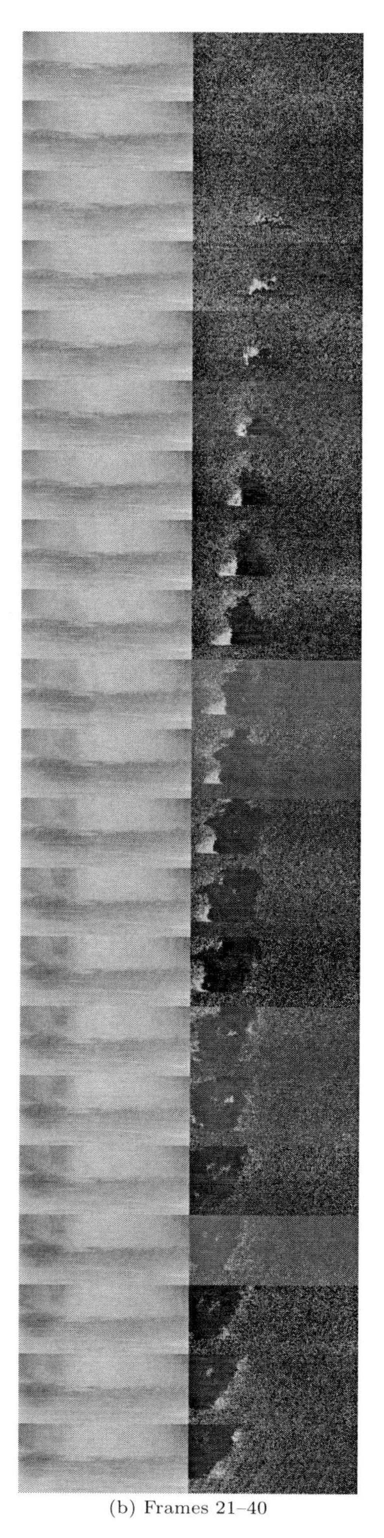

(b) Frames 21–40

FIGURE 16.5 Green component of the decomposed video sequence. (a) Frames 1–20; (b) Frames 21–40 of the green component of the decomposed video sequence. The left panel displays the low-rank approximation and the right panel shows the sparse component within each strip.

(a) Frames 1–20

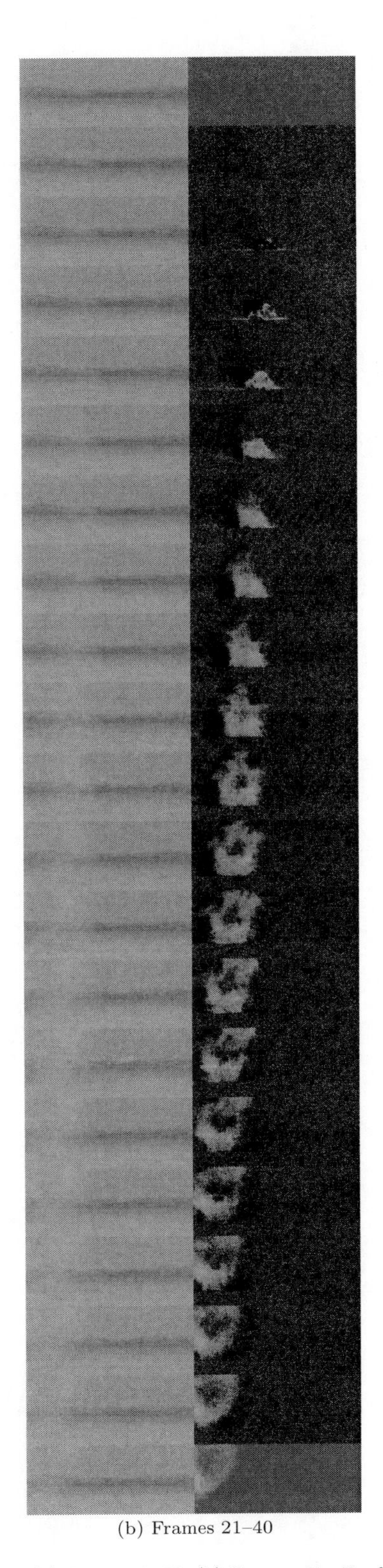

(b) Frames 21–40

FIGURE 16.6 Blue component of the decomposed video sequence.(a) Frames 1–20; (b) Frames 21–40 of the blue component of the decomposed video sequence. The left panel displays the low-rank approximation and the right panel shows the sparse component within each strip.

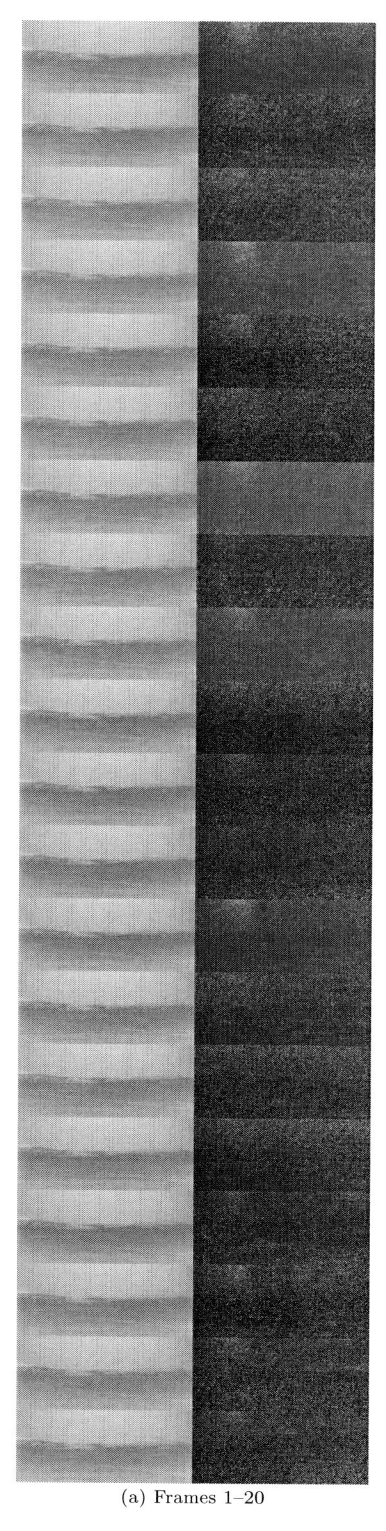

(a) Frames 1–20

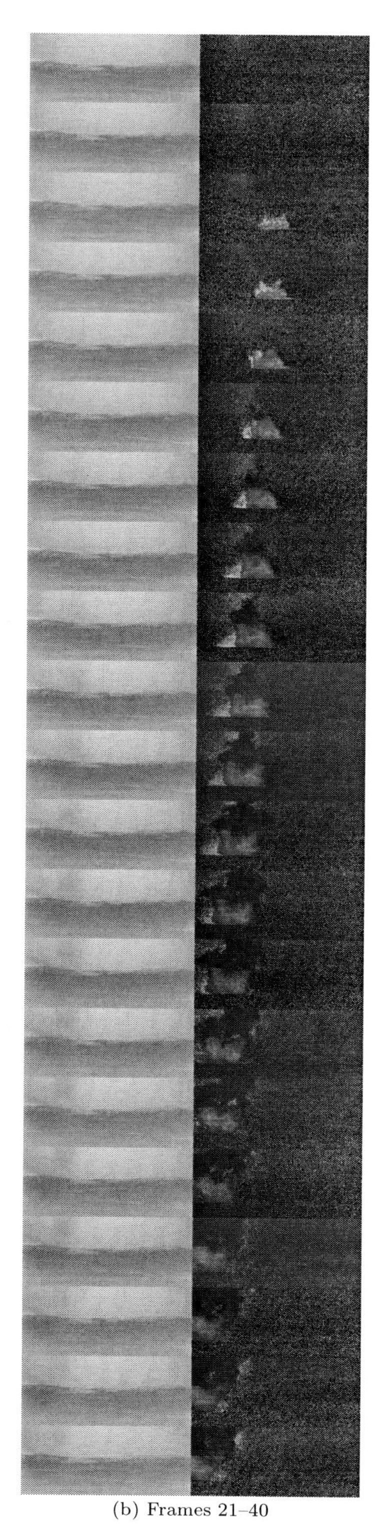

(b) Frames 21–40

FIGURE 16.7 Red, green, and blue components of the decomposed video sequence combined. (a) Frames 1–20; (b) Frames 21–40 of the red, green, and blue components of the decomposed video sequence combined.

aspects of the problem as well. A better understanding of the physical process of chemical diffusion in the atmosphere is needed, such as when detection becomes impossible due to low concentrations. A practical real-time detection and tracking system for toxic chemical releases would require special mathematical and computational considerations. In order to provide the accuracy and reliability required for such a system in the real world, many more physical and computational experiments need to be considered.

References

1. J. Broadwater, D. Limsui, and A.K. Carr. A primer for chemical plume detection using LWIR sensors. Technical report, National Security Technology Department, Johns Hopkins University, 2011.
2. E. Candès, X. Li, Y. Ma, and J. Wright. Robust principal component analysis? *Journal of ACM*, 58(1):1–37, 2009.
3. E. Candès and B. Recht. Exact matrix completion via convex optimization. *Foundations of Computational Mathematics*, 9(6):717–772, 2009.
4. E. Candès and T. Tao. The power of convex relaxation: Near-optimal matrix completion. *IEEE Transactions on Information Theory*, 56(5):2053–2080, 2010.
5. J. Ellis. Searching for oil seeps and oil-impacted soil with hyperspectral imagery. *Earth Observation Magazine*, 10(1):1058–1064, 2001.

17

Low-Rank plus Sparse Spatiotemporal MRI: Acceleration, Background Suppression, and Motion Learning

Ricardo Otazo
New York University School of Medicine

Emmanuel Candes
Stanford University

Daniel K. Sodickson
New York University School of Medicine

17.1 Introduction

Spatiotemporal MRI techniques acquire a time-series of images that encode physiological information of clinical interest, such as organ motion [2], contrast agent uptake [1, 17], signal relaxation [30], water diffusion [18], spectroscopy [6], and neural activity based on blood oxygenation [4, 5] (Figure 17.1). The rich diversity of physiological information that can be acquired using this user-defined time-dimension is a unique feature of MRI. However, the relatively limited imaging speed of conventional MRI imposes limitations on the amount of information that can be acquired per unit time. Hence, conventional methods sacrifice spatial resolution and/or volumetric coverage for an appropriate temporal resolution. In reality, because of extensive spatiotemporal correlations, the information content is much lower than the number of pixels times the number of frames. An immediate consequence is that a conventional MRI method, operating by acquiring fully sampled images at each time point, is rather wasteful. Ideally, one would just want to acquire the 'innovation' from one frame to the next.

Compressed sensing [8, 11, 21] offers an alternative approach. The existence of correlations means that the spatiotemporal signal we wish to recover is sparse in an appropriate representation space, and sparsity along with incoherent sampling can be used to reduce the number of measurements needed at each time point without information loss. This chapter, however, presents a departure from the pure sparsity model by modeling a spatiotemporal MRI dataset as the superposition of a background component (a low-rank matrix L) and of a frame-by-frame innovation component (a sparse matrix S). This low-rank plus sparse ($L + S$) decomposition [7, 10], also known as *robust principal component analysis* (robust PCA), is both intuitive and quite powerful as it often outperforms reconstructions based on sparsity or low-rank alone [14, 23, 24, 32].

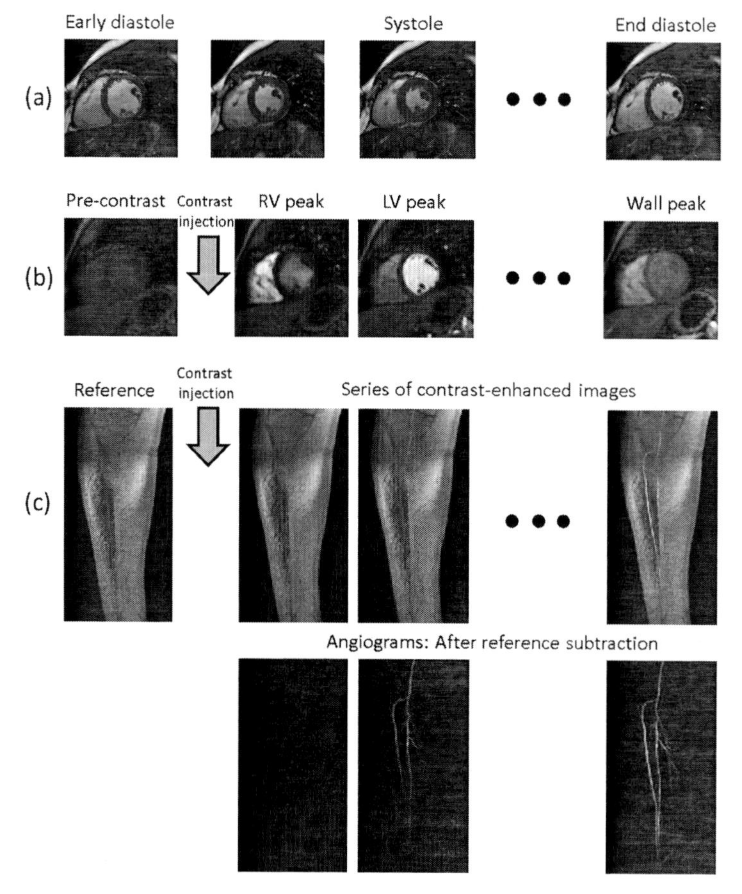

FIGURE 17.1 Examples of spatiotemporal MRI. (a) Cardiac cine: imaging of the cardiac cycle (diastolic and systolic phases) to assess cardiac function based on myocardial wall motion. (b) Cardiac perfusion: contrast agent uptake for different regions of the heart (RV: right ventricle, LV: left ventricle, myocardiac wall) to assess coronary artery disease (narrowing of the arteries). (c) Time-resolved MR angiography (MRA) of the lower extremities: contrast uptake in the arteries to assess peripheral vascular disease. Methods (b) and (c) use an intravenous contrast agent and require high temporal resolution to follow contrast enhancement over time. Conventional time-resolved MRA (c) requires explicit background suppression to subtract the non-enhanced tissue surrounding the arteries, and is sensitive to inconsistencies between the pre-contrast reference image and the post-contrast time series of images.

This chapter presents the application of the $L + S$ model (i) for reconstructing undersampled spatiotemporal MRI data with separation of background and frame-by-frame innovation information, and (ii) for automatic discovery of motion fields (motion-guided $L + S$). This second feature offers both inherent motion compensation and access to new physiological information contained in the displacement fields. We present reconstructions of clinically-relevant accelerated spatiotemporal MRI data corresponding to time-resolved angiography, cardiac perfusion, and abdominal perfusion using Cartesian and golden-angle radial sampling to demonstrate the general applicability of the $L + S$ method.

17.2 Background on Spatiotemporal MRI Techniques

MRI uses magnetic fields and radio waves to encode the Fourier transform of the image in a signal that is acquired by a detector coil or array of coils. Conventional MRI samples the Fourier domain, also called k-space, with data sufficiency determined by the Nyquist/Shannon theory, and uses an inverse Fourier transform tailored to the sampling pattern to reconstruct the image. Spatiotemporal MRI techniques acquire k-space data at each time point so that we can view the whole sampling process as operating in the higher dimensional k-t space (Figure 17.2). This requires the acquisition of each time point to be short relative to the dynamic process in order to obtain an instantaneous snapshot. The speed at which k-space can be traversed using magnetic fields does not enable us to acquire fully-sampled k-space data at each time point, and to reconstruct each time point separately. As a consequence, the amount of k-space data at each time point is usually reduced in time-sensitive acquisitions, and spatial resolution and/or volumetric coverage is decreased.

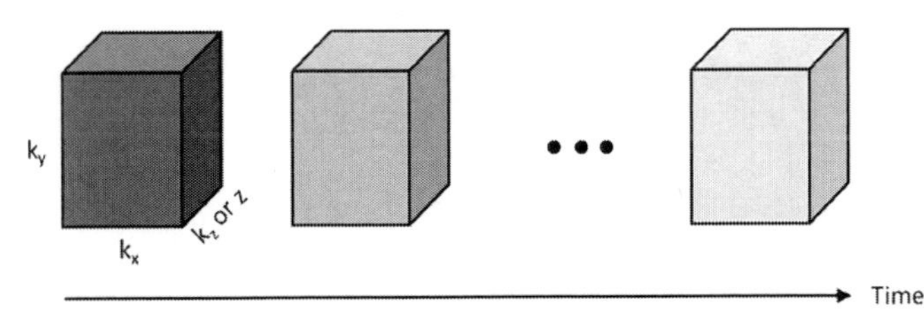

FIGURE 17.2 k-t data acquisition in spatiotemporal MRI. At each time point, 3D k-space data (k_x, k_y, k_z) or multi-slice-2D data (k_x, k_y, z) are acquired. The 3D approach offers higher SNR and the ability to reformat the volumetric data in arbitrary orientations, but it takes longer and thus is more sensitive to organ motion.

Earlier, we argued that because of spatiotemporal correlations, acquiring fully sampled images at each time point is redundant and inefficient since the information that is common to all frames is sampled over and over again. Perhaps not surprisingly, then, a number of methods have been developed to acquire undersampled k-space data at each time point and exploit spatiotemporal correlations in order to reconstruct a time-series of images without aliasing artifacts [34]. The basic idea in these methods is to reconstruct the images in an appropriate transformed domain where they are known to be sparse; i.e., in a domain where (1) most of the transformed values take on low magnitudes and can be discarded without loss of important information, and (2) only a few take on large magnitudes; these values are the pieces of information truly encoding the signal. This principle effectively reduces the number of unknowns so that reconstruction from undersampled data might be feasible. Below, we review two main categories of sparsity-based acceleration methods for spatiotemporal MRI.

- **Regular k-space undersampling and linear reconstruction**: Since sparsity reduces signal foldover that results from regular k-space undersampling, these techniques undersample k-t space in such a way that a superposition of signal coefficients is minimized in a transform domain, where the object of interest is known to be sparse; an example is the combined spatial and temporal frequency domain (x-y-f). The reconstruction first estimates the representation in the transform domain and then uses a linear filter to recover the unaliased representation from the knowledge of the sparse representation. For example, k-t BLAST [33] uses a shifted regular undersampling pattern at each time point as shown in Figure 17.2 (top row to minimize signal foldover in x-y-f space), and removes the remaining aliasing by using a previously estimated sparse representation (the latter being obtained from a series of fully sampled low-resolution images). The method k-t SENSE [33] extends this idea to include coil sensitivities in the linear filter.

- **Irregular k-space undersampling and non-linear reconstruction**: This follows the principle of compressed sensing, in which a different incoherent undersampling pattern is used at each time point in order to extend incoherence along the time domain. The reconstruction then explicitly filters out low-value coefficients (mostly aliasing artifacts) in the domain where the image is known to be sparse. For example, k-t SPARSE [22] uses a different random undersampling pattern at each time point (see Figure 17.2) and enforces sparsity by minimizing the ℓ_1-norm in the x-y-f domain, where the time-series of images is known to be sparse. A clear advantage of compressed sensing over k-t BLAST techniques is that the location of the large coefficients does not need to be known a priori. k-t SPARSE-SENSE [25] extends this idea to enforce joint multicoil sparsity, and thus exploit inter-coil correlations, when a coil array is used for data acquisition.

Since the undersampling factor, and thus the acceleration factor, is ultimately determined by sparsity, there is a high interest in developing image models that can lead to very sparse representations. As we will see, the $L + S$ model can enhance sparsity, by offering a simpler representation of spatiotemporal MRI data, and, in principle, enable higher accelerations.

17.3 $L + S$ Representation of Spatiotemporal MRI

In analogy to video sequences, the $L + S$ model can represent spatiotemporal MRI data as a superposition of a background component (L), which is slowly varying in time (e.g., underlying anatomy) or exhibits some substantial periodic structure over time (e.g., periodic organ motion), and a frame-to-frame innovation component (S), which is rapidly changing or is less structured over time (e.g., contrast uptake, signal relaxation). The innovation component can be assumed to be sparse or transform-sparse since substantial differences between consecutive temporal frames are usually limited to comparatively small numbers of pixels.

To apply the $L + S$ decomposition to a time-series of MR images, we begin to store the data into a matrix M, in which each distinct column represents a distinct frame. We then wish to express this data matrix as $M = L + S$, where L is a low-rank matrix and S a sparse matrix. The decomposition is, in general, unique and the problem well posed if L is not sparse, and, vice versa, if S does not have a low-rank representation; for example, if the singular vectors of L are not sparse and if the nonzero entries of S occur at random locations. While there are quite a few methods for finding such decompositions, we shall merely consider the solution to the following convex optimization problem:

$$[L, S] = \arg\min_{L,S} \|L\|_* + \lambda \|S\|_1 \text{ s.t. } M = L + S; \tag{17.1}$$

here, $\|L\|_*$ is the nuclear norm of the matrix L equal to the sum of its singular values, and is a convex surrogate to the rank functional. The other term $\|S\|_1 \triangleq \|\operatorname{vec}(S)\|_1$ is the ℓ_1 norm of the matrix S seen as a vector, or equivalently, the sum of the magnitudes of its entries; this is the sparsity-promoting term. The scalar λ is a tuning parameter that balances the contribution of the sparse term relative to the low-rank term.

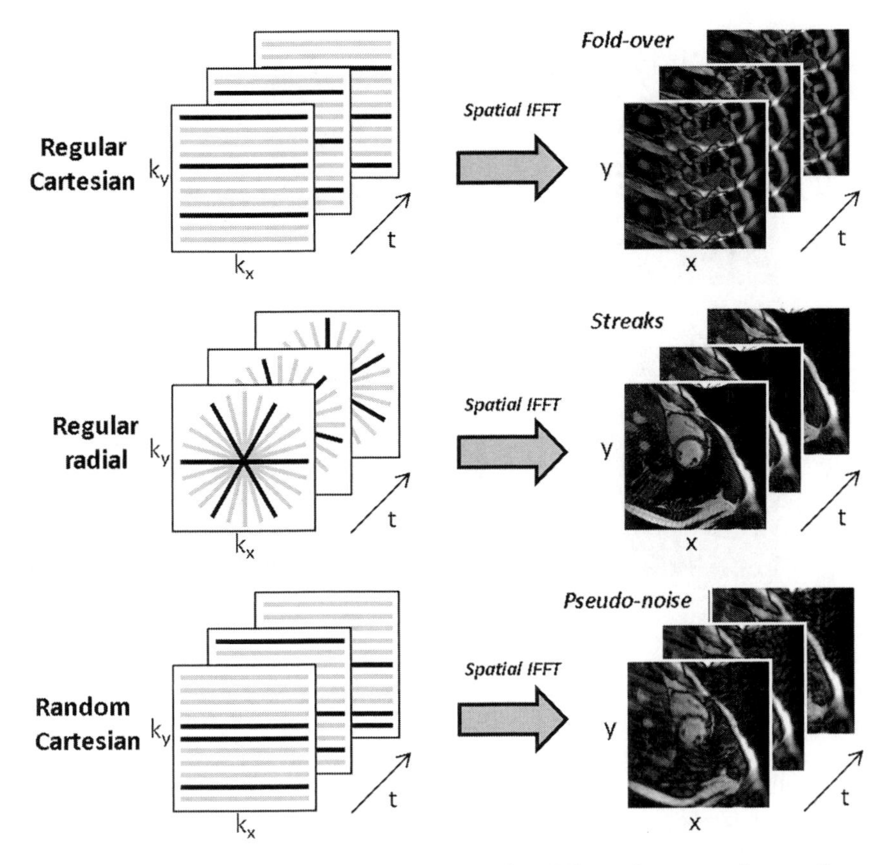

FIGURE 17.3 Typical k-t undersampling schemes, with a different k-space undersampling pattern at each time point. The top row shows a uniform Cartesian undersampling (k-t BLAST method), which involves acquiring one out of every four lines in this example, resulting in discrete fold-over artifacts. The middle row shows a uniform radial undersampling,which leads to low-value streaking artifacts. The bottom row shows random Cartesian undersampling, which leads to noise-like artifacts. The last two approaches are said to be "incoherent"; the overall image features are preserved and the undersampling interferences take on low-values and are spread out over the whole image.

Figure 17.4 shows the $L + S$ decomposition of cardiac cine and perfusion data sets, where L captures the correlated background between frames and S captures the dynamic information or frame-by-frame innovations (heart motion for cine and contrast-enhancement for perfusion). Here, the frames of L (the columns) are not constant over time but are rather slowly varying so that the background is not just estimated to be a temporal average. In fact, in the cardiac cine example, L includes periodic motion in the background, since it is highly structured over time and can be effectively represented by a single singular vector. Another important feature of the $L + S$ model is that S has a sparser representation than M since the background has been suppressed. This gain in sparsity is already obvious in the original y-t space (y is the vertical spatial axis), but it is more pronounced in an appropriate transform domain where spatiotemporal MRI is usually sparse, such as the combined vertical spatial axis (y) and temporal frequency (f) domain (y-f) that results from applying a Fourier transform along each row (rightmost column of Figure 17.4). This increase in sparsity given by the background separation will in principle enable higher acceleration factors, since fewer coefficients need to be recovered provided the load to represent the low-rank component is lower.

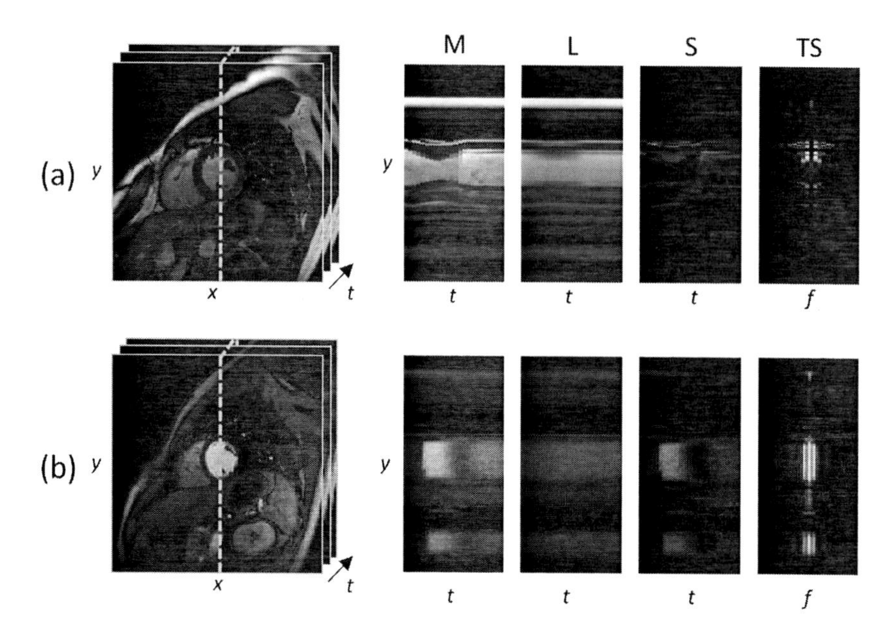

FIGURE 17.4 $L + S$ decomposition of cardiac cine (a) and perfusion (b) MRI corresponding to the central x location (dashed line). L captures the correlated background among temporal frames and S the frame-by-frame innovations that represent the physiologic information of interest (heart motion for cine and contrast-enhancement for perfusion). The L component is not static, but is rather slowly changing over time and contains the most correlated component of the cardiac motion (a) and contrast enhancement (b). The rightmost column displays the sparse component S in y-f space (Fourier transform along the rows), which shows increased sparsity compared to the original y-t domain.

17.4 $L + S$ Reconstruction of Undersampled Spatiotemporal MRI Data

As in compressed sensing, we use our $L + S$ model to reconstruct time-resolved MR images from undersampled data by solving [24]

$$[L, S] = \arg\min_{L,S} \|L\|_* + \lambda \|T(S)\|_1 \quad \text{s.t.} \quad E(L + S) = d; \tag{17.2}$$

here, T is a linear sparsifying transform for S, E is the encoding operator, and d is the undersampled k-t data. As before, L and S are defined as space-time matrices, where each column is a temporal frame, and d is defined as a stretched-out single column vector. We assume that S has a sparse representation in some transformed domain (e.g., temporal frequency domain, temporal finite differences), hence the idea of minimizing $\|T(S)\|_1$ and not $\|S\|_1$ itself. For a single-coil acquisition, the encoding operator E performs a frame-by-frame undersampled spatial Fourier transform. For acquisition with multiple receiver coils, E is given by the frame-by-frame multicoil encoding operator, which performs a multiplication by coil sensitivities followed by an undersampled Fourier transform [29]. The multicoil reconstruction case enforces a joint multicoil $L + S$ model, which presents improved performance over enforcing a coil-by-coil $L + S$ model due to the exploitation of inter-coil correlations, as demonstrated previously for the combination of compressed sensing and parallel imaging based on joint multicoil sparsity [19, 25].

$L + S$ reconstruction aims to simultaneously (a) remove aliasing artifacts in the space-time domain (or equivalently to estimate the value of nonsampled points in k-t space) and

(b) separate the resulting spatiotemporal low-rank and sparse components. Aim (a) requires incoherence between the acquisition space (k-t) and the representation space of L and S, which can be accomplished by using similar k-t undersampling patterns as in compressed sensing; e.g., by means of a different random Cartesian or radial k-space undersampling pattern at each time point (Figure 17.2). In these sampling schemes, low spatial frequencies are usually fully sampled and the undersampling factor increases as we move away from the center of k-space. First, high incoherence between k-t space and L is achieved since the column space of L cannot be approximated by a randomly selected subset of high spatial frequency Fourier modes and the row-space of L cannot be approximated by a randomly selected subset of temporal delta functions. Second, if a temporal Fourier transform is used, incoherence between k-t space and x-f space is maximal, due to their Fourier relationship. Aim (b) requires rank-sparsity incoherence; that is, L cannot be sparse in the space defined by T, and, vice versa, $T(S)$ cannot have low-rank representation. This condition is usually not fully satisfied due to the fact that dynamic information in MRI does not appear at random temporal locations and shows spatial structure. However, rank(L) is usually much lower than rank(S) and the singular values of L are much higher than the singular values of S, since most of the signal power resides in the background. Under these conditions, the highest singular values representing the background will be absorbed by L, leaving the dynamic information for inclusion in S. This approach enables an approximate separation with a small contamination from dynamic features in the background component, but removes the risk of importing the high singular values that represent the background into S.

17.4.1 Implementation Using a Proximal Gradient Algorithm

With real noisy data it is best to use a noise-aware version of the program given in Equation 17.2, which only seeks an approximate rather than an exact data matching [24]:

$$[L, S] = \arg\min_{L,S} \tfrac{1}{2} \|E(L+S) - d\|_2^2 + \lambda_L \|L\|_* + \lambda_S \|T(S)\|_1 , \tag{17.3}$$

where the parameters λ_L and λ_S weight the contribution of the nuclear- and ℓ_1-norm terms, respectively, with respect to the error term (data consistency). This optimization problem is solved using a proximal gradient algorithm, which employs soft-thresholding of the singular values of L and of the entries of $T(S)$. The soft-thresholding or shrinkage operator takes the form:

$$\Lambda_\lambda(x) = \frac{x}{|x|} \max(|x| - \lambda, 0), \tag{17.4}$$

in which x is complex valued and λ is a real-valued threshold. This is extended to matrices by applying the shrinkage operator to each entry. The singular value thresholding (SVT) operation is defined as:

$$\mathrm{SVT}_\lambda(M) = U \Lambda_\lambda(\Sigma) V^H, \tag{17.5}$$

where $M = U\Sigma V^H$ is any singular value decomposition of M. The proposed $L + S$ reconstruction is described in **Algorithm 1**,[55] where at the k-th iteration the updates for L and S are obtained by applying the SVT operator to $M_{k-1} - S_{k-1}$ and the shrinkage operator to $M_{k-1} - L_{k-1}$, respectively. Then data consistency is enforced by subtracting the aliasing artifacts corresponding to the residual in k-space from $L_k + S_k$. The algorithm iterates until the relative change in the solution is less than 10^{-5}, namely until

[55] MATLAB code that implements **Algorithm 1** and reproduces some of the examples presented in this chapter is available at `www.cai2r.net/resources/software/ls-reconstruction-matlab-code`

$\|M_k - M_{k-1}\|_2 \leq 10^{-5} \|M_{k-1}\|_2$. The multicoil encoding operator E is implemented by means of FFTs in the Cartesian sampling case and NUFFTs [13] otherwise. Coil sensitivity maps were computed from the temporal average of the accelerated data using the adaptive coil combination technique described in [37]. The SVT operator requires computing the singular value decomposition of a matrix of size $n_s \times n_t$, where n_s is the number of pixels in each temporal frame and n_t is the number of time points. Since n_t is relatively small, this is not prohibitive and can be performed very rapidly.

Algorithm 17.1 - $L + S$ reconstruction using iterative soft-thresholding

1: **input:**
2: d: multicoil undersampled k-t data
3: E: space-time multicoil encoding operator
4: T: sparsifying transform
5: λ_L: singular-value parameter
6: λ_S: sparsity parameter
7: **initialize:** $M_0 = E^{\star}d$, $S_0 = 0$
8: **while** $\|M_k - M_{k-1}\|_2 \leq 10^{-5} \|M_{k-1}\|_2$ **do**
9: $L_k = \mathrm{SVT}_{\lambda_L}(M_{k-1} - S_{k-1})$
10: $S_k = T^{-1}(\Lambda_{\lambda_S}(T(M_{k-1} - S_{k-1})))$
11: $M_k = L_k + S_k - E^{\star}(E(L_k + S_k) - d)$
12: **end while**

Algorithm 1 converges if $\|E\|^2 < 1$. The operator E is given by the multiplication of Fourier encoding elements and coil sensitivities [24]. Normalizing E by dividing the Fourier encoding elements by $\sqrt{n}$, where n is the number of pixels in the image, and the coil sensitivities by their maximum value, results in $\|E\|^2 = 1$ in the fully sampled case and $\|E\|^2 < 1$ in the undersampled case (for more details see [24]).

The parameters λ_L and λ_S can be computed from the noise level in the undersampled k-t data and the theoretical value of λ in $\|L\|_* + \lambda\|S\|_1$ in the following way:

- $\lambda_L = 2\sigma$, where σ is the noise standard deviation in the undersampled data. The noise level σ can be computed from a noise-only acquisition (no RF pulse), which is routinely performed to estimate the noise covariance matrix for a multiple coil acquisition.

- $\lambda_S = \lambda_L \cdot \sqrt{R/\max(n_s, n_t)}$, where R is the undersampling factor. Note that the ratio λ_S/λ_L corresponds to the recommended value of λ in the $L+S$ decomposition problem (Equation (17.2)), see [7].

17.4.2 Accelerated Clinical Imaging Examples

Time-resolved MR angiography

A natural application of $L+S$ is time-resolved MR angiography, which requires explicit background suppression in the time-series of images acquired after injection of an intravenous contrast agent (typically gadolinium) [17]. Contrast-enhanced time-resolved 3D MRA of the lower extremities was performed in a healthy adult volunteer using an accelerated TWIST (Time-resolved angiography WIth Stochastic Trajectories) [20, 36] pulse sequence on a 1.5T scanner (Avanto, Siemens Healthcare, Erlangen, Germany) equipped with a 12-element peripheral coil array. TWIST samples the center of k-space at the Nyquist rate and undersamples the periphery using a different pseudo-random pattern at each time point, which is suitable to obtain sufficient incoherence for the $L + S$ approach. Relevant imaging parameters are as follows: FOV = $500 \times 375 \times 115$ mm^3, acquisition matrix size = $512 \times 230 \times 42$, number of frames = 10. An acceleration factor of 7.3 is used to achieve a temporal resolution of 6.4 seconds for each 3D image. Image reconstruction was performed using the $L+S$ approach without a sparsifying transform, since angiograms are already sparse in the spatial domain. For comparison purposes, an ℓ_1-based CS reconstruction with pre-contrast data subtraction was performed. Figure 17.5 shows that the $L+S$ approach automatically separates the non-enhanced background from the enhanced vessels without the need of subtraction or modeling. At the same time, the S component provides angiograms with improved image quality as compared with the CS reconstruction with raw data subtraction. CS reconstruction results in incomplete background suppression, which might be due, in part, to inconsistencies between the time-series of contrast-enhanced images and the reference used for subtraction.

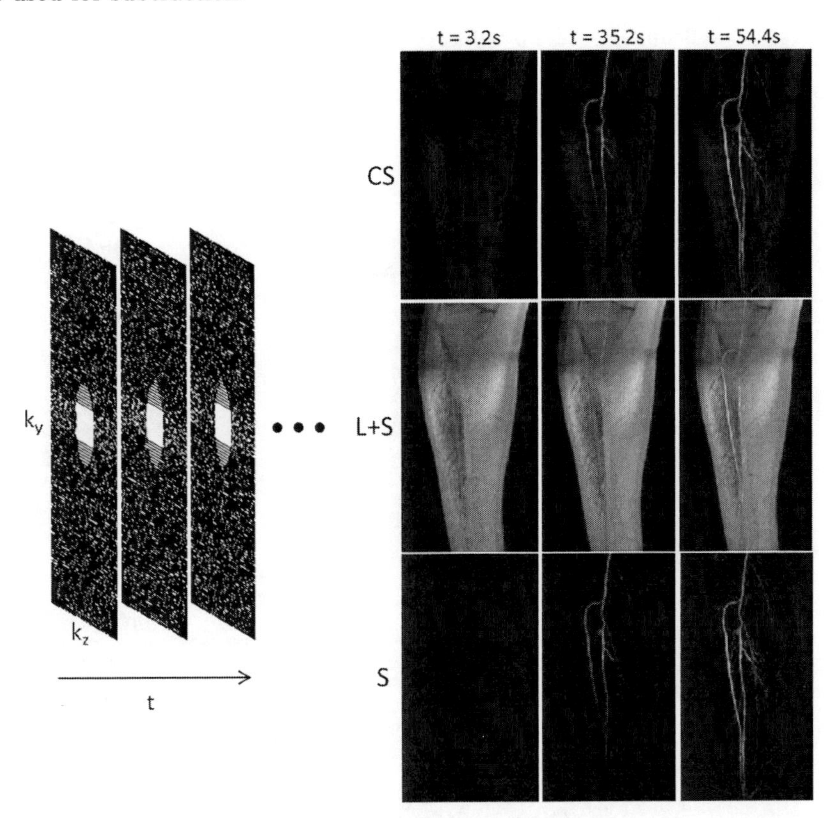

FIGURE 17.5 Accelerated time-resolved peripheral MR angiography: k_y-k_z-t undersampling pattern (white=sampled) with a reduction factor of 7.3 compared to Nyquist sampling rate (left); and maximum intensity projection (MIP, along z) maps corresponding to CS (sparsity-only) and $L+S$ reconstructions for three different contrast-enhancement phases (right). The CS approach employs raw data subtraction using a reference acquired before contrast injection. $L+S$ automatically suppresses the non-enhanced background (without the need of subtraction) in the S component, which presents improved angiograms (compared to CS). In particular, small vessels in S are better visualized.

Cardiac perfusion

Cardiac perfusion MRI is another suitable candidate for the $L+S$ model. Cardiac perfusion requires rapid imaging to follow the contrast-passage with high spatial resolution and volumetric coverage and to minimize sensitivity to cardiac motion. Moreover, it can benefit from the inherent background suppression to improve the visualization of perfusion defects.

Multislice acquisitions, where each slice is acquired independently in a very short period of time (less than 100 ms), are usually employed to freeze cardiac motion, and the goal of acceleration is to maximize spatial resolution and the number of slices per heartbeat.

Multislice first-pass cardiac perfusion data with 8-fold k_y-t acceleration were acquired during a single breath-hold on a patient with known coronary artery disease. Relevant imaging parameters are: image matrix size = 192×192, number of temporal frames = 40, spatial resolution = 1.6×1.67 mm^2 and temporal resolution = 60 ms. The $L + S$ reconstruction employed a temporal Fourier transform as sparsifying transform. For comparison purposes, a sparse-only reconstruction (CS) based on the k-t SPARSE-SENSE approach [25] was also performed using the same sparsifying transform. Figure 17.6 shows that the $L + S$ model improves performance in terms of reducing aliasing artifacts without degrading temporal fidelity. In addition, $L + S$ enhances visualization of the perfusion defect in the S component, where the background has been suppressed and better contrast is observed between the healthy portion of the myocardium and the lesion. This feature may be useful to identify lesions that are difficult to visualize in the original image.

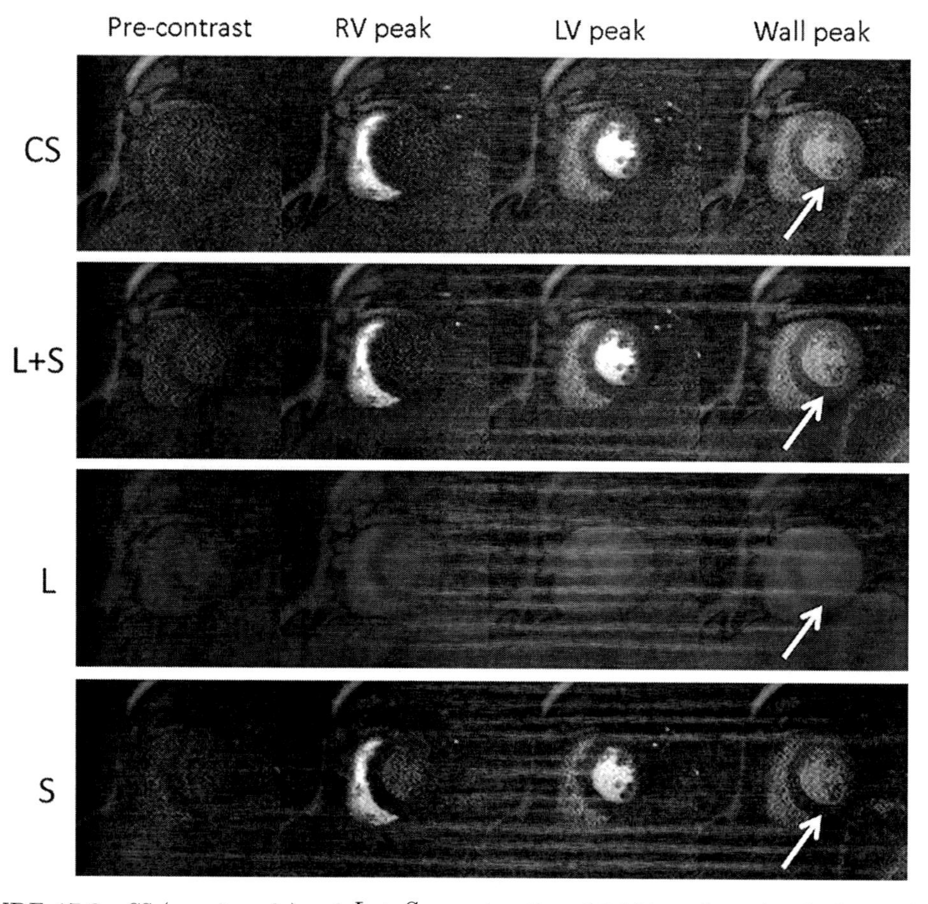

FIGURE 17.6 CS (sparsity-only) and $L + S$ reconstruction of 8-fold accelerated multislice perfusion data (in-plane spatial resolution is 1.6 by 1.67 mm and temporal resolution per slice is 60 ms) corresponding to a representative slice and three contrast-enhancement phases (RV: right ventricle, LV: left ventricle and myocardial wall). Besides improving overall image quality, the $L + S$ approach improves the visualization of the perfusion defect (white arrow) in the sparse component S, where the background has been suppressed.

Liver perfusion

Free-breathing volumetric imaging of the abdomen can be performed using non-Cartesian k-space samplings, such as radial sampling [9], which are more robust to motion than their Cartesian counterparts. For example, the repeated sampling of the center of k-space in radial sampling performs a motion-averaging operation that avoids ghosting artifacts typically seen in Cartesian sampling. However, conventional radial imaging requires more data than Cartesian imaging (about 50% more samples), which imposes challenges for dynamic imaging. On the other hand, undersampling a radial trajectory – even with a uniform pattern – inherently results in low-value aliasing artifacts that spread incoherently over the whole image space. This property makes radial k-space sampling suitable for sparsity-based reconstruction in general [12], including $L + S$ reconstruction.

Contrast-enhanced abdominal MRI data were acquired on a healthy volunteer during free breathing using a 3D stack-of-stars (radial sampling in k_y-k_x and Cartesian sampling in k_z) FLASH pulse sequence on a whole-body 3T scanner (MAGNETOM Verio, Siemens Healthcare, Erlangen, Germany) equipped with a 12-element receiver coil array. Relevant imaging parameters include: FOV is 380×380 mm^2, 384 number of points per radial spoke, and a slice thickness of 3mm. 600 spokes were continuously acquired for each of 30 slices during free breathing, to cover the entire liver (total acquisition time was 77 seconds). Radial sampling was performed using the so-called golden-angle approach [38], where consecutive spokes are separated by 111.25°. The golden-angle separation enables continuous acquisition of complementary information, which can be sorted in temporal frames of quasi-arbitrary duration. 8 consecutive spokes are employed to form each temporal frame, resulting in a temporal resolution of 0.94 sec, corresponding to an acceleration rate of 48 when compared to the Cartesian case with the same image size. Temporal resolutions of 1-2 seconds are required to capture the rapid contrast washout in small abdominal lesions. The reconstructed 4D image size is $384 \times 384 \times 30 \times 75$ with a spatial resolution of $1 \times 1 \times 3$ mm^3. The $L + S$ reconstruction employs a temporal first-order difference as a sparsifying transform. For comparison purposes, a sparse-only reconstruction (CS) is also performed using the same sparsifying transform [12].

The $L+S$ model presents improved reconstruction performance compared to the sparse-only model used in CS as indicated by a sharper representation of small structures, which appear fuzzy in the CS reconstruction; see Figure 17.7. Moreover, the intrinsic background suppression improves the visualization of contrast enhancement in the S component, which might be useful for detection of regions with low enhancement that are otherwise flooded by the background.

17.5 Motion-Guided $L + S$ Reconstruction

The superposition of organ motion with the physiological process of interest (e.g., contrast enhancement) introduces significant challenges for acceleration based on spatiotemporal sparsity [16, 26, 35] (including the $L + S$ reconstruction approach). Organ motion causes misalignment among temporal frames, which reduces the amount of temporal correlations; consequently, the low-rank and sparsity assumptions break down. Under these conditions, $L+S$ reconstruction introduces temporal blurring, where each reconstructed temporal frame is given by a superposition of adjacent temporal frames, leading to non-diagnostic information, or even worse, information that can lead to a false diagnosis.

Simultaneous alignment and $L + S$ decomposition of a series of fully-sampled images has been demonstrated in computer vision applications by means of the Robust Alignment by Sparse and Low-rank decomposition (RASL) method [28]. The idea there is to include

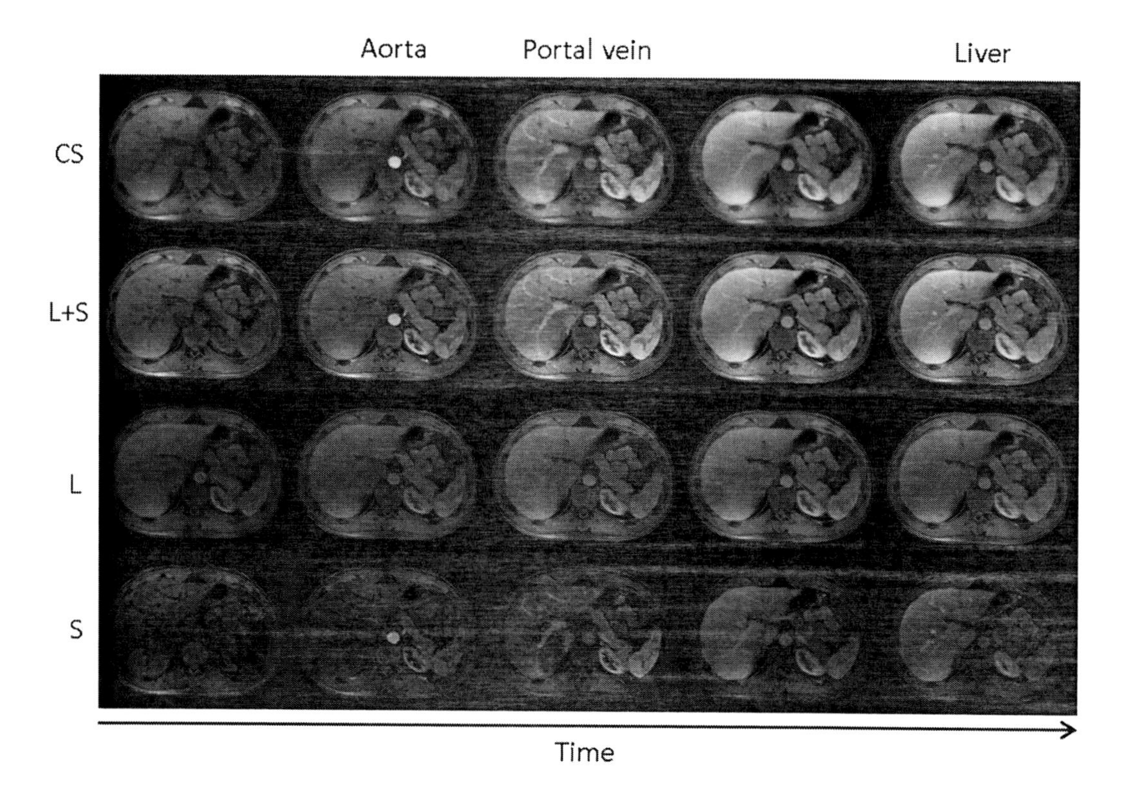

FIGURE 17.7 CS (sparsity-only) and $L+S$ reconstruction of 4D dynamic contrast-enhanced abdominal data acquired with golden-angle radial sampling (8 spokes/frame, undersampling factor is 48 and temporal resolution is 0.94 seconds per 3D volume) corresponding to a representative slice and three contrast-enhancement phases (aorta, portal vein, liver). $L + S$ compares favorably to CS, which suffers from spatiotemporal blurring. Moreover, the S component, in which the background has been suppressed, offers improved visualization of contrast-enhancement.

a frame-by-frame warping operator W in the $L + S$ decomposition problem as

$$W(L + S) = M \tag{17.6}$$

and to search jointly over the unknown W, L, and S. The idea is to find the warping that will align the images by making the image-series as low-rank + sparse as possible (so that the background stays still if you will). In 2D + time imaging, W changes the original spatial coordinates for each temporal frame separately from (x, y) to $(x, y) + v(x, y)$, where $v(x, y) \in \mathbb{R}^2$ is the displacement or motion field. RASL uses a parametric linear model for the motion operator, e.g., affine transformation, which enables users to reduce the number of unknowns. In [28], RASL was shown to align a series of images acquired under different conditions such as camera position, illumination, and partial occlusion. In MRI, organ motion usually results in non-rigid body deformations, which require more advanced models than parametric transformations.

Simultaneous estimation of L, S, and the field of motion W is feasible due to the self-consistency between image alignment (or registration) and the $L+S$ model. More explicitly, the rank of L will be minimized and the sparsity of S will be maximized (or equivalently the number of non-zero coefficients will be minimized) when the series of images is aligned. This self-consistency principle motivated us to introduce the motion-guided $L+S$ reconstruction technique [27] in order to simultaneously reconstruct undersampled k-t data and to learn the displacement fields at the same time. This method seeks to align the sequence of images given by the columns of $L + S$ as well as to perform separation by solving

$$[L, S, W] = \arg\min_{L,S,W} \ \tfrac{1}{2} \|E(W(L+S)) - d\|_2^2 + \lambda_L \|L\|_* + \lambda_S \|T(S)\|_1 . \qquad (17.7)$$

The issue is that W is unknown, which makes the problem not convex. If the deformation W were known, we could of course solve Equation (17.7) by **Algorithm 1**. With a known deformation field, the mapping $M \mapsto W(M)$ is linear and we can formally replace the operator E in the description of **Algorithm 1** with $E \circ W$ so that the update for M reads

$$M_k = L_k + S_k - W^\star \left(E^\star \left(E\left(W\left(L_k + S_k\right)\right) - d\right)\right). \qquad (17.8)$$

Of course, we do not know the motion field and our idea is to estimate it by looking at deformations between consecutive updates. As we shall see below, our motion-guided $L+S$ reconstruction method [27] uses a deformable model based on optical flows. To explain the idea, consider an image $I_0(x,y)$ and another $I_1(x,y)$. We would like to find a deformation $v(x,y)$ so that

$$I_0((x,y) + v(x,y)) \approx I_1(x,y). \qquad (17.9)$$

Then by linearizing the right-hand side [3, 15], it approximately holds that

$$I_0(x,y) + \nabla I_0(x,y) \cdot v(x,y) \approx I_1(x,y).$$

Defining $\nabla I_0(x,y) = (G_x, G_y)$, where G_x and G_y are the horizontal and vertical gradients of the image $I_0(x,y)$, we obtain at each point (x,y),

$$G_x \cdot v_x + G_y \cdot v_y \approx I_1 - I_0. \qquad (17.10)$$

This linear system is undetermined and needs additional constraints to define a meaningful solution. Here, we follow the solution proposed by Thirion [31], also known as the demons method, which corresponds to a second-order gradient descent on the sum of squares difference between I_1 and I_0:

$$v_x = \frac{G_x}{G_x^2 + G_y^2 + (I_1 - I_0)^2} \left(I_1 - I_0\right), \qquad (17.11)$$

$$v_y = \frac{G_y}{G_x^2 + G_y^2 + (I_1 - I_0)^2} \left(I_1 - I_0\right). \qquad (17.12)$$

In our setting, $I_1 = L_k + S_k$ and $I_0 = M_{k-1}$ at a given time point, and the motion field update is represented as the deformation between the solutions corresponding to the previous and current iterations,

$$(\Delta v_x, \Delta v_y) = (G_x, G_y) \cdot \frac{L_k + S_k - M_{k-1}}{G_x^2 + G_y^2 + (L_k + S_k - M_{k-1})^2} \qquad (17.13)$$

and $(G_x, G_y) = \nabla M_{k-1}$. Data consistency is then enforced using Equation 17.8, where the warping operator W is defined using the updated motion fields. The proposed motion-guided $L+S$ reconstruction is given in **Algorithm 2**, where the successive iterations search for the motion fields that optimize the $L+S$ model, and vice-versa, the optimization of the $L+S$ model results in a registered series of images.

Algorithm 17.2 - Motion-guided $L + S$ reconstruction using iterative soft-thresholding

1: **input:**
2: d: multicoil undersampled k-t data
3: E: space-time multicoil encoding operator
4: T: sparsifying transform
5: λ_L: singular-value parameter
6: λ_S: sparsity parameter
7: **initialize:** $M_0 = E^\star d$, $S_0 = 0$, $v = (0, 0)$, $k = 1$
8: **while** $\|M_k - M_{k-1}\|_2 \leq 10^{-5} \|M_{k-1}\|_2$ **do**
9: $L_k = \mathrm{SVT}_{\lambda_L}(M_{k-1} - S_{k-1})$
10: $S_k = T^{-1}(\Lambda_{\lambda_S}(T(M_{k-1} - S_{k-1})))$
11: **for** each spatiotemporal position **do**
12: $(\Delta v_x, \Delta v_y) = (G_x, G_y) \cdot \frac{L_k + S_k - M_{k-1}}{G_x^2 + G_y^2 + (L_k + S_k - M_{k-1})^2}$, where $(G_x, G_y) = \nabla M_{k-1}$
13: **end for**
14: $v_k = v_{k-1} + (\Delta v_x, \Delta v_y)$
15: $M_k = L_k + S_k - W_v^\star (E^\star (E(W_v(L_k + S_k)) - d))$
16: $k = k + 1$
17: **end while**

The motion-guided $L + S$ approach computes the motion fields in addition to L and S, which at first view would suggest that more k-space data are required since the number of unknowns has been increased. However, because of the self consistency property between the $L + S$ model and image registration, the number of unknowns can be significantly reduced since this approach enhances low-rankedness and sparsity, thereby compensating for the estimation of extra variables.

17.5.1 Free-Breathing Clinical Imaging Examples

Motion-guided $L+S$ was tested on free-breathing accelerated contrast-enhanced spatiotemporal MRI datasets in the heart (Figure 17.8) and abdomen (Figure 17.9). The case of contrast-enhanced imaging is particularly challenging due to the different signal intensity in each temporal frame, which usually causes errors in conventional image registration techniques that select one reference frame and register the other frames in the series to this reference assuming similar signal intensity.

Free-breathing cardiac perfusion MRI data were acquired using the same undersampling pattern and imaging parameters as in the breath-held case (Section 17.4.2): the acceleration or undersampling factor is 8, the image matrix is of size 192×192, the number of temporal frames is 40, the number of slices is 10, the spatial resolution is $1.6 \times 1.67\text{mm}^2$, and the temporal resolution is 60ms. This acquisition was gated using an electrocardiogram (ECG) signal, which is a common procedure in cardiac MRI to synchronize cardiac motion and enable the acquisition of different contrast phases at the same cardiac position. However, respiratory motion is not synchronized, and thus, different temporal frames will be at different respiratory motion states. Standard $L + S$ reconstruction presents temporal blurring, which results in spatial and temporal distortions seen in Figure 17.8 (top row). In this example, temporal blurring also causes a false perfusion defect in the lower part of the myocardial wall, which can be falsely diagnosed as a lesion. Motion-guided $L + S$ significantly removes temporal blurring artifacts, including the false perfusion defect. The estimated motion fields shown in Figure 17.8 (botton row) indicate the presence of non-rigid body motion, which would not have been effectively captured by rigid motion models that assume translations and rotations only.

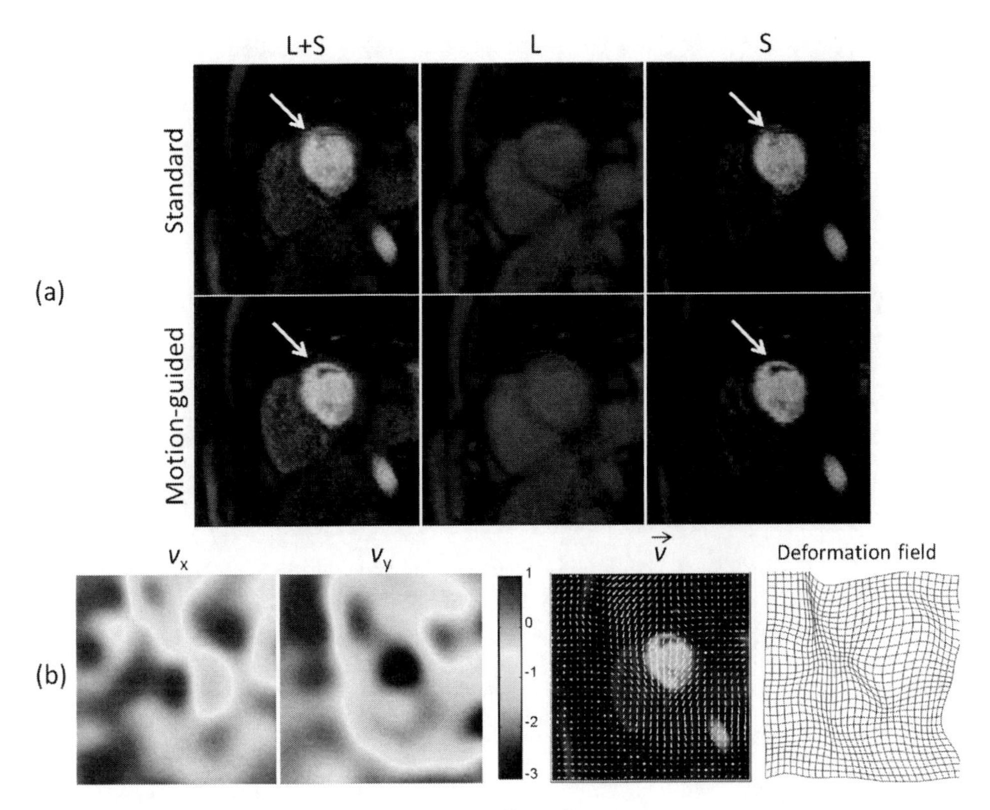

FIGURE 17.8 (a) Standard and motion-guided $L + S$ reconstruction of 8-fold accelerated cardiac perfusion data acquired during free breathing for a representative contrast phase and slice. The arrows indicate temporal blurring artifacts in the standard $L + S$ images caused by misalignment among frames, which are significantly removed by the motion-guided $L + S$ approach. (b) Estimated motion fields for the same slice and contrast phase as in (a). The motion fields clearly indicate non-rigid body motion.

Free-breathing dynamic contrast-enhanced MRI in the abdomen was performed on a 3T Siemens Verio scanner using a 15-element body matrix coil array. 4D data with whole-abdomen coverage were acquired using stack-of-stars sampling (radial in k_x-k_y and Cartesian in k_z) with 12.8-fold acceleration (20 radial lines per temporal frame). Even though radial imaging is very robust to intra-frame motion, inter-frame motion is still a problem since temporal frames can be at different respiratory motion states depending on the temporal resolution and breathing pattern of the subject being scanned. Motion-guided $L + S$ improves the visualization of contrast-enhancement of small vessels in the liver and enhances sharpness of the kidneys, which appear fuzzy in the standard $L + S$ approach due to organ motion blurring (Figure 17.9).

17.6 Conclusion

The $L+S$ decomposition proposes a novel approach for performing spatiotemporal MRI that goes beyond accelerated data acquisition, and in fact, presents new applications of sparsity including the separation of information sources (background, contrast-enhancement, organ motion, and so on) and self-discovery of motion fields. Reconstruction of highly-accelerated spatiotemporal MRI data with separation of background and dynamic information without the need for explicit modeling or data subtraction can be particularly useful for clinical

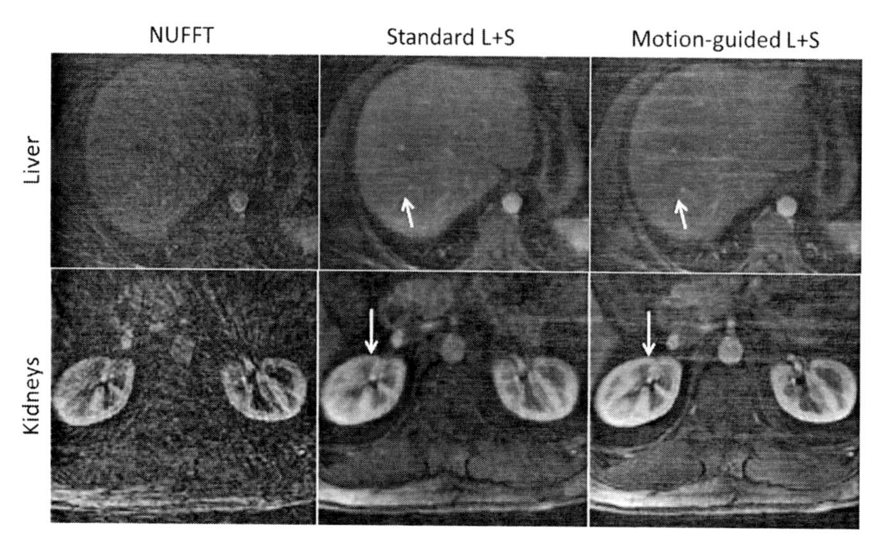

FIGURE 17.9 (a) NUFFT, Standard $L + S$ and motion-guided $L + S$ reconstructions of 12.8-fold accelerated abdominal DCE radial data acquired during free breathing for slices corresponding to liver and kidney regions. The arrows indicate temporal blurring in the standard $L + S$ images, which are removed by the motion-guided $L + S$ approach without loss of spatial resolution.

studies that require background suppression, such as contrast-enhanced angiography and free-breathing abdominal studies, where conventional data subtraction is sensitive to motion. In cases where the separation is not of interest, $L + S$ can be used as an improved compressed sensing approach with higher acceleration capabilities than can be achieved by exploiting low-rank or sparsity separately. Motion-guided $L + S$ extends the original $L + S$ approach to self-learn inter-frame motion fields without the need for a reference by exploiting self-consistency between registration and the $L + S$ decomposition. The image series registration performed by motion-guided $L + S$ is robust to signal intensity changes and enables users to extract non-rigid body motion fields. In cases where knowledge about organ motion is not of interest, motion-guided $L + S$ can be used as an accelerated imaging technique with robust motion compensation (even in cases with changes in signal intensity). The estimated motion fields can potentially add useful clinical information; for example, to analyze contrast enhancement during respiration. In summary, the general applicability of the $L + S$ technique to several problems of clinical interest has raised significant interest in the MRI community and promises new developments in the near future.

17.7 Acknowledgments

This work was supported in part by the National Institutes of Health (NIH) and was performed under the rubric of the Center for Advanced Imaging Innovation and Research (CAI^{2}R), a NIBIB Biomedical Technology Resource Center. E. C. is partially supported by a Math + X award from the Simons Foundation.

References

1. D. Atkinson, D. Burnstein, and R. Edelman. First-pass cardiac perfusion: evaluation with ultrafast MR imaging. *Radiology*, 174(3):757–62, 1991.

2. D. Atkinson and R. Edelman. Cineangiography of the heart in a single breath hold with a segmented turboflash sequence. *Radiology*, 178(2):357–60, 1991.
3. S. Baker and I. Matthews. Lucas-Kanade 20 years on: A unifying framework. *Intl J. Computer Vision*, 56:221–255, 2004.
4. J. Belliveau, D. Kennedy, R. McKinstry, B. Buchbinder, R. Weisskoff, M. Cohen, J. Vevea, T. Brady, and B. Rosen. Functional mapping of the human visual cortex by magnetic resonance imaging. *Science*, 254:716–719, 1991.
5. B. Biswal, F. Yetkin, V. Haughton, and J. Hyde. Functional connectivity in the motor cortex of resting human brain using echo-planar MRI. *Magn Reson Med.*, 34(4):537–41, 1995.
6. T.B. Brown, B.M. Kincaid, and K. Ugurbil. NMR chemical shift imaging in three dimensions. *Proc. Natl. Acad. Sci. USA*, 79(11):3523–3526, 1982.
7. E. Candes, X. Li, Y. Ma, and J. Wright. Robust principal component analysis? *Journal of the ACM*, 58(3):1–37, 2011.
8. E. Candes, J. Romberg, and T. Tao. Robust uncertainty principles: Exact signal reconstruction from highly incomplete frequency information. *IEEE Trans Inf Theory*, 52:489–509, 2006.
9. H. Chandarana, T. Block, A. Rosenkrantz, R. Lim, D. Kim, D. Mossa, J. Babb, B. Kiefer, and V. Lee. Free-breathing radial 3D fat-suppressed T1-weighted gradient echo sequence: A viable alternative for contrast-enhanced liver imaging in patients unable to suspend respiration. *Invest Radiol*, 46(10):648–53, 2011.
10. V. Chandrasekaran, S. Sanghavi, PA. Parrilo, and A. Willsky. Rank-sparsity incoherence for matrix decomposition. *Siam J Optim.*, 21(2):572–96, 2011.
11. D. Donoho. Compressed sensing. *IEEE Trans Inf Theory*, 52:1289–1306, 2006.
12. L. Feng, R. Grimm, K. Block, H. Chandarana, S. Kim, J. Xu, L. Axel, D. Sodickson, and R. Otazo. Golden-angle radial sparse parallel MRI: Combination of compressed sensing, parallel imaging, and golden-angle radial sampling for fast and flexible dynamic volumetric MRI. *Magn Reson Med.*, 72(3):707–17, 2014.
13. J. Fessler and B. Sutton. Nonuniform fast Fourier transforms using min-max interpolation. *IEEE Transactions on Signal Processing*, 51(2):560–74, 2003.
14. H. Gao, S. Rapacchi, D. Wang, J. Moriarty, C. Meehan, J. Sayre, G. Laub, P. Finn, and P. Hu. Compressed sensing using prior rank, intensity and sparsity model (prism): Applications in cardiac cine MRI. In *Proceedings of the 20th Annual Meeting of ISMRM, Melbourne*, page 2242, 2012.
15. B. Horn and B. Schunck. Determining optical flow. *Artif Intell.*, 17:185–203, 1981.
16. H. Jung and J. Ye. Motion estimated and compensated compressed sensing dynamic magnetic resonance imaging: what we can learn from video compression techniques. *Int J Imaging Syst Technol.*, 20:81–98, 2010.
17. F. Korosec, R. Frayne, T. Grist, and C. Mistretta. Time-resolved contrast-enhanced 3D MR angiography. *Magn Reson Med.*, 36(3):345–51, 1996.
18. D. Le Bihan, E. Breton, D. Lallemand, P. Grenier, E. Cabanis, and M. Laval-Jeantet. MR imaging of intravoxel incoherent motions: application to diffusion and perfusion in neurologic disorders. *Radiology*, 161(2):401–7, 1986.
19. D. Liang, B. Liu, J. Wang, and L. Ying. Accelerating sense using compressed sensing. *Magn Reson Med.*, 62(6):1574–84, 2009.
20. R. Lim, J. Jacob, E. Hecht, D. Kim, S. Huffman, S. Kim, J. Babb, G. Laub, M. Adelman, and V. Lee. Time-resolved lower extremity MRA with temporal interpolation and stochastic spiral trajectories: preliminary clinical experience. *J Magn Reson Imaging*, 31(3):663–72, 2010.
21. M. Lustig, D. Donoho, and J. Pauly. Sparse MRI: The application of compressed sensing for rapid MR imaging. *Magn Reson Med.*, 58(6):1182–95, 2007.
22. M. Lustig, J. Santos, D. Donoho, and J. Pauly. k-t SPARSE: High frame rate dynamic MRI

exploiting spatio-temporal sparsity. In *Proceedings of the 14th Annual Meeting of ISMRM, Seattle*, page 2420, 2006.

23. R. Otazo, E. Candes, and D. Sodickson. Low-rank and sparse matrix decomposition for accelerated DCE-MRI with background and contrast separation. In *Proceedings of the ISMRM Workshop on Data Sampling and Image Reconstruction, Sedona*, page 7, 2013.
24. R. Otazo, E. Candes, and D.K. Sodickson. Low-rank and sparse matrix decomposition for accelerated dynamic MRI with separation of background and dynamic components. *Magn Reson Med.*, 73(3):1125–36, 2015.
25. R. Otazo, D. Kim, L. Axel, and D. K. Sodickson. Combination of compressed sensing and parallel imaging for highly accelerated first-pass cardiac perfusion MRI. *Magn Reson Med.*, 64(3):767–76, 2010.
26. R. Otazo, D. Kim, L. Axel, and D.K. Sodickson. Combination of compressed sensing and parallel imaging with respiratory motion correction for highly-accelerated first-pass cardiac perfusion MRI. In *Proceedings of the 19th Annual Meeting of ISMRM, Montreal*, page 63, 2011.
27. R. Otazo, T. Koesters, E. Candes, and D.K. Sodickson. Motion-guided low-rank plus sparse (L+S) reconstruction for free-breathing dynamic MRI. In *Proceedings of the 22nd Annual Meeting of ISMRM, Milan*, page 742, 2014.
28. Y. Peng, A. Ganesh, J. Wright, W. Xu, and Y. Ma. Rasl: Robust alignment by sparse and low-rank decomposition for linearly correlated images. *IEEE Trans Pattern Anal Mach Intell*, 34(11):2233–46, 2012.
29. K. P. Pruessmann, M. Weiger, P. Bornert, and P. Boesiger. Advances in sensitivity encoding with arbitrary k-space trajectories. *Magn Reson Med.*, 46(4):638–51, 2001.
30. P. Schmitt, M. Griswold, P. Jakob, M. Kotas, V. Gulani, M. Flentje, and A. Haase. Inversion recovery TrueFISP: quantification of T(1), T(2), and spin density. *Magn Reson Med.*, 51(4):661–67, 2004.
31. J. Thirion. Image matching as a diffusion process: an analogy with Maxwell demons. *Medical Image Analysis*, 2(3):243–260, 1998.
32. B. Tremoulheac, N. Dikaios, D. Atkinson, and S. Arridge. Dynamic MR image reconstruction-separation from undersampled (k,t)-space via low-rank plus sparse prior. *IEEE Trans Med Imaging*, 33(8):1689–701, 2014.
33. J. Tsao, P. Boesiger, and K. P. Pruessmann. k-t BLAST and k-t SENSE: Dynamic MRI with high frame rate exploiting spatiotemporal correlations. *Magn Reson Med.*, 50(5):1031–42, 2003.
34. J. Tsao and S. Kozerke. MRI temporal acceleration techniques. *J Magn Reson Imaging*, 36(3):543–560, 2012.
35. M. Usman, D. Atkinson, F. Odille, C. Kolbitsch, G. Vaillant, T. Schaeffter, P. Batchelor, and C. Prieto. Motion corrected compressed sensing for free-breathing dynamic cardiac MRI. *Magn Reson Med.*, 70(2):504–16, 2013.
36. F. Vogt, H. Eggebrecht, G. Laub, R. Kroeker, M. Schmidt, J. Barkhausen, and S. Ladd. High spatial and temporal resolution MRA (TWIST) in acute aortic dissection. In *Proceedings of the 15th Annual Meeting of ISMRM, Berlin*, page 92, 2007.
37. D. O. Walsh, A. F. Gmitro, and M. W. Marcellin. Adaptive reconstruction of phased array MR imagery. *Magn Reson Med.*, 43(5):682–90, 2000.
38. S. Winkelmann, T. Schaeffter, T. Koehler, H. Eggers, and O. Doessel. An optimal radial profile order based on the golden ratio for time-resolved MRI. *IEEE Trans Med. Imaging*, 26(1):68–76, 2007.

V

Applications in Background/Foreground Separation for Video Surveillance

18

LRSLibrary: Low-Rank and Sparse Tools for Background Modeling and Subtraction in Videos

Andrews Sobral
Lab. L3I, University of La Rochelle, France

Thierry Bouwmans
Lab. MIA, University of La Rochelle, France

El-Hadi Zahzah
Lab. L3I, University of La Rochelle, France

18.1 Introduction

The detection of moving objects is the basic low-level operation in video analysis. This basic operation (also called "background subtraction"or BS) consists of separating the moving objects called "foreground" from the static information called "background."The background subtraction is a key step in many fields of computer vision applications, such as video surveillance to detect persons, vehicles, animals, etc., human-computer interface, motion detection, and multimedia applications. Many BS methods have been developed over the last few years [54], [7], [8] and several implementations in C++ are available in the BGSLibrary[56] [52]. The main resources can be found at the Background Subtraction Web Site.[57]

Recent research on subspace estimation by sparse representation and rank minimization represents a nice framework to separate moving objects from the background. It's based on the idea that the data matrix $\mathbf{A}$[58] can be decomposed into two components such that $\mathbf{A} = \mathbf{L} + \mathbf{S}$, where $\mathbf{L}$ is a low-rank matrix and $\mathbf{S}$ is a matrix that can be sparse or not. This decomposition can be obtained by Robust Principal Component Analysis (RPCA) solved via Principal Component Pursuit (PCP) [12], [59]. The background sequence is then modeled

[56] `https://github.com/andrewssobral/bgslibrary`

[57] `http://sites.google.com/site/backgroundsubtraction`

[58] This work follows the notation conventions in multilinear and tensor algebra as in [27], [19]. Scalars are denoted by lowercase letters, e.g., x; vectors are denoted by lowercase boldface letters, e.g., $\mathbf{x}$; matrices by uppercase boldface, e.g., $\mathbf{X}$; and tensors by calligraphic letters, e.g., $\mathcal{X}$. In this paper, only real-valued data are considered.

FIGURE 18.1 Original image, low-rank matrix **L** (background), sparse matrix **S** (foreground), foreground mask, and filtered mask.

by a low-rank subspace that can gradually change over time, while the moving foreground objects constitute the correlated sparse outliers. For example, Figure 18.1 shows the original frame of the sequence from ChangeDetection.net[59] [18] and its decomposition into the low-rank matrix **L** and sparse matrix **S**. We can see that **L** corresponds to the background whereas **S** corresponds to the foreground. The fourth image shows the foreground mask obtained by thresholding the matrix **S** and the fifth image the filtered result by 5×5 median filter.

PCP is limited to the low-rank component being exactly low-rank and the sparse component being exactly sparse, but the observations in real applications are often corrupted by noise affecting every entry of the data matrix. Therefore, Zhou et al. [73] proposed a stable PCP (SPCP) that guarantees stable and accurate recovery in the presence of entry-wise noise. SPCP assumes that the observation matrix **A** is represented as $\mathbf{A} = \mathbf{L} + \mathbf{S} + \mathbf{E}$ (also named Three-Term Decomposition), where **E** is a noise term (say i.i.d. noise on each entry of the matrix). To recover **L**, **S**, and **E**, Zhou et al. [73] proposed to solve the following optimization problem, as a relaxed version of PCP:

$$\min_{\mathbf{L},\mathbf{S},\mathbf{E}} ||\mathbf{L}||_* + \lambda||\mathbf{S}||_1 + \gamma||\mathbf{E}||_F^2, \text{ s.t. } \mathbf{A} = \mathbf{L} + \mathbf{S} + \mathbf{E} \tag{18.1}$$

where $||.||_*$, $||.||_1$, and $||.||_F$ are the nuclear norm (i.e., sum of singular values), l_1-norm (sum of matrix elements magnitude), and the Frobenius norm, respectively, while $\lambda > 0$ and $\gamma > 0$ are an arbitrary weighting parameter. This decomposition is called "stable" decomposition as it separates the outliers in **S** and the noise in **E**.

In scenes where the background is very dynamic (i.e., sea waves in maritime surveillance), the motion of the objects of interest (i.e., boats) will be mixed with the dynamic behavior of the background (i.e., waves). SPCP-based methods try to deal with this problem under the term where the multi-modality of the background (i.e., waves) is considered as a noise component (**E**), while the moving objects (i.e., boats) are considered as sparse components (**S**). The low-rank component (**L**) represents the static part of the background (uni-modality).

Several frameworks based on decomposition in low-rank and sparse matrices have been developed. For example, [36] and [37] propose a robust subspace segmentation by Low-Rank Representation (LLR). The classical RPCA assumes that the underlying data structure is a single low-rank subspace. In LRR, the data is considered as samples approximately drawn from a mixture of several low-rank subspaces (subspace clustering problem). On the other hand, it is desirable to retain the non-negative characteristics of the original data (i.e., image or video analysis and document clustering). A popular matrix factorization approach is the Non-negative Matrix Factorization (NMF) framework, first proposed by [47] and greatly popularized by [29].

[59] `http://changedetection.net/`

As being different from previous subspace learning methods, where we consider the image as a vector (each video frame is a column vector of matrix **M**), *tensor*-based approaches have been proposed to deal with this issue. Tensor decompositions have been widely studied and applied to many real-world problems [53], [27], [42], and [19]. CANDECOMP/PARAFAC(CP)-decomposition[60] and Tucker decomposition are two widely used low-rank decompositions of tensors.[61] Today, the Tucker model is better known as the Higher-order SVD (HoSVD) from the work of [28]. The HoSVD is a generalization of the well-know matrix SVD.[62]

So, these recent advances in robust matrix and tensor factorization are fundamental and can be applied to background modeling and foreground detection for video surveillance. It was for this reason that the LRSLibrary was developed. The goal is to provide an easy-to-use library to apply low-rank and sparse decomposition tools for background modeling and subtraction in videos. The library is open-source and free for academic/research purposes (non-comercial). Note that the license of the algorithms included in LRSLibrary do not necessarily have the same license as the library. Some authors do not allow their algorithms to be used for a commercial purpose, so that you will first need to contact them to ask permission.

In the next sections, we present the main features and some remarks about the LRSLibrary such as available algorithms and their respective computational cost. We also explain how to use the LRSLibrary and how to contribute. Lastly, the conclusion of the present work and future developments is shown.

18.2 The LRSLibrary

As initially presented, the LRSLibrary was designed for motion segmentation in videos. Unlike the BGSLibrary, the LRSLibrary is implemented in MATLAB and focuses on decomposition in low-rank and sparse components. The LRSLibrary provides a collection of state-of-the-art *matrix*-based and *tensor*-based factorization algorithms.

Through the graphical user interface (GUI) available in the LRSLibrary, the user can select the method type (i.e., RPCA for Robust PCA) and its respective algorithm (i.e., FPCP for Fast PCP). Next, you will need to selected the input video file and define the output video file. Then, click on the **process video** button, wait a moment and the expect results can be seen by clicking on the **display results** button. Note that in LRSLibrary GUI the blue, green, yellow, red, and dark red colors represent the computational cost of each algorithm (from very fast to very slow, respectively).[63] The LRSLibrary also disposes of an additional tool to resize and crop videos. To access this utility, please click on the **edit video** button. For large videos, it may be interesting to edit the video before processing it.

The LRSLibrary includes an additional tool to resize and crop videos. To access this utility, please click on the **edit video** button. For large videos, it may be interesting to the edit video before processing (please, see Figure 18.3).

[60]The CP model is a special case of the Tucker model, where the core tensor is superdiagonal and the number of components in the factor matrices is the same [27].

[61]Please refer to [19] for a complete review of low-rank tensor approximation techniques.

[62]The HoSVD of a tensor $\mathcal{X}$ involves the matrix SVDs of its unfolding matrices.

[63]Please refer to Section 18.4 for a detailed explanation.

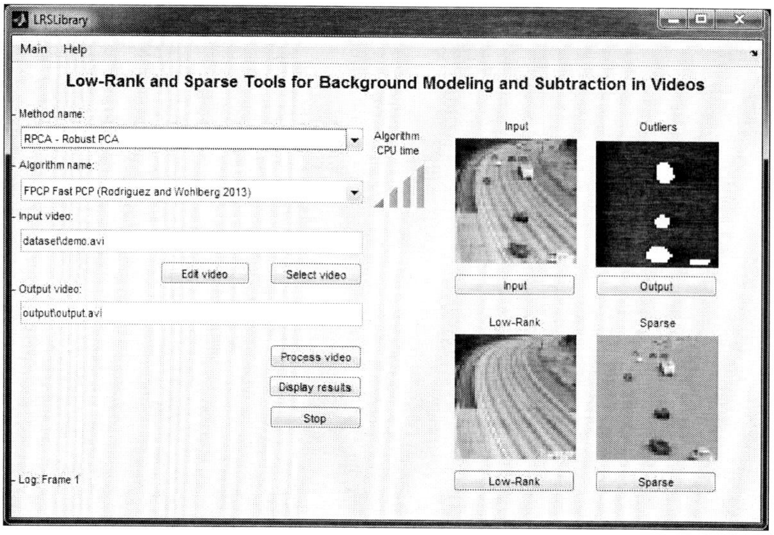

LRSLibrary

Main Help

Low-Rank and Sparse Tools for Background Modeling and Subtraction in Videos

Method name:

RPCA - Robust PCA

Algorithm CPU time

Algorithm name:

FPCP Fast PCP (Rodriguez and Wohlberg 2013)

Input video:

dataset\demo.avi

Edit video Select video

Output video:

output\output.avi

Process video

Display results

Stop

Log: Frame 1

Input Outliers

Input Output

Low-Rank Sparse

Low-Rank Sparse

FIGURE 18.2 LRSLibrary GUI.

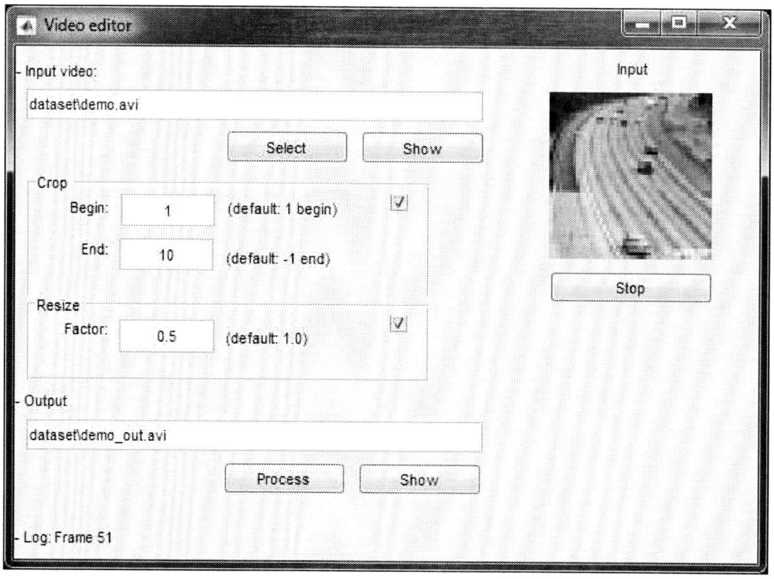

Video editor

Input video:

dataset\demo.avi

Select Show

Crop

Begin: 1 (default: 1 begin)

End: 10 (default: -1 end)

Resize

Factor: 0.5 (default: 1.0)

Output

dataset\demo_out.avi

Process Show

Log: Frame 51

Input

Stop

FIGURE 18.3 LRSLibrary video editor.

18.3 LRSLibrary Algorithms

Up to the date of writing of this paper, the LRSLibrary provided 93 algorithms for background modeling and subtraction. An updated list of available algorithms can be found on the library website.[64] The algorithms were grouped by their respective class: **RPCA** for Robust PCA, **ST** for Subspace Tracking, **MC** for Matrix Completion, **TTD** for Three-Term Decomposition, **LRR** for Low-Rank Representation, **NMF** for Non-negative Matrix Factorization, **NTF** for Non-negative Tensor Factorization, or **TD** for standard Tensor Decomposition. The mathematical models and theory of each algorithm will not be described further in this work as the reader can refer to the other chapters of this handbook for most of them. Table 18.1 and Table 18.2 show the available algorithms for each category.

For the *matrix*-based algorithms (i.e., RPCA), each video frame is a column vector of a matrix $\mathbf{M}$, while for *tensor*-based algorithms (i.e,. NTF) each video frame is a frontal slice of a third-order tensor $\mathcal{X}$. Next, depending on the selected algorithm, the matrix $\mathbf{M}$ or the tensor $\mathcal{X}$ is decomposed in its respective low-rank and sparse representation. The foreground mask (outliers) in Figure 18.2 are obtained by performing a hard-threshold (HT) function in the sparse representation, defined by:

$$HT(S) = \begin{cases} 1 & \text{if } \sqrt{S^2} > \beta \\ 0 & \text{otherwise} \end{cases}, \text{ where} \beta = \sqrt{\sigma(\boldsymbol{S})^2}. \tag{18.2}$$

In the Equation 18.2, S can be a *sparse*-matrix or a *sparse*-tensor, and $\boldsymbol{S}$ is the vectorized version of S.

18.4 Computational Cost

Many efforts have been recently concentrated to develop low-computational subspace learning algorithms. In this section, an evaluation of the computational cost of the LRSLibrary algorithms was made. Tables 18.3 and 18.4 shows the average CPU time consumption and the speed classification of each algorithm to decompose a 2304×51 matrix or $48 \times 48 \times 51$ tensor data. Both matrix and tensor data were built from **dataset/demo.avi** file.

The speed classification criterion (SCC) function was defined as:

$$SCC(\bar{t}) = \begin{cases} 1 & \text{if } \bar{t} < 1 & \text{(very fast: represented by blue color)} \\ 2 & \text{if } 1 < \bar{t} < 5 & \text{(fast: represented by green color)} \\ 3 & \text{if } 5 < \bar{t} < 20 & \text{(medium: represented by yellow color)} \\ 4 & \text{if } 20 < \bar{t} < 60 & \text{(slow: represented by red color)} \\ 5 & \text{if } \bar{t} > 60 & \text{(very slow: represented by dark red color)} \end{cases} \tag{18.3}$$

where $\bar{t}$ is the average time (in seconds) over three executions. In the LRSLibrary GUI, icons were used to represent the speed classification of each algorithm (see Figure 18.4).

The experiments were performed in an Intel Core i7-3740QM CPU 2.70GHz with 16Gb of RAM running MATLAB R2013b and Windows 7 Professional SP1 64 bits.

As can be seen in Table 18.3, FPCP and GoDec algorithms were both the fastest algorithms available in the LRSLibrary, processing data in 0.01 seconds.

[64] `https://github.com/andrewssobral/lrslibrary\#list-of-the-algorithms-available-in-lrslibrary`

TABLE 18.1 Algorithms (Part 1)

Algorithm ID	Algorithm Name
RPCA algorithms	
RPCA	Robust Principal Component Analysis [56]
PCP	Principal Component Pursuit [12]
FPCP	Fast PCP [49]
R2PCP	Riemannian Robust Principal Component Pursuit [24]
AS-RPCA	Active Subspace: Towards Scalable Low-Rank Learning [38]
ALM	Augmented Lagrange Multiplier [55]
EALM	Exact ALM [32]
IALM	Inexact ALM [32]
IALM_LMSVDS	IALM with LMSVDS [40]
IALM_BLWS	IALM with BLWS [35]
APG_PARTIAL	Partial Accelerated Proximal Gradient [33]
APG	Accelerated Proximal Gradient [33]
DUAL	Dual RPCA [33]
SVT	Singular Value Thresholding [11]
ADM	Alternating Direction Method [66]
LSADM	LSADM [43]
L1F	l_1-filtering [39]
DECOLOR	Contiguous Outliers in the Low-Rank Representation [72]
RegL1-ALM	Low-Rank Matrix Approximation under Robust l_1-Norm [69]
GA	Grassmann Average [22]
GM	Grassmann Median [22]
TGA	Trimmed Grassmann Average [22]
STOC-RPCA	Online Robust PCA via Stochastic Optimization [15]
MoG-RPCA	Mixture of Gaussians RPCA [68]
OP-RPCA	Robust PCA via Outlier Pursuit [61]
NSA1	Non-Smooth Augmented Lagrangian v1 [2]
NSA2	Non-Smooth Augmented Lagrangian v2 [2]
PSPG	Partially Smooth Proximal Gradient [3]
flip-SPCP-sum-SPG	Flip-Flop Stable PCP-sum (Spectral Projected Gradient) [1]
flip-SPCP-max-QN	Flip-Flop Stable PCP-max (Quasi-Newton) [1]
Lag-SPCP-SPG	Lagrangian SPCP (Spectral Projected Gradient) [1]
Lag-SPCP-QN	Lagrangian SPCP (Quasi-Newton) [1]
FW-T	SPCP solved by Frank-Wolfe method [45]
BRPCA-MD	Bayesian Robust PCA with Markov Dependency [14]
BRPCA-MD-NSS	BRPCA-MD with Non-Stationary Noise [14]
VBRPCA	Variational Bayesian RPCA [4]
PRMF	Probabilistic Robust Matrix Factorization [63]
OPRMF	Online PRMF [63]
MBRMF	Markov BRMF [64]
TFOCS-EC	TFOCS with equality constraints [6]
TFOCS-IC	TFOCS with inequality constraints [6]
GoDec	Go Decomposition [71]
SSGoDec	Semi-Soft GoDec [71]
Subspace Tracking (ST) algorithms	
GRASTA	Grassmannian Robust Adaptive Subspace Tracking Algorithm [23]
pROST	Robust PCA and subspace tracking using l_0-surrogates [50]
GOSUS	Grassmannian Online Subspace Updates with Structured-sparsity [62]
Matrix Completion (MC) algorithms	
LRGeomCG	Low-rank matrix completion by Riemannian optimization [58]
GROUSE	Grassmannian Rank-One Update Subspace Estimation [5]
OptSpace	A Matrix Completion Algorithm [25]
FPC	Fixed point and Bregman iterative methods [44]
SVT	A singular value thresholding algorithm for matrix completion [11]
Low-Rank Representation (LRR) algorithms	
EALM	Exact ALM [32]
IALM	Inexact ALM [32]
ADM	Alternating Direction Method [34]
LADMAP	Linearized ADM with Adaptive Penalty [34]
FastLADMAP	Fast LADMAP [34]
ROSL	Robust Orthonormal Subspace Learning [51]

TABLE 18.2 Algorithms (Part 2)

Algorithm ID	Algorithm Name
Three Terms Decomposition (TTD) algorithms	
TDD	3-Way-Decomposition [46]
MAMR	Motion-Assisted Matrix Restoration [65]
RMAMR	Robust Motion-Assisted Matrix Restoration [65]
ADMM	Alternating Direction Method of Multipliers [48]
Non-negative Matrix Factorization (NMF) algorithms	
NMF-MU	NMF solved by Multiplicative Updates
NMF-PG	NMF solved by Projected Gradient
NMF-ALS	NMF solved by Alternating Least Squares
NMF-ALS-OBS	NMF solved by Alternating Least Squares with Optimal Brain Surgeon
PNMF	Probabilistic Non-negative Matrix Factorization
ManhNMF	Manhattan NMF [20]
NeNMF	NMF via Nesterovs Optimal Gradient Method [21]
LNMF	Spatially Localized NMF [30]
ENMF	Exact NMF [16]
nmfLS2	Non-negative Matrix Factorization with sparse matrix [26]
Semi-NMF	Semi Non-negative Matrix Factorization
Deep-Semi-NMF	Deep Semi Non-negative Matrix Factorization [57]
iNMF	Incremental Subspace Learning via NMF [10]
DRMF	Direct Robust Matrix Factorization [60]
Non-negative Tensor Factorization (NTF) algorithms	
betaNTF	Simple beta-NTF implementation [41]
bcuNTD	Non-negative Tucker Decomposition by block-coordinate update [41]
bcuNCP	Non-negative CP Decomposition by block-coordinate update [41]
NTD-MU	Non-negative Tucker Decomposition solved by Multiplicative Updates [70]
NTD-APG	Non-negative Tucker Decomposition solved by Accelerated Proximal Gradient [70]
NTD-HALS	Non-negative Tucker Decomposition solved by Hierarchical ALS [70]
Tensor Decomposition (TD) algorithms	
HoSVD	Higher-order Singular Value Decomposition [Tucker Decomposition]
HoRPCA-IALM	HoRPCA solved by IALM [17]
HoRPCA-S	HoRPCA with Singleton model solved by ADAL [17]
HoRPCA-S-NCX	HoRPCA with Singleton model solved by ADAL [non-convex] [17]
Tucker-ADAL	Tucker Decomposition solved by ADAL [17]
Tucker-ALS	Tucker Decomposition solved by ALS
CP-ALS	PARAFAC/CP decomposition solved by ALS
CP-APR	PARAFAC/CP decomposition solved by Alternating Poisson Regression [13]
CP2	PARAFAC2 decomposition solved by ALS [9]
RSTD	Rank Sparsity Tensor Decomposition [31]
t-SVD	Tensor SVD in Fourier Domain [67]

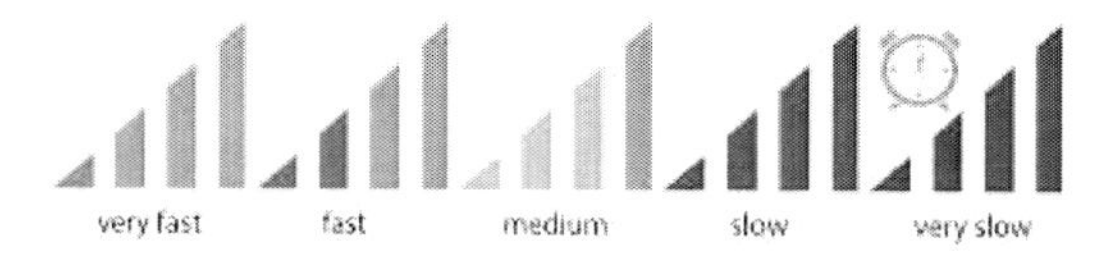

FIGURE 18.4 Icons representing the speed classification of each algorithm.

18.5 Usage example

The LRSLibrary is really easy to use, it contains several ready-to-use functions to help the user to perform video processing. Listing 18.1 demonstrates how to perform matrix and tensor factorization given an input video file. The final results are stored in the **out** variable, and the function **show_results** presents the background subtraction process, see Figure 18.5. Please, refer to **demo.m**[65] file available in the LRSLibrary package for a complete demonstration of available functions.

FIGURE 18.5 Given results from Listing 18.1.

18.6 How to Contribute

Every interested party is invited to cooperate by sending any implementation of low-rank and sparse decomposition algorithms. Two options are available:

- Sending the algorithm source code to us by email.
- Forking the LRSLibrary on GitHub,[66] pushing your changes, then sending us a pull request.

18.7 Conclusion

The LRSLibrary displays a wide variety of subspace learning algorithms that can be used via an easy-to-use graphical user interface. This library was developed to serve as a framework for detection and segmentation of moving objects using robust *matrix*-based and *tensor*-based factorization techniques. The experimental results in speed classification can further help the user to choose the best algorithm for his own situation. The library was developed in MATLAB due the majority of the researchers having more experience with scientific programming languages, but an improved version in C++ (or Python) of the LRSLibrary including only the fastest algorithm can be made for real-time or commercial purposes. The

[65] `https://github.com/andrewssobral/lrslibrary/blob/master/demo.m`
[66] `https://github.com/andrewssobral/lrslibrary`

TABLE 18.3 Computational cost of each algorithm and its respective speed classification (Part I). *In seconds.

Rank	Method	Algorithm ID	Speed classification	CPU Time*	SVD++
1	RPCA	FPCP	1	0.01	*
2	RPCA	GoDec	1	0.01	
3	RPCA	SSGoDec	1	0.02	
4	NMF	NeNMF	1	0.02	
5	RPCA	GM	1	0.03	
6	RPCA	TGA	1	0.03	
7	RPCA	GreGoDec	1	0.03	
8	RPCA	GA	1	0.04	
9	NMF	Deep-Semi-NMF	1	0.04	
10	NMF	LNMF	1	0.05	
11	NMF	iNMF	1	0.05	
12	RPCA	Lag-SPCP-QN	1	0.06	
13	LRR	ROSL	1	0.08	
14	MC	LRGeomCG	1	0.10	
15	TD	Tucker-ALS	1	0.10	
16	TD	t-SVD	1	0.14	
17	TD	CP-ALS	1	0.16	
18	RPCA	IALM	1	0.19	*
19	NMF	ManhNMF	1	0.20	
20	RPCA	FW-T	1	0.22	
21	ST	GRASTA	1	0.23	
22	NMF	Semi-NMF	1	0.24	
23	LRR	FastLADMAP	1	0.25	
24	RPCA	R2PCP	1	0.26	
25	NMF	NMF-MU	1	0.28	
26	RPCA	STOC-RPCA	1	0.30	
27	NMF	NMF-PG	1	0.30	
28	TTD	MAMR	1	0.31	
29	TTD	3WD	1	0.34	
30	RPCA	LSADM	1	0.35	
31	RPCA	PSPG	1	0.35	
32	NTF	bcuNCP	1	0.36	
33	NMF	NMF-ALS	1	0.37	
34	RPCA	AS-RPCA	1	0.42	
35	LRR	IALM	1	0.42	
36	LRR	LADMAP	1	0.43	
37	RPCA	L1F	1	0.45	*
38	NMF	NMF-ALS-OBS	1	0.48	
39	RPCA	NSA1	1	0.50	
40	RPCA	NSA2	1	0.54	
41	MC	GROUSE	1	0.54	
42	RPCA	RegL1-ALM	1	0.58	
43	NMF	DRMF	1	0.61	
44	TTD	RMAMR	1	0.62	
45	RPCA	IALM_LMSVDS	1	0.70	
46	TTD	ADMM	1	0.73	
47	LRR	ADM	1	0.74	

main goal of this work is to continuously improve the LRSLibrary by adding new features and new subspace learning methods.

18.8 Acknowledgments

First, the authors would like to thank everybody who contributed in some way to the success of this library, by making the implementation of each algorithm available on the Internet or sending it to us by email. The authors gratefully acknowledge the financial support of CAPES (Brazil) through the Brazilian Science Without Borders program (CsF) for granting a scholarship to the first author.

TABLE 18.4 Computational cost of each algorithm and its respective speed classification (Part II). *In seconds.

Rank	Method	Algorithm ID	Speed classification	CPU Time*	SVD++
48	RPCA	PCP	1	0.78	*
49	RPCA	APG_PARTIAL	1	0.92	*
50	TD	HoSVD	1	0.92	
51	RPCA	DECOLOR	1	0.93	
52	RPCA	APG	2	1.03	*
53	RPCA	VBRPCA	2	1.07	
54	RPCA	IALM_BLWS	2	1.10	
55	RPCA	PRMF	2	1.10	
56	MC	OptSpace	2	1.10	
57	NMF	nmfLS2	2	1.15	
58	NMF	PNMF	2	1.21	
59	RPCA	Lag-SPCP-SPG	2	1.50	
60	RPCA	TFOCS-IC	2	1.59	
61	RPCA	TFOCS-EC	2	1.62	
62	NMF	ENMF	2	1.80	
63	NTF	betaNTF	2	2.14	
64	RPCA	MoG-RPCA	2	2.15	
65	RPCA	EALM	2	2.21	*
66	TD	HoRPCA-IALM	2	2.42	
67	ST	GOSUS	2	2.48	
68	LRR	EALM	2	2.64	
69	TD	HoRPCA-S	2	2.92	
70	NTF	NTD-APG	2	3.31	
71	NTF	NTD-MU	2	3.50	
72	RPCA	ADM	2	3.58	
73	NTF	bcuNTD	2	3.69	
74	TD	RSTD	2	3.75	
75	RPCA	RPCA	2	3.84	
76	TD	Tucker-ADAL	2	4.45	
77	NTF	NTD-HALS	2	4.88	
78	MC	FPC	3	8.17	
79	RPCA	ALM	3	8.38	*
80	ST	pROST	3	8.94	
81	MC	SVT	3	9.68	
82	RPCA	flip-SPCP-max-QN	4	27.71	
83	TD	CP-APR	4	29.83	
84	RPCA	DUAL	4	38.41	*
85	TD	CP2	4	39.08	
86	RPCA	OPRMF	4	39.88	
87	RPCA	BRPCA-MD	4	45.90	
88	RPCA	BRPCA-MD-NSS	4	46.31	
89	RPCA	MBRMF	4	47.18	
90	TD	HoRPCA-S-NCX	4	48.43	
91	RPCA	OP-RPCA	4	54.23	
92	RPCA	flip-SPCP-sum-SPG	4	57.67	
93	RPCA	SVT	5	168.90	

Listing 18.1 demo.m

```
1  % First run the setup script
2  lrs_setup; % or run('C:/lrslibrary/lrs_setup')
3
4  % Load configuration
5  lrs_load_conf;
6
7  % Load video file
8  video = load_video_file(fullfile(lrs_conf.lrs_dir,'dataset','demo.avi'));
9
10 %%%%%%%%%%%%%%%%%%%%%%%%%%%%%%%%%%%%%%%%%%%%%%%%%%%%%%%%%%%%%%%%%%%%%%%%%%%%%%%%%%%%%%%%%%%%%%%%
11 %                    Demo: Matrix-based factorization                     %
12 %%%%%%%%%%%%%%%%%%%%%%%%%%%%%%%%%%%%%%%%%%%%%%%%%%%%%%%%%%%%%%%%%%%%%%%%%%%%%%%%%%%%%%%%%%%%%%%%
13 M = im2double(convert_video_to_2d(video));
14 m = video.height;
15 n = video.width;
16 p = video.nrFramesTotal;
17 opts.rows = m;
18 opts.cols = n;
19
20 % Robust PCA using FPCP algorithm
21 out = process_matrix('RPCA', 'FPCP', M, opts);
22 % Subspace Tracking using GRASTA algorithm
23 out = process_matrix('ST', 'GRASTA', M, opts);
24 % Matrix Completion using GROUSE algorithm
25 out = process_matrix('MC', 'GROUSE', M, opts);
26 %% Low Rank Recovery using FastLADMAP algorithm
27 out = process_matrix('LRR', 'FastLADMAP', M, opts);
28 % Three-Term Decomposition using 3WD algorithm
29 out = process_matrix('TTD', '3WD', M, opts);
30 % Non-negative Matrix Factorization using ManhNMF algorithm
31 out = process_matrix('NMF', 'ManhNMF', M, opts);
32
33 % Show results
34 show_out(M,out.L,out.S,out.O,p,m,n);
35
36 %%%%%%%%%%%%%%%%%%%%%%%%%%%%%%%%%%%%%%%%%%%%%%%%%%%%%%%%%%%%%%%%%%%%%%%%%%%%%%%%%%%%%%%%%%%%%%%%
37 %                    Demo: Tensor-based factorization                     %
38 %%%%%%%%%%%%%%%%%%%%%%%%%%%%%%%%%%%%%%%%%%%%%%%%%%%%%%%%%%%%%%%%%%%%%%%%%%%%%%%%%%%%%%%%%%%%%%%%
39 T = tensor(im2double(convert_video_to_3d(video)));
40
41 % Non-Negative Tensor Factorization using bcuNCP algorithm
42 out = process_tensor('NTF', 'bcuNCP', T);
43 % Tensor Decomposition using Tucker-ALS algorithm
44 out = process_tensor('TD', 'Tucker-ALS', T);
45
46 % Show results
47 show_3dtensors(T,out.L,out.S,out.O);
```

References

1. A. Aravkin, S. Becker, V. Cevher, and P. Olsen. A variational approach to stable principal component pursuit. *Conference on Uncertainty in Artificial Intelligence, UAI 2014*, July 2014.
2. N. Aybat, D. Goldfarb, and G. Iyengar. Fast first-order methods for stable principal component pursuit. *Preprint*, 2011.
3. N. Aybat, D. Goldfarb, and G. Iyengar. Efficient algorithms for robust and stable principal component pursuit. *Preprint*, November 2012.

4. S. Babacan, M. Luessi, R. Molina, and A. Katsaggelos. Sparse Bayesian methods for low-rank matrix estimation. *IEEE Transactions on Signal Processing*, 60(8):3964–3977, 2012.
5. L. Balzano, R. Nowak, and B. Recht. Online identification and tracking of subspaces from highly incomplete information. *Preprint*, 2010.
6. S. Becker, E. Candes, and M. Grant. TFOCS: flexible first-order methods for rank minimization. *Low-rank Matrix Optimization Symposium, SIAM Conference on Optimization*, 2011.
7. T. Bouwmans. Traditional and recent approaches in background modeling for foreground detection: An overview. In *Computer Science Review*, 2014.
8. T. Bouwmans and E. Zahzah. Robust PCA via Principal Component Pursuit: A review for a comparative evaluation in video surveillance. In *Special Issue on Background Models Challenge, Computer Vision and Image Understanding*, volume 122, pages 22–34, May 2014.
9. R. Bro. PARAFAC2. *http://onlinelibrary.wiley.com*, 2001.
10. S. Bucak, B. Gunsel, and O. Gursoy. Incremental subspace learning via non-negative matrix factorization. *Pattern Recognition*, 42(5):788–797, May 2009.
11. J. Cai, E. Candes, and Z. Shen. A singular value thresholding algorithm for matrix completion. *International Journal of ACM*, May 2008.
12. E. Candes, X. Li, Y. Ma, and J. Wright. Robust principal component analysis? *International Journal of ACM*, 58(3):117–142, May 2011.
13. E. Chi and T. Kolda. On tensors, sparsity, and non-negative factorizations. *SIAM Journal on Matrix Analysis and Applications*, 33(4):1272–1299, 2012.
14. X. Ding, L. He, and L. Carin. Bayesian robust principal component analysis. *IEEE Transaction on Image Processing*, 2011.
15. J. Feng, H. Xu, and S. Yan. Online robust PCA via stochastic optimization. *NIPS 2013*, 2013.
16. N. Gillis and F. Glineur. On the geometric interpretation of the non-negative rank. *Linear Algebra and its Applications*, pages 2685–2712, 2012.
17. D. Goldfarb and Z. Qin. Robust low-rank tensor recovery: Models and algorithms. *SIAM Journal on Matrix Analysis and Applications*, 2013.
18. N. Goyette, P. Jodoin, F. Porikli, J. Konrad, and P. Ishwar. changedetection.net: A new change detection benchmark dataset. In *IEEE Workshop on Change Detection CDW 2012 in conjunction with CVPR-2012*, pages 16–21, June 2012.
19. L. Grasedyck, D. Kressner, and C. Tobler. A literature survey of low-rank tensor approximation techniques. *arXiv/1302.7121*, 2013.
20. N. Guan, D. Tao, Z. Luo, and J. Shawe-Taylor. MahNMF: Manhattan non-negative matrix factorization. *Journal of Machine Learning Research*, 2012.
21. N. Guan, D. Tao, Z. Luo, and B. Yuan. NeNMF: an optimal gradient method for non-negative matrix factorization. *IEEE Transactions on Signal Processing*, 60(6):2882–2898, 2012.
22. S. Hauberg, A. Feragen, and M. Black. Grassmann averages for scalable robust PCA. *IEEE Conference on Computer Vision and Pattern Recognition, CVPR 2014*, June 2014.
23. J. He, L. Balzano, and A. Szlam. Incremental gradient on the Grassmannian for online foreground and background separation in subsampled video. *International on Conference on Computer Vision and Pattern Recognition, CVPR 2012*, June 2012.
24. M. Hintermuller and T. Wu. Robust principal component pursuit via inexact alternating minimization on matrix manifolds. *Journal of Mathematics and Imaging Vision*, 2014.
25. P. Jain, R. Keshavan, A. Montanari, and S. Oh. Matrix and tensor completion algorithms. *Preprint*, 2009.
26. Y. Ji and J. Eisenstein. Discriminative improvements to distributional sentence similarity. *Conference on Empirical Methods in Natural Language Processing, EMNLP 2013*,

2013.
27. T. Kolda and B. Bader. Tensor decompositions and applications. *SIAM Review*, 2008.
28. L. Lathauwer, B. Moor, and J. Vandewalle. A multilinear singular value decomposition. *SIAM Journal on Matrix Analysis and Applications*, 21(4):1253–1278, March 2000.
29. D. Lee and H. Seung. Learning the parts of objects by non-negative matrix factorization. *Nature*, 401(21):788–791, 1999.
30. S. Li, X. Hou, H. Zhang, and Q. Cheng. Learning spatially localized, parts-based representation. *IEEE Computer Society Conference on Computer Vision and Pattern Recognition, CVPR 2001*, 1:207–212, 2001.
31. Y. Li, J. Yan, Y. Zhou, and J. Yang. Optimum subspace learning and error correction for tensors. *European Conference on Computer Vision, ECCV 2010*, 2010.
32. Z. Lin, M. Chen, L. Wu, and Y. Ma. The augmented Lagrange multiplier method for exact recovery of corrupted low-rank matrices. *UIUC Technical Report*, November 2009.
33. Z. Lin, A. Ganesh, J. Wright, L. Wu, M. Chen, and Y. Ma. Fast convex optimization algorithms for exact recovery of a corrupted low-rank matrix. *UIUC Technical Report*, August 2009.
34. Z. Lin, R. Liu, and Z. Su. Linearized alternating direction method with adaptive penalty for low-rank representation. *NIPS 2011*, December 2011.
35. Z. Lin and S. Wei. A block Lanczos with warm start technique for accelerating nuclear norm minimization algorithms. *Preprint*, 2010.
36. G. Liu, Z. Lin, S. Yan, J. Sun, Y. Yu, and Y. Ma. Robust recovery of subspace structures by low-rank representation. *IEEE Transactions on Pattern Analysis and Machine Intelligence*, 35(1):171–184, January 2013.
37. G. Liu, Z. Lin, and Y. Yu. Robust subspace segmentation by low-rank representation. *International Conference on Machine Learning, ICML 2010*, 2010.
38. G. Liu and S. Yan. Active subspace: Toward scalable low-rank learning. *Neural Computation*, 24(12):3371–3394, December 2012.
39. R. Liu, Z. Lin, Z. Su, and J. Gao. Linear ime principal component pursuit and its extensions using l_1 filtering. *Neurocomputing*, 2014.
40. X. Liu, Z. Wen, and Y. Zhang. Limited memory block Krylov subspace optimization for computing dominant singular value decomposition. *Preprint*, 2012.
41. A. Liutkus. Simple to use NMF/NTF with beta divergence. *http://www.mathworks.com/*, 2012.
42. H. Lu, K. Plataniotis, and A. Venetsanopoulos. A survey of multilinear subspace learning for tensor data. *Pattern Recognition*, 44(7):1540–1551, July 2011.
43. S. Ma. Algorithms for sparse and low-rank optimization: Convergence, complexity and applications. *Thesis, Columbia University*, 2011.
44. S. Ma, D. Goldfarb, and L. Chen. Fixed point and Bregman iterative methods for matrix rank minimization. *Mathematical Programming Series A*, 128(1):321–353, 2011.
45. C. Mu, Y. Zhang, J. Wright, and D. Goldfarb. Scalable robust matrix recovery: Frank-Wolfe meets proximal methods. *Preprint*, 2014.
46. O. Oreifej, X. Li, and M. Shah. Simultaneous video stabilization and moving object detection in turbulence. *IEEE Transactions on Pattern Analysis and Machine Intelligence, PAMI 2012*, 2012.
47. P. Paatero and A. Tapper. Positive matrix factorization: a non-negative factor model with optimal utilization of error estimates of data values. *Environmetrics*, 5:111–126, 1994.
48. N. Parikh and S. Boyd. Proximal algorithms. *Foundations and Trends in Optimization*, 1(3), 2014.
49. P. Rodriguez and B. Wohlberg. Fast principal component pursuit via alternating minimization. *IEEE International Conference on Image Processing, ICIP 2013*, September 2013.

50. F. Seidel, C. Hage, and M. Kleinsteuber. pROST - a smoothed Lp-norm robust online subspace tracking method for realtime background subtraction in video. *Special Issue on Background Modeling for Foreground Detection in Real-World Dynamic Scenes, Machine Vision and Applications*, June 2014.
51. X. Shu, F. Porikli, and N. Ahuja. Robust orthonormal subspace learning: Efficient recovery of corrupted low-rank matrices. *International Conference on Computer Vision and Pattern Recognition, CVPR 2014*, June 2014.
52. A. Sobral. BGSLibrary: An OpenCV C++ Background Subtraction Library. In *Workshop on Vision Computational, WVC 2013*, Rio de Janeiro, Brazil, June 2013.
53. A. Sobral, C. Baker, T. Bouwmans, and E. Zahzah. Incremental and multi-feature tensor subspace learning applied for background modeling and subtraction. *International Conference on Image Analysis and Recognition, ICIAR 2014*, pages 94–103, 2014.
54. A. Sobral and A. Vacavant. A comprehensive review of background subtraction algorithms evaluated with synthetic and real videos. *Computer Vision and Image Understanding*, 122(1):4–21, 2014.
55. G. Tang and A. Nehorai. Robust principal component analysis based on low-rank and block-sparse matrix decomposition. *CISS 2011*, 2011.
56. F. De La Torre and M. Black. A robust principal component analysis for computer vision. *International Conference on Computer Vision*, 2001.
57. G. Trigeorgis, K. Bousmalis, S. Zafeiriou, and B. Schuller. A deep semi-NMF model for learning hidden representations. *International Conference on Machine Learning, ICML 2014*, 2014.
58. B. Vandereycken, P. Absil, and S. Vandewalle. A Riemannian geometry with complete geodesics for the set of positive semidefinite matrices of fixed rank. *IMA Journal of Numerical Analysis*, 33:481–514, 2013.
59. J. Wright, Y. Peng, Y. Ma, A. Ganesh, and S. Rao. Robust principal component analysis: Exact recovery of corrupted low-rank matrices by convex optimization. *Neural Information Processing Systems, NIPS 2009*, December 2009.
60. L. Xiong, X. Chen, and J. Schneider. Direct robust matrix factorization for anomaly detection. *International Conference on Data Mining, ICDM 2011*, 2011.
61. H. Xu, C. Caramanis, and S. Sanghavi. Robust PCA via outlier pursuit. *Annual Conference on Neural Information Processing Systems, NIPS 2010*, 2010.
62. J. Xu, V. Ithapu, L. Mukherjee, J. Rehg, and V. Singh. GOSUS: Grassmannian online subspace updates with structured-sparsity. *International Conference on Computer Vision, ICCV 2013*, September 2013.
63. N. Yang, T. Yao, J. Wang, and D. Yeung. A probabilistic approach to robust matrix factorization. *European Conference on Computer Vision, ECCV 2012*, pages 126–139, 2012.
64. N. Yang and D. Yeung. Bayesian robust matrix factorization for image and video processing. *International Conference on Computer Vision, ICCV 2013*, 2013.
65. X. Ye, J. Yang, X. Sun, K. Li, C. Hou, and Y. Wang. Foreground-background separation from video clips via motion-assisted matrix restoration. *IEEE Transactions on Circuits and Systems for Video Technology*, 2015.
66. X. Yuan and J. Yang. Sparse and low-rank matrix decomposition via alternating direction methods. *Optimization Online*, November 2009.
67. Z. Zhang, G. Ely, S. Aeron, N. Hao, and M. Kilmer. Novel factorization strategies for higher order tensors: Implications for compression and recovery of multi-linear data. *Preprint*, 2013.
68. Q. Zhao, D. Meng, Z. Xu, W. Zuo, and L. Zhang. Robust principal component analysis with complex noise. *International Conference on Machine Learning, ICML 2014*, 2014.
69. Y. Zheng, G. Liu, S. Sugimoto, S. Yan, and M. Okutomi. Practical low-rank matrix approx-

imation under robust $l1$-norm. *CVPR 2012*, 2012.
70. G. Zhou, A. Cichocki, and S. Xie. Fast non-negative matrix/tensor factorization based on low-rank approximation. *IEEE Transactions on Signal Processing*, 60(6):2928–2940, June 2012.
71. T. Zhou and D. Tao. GoDec: randomized low-rank and sparse matrix decomposition in noisy case. *International Conference on Machine Learning, ICML 2011*, 2011.
72. X. Zhou, C. Yang, and W. Yu. Moving object detection by detecting contiguous outliers in the low-rank representation. *IEEE Transactions on Pattern Analysis and Machine Intelligence*, 35:597–610, 2013.
73. Z. Zhou, X. Li, J. Wright, E. Candes, and Y. Ma. Stable principal component pursuit. *IEEE ISIT*, pages 1518–1522, June 2010.

19

Dynamic Mode Decomposition for Robust PCA with Applications to Foreground/Background Subtraction in Video Streams and Multi-Resolution Analysis

Jake Nathan Kutz
University of Washington, USA

Xing Fu
University of Washington, USA

Steven L. Brunton
University of Washington, USA

Jacob Grosek
Air Force Research Laboratories, USA

19.1 Introduction

Accurate and real-time video surveillance techniques for removing *background* variations in a video stream, which are highly correlated between frames, are at the forefront of modern data-analysis research. The objective in such algorithms is to highlight *foreground* objects of potential interest. Background/foreground separation is typically an integral step in detecting, identifying, tracking, and recognizing objects in video sequences. Most modern computer vision applications demand algorithms that can be implemented in real-time, and that are robust enough to handle diverse, complicated, and cluttered backgrounds. Competitive methods often need to be flexible enough to accommodate changes in a scene due to, for instance, illumination changes that can occur throughout the day, or location changes where the application is being implemented. Given the importance of this task, a variety of iterative techniques and methods have already been developed in order to perform background/-foreground separation [4, 8, 11, 15, 23, 24] (See also, for instance, the recent reviews by Bouwmans [2] and Benezeth et al. [1], which compare error and timing of various methods).

One potential viewpoint of this computational task is as a matrix separation problem into *low-rank* (background) and *sparse* (foreground) components. Recently, this viewpoint has been advocated by Candès et al. in the framework of *robust principal component analysis*

(RPCA) [4]. By weighting a combination of the nuclear and the L^1 norms, a convenient convex optimization problem (*principal component pursuit*) was demonstrated, under suitable assumptions, to exactly recover the low-rank and sparse components of a given data-matrix (or video for our purposes). It was also compared to the state-of-the-art computer vision procedure developed by De La Torre and Black [10]. We advocate a similar matrix separation approach, but by using the method of *dynamic mode decomposition* (DMD) [5, 17, 20–22, 26] (see also Kutz [9] for a tutorial review). This method, which essentially implements a Fourier decomposition of correlated spatial activity of the video frames in time, distinguishes the stationary background from the dynamic foreground by differentiating between the near-zero Fourier modes and the remaining modes bounded away from the origin, respectively [7]. Originally introduced in the fluid mechanics community, DMD has emerged as a powerful tool for analyzing the dynamics of nonlinear systems [5, 17, 20–22, 26].

In the application of video surveillance, the video frames can be thought of as snapshots of some underlying complex/nonlinear dynamics. The DMD decomposition yields oscillatory time components of the video frames that have contextual implications. Namely, those modes that are near the origin represent dynamics that are unchanging, or changing slowly, and can be interpreted as stationary background pixels, or low-rank components of the data matrix. In contrast, those modes bounded away from the origin are changing on $\mathcal{O}(1)$ timescales or faster, and represent the foreground motion in the video, or the sparse components of the data matrix. Thus, by simply applying the dynamical systems DMD interpretation to video frames, an approximate RPCA technique can be enacted at a fixed cost of a singular-value decomposition and a linear equation solver. The additional innovation of multi-resolution DMD (MRDMD) allows for further separation of dynamic content in the video, thus allowing for the separation of components that are happening on different time scales.

More broadly, the modeling of multi-scale systems, both in time and space, pervade modern theoretical and computational efforts across the engineering, biological, and physical sciences. Driving innovations are methods and algorithms that circumvent the significant challenges in efficiently connecting micro- to macro-scale effects that are potentially separated by orders of magnitude spatially and/or temporally. Wavelet-based methods and/or windowed Fourier Transforms are ideally structured to perform such multi-resolution analyses (MRA) as they systematically remove temporal or spatial features by a process of recursive refinement of sampling from the data of interest. Typically, MRA is performed on either space or time, but not both simultaneously. By integrating the concept of MRA with the DMD, the MRDMD naturally integrates space and time so that the multi-scale spatio-temporal features are easily separated.

19.1.1 Overview of Method

The origins of the DMD method are associated with the fluid dynamics community and the modeling of complex flows. Its growing success stems from the fact that it is an *equation-free*, data-driven method capable of providing accurate assessments of the spatial-temporal coherent structures in a complex fluid, or short-time future estimates of complex systems, thus allowing for control protocols to be enacted simply from sampling. More broadly, DMD has quickly gained popularity since it provides information about nonlinear dynamical systems. DMD analysis can be considered to be a numerical approximation to Koopman spectral analysis [16–18], and it is in this sense that DMD is applicable to nonlinear systems.

To be more precise, one may consider the DMD as a way to approximate the dynamics of a nonlinear system. Consider the dynamical nonlinear system

$$\frac{d\mathbf{x}}{dt} = f(\mathbf{x}, t)\,. \tag{19.1}$$

In addition to the governing equations, both measurements of the system

$$g(\mathbf{x}, t) = 0 \tag{19.2}$$

and initial conditions are prescribed

$$\mathbf{x}(0) = \mathbf{x}_0. \tag{19.3}$$

Typically $\mathbf{x}$ is an N-dimensional vector ($N \gg 1$) that arises from either discretization of a complex system, or in the case of video streams, it is the total number of pixels in a given frame. The governing equation and initial condition specify a well-posed initial value problem. The inclusion of measurements $g(\mathbf{x}, t)$, let's say M of them, make the system overdetermined. By including model error along with noisy measurements, one can formulate an optimal predictive strategy using a data-assimilation framework and Kalman filtering innovations [9].

Since in general the solution of governing nonlinear evolution (19.1) with (19.2) is not possible to construct, numerical solutions are used to evolve to future states. In the DMD framework, the equation-free viewpoint also assumes that the right-hand side of (19.1) governing the physics, $f(\mathbf{x}, t)$, is unknown. Thus the snapshot measurements and initial conditions alone are used to approximate the dynamics and predict the future state. The DMD procedure thus constructs the proxy, approximate linear evolution

$$\frac{d\tilde{\mathbf{x}}}{dt} = \mathbf{A}\tilde{\mathbf{x}} \tag{19.4}$$

with $\tilde{\mathbf{x}}(0) = \tilde{\mathbf{x}}_0$ and whose solution is

$$\tilde{\mathbf{x}}(t) = \sum_{k=1}^{K} b_k \psi_k \exp(\omega_k t) \tag{19.5}$$

where ψ_k and ω_k are the eigenfunctions and eigenvalues of the matrix $\mathbf{A}$. The ultimate goal in the DMD algorithm is to optimally construct the matrix $\mathbf{A}$ so that the true and approximate solution remain optimally close in a least-square sense:

$$\|\mathbf{x}(t) - \tilde{\mathbf{x}}(t)\| \ll 1. \tag{19.6}$$

Of course, the optimality of the approximation holds only over the sampling window where $\mathbf{A}$ is constructed, but the approximate solution can be used to not only make future state predictions, but also to decompose the dynamics into various time-scales since the ω_k are prescribed. Moreover, the DMD typically makes use of low-rank structure so that the total number of modes, $K \ll N$, allows for dimensionality reduction of the complex system or video stream.

At its core, the DMD method can be thought of as an ideal combination of spatial dimensionality-reduction techniques, such as the Proper Orthogonal Decomposition [9], with Fourier Transforms in time. By simply interpreting a video stream in this context, background/foreground separation can be achieved in a highly efficient manner. Moreover, our goal is to integrate the DMD decomposition with key concepts from wavelet theory and MRA. Specifically, the DMD method takes snapshots of an underlying dynamical systems to construct its decomposition. However, the frequency and duration (sampling window) of the data collection process can be adapted, much as in wavelet theory, to sift out information at different scales. Indeed, an iterative refinement of progressively shorter snapshot sampling windows and recursive extraction of DMD modes from slow to increasingly fast time scales allows for an MRDMD that allows for improved analytics. Moreover, it also allows for improved analytic predictions of the short-time future state of the system, which is of critical importance for control protocols.

19.2 Dynamic Mode Decomposition

The DMD method provides a spatio-temporal decomposition of data into a set of dynamic modes that are derived from snapshots or measurements of a given system in time. The mathematics underlying the extraction of dynamic information from time-resolved snapshots is closely related to the idea of the Arnoldi algorithm [25], one of the workhorses of fast computational solvers. The data collection process involves two parameters:

$$N = \text{number of spatial points saved per time snapshot}$$
$$M = \text{number of snapshots taken}$$

Originally the algorithm was designed to collect data at regularly spaced intervals of time. However, new innovations allow for both sparse spatial and temporal collection of data [3] as well as irregularly spaced collection times [26]. To illustrate the algorithm, we consider regularly spaced sampling in time:

$$\text{data collection times}: \quad t_{m+1} = t_m + \Delta t \tag{19.7}$$

where the collection time starts at t_1 and ends at t_M, and the interval between data collection times is Δt. In the MRDMD method [6], the total number of snapshots will vary as the algorithm extracts multi-timescale spatio-temporal structures.

The data snapshots are arranged into an $N \times M$ matrix

$$\mathbf{X} = [\mathbf{x}(t_1) \;\; \mathbf{x}(t_2) \;\; \mathbf{x}(t_3) \;\; \cdots \;\; \mathbf{x}(t_M)] \tag{19.8}$$

where the vector $\mathbf{x}$ are the N measurements of the state variable of the system of interest at the data collection points. The objective is to mine the data matrix $\mathbf{X}$ for important dynamical information. For the purposes of the DMD method, the following matrix is also defined:

$$\mathbf{X}_j^k = [\mathbf{x}(t_j) \;\; \mathbf{x}(t_{j+1}) \;\; \cdots \;\; \mathbf{x}(t_k)] \tag{19.9}$$

Thus this matrix includes columns j through k of the original data matrix.

The DMD method approximates the modes of the so-called *Koopman operator*. The Koopman operator is a linear, infinite-dimensional operator that represents nonlinear, infinite-dimensional dynamics without linearization [16–18], and is the adjoint of the Perron-Frobenius operator. The method can be viewed as computing, from the experimental data, the eigenvalues and eigenvectors (low-dimensional modes) of a linear model that approximates the underlying dynamics, even if the dynamics are nonlinear. Since the model is assumed to be linear, the decomposition gives the growth rates and frequencies associated with each mode. If the underlying model is linear, then the DMD method recovers the leading eigenvalues and eigenvectors normally computed using standard solution methods for linear differential equations.

Mathematically, the Koopman operator $\mathbf{A}$ is a linear, time-independent operator depicted in Figure 19.1 such that [17]

$$\mathbf{x}_{j+1} = \mathbf{A}\mathbf{x}_j \tag{19.10}$$

where j indicates the specific data collection time and $\mathbf{A}$ is the linear operator that maps the data from time t_j to t_{j+1}. The vector $\mathbf{x}_j$ is an N-dimensional vector of the data points collected at time j. The computation of the Koopman operator is at the heart of the DMD methodology. As already stated, the mapping over Δt is linear even though the underlying dynamics that generated $\mathbf{x}_j$ may be nonlinear. It should be noted that this is different than linearizing the dynamics.

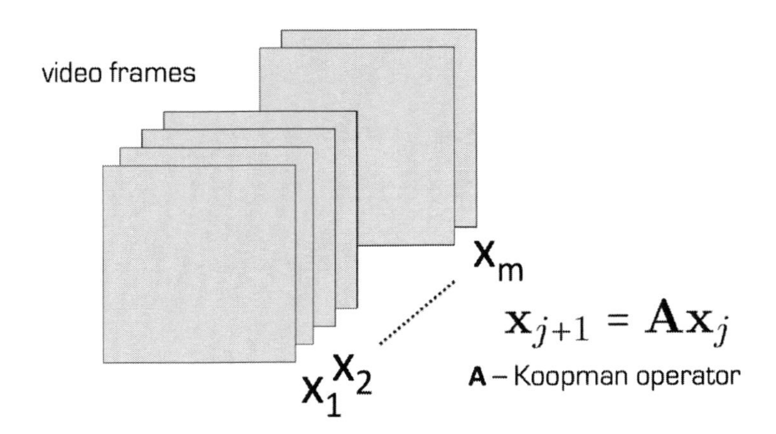

FIGURE 19.1 Illustration of the DMD method where snapshots (video frames) $\mathbf{x}_j$ are taken and a linear transformation $\mathbf{A}$ (Koopman operator) is constructed. The DMD method constructs the best matrix $\mathbf{A}$ that minimizes the least-square error for all transformations $\mathbf{x}_{j+1} = \mathbf{A}\mathbf{x}_j$ with $j = 1, 2, \cdots, m-1$.

To construct the appropriate Koopman operator that best represents the data collected, the matrix $\mathbf{X}_1^{M-1}$ is considered:

$$\mathbf{X}_1^{M-1} = [\mathbf{x}_1 \ \mathbf{x}_2 \ \mathbf{x}_3 \ \cdots \ \mathbf{x}_{M-1}] \,. \tag{19.11}$$

Making use of (19.10), this matrix reduces to

$$\mathbf{X}_1^{M-1} = [\mathbf{x}_1 \ \mathbf{A}\mathbf{x}_1 \ \mathbf{A}^2\mathbf{x}_1 \ \cdots \ \mathbf{A}^{M-2}\mathbf{x}_1] \,. \tag{19.12}$$

Here is where the DMD method connects to Krylov subspaces and the Arnoldi algorithm. Specifically, the columns of $\mathbf{X}_1^{M-1}$ are each elements in a Krylov space [25]. This matrix attempts to fit the first $M-1$ data collection points using the Koopman operator (matrix) $\mathbf{A}$. In the DMD technique, the final data point $\mathbf{x}_M$ is represented, as well as possible, in terms of this Krylov basis, thus

$$\mathbf{x}_M = \sum_{m=1}^{M-1} b_m \mathbf{x}_m + \mathbf{r} \tag{19.13}$$

where the b_m are the coefficients of the Krylov space vectors and $\mathbf{r}$ is the residual (or error) that lies outside (orthogonal to) the Krylov space. Ultimately, this best fit to the data using this DMD procedure will be done in an L^2 sense using a pseudo-inverse.

Before proceeding further, it is at this point that the data matrix $\mathbf{X}_1^{M-1}$ in (19.12) should be considered further. In particular, our dimensionality reduction methods look to take advantage of any low-dimensional structures in the data. To exploit this, the SVD of (19.12) is computed:

$$\mathbf{X}_1^{M-1} = \mathbf{U}\mathbf{\Sigma}\mathbf{V}^* \tag{19.14}$$

where $*$ denotes the conjugate transpose, $\mathbf{U} \in \mathbb{C}^{N\times K}$, $\mathbf{\Sigma} \in \mathbb{C}^{K\times K}$, and $\mathbf{V} \in \mathbb{C}^{M-1\times K}$. Here K is the reduced SVD's approximation to the rank of $\mathbf{X}_1^{M-1}$. If the data matrix is full rank and the data has no suitable low-dimensional structure, then the DMD method fails immediately. However, if the data matrix can be approximated by a low-rank matrix, then

DMD can take advantage of this low-dimensional structure to project a future state of the system. Thus once again, the SVD plays the critical role in the methodology.

With the reduction (19.14) to (19.12), we can return to the results of the Koopman operator and Krylov basis (19.13). Specifically, generalizing (19.10) to its matrix form yields

$$\mathbf{A}\mathbf{X}_1^{M-1} = \mathbf{X}_2^M . \tag{19.15}$$

But by using (19.13), the right hand-side of this equation can be written in the form

$$\mathbf{X}_2^M = \mathbf{X}_1^{M-1}\mathbf{S} + \mathbf{r}e_{M-1}^* \tag{19.16}$$

where e_{M-1} is the $(M-1)$th unit vector and

$$\mathbf{S} = \begin{bmatrix} 0 \cdots & 0 & b_1 \\ 1 \ddots & 0 & b_2 \\ 0 \ddots \ddots & & \vdots \\ \ddots \ddots & 0 & b_{M-2} \\ 0 \cdots \; 0 & 1 & b_{M-1} \end{bmatrix} . \tag{19.17}$$

Recall that the b_j are the unknown coefficients in (19.13).

The key idea now is the observation that the eigenvalues of $\mathbf{S}$ approximate some of the eigenvalues of the unknown Koopman operator $\mathbf{A}$, making the DMD method similar to the Arnoldi algorithm and its approximations to the Ritz eigenvalues. Schmid [21] showed that rather than computing the matrix $\mathbf{S}$ directly, we can instead compute the *lower-rank* matrix

$$\tilde{\mathbf{S}} = \mathbf{U}^*\mathbf{X}_2^M\mathbf{V}\mathbf{\Sigma}^{-1} \tag{19.18}$$

which is related to $\mathbf{S}$ via a similarity transformation. Recall that the matrices $\mathbf{U}$, $\mathbf{\Sigma}$, and $\mathbf{V}$ arise from the SVD reduction of $\mathbf{X}_1^{M-1}$ in (19.14).

Consider then the eigenvalue problem associated with $\tilde{\mathbf{S}}$:

$$\tilde{\mathbf{S}}\mathbf{y}_k = \mu_k\mathbf{y}_k \qquad k = 1, 2, \cdots, K \tag{19.19}$$

where K is the rank of the approximation we are choosing to make. The eigenvalues μ_k capture the time dynamics of the discrete Koopman map $\mathbf{A}$ as a Δt step is taken forward in time. These eigenvalues and eigenvectors can be related back to the similarity transformed original eigenvalues and eigenvectors of $\mathbf{S}$ in order to construct the DMD modes:

$$\psi_k = \mathbf{U}\mathbf{y}_k . \tag{19.20}$$

With the low-rank approximations of both the eigenvalues and eigenvectors in hand, the projected future solution can be constructed for all time in the future. By first rewriting for convenience $\omega_k = \ln(\mu_k)/\Delta t$ (recall that the Koopman operator time dynamics is linear), then the approximate solution at all future times, $\mathbf{x}_{\text{DMD}}(t)$, is given by

$$\mathbf{x}_{\text{DMD}}(t) = \sum_{k=1}^{K} b_k(0)\psi_k(\mathbf{x})\exp(\omega_k t) = \mathbf{\Psi}\text{diag}(\exp(\omega t)\mathbf{b} \tag{19.21}$$

where $b_k(0)$ is the initial amplitude of each mode, $\mathbf{\Psi}$ is the matrix whose columns are the eigenvectors ψ_k, diag(ωt) is a diagonal matrix whose entries are the eigenvalues $\exp(\omega_k t)$, and $\mathbf{b}$ is a vector of the coefficients b_k.

It only remains to compute the initial coefficient values $b_k(0)$. If we consider the initial snapshot $(\mathbf{x}_1)$ at time zero, let's say, then (19.21) gives $\mathbf{x}_1 = \mathbf{\Psi b}$. This generically is not a square matrix so that its solution

$$\mathbf{b} = \mathbf{\Psi}^+ \mathbf{x}_1 \tag{19.22}$$

can be found using a pseudo-inverse. Indeed, $\mathbf{\Psi}^+$ denotes the Moore-Penrose pseudo-inverse that can be accessed in MATLAB via the **pinv** command. The pseudo-inverse is equivalent to finding the best solution **b** in the least-squares (best fit) sense. This is equivalent to how DMD modes were derived originally.

Overall, then, the DMD algorithm presented here takes advantage of low dimensionality in the data in order to make a low-rank approximation of the linear mapping that best approximates the nonlinear dynamics of the data collected for the system. Once this is done, a prediction of the future state of the system is achieved for all time. Unlike the POD method, which requires solving a low-rank set of dynamical quantities to predict the future state, no additional work is required for the future state prediction outside of plugging in the desired future time into (19.21). Thus the advantages of DMD revolve around the fact that (i) no underlying governing equations are prescribed, which is ideal in the context of video processing, and (ii) the future state is known for all time (of course, provided the DMD approximation holds).

The algorithm is as follows:

(i) Sample data at N prescribed locations M times. The data snapshots should ideally be evenly spaced in time by a fixed Δt. This gives the data matrix $\mathbf{X}$.

(ii) From the data matrix $\mathbf{X}$, construct the two sub-matrices $\mathbf{X}_1^{M-1}$ and $\mathbf{X}_2^M$.

(iii) Compute the SVD decomposition of $\mathbf{X}_1^{M-1}$.

(iv) The matrix $\tilde{\mathbf{S}}$ can then be computed and its eigenvalues and eigenvectors found.

(v) Project the initial state of the system onto the DMD modes using the pseudo-inverse.

(vi) Compute the solution at any future time using the DMD modes along with their projection to the initial conditions and the time dynamics computed using the eigenvalue of $\tilde{\mathbf{S}}$.

19.3 Robust PCA with DMD

Given a collection of data from a potentially complex, nonlinear system, the RPCA method will seek out the sparse structures within the data, while simultaneously fitting the remaining entries to a low-rank basis. As long as the given data is truly of this nature, in that it lies on a low-dimensional subspace and has sparse components, then the RPCA algorithm has been proven by Candès et al. [4] to perfectly separate the given data $\mathbf{X}$ according to

$$\mathbf{X} = \mathbf{L} + \mathbf{S} \tag{19.23}$$

where

$$\begin{aligned} \mathbf{L} &\rightarrow \text{low-rank} \\ \mathbf{S} &\rightarrow \text{sparse.} \end{aligned}$$

The key to the RPCA algorithm is formulating the problem into a tractable, nonsmooth convex optimization problem known as *principal component pursuit* (PCP):

$$\begin{aligned} \arg\min \quad & \|\mathbf{L}\|_* + \lambda \|\mathbf{S}\|_1 \\ \text{subject to} \quad & \mathbf{X} = \mathbf{L} + \mathbf{S} \end{aligned} \tag{19.24}$$

Here PCP is minimizing the weighed combination of the nuclear norm: $\|\mathbf{M}\|_* := \text{trace}\left(\sqrt{\mathbf{M}^*\mathbf{M}}\right)$ and the L^1-norm: $\|\mathbf{M}\|_1 := \sum_{ij} |m_{ij}|$. The scalar regularization parameter is non-negative: $\lambda \geq 0$. From the optimization problem (19.24), it can be seen that as $\lambda \to 0$, the low-rank structure will incorporate all of the given data: $\mathbf{L} \to \mathbf{X}$, leaving the sparse structure devoid of anything. It is also true that as λ increases, the sparse structure will embody more and more of the original data matrix: $\mathbf{S} \to \mathbf{X}$, as the low-rank structure will commensurately approach the zero matrix [4].

Effectively, λ controls the dimensionality of the low-rank subspace; however, one does not need to know the rank of $\mathbf{L}$ *a priori*. Candès et al. [4] have shown that the choice

$$\lambda = \frac{1}{\sqrt{\max(n, m)}},$$

where $\mathbf{X}$ is $n \times m$, has a high probability of success at producing the correct low-rank and sparse separation provided that the matrices $\mathbf{L}$ and $\mathbf{S}$ are incoherent, which is the case for many practical applications. Some fine tuning of λ may yield slightly improved results (See Kutz [9] for a tutorial on RPCA and the effects of varying λ).

Though there are multiple methods that can solve the convex PCP problem, the *augmented Lagrange multiplier* (ALM) method stands out as a simple and stable algorithm with robust, efficient performance characteristics. The ALM method is effective because it achieves high accuracies in fewer iterations when compared against other competing methods [4]. Moreover, there is an *inexact* ALM variant [13] to the *exact* ALM method [12], which is able to converge in even fewer iterations at the cost of weaker guaranteed convergence criteria. MATLAB code that implements these methods, along with a few other algorithms, can be downloaded from the University of Illinois Perception and Decision Lab website [14]. We implement these codes in what follows.

19.3.1 Video Interpretation of the RPCA Method

In a video sequence, stationary background objects translate into highly correlated pixel regions from one frame to the next, which suggests a low-rank structure within the video data. In the case of videos, where the data in each frame is 2D by nature, frames need to be reshaped into 1D column vectors and united into a single data matrix $\mathbf{X}$. The RPCA algorithm can then implement the background/foreground separation found in $\mathbf{X} = \mathbf{L} + \mathbf{S}$, where the low-rank matrix $\mathbf{L}$ will render the video of just the background, and the sparse matrix $\mathbf{S}$ will render the complementary video of the moving foreground objects. Because the foreground objects exhibit a spatial coherency throughout the video, the RPCA method is no longer guaranteed a high probability of success; however, in practice, RPCA achieves an acceptable separation almost every time [4].

The RPCA method performs the foreground/background separation with a quality on par with the DMD method, which will be reviewed in the next section. However, as λ is decreased, the sparse reconstruction of the video, which is stored in matrix $\mathbf{S}$, starts to bring in more of the original video, including erroneous stationary pixels that should be part of the low-rank background. When λ is increased, the sparse reconstruction of the video begins to see a decrease in the pixel intensities that correspond to the moving objects, and some foreground pixels disappear altogether.

19.3.2 RPCA Interpretation of the DMD Method

The DMD algorithm can be used to produce a similar low-rank and sparse separation as in (19.23) [7]. For the DMD case, the separation relies on the interpretation of the ω_k frequencies in the solution reconstructions represented in general by (19.5), and more specifically in DMD by (19.21). In particular, low-rank features in video, for instance, are such that $|\omega_j| \approx 0$, i.e., they are slowly changing in time. Thus if one sets a threshold so as to gather all the low-rank modes where $|\omega_j| \leq \epsilon \ll 1$, then the separation can be accomplished. This reproduces a representation of the $\mathbf{L}$ and $\mathbf{S}$ matrices of the form:

$$\begin{aligned} \mathbf{L} &\approx \sum_{|\omega_k|\leq\epsilon} b_k \psi_k \exp(\omega_k t) \\ \mathbf{S} &\approx \sum_{|\omega_k|>\epsilon} b_k \psi_k \exp(\omega_k t) \end{aligned} \tag{19.25}$$

Note that the low-rank matrix $\mathbf{L}$ picks out only a small number of the total number of DMD modes to represent the *slow* oscillations or DC content in the data ($\omega_j = 0$). The DC content is exactly the background mode when interpreted in the video stream context with a fixed and stable camera.

The advantage of the DMD method and its sparse/low-rank separation is the computational efficiency of achieving (19.25), especially when compared to the optimization methods of RPCA. However, the DMD method does not do well for delta-function-like behaviors in time, i.e., a very rapid on/off behavior. This is due to the fact that such a behavior would require many Fourier modes in time to resolve, thus undermining the basic method of associating correlated spatial activity with single oscillatory behaviors in time.

19.4 DMD for Background Subtraction

A video sequence offers an appropriate application for this DMD method because the frames of the video are, by nature, equally spaced in time, and the pixel data, collected in every snapshot, can readily be vectorized. Given m frames to the video stream, the $n \times 1$ vectors $\mathbf{x}_1, \mathbf{x}_2, \ldots, \mathbf{x}_m$ can be extracted, which contain the pixel data of each frame, there being n pixels in total per frame. The DMD method can attempt to reconstruct any given frame, or even possibly future frames, by calculating $\mathbf{x}_{\text{DMD}}(t)$ at the corresponding time t. The validity of the reconstruction depends on how well the specific video sequence meets the assumptions and criteria of the DMD method.

In order to reconstruct the entire video, consider the $1 \times m$ time vector $\mathbf{t} = [t_1\ t_2\ \ldots\ t_m]$, which contains the times at which the frames were collected. If $t_j = j - 1\ \forall j$, then time becomes equivalent to the frame count, where the first frame is labelled as 0 and the jth frame is labelled as $m-1$. The video sequence $\mathbf{X}$ is reconstructed with the DMD approximation (19.21). Notice that ψ_k is an $n \times 1$ vector, that is multiplied by the $1 \times m$ vector $\mathbf{t}$, to produce the proper $n \times m$ video size. By the construction of the DMD methodology: $\mathbf{x}_1 = \mathbf{\Psi b}$, which means that $\mathbf{\Psi b}$ renders the first frame of the video with a dimensionality reduction chosen through the parameter ℓ. It becomes apparent that any portion of the first video frame that does not change in time, or changes very slowly in time, must have an associated Fourier mode (ω_j) that is located near the origin in complex space: $\|\omega_j\| \approx 0$. This fact becomes the key principle that makes possible the ability of the DMD method to separate background (approximate low-rank) information from foreground (approximate sparse) information.

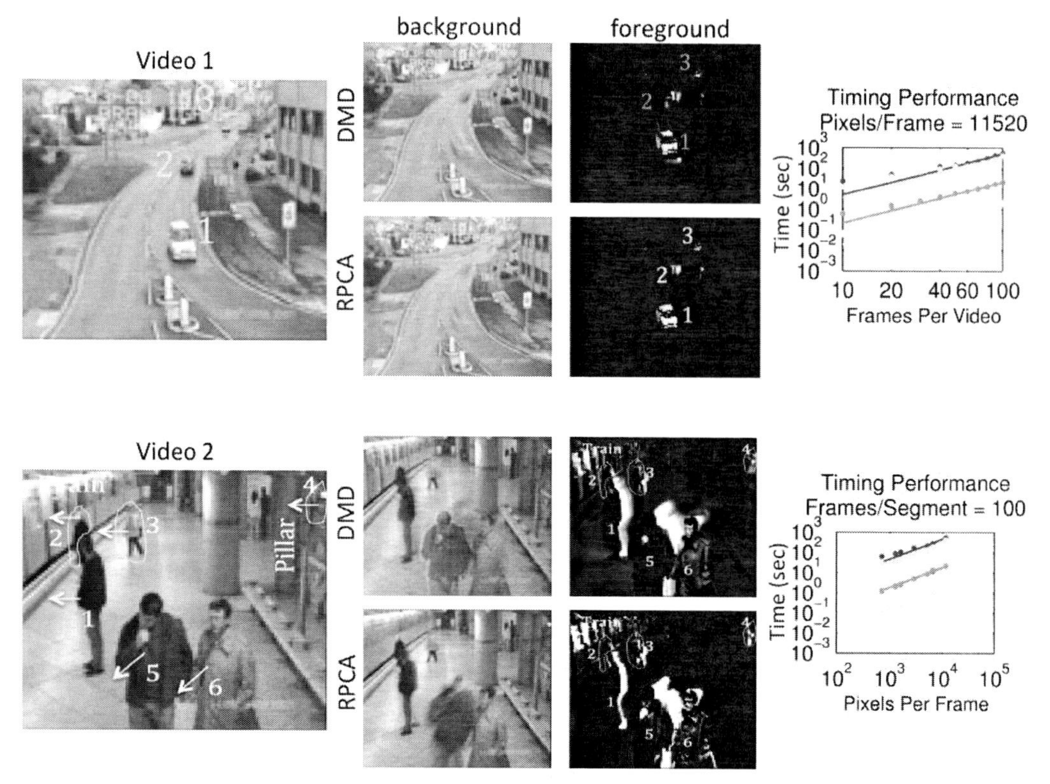

FIGURE 19.2 (See color insert.) Demonstration of foreground/background separation using the DMD method and RPCA on "Abandoned Bag" and "Parked Vehicle" videos. The far left panel shows two original clips at a given snapshot in time, or frame number. Both methods provide very similar foreground/background representations. The various items in the foreground have been labeled and numbered. In the second video, the movement has also been shown. The timing performance on a 1.86 GHz Intel Core 2 Duo processor using MATLAB is also shown on the right both as function of frames per video and pixels per video. The DMD (green and yellow, respectively), inexact ALM RPCA [14] (red and magenta, respectively), and exact ALM RPCA [14] (blue and cyan, respectively) background/foreground separation methods are graphed on a logarithmic scale. The DMD is an order-of-magnitude faster than competing RPCA methods.

Assume that ω_p, where $p \in \{1, 2, \ldots, \ell\}$, satisfies $\|\omega_p\| \approx 0$, and that $\|\omega_j\| \ \forall \ j \neq p$ is bounded away from zero. Thus,

$$\mathbf{X}_{\text{DMD}} = \underbrace{b_p \varphi_p e^{\omega_p \mathbf{t}}}_{\text{Background Video}} + \underbrace{\sum_{j \neq p} b_j \varphi_j e^{\omega_j \mathbf{t}}}_{\text{Foreground Video}} \tag{19.26}$$

Assuming that $\mathbf{X} \in \mathbb{R}^{n \times m}$, then a proper DMD reconstruction should also produce $\mathbf{X}_{\text{DMD}} \in \mathbb{R}^{n \times m}$. However, each term of the DMD reconstruction is complex: $b_j \psi_j \exp(\omega_j \mathbf{t}) \in \mathbb{C}^{n \times m} \ \forall j$, though they sum to a real-valued matrix. This poses a problem when separating the DMD terms into approximate low-rank and sparse reconstructions because real-valued outputs are desired and knowing how to handle the complex elements can make a significant difference in the accuracy of the results. Consider calculating the DMD's approximate low-rank reconstruction according to

$$\mathbf{X}_{\text{DMD}}^{\text{Low-Rank}} = b_p \psi_p e^{\omega_p \mathbf{t}}.$$

Since it should be true that

$$\mathbf{X} = \mathbf{X}_{\text{DMD}}^{\text{Low-Rank}} + \mathbf{X}_{\text{DMD}}^{\text{Sparse}},$$

then the DMD's approximate sparse reconstruction,

$$\mathbf{X}_{\text{DMD}}^{\text{Sparse}} = \sum_{j \neq p} b_j \psi_j e^{\omega_j \mathbf{t}},$$

can be calculated with real-valued elements only as follows...

$$\mathbf{X}_{\text{DMD}}^{\text{Sparse}} = \mathbf{X} - \left| \mathbf{X}_{\text{DMD}}^{\text{Low-Rank}} \right|,$$

where $|\cdot|$ yields the modulus of each element within the matrix. However, this may result in $\mathbf{X}_{\text{DMD}}^{\text{Sparse}}$ having negative values in some of its elements, which would not make sense in terms of having negative pixel intensities. These residual negative values can be put into a $n \times m$ matrix $\mathbf{R}$ and then be added back into $\mathbf{X}_{\text{DMD}}^{\text{Low-Rank}}$ as follows:

$$\mathbf{X}_{\text{DMD}}^{\text{Low-Rank}} \leftarrow \mathbf{R} + \left| \mathbf{X}_{\text{DMD}}^{\text{Low-Rank}} \right|$$
$$\mathbf{X}_{\text{DMD}}^{\text{Sparse}} \leftarrow \mathbf{X}_{\text{DMD}}^{\text{Sparse}} - \mathbf{R}$$

This way the magnitudes of the complex values from the DMD reconstruction are accounted for, while maintaining the important constraints that

$$\mathbf{X} = \mathbf{X}_{\text{DMD}}^{\text{Low-Rank}} + \mathbf{X}_{\text{DMD}}^{\text{Sparse}} = \mathbf{L} + \mathbf{S},$$

so that none of the pixel intensities are below zero, and ensuring that the approximate low-rank and sparse DMD reconstructions are real-valued. This method seems to work well empirically.

Using the Advanced Video and Signal based Surveillance (AVSS) Datasets, (*www.eecs.qmul.ac.uk/ andrea/avss2007_d.html*), specifically the "Parked Vehicle - Hard" and "Abandoned Bag - Hard" videos, the DMD separation procedure can be compared and contrasted against the RPCA procedure. The original videos are converted to grayscale and down-sampled in pixel resolution to $n = 120 \times 96 = 11520$, in order to make the computational memory requirements manageable for personal computers. Also, the introductory preambles to the surveillance videos, which constitute the first 351 frames of each video, are removed because they are irrelevant for the following illustrations. The video streams are broken into segments of $m = 30$ frames each. The results of the separation are shown in Figure 19.2 for both methods along with the timing results.

It should be noted that the RPCA method is highly sensitive to the λ parameter value. The RPCA method achieves similar results to DMD with the recommended regularization parameter value of $\lambda = 9.32 \cdot 10^{-3}$. As this parameter is changed by only ± 0.006, either extra, foreign pixels get included in the foreground ($\lambda = 3.32 \cdot 10^{-3}$), or two of the three cars begin to fade away and blend into the background ($\lambda = 1.532 \cdot 10^{-2}$). Note that the sparse results for both the DMD and RPCA methods are artificially brightened by a factor of 10 in order to illuminate all the dark extraneous pixels that would normally not be visible against the black background.

19.5 Multi-Resolution Analysis with DMD

The MRDMD [6] is inspired by the observation that the slow and fast modes can be separated for such applications as foreground/background subtraction in video feeds [7]. The

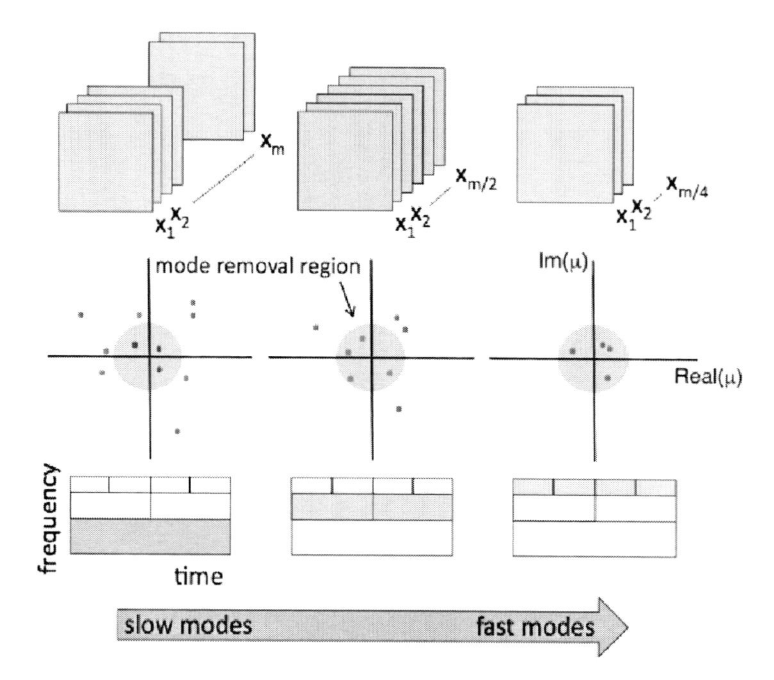

FIGURE 19.3 (See color insert.) Representation of the multi-resolution dynamic mode decomposition where successive sampling of the data, initially with M snapshots and decreasing by a factor of two at each resolution level, is shown (top figures). The DMD spectrum is shown in the middle panel where there are m_1 (blue dots) slow-dynamic modes at the slowest level, m_2 (red) modes at the next levels and m_3 (green) modes at the fastest time-scale shown. The shaded region represents the modes that are removed at that level. The bottom panels show the wavelet-like time-frequency decomposition of the data color coded with the snapshots and DMD spectral representations.

MRDMD recursively removes low-frequency content from a given collection of snapshots [6]. Typically, the number of snapshots M are chosen so that the DMD modes provide an approximately full rank approximation of the dynamics observed. Thus the M is chosen so that all high- and low-frequency content is present. In the MRDMD, M is originally chosen in the same way so that an approximate full-rank approximation can be accomplished. However, from this initial pass through the data, the slowest m_1 modes are removed and DMD is once again performed, now with only $M/2$ snapshots. Again the slowest m_2 modes are removed and the algorithm is continued until a desired termination.

Mathematically, the MRDMD separates the DMD approximate solution (19.21) in the first pass as follows:

$$\mathbf{x}_{\text{DMD}}(t) = \sum_{k=1}^{M} b_k(0)\psi_k^{(1)}(\mathbf{x}) \exp(\omega_k t) \tag{19.27}$$

$$= \underbrace{\sum_{k=1}^{m_1} b_k(0)\psi_k^{(1)}(\mathbf{x}) \exp(\omega_k t)}_{\text{(slow modes)}} + \underbrace{\sum_{k=m_1}^{M} b_k(0)\psi_k^{(1)}(\mathbf{x}) \exp(\omega_k t)}_{\text{(fast modes)}}$$

where the $\psi_k^{(1)}(\mathbf{x})$ represent the DMD modes computed from the full M snapshots.

The first sum in this expression (19.28) represents the slow-mode dynamics whereas the

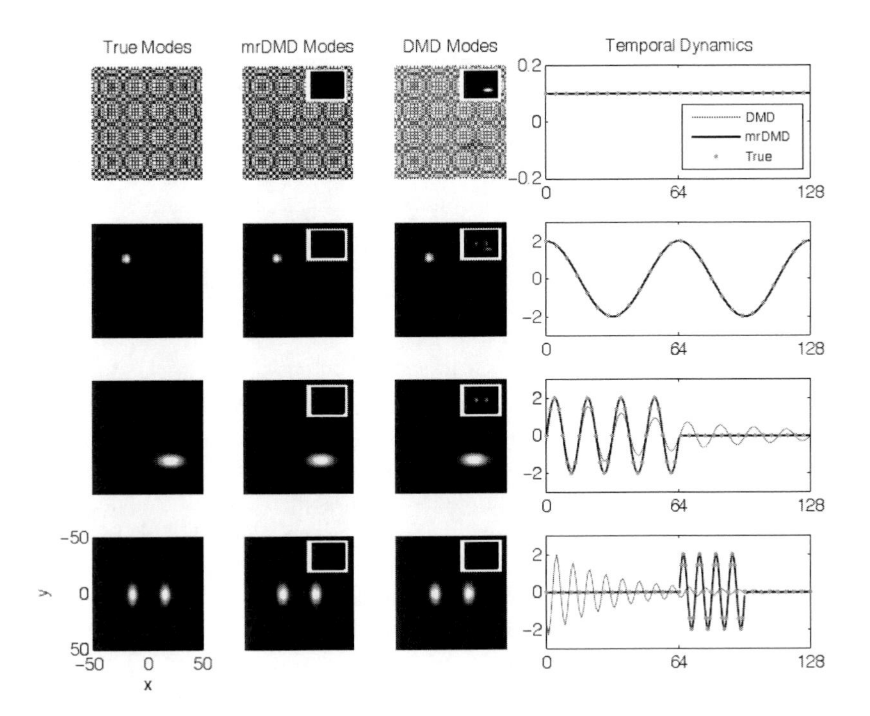

FIGURE 19.4 (See color insert.) Example of a multi-resolution analysis using the MRDMD decomposition. The left panel shows the four modes that are to be mixed. The right panel shows the true time dynamics of each mode, including some modes that turn on and off. The DMD and MRDMD reconstructions of the modes and time dynamics are both illustrated. Without the MRDMD method, the DMD fails to construct the correct dynamics, especially on those modes that turn on and off. The MRDMD works very well in getting both the spatial and temporal dynamics correctly.

second sum is everything else. Thus the second sum can be computed to yield the matrix:

$$\mathbf{X}_{M/2} = \sum_{k=m_1}^{M} b_k(0)\psi_k^{(1)}(\mathbf{x})\exp(\omega_k t)\,. \tag{19.28}$$

The DMD analysis outlined in the previous section can now be performed once again on the data matrix $\mathbf{X}_{M/2}$. However, the matrix $\mathbf{X}_{M/2}$ is now separated into two matrices

$$\mathbf{X}_{M/2} = \mathbf{X}_{M/2}^{(1)} + \mathbf{X}_{M/2}^{(2)} \tag{19.29}$$

where the first matrix contains the first $M/2$ snapshots and the second matrix contains the remaining $M/2$ snapshots. The m_2 slow-DMD modes at this level are given by $\psi_k^{(2)}$, where they are computed separately in the first of second interval of snapshots.

The iteration process works by recursively removing slow frequency components and building the new matrices $\mathbf{X}_{M/2}$, $\mathbf{X}_{M/4}, \mathbf{X}_{M/8}, \cdots$ until a desired/prescribed multi-resolution decomposition has been achieved. The approximate DMD solution can then be constructed as follows:

$$\begin{aligned}\mathbf{x}_{\text{DMD}}(t)=&\sum_{k=1}^{m_1} b_k^{(1)}\psi_k^{(1)}(\mathbf{x})\exp(\omega_k^{(1)}t)\\&+\sum_{k=1}^{m_2} b_k^{(2)}\psi_k^{(2)}(\mathbf{x})\exp(\omega_k^{(2)}t)\\&+\sum_{k=1}^{m_3} b_k^{(3)}\psi_k^{(3)}(\mathbf{x})\exp(\omega_k^{(3)}t)+\cdots\end{aligned} \tag{19.30}$$

where the $\psi_k^{(k)}$ and $\omega_k^{(k)}$ are the DMD modes and DMD eigenvalues at the kth level of decomposition, the $b_k^{(k)}$ are the initial projections of the data onto the time interval of interest, and the m_k are the number of slow-modes retained at each level. The advantage of this method is readily apparent: Different spatial-temporal DMD modes are used to represent key multi-resolution features. Thus there is not a single set of modes that dominates the SVD decomposition and potentially marginalizes features at other time scales.

Figure 19.3 illustrates the multi-resolution DMD process pictorially. In the figure, a three-level decomposition is performed with the slowest scale represented in blue (eigenvalues and snapshots), the mid-scale in red, and the fast scale in green. The connection to multi-resolution wavelet analysis is also evident from the bottom panels as one can see that the MRDMD method successively pulls out time-frequency information in a principled way. To further illustrate its usefulness, the MRDMD is applied to an example problem shown in Figure 19.4. In this example, four different spatial structures are combined, each having different temporal dynamics. In this case, some of the modes come on and off, which is a difficult situation for the DMD and/or RPCA method to handle. However, the MRDMD method is shown to adroitly decompose the mixed signals, producing a nearly perfect reconstruction of the individual modes and their time dynamics.

19.6 Conclusion

It has been demonstrated that the method of dynamic mode decomposition , typically used for evaluating the dynamics of complex systems, can be used for background/foreground separation in videos with visually appealing results and excellent computational efficiency. The separation results produced by the DMD method are on par with the quality of separation achieved with the RPCA method for realistic video scenarios. However, the results are achieved orders of magnitude faster [7]. Indeed, we demonstrate that DMD is viable as a real-time solution to foreground/background video separation tasks even with laptop-level computing platforms.

The DMD method can also be easily extended to produce a multi-resolution analysis tool for video streams or any other complex, dynamical data set [6]. The MRDMD architecture allows one to separate the data in a principled way by weighting both its spatial correlation and time-scale response. By recursively filtering out time-scales, a wavelet type analysis of a spatio-temporal data set can be produced. The method also overcomes some of the shortcomings of the DMD method by successfully handling signals that turn on or off in time, or demonstrate some translational motion. The DMD decomposition technique can also be applied to control architectures in an equation-free way [19].

As with any separation method, including RPCA and DMD, the burden of working with too much data, i.e., high-resolution images and/or many frames per video segment, can be problematic because of reduced computational speeds and limited memory sizes. Nonetheless, the DMD algorithm has shown itself to be robust and efficient enough to produce attractive results in times well below the normal frame acquisition rate of most cameras, allowing for higher pixel resolutions and video segment sizes to be used. For real-time video applications, it makes sense to break the continuous video stream into segments large enough to ensure that there is enough information to complete an adequate background/foreground separation, but small enough to keep the processing times smaller than the data acquisition times. Additionally, moving objects that turn, stop, and/or accelerate are better handled by the DMD procedure as individual actions than in one large video segment.

Finally, unlike RPCA, the analysis presented here is a simple formal procedure that is presented without proof of convergence to low-rank and sparse structures. Although the results are very promising in terms of both quality and exceptional computational costs, future work will focus on making the results rigorous under suitable conditions. Potentially, one could use the DMD technique for generating initial conditions for the convex optimization problem of RPCA, thus combining the power and speed of the method with the exact reconstruction ability of RPCA. It is also intriguing to consider the possibility of connecting the DMD methodology to the more standard L^1 convex optimization problem of RPCA and the matrix completion problem, perhaps allowing for even further improvement in background separation methods in terms of computational efficiency.

References

1. Y. Benezeth, P. Jodoin, B. Emile, H. Laurent, and C. Rosenberger. *Comparative Study of Background Subtraction Algorithms. Journal of Electronic Imaging 19 (2010)*, 19(3):033003, 2010.
2. T. Bouwmans. *Traditional and Recent Approaches in Background Modeling for Foreground Detection: An Overview. Computer Science Review*, 11:31–66, 2014.
3. S. Brunton, J. Proctor, and J. Kutz. *Compressive sampling and dynamic mode decomposition. Journal of Computational Dynamics*, Preprint, December 2015.
4. E. Candès, X. Li, Y. Ma, and J. Wright. *Robust Principal Component Analysis? Journal of ACM*, 58:1–37, 2009.
5. K. Chen, J. Tu, and C. Rowley. *Variants of Dynamic Mode Decomposition: Boundary Condition, Koopman, and Fourier Analyses. Journal of Nonlinear Science*, 22(6):887–915, 2012.
6. Xing Fu, S. L. Brunton, and J. Kutz. *Multi-Resolution Dynamic Mode Decomposition. arXiv*, page arXiv:1404.7592, 2015.
7. J. Grosek and J. Kutz. *Dynamic Mode Decomposition for Real-Time Background/Foreground Separation in Video. arXiv*, page arXiv:1404.7592, 2014.
8. J. He, L. Balzano, and A. Szlam. *Incremental Gradient on the Grassmannian for Online Foreground and Background Separation in Subsampled Video.* In *IEEE Conference on Computer Vision and Pattern Recognition, CVPR 2012*, pages 1568–1575, 2012.
9. J. Kutz. *Data-driven modeling and scientific computing: Methods for Integrating Dynamics of Complex Systems and Big Data.* Oxford Press, 2013.
10. F. De la Torre and M. Black. *A Framework for Robust Subspace Learning. International Journal of Computer Vision*, 54(1-3):117–142, 2003.
11. L. Li, W. Huang, I. Gu, and Q. Tian. *Statistical Modeling of Complex Backgrounds for Foreground Object Detection. IEEE Transactions on Image Processing*, 13(11):1459–1472, 2004.
12. Z. Lin, M. Chen, and Y. Ma. *The Augmented Lagrange Multiplier Method for Exact Recovery of Corrupted Low-Rank Matrices.* Technical Report, University of Illinois at Urbana-Champaign, 2011.
13. Z. Lin, A. Ganesh, J. Wright, L. Wu, M. Chen, and Y. Ma. *Fast convex optimization algo-*

rithms for exact recovery of a corrupted low-rank matrix. *Computational Advances in Multi-Sensor Adaptive Processing (CAMSAP)*, 61, 2009.

14. Y. Ma. Low-Rank Matrix Recovery and Completion via Convex Optimization Sample Code, 2013. Perception and Decision Lab, University of Illinois at Urbana-Champaign, and Microsoft Research Asia, Beijing, Copyright 2009.
15. L. Maddalena and A. Petrosino. A Self-Organizing Approach to Background Subtraction for Visual Surveillance Applications. *IEEE Transactions on Image Processing*, 17(7):1168–1177, 2008.
16. I. Mezić. Spectral properties of dynamical systems, model reduction and decompositions. *Nonlinear Dynamics*, 41:309–325, 2005.
17. I. Mezić. Analysis of Fluid Flows via Spectral Properties of the Koopman Operator. *Annual Review of Fluid Mechanics*, 45:357–378, 2013.
18. I. Mezić and A. Banaszuk. Comparison of systems with complex behavior. *Physica D*, 197:101–133, 2004.
19. J. Proctor, S. Brunton, and J. Kutz. Dynamic mode decomposition with control. *arXiv*, page arXiv:1409.6358, 2014.
20. C. Rowley, I. Mezić, S. Bagheri, P. Schlatter, and D. Henningson. Spectral analysis of nonlinear flows. *Journal of Fluid Mechanics*, 641:115–127, 2009.
21. P. Schmid. Dynamic mode decomposition of numerical and experimental data. *Journal of Fluid Mechanics*, 656:5–28, 2010.
22. P. Schmid, L. Li, M. Juniper, and O. Pust. Applications of the dynamic mode decomposition. *Theoretical and Computational Fluid Dynamics*, 25(1-4):249–259, 2011.
23. C. Stauffer and W. Grimson. Adaptive background mixture models for real-time trackings. In *Proceedings IEEE Conf. on Computer Vision and Pattern Recognition*, pages 246–252, 1999.
24. Y. Tian, M. Lu, and A. Hampapur. Robust and Efficient Foreground Analysis for Real-Time Video Surveillance. In *IEEE Computer Society Conference on Computer Vision and Pattern Recognition, CVPR 2005*, volume 1, pages 1182–1187, 2005.
25. L. Trefethen and D. Bau. Numerical Linear Algebra. SIAM, Philadelphia, 1997.
26. J. Tu, D. Luchtenberg, C. Rowley, S. Brunton, and J. Kutz. On Dynamic Mode Decomposition: Theory and Applications. *Journal of Computational Dynamics*, 1:391–421, 2014.

20

Stochastic RPCA for Background/Foreground Separation

Sajid Javed
School of Computer Science and Engineering, Kyungpook National University, Republic of Korea

Seon Ho Oh
School of Computer Science and Engineering, Kyungpook National University, Republic of Korea

Thierry Bouwmans
Laboratoire MIA, Universite de La Rochelle, France

Soon Ki Jung
School of Computer Science and Engineering, Kyungpook National University, Republic of Korea

20.1 Introduction

The detection of moving objects is the fundamental pre-processing step in many computer vision and image processing applications, such as video inpainting, compression, privacy, surveillance, segmentation, optical flow, and augmented reality [1], [14]. This basic step requires an accurate and efficient background subtraction (also known as foreground detection). Typically, the background subtraction process consists of isolating the moving objects called "foreground" from the static scene called "background." However, it becomes a really hard task when the background scene contains more variations such as sudden global illumination changes, waving trees, water surface, moving curtains, bootstrapping, etc. In addition, color saturation and bad weather conditions are major well-known background modeling challenges.

Many algorithms have been proposed to handle the problem of background subtraction [15] and several implementations are available in BGS[67] and LRS[68] libraries. Excellent

[67] https://github.com/andrewssobral/bgslibrary
[68] https://github.com/andrewssobral/lrslibrary

surveys on background modeling and foreground detection can be found in [10]. Especially, subspace learning model such as *Principal Component Analysis* (PCA) provides a very nice framework for moving object detection. A variety of methods have been proposed for background modeling using subspace learning models over the past few years. Among them, Oliver et al. [20] are the first authors to model the background using PCA. The foreground detection is then achieved by thresholding the difference between the reconstructed background and corresponding input image. PCA provides very robust subspace learning but it is not stable when some of the data entries are corrupted and outliers appear in the new subspace basis. Although several improvements have been proposed [31] in which the authors address the major limitations of traditional PCA with respect to outliers and noise, methods such as [25] do not provide a strong performance for background/foreground segmentation.

Therefore, Candes et al. [4] designed a very interesting scheme called *Robust Principal Component Analysis* (RPCA) to address the main limitations of PCA. RPCA basically decomposes the original data matrix A, such that $A = L + S$, where L is a *low-rank* matrix and S is a matrix that can be *sparse* or not. RPCA shows a very nice potential for many computer vision applications such as face recognition [29]. In the case of background/foreground segmentation, Candes et al. [4] proposed a convex optimization approach called *Principal Component Pursuit* (PCP). Under minimal assumptions, PCP perfectly recovers the low-dimensional subspace called *low-rank* matrix for background modeling and *sparse* error constituting the foreground mask. For example, Figure 20.1 shows encouraging results for moving object detection of real sequences taken from the I-LIDS [3] dataset.

Excellent surveys on background subtraction using RPCA can be found in [2]. All these RPCA-based approaches such as *Augmented Lagrangian Multiplier* (ALM), *Singluar Value Thresholding* (SVT), and *Linearized Alternating Direction Method with an Adaptive Penalty* (LADMAP) discussed in [2] provide a very nice framework for foreground detection, but they are not applicable for real-time systems, since some major difficulties are observed, as:

- **Computational Complexities.** Traditional RPCA-based approaches are based on Singular Value Decomposition (SVD) computation, where all the samples are required for the partial computation of SVD at each major loop with a higher cost, which is not useful for real time systems.

- **Memory Usage.** RPCA-based methodologies use batch optimization algorithms, e.g., in order to decompose the observed data into *low-rank* and *sparse* components. A number of training samples are required to be stored in a memory before any processing that needs high memory requirements and hence it is not able to process high-dimensional data. Therefore, hardly any approach is able to process more than five hundred frames due to memory issues.

- **Low-Rank Component Stability.** Although a number of existing solutions have been proposed in [19] for RPCA limitations, *low-rank* components sometimes suffer from consistency. As a result the outliers always appear in the low-dimensional subspace, which degrades the performance of real-time visual surveillance systems.

- **Lack of Multiple Features.** Previous RPCA approaches consider only one dimension or single channel of original input image for *sparse* error separation due to computation and memory issues. Therefore, the features are not integrated in conventional approaches. As a result, foreground detection is not always robust, since the pixel values are not sufficient to perform in different background scenes.

Many significant improvements have been presented in the literature to accelerate the RPCA via PCP algorithms [2]. For example, Zhou et al. [31] proposed *Go Decomposition* (GoDec), which accelerates the RPCA algorithm via PCP using a *Bilateral Random Projections* (BRP) scheme to separate the *low-rank* and *sparse* matrix. The Semi-Soft GoDec [31] method is an extension of GoDec that is four times faster than GoDec. It imposes a hard thresholding scheme in *low-rank* and *sparse* matrix entries. In [32], Zhou et al. proposed the *Detecting Contiguous Outliers in the Low-rank Representation* (DECOLOR) method, which accelerates PCP algorithm by integrating the object detection and background learning into a single process of optimization. It also adds continuity constraints on *low-rank* and *sparse* components. In [23], a fast PCP algorithm is proposed, which reduces the SVD computational time in *inexact ALM* (IALM) by adding some constants in the minimization equation of *low-rank* and *sparse*. The results in the background modeling case are very encouraging, but it is not desirable for real-time processing due to the base of PCP.

(a) (b) (c) (d)

FIGURE 20.1 An example of moving object detection using RPCA. From left to right: (a) input, (b) low-rank, (c) sparse component, and (d) foreground mask.

Incremental and *online robust* PCA methods are also developed for PCP algorithms. He et al. [11] proposed. *Grassmanian Robust Adaptive Subspace Tracking Algorithm* (GRASTA), which is an incremental gradient descent method on Grassmannian manifold for solving the RPCA problem in an online manner. At each iteration, GRASTA uses the gradient of the updated augmented Lagrangian function after revealing a new sample to perform the gradient descent. Results are encouraging for background modeling, but no theoretic guarantee of the algorithm convergence is provided. J. Mairal et al. [19] proposed an online learning method for sparse coding and dictionary learning, is which efficiently solves the smooth non-convex objective function over a convex set. A real-time processing is achieved, but it does not require learning rate tunning like regular stochastic gradient descents.

In [9], Guan et al. proposed the *Online RSA Non-negative Matrix Factorization* (NMF) method, which receives one chunk of samples per step and updates the basis accordingly. OR-NMF converges faster in each step of the basis update. But using a buffering strategy for storing a limited number of samples for memory usage remains the issue. Therefore, Feng and Xu [6] recently proposed the *Online Robust-PCA* (OR-PCA) algorithm, which processes one sample per time instance using stochastic approximations. In this approach, a nuclear norm objective function is reformulated and therefore all the samples are decoupled in an optimization process for *sparse* error separation.

Considering all of these approaches and major RPCA limitations, we present OR-PCA with its application to background/foreground segmentation in this chapter. First, we give an overview of stochastic RPCA (also known as OR-PCA), then background/foreground segmentation is presented using stochastic RPCA, which was recently introduced in [14], [16], and [16].

The rest of this chatper is organized as follows. In Section 20.2, the stochastic process for OR-PCA is reviewed. Section 20.3 describes its potential applications, such as foreground detection. Experimental results are discussed in detail in Section 20.4, and finally, conclusions are drawn in Section 20.5.

20.2 Stochastic RPCA

Feng and Xu [6] proposed an *Online Robust PCA* (also known as stochastic RPCA) method to solve the decomposition problem via online methods. In [6], it is argued that the original data matrix A can be decomposed into *low-rank* and *sparse* components using stochastic optimization . In contrast to traditional RPCA-based algorithms, where all the samples are required to be loaded in a memory for SVD computation, OR-PCA processes one sample per time instance via an iterative optimization scheme developed by Feng and Xu [6]. Therefore, its memory and computational cost are independent of the number of samples, which drastically increases its efficiency for real-time systems. The main advantage is that OR-PCA is able to efficiently handle high-dimensional data. In [6], it is proved that OR-PCA converges to a global optimal solution of the original PCP formulation [4] and the validity of this iterative optimization scheme is also provided by showing that if all the observations are bounded, then the low-dimensional subspace basis can be fully ranked.

The main goal of OR-PCA is to correctly estimate the underlying subspace basis over the noise samples. Let us say that A is the observed sample matrix, L and$S$ are its corresponding *low-rank* and *sparse* errors; PCP guarantees that these two components can be recovered by solving the convex equation given by:

$$\min_{L,S} \frac{1}{2}\|A - L - S\|_F^2 + \lambda_1\|L\|_* + \lambda_2\|S\|_1. \tag{20.1}$$

In order to solve (20.1), optimization techniques are designed, such as the Accelerated Proximal Gradient and ALM discussed in [2], but these schemes are based on batch mode and take a longer time to decompose the matrix A in to L and S. In addition, the nuclear norm function tightly couples all the samples into one vector for SVD computation, which is not suitable for memory requirements.

Therefore, Feng and Xu [6] proposed a stochastic optimization method that basically decomposes the nuclear norm of the objective function in the traditional PCP algorithms into an explicit product of two *low-rank* matrices, i.e., basis and coefficient. In other words, the nuclear norm is reformulated by eliminating the hard constraints such as $L = XR^T$ and hence OR-PCA can be presented as:

$$\min_{X\in\Re^{p\times r},R\in\Re^{n\times r},E} \frac{1}{2}\|A - XR^T - S\|_F^2 + \frac{\lambda_1}{2}\left(\|X\|_F^2 + \|R\|_F^2\right) + \lambda_2\|S\|_1, \tag{20.2}$$

where p is the number of samples, X is the optimum basis of each sample, R is a basis coefficient, and S is a *sparse* error. λ_1 controls the basis and coefficients for the *low-rank* matrix, whereas λ_2 controls the sparsity pattern, which can be tuned according to magnitude of corrupted data entries. In addition, *basis* and *coefficients* depend on the value of *rank* r, which is tuned carefully to speed up the stochastic optimization process.

In particular, the OR-PCA optimization consists of two iterative updating components. First, every incoming sample is projected onto currently initialized basis X and then the *sparse* noise component is separated, which includes the outliers contamination. Then, the basis X is updated with the incoming sample using the block-coordinate descent method [30]. The underlying low-dimensional subspace, called the *low-rank* matrix, can be recovere using $L = XR^T$. More details can be found in [6].

20.3 Application: Background/Foreground Detection

OR-PCA [6] provides a very encouraging solution for original data decompositon into *low-rank* and *sparse* components, and therefore it can be adopted for different computer vision problems. In [6], the scientific evaluations are presented only on synthetic data for optimal solution convergence, and no encouraging results have been observed for computer vision applications such as background subtraction, matrix completion, and face recognition, etc. In this section, we present some recent work on background/foreground segmentation using stochastic RPCA, which has recently been presented in [14], [16], and [13].

20.3.1 Highly Dynamic Background Subtraction Using OR-PCA with MRF

In this section, we give an overview of the robust background subtraction using stochastic RPCA, which is modified to be adapted for this application in [14]. A very nice framework was developed by Javed et al. [14] via image decomposition using OR-PCA with Markov Random Field (MRF). Figure 20.2 shows a block diagram of the complete framework.

As mentioned in Figure 20.2, first the input video frames are decomposed into Gaussian and Laplacian images using a set of two Gaussian kernels. Then, OR-PCA [6] is applied to each of the Gaussian and Laplacian images with different parameters to model two channels of background, separately. An alternative initialization scheme is proposed in [14] to speed up the stochastic optimization process. Finally, the integration stage, which combines *low-rank* and *sparse* components obtained via stochastic RPCA to recover the background model and foreground image, is performed. The reconstructed *sparse* matrix is then thresholded to get the binary foreground mask. In order to improve the foreground segmentation, an MRF is applied, which exploits structural information and similarities continuously. In the later sections, we will give the summary of each component.

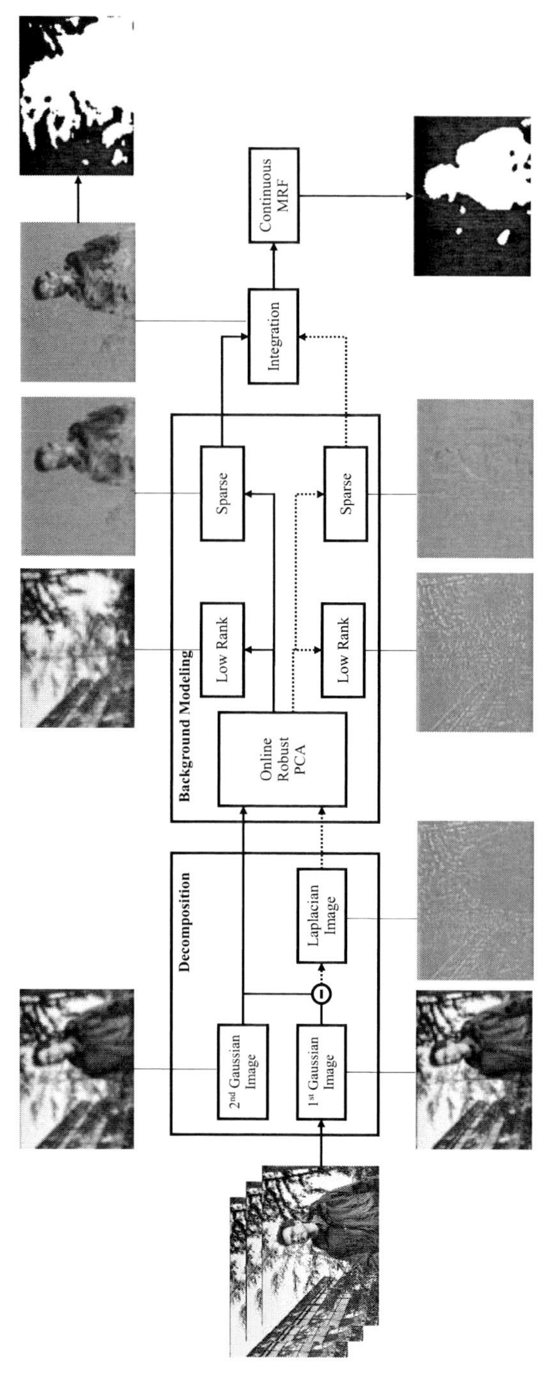

FIGURE 20.2 Overview of our background subtraction scheme in [14].

Global Pre-Processing

As mentioned in [14], two separate spatial Gaussian kernels are designed to decompose the input image into Gaussian and Laplacian images as a global pre-processing step. First, Gaussian kernels are applied on the input image to get its Gaussian images.

In the first case, we choose the standard deviation σ on the Gaussian kernel as 2 with a filter size of 5×5 to get the first Gaussian image. In the second case, we apply Gaussian kernel with the same σ value on the first blurred image due to its adequate smoothing properties. Since the difference of Gaussians is approximately the same as the Laplacian of Gaussian, the Laplacian image is obtained by the difference of two Gaussian images.

Every input video frame is decomposed into Gaussian and Laplacian images using the method discussed above. As the Gaussian image is robust against background variations, the Laplacian image provides enough edge features for small pixel variations. Therefore, the false alarms are reduced from the foreground region to some extent as a result, and our methodology provides accurate foreground detection.

Background Modeling

OR-PCA [6] is used to model two channels of background from Gaussian and Laplacian images. In this case, A is a vectorized input video frame (Guassian and Laplacian frames), X is a random initialized basis, p is the total number of pixels e.g., $width \times height$, and r is a rank in (20.2). The background sequence for each image is then modeled by a multiple of basis and its coefficient R such as $L = XR^T$, whereas the sparse component S for each image constitutes the foreground objects. More details are explored in [14].

Data Dependent Initialization Scheme

The number of subspace basis X is randomly determined using improper value of *rank*, and no initialization method is considered for OR-PCA in [6]. The rank value R is 20 and $\lambda_1 = \lambda_2 = 0.01$ in (20.2). As a result, the algorithm converges slowly to the optimal solution and outliers appear in the *low-rank* matrix, which affects the *sparse* component as well as the binary foreground mask for the background modeling case.

In order to meet the time complexity, the basis for low-dimensional subspace is initialized using first N video frames with a good selection of *rank*. Since stochastic RPCA is applied on two images in this scheme, the basis for each image is initialized using input video frames. In this case, the *rank* is a tunable parameter for each image, and a good range is provided according to different types of background scenes. By this technique, the stochastic RPCA converges to the low-dimensional subspace faster than the original one. The outliers are also reduced and good computational time is achieved without sacrificing the quality of foreground in the video surveillance case.

Foreground Detection

The *low-rank* and *sparse* components are obtained from each decomposed image after applying stochastic RPCA. Gaussian and Laplacian *low-rank* and *sparse* components are integrated in this step. The different parameters setting is adopted for OR-PCA in (20.2) on each decomposed image. λ_1 is considered as a constant 0.01 for both images. λ_2 and rank r for Laplacian, whereas λ_2' and rank r' for the Gaussian image are selected according to the background scene, for obtaining enough sparsity pattern for each decomposed image.

Since the Laplacian image provides enough edge features for small variations in background scene, λ_2 must be smaller than λ_2'. After integrating both components of each image, the binary foreground mask f is then obtained by thresholding the integrated sparse component.

(a) (b) (c) (d) (e)

FIGURE 20.3 Stochastic RPCA via image decomposition. Input, low rank, sparse, and foreground mask images are shown in each of the rows [14].

At this stage, the background subtraction scheme is good enough to deal with static and some small background dynamics such as slight illumination changes, but it fails to handle highly dynamic backgrounds, where most of the background pixels have high variations such as waving trees, water surface, rapid illumination changes, etc. For example, Figure 20.3 (a), (b), and (c) show the results of static and some small dynamic backgrounds. However, moving curtains and waving trees where most of the background pixels are moving are shown in (d) and (e), respectively. We use the best parameters as $r = 1$ and $\lambda_2 = 0.03$ for both images in (a), whereas in (b) and (c), $r = 1$, $r' = 3$, $\lambda_2 = 0.02$, and $\lambda_2' = 0.04$ are used for each decomposed image. Similarly, the best parameters are also considered for (d) and (e) as $r = 1$, $r' = 10$, $\lambda_2 = 0.02$, and $\lambda_2' = 0.06$, respectively. In order to improve the foreground detection for highly dynamic background scenes, an MRF is employed on the sparse component, which improves the quality of foreground detection for the visual surveillance system.

Markov Random Field for Improving Foreground Segmentation

Many computer vision problems have been solved using the continuous constraints as pre-processing or post-processing steps, such as image denoising and segmentation [21]. In [14], MRF is employed as a post-processing step for improving continuous foreground segmentation. Since the foreground labels of stochastic RPCA *sparse* matrix cannot be optimal, therefore it can be improved with spatio-temporal constraints. In this section, MRF is utilized to optimize the labeling field. The MRF is a set of random variables having a Markov property described by an undirected graph.

Let us consider the foreground image f as a set of pixels $\mathcal{P}$ and a set of labels $\mathcal{L} = \{0, 1\}$, such that

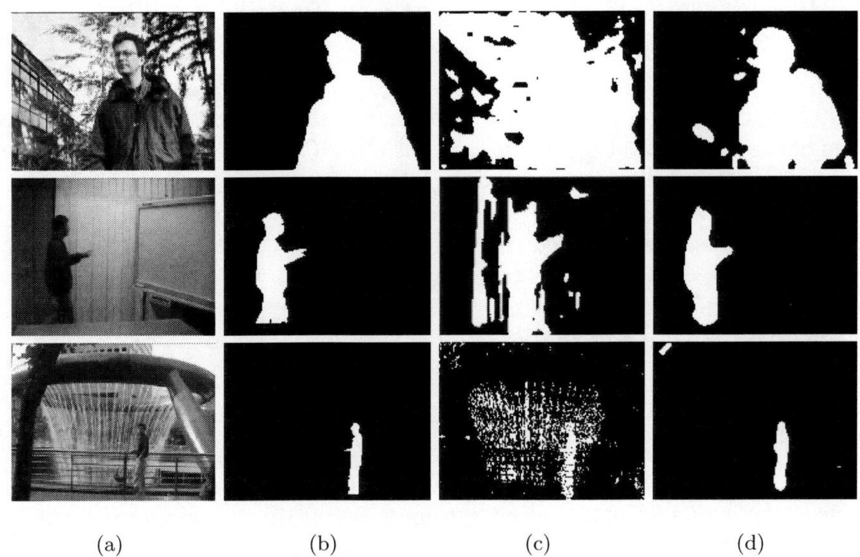

(a) (b) (c) (d)

FIGURE 20.4 Foreground detection using OR-PCA with MRF. (a) Input, (b) Groundtruth, (c) Mask without MRF, and (d) Foreground mask with MRF [14].

$$f_p = \begin{cases} 0, & \text{if } p \text{ belongs to background,} \\ 1, & \text{if } p \text{ belongs to foreground.} \end{cases} \tag{20.3}$$

The goal is to find a labeling f that minimizes the energy function:

$$E(f) = \sum_{p \in P} D_p(f_p) + \sum_{p,q \in \mathcal{N}} V_{p,q}(f_p, f_q), \tag{20.4}$$

where $\mathcal{N} \subset \mathcal{P} \times \mathcal{P}$ is a neighborhood system on pixels. $D_p(f_p)$ is a function derived from the observed data that measures the cost of assigning the label f_p to the pixel p. $V_{p,q}(f_p, f_q)$ measures the cost of assigning the labels f_p, f_q to the adjacent pixels p, q. The energy function like E is extremely difficult to minimize, however, as it is a nonconvex function in a space with many thousands of dimensions. In the last few years, however, efficient algorithms have been developed for these problems based on graph cuts.

The basic idea of graph cuts is to construct a directed graph $\mathcal{G} = (\mathcal{V}, \mathcal{E})$, where the vertices $\mathcal{V}$ stand for all pixels in image and edges $\mathcal{E}$ denote spatially neighboring pixels having non-negative weights that have two special vertices (terminals), namely, the source s and the sink t. MRF that has such a type of s-t graph is called graph-representable and can be solved in polynomial time using graph cuts [17]. In [14], the gco-v3.0 library[69] is considered for optimizing multi-label energies via the α-expansion and α-β-swap algorithms. It supports energies with any combination of unary, pairwise, and label cost terms. Figure 20.4 shows a successful foreground detection in some challenging background scenes after continuous MRF constraints.

[69]http://vision.csd.uwo.ca/code/

20.3.2 Multiple-Features-Based OR-PCA for Background/Foreground Detection

In this section, we present multi-feature-based OR-PCA presented in [16]. Previous RPCA approaches cannot process high dimensional data and they work only on a single channel due to memory and computational issues, while OR-PCA is independent of the number of samples and it can be utilized for processing high dimensional data. Therefore, multiple features are considered using OR-PCA for background subtraction in [16]. A very interesting feature selection sceheme is proposed frame by frame in [16]. The block diagram of multi-feature-based stochastic RPCA is shown in Figure 20.5.

As described in [16], the multi-feature scheme has three major components, e.g., building a feature background model, background subtraction, and, dynamic feature selection process.

Building Feature Background Model

In [16], the foreground detection process is different from the previous background modeling schemes where the model is created directly from grayscale image or color information. The background model is created using a multiple feature extraction process in [16]. The sliding block is created to store the last N frames in a data matrix e.g., $D_t \in \Re^{m \times n \times N}$, where D_t represents the input data matrix D at time t. The width and height of the frame is denoted as m and n, respectively, whereas N denotes the number of frames stored in a sliding block, e.g., ten in [16].

The matrix D_t is transformed into another matrix $A_t \in \Re^{d_1 \times N \times d_2}$ after a feature extraction process, where d_1 is the number of pixels (i.e., $m \times n$) and d_2 is the number of features. Nine different features are used in [16], such as three color channels (red, green, and blue), intensity, local binary pattern, spatial gradients in horizontal and vertical directions, and spatial gradient magnitude. In addition, Histograms of Oriented Gradients (HOG) [5], a well-known feature initially developed for human detection, is also used. In addition, a different image resolution is used according to the datasets. For example, 120×160 (19,200 pixels) are used for the wallflower dataset [26]. So, the dimension of the feature model is $A_t \in \Re^{19,200 \times 10 \times 9}$.

Once the model is created using a sliding block of N frames, the model is updated continuously when a new frame arrives. Every time a new video frame is captured, the sliding block appends (adds) the new frame and removes the old one, as with the sliding window concept. The steps described above are shown in Figure 20.5 (a), (b), (c) and (d).

Background Subtraction

The background subtraction is performed using OR-PCA [14]. In [16], OR-PCA is applied on each frame having multiple features. In this case, A is an input frame with ten sets of features, p is the number of pixels with multiple features, e.g., $(d_1 \times d_2)$, $L = XR^T$, and S represents the *low-rank* and *sparse* components having multiple features in (20.2).

A fixed setting of parameters is considered such as $\lambda_1 = 0.01$, $\lambda_2 = 0.04$ and $r = 5$ in (20.2). The *low-rank* model L_t of each feature from each frame is then obtained by a multiple of basis X and its coefficient R, whereas the sparse component S of only intensity feature is computed, which constitutes the foreground objects. The step described above is shown in Figure 20.5 (e).

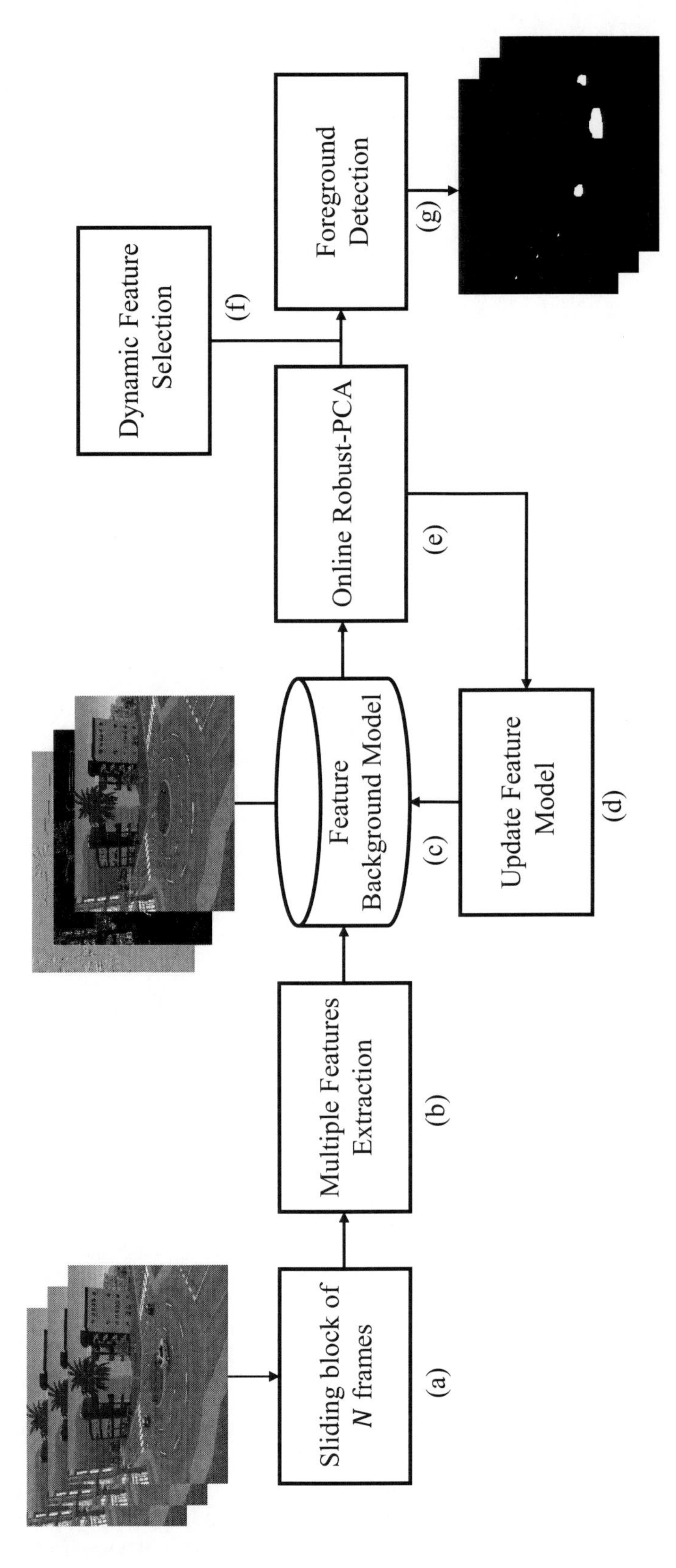

FIGURE 20.5 Multiple-features-based stochastic RPCA for background subtraction in [16].

Dynamic Feature Selection

In this section, we summarize a very encouraging scheme called dynamic feature selection (DFS), proposed by Javed et al. [16]. Let F be a set of features extracted from the input frame A_t and F' be a set of *low-rank* features extracted from the reconstructed *low-rank model* L_t. Then the similarity function SM for the kth feature at the pixel (i,j) is computed as follows:

$$SM_k(i,j) = \begin{cases} \frac{F_k(i,j)}{F'_k(i,j)}, & \text{if } F_k(i,j) < F'_k(i,j), \\ 1, & \text{if } F_k(i,j) = F'_k(i,j), \\ \frac{F'_k(i,j)}{F_k(i,j)}, & \text{if } F_k(i,j) > F'_k(i,j), \end{cases} \quad (20.5)$$

where $F_k(i,j)$ and $F'_k(i,j)$ are the feature value of pixel (i,j) for the kth feature and $SM_k(i,j)$ is between 0 and 1. In addition, since HOG features are the histogram distributions, the corresponding distance can be measured by using a well-known distance measure called Battacharya distance. Let h^{to} and h^t be two normalized histograms computed from input and reference images. Then the Battacharya distance d can be computed using two histograms whose value is in the range 0 to 1.

Next a weighted combination of similarity measures is computed as follows:

$$W(i,j) = \sum_{k=1}^{K} w_k SM_k(i,j), \quad (20.6)$$

where K is the total number of features and w_k is the weighting factor for the kth feature.

In the previous approaches, when features are extracted and matched, the weighting factor w_k of each component is chosen empirically to maximize the true pixels and minimize the false pixels in the foreground detection, which is very tedious work for large-scale video analysis. After analyzing the weighting factors in [16], the tendency is observed that the static background requires a smaller value whereas the dynamic background needs a higher value to adapt changes of the scene.

The DFS scheme selects the weight w_k for each component dynamically in [16]. The weight w_k of each component is computed frame by frame, which is the sum of the ratio of mean μ to its variance σ for each feature. It is observed experimentally that the mean μ and variance σ of dynamic backgrounds are always greater than those of the static backgrounds. Therefore, the total weighted sum of all features W_{sum} for the dynamic background is less than for the static background, as mentioned in [16]. The w_k for each feature can be computed as follows:

$$w_k = \sum_i \sum_j \frac{\mu_k(i,j)}{\sigma_k(i,j)}, \quad (20.7)$$

$$W_{sum} = \sum_{k=1}^{K} w_k, \quad (20.8)$$

where K is the number of features, and σ_k and μ_k are the variance and mean value of the kth feature, respectively. It is experimentally observed in [16] that HOG features are robust for clearly visible Human detection as shown in Figure 20.6 (b). Color information, local binary pattern, and gradient features are robust for highly dynamic backgrounds, but fail in the static case where background and foreground objects have similar features, such

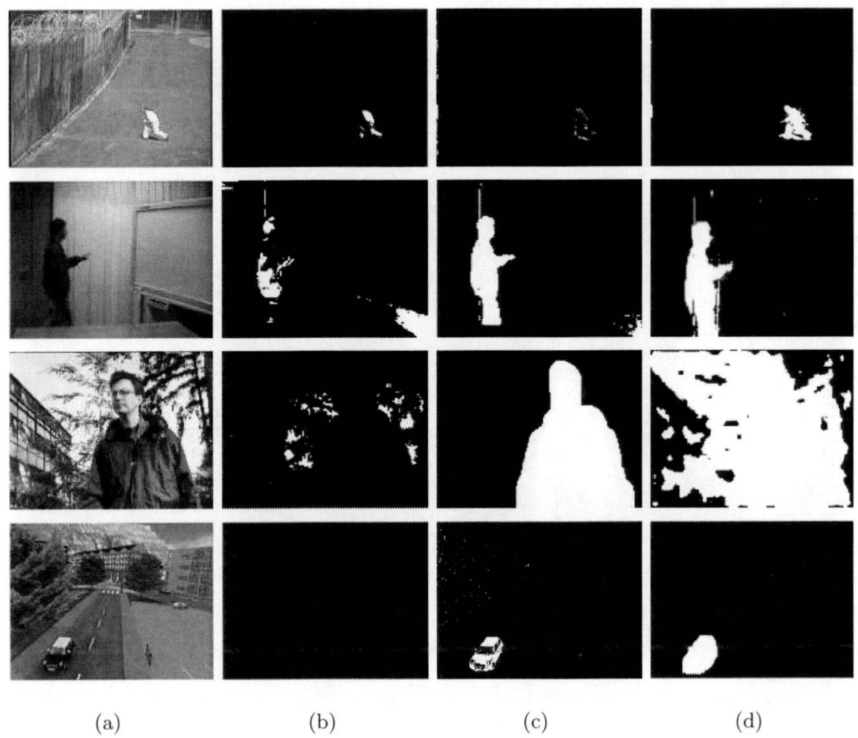

FIGURE 20.6 An example of multi-features selection. From left to right: (a) input, (b) HOG, (c) color, local binary pattern, and gradient, and (d) *sparse*-features-based background subtraction.

as in Figure 20.6 (c). Moreover, *sparse* features S in (20.2) of stochastic RPCA are very robust for static background scenes, but fail in highly dynamic backgrounds as shown in Figure 20.6 (d) and in [14].

Therefore, in [16], it is argued that the selected and useful features will participate frame by frame according to the weighted sum of all features W_{sum} in three types of background scenes: the static background scene where pixel values have no variations as shown in the first and fourth rows of Figure 20.6, the scene with little dynamic background, where some part of the background scene contains pixel variations as shown in the second row of Figure 20.6, and highly dynamic background where most of the background pixels have high variations as shown in third row of Figure 20.6. The features are contributed frame by frame according to DFS scheme, which is as under:

$$\begin{matrix} Background \\ dynamics \end{matrix} = \begin{cases} Highly\ dynamic & \text{if } W_{sum} \leq 0.2, \\ Small\ dynamic & \text{if } 0.2 < W_{sum} < 0.3, \\ Static & \text{if } W_{sum} \geq 0.3. \end{cases} \quad (20.9)$$

HOG features will participate in every situation if the foreground object is human, otherwise a minor participation will be observed. For static background scenes, only *sparse* features S will participate. All other features can be ignored in this situation. Color, local binary pattern, and gradient features will participate only for highly dynamic background scenes; the rest of the other features can be rejected. Similarly, all features will participate together in the case of a small dynamic scene. The result of these individual features are shown in Figure 20.6 with different scenarios. These different background scenes do not occur independently but simultaneously.

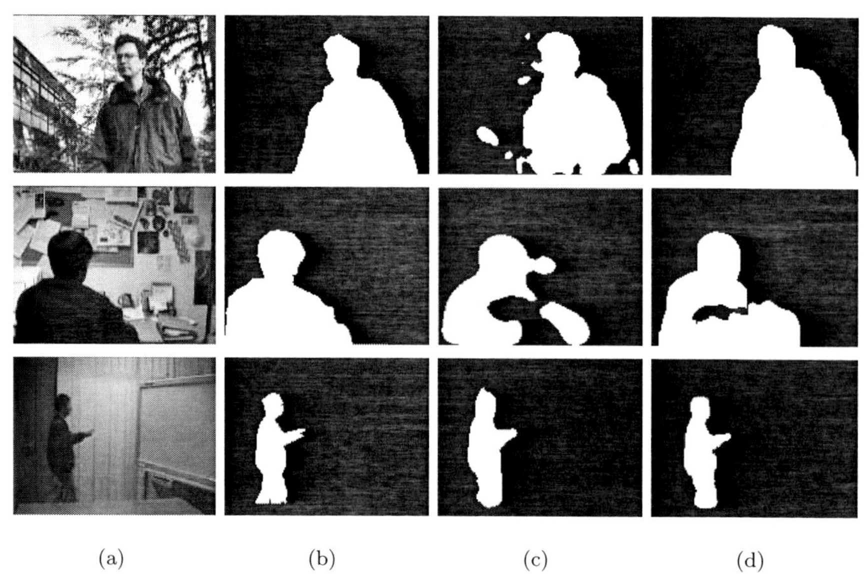

(a) (b) (c) (d)

FIGURE 20.7 Multiple-features-based stochastic RPCA with DFS. From left to right: (a) input, (b) ground truth, (c) foreground mask using [14], and (d) foreground mask using [16].

The foreground mask is obtained after the feature selection process according to W_{sum}. The stochastic RPCA sparse feature is thresholded to get the binary foreground mask when it is participating. The foreground mask is obtained from all other features by applying the following threshold function,

$$fg(i,j) = \begin{cases} 1 & \text{if } W(i,j) < t, \\ 0 & otherwise, \end{cases} \tag{20.10}$$

where t is a threshold and its value varies from 0.2 to 0.4. In [16], 0.2 is used for static, and 0.3 and 0.4 are used for little and higly dynamic backgrounds, respectively. Figure 20.6 shows a consistent improvement of foreground detection using the method OR-PCA with MRF [14] and OR-PCA with DFS [16] in (c) and (d), respectively.

20.3.3 Depth-Extended OR-PCA with Spatiotemporal Constraints for Robust Foreground Segmentation

In this section, background subtraction based on depth information using stochastic RPCA is presented. As mentioned above, due to the batch optimization methods a previous RPCA approaches are only limited to work on a single channel and they are also limited to a single camera. Therefore, Javed et al. [13] extended the main idea of OR-PCA using stereo images. In this work [13], it is shown that stochastic RPCA can be applied using the information from multiple cameras. Since range information is less affected when the background and foreground objects have similar color features, it is useful for the segmentation process to solve the color saturation problem using disparity images. Here we present the main idea of disparity-based foreground detection reported in [13]. The system diagram is shown in Figure 20.9.

As mentioned in [13], depth information provides efficient features to solve color saturation problems. First, the depth information is estimated from stereo pairs using three disparity estimation methods. Second, stochastic RPCA is applied on every color and depth frame to separate the *low-rank* and *sparse* component. Then the integration is performed on each separated *low-rank* and *sparse* matrix and a hard thresholding scheme is applied to obtain the initial foreground mask from the integrated *sparse* component. Furthermore, spatiotemporal constraints such as continuous MRF as described in Section 20.3.1 is applied on the initial foreground mask to further improve the foreground segmentation. Finally, a very nice comparison of foreground detection based on three types of disparity methods is presented. We summarize each component of Figure 20.9 in the later sections.

Disparity Estimation

Three types of depth images are computed using variational- [22], phase- [28] and SGBM- [12] based disparity algorithms in [13]. The disparity information provided as a benchmark dataset[70] is used in [13]. However, in real-time processing, the disparity map can be obtained in the same way. Figure 20.8 shows an example of input with three types of depth images taken from the dataset described in [7]. These depth images are computed based on color correspondence; therefore the noise can affect segmentation results from events such as flickering of lights, uniform regions, etc. The range is less affected when shadows appear in a background scene or foreground objects contain similar color features as background. But when the disparity is less useful, e.g., foreground objects are not very close to the camera, the method may not produce correct results, but it can be improved using spatiotemporal constraints. Therefore, both color and disparity features are useful. Overall, Depth-Extended OR-PCA (DEOR-PCA) performs accurate foreground segmentation using variational-based disparity, which is shown in a later section.

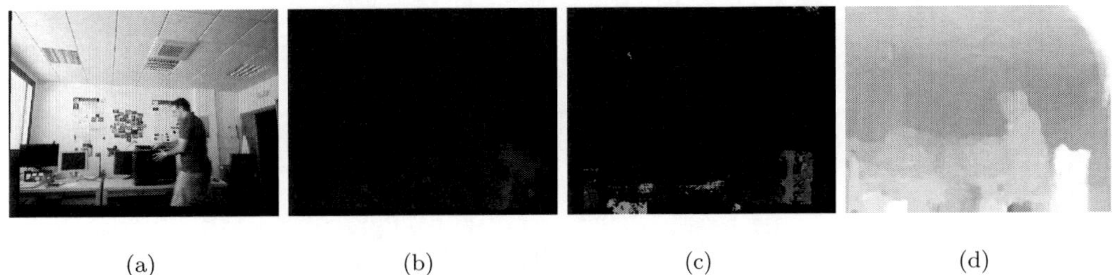

(a) (b) (c) (d)

FIGURE 20.8 Sample *LCD Screen* sequence. From left to right: (a) input, (b) Phase, (c) SGBM, and (d) variational disparity.

Background Subtraction: *Straightway Depth Extension of OR-PCA*

OR-PCA [6] is extended in [16], which is applied on each depth and color frame. In this case, A is an input data of any size (one sample), p is the number of pixels with three color features, e.g., ($width \times height \times 3$), X is a basis of each individual color channel, R is a coefficient, and S is a *sparse* error in (20.2). Then the *low-rank* and *sparse* components are obtained from each feature after decomposing original input frame A into *low-rank* L and *sparse* S components.

[70] http://atcproyectos.ugr.es/mvision

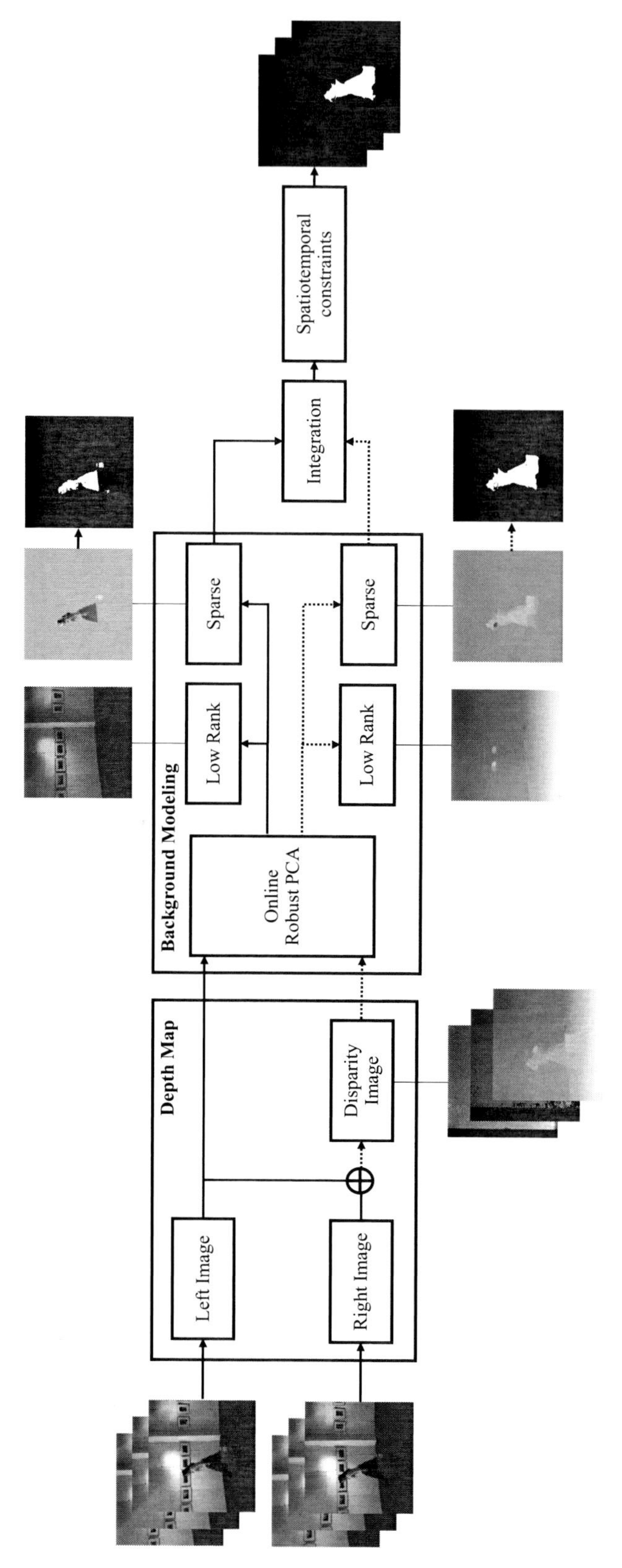

FIGURE 20.9 Block diagram of DEOR-PCA for background subtraction in [13].

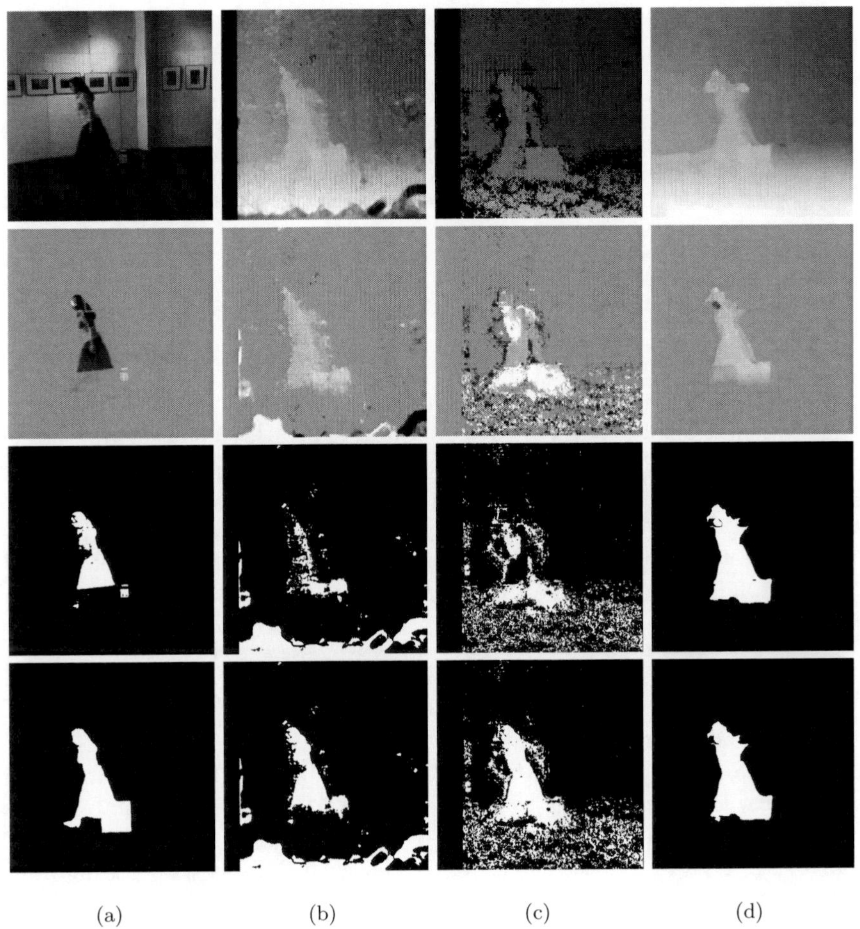

(a) (b) (c) (d)

FIGURE 20.10 *Suitecase* sequence [7]. From top to bottom, input with disparity images, *sparse* component, thresholded mask, and ground truth with integrated binary mask.

The integration is performed using color and range *low-rank* and *sparse* components in this step. The same parameter setting is used in (20.2) for each feature. λ_1 is considered a constant 0.01 for both images. λ_2 and rank r are selected for depth information, whereas λ_2' and rank r' are selected for color frame according to background scene, for obtaining enough sparsity pattern for each image. To speed up the optimization process, the paramerters $r = r' = 1$ and $\lambda_2 = \lambda_2' = 1/\sqrt{max(width, height)}$ are selected for both images. Since disparity *sparse* component provides more information in the background scene, λ_2 must be the same as λ_2'. After integrating *sparse* components of each image by adding individual features, the binary foreground mask f is then obtained by thresholding the integrated *sparse* component. A hard thresholding scheme is applied to get the initial foreground map.

At this stage, the background subtraction scheme in [13] is good enough using variational-based disparity information to deal with static and some small background dynamics as shown in Figure 20.10 (d). However, as illustrated in Figure 20.10 (b) and (c), the integrated color and phase as well as SGBM-based range information using stochastic RPCA produces false alarms. These methods are very sensitive against noise or flickering of lights, which affect the quality of depth map. In addition, as discussed above, color features do not provide enough information to detect objects that contain similar features

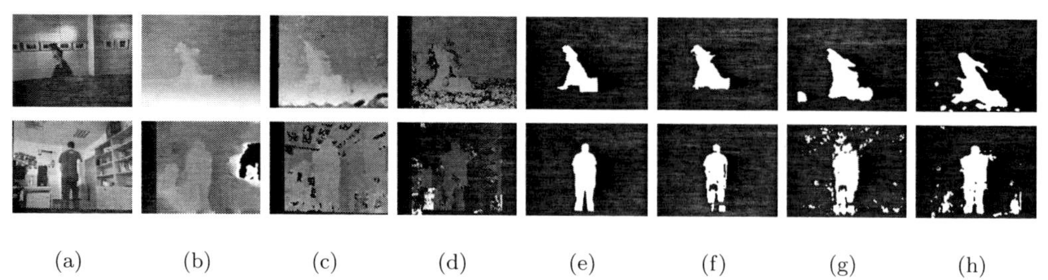

(a) (b) (c) (d) (e) (f) (g) (h)

FIGURE 20.11 Suitecase and lab door sequence. From left to right: (a) input, (b) variational, (c) phase, (d) SGBM, (e) ground truth, (f) DEOR-PCA variational, (g) DEOR-PCA phase, and (h) DEOR-PCA SGBM based foreground mask.

as in the background but depth is less effected and hence this method provides an accurate segmentation as shown in Figure 20.10 (d). Stochastic RPCA on color image with the integration of phase and SGBM-based disparity produces noise without spatiotemporal constraints. Therefore, the MRF is employed in the foreground mask to improve the quality of foreground detection and alleviates most of the noise in the depth mask as depicted in Figure 20.11 (g) and (h), respectively.

20.4 Experimental Evaluations

In this section, the experimental results of OR-PCA with MRF [14], multi-feature-based OR-PCA with DFS [16] and Depth-extended OR-PCA [13] (DEOR-PCA) are presented by comparing them with other state-of-the-art methods. All these methods are implemented in a MATLAB R2013a with 3.40 GHz Intel core i5 processor with a 4GB RAM. First we summarize some performance metrics and then each dataset with its qualitative and quantitative results are presented in detail.

In [14], [16], and [13] the ground truth-based metrics such as *F-measure*, *Precision*, and *Recall* are used for evaluation. These metrics are computed from the true positives (TP), true negatives (TN), false positives (FP), and false negatives (FN). FP and FN refer to pixels misclassified as foreground (FP) or background (FN) while TP and TN account for accurately classified pixels, respectively, as foreground and background. Recall gives the percentage of corrected pixels classified as background when compared with the total number of background pixels in the ground truth as:

$$Recall = \frac{TP}{TP + FN}. \tag{20.11}$$

Similarly, the precision metric is computed as follows:

$$Precision = \frac{TP}{TP + FP}. \tag{20.12}$$

The F-measure score is computed as:

$$F = \frac{2 \times Recall \times Precision}{Recall + Precision}. \tag{20.13}$$

The F-measure characterizes the performance of classification in Precision-Recall space. The aim is to maximize F-measure close to one.

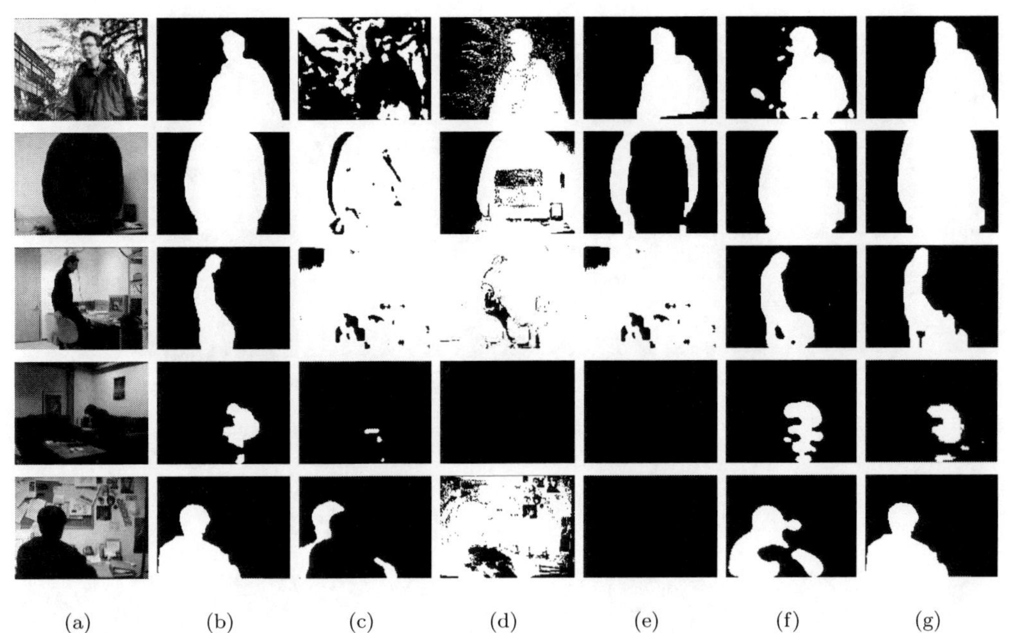

(a) (b) (c) (d) (e) (f) (g)

FIGURE 20.12 Wallflower Dataset [26]. From left to right: (a) Input, (b) Ground Truth, (c) Semi Soft GoDec [31], (d) MOG [24], (e) DECOLOR [32], (f) ORPCA-MRF [14], and (g) ORPCA-DFS [16].

20.4.1 Datasets

In this section, several available datasets are described and both quatitative and qualitative results are presented for each dataset.

Wallflower Dataset

Wallflower dataset [26] contains seven real sequences of challenging background scenes such as *Waving Trees*, *Moved Object*, *Time of Day*, *Light Switch*, *Camouflage*, *Bootstrapping*, and *Foreground Aperture*. Each sequence contains a frame size of 160×120. *F-measure* is computed with available grounds-truth data of each specific sequence. Figure 20.12 shows the visual results of five sequences using the methods ORPCA-MRF [14] and ORPCA-DFS [16] with previous methodologies.

I2R Dataset

Infocom Research (I2R) [18] dataset contains 9 statistically complex different background scenes. The ground truth is provided for some specific images for each sequence. Each sequence contains an image size of 160×128. The average *F-measure* score is computed for each sequence with its ground truth data. Figure 20.13 shows some qualitative results with other the state-of-the-art methods.

Background Models Challenge Dataset

BMC [27] dataset contains a set of both synthetic and real sequences, which are acquired from a static camera for evaluation. The image size of each sequence is 320×240 and only synthetic sequences are reported in this section using the methods described in [14] and [16]. Visual results of synthetic sequences are shown in the first two columns of Figure 20.14.

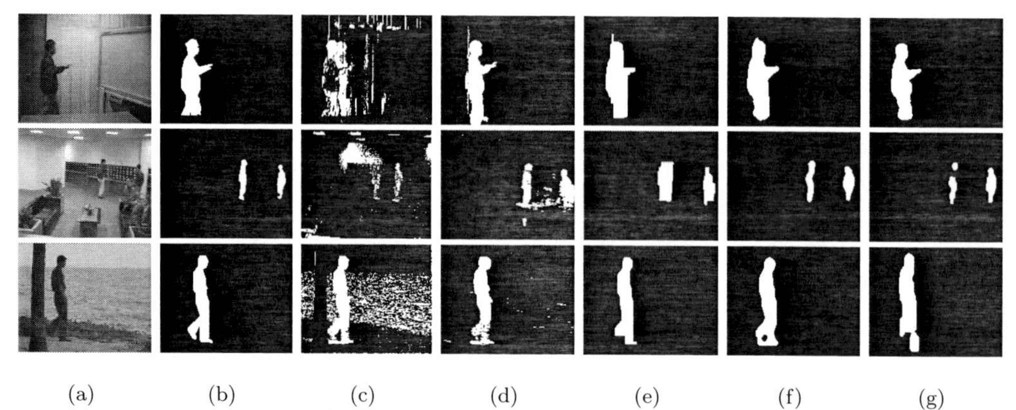

(a) (b) (c) (d) (e) (f) (g)

FIGURE 20.13 I2R Dataset [18]. From left to right: (a) Input, (b) Ground Truth, (c) Semi Soft GoDec [31], (d) MOG [24], (e) DECOLOR [32], (f) ORPCA-MRF [14], and (g) ORPCA-DFS [16].

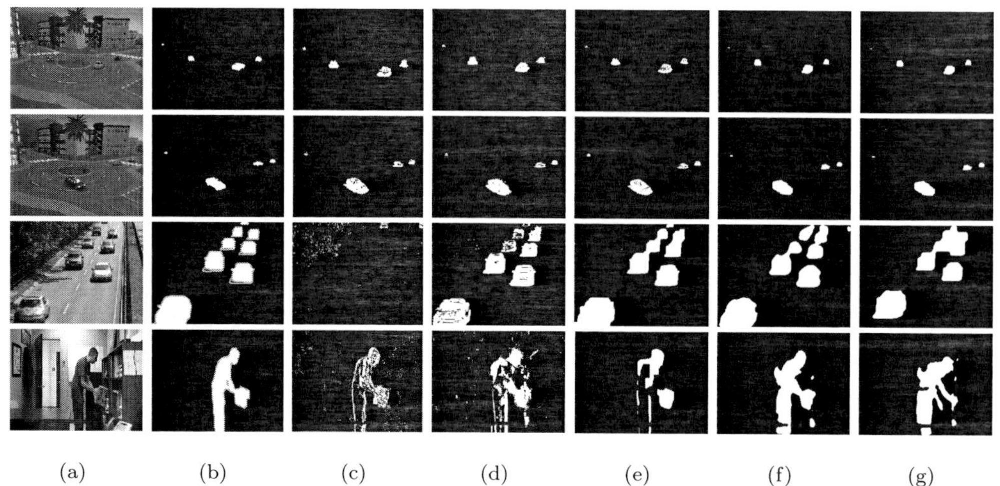

(a) (b) (c) (d) (e) (f) (g)

FIGURE 20.14 BMC [27] and CDnet [8] dataset. From left to right: (a) Input, (b) Ground Truth, (c) Semi Soft GoDec [31], (d) MOG [24], (e) DECOLOR [32], (f) ORPCA-MRF [14], and (g) ORPCA-DFS [16].

Change Detection dataset (CDnet)

Change detection [8] 2014 dataset (CDnet)[71] is the real-world region level benchmark obtained by human experts. This dataset contains about 55 video sequences that are divided into 11 encouraging video categories. The methods reported in [14] and [16] are used to test CDnet dataset using only two categories such as *baseline* and *highly dynamic backgrounds.* The image size of each sequence in each category is 320×240. The last two columns of Figure 20.14 show some of the qualitative results of the CDnet dataset.

[71] http://www.changedetection.net/

Disparity Benchmark Dataset (DBD)

This benchmark dataset [7] contains a set of real video sequences including depths information of each sequence. The dataset contains four challenging scenes called *suitecase*, *crossing*, *labdoor*, and *LCD screen*. To test any background subtraction algorithm, the main goal is to show that color information is not enough by itself but it can be enhanced with disparity images. The image size of each sequence is 1024×1024 or 640×480. Figure 20.11 shows the visual performance of *suitecase* and *labdoor* sequence using DEOR-PCA [13].

Table. 20.1 shows the quantitative performance on five datasets discussed above. The - line in the table indicates that these methods are not tested for the DBD benchmark as this dataset is designed only for depth-extended-based background subtraction algorithms. Table 20.1 depicts the average performance of full datasets such as Wallflower and I2R, whereas the *F-measure* score is computed for only two categories of CDnet and similarly only synthetic sequences of the BMC dataset as shown in Table 20.1.

Method	MOG	SSGODEC	DECOLOR	OR-PCA-MRF	OR-PCA-DFS	DEOR-PCA
Wallflower [26]	45.60	39.87	46.38	**79.81**	77.65	-
I2R [18]	58.12	38.30	65.25	**83.12**	81.21	-
CDnet [8]	59.18	37.68	54.032	**83.26**	75.63	-
BMC [27]	73.74	80.64	81.32	**85.63**	82.72	-
DBD [7]	**64.99**	-	-	-	-	**81.90**

TABLE 20.1 Comparison of average *F-measure* score on five datasets.

20.4.2 Computational Complexity

The computational time is also investigated during the experiments. The computational time is recorded frame by frame in CPU time as hh:mm:ss for 100 frames As discussed in [14] and [16], time is proportional to the value of rank. Table 20.2 shows the comparison of computational time according to different size of images.

Traditional RPCA via PCP-based algorithms either fails to load large amount of input video frames or takes a longer time for optimization, which is not useful for real-time processing. However, in these approaches, near real-time processing is achieved using an initialization scheme. Moreover, image decomposition together with continuous constraints improves the quality of foreground in [14] and multi-featured based OR-PCA [16] provides a very nice potential for background/foreground detection. Table. 20.2 indicates that OR-PCA with MRF takes less time as compare to other methods, whereas multiple features-based OR-PCA computational time is affected. The extra time is due to the larger size of the feature vector. These good experimental evaluations both qualitative and quantitative are the consequences of OR-PCA's providing a very nice framework in contrast to conventional RPCA methods for background subtraction.

20.5 Conclusion

In this chapter, a suitable application such as background/foreground segmentation using stochastic RPCA is presented in detail. In general, stochastic RPCA with MRF via image decomposition provides potential results as compared to other PCP-based strategies.

Method	120 × 160	240 × 320	576 × 720	640 × 480	1024 × 1024
OR-PCAMRF [14]	00:00:06	00:00:18	00:00:40	00:00:26	-
OR-PCADFS [16]	00:01:10	00:02:35	00:05:54	-	-
DEOR-PCA [13]	00:00:03	00:00:09	-	00:00:27	00:01:12

TABLE 20.2 Comparison of computational complexity according to varying dimension of image size.

Multiple features-based stochastic RPCA also make it worth while for robust foreground detection. Moreover, a very encouraging and consistent enhancement is presented using baseline algorithm stochastic RPCA. A rigorous experimental evaluation on each dataset guarantees that these methods can be used for real time processing. However, these good results are based on post-processing or pre-processing techniques adopted in stochastic optimization. Therefore, RPCA-based methods able to provide good results without any post-processing would be a very nice motivation for our future work. In addition, we will extend these methods for moving cameras using online tensor decomposition.

References

1. T. Bouwmans, F. El Baf, and B. Vachon. Statistical background modeling for foreground detection: A survey. *Handbook of Pattern Recognition and Computer Vision*, pages 181–199, 2010.
2. T. Bouwmans and E. Zahzah. Robust PCA via Principal Component Pursuit: A review for a comparative evaluation in video surveillance. *Computer Vision and Image Understanding*, pages 22–34, 2014.
3. Home Office Scientific Development Branch. Imagery library for intelligent detection systems I-LIDS. In *Conference on Crime and Security, 2006. The Institution of Engineering and Technology*, pages 445–448, June 2006.
4. E. Candès, X. Li, Y. Ma, and J. Wright. Robust Principal Component Analysis? *Journal of the ACM*, 58(3):11–37, 2011.
5. N. Dalal and B. Triggs. Histograms of oriented gradients for human detection. In *Conference on Computer Vision and Pattern Recognition, 2005. CVPR 2005. IEEE Computer Society*, volume 1, pages 886–893. IEEE, 2005.
6. J. Feng, H. Xu, and S. Yan. Online robust PCA via stochastic optimization. In *Advances in Neural Information Processing Systems*, pages 404–412, 2013.
7. E. Fernandez-Sanchez, L. Rubio, J. Diaz, and E. Ros. Background subtraction model based on color and depth cues. *Machine Vision and Applications*, pages 1–15, 2014.
8. N. Goyette, P. Jodoin, F. Porikli, J. Konrad, and P. Ishwar. Changedetection.net: A new change detection benchmark dataset. In *IEEE Computer Society Conference on Computer Vision and Pattern Recognition Workshops, CVPRW 2012*, pages 1–8, June 2012.
9. N. Guan, D. Tao, Z. Luo, and B. Yuan. Online non-negative matrix factorization with robust stochastic approximation. *IEEE Transactions on Neural Networks and Learning Systems*, 23(7):1087–1099, 2012.
10. C. Guyon, T. Bouwmans, and E. Zahzah. Robust principal component analysis for background subtraction: Systematic evaluation and comparative analysis. *Computer Vision and Image Understanding*, pages 223–228, 2012.
11. J. He, L. Balzano, and J. Lui. Online robust subspace tracking from partial information. *[Online]. Available: http://arxiv.org/abs/1109.3827*, 2011.
12. H. Hirschmuller. Stereo processing by semiglobal matching and mutual information. *IEEE*

Transactions on Pattern Analysis and Machine Intelligence, 30(2):328–341, 2008.
13. S. Javed, T. Bouwmans, and S. Jung. Depth Extended Online RPCA with Spatiotemporal Constraints for Robust Background Subtraction. In *21st Japan-Korea Joint Workshop on Frontiers of Computer Vision, FCV 2015.*
14. S. Javed, S. Oh, A. Sobral, T. Bouwmans, and S. Jung. OR-PCA with MRF for Robust Foreground Detection in Highly Dynamic Backgrounds. In *Asian Conference on Computer Vision, ACCV 2014.*
15. S. Javed, S. Oho, J. Heo, and S. Jung. Robust Background Subtraction via Online Robust PCA using image decomposition. In *2014 Research in Adaptive and Convergent Systems*, pages 90–96, 2014.
16. S. Javed, A. Sobral, T. Bouwmans, and S. Jung. Or-pca with Dynamic Feature Selection for Robust Background Subtraction. In *ACM Symposium on Applied Computing, SAC 2015.*
17. V. Kolmogorov and R. Zabin. What energy functions can be minimized via graph cuts? *IEEE Transactions on Pattern Analysis and Machine Intelligence*, 26(2):147–159, 2004.
18. L. Li, W. Huang, I. Gu, and Q. Tian. Statistical modeling of complex backgrounds for foreground object detection. *IEEE Transactions on Image Processing*, 13(11):1459–1472, 2004.
19. J. Mairal, F. Bach, J. Ponce, and G. Sapiro. Online learning for matrix factorization and sparse coding. *The Journal of Machine Learning Research*, 11:19–60, 2010.
20. N. Oliver, B. Rosario, and A. Pentland. A bayesian computer vision system for modeling human interactions. *IEEE Transactions on Pattern Analysis and Machine Intelligence*, 22(8):831–843, 2000.
21. D. Patra and P. Nanda. Image segmentation using Markov random field model learning feature and parallel hybrid algorithm. In *International Conference on Computational Intelligence and Multimedia Applications*, volume 3, pages 400–407. IEEE, 2007.
22. J. Ralli, J. Diaz, and E. Ros. Spatial and temporal constraints in variational correspondence methods. *Machine vision and applications*, 24(2):275–287, 2013.
23. P. Rodriguez and B. Wohlberg. Fast principal component pursuit via alternating minimization. In *IEEE International Conference on Image Processing, ICIP 2013*, pages 69–73, September 2013.
24. C. Stauffer and W. Grimson. Adaptive background mixture models for real-time tracking. In *IEEE Computer Society Conference on Computer Vision and Pattern Recognition*, volume 2, 1999.
25. F. De La Torre and M. Black. A framework for robust subspace learning. *International Journal of Computer Vision*, 54(1-3):117–142, 2003.
26. K. Toyama, J. Krumm, B. Brumitt, and B. Meyers. Wallflower: principles and practice of background maintenance. In *IEEE International Conference on Computer Vision*, pages 255–261, 1999.
27. A. Vacavant, T. Chateau, A. Wilhelm, and L. Lequièvre. A benchmark dataset for outdoor foreground/background extraction. In *Computer Vision-ACCV 2012 Workshops*, pages 291–300. Springer, 2013.
28. M. Vanegas, F. Barranco, J. Diaz, and E. Ros. Massive parallel-hardware architecture for multiscale stereo, optical flow and image-structure computation. *IEEE Transactions on Circuits and Systems for Video Technology*, 22(2):282–294, 2012.
29. Z. Wang and X. Xie. An efficient face recognition algorithm based on robust principal component analysis. In *International Conference on Internet Multimedia Computing and Service*, pages 99–102. ACM, 2010.
30. Y. Xu and W. Yin. A block coordinate descent method for regularized multiconvex optimization with applications to non-negative tensor factorization and completion. *SIAM Journal on imaging sciences*, 6(3):1758–1789, 2013.

31. T. Zhou and D. Tao. Godec: Randomized low-rank & sparse matrix decomposition in noisy case. In *International Conference on Machine Learning, ICML 2011*, pages 33–40, 2011.
32. X. Zhou, C. Yang, and W. Yu. Moving object detection by detecting contiguous outliers in the low-rank representation. *IEEE Transactions on Pattern Analysis and Machine Intelligence*, 35(3):597–610, 2013.

21

Bayesian Sparse Estimation for Background/Foreground Separation

Shinichi Nakajima
Technische Universitat Berlin, Germany

Masashi Sugiyama
University of Tokyo, Japan

S. Derin Babacan
Google Inc., USA

21.1 Introduction

Bayesian learning provides a general framework for sparse estimation, which includes non-Bayesian methods as special cases—the trace-norm regularization and the ℓ_1-norm regularization can be interpreted as *maximum a posteriori (MAP) estimation* with certain prior distributions.

A benefit of Bayesian learning is observed when unknown parameters are not point-estimated, but marginalized out—the *Bayesian estimator*, defined as the *posterior mean*, is less prone to overfitting than the MAP estimator, defined as the *posterior mode*. One can also use the *marginal likelihood* as a *model selection criterion* or an objective function for hyperparameter estimation. This enables fully-automatic sparse estimation without manual parameter tuning or cross validation.

In this chapter, we introduce a Bayesian sparse estimation method, called a *sparse additive matrix factorization (SAMF)* [19]. SAMF generalizes *probabilistic matrix factorization* [23], allows various types of sparsity design in a unified framework, and contains a Bayesian variant [1, 4] of *robust PCA* [3].

Sparsity is usually designed in the prior distribution. A popular choice in the Bayesian framework is the *automatic relevance determination (ARD)* prior [22], which automatically eliminates irrelevant model parameters through *empirical Bayesian learning*—the prior variance is estimated from observation by maximizing the marginal likelihood [6]. However, it was pointed out [18] that the structure of the likelihood function can also cause implicit regularization, called *model-induced regularization (MIR)*, and a sparse solution can be obtained even with the flat prior. Strong MIR appears when the probabilistic model is *non-identifiable* [25]—the mapping from the model parameters and the distributions is not one-to-one. Since low-rank approximation is often formulated as matrix factorization (MF), which accompanies non-identifiability, MIR should not be neglected. SAMF incorporates both of the MIR and the ARD effects. More specifically, we design the sparsity by factorizing unknown parameters and applying the ARD prior on the factors. This unifies the treatment of the low-rank term and the element-wise sparse term, and further generalizes the sparsity design.

In this chapter, we first introduce SAMF in Section 21.2, and its application to background/foreground separation. Robust PCA [1, 3, 4] can solve this task by capturing the background and the foreground with the low-rank and the element-wise sparse terms, respectively. Within the SAMF framework, we design a specialized sparse term, called a *segment-wise sparse* term, to better capture the foreground, based on side information.

We then derive an algorithm, called a *mean update (MU) algorithm*, for *variational Bayesian (VB) learning* of SAMF in Section 21.3. The MU algorithm uses the global analytic VB solution of MF [21] in each iteration, and generally gives a better local solution than the standard algorithm for VB learning. After that, we discuss a close relation between MIR and ARD, describe the standard VB algorithm (which is used as a baseline method in the experiment), and introduce related work in Section 21.4. Finally, we experimentally show the usefulness of SAMF in Section 21.5, and conclude the chapter in Section 21.6.

21.2 Formulation

In this section, we formulate the sparse additive matrix factorization (SAMF) model.

21.2.1 Matrix Factorization

In standard matrix factorization (MF), an observed matrix $\boldsymbol{A} \in \mathbb{R}^{D \times M}$ is modeled by a low-rank matrix $\boldsymbol{L} \in \mathbb{R}^{D \times M}$ contaminated with random noise $\boldsymbol{E} \in \mathbb{R}^{D \times M}$:

$$\boldsymbol{A} = \boldsymbol{L} + \boldsymbol{E}.$$

Then the low-rank matrix $\boldsymbol{L}$ is decomposed into the product of two matrices $\boldsymbol{B} \in \mathbb{R}^{D \times H}$ and $\boldsymbol{\Theta} \in \mathbb{R}^{M \times H}$:

$$\boldsymbol{L} = \boldsymbol{B}\boldsymbol{\Theta}^{\top} = \sum_{h=1}^{H} \boldsymbol{b}_h \boldsymbol{\theta}_h^{\top}, \tag{21.1}$$

where $\top$ denotes the transpose of a matrix or vector. In this chapter, we denote a column vector of a matrix by a bold small letter, and a row vector by a bold small letter with a tilde:

$$\boldsymbol{B} = (\boldsymbol{b}_1, \ldots, \boldsymbol{b}_H) = \left(\widetilde{\boldsymbol{b}}_1, \ldots, \widetilde{\boldsymbol{b}}_D\right)^{\top}, \qquad \boldsymbol{\Theta} = (\boldsymbol{\theta}_1, \ldots, \boldsymbol{\theta}_H) = \left(\widetilde{\boldsymbol{\theta}}_1, \ldots, \widetilde{\boldsymbol{\theta}}_M\right)^{\top}.$$

The last equation in Equation (21.1) implies that the matrix product $\boldsymbol{B}\boldsymbol{\Theta}^{\top}$ is the sum of rank-1 components. It was elucidated that this product induces an implicit regularization effect, called *model-induced regularization (MIR)*, and a low-rank (singular-component-wise sparse) solution is produced under the *variational Bayesian (VB) learning* [18]. Furthermore, the *empirical Bayesian procedure* with the *automatic relevance determination (ARD)* prior on the factors has been guaranteed to estimate the correct rank H under a reasonable condition [21]. In this procedure, the prior is given by:

$$p(\boldsymbol{B}) \propto \exp\left(-\frac{\mathrm{tr}(\boldsymbol{B}^\top \boldsymbol{C}_B^{-1}\boldsymbol{B})}{2}\right), \qquad p(\boldsymbol{\Theta}) \propto \exp\left(-\frac{\mathrm{tr}(\boldsymbol{\Theta}^\top \boldsymbol{C}_\Theta^{-1}\boldsymbol{\Theta})}{2}\right),$$

and the diagonal positive-definite prior covariances

$$\boldsymbol{C}_B = \mathbf{diag}(c_{b_1}^2, \ldots, c_{b_H}^2), \qquad \boldsymbol{C}_\Theta = \mathbf{diag}(c_{\theta_1}^2, \ldots, c_{\theta_H}^2)$$

are estimated by maximizing the marginal likelihood. This theoretical guarantee suggests that one does not have to tune the model rank H carefully, but can simply set it to a sufficiently large value, e.g., $H = \min(D, M)$—fully-automatic inference is realized.

Likewise, we can induce the element-wise sparsity by factorizing a matrix with the Hadamard (or element-wise) product

$$\boldsymbol{S} = \boldsymbol{B} * \boldsymbol{\Theta}, \qquad \text{where} \qquad (\boldsymbol{B} * \boldsymbol{\Theta})_{d,m} = B_{d,m}\Theta_{d,m}, \tag{21.2}$$

and applying the ARD prior

$$p(\boldsymbol{B}) \propto \exp\left(-\frac{\sum_{d=1}^{D}\sum_{h=1}^{H} B_{d,h}^2 \overline{c}_{b_{d,h}}^{-2}}{2}\right), \quad p(\boldsymbol{\Theta}) \propto \exp\left(-\frac{\sum_{m=1}^{M}\sum_{h=1}^{H} \Theta_{m,h}^2 \overline{c}_{\theta_{m,h}}^{-2}}{2}\right)$$

with the positive prior variances $\overline{c}_{b_{d,h}}^2, \overline{c}_{\theta_{m,h}}^2$ (for $d = 1, \ldots, D$, $m = 1, \ldots, M$, and $h = 1, \ldots, H$) to be estimated from observation.

21.2.2 A General Expression of Factorization

Let us generalize MF for more flexible sparsity design. We formally define a sparse matrix factorization (SMF) term, which involves partitioning, rearrangement, and factorization, by

$$\boldsymbol{U} = \boldsymbol{G}(\{\boldsymbol{U}'^{(t)}\}_{t=1}^T; \mathcal{X}), \qquad \text{where} \qquad \boldsymbol{U}'^{(t)} = \boldsymbol{B}^{(t)}\boldsymbol{\Theta}^{(t)\top}. \tag{21.3}$$

Here, $\{\boldsymbol{B}^{(t)}, \boldsymbol{\Theta}^{(t)}\}_{t=1}^T$ are parameters to be estimated, and $\boldsymbol{G}(\cdot; \mathcal{X}) : \mathbb{R}^{\prod_{t=1}^T (D'^{(t)} \times M'^{(t)})} \mapsto \mathbb{R}^{D \times M}$ is a designed function associated with an index mapping parameter $\mathcal{X}$. $\boldsymbol{G}(\cdot; \mathcal{X})$ will be explained in detail shortly.

Figure 21.1 shows how to construct an SMF term. First, we partition the entries of $\boldsymbol{U}$ into T parts. Then, by rearranging the entries in each part, we form partitioned-and-rearranged (PR) matrices $\boldsymbol{U}'^{(t)} \in \mathbb{R}^{D'^{(t)} \times M'^{(t)}}$ for $t = 1, \ldots, T$. Finally, each of $\boldsymbol{U}'^{(t)}$ is decomposed into the product of $\boldsymbol{B}^{(t)} \in \mathbb{R}^{D'^{(t)} \times H'^{(t)}}$ and $\boldsymbol{\Theta}^{(t)} \in \mathbb{R}^{M'^{(t)} \times H'^{(t)}}$ for $H'^{(t)} \leq \min(D'^{(t)}, M'^{(t)})$.[72]

In Equation (21.3), the function $\boldsymbol{G}(\cdot; \mathcal{X})$ is responsible for partitioning and rearrangement: It maps the set $\{\boldsymbol{U}'^{(t)}\}_{t=1}^T$ of the PR matrices to the target matrix $\boldsymbol{U} \in \mathbb{R}^{D \times M}$, based on the *one-to-one* mapping $\mathcal{X} : (t, d', m') \mapsto (d, m)$ from the indices of the entries in $\{\boldsymbol{U}'^{(t)}\}_{t=1}^T$ to the indices of the entries in $\boldsymbol{U}$, so that

$$\left(\boldsymbol{G}(\{\boldsymbol{U}'^{(t)}\}_{t=1}^T; \mathcal{X})\right)_{d,m} = U_{d,m} = U_{\mathcal{X}(t,d',m')} = U'^{(t)}_{d',m'}. \tag{21.4}$$

[72] For fully automatic inference, $H'^{(t)}$ should be set to the maximum possible value, i.e., $H'^{(t)} = \min(D'^{(t)}, M'^{(t)})$, unless the computational cost prohibits this choice.

$$U = \begin{pmatrix} U_{1,1} & U_{1,2} & U_{1,3} & U_{1,4} \\ U_{2,1} & U_{2,2} & U_{2,3} & U_{2,4} \\ U_{3,1} & U_{3,2} & U_{3,3} & U_{3,4} \\ U_{4,1} & U_{4,2} & U_{4,3} & U_{4,4} \end{pmatrix} \overset{G}{\longleftarrow} \begin{array}{l} U'^{(1)} = (U_{1,1} \; U_{1,2} \; U_{1,3} \; U_{1,4}) = B^{(1)}\Theta^{(1)\top} \\ U'^{(2)} = \begin{pmatrix} U_{2,1} & U_{2,2} \\ U_{3,1} & U_{3,2} \end{pmatrix} = B^{(2)}\Theta^{(2)\top} \\ U'^{(3)} = (U_{2,3} \; U_{2,4} \; U_{3,3} \; U_{3,4}) = B^{(3)}\Theta^{(3)\top} \\ U'^{(4)} = (U_{4,1} \; U_{4,2} \; U_{4,3}) = B^{(4)}\Theta^{(4)\top} \\ U'^{(5)} = (U_{4,4}) = B^{(5)}\Theta^{(5)\top} \end{array}$$

FIGURE 21.1 (See color insert.) An example of SMF-term construction. $G(\cdot; \mathcal{X})$ with $\mathcal{X}$: $(t, d', m') \mapsto (d, m)$ maps the set $\{U'^{(t)}\}_{t=1}^{T}$ of the PR matrices to the target matrix U, so that $U'^{(t)}_{d',m'} = U_{\mathcal{X}(t,d',m')} = U_{d,m}$.

As will be discussed in Section 21.4.1, the SMF-term expression (21.3) under VB learning induces low-rank sparsity in each partition. This means that partition-wise sparsity is also induced. Accordingly, partitioning, rearrangement, and factorization should be designed in the following manner.

Suppose that we are given a required sparsity structure on a matrix (sparsity structures suitable for background/foreground separation will be introduced in Section 21.2.4). We first partition the matrix, according to the required sparsity. Some partitions can be submatrices. We rearrange each of the submatrices on which the low-rank sparsity should not be imposed into a long vector ($U'^{(3)}$ in the example in Figure 21.1). We leave the other submatrices ($U'^{(2)}$) on which the low-rank sparsity should be imposed, the vectors ($U'^{(1)}$ and $U'^{(4)}$) and the scalars ($U'^{(5)}$) as they are. Finally, we factorize each of the PR matrices.

Clearly, the SMF expression (21.3) contains the plain MF (21.1) and the Hadamard product factorization (21.2) (skipping partitioning and rearrangement gives the former, while the complete partitioning into elements gives the latter), which can be used for the low-rank term L and the element-wise sparse term S, respectively.

21.2.3 Sparse Additive Matrix Factorization

We define a model as a sum of SMF terms (21.3):

$$A = \sum_{j=1}^{J} U^{(j)} + E, \quad \text{where} \quad U^{(j)} = G(\{B^{(t,j)}\Theta^{(t,j)\top}\}_{t=1}^{T^{(j)}}; \mathcal{X}^{(j)}). \tag{21.5}$$

SMF terms should be designed based on side information. In robust PCA [1, 3, 4], the observed matrix $A \in \mathbb{R}^{D\times M}$ is assumed to be a sum of a low-rank matrix $U^{\text{low-rank}}$ and an element-wise sparse matrix U^{element}, contaminated with Gaussian noise E:

$$A = U^{\text{low-rank}} + U^{\text{element}} + E. \tag{21.6}$$

Here and hereafter in this chapter, in order to treat various types of sparsity in a unified manner, we denote any type of sparse term by U with a superscript specifying the type of sparsity, e.g., $L = U^{\text{low-rank}}$ and $S = U^{\text{element}}$.

Robust PCA was applied to background/foreground separation, where *moving* objects (such as a person in Figure 21.2) are considered to belong to the foreground [1, 3, 4]. The observed matrix A is constructed by stacking all pixels in each frame into each column (Figure 21.3), and fitted to the robust PCA model (21.6). The low-rank term and the element-wise sparse term are expected to capture the *static* background (BG) and the *moving* foreground (FG), respectively.

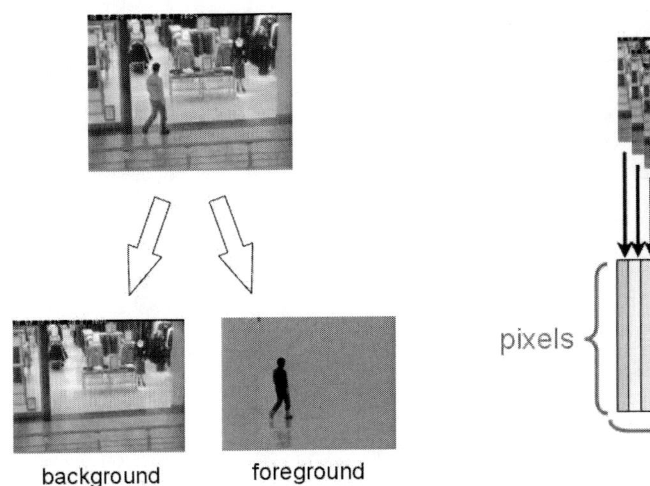

FIGURE 21.2 Background/foreground separation task.

FIGURE 21.3 (See color insert.) The observation matrix $\boldsymbol{A}$ is constructed by stacking all pixels in each frame into each column.

21.2.4 Segmentation-Based SAMF

The SMF expression (21.3) enables us to use side information in a more flexible way. For BG/FG separation, we propose a *segmentation-based SAMF (sSAMF)*, which relies on a natural assumption that a pixel segment with small color diversity tends to belong to a single object.

Assuming that the pixels in an image segment with similar colors or intensities tend to share the same label (i.e., BG or FG), we form a *segment-wise sparse* term, and use it to capture the FG, i.e.,

$$\boldsymbol{A} = \boldsymbol{U}^{\text{low-rank}} + \boldsymbol{U}^{\text{segment}} + \boldsymbol{E}. \tag{21.7}$$

The segment-wise sparse term $\boldsymbol{U}^{\text{segment}}$ is constructed based on an over-segmented image: $\boldsymbol{U}'^{(t)}$ for each t is a column vector consisting of all pixels in each segment. We use the efficient graph-based segmentation (EGS) algorithm [7], which is computationally very efficient, to produce an over-segmented image from each frame (see Figure 21.4). We will show in Section 21.5 that the segment-wise sparse term captures the FG more accurately than the element-wise sparse term.

21.3 Inference

In this section, we introduce the variational Bayesian learning for SAMF, and then derive an inference algorithm, called a mean update algorithm.

21.3.1 Probabilistic Formulation of SAMF

Let

$$\boldsymbol{w} = \{\boldsymbol{w}_{\mathrm{B}}^{(j)}, \boldsymbol{w}_{\Theta}^{(j)}\}_{j=1}^{J}, \qquad \text{where} \qquad \boldsymbol{w}_{\mathrm{B}}^{(j)} = \{\boldsymbol{B}^{(t,j)}\}_{t=1}^{T^{(j)}}, \qquad \boldsymbol{w}_{\Theta}^{(j)} = \{\boldsymbol{\Theta}^{(t,j)}\}_{t=1}^{T^{(j)}},$$

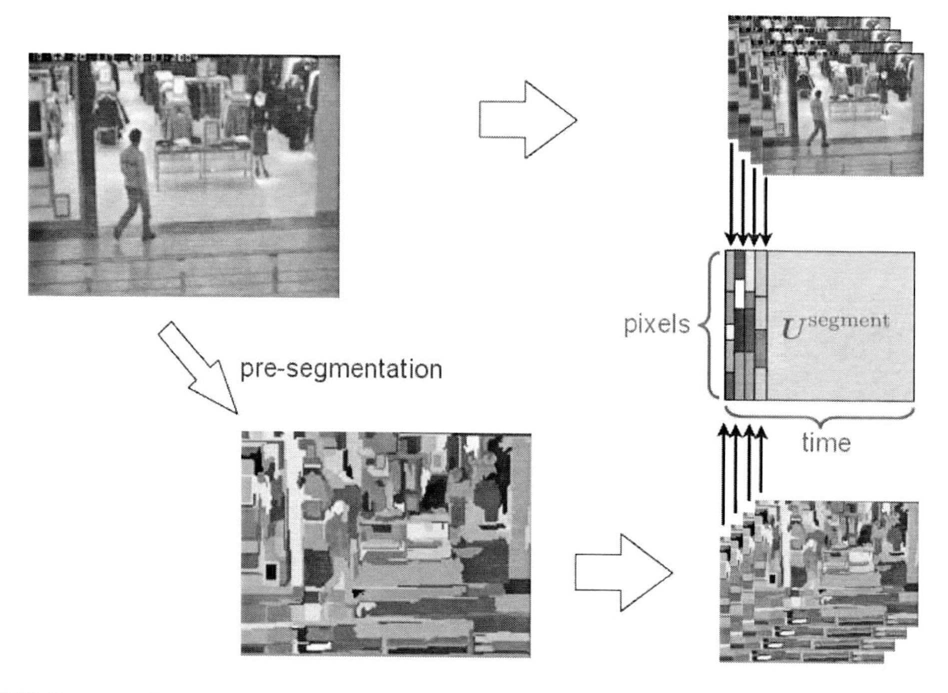

FIGURE 21.4 (See color insert.) Construction of a segment-wise sparse term. The original frame is pre-segmented and the sparsity is induced segment-wise.

summarize the parameters of the SAMF model (21.5). As in the probabilistic MF [23], we assume independent Gaussian noise and priors, i.e., the likelihood and the priors are written as

$$p(\boldsymbol{A}|\boldsymbol{w}) \propto \exp\left(-\tfrac{1}{2\sigma^2}\left\|\boldsymbol{A} - \textstyle\sum_{j=1}^{J}\boldsymbol{U}^{(j)}\right\|_{\text{Fro}}^2\right), \tag{21.8}$$

$$p(\{\boldsymbol{w}_{\text{B}}^{(j)}\}_{j=1}^{J}) \propto \exp\left(-\tfrac{1}{2}\textstyle\sum_{j=1}^{J}\sum_{t=1}^{T^{(j)}}\operatorname{tr}\left(\boldsymbol{B}^{(t,j)}\boldsymbol{C}_B^{(t,j)\,-1}\boldsymbol{B}^{(t,j)\top}\right)\right), \tag{21.9}$$

$$p(\{\boldsymbol{w}_{\Theta}^{(j)}\}_{j=1}^{J}) \propto \exp\left(-\tfrac{1}{2}\textstyle\sum_{j=1}^{J}\sum_{t=1}^{T^{(j)}}\operatorname{tr}\left(\boldsymbol{\Theta}^{(t,j)}\boldsymbol{C}_{\Theta}^{(t,j)\,-1}\boldsymbol{\Theta}^{(t,j)\top}\right)\right), \tag{21.10}$$

where $\|\cdot\|_{\text{Fro}}$ and $\operatorname{tr}(\cdot)$ denote the Frobenius norm and the trace of a matrix, respectively. We assume that the prior covariances of $\boldsymbol{B}^{(t,j)}$ and $\boldsymbol{\Theta}^{(t,j)}$ are diagonal:

$$\boldsymbol{C}_B^{(t,j)} = \mathbf{diag}\left(c_{b_1}^{(t,j)\,2}, \ldots, c_{b_H}^{(t,j)\,2}\right), \qquad \boldsymbol{C}_{\Theta}^{(t,j)} = \mathbf{diag}\left(c_{\theta_1}^{(t,j)\,2}, \ldots, c_{\theta_H}^{(t,j)\,2}\right).$$

Without loss of generality, we assume that the diagonal entries of $\boldsymbol{C}_B^{(t,j)}\boldsymbol{C}_{\Theta}^{(t,j)}$ are arranged in the non-increasing order, i.e., $c_{b_h}^{(t,j)}c_{\theta_h}^{(t,j)} \geq c_{b_{h'}}^{(t,j)}c_{\theta_{h'}}^{(t,j)}$ for any pair $h < h'$.

21.3.2 Variational Bayesian Learning

The *Bayes posterior* is written as follows:

$$p(\boldsymbol{w}|\boldsymbol{A}) = \frac{p(\boldsymbol{A}|\boldsymbol{w})p(\boldsymbol{w})}{p(\boldsymbol{A})}, \tag{21.11}$$

where $p(\boldsymbol{A}) = \langle p(\boldsymbol{A}|\boldsymbol{w})\rangle_{p(\boldsymbol{w})}$ is the *marginal likelihood*. Here, $\langle\cdot\rangle_p$ denotes the expectation over the distribution p. Since the Bayes posterior (21.11) for matrix factorization is computationally intractable, the *variational Bayesian (VB) learning* was proposed [1, 2, 12, 15].

Let $r(\boldsymbol{w})$, or r for short, be a trial distribution. The following functional with respect to r is called the *free energy*:

$$F(r|\boldsymbol{A}) = \left\langle \log \frac{r(\boldsymbol{w})}{p(\boldsymbol{A}|\boldsymbol{w})p(\boldsymbol{w})} \right\rangle_{r(\boldsymbol{w})} = \left\langle \log \frac{r(\boldsymbol{w})}{p(\boldsymbol{w}|\boldsymbol{A})} \right\rangle_{r(\boldsymbol{w})} - \log p(\boldsymbol{A}). \tag{21.12}$$

The first term is the *Kullback-Leibler (KL) distance* from the trial distribution to the Bayes posterior, and the second term is constant with respect to r. Therefore, minimizing the free energy (21.12) amounts to finding a distribution closest to the Bayes posterior (21.11) in the sense of the KL distance. In the VB learning, the free energy (21.12) is minimized over some restricted function space, which gives an approximation to the Bayes posterior. The hyperparameters $\{\boldsymbol{C}_B^{(t,j)}, \boldsymbol{C}_\Theta^{(t,j)}\}$ can also be estimated by minimizing the free energy (21.12). This procedure is called *empirical Bayesian learning*.

Following the standard derivation of VB learning [1, 2, 15], we impose the following decomposability constraint on the posterior:

$$r(\boldsymbol{w}) = \prod_{j=1}^{J} r_{\mathrm{B}}^{(j)}(\boldsymbol{w}_{\mathrm{B}}^{(j)}) \cdot r_{\Theta}^{(j)}(\boldsymbol{w}_{\Theta}^{(j)}). \tag{21.13}$$

Under this constraint, it is easy to show that the VB posterior minimizing the free energy (21.12) is written as

$$r(\boldsymbol{w}) = \prod_{j=1}^{J}\prod_{t=1}^{T^{(j)}} \left(\prod_{d'=1}^{D'^{(t,j)}} \mathcal{N}_{H'^{(t,j)}}(\tilde{\boldsymbol{b}}_{d'}^{(t,j)}; \widehat{\tilde{\boldsymbol{b}}}_{d'}^{(t,j)}, \boldsymbol{\Sigma}_B^{(t,j)}) \cdot \prod_{m'=1}^{M'^{(t,j)}} \mathcal{N}_{H'^{(t,j)}}(\tilde{\boldsymbol{\theta}}_{m'}^{(t,j)}; \widehat{\tilde{\boldsymbol{\theta}}}_{m'}^{(t,j)}, \boldsymbol{\Sigma}_\Theta^{(t,j)}) \right), \tag{21.14}$$

where $\mathcal{N}_D(\cdot; \boldsymbol{\mu}, \boldsymbol{\Sigma})$ denotes the D-dimensional Gaussian distribution with mean $\boldsymbol{\mu}$ and covariance $\boldsymbol{\Sigma}$.

21.3.3 Mean Update Algorithm

Based on the form of the posterior (21.14), one can easily derive a coordinate descent algorithm, which we call the standard VB iteration and introduce in Section 21.4.2. However, the unified framework of SAMF as a sum of MF terms enables us to derive a more sophisticated algorithm, called the *mean update (MU) algorithm*.

Let us denote the mean of $\boldsymbol{U}^{(j)}$, defined in Equation (21.5), over the VB posterior by

$$\widehat{\boldsymbol{U}}^{(j)} = \langle \boldsymbol{U}^{(j)} \rangle_{r_{\mathrm{B}}^{(j)}(\boldsymbol{w}_{\mathrm{B}}^{(j)}) r_{\Theta}^{(j)}(\boldsymbol{w}_{\Theta}^{(j)})} = \boldsymbol{G}\left(\{\widehat{\boldsymbol{B}}^{(t,j)} \widehat{\boldsymbol{\Theta}}^{(t,j)\top}\}_{t=1}^{T^{(j)}}; \mathcal{X}^{(j)} \right). \tag{21.15}$$

Then we can prove the following theorem (see [19] for the proof):

THEOREM 21.1 *Given $\{\widehat{\boldsymbol{U}}^{(j')}\}_{j' \neq j}$ and the noise variance σ^2, the VB posterior of $(\boldsymbol{w}_{\mathrm{B}}^{(j)}, \boldsymbol{w}_{\Theta}^{(j)}) = \{\boldsymbol{B}^{(t,j)}, \boldsymbol{\Theta}^{(t,j)}\}_{t=1}^{T^{(j)}}$ coincides with the VB posterior of the following MF model:*

$$p(\boldsymbol{Z}'^{(t,j)}|\boldsymbol{B}^{(t,j)}, \boldsymbol{\Theta}^{(t,j)}) \propto \exp\left(-\frac{1}{2\sigma^2} \left\| \boldsymbol{Z}'^{(t,j)} - \boldsymbol{B}^{(t,j)}\boldsymbol{\Theta}^{(t,j)\top} \right\|_{Fro}^2 \right), \tag{21.16}$$

$$p(\boldsymbol{B}^{(t,j)}) \propto \exp\left(-\frac{1}{2} tr\left(\boldsymbol{B}^{(t,j)} \boldsymbol{C}_B^{(t,j)\,-1} \boldsymbol{B}^{(t,j)\top} \right) \right), \tag{21.17}$$

$$p(\boldsymbol{\Theta}^{(t,j)}) \propto \exp\left(-\frac{1}{2} tr\left(\boldsymbol{\Theta}^{(t,j)} \boldsymbol{C}_\Theta^{(t,j)\,-1} \boldsymbol{\Theta}^{(t,j)\top} \right) \right), \tag{21.18}$$

for each $t = 1, \ldots, T^{(j)}$. *Here,* $\boldsymbol{Z}'^{(t,j)} \in \mathbb{R}^{D'^{(t,j)} \times M'^{(t,j)}}$ *is defined as*

$$Z'^{(t,j)}_{d',m'} = Z^{(j)}_{\mathcal{X}^{(j)}(t,d',m')}, \qquad \text{where} \qquad \boldsymbol{Z}^{(j)} = \boldsymbol{A} - \sum_{j' \neq j} \widehat{\boldsymbol{U}}^{(j)}. \tag{21.19}$$

The left formula in Equation (21.19) relates the entries of $\boldsymbol{Z}^{(j)} \in \mathbb{R}^{D \times M}$ to the entries of $\{\boldsymbol{Z}'^{(t,j)} \in \mathbb{R}^{D'^{(t,j)} \times M'^{(t,j)}}\}_{t=1}^{T^{(j)}}$ by using the map $\mathcal{X}^{(j)} : (t, d', m') \mapsto (d, m)$ (see Equation (21.4) and Figure 21.1).

Theorem 21.1 states that a partial problem of SAMF—finding the posterior of $(\boldsymbol{B}^{(t,j)}, \boldsymbol{\Theta}^{(t,j)})$ for each $t = 1, \ldots, T^{(j)}$, given $\{\widehat{\boldsymbol{U}}^{(j')}\}_{j' \neq j}$ and σ^2— can be solved in the same way as the standard MF.

The noise variance σ^2 is also unknown in many applications. To estimate σ^2, we can use the following lemma:

LEMMA 21.1 Given the VB posterior for $\{\boldsymbol{w}_{\mathrm{B}}^{(j)}, \boldsymbol{w}_{\Theta}^{(j)}\}_{j=1}^{J}$, the noise variance σ^2 minimizing the free energy (21.12) is given by

$$\sigma^2 = \frac{1}{DM}\Bigg\{ \|\boldsymbol{A}\|^2_{\mathrm{Fro}} - 2\sum_{j=1}^{J} \mathrm{tr}\left(\widehat{\boldsymbol{U}}^{(j)\top}\left(\boldsymbol{A} - \sum_{j'=j+1}^{J} \widehat{\boldsymbol{U}}^{(j')}\right)\right)$$

$$+ \sum_{j=1}^{J}\sum_{t=1}^{T^{(j)}} \mathrm{tr}\left((\widehat{\boldsymbol{B}}^{(t,j)\top}\widehat{\boldsymbol{B}}^{(t,j)} + D'^{(t,j)}\boldsymbol{\Sigma}_{\mathrm{B}}^{(t,j)}) \cdot (\widehat{\boldsymbol{\Theta}}^{(t,j)\top}\widehat{\boldsymbol{\Theta}}^{(t,j)} + M'^{(t,j)}\boldsymbol{\Sigma}_{\Theta}^{(t,j)})\right)\Bigg\}. \tag{21.20}$$

Theorem 21.1 allows us to use the results given in [21], where the global analytic VB solution for MF was derived, despite the non-convexity of the problem. Thus, we can obtain the following corollaries:

COROLLARY 21.1 Assume that $D'^{(t,j)} \leq M'^{(t,j)}$ for all (t, j), and that $\{\widehat{\boldsymbol{U}}^{(j')}\}_{j' \neq j}$ and the noise variance σ^2 are given. Let $\gamma_h^{(t,j)}$ (≥ 0) be the h-th largest singular value of $\boldsymbol{Z}'^{(t,j)}$, and let $\boldsymbol{\omega}_{b_h}^{(t,j)}$ and $\boldsymbol{\omega}_{\theta_h}^{(t,j)}$ be the associated left and right singular vectors:

$$\boldsymbol{Z}'^{(t,j)} = \sum_{h=1}^{D'^{(t,j)}} \gamma_h^{(t,j)} \boldsymbol{\omega}_{b_h}^{(t,j)} \boldsymbol{\omega}_{\theta_h}^{(t,j)\top}. \tag{21.21}$$

Let $\alpha^{(t,j)} = \frac{D'^{(t,j)}}{M'^{(t,j)}}$, and let $\underline{\tau}^{(t,j)} = \underline{\tau}(\alpha^{(t,j)})$ be the unique zero-cross point of the following decreasing function:

$$\Xi(\tau; \alpha) = \Phi(\tau) + \Phi\left(\frac{\tau}{\alpha}\right), \qquad \text{where} \qquad \Phi(x) = \frac{\log(x+1)}{x} - \frac{1}{2}. \tag{21.22}$$

Then, the global empirical VB solution for the j-th SMF term $\{\boldsymbol{B}^{(t,j)}, \boldsymbol{\Theta}^{(t,j)}\}_{t=1}^{T^{(j)}}$ (where the hyperparameters $\{\boldsymbol{C}_B^{(t,j)}, \boldsymbol{C}_\Theta^{(t,j)}\}_{t=1}^{T^{(j)}}$ are also estimated from observation) is given by

$$\widehat{\boldsymbol{U}}'^{(t,j)\,\mathrm{EVB}} = \left\langle \boldsymbol{B}^{(t,j)}\boldsymbol{\Theta}^{(t,j)\top}\right\rangle_{r^{(t,j)}(\boldsymbol{B}^{(t,j)},\boldsymbol{\Theta}^{(t,j)})} = \sum_{h=1}^{H'^{(t,j)}} \widehat{\gamma}_h^{(t,j)\,\mathrm{EVB}} \boldsymbol{\omega}_{b_h}^{(t,j)} \boldsymbol{\omega}_{\theta_h}^{(t,j)\top},$$

$$\text{where} \qquad \widehat{\gamma}_h^{(t,j)\,\mathrm{EVB}} = \begin{cases} \breve{\gamma}_h^{(t,j)\,\mathrm{EVB}} & \text{if } \gamma_h^{(t,j)} > \underline{\gamma}_h^{(t,j)\mathrm{EVB}}, \\ 0 & \text{otherwise.} \end{cases} \tag{21.23}$$

Here, the threshold and the amplitude are given by

$$\underline{\gamma}_h^{(t,j)\,\mathrm{EVB}} = \sigma\sqrt{M'^{(t,j)}\left(1+\underline{\tau}^{(t,j)}\right)\left(1+\frac{\alpha^{(t,j)}}{\underline{\tau}^{(t,j)}}\right)}, \tag{21.24}$$

$$\breve{\gamma}_h^{(t,j)\,\mathrm{EVB}} = \frac{\gamma_h^{(t,j)}}{2}\left(1-\frac{(D'^{(t,j)}+M'^{(t,j)})\sigma^2}{\gamma_h^{(t,j)\,2}} + \sqrt{\left(1-\frac{(D'^{(t,j)}+M'^{(t,j)})\sigma^2}{\gamma_h^{(t,j)\,2}}\right)^2 - \frac{4D'^{(t,j)}M'^{(t,j)}\sigma^4}{\gamma_h^{(t,j)\,4}}}\right). \tag{21.25}$$

COROLLARY 21.2 Assume that $D'^{(t,j)} \leq M'^{(t,j)}$ for all (t,j), and let $\boldsymbol{I}_D$ be the D-dimensional identity matrix. Given $\{\widehat{\boldsymbol{U}}^{(j')}\}_{j'\neq j}$ and the noise variance σ^2, the VB posterior for the j-th SMF term is given by

$$r_{\mathrm{B}}^{(t,j)}(\boldsymbol{B}^{(t,j)}) = \prod_{h=1}^{H'^{(t,j)}} \mathcal{N}_{D'^{(t,j)}}(\boldsymbol{b}_h^{(t,j)}; \widehat{\boldsymbol{b}}_h^{(t,j)}, \sigma_{b_h}^{(t,j)\,2}\boldsymbol{I}_{D'^{(t,j)}}),$$

$$r_{\Theta}^{(t,j)}(\boldsymbol{\Theta}^{(t,j)}) = \prod_{h=1}^{H'^{(t,j)}} \mathcal{N}_{M'^{(t,j)}}(\boldsymbol{\theta}_h^{(t,j)}; \widehat{\boldsymbol{\theta}}_h^{(t,j)}, \sigma_{\theta_h}^{(t,j)\,2}\boldsymbol{I}_{M'^{(t,j)}}),$$

where, for (h,t) such that $\gamma_h^{(t,j)} \geq \underline{\gamma}_h^{(t,j)\,\mathrm{EVB}}$,

$$\widehat{\boldsymbol{b}}_h^{(t,j)} = \pm\sqrt{\frac{\breve{\gamma}_h^{(t,j)\,\mathrm{EVB}}}{\delta_h^{(t,j)}}}\cdot\boldsymbol{\omega}_{b_h}^{(t,j)}, \qquad \sigma_{b_h}^{(t,j)\,2} = \frac{\sigma^2}{\gamma_h^{(t,j)}\delta_h^{(t,j)}},$$

$$\widehat{\boldsymbol{\theta}}_h^{(t,j)} = \pm\sqrt{\breve{\gamma}_h^{(t,j)\,\mathrm{EVB}}\delta_h^{(t,j)}}\cdot\boldsymbol{\omega}_{\theta_h}^{(t,j)}, \qquad \sigma_{\theta_h}^{(t,j)\,2} = \frac{\sigma^2\delta_h^{(t,j)}}{\gamma_h^{(t,j)}},$$

$$\delta_h^{(t,j)} = \sqrt{\frac{M'\breve{\gamma}_h^{(t,j)\,\mathrm{EVB}}}{D'\gamma_h^{(t,j)}}}\left(1+\frac{D'\sigma^2}{\gamma_h^{(t,j)}\breve{\gamma}_h^{(t,j)\,\mathrm{EVB}}}\right),$$

and, for (h,t) such that $\gamma_h^{(t,j)} < \underline{\gamma}_h^{(t,j)\,\mathrm{EVB}}$,

$$\widehat{\boldsymbol{b}}_h^{(t,j)} = \boldsymbol{0}, \qquad \sigma_{b_h}^{(t,j)\,2} \to +0, \qquad \widehat{\boldsymbol{\theta}}_h^{(t,j)} = \boldsymbol{0}, \qquad \sigma_{\theta_h}^{(t,j)\,2} \to +0.$$

Note that the corollaries above assume that $D'^{(t,j)} \leq M'^{(t,j)}$ for all (t,j). However, we can easily obtain the result for the case when $D'^{(t,j)} > M'^{(t,j)}$ by considering the transpose $\widehat{U}'^{(t,j)\top}$ of the solution. Also, we can always take the mapping $\mathcal{X}^{(j)}$ so that $D'^{(t,j)} \leq M'^{(t,j)}$ holds for all (t,j) without any practical restriction. This eases the implementation of the algorithm.

Using Corollary 21.1, Corollary 21.2, and Lemma 21.1, we can build a local search algorithm, which we call a *mean update (MU) algorithm*. A pseudo-code is given in Algorithm 21.1, where $\boldsymbol{0}_{(D_1,D_2)}$ denotes the $D_1 \times D_2$ matrix with all entries equal to zero. Note that, under the empirical Bayesian procedure, all unknown parameters are estimated from observation, which allows fully-automatic inference without manual parameter tuning.

The MU algorithm is similar in spirit to the backfitting algorithm [5, 10], where each additive term is updated to fit a dummy target. In the MU algorithm, $\boldsymbol{Z}^{(j)}$ defined in Equation (21.19) corresponds to the dummy target. Although each of the corollaries and the lemma above guarantee the global optimality for each step, the MU algorithm does not generally guarantee the global optimality over the entire parameter space. Nevertheless, experimental results in Section 21.5 show that the MU algorithm performs well in practice.

Algorithm 21.1 - Mean update (MU) algorithm

1: Initialization: $\widehat{\boldsymbol{U}}^{(j)} \leftarrow \boldsymbol{0}_{(D,M)}$ for $j = 1, \ldots, J$, $\sigma^2 \leftarrow \|\boldsymbol{A}\|_{\mathrm{Fro}}^2/(DM)$.
2: **for** $j = 1$ to J **do**
3: The empirical VB solution of $\boldsymbol{U}'^{(t,j)} = \boldsymbol{B}^{(t,j)}\boldsymbol{\Theta}^{(t,j)\top}$ for each $t = 1, \ldots, T^{(j)}$, given $\{\widehat{\boldsymbol{U}}^{(j')}\}_{j' \neq j}$ and σ^2, is computed by Corollary 21.1.
4: $\widehat{\boldsymbol{U}}^{(j)} \leftarrow \boldsymbol{G}(\{\widehat{\boldsymbol{U}}'^{(t,j)}\}_{t=1}^{T^{(j)}}; \mathcal{X}^{(j)})$.
5: **end for**
6: The noise variance σ^2 is estimated by Lemma 21.1, given the empirical VB posterior on $\{\boldsymbol{w}_{\boldsymbol{\Theta}}^{(j)}, \boldsymbol{w}_{\boldsymbol{B}}^{(j)}\}_{j=1}^{J}$ (computed by Corollary 21.2).
7: Repeat 2 to 6 until convergence.

When Corollary 21.1 is applied in Step 3 of Algorithm 21.1, a singular value decomposition (21.21) of $\boldsymbol{Z}'^{(t,j)}$, defined in Equation (21.19), is required. However, for many practical SMF terms, including the element-wise sparse term and the segment-wise sparse term, $\boldsymbol{Z}'^{(t,j)} \in \mathbb{R}^{D'^{(t,j)} \times M'^{(t,j)}}$ is a vector or scalar, i.e., $D'^{(t,j)} = 1$ or $M'^{(t,j)} = 1$. In such cases, the singular value and the singular vectors are given simply by

$$\gamma_1^{(t,j)} = \|\boldsymbol{Z}'^{(t,j)}\|, \quad \boldsymbol{\omega}_{b_1}^{(t,j)} = 1, \quad \boldsymbol{\omega}_{\theta_1}^{(t,j)} = \boldsymbol{Z}'^{(t,j)}/\|\boldsymbol{Z}'^{(t,j)}\| \qquad \text{if } D'^{(t,j)} = 1,$$
$$\gamma_1^{(t,j)} = \|\boldsymbol{Z}'^{(t,j)}\|, \quad \boldsymbol{\omega}_{b_1}^{(t,j)} = \boldsymbol{Z}'^{(t,j)}/\|\boldsymbol{Z}'^{(t,j)}\|, \quad \boldsymbol{\omega}_{\theta_1}^{(t,j)} = 1 \qquad \text{if } M'^{(t,j)} = 1.$$

21.4 Discussion

In this section, we first discuss a close relation between MIR and ARD. Then, we describe the standard VB iteration for SAMF, which is used as a baseline in the experiment. After that, we introduce related previous work.

21.4.1 MIR and ARD

The *model-induced regularization (MIR)* effect [18] induced by *factorization* actually has a close connection to the *automatic relevance determination (ARD)* effect [22]. Assume $\boldsymbol{C}_{\Theta} = \boldsymbol{I}_H$ in the *plain* MF model (21.16)–(21.18) (here we omit the suffixes t and j for brevity), and consider the following transformation: $\boldsymbol{B}\boldsymbol{\Theta}^\top \mapsto \boldsymbol{L} \in \mathbb{R}^{D \times M}$. Then, the likelihood (21.16) and the prior (21.18) on $\boldsymbol{\Theta}$ are rewritten as

$$p(\boldsymbol{Z}'|\boldsymbol{L}) \propto \exp\left(-\frac{1}{2\sigma^2}\|\boldsymbol{Z}' - \boldsymbol{L}\|_{\mathrm{Fro}}^2\right), \tag{21.26}$$

$$p(\boldsymbol{L}|\boldsymbol{B}) \propto \exp\left(-\frac{1}{2}\mathrm{tr}\left(\boldsymbol{L}^\top(\boldsymbol{B}\boldsymbol{B}^\top)^\dagger\boldsymbol{L}\right)\right), \tag{21.27}$$

where † denotes the Moore-Penrose generalized inverse of a matrix. The prior (21.17) on $\boldsymbol{B}$ is kept unchanged. $p(\boldsymbol{L}|\boldsymbol{B})$ in Equation (21.27) is the so-called ARD prior with the covariance hyperparameter $\boldsymbol{B}\boldsymbol{B}^\top \in \mathbb{R}^{D \times D}$. It is known that this induces the ARD effect, i.e., the *empirical Bayesian learning*, where the prior covariance hyperparameter $\boldsymbol{B}\boldsymbol{B}^\top$ is estimated from observation by maximizing the marginal likelihood (or minimizing the free energy), inducing strong regularization and sparsity [6, 22].

In the current context, the ARD prior (21.27) induces low-rank sparsity on $\boldsymbol{L}$. Similarly, we can show that $B_{d,m}^2$ in Equation (21.2) acts as the prior variance for $S_{d,m} \in \mathbb{R}$. This explains why the factorization forms in Equations (21.1) and (21.2) induce low-rank and element-wise sparsity, respectively.

When we employ the SMF-term expression (21.3), MIR occurs in each partition. Therefore, partition-wise sparsity and low-rank sparsity in each partition are induced. Corollary 21.1 theoretically supports this fact: Small singular values are discarded by thresholding in Equation (21.23).

21.4.2 Standard VB Iteration

Following the standard procedure for the VB learning [1, 2, 15], we can derive the following algorithm, which we call the *standard VB iteration*:

$$\widehat{\boldsymbol{B}}^{(t,j)} = \sigma^{-2} \boldsymbol{Z}'^{(t,j)} \widehat{\boldsymbol{\Theta}}^{(t,j)} \boldsymbol{\Sigma}_{\mathrm{B}}^{(t,j)}, \tag{21.28}$$

$$\boldsymbol{\Sigma}_{\mathrm{B}}^{(t,j)} = \sigma^2 \left(\widehat{\boldsymbol{\Theta}}^{(t,j)\top} \widehat{\boldsymbol{\Theta}}^{(t,j)} + M'^{(t,j)} \boldsymbol{\Sigma}_{\Theta}^{(t,j)} + \sigma^2 \boldsymbol{C}_{B}^{(t,j)\ -1} \right)^{-1}, \tag{21.29}$$

$$\widehat{\boldsymbol{\Theta}}^{(t,j)} = \sigma^{-2} \boldsymbol{Z}'^{(t,j)\top} \widehat{\boldsymbol{B}}^{(t,j)} \boldsymbol{\Sigma}_{\Theta}^{(t,j)}, \tag{21.30}$$

$$\boldsymbol{\Sigma}_{\Theta}^{(t,j)} = \sigma^2 \left(\widehat{\boldsymbol{B}}^{(t,j)\top} \widehat{\boldsymbol{B}}^{(t,j)} + D'^{(t,j)} \boldsymbol{\Sigma}_{\mathrm{B}}^{(t,j)} + \sigma^2 \boldsymbol{C}_{\Theta}^{(t,j)\ -1} \right)^{-1}. \tag{21.31}$$

Iterating Equations (21.28)–(21.31) for each (t, j) in turn until convergence gives a local minimum of the free energy (21.12).

In the empirical Bayesian procedure, the hyperparameters $\{\boldsymbol{C}_{\Theta}^{(t,j)}, \boldsymbol{C}_{B}^{(t,j)}\}_{t=1,j=1}^{T^{(j)}\ J}$ are also estimated from observation. The following update rules give a local minimum of the free energy:

$$c_{b_h}^{(t,j)\ 2} = \|\widehat{\boldsymbol{b}}_h^{(t,j)}\|^2 / D'^{(t,j)} + (\boldsymbol{\Sigma}_{\mathrm{B}}^{(t,j)})_{hh}, \tag{21.32}$$

$$c_{\theta_h}^{(t,j)\ 2} = \|\widehat{\boldsymbol{\theta}}_h^{(t,j)}\|^2 / M'^{(t,j)} + (\boldsymbol{\Sigma}_{\Theta}^{(t,j)})_{hh}. \tag{21.33}$$

When the noise variance σ^2 is unknown, it is estimated by Equation (21.20) in each iteration.

The standard VB iteration is computationally efficient since only a single parameter in $\{\widehat{\boldsymbol{B}}^{(t,j)}, \boldsymbol{\Sigma}_{\mathrm{B}}^{(t,j)}, \widehat{\boldsymbol{\Theta}}^{(t,j)}, \boldsymbol{\Sigma}_{\Theta}^{(t,j)}, c_{b_h}^{(t,j)\ 2}, c_{\theta_h}^{(t,j)\ 2}\}_{t=1,j=1}^{T^{(j)}\ J}$ is updated in each step. However, it is known that the standard VB iteration is prone to suffer from the local minima problem [20]. On the other hand, although the MU algorithm is also not guaranteed to give the global solution as a whole, it gives the global optimal solution of the set $\{\widehat{\boldsymbol{B}}^{(t,j)}, \boldsymbol{\Sigma}_{\mathrm{B}}^{(t,j)}, \widehat{\boldsymbol{\Theta}}^{(t,j)}, \boldsymbol{\Sigma}_{\Theta}^{(t,j)}, c_{b_h}^{(t,j)\ 2}, c_{\theta_h}^{(t,j)\ 2}\}_{t=1}^{T^{(j)}}$ for each j in each step. Experimental results in Section 21.5 show that the MU algorithm tends to give a better solution with smaller free energy and reconstruction error than the standard VB iteration.

21.4.3 Related Work

As is widely known, traditional PCA is sensitive to outliers in data and generally fails in their presence. Robust PCA [3] was developed to cope with large outliers that are not modeled within the traditional PCA. Unlike the methods based on robust statistics [8, 9, 11, 13, 14, 17, 24], robust PCA explicitly captures the spiky noise by an additional element-wise sparse term (see Equation (21.6)). Accordingly, robust PCA can also be used in applications where the task is to estimate the element-wise sparse term itself (as opposed to discarding it as noise). A typical such application is background/foreground video separation (Figure 21.2).

The original formulation [3] of robust PCA is non-Bayesian, and the sparsity is induced by the ℓ_1-norm regularization. Although its solution can be efficiently obtained via the augmented Lagrange multiplier (ALM) method [16], there are unknown algorithmic parameters that should be carefully tuned to obtain its best performance. Employing a Bayesian formulation addresses this issue: A sampling-based method [4] and a VB method [1] were proposed, where all unknown parameters are estimated from the observation.

An extensive experimental comparison was conducted [1] between the VB method, called the VB robust PCA, and other methods. It was reported that the ALM method [16] requires careful tuning of its algorithmic parameters, and the Bayesian sampling method [4] has high computational complexity that can be prohibitive in large-scale applications. Compared to these methods, the VB robust PCA is favorable both in terms of computational complexity and estimation performance.

Our SAMF framework contains the robust PCA model as a special case where the observed matrix is modeled as the sum of a low-rank and an element-wise sparse term. The VB algorithm used in [1] is the same as the standard VB iteration introduced in Section 21.4.2, except for a slight difference in the hyperprior setting. Accordingly, the method introduced in this chapter is an extension of the VB robust PCA in two ways—more variation in sparsity with different types of factorization and higher accuracy with the MU algorithm. In Section 21.5, we experimentally show advantages of these extensions, by comparing them with a SAMF counterpart of the VB robust PCA and the standard VB iteration as baseline methods.

Group LASSO [26] also provides a framework for arbitrary sparsity design, where the sparsity is induced by the ℓ_1-regularization. Although the convexity of the group LASSO problem is attractive, it typically requires careful tuning of regularization parameters, as the ALM method for robust PCA. On the other hand, group-sparsity is induced by MIR and ARD in SAMF, and all unknown parameters can be estimated within the Bayesian framework.

21.5 Experiment

In this section, we first conduct an experiment on artificial data, and show the superiority of the MU algorithm over the standard VB iteration. After that, we demonstrate the usefulness of SAMF in background/foreground separation.

21.5.1 Artificial Data

We assess the quality of the solution in the robust PCA model:

$$\boldsymbol{A} = \boldsymbol{U}^{\text{low-rank}} + \boldsymbol{U}^{\text{element}} + \boldsymbol{E}.$$

We assume the empirical VB scenario with unknown noise variance, i.e., the hyperparameters $\{\boldsymbol{C}_B^{(t,j)}, \boldsymbol{C}_\Theta^{(t,j)}\}_{t=1,j=1}^{T^{(j)}\ J}$ and the noise variance σ^2 are also estimated from observation. We use the full-rank model ($H = \min(D, M)$) for the low-rank term $\boldsymbol{U}^{\text{low-rank}}$, and expect the MIR effect to find the true rank of $\boldsymbol{U}^{\text{low-rank}}$, as well as the true non-zero entries in $\boldsymbol{U}^{\text{element}}$.

We created an artificial dataset with the data matrix size $D = 40$ and $M = 100$, and the rank $H^* = 10$ for a *true* low-rank matrix $\boldsymbol{U}^{\text{low-rank}*} = \boldsymbol{B}^*\boldsymbol{\Theta}^{*\top}$. Each entry in $\boldsymbol{B}^* \in \mathbb{R}^{D\times H^*}$ and $\boldsymbol{\Theta}^* \in \mathbb{R}^{M\times H^*}$ was drawn from $\mathcal{N}_1(0,1)$. A *true* element-wise sparse term $\boldsymbol{U}^{\text{element}*}$ was created by first randomly selecting ρDM entries for $\rho = 0.05$, and then adding noise subject to $\mathcal{N}_1(0,\kappa)$ for $\kappa = 100$ to each of the selected entries. Finally, an observed matrix $\boldsymbol{A}$ was created by adding a dense noise matrix $\boldsymbol{E}$, each entry of which is subject to $\mathcal{N}_1(0,1)$.

It is known that the standard VB iteration (reviewed in Section 21.4.2) is sensitive to initialization [20]. We set the initial values in the following way: The mean parameters $\{\widehat{\boldsymbol{B}}^{(t,j)}, \widehat{\boldsymbol{\Theta}}^{(t,j)}\}_{t=1,j=1}^{T^{(j)}\ J}$ were randomly created so that each entry follows $\mathcal{N}_1(0,1)$. The covariances $\{\boldsymbol{\Sigma}_{\text{B}}^{(t,j)}, \boldsymbol{\Sigma}_\Theta^{(t,j)}\}_{t=1,j=1}^{T^{(j)}\ J}$ and the hyperparameters $\{\boldsymbol{C}_B^{(t,j)}, \boldsymbol{C}_\Theta^{(t,j)}\}_{t=1,j=1}^{T^{(j)}\ J}$ were set to

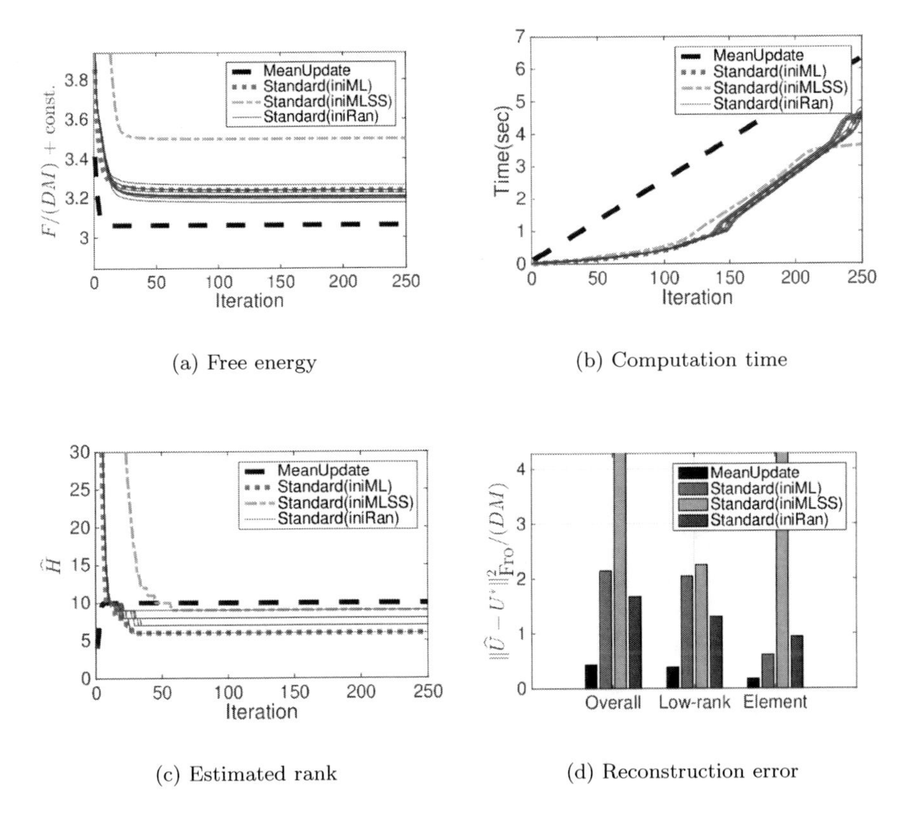

(a) Free energy

(b) Computation time

(c) Estimated rank

(d) Reconstruction error

FIGURE 21.5 (See color insert.) Experimental results in robust PCA on an artificial dataset ($D = 40, M = 100, H^* = 10, \rho = 0.05$).

be the identity matrix. The initial noise variance was set to $\sigma^2 = 1$. Note that we rescaled $\boldsymbol{A}$ so that $\|\boldsymbol{A}\|^2_{\mathrm{Fro}}/(DM) = 1$, before starting iteration. We ran the standard VB iteration algroithm 10 times, starting from different initial points, and each trial is plotted by a solid line (labeled as 'Standard(iniRan)') in Figure 21.5.

Initialization of the MU algorithm (described in Algorithm 21.1) is simple: We just set $\widehat{\boldsymbol{U}}^{(j)} = \boldsymbol{0}_{(D,M)}$ for $j = 1, \ldots, J$, and $\sigma^2 = 1$. Initialization of all other variables is not necessary. Furthermore, we empirically observed that the initial value for σ^2 has little effect on the result, unless it is too small. Note that, in the MU algorithm, initializing σ^2 to a large value is not harmful, because it is set to an adequate value after the first iteration with the mean parameters kept $\widehat{\boldsymbol{U}}^{(j)} = \boldsymbol{0}_{(D,M)}$. The result with the MU algorithm is plotted by the dashed line in Figure 21.5.

Figures 21.5(a)–21.5(c) show the free energy, the computation time, and the estimated rank, respectively, over iterations, and Figure 21.5(d) shows the reconstruction errors after 250 iterations. The reconstruction errors consist of the *overall* error $\|(\widehat{\boldsymbol{U}}^{\mathrm{low-rank}} + \widehat{\boldsymbol{U}}^{\mathrm{element}}) - (\boldsymbol{U}^{\mathrm{low-rank}*} + \boldsymbol{U}^{\mathrm{element}*})\|_{\mathrm{Fro}}/(DM)$, the *low-rank term* error $\|\widehat{\boldsymbol{U}}^{\mathrm{low-rank}} - \boldsymbol{U}^{\mathrm{low-rank}*}\|_{\mathrm{Fro}}/(DM)$, and the *element-wise sparse term* error $\|\widehat{\boldsymbol{U}}^{\mathrm{element}} - \boldsymbol{U}^{\mathrm{element}*}\|_{\mathrm{Fro}}/(DM)$. The graphs show that the MU algorithm, of which iteration is computationally slightly more expensive than the standard VB iteration, immediately converges to a local minimum with the free energy substantially lower than the standard VB iteration. The estimated rank agrees with the true rank $\widehat{H} = H^* = 10$, while all 10 trials of the standard VB iteration failed to estimate the true rank. It is also observed that the MU algorithm reconstructs the true terms well.

We also tested different initialization schemes for the standard VB iteration. The line labeled "Standard(iniML)" in Figure 21.5 indicates the maximum likelihood (ML) initialization, i.e., $(\widehat{\boldsymbol{b}}_h^{(t,j)}, \widehat{\boldsymbol{\theta}}_h^{(t,j)}) = (\gamma_h^{(t,j)1/2}\boldsymbol{\omega}_{b_h}^{(t,j)}, \gamma_h^{(t,j)1/2}\boldsymbol{\omega}_{\theta_h}^{(t,j)})$. Here, $\gamma_h^{(t,j)}$ is the h-th largest singular value of the (t,j)-th PR matrix $\boldsymbol{A}'^{(t,j)}$ of $\boldsymbol{A}$ (such that $A'^{(t,j)}_{d',m'} = A_{\mathcal{X}^{(j)}(t,d',m')}$), and $\boldsymbol{\omega}_{b_h}^{(t,j)}$ and $\boldsymbol{\omega}_{\theta_h}^{(t,j)}$ are the associated left and right singular vectors. Starting the initialization from small σ^2 was reported to alleviate the local minima problem in MF [20]. The line labeled "Standard(iniMLSS)" indicates the ML initialization with $\sigma^2 = 0.0001$. Although this scheme slightly improved the rank estimation performance, the standard VB iteration with any initialization scheme gave substantially worse free energy and reconstruction error than the MU algorithm.

Similar tendencies were observed in different SAMF models with different sizes of matrices [19]. Thus, we conclude that the MU algorithm generally outperforms the standard VB iteration.

21.5.2 Background/Foreground Separation

Finally, we demonstrate the usefulness of SAMF in a background/foreground video separation problem (Figure 21.2). As explained in Section 21.2.3, we form the observed matrix $\boldsymbol{A}$ by stacking all pixels in each frame into each column (Figure 21.3). As a baseline, we applied the robust PCA model (21.6), where the low-rank term captures the *static* background (BG) and the element-wise (or pixel-wise) term captures the *moving* foreground (FG), e.g., people walking through.

To show the advantage of SAMF, we evaluate the performance of the segmentation-based SAMF (sSAMF) model (21.7), where the segment-wise sparse term is used to capture the FG. We produced an over-segmented image from each frame by using the efficient graph-based segmentation (EGS) algorithm [7]. Note that EGS is computationally very efficient: It takes less than 0.05 second on a normal laptop to segment a 192×144 grey image. EGS has several tuning parameters, and the obtained segmentation is sensitive to some of them. However, we confirmed that sSAMF performs similarly with visually different segmentations obtained over a wide range of tuning parameters (see detailed information below on the segmentation algorithm). Therefore, careful parameter tuning of EGS is not necessary for our purpose.

We compared sSAMF with robust PCA on the "WalkByShop1front" video from the *Caviar dataset.*[73] Thanks to the empirical Bayesian procedure, all unknown parameters (except the ones for segmentation) are estimated without manual tuning. For both models (robust PCA and sSAMF), we used the MU algorithm, which has been shown to outperform the standard VB iteration (see Section 21.5.1). The original video consists of 2360 frames, each of which is a color image with 384×288 pixels. We resized each image into 192×144 pixels, averaged over the color channels, and sub-sampled every 15 frames (the frame IDs are $0, 15, 30, \ldots, 2355$). Thus, $\boldsymbol{A}$ is of the size 27,684 (pixels) $\times$ 158 (frames). We evaluated robust PCA and sSAMF on this video, and found that both models perform well (although robust PCA failed in a few frames).

To contrast robust PCA and sSAMF more clearly, we created a more *difficult* video by sub-sampling every 5 frames from 1501 to 2000 (the frame IDs are $1501, 1506, \ldots, 1996$ and $\boldsymbol{A}$ is of the size 27,684 (pixels) $\times$ 100 (frames)). Since more people walked through in this

[73] `http://groups.inf.ed.ac.uk/vision/CAVIAR/CAVIARDATA1/`

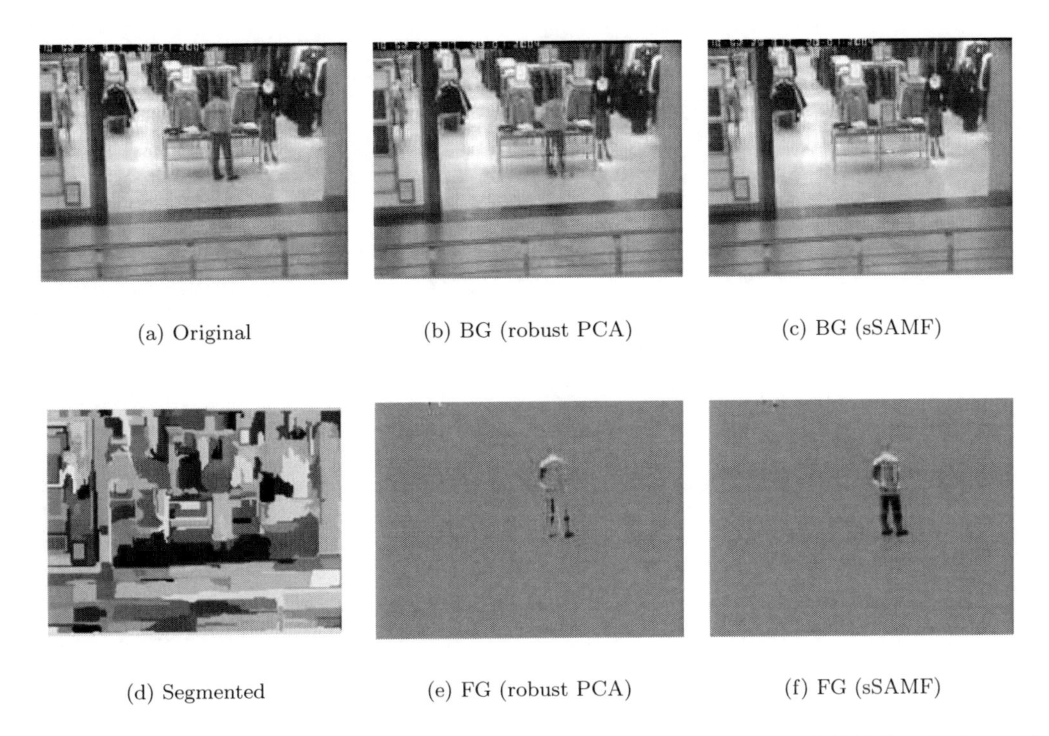

(a) Original (b) BG (robust PCA) (c) BG (sSAMF)

(d) Segmented (e) FG (robust PCA) (f) FG (sSAMF)

FIGURE 21.6 (See color insert.) Robust PCA vs. segmentation-based SAMF (sSAMF) in background (BG)/foreground (FG) video separation.

period, BG estimation is more challenging. Figure 21.6 shows results at one frame.

Figure 21.6(a) shows an original frame. This is a difficult snapshot, because a person stayed at the same position for a while, which confuses separation. Figures 21.6(b) and 21.6(e) show the BG term and the FG term, respectively, obtained by robust PCA. We can see that robust PCA failed to separate the person from BG (the person is partly captured in the BG term). On the other hand, Figures 21.6(c) and 21.6(f) show the BG term and the FG term, respectively, obtained by sSAMF, based on the segmented image shown in Figure 21.6(d). We can see that sSAMF successfully separated the person from BG in this difficult frame. A careful look at the legs of the person explains how the segment-wise sparsity helps separation—the legs form a single segment in Figure 21.6(d), and the segment-wise sparse term (Figure 21.6(f)) captures all pixels on the legs, while the pixel-wise sparse term (Figure 21.6(e)) captures only a part of those pixels. We observed that, in all frames of the *difficult* video, as well as the *easier* one, sSAMF gave good separation, while robust PCA failed in several frames.

For reference, we applied the convex optimization approach [3], which solves the minimization problem

$$\min_{\boldsymbol{L},\boldsymbol{S}} \|\boldsymbol{L}\|_{\mathrm{tr}} + \lambda \|\boldsymbol{S}\|_1 \qquad \text{s.t.} \qquad \boldsymbol{A} = \boldsymbol{L} + \boldsymbol{S},$$

where $\|\cdot\|_{\mathrm{tr}}$ and $\|\cdot\|_1$ denote the trace norm and the ℓ_1-norm of a matrix, respectively, by the inexact ALM algorithm [16]. Figure 21.7 shows the obtained BG term $\boldsymbol{L}$ and the FG term $\boldsymbol{S}$ at the same frame as in Figure 21.6 for $\lambda = 0.001, 0.005, 0.025$. We see that the performance strongly depends on the parameter λ, and that sSAMF gives an almost identical result (right column in Figure 21.6) to the best ALM result with $\lambda = 0.005$ (middle column in Figure 21.7) without any manual parameter tuning.

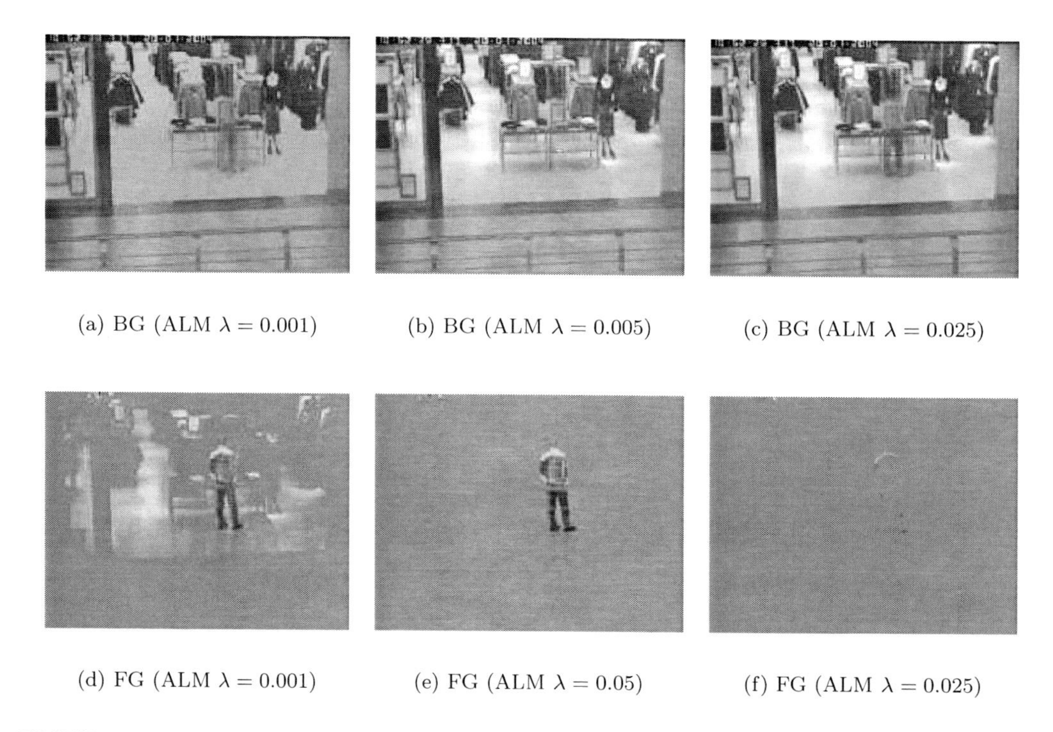

(a) BG (ALM $\lambda = 0.001$) (b) BG (ALM $\lambda = 0.005$) (c) BG (ALM $\lambda = 0.025$)

(d) FG (ALM $\lambda = 0.001$) (e) FG (ALM $\lambda = 0.05$) (f) FG (ALM $\lambda = 0.025$)

FIGURE 21.7 Results with the inexact ALM algorithm [16] for $\lambda = 0.001$ (left column), $\lambda = 0.005$ (middle column), and $\lambda = 0.025$ (right column).

Below, we give detailed information on the segmentation algorithm and the computation time.

Segmentation Algorithm

For the efficient graph-based segmentation (EGS) algorithm [7], we used the code publicly available from the authors' homepage.[74] EGS has three tuning parameters: the smoothing parameter *sigma*, the threshold parameter t, and the minimum segment size *minc*. Among them, t dominantly determines the typical size of segments (larger t leads to larger segments). To obtain over-segmented images for sSAMF in our experiment, we chose $t = 50$, and the other parameters are set to $sigma = 0.5$ and $minc = 20$ as recommended by the authors. We also tested other parameter settings, and observed that BG/FG separation by sSAMF performed almost equally for $1 \leq t \leq 100$, despite the visual variation of segmented images (see Figure 21.8). Overall, we empirically observed that the performance of sSAMF is not very sensitive to the selection of segmented images, unless it is highly under-segmented.

Computation Time

The computation time for segmentation by EGS was less than 10 sec (for 100 frames). Forming the one-to-one mapping $\mathcal{X}$ took more than 80 sec (which is expected to be improved). In total, sSAMF took 600 sec on a Linux machine with Xeon X5570(2.93GHz),

[74] `http://www.cs.brown.edu/~pff/`

(a) Segmented ($t = 1$)

(b) Segmented ($t = 10$)

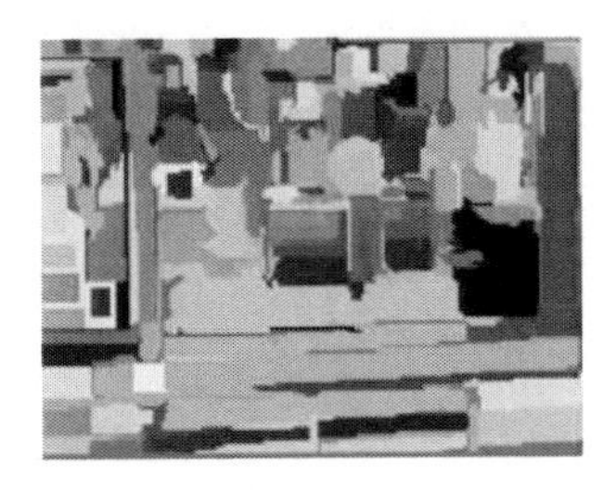

(c) Segmented ($t = 100$)

FIGURE 21.8 (See color insert.) Images segmented by the efficient graph-based segmentation (EGS) algorithm with different t values. They are visually different, but with all these segmentations, the BG/FG separation results by sSAMF were almost identical. The original image is shown in Figure 21.6(a).

while robust PCA took 700 sec. This slight reduction in computation time comes from the reduction in the number T of partitions for the FG term, and hence the number of computations of partial analytic solutions.

21.6 Conclusion

In this chapter, we introduced a Bayesian sparse estimation method, called a sparse additive matrix factorization (SAMF), and its application to background/foreground separation. SAMF allows flexible design of matrix factorization to induce various types of sparsity, in order to incorporate side information. We also introduced a sophisticated algorithm, called a mean update (MU) algorithm, for variational Bayesian (VB) learning in SAMF. Experiments demonstrated the usefulness of SAMF.

Future work includes analysis of convergence properties of the MU algorithm, theoretical elucidation of the reason why the MU algorithm tends to give a better solution than the standard VB algorithm, and further exploration of application-specific sparsity design.

References

1. S. Babacan, M. Luessi, R. Molina, and A. Katsaggelos. Sparse Bayesian methods for low-rank matrix estimation. *IEEE Transactions on Signal Processing*, 60(8):3964–3977, 2012.
2. C. Bishop. Variational principal components. *International Conference on Artificial Neural Networks*, 1:514–509, 1999.
3. E. Candès, X. Li, Y. Ma, and J. Wright. Robust principal component analysis? *Journal of the ACM*, 58(3):1–37, May 2011.
4. X. Ding, L. He, and L. Carin. Bayesian robust principal component analysis. *IEEE Transactions on Image Processing*, 20(12):3419–3430, 2011.
5. A. D'Souza, S. Vijayakumar, and S. Schaal. The Bayesian backfitting relevance vector machine. In *Proceedings of the 21st International Conference on Machine Learning*, 2004.
6. B. Efron and C. Morris. Stein's estimation rule and its competitors–an empirical Bayes approach. *Journal of the American Statistical Association*, 68:117–130, 1973.
7. P. Felzenszwalb and D. Huttenlocher. Efficient graph-based image segmentation. *International Journal of Computer Vision*, 59(2):167–181, 2004.
8. M. Fischler and R. Bolles. Random sample consensus: a paradigm for model fitting with

applications to image analysis and automated cartography. *Communications of the ACM*, 24:381–385, 1981.
9. J. Gao. Robust l_1 principal component analysis and its Bayesian variational inference. *Neural Computation*, 20:555–578, 2008.
10. T. Hastie and R. Tibshirani. Generalized additive models. *Statistical Science*, 1(3):297–318, 1986.
11. P. Huber and E. Ronchetti. *Robust Statistics*. Wiley, 2009.
12. A. Ilin and T. Raiko. Practical approaches to principal component analysis in the presence of missing values. *Journal of Machine Learning Research*, 11:1957–2000, 2010.
13. Q. Ke and T. Kanade. Robust l_1 norm factorization in the presence of outliers and missing data by alternative convex programming. In *IEEE Conference on Computer Vision and Pattern Recognition, CVPR 2005*, 2005.
14. B. Lakshminarayanan, G. Bouchard, and C. Archambeau. Robust Bayesian matrix factorisation. In *International Conference on Artificial Intelligence and Statistics, AISTATS 2011*, volume 15, 2011.
15. Y. Lim and Y. Teh. Variational Bayesian approach to movie rating prediction. In *Proc. of KDD Cup and Workshop*, 2007.
16. Z. Lin, M. Chen, and Y. Ma. The augmented Lagrange multiplier method for exact recovery of corrupted low-rank matrices. *UIUC Technical Report UILU-ENG-09-2215*, 2009.
17. J. Luttinen, A. Ilin, and J. Karhunen. Bayesian robust pca for incomplete data. In *International Conference on Independent Component Analysis and Signal Separation*, 2009.
18. S. Nakajima and M. Sugiyama. Theoretical analysis of Bayesian matrix factorization. *Journal of Machine Learning Research*, 12:2579–2644, 2011.
19. S. Nakajima, M. Sugiyama, and S. Babacan. Variational Bayesian sparse additive matrix factorization. *Machine Learning*, 92:319–1347, 2013.
20. S. Nakajima, M. Sugiyama, S. D. Babacan, and R. Tomioka. Global analytic solution of fully-observed variational Bayesian matrix factorization. *Journal of Machine Learning Research*, 14:1–37, 2013.
21. S. Nakajima, R. Tomioka, M. Sugiyama, and S. Babacan. Condition for perfect dimensionality recovery by variational Bayesian PCA. *To appear in Journal of Machine Learning Research*, December 2015.
22. R. Neal. *Bayesian Learning for Neural Networks*. Springer, 1996.
23. R. Salakhutdinov and A. Mnih. Bayesian probabilistic matrix factorization using Markov chain Monte Carlo. In *International Conference on Machine Learning, ICML 2008*, pages 1257–1264, Cambridge, MA, 2008. MIT Press.
24. F. De La Torre and M. Black. A framework for robust subspace learning. *International Journal of Computer Vision*, 54:117–142, 2003.
25. S. Watanabe. *Algebraic Geometry and Statistical Learning*. Cambridge University Press, Cambridge, UK, 2009.
26. M. Yuan and Y. Lin. Model selection and estimation in regression with grouped variables. *Journal of the Royal Statistical Society B*, 68(1):49–67, 2006.

Index